Kindly Donated
by
Mr Lewis

Clinical Cytotechnology

To our husbands Jack and Brian for their constant support and encouragement during the preparation of this book

Clinical Cytotechnology

Dulcie V. Coleman MD MRCPath FIAC
Professor of Cell Pathology and Consultant Cytopathologist, St Mary's Hospital Medical School, London

Patricia A. Chapman FIMLS CFIAC FRSH
Chief Medical Laboratory Scientific Officer, Department of Cytology, Watford General Hospital, UK

*For invaluable help and advice over the years.
Best wishes
Pat*

MEDICAL LIBRARY
WATFORD POSTGRADUATE
MEDICAL CENTRE
WATFORD GENERAL HOSPITAL
VICARAGE ROAD
WATFORD WD1 8HB

Butterworths

London   Boston   Singapore   Sydney   Toronto   Wellington

PART OF REED INTERNATIONAL P.L.C.

All rights reserved. No part of this publication may be reproduced or transmitted in any form or by any means (including photocopying and recording) without the written permission of the copyright holder except in accordance with the provisions of the Copyright Act 1956 (as amended) or under the terms of a licence issued by the Copyright Licensing Agency Ltd, 33–34 Alfred Place, London, England, WC1E 7DP. The written permission of the copyright holder must also be obtained before any part of this publication is stored in a retrieval system of any nature. Applications for the copyright holder's written permission to reproduce, transmit or store in a retrieval system any part of this publication should be addressed to the Publishers.

Warning: The doing of an unauthorised act in relation to a copyright work may result in both a civil claim for damages and criminal prosecution.

This book is sold subject to the Standard Conditions of Sale of Net Books and may not be re-sold in the UK below the net price given by the Publishers in their current price list.

First published 1989

© Butterworth & Co. (Publishers) Ltd, 1989

British Library Cataloguing in Publication Data

Clinical cytotechnology.
  1. Man. Diagnosis. Applications of cytology.
  I. Coleman, Dulcie V. II. Chapman, Patricia A.
616.07′582

ISBN 0-407-00176-X

Library of Congress Cataloging-in-Publication Data

Clinical cytotechnology / [edited by] Dulcie V. Coleman,
  Patricia A. Chapman.
    p. cm.
    Bibliography: p.
    Includes index.
    ISBN 0-407-00176-X:
    1. Diagnosis, Cytologic. I. Coleman, Dulcie V.
  II. Chapman, Patricia A.
  RB43.C56 1989
  616.07′582—dc19

Typeset by Bath Typesetting Ltd, Bath
Printed and bound in Great Britain by Courier International Ltd, Tiptree, Essex

Preface

Clinical cytology is one of the most rapidly developing areas of pathology. All pathology laboratories will confirm that there has been a significant increase in demand for cytology services in the last decade, reflecting an increased awareness of the value of cytology for the investigation of the patient with symptoms and signs suggestive of malignant disease and a better understanding by the public of the importance of cancer prevention as a method of primary health care.

While this increase in demand for cytology services is gratifying to those of us who have worked in this field for many years, it brings with it an urgent need for properly trained pathologists and cytotechnologists to provide the service and who, as a result of their training, are fully aware of the scope of cytological techniques. This book has been written with a view to meeting this need.

The purpose of this book is to explain the principles and practice of clinical cytology to medical and non-medical scientists who have responsibility for the provision of a cytology service or who have a major interest in it. The text covers all aspects of gynaecological and non-gynaecological cytology. As cervical cytology provides the major workload in most laboratories, special emphasis is given to the cytology of the female genital tract. Four chapters are devoted to this theme and screening strategy, reporting and management are included among the topics discussed. Other chapters cover the cytology of the respiratory, urinary and gastrointestinal tracts, serous effusions, cerebrospinal fluid, and synovial fluid. Semen analysis, basic cell biology and parasitology have also been identified as important areas meriting the special attention of the reader. A large section of this book deals with practical aspects of cytology including specimen preparation and staining, laboratory management (including quality control) and microscope technique. The incorporation of chapters on immunocytochemistry, fine needle aspiration cytology, iatrogenic changes and automation make this book fully up-to-date.

As stated above, this book is intended for the education and training of all medical and non-medical personnel responsible for providing a clinical cytology service. It is, however, written with special attention to cytotechnologists and cytopathologists preparing for national and international examinations and is intended as a reference book for this group of students. The range of contents is such that it will also be useful to histologists, research workers, students of biological science, nurses and health care workers whose work demands a knowledge of some, if not all, aspects of clinical cytology.

The book is well illustrated and besides the many carefully selected high quality monochrome illustrations there are numerous diagrams, helpful tables and a collection of unique colour plates. The chapters are written by a team of authors who are experts in cytology, many of whom have written authoritative articles or books on topics which they have made their special subject. Each chapter is written in a clear and structured way. The collection and preparation of specimens is discussed in detail, and information about the anatomy, physiology, histology and histopathology of individual organs or tissue is provided so that the reader can assemble all the scientific information necessary for the correct light microscope interpretation of a smear. All authors are fully aware of the pitfalls of diagnosis which are discussed in their individual chapters. Most of the contributors are closely involved in the teaching of cytology and some are examiners at national and international level.

The early pioneers of clinical cytology have left a legacy which, by continued research and development over the years, has become a powerful tool for cancer diagnosis and cancer prevention. We hope this book will make readers aware of the contribution they have made to the field of clinical pathology and to medical practice in general. Dr George N. Papanicolaou likened cytology to 'a new tributary flowing into an old stream'. We hope this book will help cytology gather momentum and bring fresh ideas to the mainstream of pathology.

We would like to acknowledge the following for their sponsorship contributions:

BDH Limited, Diagnostics Division
DAKO Limited
Histopath Limited
Raymond Lamb
Shandon Southern Properties Limited

Patricia A. Chapman
Dulcie V. Coleman

Contributors

W. K. Blenkinsopp MA, MD, PhD, FRCPath
Consultant Pathologist, Watford General Hospital, Watford, UK

Michael M. Boddington MA, MSc(Oxon.), FRCPath
Consultant Cytopathologist, The Royal Berkshire Hospital, Reading, UK

E. Blanche Butler MD, FRCOG, MRCPath
Early Diagnostic Unit, Elizabeth Garrett Anderson Hospital, London, UK

Phillipa C. Carroll PhD, FIMLS, CMIAC
Formerly Senior Medical Laboratory Scientific Officer, Department of Cytology, St Mary's Hospital, London, UK

Gordon M. Campbell FIMLS
Senior Chief Medical Laboratory Scientific Officer, Royal Infirmary, Glasgow; Specialist Visiting Lecturer, Glasgow College of Technology, Glasgow, UK

Patricia A. Chapman FIMLS, CFIAC, FRSH
Chief Medical Laboratory Scientific Officer, Department of Cytology, Watford General Hospital, UK

Dulcie V. Coleman MD, MRCPath, FIAC
Professor of Cell Pathology and Consultant Cytopathologist, St Mary's Hospital Medical School, London, UK

Michael Drake MB, BS, FRCPA, FRCPath, FRACP, FIAC
Consultant Pathologist, Prince Henry's Hospital, Melbourne, Australia

Janice I. Guerra FIMLS
Senior Medical Laboratory Scientific Officer, Cytology Department, St Mary's Hospital Medical School, London, UK

Margaret Haddon FIMLS CTIAC
Three Legged Cross, Dorset, UK

Amanda Herbert MB, BS, FRCPath
Consultant Pathologist, Southampton General Hospital, Southampton, UK

Anthony D. Hoyes BSc, MB, BS, PhD
Formerly Professor of Neuroanatomy, St Mary's Hospital Medical School, London, UK

Elizabeth Hudson MD, MRCPath
Consultant Cytopathologist, Northwick Park Hospital and Clinical Research Centre, Harrow, UK

O. A. N. Husain MD, FRCPath, FRCOG
Consultant Pathologist, Head of Departments of Cytopathology of Charing Cross and St Stephen's Hospitals, London, UK

Jocelyn E. A. Imrie MB, ChB, MRCPath
Consultant Cytopathologist, Monklands District General Hospital, Airdrie, UK

Tadao K. Kobayashi MT, CFIAC
Head, Section of Cytopathology; Chief Medical Laboratory Scientist in Laboratory Medicine, Saiseikai Shiga Hospital, Imperial Gift Foundation Inc., Shiga, Japan

David Melcher MA (Cantab.), MB, ChB (Capetown), FRCPath, FIAC
Consultant Pathologist, Brighton General Hospital, Brighton; Visiting Research Fellow, University of Sussex, UK

Patricia M. Norman BA, MB, BCh, BAO, FRCPath
Clinical Pathologist, National Hospitals for Nervous Diseases, London, UK

Noel R. Padley MB, ChB, FRCPath
Consultant Pathologist, Department of Pathology, South East Kent Health Authority, William Harvey Hospital, Ashford, UK

Alan R. Potter MPhil, FIMLS
Principal Scientific Officer – Swindon Health District, Princess Margaret Hospital, Swindon, UK

David T. Proctor FIMLS, CMIAC
Senior Chief Medical Laboratory Scientific Officer, Cytology Department, St Mary's Hospital, Manchester, UK

Michael Reeves MRCP, FRCR
Consultant Radiologist, Bondi Junction, New South Wales, Australia

A. J. Salsbury MD, MB, BChir, MRCPath
Late Consultant Haematologist, Brompton Hospital, London, UK

Rod Setterington
Carl Zeiss (Oberkochen) Ltd, Welwyn Garden City, UK

John Sims FIMLS, CMS
Senior Chief Medical Laboratory Scientific Officer, Cytology Department, Southmead Hospital, Bristol, UK

Russell Smith FIMLS, CMIAC
Senior Chief Medical Laboratory Scientific Officer, Department of Clinical Cytopathology, Brighton General Hospital, Brighton, UK

Keith C. Watts CMIAC, FIMLS, PhD (Lond.)
Formerly Chief Medical Laboratory Scientific Officer, Department of Cytopathology, Charing Cross Hospital, London; Currently Preclinical undergraduate, University College and Middlesex School of Medicine, London

George W. Wikeley CFIAC
Chairman South African Society of Clinical Cytology, Durban, South Africa

Elizabeth M. H. Wilson BSc, MPhil, FIMLS, CTIAC
Extraordinary member of the BSCC; Cytology Department, Dudley Road Hospital, Birmingham, UK

Conventions used in the text

The following conventions are used throughout this text

Ethanol (C_2H_5OH)
syn: ethyl alcohol, absolute alcohol

The term 'ethanol' is used throughout this book when pure absolute alcohol (ethyl alcohol) is necessary. Use of ethanol in all forms is subject to control by Customs and Excise.

Alcohol (C_2H_5OH + denaturant)
syn: industrial methylated spirits, 'meths'

The term 'alcohol' is used through this book as an abbreviation for industrial methylated spirits. The reagent is obtained by denaturing 95 volumes of synthesized ethanol with 5 volumes of wood naphtha, a wood distillate containing methanol. Four grades are available, differing only in alcohol and water content. In this book the term *alcohol* refers to 74 overproof (OP), which is the grade most commonly used in cytology laboratories. 74 OP has a total alcohol content of 99.2% v/v.

Isopropanol (($CH_3)_2CHOH$)
syn: propan-2-ol, isopropyl alcohol

This has similar uses to ethanol. It has a strong characteristic odour, and is not generally subject to customs regulations although some countries require denaturants to be added.

Methanol (CH_3OH)
syn: methyl alcohol

This is used principally undiluted as a fixative and solvent in Romanowsky methods. For this purpose methanol (methyl alcohol) must be acetone free. Care must be taken when handling methanol because of toxicity when ingested or inhaled.

Mixtures of fluids

Percentages refer to the volumes of the fluids mixed (v/v), not the weights.

Revolutions per minute

Centrifugation in this book is defined as revolutions per minute for convenience of use with the standard bench centrifuge or cytocentrifuge. However, the relative centrifugal force (RCF) in gravitational units can be calculated by using the following formula:

$$RCF = 1.118 \times 10^{-5} \times r \times N^2$$

where
r equals the radius of the centrifuge arm in centimetres
N equals the speed in rev/min.

Experience at St Mary's Hospital indicates that most cytological specimens (other than delicate samples such as CSF) will tolerate a range of RCF from 400–1000 G.

Saline

This term refers to 0.85% isotonic sodium chloride solution.

Solutions

Percentages refer to the weights of solids and the volumes of liquids (w/v), e.g. 1% is equal to 1 g per 100 ml

Contents

Preface   v

List of Contributors   vii

Conventions used in the text   x

1   Introduction   1
    Dulcie V. Coleman

2   The cell   7
    A. D. Hoyes

3   Disease at the cellular level   29
    Amanda Herbert

4   Preparatory techniques   52
    David T. Proctor

5   Staining techniques   79
    David T. Proctor

6   Immunocytochemical methods   106
    Alan R. Potter

7   Microscopy   125
    Rod Setterington

8   Female genital tract: normal cytology   137
    Dulcie V. Coleman

9   Female genital tract: infection, inflammation and repair   167
    Patricia A. Chapman

10 Intraepithelial and invasive carcinoma of the cervix   195
    E. Blanche Butler

11 The corpus uteri   220
    Elizabeth Hudson and W. K. Blenkinsopp

12 The respiratory tract: normal structure, function and methods of investigation   235
    Gordon M. Campbell and Jocelyn E. A. Imrie

13 The respiratory tract: pathology and cytology of respiratory tract diseases   248
    Jocelyn E. A. Imrie and Gordon M. Campbell

14 **Serous effusions**   271
Michael M. Boddington

15 **The place of special techniques in the investigation of serous effusions**   283
Janice I. Guerra

16 **Cerebrospinal fluids**   293
Patricia M. Norman

17 **The urinary tract**   302
Elizabeth M. H. Wilson

18 **Semen analysis, and the cytology of hydrocoeles and spermatocoeles, and the prostatic gland**   327
Margaret Haddon

19 **Fine needle aspiration biopsy**   339
David Melcher, Michael Reeves and Russell Smith

20 **Gastrointestinal tract**   368
Michael Drake

21 **Synovial fluids**   384
Margaret Haddon

22 **Tumour cells in blood and bone marrow**   395
A. J. Salsbury

23 **Parasitic infections**   408
George W. Wikeley

24 **Iatrogenic changes**   425
Tadao K. Kobayashi

25 **Automation**   441
Keith C. Watts and O. A. N. Husain

26 **Laboratory organization and management**   455
John Sims and Phillipa C. Carroll

27 **Computerization**   467
Noel R. Padley

Index   471

1

Introduction

Dulcie V. Coleman

For someone to understand his science, he must first know its history.
Auguste Compte (1798–1857)

The aim of this introductory chapter is to provide the reader with an historical perspective of the development of clinical cytology so that he/she can appreciate the skill and dedication of the scientists who pioneered the development of this new specialty of pathology and realize the advances that have been made in the past few years. The list of pioneers mentioned in the chapter is far from complete, but it is hoped that the reader will be sufficiently stimulated to delve more deeply into the subject. The books by Grunze and Spriggs (1980) and Carmichael (1973) are recommended for further reading.

The foundations of clinical cytology as we know it today were laid in the first half of the nineteenth century at a time when technical improvements in the microscope and intellectual acceptance of cellular theory were developing hand in hand. At this time Joseph Lister (1829) was developing a microscope that was free from spherical and chromatographic aberration and the physiologists Schleiden (1804–1881) and Schwann (1810–1882) were promoting the concept of the cell as the basic unit of living and growing organisms. The effect of these two developments was to create a climate of exploration in biology and medicine which led to the study of tissue fragments and body secretions in the light microscope with a view to understanding the pathological basis of disease. In short, medical microscopy was born in the nineteenth century and the scientific literature of the time was swarming with drawings and descriptions of the microscopic appearances of a wide variety of benign and malignant lesions as they appeared in histological section or cytological smear.

Scientists from all nations contributed to this fund of scientific knowledge. Among those who undertook some of the earliest cytological studies were the eminent German physiologist Johannes Müller (1801–1858), Alfred Donne (1801–1878), a former French lawyer with an interest in microscopy, and Lionel S. Beale (1828–1906), Emeritus Professor of Medicine, King's College London. These scientists published descriptions of the microscopic appearances of the cells in blood, pus, ascitic fluid and urine which correspond very closely to our concepts of these body fluids today. One of the first Americans with an interest in cytology was a physician from Philadelphia, Dr F. Donaldson (1853) who made impression smears from the cut surfaces of tumours. Another pioneer was Henri Lebert (1845), a French pathologist who aspirated cells from solid and cystic tumours. The foundations of gastric cytology (Rosenbach, 1882) and the cytology of effusions and CSF (Bennett, 1849) were also laid at this time. The careful scientific drawings produced during this period show a remarkable similarity to the photographic illustrations that are found in cytological textbooks and published reports today.

Some of these authors made astute observations about the nature of cancer cells. They accurately described their abnormal nuclei, structural pleomorphism and altered nucleo-cytoplasmic ratio. There was much confusion about the origin of the cancer cell as the mechanisms of cell division and replication were not fully understood. One of the key contributors to the development of cytology during this period was Johannes Müller, a German physiologist who was one of the first to bring some order into the chaos of descriptive pathology in the nineteenth century. In 1838, 20 years before Virchow's treatise on cellular pathology, Muller published a monograph *On the Nature and Structural Characteristics of Cancer and those Morbid Growths that may be Confounded with it*. This book described the microscopic criteria that distinguished malignant from benign tumours and the text contains some illuminating statements that are still true today (Grunze and Spriggs, 1980). Muller argued that all malignant tumours are closely related physiologically and supported his argument by demonstrating that tumour cells exfoliate more readily than normal cells. He classified tumours into two groups—those with abundant connective tissue and those with scanty connective tissue—which roughly correspond to the classification of sarcomas and carcinomas that we use today.

Among the medical literature published at this time are case reports describing how the study of cells could be applied to clinical diagnosis. Sir Julius Vogel (1843) is credited with being the first to describe the technique which a century later became known as exfoliative cytology. A patient was referred to him with a large tumour set deeply behind the angle of the mandible. A fistula led from the tumour to the skin below the ear. Vogel examined the fluid issuing from the fistula under the microscope and reported the presence of malignant cells. As he predicted in his report the patient died shortly from her tumour.

Another pioneer of exfoliative cytology was Professor Lionel S. Beale who drew attention to the value of cytological diagnosis in clinical practice when he published a detailed report of a case of pharyngeal cancer diagnosed by sputum cytology in 1860. The first clinical diagnosis of bladder cancer by the cytological examination of tumour cells in urinary sediment is attributed to Dr Lambl of Prague who reported eight cases in 1856. Lambl's paper is particularly interesting as he defines the cytological examination of urine as a bedside test and describes the problems of diagnosis which occur when the urine is infected.

Unfortunately the enthusiasm for exfoliative cytology which was generated by Beale and Lambl and others at the turn of the century was not sustained in the years that followed. The reasons are two-fold. Firstly, as the techniques of histological section and staining were perfected and the diagnostic skills of the histopathologists improved, clinicians became aware of the limitations of cytological diagnosis. Tumour typing of exfoliated cells was not as accurate as histological studies of biopsy material and the presence of inflammation often rendered a sample unsuitable for diagnosis just as it does today. Even the speed and simplicity of cytological processing which was its major advantage was soon to be matched by histology with the introduction of frozen sections.

In the opening years of the twentieth century only a few pathologists advocated the use of exfoliative cytology in clinical practice and many even denied it a place in their diagnostic methods (Bamforth and Osborn, 1958). Among the few who pursued the techniques of clinical cytology at this time were Professor Dudgeon, pathologist at St Thomas's Hospital, London whose monograph on sputum cytology is now a classic (Dudgeon and Wrigley, 1935); Dr U. Quensel, University of Uppsala, Sweden who devised a method of supravital staining which was widely used for the diagnosis of effusions (Quensel, 1928); and Drs Martin and Ellis (1930, 1934) at the Memorial Hospital, New York who pioneered the techniques of fine needle aspiration.

Despite these efforts, many years were to pass before exfoliative cytology was accepted as a valid method of scientific investigation. Much of the credit for the emergence of cytology as a recognized specialty must surely go to Dr George Papanicolaou who, through teaching and research, convinced his medical colleagues of the clinical value of cytology with such success that the technique is now used in clinics and hospitals throughout the world as a first line of investigation of the patient suspected of having malignant disease.

George N. Papanicolaou (1883–1962)

George Papanicolaou was born in Kymi, the capital of Euboea, the largest of the many islands that surround the Greek mainland. His father Nicholas Papanicolaou was a general medical practitioner who was keen for his son to follow in his footsteps and become a doctor. George Papanicolaou duly received a medical education in the University of Athens and graduated at the age of 21 as an 'A' grade student. After a short period of military service

George Papanicolaou had to make up his mind about his future career. He was keen to further his education by study abroad whereas his father wanted his son to enter private practice in Kymi. George refused to do so, causing a rift in the family that never really healed. Instead he pursued a career in biological research, first in Munich where he studied under Professor Richard Hertwig who was renowned for his work on fertilization of the ovum, and later in Monaco at the Oceanic Institute. He returned to Greece in 1912 but the opportunity for advancing his career in this war-torn country was poor and in October 1913 he and his wife departed for the USA, full of hope and expectation for his future in research.

One of his first research appointments in America was in the Department of Anatomy at Cornell Medical College, New York as an assistant to Dr Stockard, the departmental Chairman. Here he continued the work he had started in Munich on fertilization and sex determination in the guinea pig and it was at this time that he made the important discovery that the oestrous cycle in guinea pigs could be monitored by vaginal smears. Subsequent studies on other laboratory mammals revealed that their sexual cycles could be monitored in a similar way.

Papanicolaou recognized the potential clinical value of this approach in the human female and soon obtained financial help and clinical support to extend his research to a study of vaginal aspirates from women. Although the aim of the study was to relate ovarian hormone production to the cellular changes in the smear, it was not long before Papanicolaou encountered his first case of cervical carcinoma detected by this technique. He reported this finding in 1928 at a meeting in Battle Creek, Michigan where he presented a paper entitled 'New cancer diagnosis', but his report was received with little enthusiasm even though he pointed out to the clinicians in the audience that the test would be valuable in detecting and treating cervical cancer at its earliest stages when the lesions would be most amenable to treatment. Unfortunately the prevailing opinion of the time was that 'since the uterus was accessible to diagnostic exploration by biopsy cytological examination of vaginal smears appeared to be superfluous', and Papanicolaou was discouraged from pursuing this line of research any further at this time.

In the years that followed Papanicolaou consolidated his work on endocrine cytology. He made a meticulous study of vaginal cytology in the human female and reported on the smear patterns in pregnancy and the menopause. This approach proved to be of immense value to his contemporaries who were elucidating the ovarian hormones and their role in human reproduction.

The direction of his research changed dramatically in 1939 with the appointment of a new Chairman to the Department of Anatomy at Cornell University. The new Chairman, Dr Joseph Hinsey, had been one of the few people to realize the potential of Papanicolaou's original report on cancer diagnosis by the vaginal smear and advised him to return to this work and explore its clinical potential more thoroughly. He arranged for Papanicolaou to collaborate with another member of his staff at Cornell, Dr Herbert Traut, a pathologist with a special interest in gynaecology, and such was their enthusiasm for their work that within a year they were to report their findings to the New York Obstetrical Society in a paper entitled 'The diagnostic value of vaginal smears in carcinoma of the uterus' (Papanicolaou and Traut, 1941). This time the report was favourably received and Papanicolaou and Traut proceeded to prepare their now classic monograph *Diagnosis of Uterine Cancer by the Vaginal Smear* which was published by the Commonwealth Fund in 1943.

Confirmation of the findings of Papanicolaou and Traut by other scientists led the American Cancer Society to take up the challenge of cervical screening. They accepted that the vaginal smear was unique in that it was capable of detecting uterine cancer before it became visible to the naked eye and before it produced the danger signals of cancer, and undertook to popularize the vaginal smear test and to educate the public and the medical profession of its value. In this they had no easy task especially as many pathologists were reluctant to accept this new technique. Papanicolaou was encouraged to set up training courses at Cornell, the first of which was held in 1947 and attracted 70 students, 45 of whom were pathologists from various states in the nation. The American Cancer Society also sponsored the first National Cytology Conference in Boston in 1948 which was attended by 90 persons including gynaecologists and pathologists from Europe and the UK.

Many came away convinced of the value of the vaginal smear as a method of detecting preinvasive cancer and the first cervical cancer screening clinics were established in the USA, UK and Europe about this time. One of the first in the UK was founded by the gynaecologist Mr Stanley Way in Gateshead in 1948; in North America the first well woman clinic to be opened was in Massachusetts in 1945 (McSweeney and McKay, 1948).

Papanicolaou continued to explore the potential of cytology as a method of diagnosing other gynaecological cancers and applied the wet smear technique that he had developed for vaginal cytology to the investigation of urine, sputum, gastric washings, ascites, prostatic secretions, nipple discharges and spinal fluid. In 1954 the *Atlas of Exfoliative Cytology* was published which contained the remarkable drawings of Hashime Murayama who did the original illustrations for Papanicolaou's earlier monograph on the diagnosis of cervical cancer by the vaginal smear. Papanicolaou was acknowledged to be the founder of a new medical specialty and many honours were heaped on him, one of the most enduring being the establishment of the Papanicolaou Cancer Research Institute of Miami of which he became director in 1961. He died suddenly in February 1962, secure in the knowledge that the techniques he had developed were now accepted worldwide as a valid method of early cancer diagnosis.

After Papanicolaou

Papanicolaou's achievements for exfoliative cytology were many. He convinced pathologists and clinicians alike that exfoliative cytology was a quick, safe, simple and efficient way of investigating a patient with a clinically suspicious lesion without resorting to surgery. He also showed that exfoliative cytology was a suitable non-invasive tool for screening apparently healthy individuals for precancerous changes. He perfected the staining protocol which is associated with his name so that the morphology of the cell was preserved as perfectly as possible and he described and classified a multitude of cellular changes found in benign and malignant disease.

Because of the secure foundations which Papanicolaou provided for the new science of exfoliative cytology, the way was open for changes in cytological practice. Some of these changes occurred in Papanicolaou's lifetime while others occurred 10 or 20 years after his death. For example, it soon became apparent that a cervical scrape was more efficient than a vaginal aspirate for detecting cervical cancer and a spatula was designed for this purpose (Ayre, 1944). It also became evident that the scope for cytological diagnosis was not limited to the recognition of tumour cells or hormonal changes but could be extended to the diagnosis of viruses, e.g. herpes and papovaviruses and other specific infectious agents, e.g. *Trichomonas vaginalis* and *Actinomyces* sp. As our understanding of the physiology and pathobiology of the cervix improved, Papanicolaou's original system of classifying vaginal smears into five categories was challenged for its lack of precision and was abandoned in many centres. The tendency today is to construct a narrative report using a series of clearly defined terms to describe the cytological findings which are then related to the underlying histology. This approach contrasts sharply with the rather ill-defined terminology used by Papanicolaou and his contemporaries to describe the cellular changes they observed.

However, perhaps the most remarkable change of all that has occurred in cytology since Papanicolaou's demise is the resurgence of interest in fine needle aspiration (FNA) as a method of cytological examination.

Fine needle aspiration

Early records show that fine needle aspiration was used as a tool for the diagnosis of palpable tumours in a variety of body sites in the latter half of the nineteenth century. In 1853 Sir James Paget, an eminent surgeon at St Bartholomew's Hospital, London reported the value of this approach for the diagnosis of breast lumps (Webb, 1974). Several years later Menetrier (1886) and Kronig (1887) recorded their attempts to use FNA for the diagnosis of lung tumours. However in the years that followed it would seem that the technique was in regular use in only a few centres. One of these centres was the Memorial Hospital for Cancer and Allied Diseases, New York where, in 1925, two surgeons Martin and Coley began experimenting with the technique for the preoperative diagnosis of palpable tumours (Koss, Woyke and

Olszewski, 1984). Within a short time they had gained experience of over 1000 cases including tumours of breast, lymph node, lung and bone and reported their evaluation of the technique in a series of major papers (Coley, Sharp and Ellis, 1931; Martin and Ellis, 1930, 1934; Martin and Stewart, 1936). They demonstrated that FNA was a safe and accurate method of diagnosis which in many cases obviated the need for incisional biopsy.

Although Martin and Coley had few disciples in America at this time, FNA was put on a firm footing in Sweden by Nils Soderstrom and Josef Zajicek. These pioneers wrote extensively of their experience with FNA (Soderstrom, 1966; Zajicek, 1974, 1979), and with their colleagues at the Karolinska hospital were responsible for training many hundreds of cytologists and surgeons from all nations in the technique. It is largely as a result of their efforts that FNA is so widely used today.

The uptake of FNA in the last decade has been accelerated by the recent improvement in roentgenological and ultrasound imaging techniques. Once it was clear that internal organs could be sampled safely and swiftly under ultrasound or radiological control, radiologists and ultrasonographers joined the ranks of the practitioners of FNA and started to use the technique for the diagnosis of tumours at almost any body site.

A look into the future

Although there have been important advances in cytology in the last 30 years there is no room for complacency.

In the first place there is a need to develop techniques which will improve the accuracy of cytodiagnosis. This is still an area of concern among those who provide a cervical screening service as the false negative rate of cervical cytology is estimated to lie between 7 and 17%.

Improved performance could be achieved in several ways; for example, by better methods of sample collection, more efficient quality control and the introduction of structured training programmes which are lacking in many countries. Diagnostic accuracy could also be improved by the development of new methods of staining and preparing specimens which will focus attention on tumour cells. This last goal could be achieved by developing antibodies or probes specific for dyskaryotic cells and is a research exercise well worth pursuing. Not only would this approach facilitate distinction between benign and malignant cells at the level of the light microscope, but it would also provide scope for a truly effective automated screening system.

Another area where the accuracy of cytological techniques could be improved is in the field of tumour typing. Many cytologists base their diagnosis of fine needle aspirates and serous effusions entirely on the cell morphology afforded by a Papanicolaou-stained smear, so that the opportunity for precise tumour typing is limited. More use could be made of special staining techniques in routine cytology including immunocytochemical staining methods, an aspect that is discussed more extensively in Chapter 5. The recent application of cytomorphometry to cytological specimens indicates that this approach may also prove to be of value for tumour grading where there is a large element of subjectivity in the cytological interpretation of the specimen.

Finally, it is not always appreciated that cytology has an important role to play in basic research. It provides research scientists with unrivalled opportunities to repeatedly sample tumours at any site *in vivo* over a period of time without interfering with the biological growth pattern of the tumour and with minimal inconvenience to the patient. Thus it provides exceptional opportunities to monitor tumour response to therapy, drug uptake and tumour targetting. There is also scope for using cytology to advance our understanding of the basic mechanisms of the inflammatory response and for expanding our knowledge of the pathobiology of viruses and other infectious agents. The potential of cytology in pure and applied research has not yet been fully realized and it may be in this area that the future of cytology will lie.

References

AYRE, J. E. (1944). A simple office test for uterine cancer diagnosis. *Can. Med. Ass. J.*, **51**, 17

BAMFORTH, J. and OSBORN, G. R. (1958). Diagnosis from cells. *J. Clin. Pathol.*, **11**, 473

BEALE, L.S. (1860). Results of the chemical and microscopical examination of solid organs and secretions. Examination of sputum from a case of cancer of the pharynx and the adjacent parts. *Arch. Med. Lond.*, **2**, 44

BENNETT, J. H. (1849). *On Cancerous and Cancroid Growths* Edinburgh: Southerland and Knox

CARMICHAEL, D. E. (1973). *The Papanicolaou Smear; Life of George N. Papanicolaou.* Springfield, Illinois: Charles C. Thomas

COLEY, B. L., SHARP, G. S. and ELLIS, E. B. (1931). Diagnosis of bone tumours by aspiration. *Am. J. Surg. (NS)*, **13**, 215

DONALDSON, F. (1853). The practical application of the microscope to the diagnosis of cancer. *Am. J. Med. Sci.*, **25**, 43

DUDGEON, L. S. and WRIGLEY, C. H. (1935). On the demonstration of particles of malignant growth in the sputum by means of the wet film method. *J. Laryngol. Otol.*, **50**, 752

GRUNZE, H. and SPRIGGS, A. I. (1980). *History of Clinical Cytology. A Selection of Documents.* Darmstadt: Verlag Ernst Giebeler

KOSS, L. G., WOYKE, S. and OLSZEWSKI, W. (1984). *Aspiration Biopsy, Cytologic Interpretation and Histologic Bases.* New York and Tokyo: Igaken Shoin

KRONIG, G. (1887). Diagnostischer Beitrag zur Hertz-Und Lungenpathologie. *Berlin Klin. Wschr.*, **24**, 961

LAMBL (1856). Uber Harnblasenkrebs. *Prag. Viertelj. Schr. Heilkunde*, **49**, 1

LEBERT, H. (1845). *Physiologie, Pathologie ou Recherches Clinique, Experimentales et Microscopiques.* Paris: J. B. Baillière

LISTER, J. J. (1829). On some properties in achromatic object glasses applicable to the improvement of the microscope. *Phil. Trans.*, **130**, 187

MARTIN, H. E. and ELLIS, E. B. (1930). Biopsy by needle puncture and aspiration. *Ann. Surg.*, **92**, 169

MARTIN, H. E. and ELLIS, E. B. (1934). Aspiration biopsy. *Surg. Gynecol. Obstet.*, **59**, 578

MARTIN, H. E. and STEWART, F. W. (1936). The advantages and limitations of aspiration biopsy. *Am. J. Roentgenol.*, **35**, 245

MCSWEENEY, D. J. and MCKAY, D. (1948). Uterine cancer: its detection by simple screening methods. *N. Engl. J. Med.*, **238**, 867–870

MENETRIER, P. (1886). Cancer primitif du poumon. *Bull. Soc. Anat. Paris*, **11**, 643

MULLER, J. (1840). *On the Nature and Structural Characteristics of Cancer and those Morbid Growths which may be Confounded with it* (translated by C. West). London: Sherwood, Gilbert and Piper

PAPANICOLAOU, G. N. (1928). New cancer diagnosis. In: *Proceedings of 3rd Race Betterment Conference, Battle Creek, Michigan,* Race Betterment Fdn., p. 528

PAPANICOLAOU, G. N. (1954). *Atlas of Exfoliative Cytology.* Cambridge, Massachusetts: Harvard University Press

PAPANICOLAOU, G. N. and TRAUT, H. F. (1941). The diagnostic value of vaginal smears in carcinoma of the uterus. *Am. J. Obstet. Gynecol.*, **42**, 193

PAPANICOLAOU, G. N. and TRAUT, H. F. (1943). *Diagnosis of Uterine Cancer by the Vaginal Smear.* New York: Commonwealth Fund

QUENSEL, U. (1928). Zur Frage der Zytodiagnostik der Ergusse seroser Hohlen. *Acta Med. Scand.*, **68**, 427

ROSENBACH, O. (1882). Uber die Arwesenheit von Geschwulstpartiken in dem urch die Magenpumpe entleerten Mageninhalte bei Carcinoma Ventriculi. *Deutsch Med. Wschr.*, **8**, 452

SCHWANN, T. H. (1839). *The Microscopical Researches into the Accordance in the Structure and Growth of Animals and Plants.* Berlin: G. D. Reiner

SODERSTROM, N. (1966). *Fine Needle Aspiration Biopsy.* New York: Grune and Stratton

VOGEL, J. (1843). *The Pathology of the Human Body* (translated with additions by G. E. Day). London: H. Baillière

WEBB, A. J. (1974). Through a glass darkly. (The development of needle aspiration biopsy). *Bristol Med. Chir. J.*, **89**, 59–68

ZAJICEK, J. (1974). *Aspiration Biopsy Cytology Part I: Cytology of Supradiaphragmatic Organs.* Monographs in Clinical Cytology, Vol. 4. Basel and New York: S. Karger

ZAJICEK, J. (1979). *Aspiration Biopsy Cytology Part II: Cytology of Infradiaphragmatic Organs.* Monographs in Clinical Cytology, Vol. 7. Basel and New York: S. Karger

2

The cell

A. D. Hoyes

Cells are the fundamental structural units of all living organisms. They exhibit all of the functional activities of such structures including absorption, synthesis, respiration and excretion. In multicellular organisms, distinct populations of cells become specialized to perform particular functions. In such cases, one or more of the basic functions of the cells may become accentuated or modified, as in nerve and muscle cells, in which the property of response to a stimulus is developed in high degree to provide, on the one hand, for transmission of information and, on the other, for organized contractile responses. Cells which assume highly specialized functions of this type frequently lose one of the other fundamental attributes of the cells—the ability to divide.

Specialization of cells in respect of their functional activity is often accompanied by the adoption by the cells of particular shapes and by the development of particular features, whether at the surface or in the interior. There are, nevertheless, a number of features which are common to all cells. Thus, with the exception of the red blood cell or erythrocyte and platelets, which are in fact nothing more than small segments of cytoplasm, the cells of the body all possess at least one *nucleus*. The nucleus represents the site of storage of DNA, the protein which provides, through messages relayed to the rest of the cell, overall control of the functional activity of the cell. The *cytoplasm* is a semi-fluid gel within which are embedded a variety of discrete structures collectively known as *organelles*. The cell is separated from the exterior by the cell or *plasma membrane*. It is through this membrane that the materials required by or formed during the activity of the cell are taken up or released into the surrounding tissue.

The plasma membrane

The plasma membrane is normally approximately 7–8 nm in diameter. In some cells it is much thicker, but rarely exceeds 15 nm in diameter. This is well below the resolving power of the light microscope which is, at best, in the order of 180 nm. For this reason, what is often identified as the plasma membrane in light microscopical preparations must be either extended areas of membrane viewed obliquely or aggregations of stain on the membrane and its associated structures. A particular example of this kind of phenomenon is the neurilemma. This is an apparently membranous structure which can be seen in light microscopical preparations around the outside of myelinated nerve fibres. Examination of such fibres has shown that it cannot be a single structure and suggests that it may represent aggregates of stain on structures such as the plasma membrane and the associated basement membrane.

It was shown in the early years of the present century that the plasma membrane consists of two layers of lipid. In each layer the lipid molecules are arranged with their polar or hydrophobic bonds facing inwards and their non-polar or hydrophilic bonds facing outwards. This highly regular arrangement is responsible for the birefringence shown by membranes and, in particular, membrane derivatives such as myelin in polarized light. More recently, it was recognized that substantial amounts of protein are associated with plasma membranes. In the model proposed by Davson and Danielli (1952) the protein was considered to be arranged as a layer on the inner and outer surfaces of the membrane. In the more recent model proposed by Singer and Nicholson (1972) the protein molecules are considered to lie embedded on the inner or outer aspects of the lipid layers or to traverse the layers (*Figure 2.1*). Because the protein molecules are thought to be highly mobile this model is known as the fluid mosaic model.

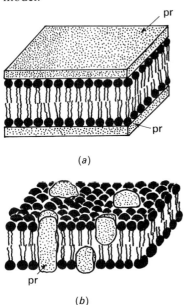

Figure 2.1 Diagrams to show the essential features of the plasma membrane models proposed by (*a*) Davson and Danielli (1952) and (*b*) Singer and Nicholson (1972). The non-polar regions of the lipid molecules which, in both models, form two layers are represented by black circles and the polar regions by lines. In the unit membrane model of Davson and Danielli (1952) protein forms a layer on both surfaces of the lipid layers (pr). In the fluid mosaic model of Singer and Nicholson (1972) the proteins integral to the membrane are embedded in the inner or outer surfaces of the membrane or extend through the membrane (pr).

The membrane protein molecules not only include enzymes but also probably contain the receptors for chemicals which are bound to the membrane. These chemicals include substances released from nerves, the so-called neurohumoral transmitters, and hormones which act upon cells to produce particular responses such as the release of secretory products. The protein molecules which cross the membrane may also contain the pores through which many substances are considered to be transported across the membrane. Transport across membranes is an important and very widely studied phenomenon. It may be active or passive.

Active transport is envisaged as involving some sort of carrier mechanism which transports the substances from one surface of the membrane to the other. It requires the utilization of energy and can be inhibited by substances such as ouabain which block the metabolic processes providing the energy. Active transport can also occur against a concentration gradient. It is through the existence of active transport mechanisms for sodium and, to a lesser extent, potassium across plasma membranes that the concentrations of sodium are maintained at a much lower level in the cytoplasm of the cell than in the surrounding tissue, and potassium occurs in much higher concentrations within than outside the cell. These mechanisms also play a major part in maintaining the difference in electrical potential which exists across the plasma membrane between the interior and the exterior of the cell.

Passive transport occurs by diffusion across the membrane and with small molecules such as water molecules the amount and direction are governed by the osmotic and hydrostatic gradients across the membrane. Another form of transport known as facilitated diffusion has also been recognized. This has the features of active transport in that it involves the use of carriers, but is limited in degree because the carriers are considered to be readily saturated.

All cells possess on their outer surfaces a layer of carbohydrate. This can be stained, for example with silver, and under the electron microscope it appears as a thin continuous layer of material. Current evidence suggests that the surface coat consists of the carbohydrate components of the membrane proteins. The basement membrane, or basal lamina, which separates the basal surface of most epithelial and many other types of cell from the surrounding tissue, is a much thicker layer of material. It also contains carbohydrate and can be stained with carbohydrate stains such as PAS. Although under the electron microscope the basement membrane appears to be amorphous, it has a filamentous component. While there is evidence that the membrane is formed largely by the cells with which it is associated, it is probable that it also contains material incorporated from the surrounding tissue. Most basement membranes are highly porous and allow the passage of all but very large molecules. In some cases, notably the renal corpuscle, the basement membrane is highly specialized. In the renal corpuscle the membrane is very thick and normally only allows passage of molecules up to the size of the plasma protein, albumin.

Figure 2.2 Microvilli on the surface of an absorptive cell of the ileum. The microvilli contain fine filaments which extend into the underlying cytoplasm (f). Between the cells the various components of the terminal bar including a desmosome (d) can be seen. Stain: uranyl acetate/lead citrate. Magnification × 57 500, reduced to 70% in reproduction.

It has recently been shown that on the inner or cytoplasmic surface of the plasma membrane of many cell types there is an accumulation of very fine filaments. There is now good evidence that these filaments consist of the contractile protein, actin. The filaments are apparently attached to the membrane and it is probable that they play a major part in the movements of the membrane. These movements contribute, not only to the formation of the cellular processes which occur in motile cells such as macrophages, but also to the processes of exocytosis and endocytosis.

In most cells the plasma membrane shows a number of special features. These include microvilli and other similar processes, cilia and flagella, and a number of different types of intercellular junctions. The plasma membrane also plays an integral part in the processes of exocytosis and endocytosis, processes through which many materials, including large molecules and particles, are taken into or eliminated from the cell.

Microvilli

These are elongated folds of the plasma membrane. They are highly developed on the surfaces of the epithelial cells of the gastrointestinal tract (*Figure 2.2*) where their regular arrangement and close packing gives rise to the structure classified as the brush border. In this situation the microvilli are approximately cylindrical and contain an orderly series of regularly arranged longitudinal filaments. At the base of the microvilli the filaments terminate in a transversely orientated band of similar filaments. This band of filaments comprises the terminal web, a structure which can be recognized in intestinal epithelial cells with the light microscope. The filaments in the cytoplasmic core of the microvilli also consist of actin and are probably responsible for the movements of the microvilli. Villous folds which resemble the microvilli of the intestinal epithelial cells are found on the surfaces of many other types of cell. Such processes are also normally called microvilli, but they do not always show the highly developed internal structure seen in the microvilli of the intestinal epithelial cell. They typically occur on surfaces of cells bathed in an aqueous medium and on these and other surfaces of cells involved in the transport of materials. They are, however, not normally found on the basal surfaces of epithelial cells. In such situations the plasma membrane may be

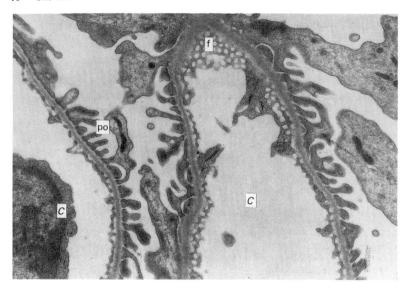

Figure 2.3 Processes of podocytes (po) applied to the basement membrane of the capillaries (C) of a renal glomerulus. Note the fenestrations in the capillary endothelium, in particular where the endothelial cells are cut obliquely (f). Stain: uranyl acetate/lead citrate. Magnification ×26 000, reduced to 70% in reproduction.

thrown into a series of interdigitating processes. Such processes are found in the epithelial cells of the kidney tubules. A somewhat unusual instance of the formation of this kind of process is seen in the podocytes of the renal corpuscle, the foot processes of which form a layer on the outer surface of the glomerular basement membrane (*Figure 2.3*).

Cilia

These are found on the surfaces of epithelia in which there is movement of materials associated with the surface, such as in the respiratory tract (*Figure 2.4*) where they are responsible for the movement of the layer of mucus secreted by the mucous cells and glands either upwards towards the larynx or, in the nose, backwards towards the pharynx. They are elongated, hair-like structures and differ from microvilli and from stereocilia, which are really nothing more than long microvillous-like processes, in that they contain microtubules. The microtubules are arranged in a regular manner and consist of nine outer pairs and two central tubules. At the base of the cilium the microtubules are attached to a structure known as the basal body. This is in many ways very similar to a centriole, and is the structure from which the microtubules are thought to be formed during development.

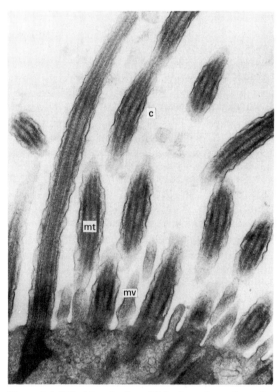

Figure 2.4 Cilia (c) and microvilli (mv) on the surface of a cell of the epithelium of the trachea. Note the presence in the cilia of microtubules (mt). Stain: uranyl acetate/lead citrate. Magnification ×117 000, reduced to 50% in reproduction.

Cilia move by beating and their movement is highly organized so that waves of beats pass across the surface of the epithelium. The beats of an individual cilium can be resolved into an active effective stroke and a passive recovery stroke. Quite how the microtubules and their associated proteins are involved in producing these movements is not yet clear. In some structures there are cilia which lack the central pair of microtubules. These cilia are non-motile and are called kinocilia.

Flagella

Flagella are essentially similar structures to cilia, but they are much longer. They occur in cells such as spermatocytes in which they are responsible, as in many unicellular organisms, for the movement of the cells.

Cell junctions

These are structures through which one cell is connected to another. The terminal bar, to which the terminal web of the intestinal epithelial cell is attached, represents a complex series of different types of junction (*Figure 2.2*). The principal types of junction which are now recognized between cells are the *tight junction*, the *gap junction* and the *desmosome*. Analysis of the membranes at the sites of junctions using the technique of freeze fracture have shown that there are special features of the membrane in these regions.

The tight junction often forms one or a series of rings around the cells of epithelia near their surfaces. At this form of junction the membranes of apposing cells fuse to eliminate the gap between the cells. Tight junctions therefore form a seal between the cells. Although the presence of such a seal would be expected to prevent passage of materials through the intercellular spaces, it is now recognized that some tight junctions, classified as leaky tight junctions, do allow passage of some materials between the cells at a slow rate.

Gap junctions occur on small areas of the membranes of apposing cells. At these sites the membranes become closely approximated. They do not fuse as in the tight junction and a gap of 2–3 nm, about one-tenth of the normal gap between cells, remains between them. This gap is sufficient to allow fairly free passage of materials across the junction. It has recently been recognized that at the gap junction there are pores which pass from the cytoplasm of one cell to the cytoplasm of the other. These pores are very small and allow passage only of small molecules between the cytoplasm of the cells. The presence of the pores can be demonstrated with the light microscope by injecting dyes such as fluorescein into one cell and tracing its passage into neighbouring cells. Although such experiments indicate that the gap junction permits some form of communication between cells, it is probable that the principal function of gap junctions is to transmit changes in the electrical potential of the membrane of one cell to that of another. Gap junctions are therefore considered to play an important part in the electrical coupling which occurs between cardiac muscle cells at the intercalated disc where the junctions are numerous, and in other cells such as smooth muscle cells. In such cells the junctions are frequently called *nexuses*.

Desmosomes also normally occupy only small areas of the membranes of apposing cells. An exception is the terminal bar of intestinal and other epithelial cells in which the desmosome forms a band around the cell. At the desmosomes the membranes remain separated by the normal intercellular interval of 20 nm. Sometimes a dense line can be demonstrated in the gap between the membranes. The gap contains glycoprotein through which the membranes are firmly attached to one another. On the inner surface of the membranes there is a layer of dense material. In cells such as those of the epidermis, in which desmosomes are particularly numerous, the structural filaments in the cytoplasm loop into this material. In the epidermis in particular the basal surface of the basal cells shows a similar specialization, even though it approximates only to the basement membrane. Such structures are known as hemidesmosomes. As is suggested by their frequency in the highly mobile epithelium which comprises the epidermis, the function of desmosomes is to provide for intercellular attachment.

Endocytosis and exocytosis

These are processes which involve the formation of various forms of vesicle and vacuole from the plasma membrane, and their fusion with and incorporation into the membrane. In endocytosis an area of the plasma membrane folds into the cytoplasm and then becomes detached from the rest of the membrane to pass into the cytoplasm as a membrane-bounded structure. Two forms of endocytosis are recognized. These are *pinocytosis*, sometimes also called micropinocytosis, and *phagocytosis* which is sometimes defined as macropinocytosis. Micropinocytosis results in the formation of micropinocytotic vesicles which are usually not more than 100 nm in diameter. Two kinds of micropinocytotic vesicle exist. These are coated and smooth vesicles. The membrane of coated vesicles is invested on its cytoplasmic surface by a series of radiating spikes. Because coated vesicles are formed in large numbers at the surface of oocytes, which incorporate substantial amounts of protein from the exterior, it is thought that the primary function of such vesicles is in the transport of protein across the membrane. Other functions of coated vesicles have been proposed, and there are situations in which the endocytosis of protein does not involve these vesicles. Smooth vesicles have no coat of spikes. They are especially numerous in the endothelial cells of blood vessels where it is thought that they are involved in the transport of materials from one surface of the cell to the other. Some support for this view is provided by the fact that the vesicles are rarely seen in the endothelium of capillary vessels in the brain in which there is a distinct barrier between the blood in the vessels and the brain tissue.

Macropinocytosis or phagocytosis is seen in many types of cell, but notably in the macrophage. Macrophages ingest particulate material such as the carbon particles which were once used to demonstrate their distribution in the body, as well as debris derived from the breakdown or activity of other cells and structures such as bacteria.

Exocytosis is essentially the reverse process to endocytosis. It occurs with micropinocytotic vesicles and larger vacuolar structures such as secretory vacuoles released from exocrine secretory cells and the residual bodies formed as a result of the activity of lysosomes. In exocytosis the membrane of the vacuole fuses with the plasma membrane and the contents are expelled to the exterior of the cell. The process involves the incorporation of the membrane into the plasma membrane, and much attention has recently been given to the process by which this additional membrane is retrieved by the cell. In neurons it has been postulated that the retrieval process involves coated vesicle formation, but it is probable that other processes are involved as well. The mechanism by which exocytotic vesicles move to the surface has not been worked out completely, but microtubules seem to play some part in the process. The actual release mechanism probably involves the contractile proteins associated with the membrane and, in secretory cells at least, may be initiated by changes in the electrical potential of the membrane, similar to those known to exist in muscle cells. Another aspect of exocytosis is that it occurs much more rapidly than endocytosis, and the actual process of release of the exocytotic vesicle or vacuole is rarely seen.

The cytoplasm

The cytoplasm is the part of the cell which surrounds the nucleus and which is bounded at the surface by the plasma membrane. It consists of a sometimes gel-like and, in the sections used for light and normal forms of electron microscopy, apparently structureless matrix and the organelles. With high voltage electron microscopy the matrix or hyaloplasm, as it is sometimes called, can be seen to consist of a network of material which is largely protein in nature, but may include other components such as fat micelles.

The cytoplasm is sometimes divided into several zones. These are the zone immediately surrounding the nucleus, the cytocentrum, the endoplasm which contains the cell organelles and in which most of the movement of the organelles and other forms of movement such as streaming occur, and the ectoplasm. The ectoplasm is the thin zone beneath the plasma membrane. It contains the actin filaments which are involved in movements of the plasma membrane and in functions such as endocytosis.

Different types of cell contain different amounts of cytoplasm. Some cells such as small lymphocytes possess very little cytoplasm. In most other types of cell the cytoplasm is much more abundant. The volume of the cytoplasm is usually expressed as a proportion of that of the nucleus. This is the nucleo-cytoplasmic ratio.

The cell organelles

The cell organelles are the particulate components of the cytoplasm. Each type of organelle performs particular functions and it is through the integration of these functions that the normal activity of the cell is expressed. Except in cells such as the red cell, most of the organelles are present if only in small numbers or amounts. Some organelles such as centrioles and glycogen, are, however, not infrequently absent.

The amount and number of individual organelles varies widely in different types of cell. This is primarily a reflection of the function of the cells. Thus, cells which are involved in the manufacture of protein for secretion typically contain large amounts of granular endoplasmic reticulum and numerous secretory granules and cells with high rates of metabolic activity such as neurons contain numerous mitochondria.

Mitochondria

The energy required for cellular metabolism is provided by oxidation of substances derived from the food. Most cells use glucose for this process but some cells also use fatty acids. The oxidation of glucose to provide energy takes place in two stages. The first stage known as glycolysis occurs in the cytoplasm. The second stage takes place in the mitochondria. It results in the oxidation of pyruvate produced in the first stage to carbon dioxide and water. Most of the energy required for cell metabolism is produced at this stage. This energy is stored in the form of ATP.

Mitochondria are rod or sausage-shaped structures. They are normally 1–2 μm long and 0.5 μm in diameter. Because their diameter exceeds the resolving power of the light microscope they cannot be visualized by light microscopy. They can be demonstrated in living cells by phase contrast microscopy or after staining with vital dyes such as Janus Green B. Because they contain large amounts of phospholipid, they can be demonstrated in fixed cells with iron haematoxylin. They can also be demonstrated with histochemical techniques for oxidative enzymes involved in the Krebs cycle, the chain of reactions which occurs during the second or mitochondrial phase of glucose oxidation.

With the electron microscope (*Figures 2.5* and *2.6*) mitochondria are seen to possess two unit membranes similar to the plasma membrane. Of these, the inner membrane is folded into the interior of the mitochondrion in a number of places to form a series of structures known as

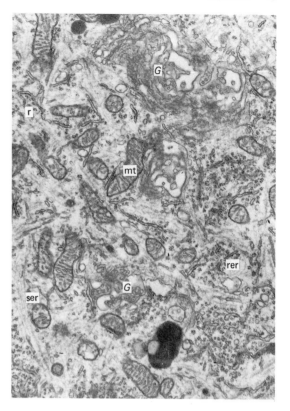

Figure 2.5 Cytoplasm of a neuronal cell body, showing mitochondria (mt), free ribosomes mostly in the form of polyribosomes (r), sacs of granular endoplasmic reticulum only partly covered by ribosomes (rer), smooth endoplasmic reticulum (ser) and an obliquely cut Golgi apparatus (G). Stain: uranyl acetate/lead citrate. Magnification × 51 500, reduced to 50% in reproduction.

cristae. The gap between the inner and outer membranes is usually small, but can be greatly increased by adding ADP to the tissue prior to fixation. When mitochondria are subjected to osmotic shock, numerous small particles, the elementary particles, can be seen attached to the inner surface of the inner membrane. The space enclosed by the inner membrane, the inner chamber, contains the mitochondrial matrix. This contains proteins, small amounts of DNA and, in many cells, electron-dense granules (*Figure 2.6*). The DNA may be involved in the synthesis of some of the mitochondrial enzymes, and it has been argued that its presence in these structures is related to the fact that they represent independent structures living in symbiosis with the rest of the cell. The dense granules in the matrix have been regarded by some as being involved in the metabolism of calcium. Although mitochondria play an im-

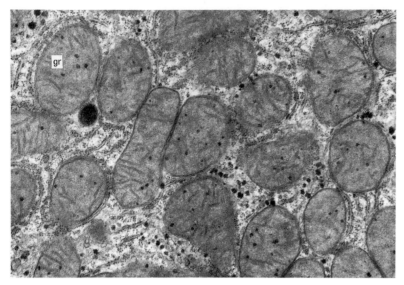

Figure 2.6 Mitochondria in a hepatocyte. Note the presence within the mitochondria of numerous dense granules (gr). Stain: uranyl acetate/lead citrate. Magnification × 29 000, reduced to 70% in reproduction.

portant part in the storage of intracellular calcium and it is probable that pathological calcification of cells commences in the matrix, there is biochemical and some structural evidence to suggest that the calcium is stored mainly in the outer chamber, the gap between the inner and outer membranes.

The number and form of the mitochondria varies widely in different cell types. They are particularly numerous in cells such as those of the renal tubules in which they are also elongated and are specifically associated with the infoldings of the basal plasma membrane. In such cells, the cristae tend to cross the inner chamber and fuse with both sides of the inner membrane. In most cells the cristae are shelflike in form but in cells involved in steroid synthesis, such as the cells of the adrenal cortex, they are frequently tubular, attaching to the inner membrane only at their ends.

Ribosomes

Ribosomes are small particles which are generally not more than 25 nm in diameter. They consist of protein and RNA with its associated negatively-charged, acidic, phosphate groups. Although individual ribosomes can be visualized only with the electron microscope, when they occur in large numbers in localized areas of cells, the cytoplasm in these regions is basophilic and stains with cationic or basic dyes such as haematoxylin.

Ribosomes occur either free in the cytoplasm or attached to membranes. Free ribosomes may either be single or occur in groups (*Figures 2.5* and *2.7*). Groups of ribosomes are called polyribosomes (*Figure 2.7*) and may appear as chains, whorls or in other configurations. Ribosomes act as the templates for the formation of proteins in the cytoplasm. Polyribosomes are probably involved primarily in the formation of the structural proteins of the cell; they occur in large numbers in cells forming large amounts of such protein, e.g. embryonic cells, and the early precursors of cells formed in the adult such as the so-called haemocytoblasts, the cells which represent the precursor of red cells if not of other types of blood cell.

Endoplasmic reticulum

The endoplasmic reticulum is formed by membrane bounded structures which may occur in the form of tubules, vacuoles or flattened sacs or cisternae. The sacs of the reticulum can often be shown to be continuous with the outer membrane of the nucleus, but the communications with other membrane systems of the cell, notably the plasma membrane which were

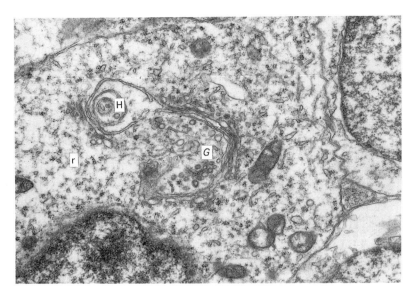

Figure 2.7 Part of a cell in the fetal cerebral hemisphere showing the Golgi apparatus (G) GERL membranes (H) and numerous polyribosomes (r). Stain: uranyl acetate/lead citrate. Magnification ×23 000, reduced to 70% in reproduction.

postulated in the early days of electron microscopy, probably occur rarely if at all.

There are two types of endoplasmic reticulum. Rough or granular endoplasmic reticulum has numerous ribosomes attached to the outer surfaces of the membranes. Smooth endoplasmic reticulum is devoid of attached ribosomes. Because it is also involved in the synthesis of protein, granular endoplasmic reticulum occurs in large amounts in cells active in protein production, such as neurons and the cells of exocrine glands such as the pancreas. In such cells the cisternae of the reticulum are frequently arranged in an approximately parallel fashion and in the form of stacks. In the pancreatic acinar cell (*Figure 2.8*) the ribosomes are almost all attached to the stacks. They also cover the external surface of the sacs almost completely. Although in such cases the arrangement of the ribosomes appears to be highly regular, examination of sections which graze the surface of the sacs indicates that the ribosomes occur in the same kinds of groups as in polyribosomes.

In neurons (*Figure 2.5*) the sacs are only partially covered on their external surfaces by ribosomes and free ribosomes are numerous. The difference in the arrangement of the ribosomes on the sacs of the endoplasmic reticulum in neurons and pancreatic acinar cells may be related to the fact that, whereas in the pancreatic cell the protein is mainly synthesized for secretion, in the neuron a high proportion of the protein is utilized to replace structural protein degraded as a result of the metabolic activity of the cell.

The protein produced on the ribosomes attached to the sacs of the reticulum is probably released, not into the cytoplasm but into the lumen of the sacs of the reticulum. It is then transported, probably in the form of small vesicles which bud off the ends of the sacs, to the Golgi apparatus.

Because of its associated ribosomes, accumulations of endoplasmic reticulum are also basophilic. The distribution of basophilia in different types of cell can therefore be used to provide an indication of the amount and distribution of the reticulum. Thus, in pancreatic acinar cells in which there is a high degree of polarity of the organelles and the granular endoplasmic reticulum is located near the base of the cell, the cytoplasm is intensely basophilic in this region. In neurons, accumulations of granular endoplasmic reticulum and its associated ribosomes occur in fairly discrete areas of the cytoplasm. These areas stain with basic dyes such as those in the Nissl stain, and are recognized with the light microscope as localized bodies, the Nissl bodies.

Smooth endoplasmic reticulum consists of

16 The cell

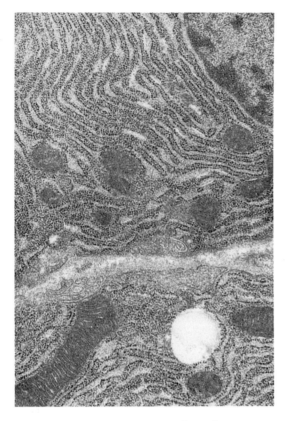

Figure 2.8 Regularly arranged sacs of granular endoplasmic reticulum in the basal regions of cells of the exocrine pancreas. Note that the surfaces of the sacs are almost entirely covered by ribosomes. Stain: uranyl acetate/lead citrate. Magnification × 28 500, reduced to 70% in reproduction.

tubules or sacs similar to those of the granular reticulum. The sacs only sometimes show the stacked form exhibited by the sacs of the granular endoplasmic reticulum and generally occur in the form of branching and anastomosing networks (*Figure 2.5*). It is particularly abundant in the cells of the adrenal cortex which are active in the secretion of steroid hormones. It is also abundant in the hepatocyte, and the massive increase in the amount of the reticulum in such cells after injection of substances such as phenobarbitone, which are inactivated in the liver, suggests that it may also play some part in this process. In smooth muscle cells, short sacs of smooth endoplasmic reticulum are found in association both with the mitochondria and the plasma membrane, and may play a part in the mechanism of release into and removal from the cytoplasm of the calcium upon which the process of coupling between the stimulus and the contraction of the cell depends. In neurons the elements of the reticulum also probably play a part in the transport of materials and in the formation of the small vesicles which contain the transmitters released at the endings of the cells.

Apart from the structures which can clearly be defined as components of the smooth endoplasmic reticulum, many cells contain membranous sacs which are structurally similar to the elements of the reticulum. The T-system of the striated muscle cells, which as in the smooth muscle cell is probably involved in calcium metabolism associated with contraction, is a particular example of these membranous sacs. Apart from this, in megakaryocytes, the precursors of platelets, a system of membranes known as demarcation membranes separate from one another the small areas of cytoplasm destined to become individual platelets. Another system of membranes, the GERL, probably plays some part in the formation of secondary lysosomes. The multiplicity of functions proposed for these and the various smooth surfaced sacs normally defined as those of the smooth endoplasmic reticulum raises some questions as to whether all of these types of sac can be regarded as a single structural unit.

The Golgi apparatus

Substances synthesized in the endoplasmic reticulum which are intended for packaging in secretory vesicles or vacuoles and other structures such as lysosomes are processed by the Golgi apparatus. The apparatus, which is sometimes also called the Golgi body, is normally located close to the nucleus.

Like the endoplasmic reticulum, this also consists of a series of flattened sacs arranged in stacks (*Figures 2.5* and *2.7*). The sacs curve towards the nucleus and frequently show discontinuities or fenestrations. The sacs near the nucleus show clearcut differences in both diameter and cytochemical characteristics from those further away from the nucleus. By analysing the fate of radioactively labelled substances incorporated into the secretory products, it has been possible to show that the proteins formed in the endoplasmic reticulum enter the sacs of the Golgi apparatus closest to the nucleus. This region is defined as the forming face of the apparatus. It is thought that as new sacs are

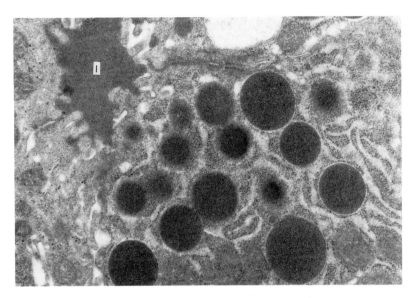

Figure 2.9 Secretory granules in the apical part of an acinar cell of the exocrine pancreas. Exocytosed secretory material is present in the acinar lumen (l). Stain: uranyl acetate/lead citrate. Magnification × 28 500, reduced to 70% in reproduction.

formed by incorporation of the transport vesicles derived from the endoplasmic reticulum, the previously formed sacs move outwards to the opposite surface. There is evidence that during this process carbohydrate groups are added to the proteins and the secretory product is concentrated. When the sacs reach the outer or maturing face of the apparatus, they break up into small vesicles or vacuoles. Some of these are lysosomes and others form secretory vacuoles.

Because the Golgi apparatus reduces heavy metals such as osmium or silver, it can readily be demonstrated by light microscopy. The apparently discrete nature of the apparatus in such preparations is the reason why it was often classified as the Golgi body. Examination of light microscopical preparations gives a clear idea of the size and distribution of the apparatus. Cells in which there is only limited synthetic activity generally have only a small Golgi apparatus. In cells involved in secretion it is generally very well developed, and in exocrine gland cells and other cells which exhibit some degree of polarity in respect of their functional activity it typically lies near one pole of the nucleus. In neurons, in which it is also well developed, it forms a coarse network which surrounds the nucleus.

Secretory vesicles and vacuoles

Secretory vesicles and vacuoles vary considerably in size and in the density of their contents. In some situations the vesicles are little more than 50–100 nm in diameter and in others they are 1 μm or more in diameter. In many vesicles, especially those in which the secretory material appears highly electron-dense under the electron microscope, there is a clear interval or halo between the material and the membrane which bounds the vesicle. Although it is sometimes argued that the halo and the secretory material or granule as it is often called consist of different components, it is probable that the halo is formed by condensation of the material during fixation and processing of the tissue. In cells such as those of the exocrine pancreas, the vesicles formed from the Golgi apparatus fuse to form large membrane-bounded structures known as condensing vacuoles. These probably then become transformed into definitive secretory vacuoles by further condensation of their contents.

The distribution of secretory vacuoles differs considerably in different types of cells. In cells in which the organelles are polarized and the secretion is released at a particular part of the surface, the vacuoles accumulate in the region of the cell close to this surface (*Figure 2.9*). Such is the case in exocrine gland cells where the vacuoles are found in the apical cytoplasm. In other cells in which the vacuoles can be released at any part of the surface, such as the cells of most endocrine glands, the granules are often more or less randomly distributed in the cytoplasm. The contents of secretory vesicles and vacuoles are released from the cells by the process of exocytosis. The release of secretory material may be continuous or intermittent. In cells in which the release of secretory material is intermittent, the secretory vacuoles often accumulate in large numbers in the cell (*Figure 2.7*). In cells in which the secretion is continuous, there may be very few secretory vacuoles in the cytoplasm.

Many cells secrete more than one substance. In such cells the secretory vacuoles may contain a mixture of different substances. Alternatively, as appears to be the case in the thyroid gland, different substances may be contained in different vacuoles. In some cases the secretory vesicles or vacuoles contain materials which appear to play no specific functional role after release. Such is the case in the secretory vesicles in neurons, and in particular in adrenergic neurons, where the specific secretory product is associated in the vesicles with protein and the enzyme dopamine β-hydroxylase.

The widely different chemical composition of the substances contained in secretory vesicles and vacuoles often endows the vacuoles with different staining properties in material processed for light microscopy. Thus by using appropriate stains or combinations of stains it is possible to differentiate between otherwise apparently similar cells such as mucus-secreting cells, the various form of granulocyte in the blood and the endocrine cells of the anterior pituitary. In situations such as the pituitary there are also differences in the number, size and electron density of the contents of the vacuoles, and these can be used to differentiate the various types of cell from one another. Other more sophisticated techniques, including cytochemical techniques based on enzyme staining and immunological techniques, can also be applied to the demonstration of specific substances within secretory vacuoles. Immunological techniques have, for example, been used to show that in the gut there are cells which contain a wide variety of different forms of peptide.

Lysosomes and associated structures

In the course of their metabolic activity cells inevitably accumulate material which needs to be degraded and removed. Such materials in-

Figure 2.10 Myofilaments in a striated muscle cell. The regular arrangement of myosin (m) and actin (a) filaments within the myofilaments and the alignment of similar regions within adjacent myofilaments and myofibrils is responsible for the crossbanded or striated appearance of this type of muscle cell as seen with the light microscope. Stain: uranyl acetate/lead citrate. Magnification × 22 000, reduced to 70% in reproduction.

clude worn out organelles, excess secretory material and material ingested into the cell by the process of phagocytosis. A particular example of such processes is the change in function of the Schwann cell, the cell which surrounds nerve fibres in the periphery, where degeneration of the nerve fibres results in the redundancy of the myelin, one of the major components of the cell. Lysosomes play a major role in this process of elimination.

Two major forms of lysosome are now recognized. Primary lysosomes are small vesicular structures with a highly electron-dense core. They are produced from the Golgi apparatus and contain a wide variety of enzymes active in hydrolysis. The best known of these enzymes is acid phosphatase, and cytochemical reactions for this enzyme have been used very widely to differentiate lysosomes from other similar cellular organelles. Primary lysosomes fuse with other membrane-bounded vesicles or vacuoles and empty their contents into these structures by a process which closely resembles exocytosis. The resulting structure is known as a secondary lysosome. The structures with which primary lysosomes fuse are of several types. They include the vacuoles containing material ingested by phagocytosis. They also include secretory vacuoles present in excess in the cells, multivesicular bodies and vacuoles derived from the GERL membranes. Multivesicular bodies are vacuoles which contain small vesicles. These small vesicles are derived from vesicles which fuse with the membrane of the vacuole, but which then, instead of merely becoming incorporated into the membrane, become invaginated and pinched off into the interior of the vacuole by a process akin to micropinocytosis. GERL membranes are elements of smooth endoplasmic reticulum which occur near the Golgi apparatus and which appear to produce various types of lysosomes (Novikoff, 1976). They enclose small segments of cytoplasm, wrapping around them to form a vacuole which initially possesses not one, but two, bounding membranes (*Figure 2.7*).

In the secondary lysosomes the enzymes derived from the primary lysosomes begin to break down the contents of the vacuole. In cases where not all of the material can be degraded, it remains in the vacuole. Such structures are called residual bodies. Although there is some doubt about their subsequent fate, it is possible that some residual bodies are eliminated from the cell by exocytosis, following which the material contained by the bodies is presumably eliminated from the tissue by macrophages.

Microtubules and filaments

Many cells contain a variety of types of filament and elongated tubular structures known as microtubules. Three types of cytoplasmic filament have been defined. Of these, two are found in abundance in muscle cells. Fine filaments, approximately 5 nm in diameter, consist of the contractile protein actin. These are essentially the same as the filaments located beneath the plasma membrane and in microvilli. Thick filaments, approximately 15 nm in diameter, consist of the contractile protein myosin. They are normally readily visible only in cardiac and striated muscle cells (*Figure 2.10*) where they occur in a highly regular arrangement in the myofibrils with actin filaments. It is through the interaction between actin and myosin that the contraction of the cells is brought about. Myosin also occurs in smooth muscle cells and probably in other types of cell in which contraction and movement takes place. In such cells no thick filaments can normally be demonstrated. It is thought that this is due to the fact that the myosin exists in a disaggregated state or becomes disaggregated during fixation and processing.

The third type of filament, the *intermediate filament*, occurs in many types of cell. It is usually approximately 10 nm in diameter. Intermediate filaments are considered to provide structural support and to act as a kind of cytoskeleton. They are particularly numerous in the cells of the epidermis, in which they consist of the fibrous protein keratin. Neurons, and in particular the processes of neurons, also contain large numbers of filaments. In such cells they are usually classified as neurofilaments. They also occur in erythrocytes in which they consists of the protein spectrin, and not only help to maintain the distinctive shape of the cells but also probably protect them from shearing forces during their flow through the blood. Intermediate filaments composed of the protein desmin are found in muscle cells; in such cells, they are thought to contribute to holding the contractile filaments to the plasma membrane, and in striated muscle cells they are located in particular regions of the myofibrils. Intermediate filaments consisting of vimentin surround the nucleus and may be involved in maintaining the position of the nucleus within the cytoplasm.

Microtubules can also be identified in most types of cell. They consist of hollow cylinders approximately 22 nm in diameter (*Figure 2.4*). They are frequently very long and often occur in groups. They consist of the protein tubulin and their wall has a highly ordered helical structure. Microtubules not only occur in cilia and flagella, but also form a major constituent of the mitotic apparatus which is formed during cell division. They occur in cells or cellular elements such as red cells and platelets in which, together with intermediate filaments, they probably play a part in the maintenance of cell shape. In neurons in which they are also abundant, and in which they also probably play a part in the maintenance of the shape of the cell, they probably perform additional roles, notably in the guiding of movement of structures such as vesicles within the cell.

Centrioles

Centrioles are cylindrical structures which are found normally in pairs in cells which retain the capacity for division. They are closely similar to the basal bodies of cilia. They consist of nine groups of three microtubules arranged in a circle, each group of microtubules being orientated in a manner similar to the vanes of a water wheel. A variety of types of pericentriolar body are found in association with centrioles, notably centriolar satellites which possibly represent aggregates of the subunits present in microtubules.

The association of centrioles with cilia and their participation in the formation of the microtubules comprising the mitotic spindle suggests that they play a part in the genesis of cellular microtubules.

Lipid droplets

Although the hyaloplasm contains substantial amounts of lipid, it is only in some circumstances and in some types of cell that the lipid accumulates in sufficient amounts to form distinct lipid droplets. Lipid droplets differ from other recognizable cell organelles in that they are not bounded by membranes. They can be stained in light microscopical preparations by dyes which dissolve in the lipid, such as the Sudan dyes. They frequently accumulate in ageing cells and are found in cells in a variety of pathological conditions, notably atheroma; but, except in cells which secrete lipid materials such as those of the adrenal cortex, they are large only in the fat cells which comprise the principal cellular component of adipose tissue.

Other cell organelles

A variety of other types of organelle are found in different types of cell. These include glycogen granules and ferritin granules. Glycogen granules appear in electron microscopical preparations as electron-dense granules which are much larger than ribosomes. In cells such as those of the liver they frequently occur in large groups. Ferritin granules are very small electron-dense granules which can readily be mistaken in electron microscopical preparations for stain deposit. They represent the storage form of iron and are found in substantial numbers in the cells of the liver.

The nucleus

Most cells contain a single nucleus. In the red blood cell the nucleus is extruded during the later stages of development and prior to release of the cell from the bone marrow into the blood. In other cells, such as striated muscle cells, the osteoclasts which are found in bone and which are responsible for its resorption and the foreign body giant cells which occur in certain pathological conditions, there are large numbers of nuclei.

At the light microscopical level the nuclei of most cells appear approximately spherical. The nuclei may, however, be deformed, as during contraction of smooth muscle cells when they become elongated, and in macrophages during their passage through narrow spaces between other types of cell. When examined with the electron microscope, nuclei often show one or more infoldings or indentations of the nuclear envelope. In some cells, notably the neutrophils and other granular white cells of the blood, the nuclei are divided into a series of lobes connected by thin strands. Typically, nuclei are approximately the same size as red blood cells, that is about 8 µm in diameter.

The nuclear contents are separated from the cytoplasm by the nuclear envelope or mem-

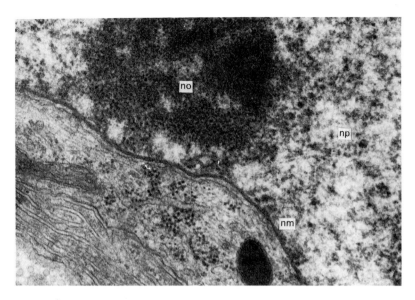

Figure 2.11 Part of the nucleus of a supporting cell in a neuronal ganglion showing the components of the nuclear membrane (nm), the nucleolus (no) and the nucleoplasm (np). Stain: uranyl acetate/lead citrate. Magnification × 66 500, reduced to 70% in reproduction.

brane (*Figure 2.11*). This consists of two unit membranes. Of these, the outer membrane is often studded on its outer surface by ribosomes. On the internal aspect of the inner membrane a thin layer of granular material is normally present. This is the nuclear lamina. The nuclear envelope is perforated by numerous nuclear pores. These are regions where the two membranes comprising the envelope fuse with one another to leave a circular aperture allowing communication between the interior of the nucleus and the cytoplasm. The diameter of the nuclear pores is in the order of 60 nm, but the pore is filled with a complex structure consisting of filaments and granules. This pore complex clearly reduces considerably the size of the spaces available for communication between the nuclear interior and the cytoplasm. Nuclear pores occur in varying number in different types of cell. In some cells they are infrequent, but in others they occupy as much as one-third of the surface.

The nuclear interior, the nucleoplasm, is occupied by the nuclear chromatin and the nucleolus. The chromatin occurs in two forms, euchromatin and heterochromatin. Euchromatin cannot normally be shown to possess a definitive structure with the light microscope and is represented by the relatively unstained areas of the nucleoplasm. At the electron microscopical level it has an apparently granular structure, but in fact consists of fine filamentous strands. Heterochromatin occurs in irregular masses which can be defined readily with the light microscope. The distribution and amount of heterochromatin in nuclei varies with the type of cell and its arrangement within the nuclei is sometimes highly characteristic. For example, in plasma cells, the cells of the lymphoid system which secrete the antibodies responsible for humoral immunity, the heterochromatin is arranged in a fashion which resembles the spokes of a wheel.

The nuclear chromatin is formed by the chromosomes. These consist of DNA and histones and variety of other proteins. Because, in common with RNA, DNA possesses numerous acidic phosphate groups, it can also be stained with basic dyes such as haematoxylin. It can also be stained by histochemical techniques specific for DNA, such as the Feulgen reaction. The two forms of chromatin both contain components of the chromosomes. In the heterochromatin the chromosomes are tightly coiled, whereas in the euchromatin they are uncoiled. The chromosomes carry the genes, each gene being a component of the chromosome which determines the formation of a particular form of

RNA. It does this by acting as a template for the formation of RNA molecules. It is from the RNA molecules that the information required for the synthesis of particular substances, especially proteins such as enzymes, is derived. Several different forms of RNA are now recognized. They are messenger RNA (mRNA), transcription RNA (tRNA) and ribosomal RNA (rRNA). DNA acts as the template, or encodes each of these forms of RNA, but there is evidence that tRNA and rRNA are formed in the nucleolus, and that mRNA is formed directly on the chromosomes. All three types of RNA pass into the cytoplasm where they interact in complex but highly specific ways to form proteins and other cell components. Because uncoiled chromatin is active in the production of mRNA, cells which produce large amounts of protein, whether for secretion or for cellular growth and maintenance, contain large amounts of this type of chromatin. Conversely, in cells which are relatively inactive, the chromatin is largely in the form of heterochromatin. In the cells of females one of the sex chromosomes remains tightly coiled. The mass of heterochromatin represented by this chromosome lies against the nuclear membrane and is sufficiently characteristic to enable its presence to be used in the differentiation of female from male cells.

Apart from the chromatin, the nucleus contains one or more nucleoli. These are usually about 1 µm in diameter and can be demonstrated without difficulty with the light microscope. They contain RNA and protein and can be stained with stains specific for RNA such as pyronin. With the electron microscope nucleoli can be shown to possess two components, one of which is granular and the second fibrillar in nature.

In common with many other cellular components, the number and size of the nucleoli is related to the activity of the cells. In cells active in protein synthesis the nucleoli are prominent and are often multiple. In inactive cells they are often very difficult to demonstrate.

The cell cycle

The formation of new cells, whether during the early stages of development, during growth, for replacement of cells which are lost or die during later life or during the process of tissue repair, is accomplished by the process of division of pre-existing cells. Although in most cases this is accomplished by the process of mitosis, in some pathological situations cell division occurs amitotically by a simple process of separation of the cell into two random halves. Mitosis is divided into four stages known as prophase, metaphase, anaphase and telophase, and represents the second of the two major phases in the cell cycle. The other phase of the cell cycle is known as interphase (*Figure 2.12*). Most cells in the body spend the majority of their time in interphase, and the cell structure described above is that of the interphase cell.

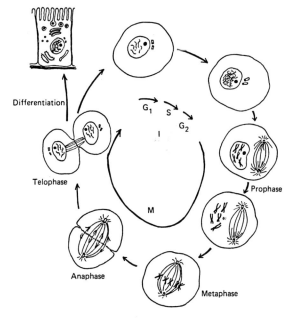

Figure 2.12 Diagram to show the various events of the cell cycle. In the first stage of mitosis (M), prophase, the mitotic spindle forms and the chromosomes become visible in the nucleus as discrete structures. At the end of prophase the nuclear membrane disappears. In metaphase the chromosomes arrange themselves at the equator of the spindle. In anaphase the new chromosomes formed by separation of the chromosome pairs move to the end of the spindle and the cleavage furrow develops in the equatorial region. In telophase the nuclear envelope is reformed around the chromosomes and the cell ultimately divides into two by deepening of the cleavage furrow. During interphase (I), the cells either undergo differentiation or pass through a series of stages, G_1, S (during which there is synthesis of DNA) and G_2, which is followed by further division by mitosis.

Since all of the cells produced by division must contain the full complement of the genes carried on the chromosomes (except germ cells which undergo a special form of division known as meiosis), division must be preceded by duplication of the chromosomes. This process of chromosome duplication occurs during interphase. Since chromosome duplication involves the synthesis of new DNA, it can be studied by analysing the incorporation into the cells of radioactively labelled components of DNA such as tritiated thymidine. Studies of this kind have shown that interphase can be divided into three separate periods. These are the period which precedes DNA synthesis classified as G_1, the period of DNA synthesis classified as S and the period which follows DNA synthesis and precedes division, classified as G_2. Using tritiated thymidine it has been possible to determine the length of each of these periods in cells undergoing regular division, such as cultured cells and intestinal epithelial cells.

Mitosis is accompanied by separation of the chromosome pairs formed during interphase and their partition equally between the two newly formed cells. In prophase the chromosomes become visible as rod-like structures. The duplicated chromosomes are joined to one another at distinctive sites along their length, known as kinetochores or centromeres. At this stage the members of each pair are called chromatids. In the cytoplasm microtubules are generated from the centrioles, which move to opposite ends of the cells. The microtubules radiate from the centrioles forming asters and also connect the two centrioles to form the mitotic spindle. At the end of prophase the nuclear envelope breaks down into membrane-bounded vesicular structures resembling endoplasmic reticulum and the chromosomes are released into the cytoplasm.

In metaphase the spindle formed by the microtubules connecting the two centrioles moves to the centre of the cell and the chromosomes move to the centre of the spindle where they attach to the microtubules by the centromeres. This attachment occurs in a plane equidistant between the centrioles known as the equatorial plane. Drugs such as colchicine which have a marked effect on microtubules, are often used to arrest the process of mitosis at this stage. Such techniques have been used very widely for studying chromosomes and to determine the rate at which particular populations of cells divide. It is through examination of the chromosomes of cells in metaphase that the number of chromosomes in different animal cells has been determined. The normal chromosomal number in human cells is 46. The chromosomes occur in pairs, and each pair can be differentiated from the rest by the fact that they differ in length and in the distance along the chromosomes of the centromeres. One pair of chromosomes are the sex chromosomes. In females these consist of two structurally similar chromosomes, the X chromosomes. In males the cells contain one X chromosome and a much smaller Y chromosome. For this reason it is possible by examining the chromosomes in cells in metaphase to determine the sex of the individual. It is also possible to demonstrate the existence of differences in chromosome number in the cells of individuals with congenital abnormalities such as Down's syndrome.

Anaphase is preceded by division of the centromere into two. Each chromatid now comprises a separate new chromosome. In anaphase the new chromosomes separate and each moves towards one end of the spindle. At the same time the first stage in the separation of the cell into two occurs by the formation of a furrow, the cleavage furrow, in the region of the equator.

In telophase the chromosome movement along the spindle is completed. The cleavage furrow deepens, probably by addition to the surface of membrane derived from small vacuoles which appear in the region of the furrow, and the cell becomes divided into two. During this phase the nuclear envelope is reformed, probably from the endoplasmic reticulum. It is possible that the formation of excess amounts of envelope is responsible for the presence in the cytoplasm of some interphase cells of the structures known as annulate lamellae. The nucleoli, which disappeared during prophase, also reappear. The two cells formed during mitosis each now enters the first period of interphase, G_1.

Meiosis, the special form of division which occurs in germ cells in the ovary and testis, is a process of division in which the chromosome number is reduced to half the normal complement (*Figure 2.13*). In effect it consists of two separate divisions. In the prophase of the first division the chromosomes of each pair join together and the chromatids appear. During this stage there is a certain amount of interchange of genetic material between the elements of each pair, a process known as crossing-over. In metaphase the paired chromosomes, each

24 *The cell*

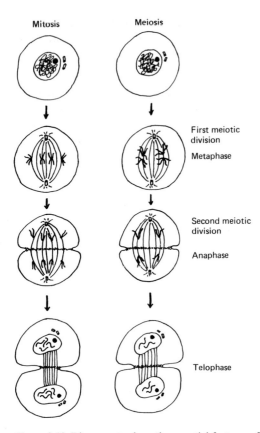

Figure 2.13 Diagrams to show the essential features of meiosis as compared with mitosis. Meiosis involves two separate divisions. In the diagrams, phases of these divisions are compared with equivalent phases of two mitotic divisions. In the first meiotic division, at metaphase, homologous pairs of chromosomes, rather than single pairs of chromosomes attach to the mitotic spindle at its equator. Individual pairs of chromosomes then become segregated into the two daughter cells, thereby reducing the chromosome number by half. In the second meiotic division the pairs of chromosomes attach to the equator during the metaphase as in mitosis, and at anaphase one chromosome of each pair moves to each end of the spindle so that when the nucleus of the daughter cells is reformed during telophase, each contains only half the normal number of chromosomes.

still consisting of two chromatids, attach to the equator. In anaphase each pair of chromatids moves along the spindle. The effect of this is that the new cells formed during telophase contain only half the normal number of chromosomes. The second meiotic division follows without an intervening stage of DNA synthesis and duplication of the chromosomes and results, as in mitosis, in the separation of the already formed chromatids from one another. As a result the cells retain only half the normal number of chromosomes. Restoration of the normal chromosome number results after fertilization and with fusion of the sperm with the ovum. It is through abnormalities of meiosis that differences in the normal chromosome number of the kind seen in Down's syndrome occur.

Mitotic division occurring in early development is normally controlled by substances derived from other cells in the vicinity. In later life a variety of other mechanisms appear to control the division of cells. These include a variety of hormones and other substances classified as chalones which are produced in the tissues themselves. It is thought that chalones normally inhibit or limit the rate of production of new cells, and that the reduction in the concentration of these substances which occurs in, for example, tissue damage is an important factor in stimulating division to produce the new cells required for repair of the tissue.

Cell differentiation and control mechanisms

Differentiation is the process by which cells assume their particular structural and functional characteristics. It often occurs in a series of stages, and is probably primarily due to changes in the activity of particular genes or gene groups. These changes are brought about by a number of processes and through the action of a variety of substances, and include activation, repression and derepression. In early development these are brought about by the action on the cells of diffusable substances known as inducers. At this stage, as well as possibly at much later stages, growth and development are also affected by interactions between cells. Such processes include the growth of nerve fibres along preferential pathways formed by Schwann cells. Contact between cells may also influence both their movement and the patterns which they adopt during development. Later in development and in the adult, hormones play a considerable role in controlling differentiation and growth. These include somatotrophic hormone produced from the anterior part of the pituitary gland, and the chalones.

Cells undergoing differentiation typically undergo a regular series of structural changes. The existence of these changes often enables cells in particular phases of the process of differentiation to be defined. Thus, in the process of formation of red cells in the bone marrow, it is possible to define the early precursor, the haemocytoblast, which contains large numbers of polyribosomes responsible for the formation of haemoglobin. It is also possible to define various forms of normoblast, in which there is a series of changes which is accompanied eventually by the extrusion of the nucleus of the cell and which result in the formation of the reticulocyte and the mature erythrocyte, in which the remnants of cell organelles present in the reticulocyte have finally been eliminated. A similar series of changes occurs in the epidermis, in which, as the cells move upwards through the layers of the epidermis, the cells formed by mitosis in the basal layer undergo a characteristic series of changes resulting in their transformation into the dead and cornified squames which are present in the surface. In the formation of red cells and of many other types of cell in the adult, the process of differentiation is also accompanied by the loss of the ability of the cells to divide. In such tissues the initial stages of formation of the cells also results from the division of immature or stem cells. It is believed that, to retain the population of such cells, only some of the newly formed cells undergo differentiation, the rest remaining undifferentiated to preserve the stock of precursor cells.

Tissues

As they are classically defined, tissues consist of aggregates of a particular type or particular types of cell. Because the cells of each tissue have particular functions, they all possess a basically similar structure. The major tissues are classically considered to be four in number, namely epithelial tissue or epithelia, connective tissue, muscle and nervous tissue. Each of these include several sub-types. Thus connective tissue includes fibrous connective tissue, of which there are several varieties, cartilage, bone and blood. Organs are formed by two or more tissues arranged in specific and generally highly specific ways. Unfortunately this classical concept of four basic tissues has a number of disadvantages. Thus, while it could be argued that they comprise organs rather than tissues, it is difficult to avoid classifying some structures, such as bone marrow and lymphoid structures, as tissues. In an attempt to eliminate some of the problems arising from such terminology, many histologists have expanded the definition of tissue types to include some of these structures.

The tissues of the body are all derived from three cell layers, the germ layers, which are formed at the stage of embryonic development in which the embryo consists of a flattened disc. The ectoderm is the outermost of these layers and gives rise, amongst other things, to the epidermis and a large part of the nervous system. The intermediate layer comprises the mesoderm. This gives rise to the connective tissues and muscle. The inner layer, the endoderm, forms the epithelial lining of the alimentary tract. A column of cells which is formed from the ectoderm in the region of the developing nervous system, the neural crest, is often regarded as the fourth germ layer. In common with the other germ layers, it also gives rise to a variety of types of cell. These include many of the neurons in the peripheral nervous system and the melanocytes. Recent work has shown that the cells derived from each of the germ layers probably possess the potential for much more varied patterns of development than was originally supposed.

Epithelia

Epithelia are formed by sheets of cells, the components of which are usually separated by an intercellular gap of approximately 20 nm. Epithelia line the surface of the body and internal cavities, including the alimentary and respiratory tracts, and many glands are formed during the development from surface or internal epithelia. In epithelial tissues the cells are typically joined by cell junctions and often interlock in complex ways. Except in rare instances, they are separated from the underlying tissue by a basement membrane.

Unilaminar epithelia consist of several types. They may be squamous, cuboidal or columnar, depending upon the shape of the cells of which they are principally composed. Squamous epithelia consist of flattened cells and include the mesothelia which line the serous cavities (pleural, pericardial and abdomino-pelvic), the endothelium of the cardiovascular system and the epithelium of the alveoli of the lung in which, in association with the necessity for rapid exchange of gases, the peripheral parts of the cells may be as little as 25 nm thick. Cuboidal or columnar epithelia are found in structures such as the alimentary tract. Although they may not be strictly classified as unilaminar, the epithelium of glands, whether exocrine or endocrine, is usually composed of cuboidal or columnar cells. In epithelia of this type the cells often exhibit a variety of surface or other specializations. In the alimentary tract the absorptive cells of the intestine possess a distinctive brush border composed of microvilli. Other cells found in the intestinal epithelium include goblet cells, which typically contain numerous apical granules of mucus, and various other forms of secretory cell. Most of the latter types of cell are similar in their structure to the protein-secreting cells found in endocrine and exocrine glands, but in the highly distinctive oxyntic cells of the glands of the stomach, which secrete hydrochloric acid, the cells are invaginated by a series of membrane-bounded, microvillus-lined channels. Microvilli are also found on the surfaces of other cells apposed to a fluid environment and involved in transport functions, such as those comprising the epithelia of the tubules of the kidney. In such epithelia, the lateral and basal regions of the cells may also exhibit villous or other types of folding of the plasma membrane. In the tubules of the kidney, the presence of such folds and the vertical orientation of the elongated mitochondria present within them are responsible for the striated appearance of the basal parts of the cell which can be seen with the light microscope.

Pseudostratified epithelium lines much of the respiratory tract. In addition to columnar cells, respiratory epithelium is characterized by small cells located near the basement membrane which form a further, but only partial, layer. At least some of these cells are the source of new cells required for replacement of worn out or effete columnar cells. In specialized structures such as the taste buds they form new receptor cells, and in the olfactory epithelium they are probably the source of new neurons, a phenomenon which is unique in the mammal, in which neuronal replacement as opposed to regeneration does not occur postnatally.

Stratified epithelia consist of several layers of cells. In such epithelia, instead of moving laterally, new cells are formed in the layer next to the basement membrane and move upwards through the other layers to the surface, undergoing characteristic types of differentiation as they do so. The two principal types of stratified epithelium are transitional epithelium and the so-called stratified squamous epithelium which forms the epidermis.

Transitional epithelium lines part of the urinary tract, such as the ureter and bladder. It is classified as transitional because when the structures are empty or collapsed, it appears to consist of several layers of cells, but when the organs are distended, it consists of as little as two layers of cells. In this kind of epithelium, the superficial cells possess a relatively impermeable surface membrane, which is both thicker than most other plasma membranes and rather stiff, so that in the collapsed state the surface membrane folds in a plate-like fashion into the cell.

Stratified squamous epithelium varies in the number of layers it possesses, and is usually described as consisting of a number of strata. The stratum basale is the layer apposed to the basement membrane and is the source of new cells for replacement of those shed from the surface. The next most superficial layer is the stratum spinosum, in which in light microscopical preparations the cells show a spiny appearance, primarily due to shrinkage during processing and the formation of spikes or spines at the points where the cells are held together by desmosomes. Superficial to that is the stratum granulosum in which the cytoplasm contains distinctive granules of keratohyalin. Above this are layers of dead cells which comprise the stratum corneum, and in which the cells have assumed the squamous appearance from which the epithelium derives its name.

Connective tissue

Connective tissue consists of cells embedded in a matrix which is normally secreted by the cells. In the true connective tissues the matrix includes fibres, notably collagen and elastin, and the ground substance. In blood the matrix comprises the plasma.

Fibrous connective tissue (*Figure 2.14*)

This occurs in several forms. The cell normally considered to form the matrix is the fibroblast or, more properly, the fibrocyte. The matrix consists of varying amounts of collagen and elastin and the ground substance. The ground substance contains acid mucopolysaccharides.

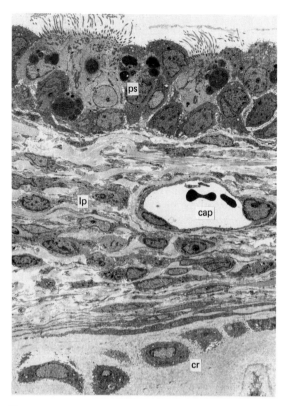

Figure 2.14 Low power electron micrograph of the trachea showing the pseudostratified epithelium (ps), the underlying lamina propria (lp) composed of fibrous connective tissue and containing numerous cells of various types and a capillary (cap) lined by endothelium, and part of one of the cartilage rings with its chondrocytes (cr). Stain: uranyl acetate/lead citrate. Magnification ×2500, reduced to 50% in reproduction.

Although the fibrocyte probably forms collagen and other components of the matrix, recent evidence has suggested that elastin is formed by smooth muscle cells or by cells very similar to them. Other cells commonly found in fibrous connective tissue are mast cells and macrophages.

In loose or areolar fibrous connective tissue the fibre is small in amount. Such tissue is found in places where there is a necessity for considerable mobility. In dense fibrous connective tissue there are large amounts of fibre. Such tissue comprises tendons and ligaments. The fibre in tendons is collagen, but in ligaments it may consist largely of elastin. In adipose tissue, such as that found beneath the skin, the tissue contains large numbers of fat cells as well as the other cells and fibre characteristic of fibrous connective tissue.

Cartilage

Cartilage contains cells known as chondrocytes (*Figure 2.14*). In cartilage the intercellular substance, which also consists of mucopolysaccharide, is present in large amounts. In hyaline cartilage, which lines the articular surfaces at synovial joints, the fibre is present in such small amounts that it cannot normally be seen. In fibrocartilage, the fibre is present in much larger amounts and may consist mainly of collagen or, in elastic cartilage, mainly of elastin. Cartilage has no direct blood supply and this is probably related to the fact that, when it is damaged or worn away, it cannot normally be repaired.

Bone

Bone contains two types of cell. The osteocyte forms the matrix and the multinucleate osteoclast breaks it down. The matrix contains large amounts of collagen around which is deposited the calcium-containing mineral, hydroxyapatite. In compact bone, the matrix is arranged in a series of circular layers comprising the Haversian systems. In cancellous or trabecular bone, the matrix forms a series of trabeculae between which there is a series of intercommunicating spaces.

Blood

There are various forms of blood cell, the erythrocytes and the white cells. The white cells include the granulocytes or polymorphonuclear leucocytes, lymphocytes which comprise the small and large lymphocytes and the platelets. The matrix, the plasma, contains the plasma proteins, but is not formed in any great degree by the cells. Blood cannot therefore be regarded as a true connective tissue. The erythrocytes, granulocytes and platelets are formed in the bone marrow from stem cells. It has recently been shown that these cells have many of the structural characteristics of small lymphocytes. Many small lymphocytes are formed in the germinal centres in lymphoid organs such as the spleen. Such structures also occur in lymph nodes and in the wall of organs such as the gut. Collections of lymphocytes of this kind are often classified as lymphatic tissue.

Muscle

There are three types of muscle, namely striated,

cardiac and smooth muscle. Striated muscle (sometimes also called skeletal or voluntary muscle) comprises the muscles responsible for most of the movements of the body. These cells are elongated cylinders, and the regular arrangement of the contractile proteins actin and myosin in the cells (*Figure 2.10*) gives them a striated or cross-banded appearance. Striated muscle cells possess large numbers of nuclei and normally contract only when they are stimulated to do so by nerve fibres. Cardiac muscle cells are also striated, but often branch and only contain one nucleus. They also possess the property of contracting independently. Cardiac muscle cells are coupled electrically to one another by gap junctions in the region of the intercalated discs. Smooth muscle cells are also often called visceral or involuntary. They differ from other forms of muscle cell in that they show no striations. They also possess a single nucleus and are sometimes capable of independent rhythmic contractions.

Nervous tissue

Nervous tissue contains neurons and various forms of supporting cell. The neurons are specialized for transmission of information, often over long distances, and possess elongated processes. These are the dendrites and the axons. Information transmission along the processes occurs as a wave of depolarization of the plasma membrane. This constitutes the nerve impulse. In most cases, transmission of the impulse to other neurons or to effector cells such as muscle cells is achieved by release of chemical messengers, the neurotransmitters. These include acetylcholine and noradrenaline.

The supporting cells in the central nervous system comprise the neuroglia. They include the star-shaped astrocytes, cells with few processes, the oligodendrocytes and the potentially phagocytic microglial cells. In the peripheral nervous system, the principal type of supporting cell is the Schwann cell. This surrounds the axons in peripheral nerves and forms the myelin around the myelinated nerve fibres.

Acknowledgements

I am indebted to Dr D. R. Kershaw, who drew the diagrams in *Figures 2.1, 2.12* and *2.13* and to Mr H. Jagessar, who processed much of the material from which the electron micrographs in the remaining figures were taken. Mr Jagessar also took the micrograph reproduced in *Figure 2.14*, which is reproduced with his kind permission.

References and further reading

BLOOM, W. and FAWCETT, D. W. (1986). *A Textbook of Histology*, 11th Edition. Philadelphia: Saunders

DAVSON, H. and DANIELLI, J. F. (1952). *The Permeability of Natural Membranes*. Cambridge: Oxford University Press

NOVIKOFF, A. B. (1976). The endoplasmic reticulum: a cytochemist's view (a review). *Proc. Natl Acad. Sci., USA*, **73**, 2781–2787

SINGER, S. J. and NICHOLSON, G. L. (1972). The fluid mosaic model of the structure of the cell membrane. *Science*, **175**, 720–731

WEISS, L. (1983). *Histology. Cell and Tissue Biology*, 5th Edition. New York: Elsevier

WILLIAMS, P. L. and WARWICK, R. (1980). *Gray's Anatomy*, 36th Edition. Edinburgh: Churchill Livingstone

3

Disease at the cellular level

Amanda Herbert

Introduction

The abnormal cellular changes seen in cytological preparations reflect the pathological changes in the tissue from which the samples have been taken. These cellular changes represent the combined effect of damage to cells and tissues, together with the host response, and are the manifestation of disease in the tissue of origin. In this chapter the genetic and environmental factors which cause cell damage are described and host defence mechanisms discussed. The cellular and tissue manifestations of the major disease processes resulting from these interrelated factors are also considered, grouped as inflammatory disease, immunological defects, vascular disease and disorders of growth. It is hoped that the information contained in this chapter will help the cytotechnologist to understand the pathobiological processes which underlie the variety of cellular patterns which can be found in routine cytological smears.

Genetically determined disease

Genetically determined diseases are of two main types: those which are inherited in a Mendelian fashion, and those which occur more often in families than would be expected from the frequency of the condition in the population at large. The former group are characterized either by the presence of a demonstrable chromosomal abnormality or by an inborn error of metabolism which can be attributed to a single mutant gene of large effect: environmental factors play little if any role in the pathogenesis of these diseases. In contrast, the group of diseases which show a familial tendency are believed to reflect a pattern of inheritance in which multiple genes act together with environmental factors: this group of diseases is more common than the former and often shows a sex predilection.

In many other diseases there is a genetic component which is more difficult to quantify. For example, genetic factors may determine the effectiveness of the host response to a particular infection or the susceptibility of an individual to tissue damage by a specific environmental agent. There is a variation in the susceptibility of different individuals to carcinogenic agents, such as those in cigarette smoke, which may well be genetic in origin. Carcinogenic agents are thought to act directly on the cellular genome, ultimately giving rise to a malignant clone: thus acquired genetic defects are also important in the development of cancer. The role of inherited and acquired genetic defects in carcinogenesis will be discussed in more detail in the section on neoplasia.

Chromosomal abnormalities

These are found in approximately 0.5% of live births and are characterized by aberration of chromosome number or structure. Numerical abnormalities are usually due to non-disjunction of a chromosome pair at the first meiotic division. Structural abnormalities include isochromosome formation, translocations, deletions and chromosome breaks. Both numerical and structural abnormalities can be recognized by light microscope examination of somatic cells which have been stimulated to undergo mitosis. Improved preparatory techniques make it possible to detect a chromosome abnormality involving a single chromosome band—representing approximately 6000 kilobase pairs of DNA and containing many hundreds of genes. The majority of chromosome abnormalities occurring during embryogenesis are not compatible with life. Up to 50% of conceptions which abort spontaneously in the first trimester of pregnancy are found to have a chromosome aberration involving the autosomes or sex chromosomes. The tragedy associated with chromosomal abnormalities which are not life threatening is that they are almost always associated with severe mental retardation. An example of this is Down's syndrome (trisomy 21). These individuals constitute a third of the residents of UK mental institutions at this time. Fortunately, the severity of mental retardation may be modified by the presence of a normal cell line in addition to the abnormal one. (The presence of more than one cell line in a karyotype is termed 'mosaicism'.)

Chromosomal abnormalities involving the sex chromosomes are usually compatible with life, although the reproductive capacity of the individual is usually seriously impaired. Thus Turner's syndrome (45XO) is characterized by a female phenotype with various physical abnormalities including short stature and hypoplastic gonads, and Klinefelter's syndrome (47XXY) is characterized by a male phenotype, gynaecomastia and testicular atrophy. Mental retardation is less likely to be found in patients with sex chromosome anomalies than in patients with autosomal aberrations.

Mutant genes of large effect

Mutation of a single base in a DNA sequence, causing replacement of one amino acid by another in the protein for which it is coding, may produce a major functional alteration and lead to an inborn error of metabolism. Most of these errors of metabolism are characterized by a deficiency in an enzyme essential for the normal function of a particular type of cell. A group of glycogen and lipid storage diseases falls into this category, each with a specific enzyme deficiency, for example Von Gierke's disease in which a deficiency in glucose-6-phosphatase results in the accumulation of glycogen in the liver and kidney, leading to hepatomegaly, hypoglycaemia and failure to thrive.

Although individually each of these diseases is rare (occurring in 1 in 2000 to 1 in 10 000 births), there are over 2000 diseases which can be attributed to single gene mutations, so that as a group they may be found in 0.2% of live births. Consanguinity of the parents increases the risk of having an affected fetus. Advances in biochemistry and DNA hybridization techniques have enabled the mutant gene to be identified in over 100 of these diseases, and this technology has made pre-natal diagnosis possible in some instances. Most of these mutant genes show a recessive mode of inheritance, and only cause disease if the abnormal gene is inherited from both parents. Typical examples are sickle cell disease and cystic fibrosis. Diseases with a dominant form of inheritance are rarer, and may occur as a result of mutations arising *de novo* during gametogenesis. Examples are achondroplastic dwarfism and familial polyposis coli.

When the abnormal gene is carried on the X chromosome, giving rise to a sex-linked recessive mode of inheritance, the disease will occur in all males who inherit the gene but only in female homozygotes who have inherited the abnormal gene from both parents. Female heterozygotes will only have one affected X chromosome, and will be carriers. A typical example of this is haemophilia.

Many genetic diseases due to mutation of a single gene show racial and geographical variation in their incidence.

Multifactorial inheritances

There are many diseases such as systemic hypertension, atopic asthma, pyloric stenosis, cleft palate, insulin-dependent diabetes and certain forms of cancer in which there is a strong familial element. Most of these diseases are considered to be polygenic in origin. The nature of the underlying genetic changes are mostly unknown, although they may reflect the cumulative effect of a large number of 'weak' genes of small effect interacting with environmental factors. In some instances, such as ankylosing spondylitis and rheumatoid arthritis, predisposition to disease appears to be related to the histocompatibility complex whose gene loci are closely related to those which determine immune reactivity.

Association of disease with the histocompatibility complex

The histocompatibility complex is a closely linked group of gene loci on chromosome 6 which code for surface antigens which can be recognized on leucocytes, but are also present in varying amounts on all nucleated cells. There are four gene loci, named human leucocyte antigen (HLA) A, B, C and D, each of which have multiple allelic forms. Certain alleles, for example HLA A1 and B8, tend to be linked more often than accounted for by chance (linkage disequilibrium). These linkages show racial and geographical variation and are thought to confer some biological advantage, possibly involving resistance to infections. The four gene loci are inherited as a linked set and are closely connected with gene loci determining immune reactivity and also some of the components of complement. Certain diseases involving chronic inflammation and disorders of immunity tend to be associated with particular HLA sub-types. HLA B27 is strongly associated with ankylosing spondylitis and its related forms of arthritis, and HLA B8 linked to HLA DR3 is associated with autoimmune disease.

Environmental causes of disease

There is a wide variety of environmental agents causing disease, which includes infectious, chemical and physical agents as well as nutritional factors all of which may be modified by the effects of cell aging. These agents exert their effect by damage to cell membranes, cellular enzyme systems or nucleic acid. The damage may be acute and localized, such as can be seen in patients with hepatic necrosis caused by carbon tetrachloride, or may remain dormant, as during the long latent period before the development of malignant mesothelioma many years after a brief episode of exposure to asbestos.

In some instances, e.g. an acute staphylococcal abscess, the causative agent is highly specific and readily demonstrable, whereas in others a variety of agents can produce an identical pathological process. For example, the complex pathological pattern of cell damage, regeneration and repair characterized by diffuse interstitial pulmonary fibrosis can be produced by viral infection oxygen toxicity, chemotherapeutic drugs or irradiation. To complicate the picture further, the same pathological pattern can occur as an idiopathic disease without a recognizable cause.

Infectious agents

Included in this group of environmental pathogens are a multitude of bacterial species, viruses, fungi, protozoa and other multicellular organisms which parasitize the human body. In order for the organism to cause disease it must first breach the body defences and multiply within the host tissue. Damage to the host tissue by the invading organism is effected in different ways. For example, damage due to bacterial infection is mediated by endotoxins which are components of the organism itself, or by exotoxins which are secreted by the organism and may cause generalized damage far from the site of infection. Damage by viruses may be caused by complete cell lysis and cell death during the process of viral replication. This is usually a short-lived and localized process, rapidly overcome by the host defences. Overwhelming life-threatening viral infections such as yellow fever and Lassa fever are, fortunately, rare. Viral infections such as measles which only cause a mild disease in Western communities may cause overwhelming infection in communities which have never been previously exposed to the virus. In most cases of virus infections, recovery from the disease is associated with elimination of the infectious organism from the body. In some viral infections, however, e.g. herpes virus and papillomavirus, the organism may remain latent in the host cell, often integrated into the genome of the host. Latent infections may reactivate at any time to cause acute infection. This is particularly likely to occur in the immunocompromised patient. Latent infection has been implicated in the development of certain inflammatory, immunological and neoplastic diseases including rheumatoid disease, AIDS and cervical cancer.

It is important to remember that there are large numbers of microorganisms which inhabit the human body on the skin, the genital tract and in the gastrointestinal tract which are not normally pathogenic. They are regarded as harmless and even useful passengers as there is evidence that they keep pathogenic bacteria at bay. They are often referred to as the normal flora or commensals. This symbiotic relationship may break down if the host defence system is impaired by therapy or disease; under these conditions some of the commensal organisms may assume a pathogenic role.

Chemicals and drugs

Many naturally occurring and synthetic chemicals and drugs can cause direct damage to cells and enzyme systems. As with microorganisms, the damage can be acute or chronic. Chemical agents may be modified within the host, particularly by enzymic metabolism and detoxification in the liver. These enzymes may be overwhelmed by an excessive dose of the toxic chemical or may occasionally convert a harmless chemical into a more active form. Thus 2-naphthylamine is activated to a carcinogen by hydroxylation in the liver. This is conjugated to an inactive glucuronidase, but released as a carcinogen in the bladder if host urothelial cells secrete the genetically determined enzyme β-glucuronidase.

Physical agents

Direct mechanical trauma, heat, ultraviolet radiation and ionizing radiation cause a spectrum of disease ranging from localized cell death to long-term cell damage, the latter usually mediated by damage to DNA. Mechanical trauma and heat cause direct localized cell damage whereas ultraviolet light and ionizing radiation also cause sub-clinical damage to DNA which may result in neoplasia many years later, or teratogenic effects in the next generation. Thus laser or heat coagulation of an area of cervical neoplasia produces total destruction of all the affected cells, with little if any damage to neighbouring cells. Conversely, radiotherapy causes lethal damage restricted to dividing cells, but also results in secondary effects on adjacent blood vessels, regenerating fibroblasts and epithelial cells which may lead to irradiation fibrosis or even neoplasia if the doses are excessive.

Effects of environmental agents on the cell

The physical, chemical and microbiological agents described in the previous section affect the cell in several different ways. They may act by damaging the cell membrane, releasing lysosomal enzymes, deactivating essential cell enzymes or by damaging the DNA. The damage caused by these agents may be so severe that the cell may die. If exposure to injury is brief or the injury mild, the cell may survive and return to normal function. Damage to DNA may permit cell survival yet result in the development of an abnormal clone. The mode of action of some physical, chemical and microbiological agents is discussed below and the pathological changes associated with them are described.

Damage to the cell membrane

Cell membranes may be damaged by activation of complement on the cell surface by antigen–antibody complexes: this is the mechanism of cell lysis in ABO incompatibility. Many forms of membrane injury are mediated by the formation of superoxides and free radicals: this is the mechanism of oxygen toxicity, paraquat and carbon tetrachloride poisoning. Membrane damage can be caused indirectly by loss of ATP production by the cytochrome enzyme system in mitochondria: this is the mechanism by which hypoxia and cyanide poisoning cause their effect. Loss of energy from ATP leads to failure of the sodium pump mechanism, which exchanges sodium for potassium at the cell membrane in order to maintain the high concentration of potassium within the cell relative to extracellular fluid.

Loss of integrity of cell membranes results in cellular oedema, swelling of mitochondria and leakage of cellular enzymes into the extracellular fluid. These degenerative changes may be apparent by light microscopy as a granular or hyaline appearance of the cytoplasm.

Damage to DNA

Damage to DNA can be caused by many environmental agents. Ultraviolet light, ionizing radiation and cytotoxic drugs are but a few examples. Ultraviolet light acts by interfering with DNA repair mechanism, causing chromosomal damage and gene mutation. Some cytotoxic drugs act by interfering with the synthesis of nucleic acid, or the formation of the spindle. Thus they exert a maximum effect on cells that are undergoing mitosis, e.g. germ cells, haemopoietic tissue and tumour cells, which makes them particularly useful anticancer agents. They also have an important role to play in preventing rejection of the allograft in organ transplantation.

Given in the appropriate dose, ionizing radiation or cytotoxic agents cause cell death. However, they also have the potential for mutagenesis and the cells, should they survive, may have undergone transformation to a potentially malignant clone. Thus cytotoxic therapy administered for the destruction of cancer cells may under certain conditions actually initiate neoplastic change.

Enzyme deactivation

Severe heat or cold may deactivate enzymes and lead to impairment of cell function. Disturbance of the enzymes involved in lipid metabolism may result in the accumulation of fat in liver cells, recognized histologically as fatty change. Fatty change can also result from failure of enzyme synthesis in malnutrition or from enzyme poisoning by alcohol.

Release of lysosomal enzymes

Proteolytic enzymes are normally bound in lysosomes within the cytoplasm and play a physiological role in phagocytosis and the inflammatory response. When lysosomes rupture, their enzymes are released, causing damage to the cell or surrounding tissues. Release of lysosomal enzymes has been observed in gout and silicosis when ingested crystals fuse with the lysosomal membranes of neutrophils and macrophages causing them to rupture.

Cell death results in rupture of lysosomal membranes with release of lysosomal enzymes and autodigestion (autolysis) of both nucleus and cytoplasmic proteins. Microscopically, the nuclear mass shrinks and becomes hyperchromatic (pyknosis) and may fragment (karyorrhexis). The cytoplasm may become intensely eosinophilic. Autolysis of a large number of cells in a tissue or organ is termed *necrosis*, which may be coagulative or colliquative, depending on whether or not the underlying tissue has a fibrous reticulin framework. A particular form of amorphous necrosis, known as *caseation*, is seen in tuberculosis. Secondary infection of necrotic tissue by putrefying organisms is termed *gangrene*.

Damage due to microbiological agents

Although it is obvious that bacterial toxins damage and kill body cells little is known about how they do so. Cell membranes are particularly susceptible to damage by bacterial phospholipases but some bacteria appear to act by inhibiting protein synthesis (e.g. diphtheria toxin) while others act by stimulating enzyme synthesis (e.g. cholera toxin).

The mode of attack of viruses is better understood than that of bacteria. Viruses attach themselves to a cell by absorption and penetrate the cell by a process of pinocytosis. The protein capsid is stripped by host cell enzymes and the viral nucleic acid freed to take over the cell function. The host cell proceeds to synthesize viral DNA, RNA, protein and lipid and to assemble new virus particles, bringing to a halt normal host cell activities. Thus a virus may cause cell damage or cell death by a variety of methods.

The distinctive changes induced in the cell by the virus are described as the *cytopathic effect* and may take the form of intranuclear or intracytoplasmic inclusions and degenerative changes in cytoplasm and nucleus. They may also result in cell fusion. Occasionally, the effect of the virus is to alter the host genome in such a way that the cell may be transformed and acquire neoplastic characteristics.

Host responses to disease

The main components of the host response to cell and tissue damage due to physical, chemical and infectious agents are the inflammatory and the immune responses. As will emerge during consideration of the inflammatory diseases, these are frequently modified by genetic factors as well as by additional host factors such as ageing and nutrition. Essentially, the inflammatory response is a reaction to cell damage, irrespective of its cause, whereas the immune response is a specific response directed against foreign cells or antigens recognized as 'non-self'.

The inflammatory response

The clinical features of the acute inflammatory response are redness, heat, swelling, pain and loss of function. The redness and heat are due to dilatation of capillaries, bringing blood to the affected area. Only in the skin will this raise the temperature because it is normally below that of blood. Later in the response there is slowing and stasis of blood, following escape of plasma from the vessels. The swelling results from increased vascular permeability, leading to tissue oedema due to exudation of protein-rich fluid, including fibrinogen, through gaps in the endothelial lining of the venules. This increase in vascular permeability is mediated by chemicals such as histamine. In severe injury vascular permeability is further enhanced by direct endothelial cell damage. Extravascular accumulation of leucocytes also contributes to the swelling. The third component of the acute inflammatory response is pain. The degree of pain experienced is related to increased tissue tension, but may also be caused by some of the chemical mediators such as the kinins and prostaglandins. Loss of function, except in the case of extensive tissue necrosis, is mainly a reflex response to pain and swelling.

From this description it is clear that there is both a cellular and a chemical component to the acute inflammatory response. The effect of chemical mediators on vascular permeability and the movement of leucocytes into the affected area result in the physical signs which we associate with this response.

Cellular components of the inflammatory response

The main cells involved in the acute inflammatory response are neutrophil polymorphonuclear leucocytes and macrophages which accumulate in the later stages of the process. Both are actively phagocytic cells, which are attracted by chemotaxis to sites of cell damage or invasion by microorganisms. Neutrophils are the main cells in an acute inflammatory exudate.

Neutrophil polymorphonuclear leucocytes

Early in the acute response, when blood flow slows down, neutrophils marginate and adhere to the endothelial surface of small blood vessels. They insinuate themselves between endothelial cells by putting out pseudopodia, and thus reach the extracellular space. The prime function of neutrophils is to phagocytose and destroy invading microorganisms. This is facilitated if the organisms are coated either by IgG or the C3b component of complement (opsonization). Receptors for C3b are particularly important because they enable neutrophils to engulf organisms which are not yet recognized by the immune system. Neutrophils are short-lived phagocytes, which either kill and digest the engulfed material, or die.

Eosinophils

These are also involved in the inflammatory response, are less actively phagocytic, and are attracted by factors released by mast cells. Their main role lies in controlling immediate hypersensitivity (allergy) and in resistance to parasites.

Macrophages

These cells are derived from maturation of blood monocytes originating in the bone marrow, and have similar receptors and phagocytic properties to neutrophils. They are longer lived, appear later in the response, and play a more important role in the uptake and clearance of cell debris. Microorganisms may survive for long periods within macrophages, and their destruction depends on interaction with lymphocytes. Macrophages secrete a wide range of enzymes and chemical mediators involved in chronic inflammation, and are involved in processing antigen for the immune response.

Chemical mediators of inflammation

The increased vascular permeability and accumulation of leucocytes at the site of inflammation are largely mediated by endogenous chemicals derived from the plasma cascade and complement systems or from damaged cells. The plasma cascade systems which are involved in the inflammatory response include the clotting, kinin and fibrinolytic systems. The stimulus for their activation is believed to be exposure of collagen and its contact with Hageman factor (clotting factor). Thus the mediators of the inflammatory response and the mediators of haemostasis and intravascular clotting are intimately related. Among the important chemical mediators of the kinin system is bradykinin which is involved in vasodilatation and vascular permeability and in the production of pain. Also important are the components of complement C3a and C5a which are chemotactic for leucocytes.

As stated above, chemical mediators of the inflammatory response can also be released by damage to cells. The most powerful mediator released in this way is histamine, which is derived from mast cells. The release of histamine is achieved in several different ways. It can be released by the action of physical and chemical agents, by the components of complement C3a and C5a, and by the action of neutrophil lysosomal enzymes on the mast cell. It can also be released by antigen binding to mast cell-bound IgE.

Prostaglandins derived from lipid in cell membranes, leucotrienes secreted from polymorphs and lysosomal proteolytic enzymes mainly derived from neutrophils are other chemical mediators of inflammation which, like histamine, are also derived from cells.

The immune response

Cells of the lymphocyte series recognize foreign antigens which they must distinguish from 'self' antigens. They proliferate to form clones of cells which react against the antigens, either through antibody (immunoglobulin) production or by cell-mediated immunity. They also provide a circulating pool of memory cells which recognize the antigens on subsequent encounters. Two lymphocyte populations have been recognized: B lymphocytes and T lymphocytes.

Circulating antibodies and the B lymphocyte system

B lymphocytes in man probably mature in the bone marrow, but derive their name from avian 'bursa-dependent' lymphocytes which mature and develop immunological competence in the bursa of Fabricius. During lymphocyte differentiation in fetal life, cells of the B series undergo gene rearrangements involving translocations of specific gene loci on different chromosomes, so that individual B lymphocytes have a heritable ability to code for one of millions of possible immunoglobulin variable regions, each of which is specific for a particular antigen. Unstimulated B lymphocytes synthesize immunoglobulins which float in the lipid cell membrane with their antigen binding sites exposed.

The binding of antigen to surface immunoglobulin on B lymphocytes stimulates them to undergo blast transformation and secrete immunoglobulin into the plasma. These immunoglobulin molecules have an identical specificity to that of the surface immunoglobulins on the original unstimulated lymphocyte. The transformed B lymphocytes are known as plasma cells. B memory cells are also produced with the same specificity, which join the circulating pool of lymphocytes and mount a greater and more rapid response on subsequent encounter with the same antigen.

Immunoglobulins

Immunoglobulins (Ig) are plasma proteins with a molecular weight of 500 000–960 000 daltons. Each molecule is composed of two heavy chains, which are antigenically different for the five classes of Ig (IgG, IgA, IgM, IgE, IgD), and two light chains which must be either kappa or lambda in any one lymphocyte. The molecule can be cleaved by papain into three fragments: one Fc fragment and two identical Fab fragments (*Figure 3.1*). The Fc fragment, composed entirely of heavy chain, binds to cell surface receptors and activates complement. The Fab fragments contain the variable regions which comprise the antigen binding sites.

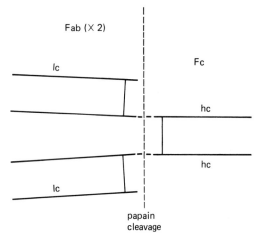

Figure 3.1 Diagram of immunoglobulin molecule to show papain cleavage fragments. lc = light chain; hc = heavy chain

The functions of the main classes of Ig are as follows:

(1) IgG. The most abundant Ig in plasma, IgG readily diffuses into extracellular fluid and crosses the placenta. It neutralizes bacterial toxins, opsonizes bacteria facilitating

phagocytosis, and binds complement when combined with antigen.
(2) IgA. Dimeric IgA is secreted into the gut or glandular lumen from mucosa-associated lymphoid tissue in the Peyer's patches of the small intestine, breast and many other sites.
(3) IgM. Pentameric IgM is mainly confined to the plasma, binding to and lysing circulating microorganisms and virus infected cells. IgM is the predominant Ig type involved in antigen binding and is secreted early in the immune response; later there is a switch, probably in the same cells, to IgG or IgA synthesis.
(4) IgE. Reaginic monomeric IgE antibody binds to Fc receptors on mast cells. Binding of specific antigen causes release of the mediators of immediate hypersensitivity.
(5) IgD. This is involved as a surface Ig binding antigen. Its function is obscure.

Cell-mediated immunity and the role of T lymphocytes

T lymphocytes form a separate series of cells which mature and acquire immunological competence in the thymus during fetal life. Unstimulated immunologically competent T cells have receptor sites which bind specific antigen. They differ from B cells in that the T cell receptors only recognize cell-bound antigens linked to histocompatibility antigens which are present on macrophages, transplanted cells and certain tumour cells: hence T cells have an important role to play in the control of virus infection, in organ transplantation and in neoplasia. They are also responsible for delayed hypersensitivity reactions. T lymphocytes activated by binding with a specific bound antigen produce clones of effector cells against that antigen with the following functions:

(1) Cytotoxicity: direct lysis of target cells.
(2) Lymphokine secretion: soluble factors which are chemotactic to macrophages and facilitate macrophage destruction of ingested microorganisms.
(3) Helper function for Ig synthesis: soluble factors from T lymphocyte helper sub-sets stimulate proliferation and maturation of B lymphocytes with receptors for the same antigen.
(4) Memory cells: circulating T memory cells are produced during clonal proliferation for a specific antigen.
(5) Suppression of immunity: T lymphocyte suppressor sub-sets exert control over both B and T cell-mediated immunity.

Inflammatory and immunological disease processes

The cell and tissue changes seen in this group of diseases reflect the balance between the damaging action of toxic and infectious agents on the cells of the body and the host response. The histopathological changes visible in the light microscope depend on the intensity and duration of exposure of the cells and tissue to injury, and on the efficiency of the inflammatory and immune responses. In those instances where the host responses are normal and able to contain the infection within a short space of time, the pattern of acute inflammatory disease may be seen. Incomplete resolution of the infection or continuous irritation by physical or chemical agents may result in chronic inflammatory disease. Abnormal, deficient or hyperactive inflammatory or immune response may result in autoimmune disease, immunodeficiency disease, hypersensitivity reactions and granuloma formation. These various patterns of inflammatory and immunological disease and their sequelae are discussed in this section.

Acute inflammatory disease

The pathological features of acute inflammatory disease vary in different sites and with the severity of infection. Some acute infections are more likely to occur when host resistance is lowered. Thus bronchopneumonia is more common at the extremes of life, when the immune and inflammatory responses are less efficient, and also during debilitating diseases such as advanced malignancy. Impairment of the cough reflex and acute viral damage to the ciliated bronchial epithelium also predispose to the development of bronchopneumonia.

The histological changes of acute inflammation are manifest in histological section by tissue damage and the presence of an exudate. Tissue oedema, cellular degeneration and cell death occur together with a polymorphonuclear leucocyte infiltrate. Several types of exudate may be seen.

Mucinous exudate

Low grade infection of mucosal surfaces produces a hypersecretion of mucus accompanied by an exudate of neutrophils. Typical examples are the sticky neutrophil-rich exudates seen in nasopharyngeal and endocervical secretions associated with relatively little evidence of inflammation in the underlying tissue. This pattern of response is often described as catarrhal inflammation.

Fibrinous exudate

In reactive exudates in which there is no bacterial growth, there is often a rough shaggy exudate which is rich in fibrin. This is typically seen in reactive serous effusions overlying pneumonia or non-infective lesions such as infarcts.

Pseudomembranous exudate

This occurs when an exudate is accompanied by necrosis of the superficial epithelium, and is typical of diphtheria, colitis due to certain bacterial endotoxins, and ulcers.

Purulent exudates

These may be mucopurulent of fibrinopurulent, depending on whether a mucous membrane is involved or a serosal surface or visceral organ. A typical example of the latter is seen in acute lobar pneumonia. An exudate composed of neutrophils and fibrin-rich fluid fills the alveolar spaces, giving the lung a consolidated airless appearance. This is red during the early stage of vasodilation and becomes grey when alveolar capillaries constrict to divert blood away from non-aerated lung.

Pus

Pus is an exudate containing a heavy growth of pyogenic bacteria, which is rich in dead and dying neutrophils, and tissue liquefied by lysosomal enzymes. In enclosed tissue spaces this will form an abscess, and the increased tissue pressure will inhibit the further formation of exudate. This will become walled off by fibrous tissue unless the pus discharges spontaneously, or is drained surgically onto the nearest surface.

Sequelae of acute inflammation

Acute inflammation can result in complete resolution, regeneration of damaged tissue or organization by fibrous scar tissue. Under some circumstances the inflammatory process does not resolve and becomes chronic.

Resolution and regeneration

Acute inflammation without tissue necrosis, such as in acute lobar pneumonia, can undergo complete resolution. This process relies on effective fibrinolysis and macrophage uptake of cellular and exudate debris. If cells have been damaged the possibility of regeneration depends on whether the cells are labile, stable or permanent.

Labile cells

These multiply and regenerate throughout life, and include haemopoietic cells, germ cells, the cells of the basal layer of the squamous epithelium and cells of serosal surfaces. Unless very extensively damaged, these will regenerate, even when there is fibrous scarring of the underlying tissue.

Stable cells

These do not normally multiply during adult life, but retain the ability to do so following tissue damage. Examples are hepatocytes and renal tubular cells. Functional regeneration of these cells often depends on whether or not the underlying connective tissue framework has been damaged. If the reticulin framework is intact, the hepatocytes or renal tubular cells will proliferate. If the framework is damaged, repair will be by fibrosis.

Permanent cells

These comprise neurones and voluntary muscle cells which are unable to multiply in adult life. In every case repair is effected by fibrosis.

Wound healing and repair by fibrosis

Repair by fibrosis will occur whenever there has been extensive tissue damage or a fibrinous exudate has overloaded the fibrinolytic system.

Healing of a skin wound involves regeneration

of surface epithelium, the formation of granulation tissue and fibrosis. The difference between healing of an incised wound (primary intention) and an open wound (secondary intention) are only differences of degree, and the same processes are involved in both. Similar processes are involved in healing of visceral surfaces, although a slightly different mechanism of regeneration exists for serosal surfaces.

Epithelial regeneration

Migration of epithelial cells over the wound surface is accompanied by increased mitotic activity at the wound edge. The regenerating epithelium spreads onto the exposed surface and temporarily down suture tracks. This process may in part be mediated by loss of feedback from normal inhibitory factors, known as chalones, which are thought to be produced by mature epithelial cells and lost when these cells are damaged. Active growth factors such as epithelial growth factor and platelet-derived growth factor may also be involved.

Granulation tissue

Within a few days of injury small capillaries grow as buds from dermal capillaries. These capillary loops, which bleed easily, form a pink granular surface on an exposed wound giving rise to the term 'granulation tissue'. Proliferating alongside these vessels are fibroblasts and myofibroblasts. The latter contain actin and myosin which contract during early wound healing, considerably reducing the size of the wound.

Fibrous scar tissue

Fibroblasts migrate with the small vessels and become orientated parallel to the wound surface. These cells synthesize collagen which provides the structure and tensile strength of scar tissue. The production of collagen is inhibited when the overlying epithelial surface is replaced. In some circumstances, particularly when the overlying surface is extensively damaged, the process of fibrosis may be excessive and lead to the formation of a fibrous scar. In some instances this can result in severe restriction of movement and function. An example is the constrictive pericarditis which may follow tuberculous infection.

Wound healing at any site requires vitamin C, which is involved in collagen synthesis, and is accelerated by zinc. The deposition of fibronectin, a connective tissue glycoprotein, from soluble precursors in plasma, extracellular fluid and cell surfaces, facilitates cell migration and may potentiate the activity of macrophages and fibroblasts in wound healing. Corticosteroids delay wound healing as does an impaired blood supply, such as occurs with atherosclerosis.

Chronic inflammatory disease

Inflammation may persist and become chronic if drainage of an exudate is impaired, if the vascular supply to the area is reduced or if foreign material is present in the lesion. The histological appearance is characterized by a combination of unresolved inflammation and an immune response. There is likely to be an exudate, tissue necrosis, granulation tissue, fibrosis and a mixture of inflammatory cells almost always including **plasma cells**.

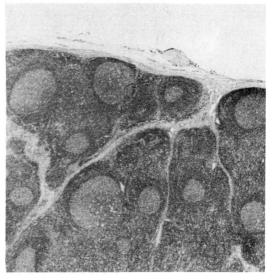

Figure 3.2 Low power photomicrograph of a lymph node showing reactive follicular hyperplasia. Stain: haematoxylin and eosin. Magnification × 60

Most forms of localized chronic inflammation will show evidence of a reactive follicular hyperplasia in lymph nodes draining the area and in secondary lymphoid follicles which develop at the site of inflammation (*Figure 3.2*). The reactive follicles, or germinal centres, contain macrophages which have ingested antigen and B lymphocytes undergoing blast transformation and maturation to antibody-producing cells.

A typical example of chronic inflammatory disease is bronchiectasis. This is a persistent non-specific bacterial infection of the major bronchi which results in the production of copious purulent sputum. The damaged bronchi become grossly dilated and are lined by chronically inflamed granulation tissue, in which lymphoid follicles and plasma cells are both prominent. The inflammation is always associated with some degree of impairment of bronchial clearance, either due to mechanical bronchial obstruction by tumour or foreign body or as a result of a severe acute infection such as whooping cough, in which bronchial ciliary clearance is impaired. This sets up a vicious circle of infection and progressive loss of clearance. In the example illustrated in *Figure 3.3*, bronchiectasis has been caused by a central bronchial carcinoma.

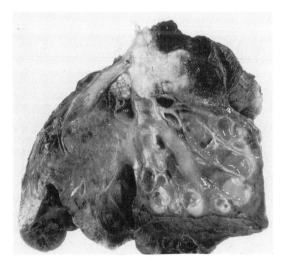

Figure 3.3 Surgical specimen of the lower lobe of a lung, demonstrating saccular bronchiectasis behind an obstructing bronchial carcinoma

Hypersensitivity reactions

In some cases of host response to infection and injury the immune response is excessive, and results in damage to the host rather than elimination of infection. This is known as hypersensitivity. In some infections which are characterized by hypersensitivity reactions, the altered immune response is as important as the microbiological organism as a cause of cell and tissue damage. Four types of hypersensitivity reactions are recognized.

Type I, immediate hypersensitivity

In certain individuals there is an inherited tendency to produce IgE antibodies to a wide range of non-infective antigens, including pollen, house mites and animal dander. This produces a variety of conditions including eczema, hay fever and asthma.

Type II, cytotoxic hypersensitivity

This involves circulating antibodies reactive against surface antigens on host cells, resulting in their lysis through the mechanisms described above. Examples are immune haemolysis, as in incompatible blood transfusions, and drug-induced haemolytic anaemia, in which non-protein drugs form antigenic 'haptens' with host cells.

Type III, immune complex-induced hypersensitivity

Under some circumstances antigen/antibody complexes or 'immune' complexes are formed which are not phagocytosed and eliminated in the normal way. These complexes usually have an excess of antigen and cause a local inflammatory reaction mediated by the activation of complement when they are deposited in tissues. Typically these local reactions may be in vessel walls, as in the necrotizing vasculitis of the Arthus reaction, or in glomeruli when deposited complexes produce a focal glomerulonephritis.

Type IV, delayed hypersensitivity

A typical example of this is contact dermatitis, in which host cells are altered by exogenous agents such as certain chemicals or metals, to render them antigenic to T lymphocytes. These antigens are presented to the immune system by specialized Langerhan's cells in the epidermis, and mount a cell-mediated immune response, characterized by oedema, blisters and lymphocyte infiltration. Delayed hypersensitivity also plays an important part in chronic granulomatous infections and the tuberculin reaction.

Chronic granulomatous disease

This group of diseases involves both chronic inflammation and delayed type, T lymphocyte-mediated hypersensitivity. They are characterized by the persistence of microorganisms or other particles within macrophages which become converted to secretory rather than phagocytic forms. Because of their appearance these are described as 'epithelioid' cells, and they frequently fuse to form various morphological types of giant cells. These converted macrophages contain lysosomal enzymes, continue to react with T lymphocytes and secrete factors which stimulate fibroblasts.

Chronic granulomatous infections are typified by tuberculosis, leprosy and schistosomiasis. Foreign material such as talc, surgical sutures and asbestos can also give rise to granulomatous reactions. Many conditions exist in which the causative agent is unknown. These may form necrotizing granulomata, as in Wegener's granulomatosis, or non-caseating granulomatosis as in sarcoidosis.

Tuberculosis

The uptake of mycobacteria by macrophages, their transformation to epithelioid cells and Langhans-type giant cells and the subsequent interaction of the altered macrophages with T lymphocytes leads to the caseation necrosis which is so typical of tuberculosis. Identical caseating granulomata are formed in BCG immunization suggesting that hypersensitivity to some component of the organism is involved. Good cell-mediated immunity leads to eventual killing of the organisms by macrophages and the lesions become progressively organized by fibrosis. Poor immunity leads to the formation of widespread small poorly developed miliary granulomata with little necrosis or fibrosis.

Sarcoidosis

This obscure condition is characterized by the formation of non-caseating granulomata, consisting of epithelioid cells, Langhans-type giant cells and T lymphocytes, with a marked tendency to undergo fibrous healing (*Figure 3.4*). These can be widespread in the body, but most often involve hilar lymph nodes and lung. This appears to be a classic example of cell-mediated hypersensitivity, but the causative agent is unknown. It appears unlikely to be an infective organism, since the condition responds well to corticosteroids.

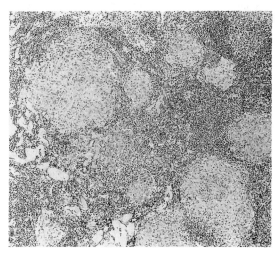

Figure 3.4 Photomicrograph of non-caseating granulomata in a lymph node from a patient with sarcoidosis. Stain: haematoxylin and eosin. Magnification × 150

Autoimmune disease

In this group of conditions circulating antibodies are directed against components of normal host cells. There are two main groups of autoimmune diseases. The first involves organ-specific antibodies, and includes Addison's disease, chronic autoimmune thyroiditis and pernicious anaemia. The cause is obscure, but in the case of autoimmune thyroiditis inappropriate expression of histocompatibility antigens has been demonstrated on epithelial cells. The second group includes the connective tissue diseases such as systemic lupus erythematosis in which anti-DNA antibodies are formed, and rheumatoid arthritis in which antibodies are formed against immunoglobulin. This latter group is associated with certain inherited combinations of histocompatibility antigens.

Immune deficiency disease

There are numerous syndromes of inherited deficiencies of specific components of B and T cell-mediated immunity. Information about these rare conditions can be found in the references listed at the end of this chapter. In clinical practice the acquired immune deficiency syndromes are more important. These usually result from immune suppresion by irradiation or chemotherapeutic anticancer drugs, but may result from infection by the human immunodeficiency virus (HIV). Other viruses, including measles, and chronic infections such as malaria also cause some degree of immunosuppression.

Acquired immune deficiency syndrome (AIDS)

This is a sexually transmitted disease with a high incidence in homosexual men, prostitutes and drug addicts. It is caused by a virus which acts directly on T lymphocyte helper cells, resulting in suppression of the immune response and the development of opportunistic infections, many of which are fatal. People with overt disease, an intermediate group with persistent lymphadenopathy, and many more who have been exposed to the virus, have antibodies to HIV in their serum.

Vascular disease processes

In this group of diseases the normal haemostatic mechanisms responsible for the repair of damaged and inflamed blood vessels are deranged by multiple environmental and host-related factors. Thrombosis and atherosclerosis are the commonest causes of death in the western world.

Thrombosis

Thrombosis refers to intravascular coagulation of red blood cells, leucocytes and platelets, bound together by fibrin. Unlike extravascular clotting, thrombosis is initiated by the aggregation of platelets and subsequent release of mediators. Contact with collagen exposed under damaged endothelium, release of ADP from damaged cells, and thrombin all stimulate platelet aggregation.

Factors causing thrombosis

Thrombosis can result from changes in blood flow, damage to vascular endothelium or changes in coagulability of blood. More than one mechanism is usually involved in the initiation and development of thrombosis at any site:

(1) Alterations in blood flow. Stasis of blood is the main mechanism involved in the developments of deep vein thrombosis which may occur after surgical operations and prolonged bed rest. Turbulence and loss of laminar flow are more often involved in arterial and intracardiac thrombosis.

(2) Damage to vascular endothelium. The most important of these is the damage to endothelium which is seen overlying an atheromatous plaque.

(3) Changes in coagulability of blood. There is an increased tendency to thrombosis following severe trauma, major abdominal operations and childbirth. Pregnancy and certain oral contraceptives also increase the coagulability of blood. Multiple factors are involved, including alterations in fibrinolysis and platelet aggregation. Inflammation, neoplastic infiltration and trauma can also result in local thrombosis.

Complications of thrombosis

Complications of thrombosis may occur depending on the degree and site of vascular obstruction, and whether arteries or veins are involved. They result from increased hydrostatic pressure, ischaemia and thromboembolism formation.

Venous congestion and oedema

The increased hydrostatic pressure caused by venous thrombosis results in peripheral oedema only if there is no alternative venous drainage. Thus there is little oedema in cases of deep vein thrombosis as the superficial veins provide an alternative pathway.

Where there is no alternative drainage, there may be intense congestion and haemorrhage, resulting in necrosis of the affected area. This is referred to as infarction, and is seen in strangulated hernias and mesenteric venous thrombosis.

Ischaemic infarction

Thrombotic occlusion of an artery, again where there is no collateral blood supply, will result in ischaemic necrosis of the affected part. This necrotic tissue will elicit an inflammatory responce and ultimately be organized by fibrous scar tissue.

Embolism

Thrombi can detach from their site of origin and become impacted in a vessel at a distant site. The commonest example of this is pulmonary embolism, usually derived from deep vein thrombosis in the calf muscles, which may cause major circulatory disturbances or sudden death. Embolism is most commonly due to thrombosis, but can result from any space-occupying substance carried in the bloodstream. This can be air introduced during surgical procedures, nitrogen released from tissues during decompression, or fat and fragments of bone marrow introduced into the veins following fractures of bones.

Atherosclerosis

The terms atheroma and atherosclerosis tend to be used synonymously and refer to the deposition of lipid-rich material in the intima and media, leading to disruption of the vessel wall, connective tissue proliferation and narrowing of the lumen. There is no single aetiological factor and the pathogenesis must be considered in the light of the morphology and distribution of the lesions, and the aetiological risk factors.

Morphology, distribution and complications of atheroma

A spectrum of lipid-rich arterial lesions occur during life, and the early ones may not be the precursors of the fully developed lesions. The latter consist of raised atheromatous plaques, composed of cholesterol-rich material deposited in the intima and media often disrupting the internal elastic lamina. The media is often thinned and the luminal surface capped by connective tissue under the endothelium.

Fully developed atheromatous plaques are mainly seen in the abdominal aorta, particularly around the ostia of its branches. Coronary, iliac, carotid and cerebral arteries are frequently affected: coronary arteries tend to be affected at an earlier age. Complication may occur if the connective tissue cap ulcerates, exposing collagen to platelets and leading to thrombosis. Haemorrhage and calcification may occur within the plaque. Haemorrhage and thrombosis may both result in sudden vascular occlusion, particularly in coronary arteries. The thinned media may undergo aneurysmal dilatation, particularly in the abdominal aorta, and this may ultimately rupture causing catastrophic haemorrhage.

Risk factors and pathogenesis of atherosclerosis

The most important risk factors are hyperlipidaemia, hypertension and cigarette smoking. Stress, obesity, diet, oral contraceptives, lack of exercise, diabetes mellitus and genetic factors may also be involved, possibly mediated by an effect on plasma lipids.

The precise pathogenetic mechanism of atheroma is unknown, but the evidence suggests that the process involves firstly damage to endothelium, mediated by cigarette smoking and/or hypertension, secondly the increased uptake of plasma lipids into the vessel wall, possibly related to increased permeability of the endothelium, and thirdly the release of platelet derived growth factor from the aggregation of platelets on a functionally damaged or interrupted endothelial surface, which results in the proliferative connective tissue changes.

Disorders of growth

Many cells in the body continue to undergo cell division throughout life: these are the labile cells. The dividing cells give rise to daughter cells which either become reserve cells which maintain the ability to divide, thereby ensuring the continuation of that particular cell line, or become differentiated cells which progressively mature and die. An example is provided by the cells lining the crypts and villi of the small intestine. The reserve cells in the crypts divide to produce the epithelial cells which advance up the villi. These epithelial cells differentiate, mature and die, eventually to be exfoliated into the lumen of the gut while the reserve cells continue to proliferate.

Cell division and differentiation is normally controlled so that the structure and function of a tissue is maintained to meet the body's demand and an equilibrium is reached between cell gain and cell loss. Cell division can be stimulated by an increase in functional demand. Increase in the number of cells as a response to functional demand is known as *hyperplasia*. Cells which do not usually divide, i.e. stable and permanent cells, respond to increased demand by an increase in size of the individual cells. This is known as *hypertrophy*. Both these processes must be distinguished from *neoplasia* which describes the uncontrolled growth of one of the cell types in a tissue or organ.

Hypertrophy

This describes the increase in bulk of an organ due to increase in cell size as a response to functional demand. Most examples of hypertrophy are found in voluntary and smooth muscle, although visceral organs such as liver and kidney can hypertrophy, for example in the residual kidney as a response to nephrectomy. Hypertrophy may be found in the heart and blood vessels. Left ventricular hypertrophy results from systemic hypertension, aortic valvular stenosis and mitral or aortic incompetence. Increase in pulmonary vascular resistance (pulmonary hypertension), most commonly resulting from hypoxia, causes right ventricular hypertrophy. Increase in vascular resistance is associated with medial hypertrophy of the arterioles: this is the hallmark of hypertension, whether systemic or pulmonary. Secondary changes, such as intimal proliferation or fibrinoid necrosis, may be present in severe hypertension, but the underlying pathological change is hypertrophy of smooth muscle in the media.

Hyperplasia

In hyperplasia the increase in organ bulk results from an increase in the number of cells, although there may be simultaneous hypertrophy. Physiological hyperplasia is seen in the breast and thyroid in pregnancy and lactation. The best examples of pathological hyperplasia occur in the endocrine glands, in which hyperplasia may result from an increase in stimulatory hormone production or from a breakdown in the regulatory negative feedback system.

Cushing's syndrome, which is characterized by hypersecretion of corticosteroids, may occur as a result of either of these mechanisms. The adrenal cortex secretes corticosteroids as a response to adrenal cortical stimulating hormone (ACTH), normally secreted by the anterior pituitary. ACTH secretion is modulated by the level of cortisol in the plasma. Adrenal cortical hyperplasia may be produced by excessive ACTH secretion from a pituitary adenoma or from an extrapituitary tumour such as an oat-cell carcinoma of bronchus. In some instances of Cushing's syndrome with adrenal cortical hyperplasia there is increased secretion of ACTH by the pituitary but no tumour is found. The mechanism whereby the plasma cortisol has failed to switch off pituitary secretion of ACTH is incompletely understood. Cushing's syndrome can be caused by a functioning adrenal neoplasm as well as by adrenal hyperplasia, in which case the surrounding adrenal cortex will be atrophic.

Other examples of hyperplasia reflect a mixture of physiological and pathological conditions. Cyclical hyperplasia and involution occur in the breast and thyroid during reproductive life in the female, and may result in conditions of nodular hyperplasia. A relative iodine deficiency may contribute to the development of a colloid nodular goitre. Fibrocystic disease of the breast is characterized by features of ductal and lobular epithelium coexisting with involution and fibrosis. A rather similar form of nodular hyperplasia is seen in the prostate in which hyperplasia coexists with smooth muscle hypertrophy.

Atrophy

Atrophy may reflect lack of cell division or reduction in cell size. An example of the former is seen in testicular atrophy. This occurs in cases of undescended testes, in pituitary hypoplasia and following oestrogen therapy for carcinoma of the prostate.

Muscular atrophy follows inactivation, ischaemia or denervation.

Metaplasia

Labile epithelial cells at any site may differentiate towards a different cell type as a response to irritant stimuli. Several different types of metaplastic change have been described:

(1) *Squamous metaplasia*. This is particularly common in glandular epithelium, which undergoes metaplasia to the more resistant squamous epithelium. This occurs as a physiological event in the epidermidization of endocervical glandular epithelium, particularly at puberty and after pregnancy. Bronchial epithelial cells frequently undergo squamous metaplasia as a reaction to irritants in cigarette smoke.

(2) *Mucous metaplasia.* Bronchial epithelial cells also undergo mucous metaplasia to goblet cells as a reaction to a variety of irritants, including those in cigarette smoke. This, in addition to hyperplasia of bronchial mucous glands, results in the increased sputum production which characterizes chronic bronchitis.

(3) *Intestinal metaplasia.* Gastric epithelium may undergo intestinal metaplasia to the type of mucus-secreting epithelium normally seen in the colon. This occurs as a response to irritation, and is a particular feature of atrophic gastritis in pernicious anaemia.

Metaplasia should not be regarded as a premalignant change, but it can be associated with the development of tumours. Bronchial carcinomas frequently develop in areas of squamous metaplasia and cervical carcinoma tends to occur in the transformation zone, which is the area in which squamous metaplasia takes place. Similarly intestinal metaplasia is frequently found adjacent to gastric carcinomas. It is also associated with atrophic gastritis which is a precancerous condition.

Neoplasia

A neoplasm is a tumour mass derived from the uncontrolled growth of cells. Most neoplasms arise from a single abnormal cell, although the abnormal clone may undergo further mutations as the tumour develops. The resultant tumour mass lacks the structure, organization, and consequently most of the function of the organ in which it has arisen, even though the individual neoplastic cells may closely resemble the equivalent normal cells.

The structure, function and behaviour of neoplasms is related to their potential for growth, differentiation, invasion and metastasis. They are classifed as benign or malignant, according to their growth characteristics.

Tumour growth

As the term neoplasia implies, growth is a function of all tumours, whether benign or malignant, and in general the microscopic measurement of the number of cells in mitosis per unit area of a tumour is an indication of its degree of malignancy. It must be remembered, however, that few tumours have a growth rate comparable to that of a developing fetus. Similarly, the mitotic rate in glands of proliferative phase endometrium, or regenerating epithelium adjacent to a wound or ulcer, may be equal to or greater than that of a malignant tumour.

There is evidence that the inappropriate growth of tumours is frequently related to the presence or absence of growth factors or their receptors. Normal squamous cells can only proliferate in the presence of epidermal growth factor (EGF), which they themselves secrete, and as they differentiate and mature so they lose their ability to respond to this stimulus. It has been suggested that overproduction of EGF receptor may be involved in the growth of squamous carcinomas and other tumours. Some tumours produce abnormal growth factor; the best example of this is bombesin, synthesized by oat-cell carcinoma.

Differentiation of tumours

When normal cells differentiate they lose the ability to divide, while undifferentiated stem cells retain this ability. Similarly, highly differentiated tumours, which closely resemble the equivalent normal cell, tend to grow slowly, whereas poorly differentiated tumours grow rapidly. The degree of differentiation of malignant tumours can be used to assess their potential aggressiveness, and is the basis for grading. Thus Grade I tumours are well differentiated, Grade II intermediate and Grade III poorly differentiated. Increasing grade tends to parallel the mitotic rate and also the degree of nuclear pleomorphism.

The spread of tumours

Malignant tumours can be distinguished from benign ones by their ability to infiltrate adjacent normal tissues and set up metastatic foci of growth at distant sites. Both these features are responsible for the life-threatening nature of malignant disease. Benign tumours expand by compressing rather than infiltrating the adjacent tissue and do not have the potential for metastasis: their clinical significance depends entirely on their location and function. Thus a benign intracranial tumour may be life-threatening, as may a functioning pituitary adenoma, due to the excessive production of ACTH. Malignant tumours spread in a number of different ways:

(1) *Local spread.* Infiltrating tumours tend to spread preferentially along pre-existing loose connective tissue planes. Infiltration may be related to the loss of cohesiveness of malignant cells and to their ability to break down collagen, particularly basement membrane. Certain tissues, notably cartilage, are more resistant than others to infiltration by malignant cells. Also arteries are less likely to be infiltrated than veins.

(2) *Lymphatic spread.* Carcinomas readily infiltrate lymphatics, and initially metastasize to the lymph nodes draining the primary site (*Figure 3.5*). This may cause obstruction to lymphatic flow, resulting in lymphoedema, serous effusions or retrograde spread of tumour cells in lymphatics (e.g. pulmonary lymphangitis carcinomatosis). Sarcomas are less likely to spread through the lymphatics, but may give rise to blood-borne metastases.

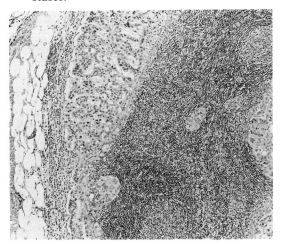

Figure 3.5 Photomicrograph of a lymph node showing infiltration of the peripheral sinus and cortex by metastatic adenocarcinoma. Stain: haematoxylin and eosin. Magnification ×150, reduced to 80% in reproduction

(3) *Blood-borne metastases.* Tumour cells reach the circulation through venous invasion and through lymphatic drainage into the thoracic duct. Many of the tumour cells which reach the circulation fail to form viable metastases. The development of metastases depends on characteristics of the tumour cells themselves and of the anatomical site. Thus malignant melanoma and small-cell carcinoma of bronchus have a striking ability for widespread metastasis early in the course of the disease, whereas squamous cell carcinomas of bronchus can grow to several centimetres in size over a period of years before they metastasize. Tumours in general metastasize with far greater frequency to liver, lung, bones and adrenal than to spleen, skin or muscle. The reasons for this are obscure, but are not related to blood flow.

(4) *Trans-coelomic spread.* Malignant infiltration of serous membranes frequently causes an effusion which provides a fluid medium in which exfoliated tumour cells can proliferate. These cells may seed to the serosal surface.

Staging of tumours

The possibility of surgical resection of a malignant tumour, and the likelihood of it having already given rise to distant metastases, depend largely on the stage of growth which the primary tumour has reached. The stage reflects tumour size and extent of direct spread. Both tumour size and spread vary in significance for different tumours. In cervical cancer it is the extent of invasion which is important, whereas in breast carcinoma it is tumour size. The TNM system of staging allows the precise criteria for staging a primary tumour at its site of origin (T) to be combined with an assessment of lymph node involvement (N) and distant metastases (M).

Intraepithelial neoplasia

Histological sections from early invasive squamous carcinomas may show changes in adjacent epithelium in which the normal stratified epithelium is replaced by abnormal cells showing features of malignancy, without invasion of the underlying stroma. Although first observed in biopsies adjacent to invasive squamous carcinoma of the uterine cervix, a similar histological appearance was later found at many different sites in the absence of an invasive lesion. This is now recognized as intraepithelial neoplasia and is believed to represent an early preinvasive stage in the development of cancer.

Intraepithelial neoplasia may be found in cervical, oesophageal and vulval squamous epithelium, in metaplastic squamous epithelium in the bronchus, in transitional epithelium in the bladder and in glandular epithelium of the endocervix, endometrium and gastrointestinal tract: adenomatous polyps of the colon represent the full spectrum of intraepithelial neoplasia. In each situation this is thought to represent a lesion which may progress to invasive cancer if left untreated. This is supported by long term follow-up studies of untreated patients, but by no means all cases progress to cancer. The preinvasive lesion itself can be treated effectively by local ablation or resection because it has no potential for metastasis in the absence of stromal invasion.

In most instances there are no symptoms from intraepithelial neoplasia and at present the only practical way of detecting these lesions is cytological screening. Cytological screening for the detection of preinvasive lesions has proved to be effective in preventing invasive cancer of the cervix, but has not been found practical for other types of cancer, except in certain high risk groups.

There is a spectrum of change constituting intraepithelial neoplasia which can be graded according to the degree of nuclear atypia and cytoplasmic maturation. In the uterine cervix the spectrum from mild to severe is designated cervical intraepithelial neoplasia grade 1–3 (CIN 1–3). This has replaced the old terminology which attempted to distinguish carcinoma-in-situ (CIN 3) from dysplasia (CIN 1–2). At the severe end of the spectrum (CIN 3) undifferentiated, mitotically active cells with scanty cytoplasm are found in all layers of the epithelium, whereas at the mild end of the spectrum (CIN 1) these cells are present in the deeper layers but show differentiation with cytoplasmic maturation and loss of mitotic activity towards the surface. CIN 2 represents an intermediate grade of differentiation.

Although there is little controversy about the criteria for diagnosis and the clinical significance of the more severe lesions, the classification of CIN 1, 2 and 3 is highly subjective and there can be difficulty in distinguishing the milder grades from certain reactive, potentially reversible lesions. In particular, there is a considerable overlap between the histological appearance of CIN 1–2 and human papillomavirus infection. Long term follow-up studies of untreated patients are rare and criteria for classification have changed with time, leaving some doubt as to the clinical significance of the milder lesions. In general it can be said that regression is less likely and progression to invasion more likely with increasing grades of CIN, and management of patients continues to be based on a histological assessment of this grade.

The morphology of tumours

The relationship of stroma and tumour cells largely determines the gross structure of a tumour, whereas degree of diffferentiation determines the microscopic appearance. Differentiation of tumours allows the site of origin to be recognized and forms the basis for the histological classification of tumours.

Gross structure of tumours

In general, benign tumours form encapsulated masses which can often be enucleated, whereas malignant ones form structureless lobulated masses, diffusely infiltrating and replacing normal tissues. Where they infiltrate external, mucosal or serosal surfaces they may form exophytic growths. When ulcerated they may be depressed below the surface. Some patterns of gross structure can be recognized:

(1) *Encephaloid tumours.* Non-epithelial tumours such as malignant lymphomas and sarcomas do not usually stimulate fibroblastic proliferation and tend to form soft, fleshy 'encephaloid' masses. Carcinomas, particularly undifferentiated ones, may have similar features but more often have a fibrous stroma. Medullary carcinoma of breast is a typical example of an 'encephaloid' carcinoma and is illustrated in *Figure 3.6.*

(2) *Scirrhous and sclerosing tumours.* These are carcinomas in which the infiltrating cells stimulate a dense desmoplastic stroma and form poorly circumscribed, often stellate lesions, typified by scirrhous breast carcinomas. The word cancer, meaning crab, is particularly appropriate for the scirrhous breast carcinoma illustrated in *Figure 3.7.* Lobulated masses of tumour may be separated by fibrous bands: this is not only seen in carcinomas as it is characteristic of nodular sclerosing Hodgkin's disease.

(3) *Polyps, papillomas and papillary carcinomas.* The growth of neoplastic epithelial cells may draw the stroma out into frond-like

The morphology of tumours 47

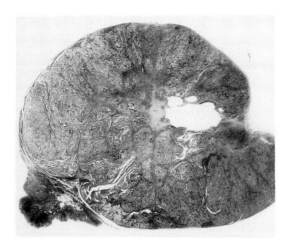

Figure 3.6 Whole section of a medullary carcinoma of breast (tumour 2 × 2.5 cm). Stain: haematoxylin and eosin.

Figure 3.7 Whole section of a scirrhous carcinoma of breast (tumour 2 × 2.5 cm). Stain: haematoxylin and eosin.

fluid. Similar spaces may form in enclosed sheets of stratified squamous epithelium, into which the more differentiated superficial cells degenerate and exfoliate. This may form cysts in a benign teratoma, or extensive cavitation in bronchogenic squamous carcinomas. A benign cystic teratoma is illustrated in *Figure 3.9*.

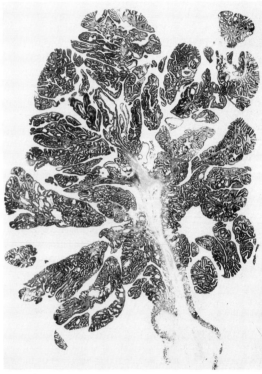

Figure 3.8 Whole section of a benign adenomatous colonic polyp (polyp 2 × 2.5 cm). Stain: haematoxylin and eosin.

papillary structures. These range from benign papillomas, as seen in bladder or mammary ducts, to highly malignant tumours such as serous and mucinous papillary cystadenocarcinomas of ovary, which are characterized by the growth of papillary fronds on mucosal or serosal surfaces and within enclosed neoplastic cysts. A papillary adenomatous polyp from the colon is illustrated in *Figure 3.8*.

(4) *Cystic and cavitating tumours.* Areas of necrosis, haemorrhage and mucoid degeneration may give rise to apparent cystic spaces in any type of tumour. True neoplastic cysts result from the growth of enclosed epithelium, usually secreting mucous or serous

Figure 3.9 Surgical specimen of a benign cystic teratoma. Scale given in centimetres

Microscopic structure of tumours

Microscopy of tumours reveals details of invasion, differentiation and grading which form the basis for classification and allow for prediction of the behaviour of tumours and selection of appropriate treatment or management of the patients. Microscopy is essential for recognition of the malignant potential of a tumour, and is particularly important in the recognition and classification of pre-invasive stages of neoplasia. Accurate classification also permits recognition of a large number of tumours which show features such as invasion, nuclear pleomorphism and even mitotic activity, but either do not have the potential for metastasis or are not true neoplasms. Examples abound in breast, lymph node and soft tissue pathology, as well as at many other sites.

Tumours can be divided broadly into those arising from the main types of normal cells.

(1) *Epithelial tumours.* These are derived from skin and its glandular appendages (breast, hair follicles, sweat glands), the gastrointestinal tract and the exocrine or visceral glands derived from it (pancreas, lung, liver) and the specialized epithelium, embryologically derived from mesodermal structures, which have differentiated into the genitourinary organs.

Benign epithelial tumours are referred to as adenomas or papillomas and malignant epithelial tumours as carcinomas. The main sub-types at any site are adenocarcinomas which form glands and may secrete mucus, transitional cell carcinomas and squamous cell carcinomas which are stratified, form intercellular bridges and may undergo keratinization.

Apart from simple differentiation at a cellular level, microscopy shows patterns of growth which may allow for more precise classification: ductal and lobular invasive breast carcinomas, renal, hepatic and certain bronchogenic carcinomas may be recognizable, even in metastases.

(2) *Connective tissue tumours.* Tumours may arise from any type of mesenchyme including fibroblastic tissue, fat, muscle, bone and vascular endothelium. Although some benign tumours, e.g. uterine leiomyomas, are common, malignant ones are rare. Malignant connective tissue tumours are known as sarcomas, and are classified wherever possible according to the type of differentiation. There are many benign connective tissue tumours which mimic malignancy. Classification, according to reproducible criteria and with recognition of variant growth patterns, requires specialist knowledge and is essential for selection of appropriate treatment and prediction of prognosis.

Specialized connective tissue in the central nervous system gives rise to the commonest group of brain tumours, the gliomas, which show differentiation towards astrocytes. Specialized connective tissue in serous cavities gives rise to mesotheliomas, most of which are malignant and which may show biphasic differentiation towards surface mesothelium and mesenchyme.

(3) *Reticuloendothelial tumours.* These include malignant lymphomas, which commonly arise in the lymph nodes and far less often in mucosa-associated lymphoid tissue. They also include malignant change in haemopoietic tissue (leukaemia) and histiocytic lymphomas. Classification on the basis of morphology alone is limited. Immunocytochemical techniques for determining cell type are usually necessary for accurate classification and patient management.

(4) *Neuroectodermal and neuroendocrine tumours.* In addition to tumours of the central and peripheral nervous system, neuroectoderm gives rise to benign naevi and malignant melanomas. These arise from melanocytes which migrate from the neural crest during fetal life.

Endocrine and para-endocrine tumours at different sites have many features in common and, although there are recognizable benign and malignant sub-types, it is frequently difficult to predict the behaviour of these tumours on the basis of their histological appearances.

(5) *Tumours arising from germ cells and fetal structures.* Germ cell tumours include seminomas and dysgerminomas, which closely resemble sperm precursor cells. Teratomas arise from totipotential cells, which are capable of differentiation towards mesenchymal, epithelial, neural, placental and fetal structures. These vary from benign cystic tumours, which show mature but disorganized differentiation towards cells of

all germ layers, to highly undifferentiated tumours which may show partial differentiation towards yolk sac or trophoblast. Teratomas and seminomas most commonly arise in the gonads, but may also arise in the thymus.

There is a group of tumours, predominantly occurring in childhood, which are derived from immature fetal structures. These include nephroblastoma, neuroblastoma and hepatoblastoma.

The aetiology of tumours

Neoplasia is a prime example of a disease process in which multiple genetic and environmental factors are involved. In those instances where tumour development has been studied there appear to be several stages between the initial mutation and the development of a malignant tumour. Not all of these stages are irreversible, and they frequently include a stage of benign proliferation.

Genetic factors in tumour development

Central to the pathogenesis of benign or malignant tumours is the acquisition of heritable genetic defects in previously normal cells which have the potential for cell division. The effect of these defects is to allow uncontrolled cell division and, in 'malignant' tumours, to confer upon a population of cells the ability to migrate and invade tissues. In most instances the primary cause and nature of the genetic mutation from which a tumour develops is unknown, but mutagenic agents such as chemical carcinogens and ionizing radiation, which are known to act directly on DNA, have been implicated in the development of certain human tumours. Research into the clonal origin of tumours suggests that most tumours are monoclonal and arise from a single mutant cell.

In a minority of instances characteristic genetic defects can be detected in successive generations of tumour cell lines: examples of this are the 8/14 translocation seen in Burkitt's lymphoma and the 9/22 translocation in chronic myeloid leukaemia (Philadelphia chromosome). In most instances the changes are more subtle, and can only be detected by sophisticated molecular techniques. Malignant cells harvested from solid tumours or malignant serous effusion may show gross karyotypic abnormalities in successive generations of cells: these represent a late stage in the evolution of tumours as the potential for error in the mitotic cycle increases after repeated cycles of replication. Bizarre changes such as aneuploidy, ring chromosomes and tripolar mitoses probably represent an end stage in tumour development: such cells are unlikely to continue to propagate.

One of the most important advances in our knowledge of the mechanism of oncogenesis has been the discovery of a group of genes known as 'oncogenes'. These were first identified in a group of RNA viruses, the retroviruses, which infect certain birds and rodents. These viral oncogenes were not only shown to be responsible for the development of tumours in their natural hosts but also to be capable of malignant transformation of cells in culture. Subsequent studies on human tissue have shown sequences of DNA in normal cells which show a remarkable homology with the viral oncogenes. These human 'cellular-' or 'proto-' oncogenes are not in themselves oncogenic, but have been shown to code for proteins involved in cell proliferation and to be activated in many common human cancers. It appears probable that activation occurs as a result of a number of different mechanisms, including point mutation in the cellular- or proto- oncogene itself or in a sequence of DNA which controls its activity. The chromosomal translocation which has been described in Burkitt's lymphoma has been shown to approximate the oncogene *c-myc* to the site for immunoglobulin synthesis which is being actively transcribed in this cell line. This may represent another mechanism for activation of cellular oncogenes.

There are some forms of cancer, particularly breast cancer, which show a tendency to occur in families. This suggests a genetic predisposition which may be activated by other factors. There are certain rare forms of cancer, such as retinoblastoma, in which a genetic abnormality is directly related to the development of the malignant tumour. There are also inherited and congenital conditions, such as Down's syndrome and ataxia telangiectasia which are associated with certain forms of lymphoreticular malignancy. In some of these instances the genetic defect has been identified but in most of them the mechanism remains obscure.

Chemical carcinogenesis

Chemicals were the first environmental factors recognized as causing malignant tumours, dating from Percivall Pott's observation of scrotal cancer in chimney sweeps in 1775. The main carcinogens in soot and tar are polycyclic hydrocarbons, such as 3,4-benzpyrene. Many other hydrocarbons and also aromatic amines, azo-dyes and nitrosamines have since been demonstrated to be carcinogenic. There are certain characteristics of chemical carcinogens which throw some light on the pathogenetic mechanisms involved in the development of tumours in general.

There is often a latent period, which may be dose-related, between the application of an experimental carcinogenic 'initiator' and the development of a tumour. The process can be enhanced by the application of a variety of agents which are not themselves carcinogenic, but which act as 'promotors'. Most initiators act by direct interaction with DNA, causing a genetic defect which is inheritable if the affected cell is one that is capable of mitosis. Promotion is also a multi-step process, often involving the stimulation of growth factors and enzymes which allow successive generations of mutant clones to proliferate at the expense of normal ones.

Viruses and neoplasia

Although a number of RNA retroviruses are known to be oncogenic in animals, there is no direct evidence for oncogenic RNA viruses causing tumours in man. There is, however, some evidence that DNA viruses may be causative agents in certain forms of human cancer, under conditions in which viral DNA becomes integrated into the host genome. Certain sub-types of human papillomavirus have been implicated in cervical and skin carcinomas, hepatitis B virus in hepatocellular carcinoma and Epstein–Barr virus in Burkitt's lymphoma.

Integration of viral DNA is probably only one step in the process of carcinogenesis, which is well shown by the development of skin cancer in the condition epidermodysplasia verruciformis. In this condition there is a predisposition to infections by many sub-types of human papillomavirus (HPV), probably mediated by an alteration in immunity, but only the lesions infected with HPV sub-type 5, and only those exposed to sunlight, develop into squamous carcinomas. There may be a similar relationship between HPV sub-types 16 and 18, immunosuppression and the chemical carcinogens of cigarette smoke in the development of cervical cancer.

Physical agents

There is a strong association between exposure to ionizing radiation and the development of malignant tumours. This is probably directly related to damage to DNA, resulting from an increased rate of mutation. Examples of this association were seen in the increased risk of leukaemia in those exposed to the atomic bomb in Hiroshima, the development of skin cancers in early radiologists and lung cancer in cobalt and uranium miners.

Prolonged exposure to ultraviolet light, particularly in fair-skinned people, causes an increased risk of skin cancers, including malignant melanoma. This risk is exaggerated in the condition xeroderma pigmentosa, in which there is an inherited deficiency of an enzyme involved in the repair of damaged DNA.

Hormones and cancer

There is an increased risk of breast and endometrial cancer in conditions associated with unopposed secretion of oestrogen, and both types of cancer cells may have oestrogen receptor sites. The action of hormones in carcinogenesis is analogous to promotion, and initiating carcinogenic agents are probably also involved.

Immunological surveillance

There is an increased risk of the development of malignant tumours in individuals who are immunosuppressed, but the tumours are almost all either lymphoreticular tumours or skin tumours which may be associated with viral infections. There is some evidence that cervical cancers may be related to immunosuppression. There is no evidence for an increased risk for common epithelial cancers such as those arising in the lung, breast or colon. There is no doubt that an immune response is directed against tumour cells in some instances, but there remains doubt as to the role of immune surveillance in the prevention and control of neoplasia.

Conclusion

There is evidence to suggest that the malignant transformation of cells is the result of interaction between viruses, chemical and physical agents on one hand and genetically determined growth factors, physiological DNA repair mechanisms and immunological surveillance on the other. Most of the time cells are maintained in a normal physiological state and the development of a malignant clone is a rare event.

References and further reading

ANDERSON, J. R. (ed.) (1985). *Muir's Textbook of Pathology*, 12th Edition. London: Edward Arnold

BOWRY, T. R. (1984). *Immunology Simplified*, 2nd Edition. Oxford: Oxford Medical Publications

CONNOR, J. M. and FERGUSON SMITH, M. A. (1984). *Essential Medical Genetics*. London: Blackwell Scientific Publications

HURLY, J. V. (1983). *Acute Inflammation*, 2nd Edition. Edinburgh: Churchill Livingstone

ROBBINS, S., COTRAN, R. and KUMAR, V. (1984). *Pathologic Basis of Disease*. Philadelphia: W. B. Saunders

ROONEY, D. E. and CZEPULKOWSKI, B. H. (eds) (1986). *Human Cytogenetics*. London: IRL Press

WOOLF, N. (1986). *Cell, Tissue and Disease: The Basis of Pathology*, 2nd Edition. London: Baillière Tindall

4

Preparatory techniques

David T. Proctor

A thorough understanding of the basic principles of fixation, preparation and staining are fundamental for good laboratory practice and an awareness of new developments in these fields is necessary to ensure standards are maintained. In this chapter preparatory techniques which are in common use and which have been found to give consistently good results are described. Many laboratories have their own modification of these procedures which are quite acceptable providing standards are not compromised.

Fixatives in cytology

The main objective of fixation in cytology is to preserve the morphological detail of the cell in as perfect condition as possible. This can be achieved by rapid inactivation of cellular enzymes and coagulation of cell protein, which is the mode of action of most fixatives. A good fixative will also prevent cell lysis by bacterial action. All fixatives must be capable of penetrating the cell membrane, but some act more quickly than others and several fixing agents are often combined to obtain a balance between rapid penetration of the cell and preservation of the structure.

Regretfully, there is no such thing as the perfect fixative; all fixation causes intracellular disruption and it is this 'fixation artefact' which is recognized by the microscopist. A working knowledge of the action of fixatives will enable the cytotechnologist to choose the optimal fixative for the specimen under investigation. Compatibility with subsequent processing and staining procedures must be a prime consideration when selecting a fixative. In cytological practice, alcoholic fixatives are the fixatives of choice; other fixatives may be needed where special staining of cytological material is indicated.

Alcoholic fixatives

Alcohol fixation is essential whenever Papanicolaou staining is contemplated. Alcoholic fixatives act by coagulating cell protein and dehydrating the cell, removing about 80% of the water chiefly from the protein ultrastructure. The degree of shrinkage and chromatin condensation depends on the grade of alcohol used. A major advantage of alcoholic fixatives is that fixation is quick and can be accomplished in 10–15 minutes. Their major disadvantage is that they are flammable. Alcoholic fixatives can be used in solution or as coating fixatives. The latter must be used if transport of the smears is anticipated.

The alcoholic fixatives in common use are ethanol, industrial methylated spirit, methanol, isopropanol and coating fixative.

Ethanol (C_2H_5OH), syn. ethyl alcohol, absolute alcohol

This fixative provides excellent preservation of cytoplasmic and nuclear detail without excessive cell shrinkage or condensation of chromatin. As a 95% solution, it is the fixative of choice for

Papanicolaou-stained smears. However, its high cost prohibits its widespread use.

Industrial methylated spirit (C_2H_5OH + denaturant), syn. 'alcohol', 'meths' (see Conventions)

Four grades are available differing only in alcohol and water content. The grade most commonly used in cytology is 74 overproof (O.P.), the total alcohol content equalling 99.27 v/v. In some centres 66 O.P. alcohol is used. Industrial methylated spirit is a very acceptable alternative to ethanol for fixation of Papanicolaou smears and is less expensive.

Methanol (CH_3OH), syn. methyl alcohol

This is used principally for the fixation of air-dried smears for Romanowsky staining but for this purpose must be acetone-free. It is toxic when ingested or inhaled.

Isopropanol (($CH_3)_2CHOH$), syn. propan-2-ol, isopropyl alcohol

This fixative causes more cell shrinkage than either methanol or ethanol. An 80% solution is often recommended for fixation of smears for Papanicolaou staining as the addition of water counteracts the shrinkage effect of the isopropanol. It is sometimes used as a fixative for membrane filters.

Coating fixatives (PEG)

Coating fixatives protect the cells in the smear from damage by providing a waxy cover. They are usually composed of polyethylene glycol in an alcohol base and are usually applied to the slide as a spray or in dropper form. Care must be taken when applying the fixative for, if it is used too vigorously, it may sweep the cells off the smear. In either case the fixative must be administered before drying occurs. As the alcohol which fixes the cells evaporates, they are left covered with a waxy protective layer giving the surface of the slide a greasy appearance. It is important to remove this coating before staining as the penetration of dyes into the cells is impaired. This is achieved by immersing the slide in 50% alcohol for 5 min and rinsing in water. Coated smears can be stored for a week or more without deteriorating but should be kept in an airtight box to prevent air-borne contaminants adhering to the surface. Slides fixed in this way can be safely sent through the post in suitable containers. It is important to remember that some commercially available spray fixatives are prepared with 80% isopropanol with lanolin. These are not recommended as cytological fixatives.

Polyethylene glycol coating fixative (PEG) can be prepared in the cytology laboratory using the following formula which is a modification of that recommended by Higgins and Smith (1963):

Industrial methylated spirit	100 ml
Polyethylene glycol 400	11 ml
Glacial acetic acid	0.5 ml

The final concentration of alcohol is 90%. The quality of fixation is comparable with that of 95% ethanol.

Special purpose fixatives

The following fixatives are not used routinely in cytology but have a place when special staining is required, or when a clot or cell button is being prepared for histology. They may also be useful for the processing of high risk samples from patients with hepatitis or tuberculosis, for demonstrating lipids or for the processing of bloodstained specimens.

Carnoy's fluid

Chloroform	300 ml
Alcohol	600 ml
Glacial acetic acid	100 ml

Suitable for fixation of bloodstained smears. Achieves rapid fixation and dehydration; haemolyses red blood cells; causes excessive shrinkage and coarse chromatin clumping; may dissolve cytoplasmic elements; denatures on storage to form hydrochloric acid. It is sometimes classified as an alcoholic fixative.

Formal saline

Formaldehyde (37–40%)	100 ml
Sodium chloride	9 g
Tap water	900 ml

Widely used for rapid fixation of tissue fragments.

10% neutral formal saline

Formaldehyde (37–40%)	100 ml
Distilled water	900 ml
Sodium dihydrogen phosphate (anhydrous)	3.5 g
Disodium hydrogen phosphate (anhydrous)	6.5 g

Suitable for fixation of cell blocks and clots. Fixation is too slow for routine cytology; protein is not coagulated but is structurally modified; cells tend to float off the slide surface. One gram of eosin, light green or orange G may be added to the above recipe to colour cell blocks or clots to facilitate microtomy.

Alcoholic formaldehyde

Formaldehyde (37–40%)	100 ml
Alcohol	900 ml
Glacial acetic acid (optional)	5 ml

A good combination of fixatives which achieves rapid coagulation of protein together with dehydration; good morphological detail. Used for secondary fixation and preservation of glycogen.

Formal vapour fixation

This is a particularly useful fixative for demonstrating lipids. Place a small ball of cotton wool in a coplin jar. Soak with a few drops of concentrated formalin (40% formaldehyde). Immediately drop slide into the coplin jar and replace the lid with a tight seal. Leave in an exhaust cabinet at room temperature for 30 min.

Formal–calcium

Formaldehyde (37–40%)	100 ml
Distilled water	900 ml
10% Calcium chloride	100 ml

Used for the preservation of lipids.

Saccomanno's fixative

| 50% Alcohol | 100 ml |
| Carbowax | 2 g |

Used for sputum blended by the Saccomanno technique (see p. 58).

Dubosq Brazil sputum fixative

Alcohol	2880 ml
Acetic acid	360 ml
Formalin	144 ml
Water	2016 ml
Picric acid	24 g

Fixative of choice when preparing cell blocks from sputum (see p. 24).

Husain's preservation fluid for urine samples

Monoethylene glycol	3446 ml
Diethylene glycol	181 ml
Borax pentahydrate	363 ml
Glacial acetic acid	500 ml
Distilled water	5667 ml

Add 10 ml of fixative to each 100 ml container. Urine should be placed in this fluid if a delay in processing is anticipated. It will reduce intracellular enzyme activity and is useful for preserving urine taken as part of a bladder cancer screening programme.

Esposti's fluid for fixation of urine

Methanol	225 ml
Distilled water	225 ml
Glacial acetic acid	50 ml

An equal volume of this fixative should be added to urine to achieve rapid fixation.

Bouin's fluid

Saturated aqueous picric acid solution	75 ml
Formaldehyde (37–40%)	25 ml
Glacial acetic acid	5 ml

This fixative is recommended for bone marrow, endometrial fragments and other small biopsies. After fixation washing with 50% alcohol will remove most of the yellow colour (picrates).

Preparation of cytological specimens

Specimens received by the cytology laboratory can be broadly categorized into three groups.

Group 1

This group comprises smears that have been prepared by the clinician and fixed immediately in alcohol solution or a coating fixative (PEG) before air drying occurs. This group comprises a major part of the workload in the vast majority of cytology laboratories. It includes all smears taken as part of the cervical cancer screening programme as well as smears from other sites. Thus it includes cervical smears, smears from the vulva, oral and nasal cavities, skin and rectum (Linehan, Melcher and Strachan, 1983); smears prepared from brushings from gastrointestinal, respiratory and urinary tracts; and smears prepared from tumour imprints, endometrial aspirates, nipple discharge and fine needle aspiration biopsy specimens.

Group 2

This group comprises smears that have been prepared by the clinician and allowed to air dry. This group consists mainly of smears prepared from fine needle aspiration biopsy specimens and tumour imprints but may include other examples from Group 1.

Group 3

This group comprises specimens that are unfixed. These may be in semi-solid fluid form and include sputum, urine, serous effusions, cyst fluids, bronchial and gastric washings, peritoneal lavage specimens and cell suspensions prepared from fine needle aspiration biopsy specimens or brushings from the gastrointestinal and respiratory tract.

The principles of processing these different types of samples are discussed in this section. Smears that have been fixed or air-dried before dispatch must be stained and mounted in the laboratory. Fresh specimens make greater demand on the skill of the cytotechnologist who has prime responsibility for the preparation of adequate smears from the sample.

Alcohol-fixed smears

The majority of smears received by the laboratory are already prepared and fixed by the clinician who has taken the sample. Providing the smears have been fixed immediately in alcohol or a coating fixative (PEG) before air drying occurs they are suitable for staining by the Papanicolaou method which is the stain of choice in cytology. Smears should be clearly labelled with the patient's name and the material should be evenly spread on the slide. If the smears are heavily bloodstained, steps can be taken to lyse the red blood cells before staining (p. 66). If several smears are received from the same site, it is wise to save one or two for immunocytochemical staining or other 'special' staining procedures should these be required. However, all smears should be stained and screened ultimately to ensure a definitive report is issued. A technique has been developed for reprocessing alcohol-fixed Papanicolaou-stained smears for electron microscopy should ultrastructural studies be indicated (p. 71).

Air-dried smears

These smears are suitable for Romanowsky staining only. Many clinicians prepare air-dried smears from fine needle aspiration biopsy specimens routinely as the amount of material aspirated is so small that it is almost impossible to fix the smear before air drying occurs. Romanowksy stains are particularly useful for displaying the non-epithelial elements of the smear and air-dried smears should always be prepared when a non-epithelial tumour e.g. lymphoma is suspected.

Unfixed samples

All unfixed specimens present a potential biological hazard and must be handled with caution. A Class I microbiological exhaust protective cabinet sited within the cytology laboratory, but separate from the screening area, is mandatory and must always be used when handling unfixed specimens. Centrifuges must not be used inside these cabinets so extreme care must be taken when transporting material from bench to cabinet. Rubber gloves and a plastic disposable apron worn over a laboratory coat give added protection. All equipment must be sterilized after use in an approved disinfectant and working surfaces must be swabbed daily. Those specimens known to contain harmful pathogens (e.g. *M. tuberculosis*, hepatitis virus or human immunodeficiency virus) require special handling (*see* Chapter 26).

Table 4.1 Preparatory techniques for cytological specimens

Type of specimen	Consistency	Preferred procedure	Technical note
Sputum, bronchial aspirates, gastric aspirates	Mucinous, mucopurulent	Direct smears	Liquefaction and cell block method used in some centres
Bronchial washings, bronchoalveolar lavage, gastric washings	Watery, mucoid	Centrifugation	Solid fragments should be picked out before centrifugation and direct smears or cell blocks prepared from them
Serous effusions (pleural, pericardial, peritoneal fluid), cyst aspirates, wound drainage fluid, peritoneal lavage specimens	Serous	(1) Centrifugation (2) Cell block	Suitable for large volumes. Small volumes should be processed by cytocentrifugation (p. 59) or membrane filtration (p. 63)
CSF	Serous	Cytocentrifugation	Sedimentation methods (Sayk and Sörnäs)
Urine	—	(1) Centrifugation (2) Membrane filtration (3) Cytocentrifugation	Special techniques may be needed to ensure cells adhere to slides. pH correction may be necessary before processing
Salinated washings from fine needle aspiration biopsy specimens and endoscopic brush specimens (gastric and bronchial)	Watery	(1) Cytocentrifugation (2) Positive pressure filtration (3) Cell block	Special techniques may be needed to ensure cells adhere to slide. Nylon brushes must be shaken vigorously in saline to release trapped material before processing
Bloodstained effusions	—	Lyse red blood cells by (1) Buffy layer techniques (2) Microhaematocrit method (3) Flotation technique (4) Enzymolysis	
Bloodstained smears	—	Haemolysis by glacial acetic acid, 2 M urea or Carnoy's fixative	Carnoy's fixative can be used with caution to haemolyse red blood cells in both fixed and unfixed smears
Fibrinous clot	Solid material	Tease out clot and prepare paraffin block	
Tissue fragment	Solid material	(1) Paraffin block (2) Imprint smear (3) Crush smear (4) Frozen sections	
All deposits from salinated washings, effusions, liquefied sputum, etc.	Solid material	(1) Direct smear (2) Cell block	

Fresh specimens present a special challenge to the cytotechnologist responsible for preparing the specimen for light microscopy. It is his or her job to select a preparatory method which will optimize the chances of detecting cellular elements in the sample if they are present. Many techniques have been developed for this purpose and the skill of the cytotechnologist lies in selecting the most appropriate technique for each specimen (*see Table 4.1*). Of course skill and experience develop hand in hand; however, in this chapter guidelines have been set out which will help with the decision-making process.

As soon as the specimen is received in the laboratory a note should be made of the quantity, colour and consistency, e.g. whether it is mucinous, serous or watery. The pH should be recorded if appropriate and the specimen neutralized if necessary. The presence of clots, blood staining, or tissue fragments should be noted. This information will help the cytotechnologist in selecting the appropriate procedure for each sample. Generally speaking, smears can be prepared directly from mucinous samples, whereas cell concentration techniques (centrifugation or membrane filtration) are needed for large volumes of serous or watery fluid. Small volumes such as CSF must be handled with care to minimize cell loss and special cell concentration techniques such as cytocentrifugation must be used. Bloodstained specimens, clotted specimens, and tissue fragments all need special techniques. The processing techniques in common use are described in the following pages. Technical aspects of smear preparation are discussed in detail on page 72.

Preparation of mucinous samples by the direct smear technique

Sputum samples, bronchial and gastric aspirates are included in this group. Mucin is an excellent adhesive and slide preparations are made by spreading the material directly onto the slides.

Procedure

(1) Label four slides with the patient's name and specimen number using a pencil (frost-ended slide) or diamond writer.
(2) Carefully tip the specimen into a petri dish. Tease it out slowly using a wooden 'orange' stick.
(3) View the specimen against light and dark backgrounds and note the macroscopic appearance (e.g. fluidity, colour, blood, flecks, food).
(4) Careful scrutiny is needed to select 'suspicious' areas for sampling. Look for altered blood (brown flecks) rather than fresh blood. If there are a few blood-flecked areas use them all. Odd-looking discolouration, white flecks and streaks should be sampled. Remember that cells from an oat cell carcinoma may be present in watery sputum samples.
(5) Break an 'orange' stick in half; the jagged ends are used for teasing the selected areas away from the main specimen.
(6) Pick up the selected material by twisting the stick. Place material on to the slide and add to it until the volume is similar to that of half a peanut. Make smear by one of the methods shown in *Figure 4.1*. Tease out large lumps into flatter streaks quickly and fix immediately. Make four smears from each sample.
(7) Immerse three slides in alcohol for at least 30 min or apply coating fixative (PEG) and allow one to air dry.
(8) Pour the specimen back into the container. Keep specimen at 4 °C until report is dispatched in case additional smears are needed.
(9) Discard used sticks into an approved disinfectant and swab the working area. Dispose of specimen appropriately when no longer needed.
(10) Stain wet-fixed smears by the Papanicolaou technique and the air-dried smear by Romanowsky technique.

The sampled areas, of course, represent only a small proportion of the total specimen and some laboratories routinely prepare cell blocks from the residual specimen. Many argue that provided all suspicious areas of the specimen are sampled, routine cell blocks are unnecessary. However, cell blocks may be of value in the

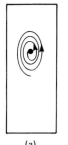 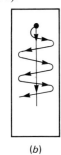  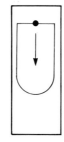 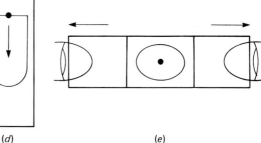

(a) (b) (c) (d) (e)

Figure 4.1 Methods of making direct smears: (a) large lake; (b) criss-cross; (c) fish-tail; (d) blood smear; (e) squash-spread

search for asbestos bodies. The preparation of cell blocks from sputum requires the liquefaction of mucus and four methods are described below.

Saccomanno's technique (Saccomanno et al., 1963)

Add 50% alcohol containing 2% carbowax to the sputum to bring the volume to 100 ml. Leave for half an hour to fix. Pour the specimen into a high speed blender and mix for 5–10 s, depending on the mucin content. When the specimen is homogenized, pour into sealed centrifuge tubes and centrifuge at 2000 rev/min for 5 min. Discard the supernatant, fix the deposit in 10% neutral formal saline overnight and prepare a cell block (see p. 69).

Tween 80 technique

Add Tween 80 together with glass beads to the unfixed sputum in a capped container. Place on a mechanical shaker for 5–10 min until the mucus is liquefied. Decant into sealed centrifuge tubes, discarding the glass beads into a suitable disinfectant. Rinse the container with a small amount of saline and add to the centrifuge tube. Centrifuge at 2000 rev/min for 5 min, discard the supernatant and fix the deposit in 10% coloured neutral formal saline. Prepare a cell block (see p. 69).

Enzyme digestion

Cytoclair* is widely used for this purpose. Add an equal amount of 1–2% Cytoclair powder dissolved in saline to the specimen. Incubate at 37 °C for up to 2 h. When liquefaction is complete centrifuge at 2000 rev/min for 5 min. Discard the supernatant, fix deposit in 10% coloured formal saline and prepare a cell block (see p. 69).

Ultrasonic technique

Add an equal amount of 10% formal saline to the sputum sample and allow to fix overnight. Position the tip of the ultrasonic disintegrator probe in the top of the liquid. Apply 60 W power at a frequency of 20 kHz for 1 min (or until the audible noise level reaches a steady

---

* Sinclair Pharmaceuticals Ltd, Ockford Road, Godalming, Surrey.

high pitch). Pour into sealed centrifuge tubes and centrifuge at 2500 rev/min for 5 min; discard the supernatant and prepare a cell block (see p. 69).

Technical note

Smears can be made directly from the centrifuged deposits but it must be remembered that the adhesive qualities of mucus have been destroyed and a coating adhesive should be used to ensure the cells adhere to the slide (p. 77).

Centrifugation

This technique is suitable for large volumes (> 50 ml) of serous fluids, e.g. effusions, cyst aspirates, and urine or salinated washings. Often several litres of pleural or peritoneal fluid are sent to the laboratory. If a large volume is received it should be left to sediment for an hour at 5 °C. Decant all but 200 ml of the supernatant into approved disinfectant and mix the remaining 200 ml to resuspend the cells before processing.

Serous fluids may often be clotted or bloodstained. The special procedures for processing bloodstained samples are described on p. 66. The deposits from transudates, salinated washings and urine may not adhere to the slides. The special techniques for improving their adhesive qualities are described on p. 77.

Procedure for direct smears

(1) Record the macroscopic appearance and volume of the fluid sample.
(2) Agitate the specimen and add equal amounts (50 ml) to two balanced capped centrifuge tubes. Use more tubes for large volumes.
(3) Centrifuge at 2500 rev/min for 5 min. Decant the supernatant into disinfectant, add more specimen to the tubes and centrifuge again.
(4) Decant the supernatant carefully and keep the tubes upside down to drain for at least 10 s.
(5) Mix the cell deposit with the remaining drops of supernatant by flicking the tube with a finger.
(6) Using a wire loop, pipette or micropipette, remove enough cell deposit to make a small drop on pre-labelled slides and prepare

smears immediately by one of the methods shown in *Figure 4.1*.

(7) Two smears should be fixed immediately for Papanicolaou staining. Plunge slides quickly into alcoholic fixative to avoid 'tidal wave' effect (*see* p. 76). Sometimes, if the surface of the preparation is very wet, it is advisable to let the smears dry slightly at the edges before fixing.

(8) Allow two smears to air dry before fixing in methanol for Romanowsky staining. More preparations may be made and set aside for further studies later.

Cytocentrifugation

First described by Watson in 1966, the Cytospin* (*Figure 4.2*) has undergone very little modification since then. The principle is both simple and efficient. Small aliquots of the fluid specimen under investigation are spun laterally onto microscope slides to form an even monolayer of cells. The cells are fixed and stained in the usual way for light microscopy. Smears are particularly suitable for Romanowsky staining. The advantages of this technique are:

(a) Suitability for processing small fluid samples containing few cells, e.g. CSF, salinated washings from fine needle aspiration biopsy specimens.
(b) Rapidity of preparation; specimens are prepared directly onto slides.

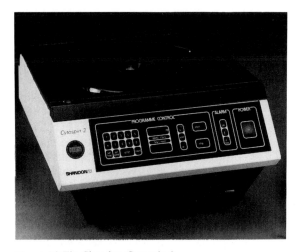

Figure 4.2 The Shandon Cytospin 2

* Shandon Southern Products, 93/96 Chadwick Road, Astmoor, Runcorn, Cheshire AW7 1PR, UK.

(c) Ease of staining and screening; any stain can be used and monolayers are easy to screen.
(d) Small defined area for screening (6.5 mm diameter).
(e) Relatively inexpensive.
(f) Safe and efficient. Enclosed system reduces health risks and apparatus is autoclavable if spillage of infected material occurs.
(g) Cytospin 2 must be loaded and unloaded in a microbiological safety cabinet.

The disadvantages of the technique are few, namely:

(a) There is always some cell loss due to absorption on the filter card.
(b) The method is not suitable for very cellular, mucinous samples.

Procedure

(1) Describe macroscopic appearance and volume of specimen.
(2) Centrifuge fluid specimen to obtain a cell deposit and discard supernatant into disinfectant.
(3) Place the Cytospin sample chambers, filter cards and labelled slides in the centrifuge head (*Figure 4.3*).
(4) Resuspend cells to optimal cellularity in saline.
(5) Add an aliquot of specimen preferably not exceeding 0.5 ml to each chamber (*Figure 4.4*).
(6) Centrifuge for optimal time using the high acceleration mode.
(7) Remove slides and fix immediately or air dry.

Technical notes

Step 2: The Cytospin is not suitable for mucinous fluids. Heavily bloodstained specimens should be treated as described on p. 66 before cytocentrifugation. One or two drops of albumin should be added to non-serous fluid cell suspensions to improve cell adhesion. Up to 12 samples can be processed simultaneously, as long as they are balanced.

Step 3: Cytospin sample chambers have undergone the modifications shown in *Figure 4.5*. The early Cytospin used a chamber (*Figure 4.5a*) that was designed in such

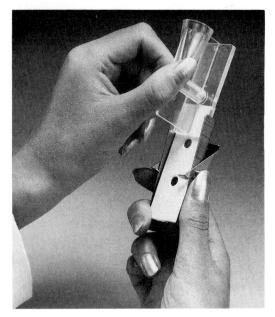

Figure 4.3 Cytospin 2: sample chamber being fitted into the special clip. The slide and filter card are already in position

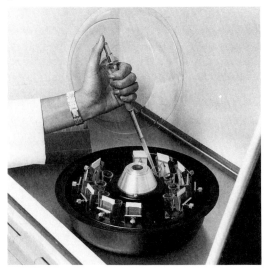

Figure 4.4 Cytospin 2: cell suspension being pipetted into the sample chamber

a way that the cell suspension was in immediate contact with the absorbent filter card. Cell losses were appreciable with this type of chamber and it has been replaced. A later modification (*Figure 4.5b*) was intended to produce an air bubble between the sample and the filter card. In practice, the porous filter card allowed air to leak away before centrifugation, and cells were lost with it. Evans, O'Rourke and Morris Jones (1974) found that this could be prevented by moistening the filter around the area of the hole with oil. This sealed the filter pores. However, fluid remained in the chamber and the oil contaminated the preparation. A recent study (Beyer-Boon, Wickel and Davoren, 1983) showed that excellent results could be obtained if the filter card is pre-moistened with saline and the chamber compression lip compresses the card. Results could also be improved if no more than 0.5 ml of sample was processed at any one time. Cytospin 2 machines are fitted with metal clips which allow the sample assemblies to rest at an angle (*Figure 4.5c*) so that an air bubble is automatically formed.

Step 4: Optimal cellularity and volume of the sample relative to the finished preparation must be assessed. A simple method (O'Sullivan and Proctor, 1982) is given at the end of this section. It may be advisable to use albuminized saline when resuspending cells. Superior morphology of cells from CSF has been reported by using dextran in the suspension fluid (Pelc, 1982).

Step 5: Small aliquots are to be preferred, usually 0.5 ml. CSF may be added directly to the sample chamber unless it is hypercellular. A Japanese cytocentrifuge model based on the Mark I Shandon machine has advantages in that it can accommodate larger volumes of fluid; it is therefore particularly suitable for urine samples.

Step 6: Centrifugation time is important. Safety locks are now fitted so take this delay time into consideration. The cell surface of wet-fix preparations must still be moist when removed for fixation if Papanicolaou staining is planned. Ideally a small amount of fluid should remain in the sample chamber of Cytospin 2 models. The optimal centrifugation time for a given volume is quickly determined by trial and error; however, CSF samples should be spun more slowly otherwise cell disintegration may occur.

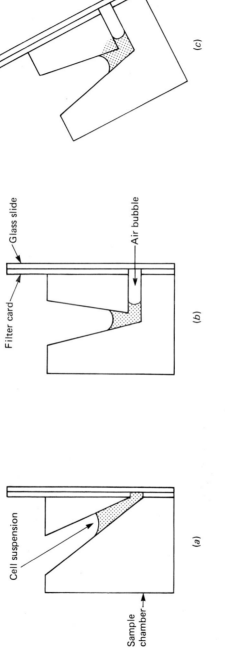

Figure 4.5 Diagrammatic view of modifications to sample chamber assemblies

Step 7: On removal of the sample, residual supernatant may be present.
With Cytospin 1 models, the fluid should be tilted away from the slide before lifting the preparation from the filter card (*Figure 4.5c*).

Health and safety note

Discard all used chambers and filter cards to an approved disinfectant. Early Cytospin models were fitted with a large perspex hood. In an infection control trial, open nutrient agar petri dishes were placed inside the hood when known bacterially infected urinary tract specimens were being processed. Air-borne contaminants were the only organisms grown (unpublished observation, St Mary's Hospital, Manchester).

Quantitative cell suspension for cytocentrifugation

Experienced laboratory workers usually prepare cell suspensions for cytocentrifugation by assessing the opacity on a trial and error basis. A simple and quick technique for quantitating cell suspensions has been described by O'Sullivan and Proctor (1982). The technique makes use of a simple optical illusion which is described below.

Calibration of Cellularity Comparison Card (CCC)

(1) Construct a Cellularity Comparison Card (*Figure 4.6*) by drawing circles approximately 2.5 inches in diameter on a stiff white card. Allow 0.75 inch between each circle.
(2) Prepare five cell suspensions (CS) by trial and error using fixed aliquots of each so that:
one CS has too many cells per microscope field;
one CS has an upper optimal limit of cells per microscope field;
one CS has the optimal number of cells per microscope field;
one CS has a lower optimal limit of cells per microscope field;
one CS has too few cells per microscope field.
Retain the aliquots of each CS.
(3) Place one drop of each CS (in turn) on a clean glass slide and view at ×100 magnification through a monocular microscope or one eyepiece of a binocular microscope.
(4) Hold the prepared CCC at an angle of 45° to the bench in line with the eyepiece. Adjust the light source until the illumination equals that of the room lighting. This is done by observing the microscope field with one eye and the CCC with the other.
(5) Position the CCC so that one of the outer circles is superimposed within the microscope field. The optical illusion should now be apparent; the cells are seen superimposed on the empty white circle. (Many people have difficulty with this phenomenon. As an experiment, place your hand six inches in front of one eye while observing some distant object with the other. The object can be seen superimposed upon the hand.)
(6) Plot the cells seen superimposed within the CCC circle onto the card, identify it and repeat for the other four cell suspensions.
(7) Add guidelines on the face or back of the card for each circle (e.g. dilute ×2, concentrate ×2, 10 drops).

Because workers have their own preferences regarding optimal cellularity, amounts of fluid and numbers of cells are not furnished but are available from the authors on request.

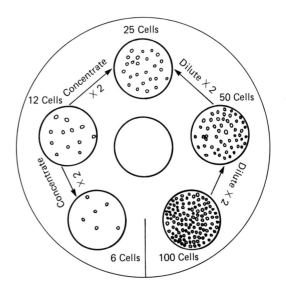

Figure 4.6 An example of a Cellularity Comparison Card. In this case, 25 cells per microscope field would produce an optimal monolayered cytospin preparation

Method of assessing cellularity of sample

Before cytocentrifugation, make a cellularity check using the calibrated CCC.

(1) After centrifugation of the fluid specimen and reconstitution of the cells in saline, place a drop of the suspension on a slide and focus on the cells using ×100 magnification.
(2) Position the CCC in line with the eyepiece as before.
(3) Equalize the room lighting and microscope illumination.
(4) Move the CCC until the centre (empty) white circle is apparently within the microscope field of view. Again the cells will appear superimposed on the card.
(5) Compare the cellularity with those cells already plotted on the outer circles and follow the guidelines appended.
(6) Make the preparations in the usual way.

With practice, each cellularity assessment can be made in less than 30 s and only one drop of the sample is used.

Membrane filtration

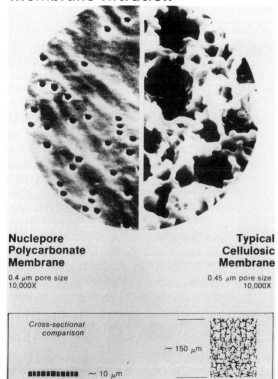

Figure 4.7 Different types of membrane filters

Vacuum filtration

Introduced initially for environmental and pollution studies, the potential benefit of vacuum filtration for the cytodiagnosis of fluid samples was quickly realized (Seal, 1956). In principle, fluid specimens are drawn through a membrane filter of known pore size under negative pressure (*Figure 4.7*). Cellular material is retained on the filter for staining and light microscope analysis.

The advantages of the technique are:

(a) Large volumes of fluid sample can be processed.
(b) More cells are captured for cytodiagnosis than by centrifugation.
(c) The technique is suitable for fixed as well as unfixed cell samples, thereby reducing the health hazard or problems of delay in transport.

The disadvantages of the technique are:

(a) The equipment is cumbersome.
(b) Filters become clogged rather quickly and the cells distorted.
(c) There is a health and safety hazard.
(d) Filters cannot be stained by the Romanowsky method and therefore are particularly unsuitable for the routine investigation of serous effusions or CSF where the non-epithelial content of the sample is often of considerable diagnostic significance.

Types of filter

Millipore filters*

Made from cellulose acetate, the filters are 150 μm thick, white in colour, with tortuous pores covering some 85% of the surface area. Intricate channels through the membrane form a mesh-like matrix. A pore size of 5 μm is recommended for cytology. The filters must be immersed in 70% alcohol before use to allow them to expand otherwise they will crinkle and curl up.

Gelman filters**

Made of cellulose triacetate, the filters are 150 μm thick, white in colour, with tortuous

* Millipore (UK) Ltd, Millipore House, The Boulevard, Ascot Road, Croxley Green, Watford, Herts.
** Gelman Hawksley Ltd, 10 Harrowden Road, Brackmills, Northampton.

pores covering 80% of the surface area. The filters have a mesh-like structure and must be pre-expanded in 70% alcohol before use. A pore size of 5 µm is recommended for cytology.

Nuclepore filters*

Made from polycarbonate, these filters are only 10 µm thick and are transparent. They have randomly positioned pores which occupy only 2% of the surface area and pass straight through the filter. A pore size of 5 µm is recommended. The filter should be moistened with saline before use. Do not allow the filters to remain in xylene longer than 5 min because they curl and become difficult to handle.

Cell recovery varies according to the type of filter used. A cell recovery rate of 81% was recorded when CSF was filtered through a Millipore filter. The recovery rate for the same specimen was 64% with a Gelman filter and 59% with a Nuclepore filter (Barratt and King, 1976).

Procedure

The apparatus required for vacuum filtration comprises sample funnel, clamp, membrane filter, porous membrane support, vacuum pump and fluid reservoir (*Figure 4.8*).

(1) Describe volume and macroscopic appearance of the fluid.
(2) Label the filter with patient's name and laboratory number using an indelible ink pen; pre-expand filter if necessary and place on the salinated membrane support. Secure the sample funnel in position over the filter with the clamp.
(3) Pour 10–20 ml of saline into the sample funnel.
(4) Add the cell sample and apply suction which must not exceed 50 mmHg.
(5) Add 20 ml of alcohol.
(6) Stop the suction, dismantle the apparatus and post-fix the filter preparation in alcohol for 10–15 min.
(7) Stain the preparation.

Technical notes

Step 2: Large filters are often used and cut in half. Care should be taken to label both

* Sterilin Ltd, Lampton House, Lampton Road, Hounslow, Middlesex.

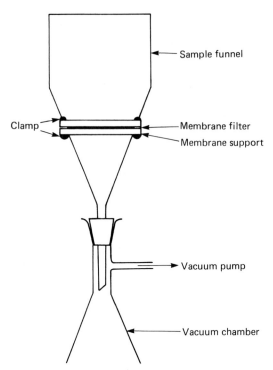

Figure 4.8 Apparatus for vacuum filtration

halves with an indelible ball point marker. Filters should be handled with flat-nose forceps.

Step 3: A squeeze bottle should be used to pour the saline down the sides of the funnel. Gravitational filtration will commence at once so have the cell sample ready. Never allow filter to dry.

Step 4: Caution—from here on a health hazard exists so extra care should be taken to avoid aerosols and spillages. Suction should be applied gradually to speed up filtration but should not exceed 50 mmHg. As the pores become clogged with cells, the filtrate slows to a dribble. Do not increase suction in an attempt to continue filtration as this will cause multilayering and the cells will be drawn into or through the pores and become distorted. A more prudent approach is to dilute the cell sample with saline.

Red cells can by lysed at this stage by adding 50% alcohol. The red colour should disappear from the membrane surface if lysis is complete. However, should lysis not occur and the red blood corpuscles coat the membrane, the filter should be discarded and an alternative method of processing the specimen used.

Step 5: Allow cells to fix for 2 min then re-apply suction. When the fixative is just disappearing, stop the suction. Never allow the surface of the membrane filters to dry as cell distortion and air-drying artefact are quick to occur.

Step 6: Remove the filter and place in fixative. Disinfect the apparatus. Red cell lysis is possible at this stage by placing the filter in 10% aqueous glacial acetic acid for 5 min prior to fixation. Carnoy's fluid is not recommended for Gelman filters and Nuclepore filters.

Step 7: It is important when using cellulose acetate filters to allow the stain to penetrate from both sides; this is facilitated by suspending them in the staining troughs. Cellulose acetate filters retain some stain, which is particularly noticeable with urine samples when phosphates and urates form lakes with haematoxylin to produce a high level of background staining. Nuclepore filters can be attached to glass slides with stainless steel clips. The use of Nuclepore filters reduces the stained background but the pores are visible on screening. Green and Wagstaffe (1973) recommend staining of the filtered cells by passing dyes through the apparatus.

Alternative procedure for membrane filtration

As there is a health risk in putting unfixed material through filtration equipment, some laboratories prefer to fix the sample before filtration.

(1) Describe the volume and macroscopic apearance of the fluid.
(2) Centrifuge the sample in a capped tube(s) at 2500 rev/min for 3 min.
(3) Decant supernatant into disinfectant.
(4) Resuspend deposit, or if heavily blood-stained, the buffy layer, in 10 ml of 50% alcohol for 30 min. The weak solution of alcohol does not distort the cells and allows them to flatten onto the filter.

(5) Proceed as described in the basic procedure from Step 2, but post-fixation in Step 6 is unnecessary.

Technical note

Step 5: Urine specimens may be fixed before filtration with Esposti's fluid fixative (p. 54). The weak solution of alcohol in this fixative does not distort the cells.

Special notes

It is possible to transfer cells from membrane filters to glass slides by an imprint method (Neilson et al., 1983; Oud 1984).

Alternatively, a technique for eliminating Nuclepore outlines is described on page 102.

Positive pressure filtration

Plastic or stainless steel capsules which have a fitting for a luer-lock syringe on one side and a hypodermic needle on the other are used for positive pressure filtration (*Figure 4.9*). Examples of these are the Swinnex (Millipore Ltd), the Cytoseive (Gelman Hawksley Ltd) and the Swin-lok (Nuclepore/Sterilin Ltd). The principles, advantages and disadvantages are similar to those of vacuum filtration. Positive pressure filtration is particularly suitable for the filtration of small volumes of fluid and hypocellular samples of CSF.

Procedure

(1) Label the filter, open the capsule, place the filter on the plastic support grid and reassemble the capsule.
(2) Fill the capsule chamber with saline from a 10 ml syringe and attach a fine gauge needle (22 gauge).
(3) Remove the syringe, draw the cell sample into it and re-attach to the capsule.
(4) Apply positive pressure to the syringe and allow the filtrate to discharge into disinfectant.
(5) Load the syringe with alcoholic fixative and continue filtration.

Technical notes

Step 1: Remember to expand Millipore and Gelman filters in 70% alcohol before

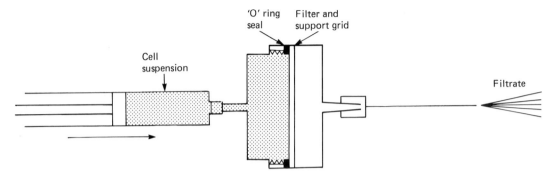

Figure 4.9 Diagrammatic view of filtration capsule in use

use. Check that the 'O' ring seal is uniform and seated correctly above the filter to secure it in position.

Step 2: As with vacuum filtration, a large volume of saline helps to maintain a steady flow. Place the capsule (needle side down) on the rim of a test tube while the syringe is recharged with fluid.

Step 3: Use a wide bore needle to draw an aliquot of specimen into the syringe. Avoid air bubbles.

Step 4: The fine bore needle will control the amount of pressure exerted. Red blood cells may be lysed at this stage as described on p. 65. Diminished flow of filtrate is indicative of clogging of the pores and overloading of filters. Do not try to force more fluid through or mechanical damage will occur to the cells.

Step 5: Fluid always remains above the membrane surface so dismantle the capsule carefully and place the filter immediately in fresh fixative. Sterilize the apparatus in an approved disinfectant.

Special preparatory techniques

Bloodstained smears

The presence of blood cells in fixed smears may cause problems during microscopy as they will mask the nucleated cells and diminish the number of cells available for diagnosis. Lysing of blood cells is accomplished by exposure to glacial acetic acid or 2 M urea, or by treatment with Carnoy's fixative. All these techniques are appropriate for achieving haemolysis of red blood corpuscles in fixed smears. Carnoy's fixative can also be used to haemolyse red blood corpuscles in unfixed smears.

Glacial acetic acid or 2 M urea method

(1) Immerse slide in 50% alcohol for 5 min. to remove coating fixatives.
(2) Rinse in water.
(3) Place in 5% glacial acetic acid for 10 min or 2 M urea (Pieslor, Oertel and Mendoza, 1979) for 1 min. With gentle agitation, haemoglobin will be seen diffusing from the slide surface.
(4) Rinse in several changes of water.
(5) Proceed with the staining sequence.

Technical notes

Step 3: Great care must be taken during the lysing of red blood corpuscles, particularly with acetic acid, as the whole smear is apt to lift off the slide surface. If this does happen, the smear must be removed from the acid quickly, and blotted flat again with fine filter paper moistened with the acid solution. The smear must thereafter be processed alone on a slide rack to avoid cross contamination by 'floaters'. This problem is usually encountered with thick mucinous bloodstained smears. Cell loss can sometimes be minimized by coating the slide with celloidin before attempting to lyse the red blood cells (p. 77).

Carnoy's method of haemolysis

Carnoy's fluid has long been advocated for removing red blood cells from unfixed smears since fixation and haemolysis are achieved simultaneously. It can also be used for lysing red cells on fixed smears. Carnoy's fluid must be freshly prepared as hydrochloric acid is formed on long storage. Optimum effects are achieved in 3 min, after which time it must be replaced by alcohol or PEG fixative. As Carnoy's causes excessive shrinkage and condensation of nuclear chromatin, it is necessary to reduce haematoxylin staining time. In addition, Carnoy's may destroy acid-soluble granules and pigments essential for diagnosis. As an alternative for unfixed smears, 3% acetic acid in alcohol is recommended.

Technical note

Haemolysis of cytocentrifuged smears is particularly hazardous. Because cells are deposited directly onto the slide, the red cells will form a supporting layer for the buffy (nucleate) cells (*Figure 4.10a*). Application of the lysing agent will swell the red cells causing them to rupture before they collapse like a deflated balloon. Many nucleate cells will be lost and indeed may continue to be shed in subsequent staining procedures (*Figure 4.10b*).

Bloodstained fluids

There are many methods for removing red cells from bloodstained fluids. Those in common use include the buffy layer technique, microhaematocrit technique, flotation technique, enzymolysis and acid haemolysis.

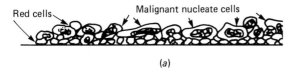

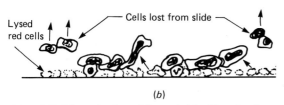

Figure 4.10 Schematic view of the probable effect of red cell lysis on a Cytospin preparation: (*a*) before lysis; (*b*) after lysis

Buffy layer technique

(1) Centrifuge the fluid at 2500 rev/min for 15 min to produce a layer of cells as shown in *Figure 4.11a*.
(2) Decant or draw off the supernatant.
(3) Remove the buffy layer with a pipette and prepare smears by one of the methods shown in *Figure 4.1* (p. 57).
(4) Fix two smears immediately in alcohol or PEG fixative and allow two smears to air dry for Romanowsky staining.
(5) Alternatively the cells in the buffy coat may be resuspended in saline for cytocentrifugation or membrane filtration.

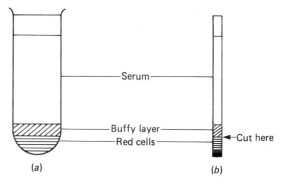

Figure 4.11 Separation of cells by (*a*) centrifugation; (*b*) microhaematocrit

Microhaematocrit technique

This technique is suitable for small heavily bloodstained samples or bloodstained deposits after centrifugation (To *et al.*, 1983; Yam and Janckila, 1983).

(1) Draw the sample into several capillary tubes 5 cm in length.
(2) Seal the clean end by flaming or with plasticine and centrifuge the tubes for 10 min.
(3) Identify the buffy coat (*Figure 4.11b*), scribe the tube and snap it at this point. Allow buffy coat to run onto a slide and prepare a smear quickly by one of the methods shown in *Figure 4.1* (p. 57).
(4) Prepare more smears and fix two immediately in alcohol or PEG fixative and allow two smears to air dry for Romanowsky staining.

Flotation technique

This technique involves the use of density

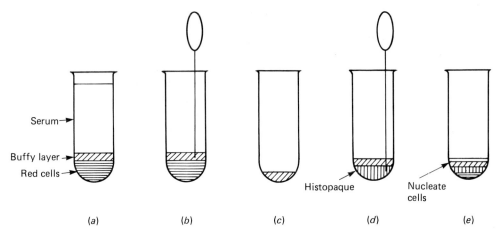

Figure 4.12 Sequential steps in Histopaque flotation technique

gradient columns or macromolecular solutions of varying density. These solutions permit passage of red blood cells while nucleated cells are retained higher in the column of fluid. Gradient columns can be used to separate the various nucleate cell lines. Those used in cytology are selected for their ability to separate red cells only. Macromolecular solutions in common use are albumin (McGrew and Nanos, 1975), Ficoll (Elequin *et al.*, 1977), Hypaque (Spriggs, 1975) Lymphoprep (To *et al.*, 1983), Percoll (Nagasawa and Nagasawa, 1983) and Histopaque. Histopaque contains Ficoll (5.7 g/dl) and sodium diatrizoate (9.0 g/dl).

The Histopaque 1077* technique

(1) Centrifuge the bloodstained fluid at 2500 rev/min for 15 min (*Figure 4.12a*). The buffy layer may be discernible.
(2) Remove the superantant and discard (*Figure 4.12b*).
(3) Pipette off the buffy area and top layer of red cells and transfer to a clean centrifuge tube (*Figure 4.12c*).
(4) Introduce Histopaque carefully underneath the cells so that they rise above it (*Figure 4.12d*).
(5) Centrifuge at 2000 rev/min for 20 min. At the interface between Histopaque and residual supernatant there will be an opaque band of white cells (*Figure 4.12e*)—the buffy layer.

---
* Sigma Chemical Co. Ltd, Fancy Road, Poole, Dorset BH17 7NH, UK.

(6) Remove the buffy cells to a clean tube, rinse three times in saline and centrifuge at 1000 rev/min for 10 min in order to remove the macromolecular salts. Retention of the salts will impair the adhesive properties of the cells.
(7) Remove the supernatant and make direct spreads, fixing or air drying as appropriate, or resuspend in saline for cytocentrifugation or membrane filtration.

Enzyme lysing agents

Enzyme lysing agents such as Streptolysin, Saponin and Zaponin are rarely used in cytology. Haemolysis is achieved rapidly but a gelatinous deposit frequently forms, particularly in heavily bloodstained fluids, which has to be dispersed with copious saline washing. This diminishes the cellular protein coat which may have been modified anyway by the enzyme used. Slight morphological changes may occasionally be noted.

Acid haemolysis

Strong acids such as glacial acetic acid (in Carnoy's fluid) and hydrochloric acid have the ability to lyse red blood cells but they are not recommended for haemolysis of bloodstained fluids since they fix the cells in solution and impair their ability to adhere to glass slides.

Aqueous formaldehyde has been recommended (Masin and Masin, 1974), but again this causes problems of cell adhesion.

Fibrinous clots

Serous fluids often arrive in the laboratory containing one or more clots.

Procedure

(1) Express fluid from the clot by gently squeezing it against the side of the specimen container. Remove clot and fix overnight in 10% neutral formal saline to which a dye has been added (white fibrin is difficult to visualize during sectioning).
(2) Process histologically and prepare a paraffin block.
(3) Trim off the surface wax until just beginning to cut into the clot.
(4) Cut a ribbon of 45 sections at 5 µm thickness.
(5) Mount on a slide(s) the 1st, 15th, 30th and 45th sections and stain routinely with haematoxylin and eosin.
(6) Retain the ribbon for special staining as required.

Cell block method

Cell deposits from all fluid specimens (including salinated washings) can be processed as a cell block. Sputum samples can also be processed in this way (*see* p. 240).

2% Agar

Powdered agar	2 g
Distilled water	100 ml

Agar is a carbohydrate derived from seaweed. The melting point is 98 °C and it is cooled to 50 °C before use. The setting point is 42 °C.

Procedure

(1) Place the fixed cell deposit (achieved by centrifuging at 2500 rev/min for 5 min and decanting supernatant fluid) into a capped conical centrifuge tube; resuspend cells in saline; centrifuge at 2500 rev/min for 5 min; decant supernatant. Leave tube upside down to drain for a few seconds.
(2) Warm the tube containing the deposit to about 45 °C under a hot water tap.
(3) Add an equal volume of 2% agar cooled to 50 °C to the cell deposit. Mix thoroughly.
(4) Immediately centrifuge at 1000 rev/min for 2 min.
(5) Examine the tube. If the agar is not solid, allow it to set.
(6) Tease out the agar cone with a sharpened stick.
(7) Place in 10% formal saline to which a few drops of eosin have been added. Leave at least one hour. This will convert the agar to an irreversible gel.
(8) Trim away excess agar and process by routine histological methods (see 2–6 in above procedure).

Technical notes

Step 3: It is very important that the deposit is mixed with the agar to ensure cells are fixed in the gel matrix, otherwise the cell block will disintegrate during processing.

Step 6: An alternative method of removing the agar cone may prove helpful. Place the tube in a freezer for 10–15 min. The agar tends to contract away from the glass. Handwarm the bottom of the tube for about 15 s and tap the rim smartly against the palm of the other hand; the agar cone should drop out.

Frozen cell blocks

This cryostat technique can be used to process unfixed or fixed cell deposits for enzyme or lipid studies.

Procedure

(1) Centrifuge the cells to a tightly packed button and invert the tube to drain.
(2) Add one drop of OCT* compound to the button and mix with a Pasteur pipette to produce a smooth paste.
(3) Using a suitable holder, place the cryostat chuck in liquid nitrogen for 15 s. Remove, but leave the chuck in the holder.
(4) With the pipette, add the OCT/cell mixture slowly to the supercooled chuck surface. The OCT will freeze solid very quickly and a suitable sized block is made by adding more specimen mixture.
(5) Quench in liquid nitrogen and prepare cryostat sections in the normal way.

*Lab-Tek Divison, Miles Laboratories Ltd, P.O. Box 37, Stoke Court, Stoke Poges, Slough, Berkshire, UK.

Tissue fragments

Unfixed tissue fragments may be prepared for cytoscreening by the squash-smear technique (*Figure 4.1e*) or fixed and made into cell blocks (p. 69). Because of the risk of damage to delicate tissues and cells, the fragments should be squashed gently and the slides pulled apart without added pressure. The slides may also be lifted apart after the tissue has been squeezed to a reasonable thickness. If speed is important, squash preparations are faster than frozen section or paraffin blocks and are particularly useful in neuropathology (Gandolfi, Tedeschi and Brizzi, 1983). Fixed tissue fragments are better made into cell blocks.

Endoscopic brushes

Brushes from any site are sometimes received in 10% formal saline and cellular material is always present among the fibres even after the most careful smearing and rinsing (Smith, Kini and Watson, 1980).

Procedure for nylon fibre brushes

(1) Shake the brush vigorously in the fixative or comb against another brush to release trapped material.
(2) Centrifuge at 2500 rev/min for 5 min. Decant the fixative.
(3) Prepare a cell block (p. 69).

Procedure for acetate fibre brushes
(Kuper *et al.*, 1966)

(1) Carefully decant most of the fixative and transfer material, including the brush, to a glass conical test tube.
(2) Fill with acetone, mix the contents, cap the tube and leave until the fibres are dissolved.
(3) Centrifuge at 2500 rev/min for 5 min. Decant the acetone.
(4) Prepare a cell block (p. 69).

Imprint preparations

These are chiefly used in neuropathology for rapid diagnosis of brain lesions but may be applied to any fresh tissue. For example, imprints of excised lymph nodes may be prepared for the preliminary diagnosis of lymphoma, and imprints of bronchial biopsy material prepared for the detection of *Pneumocystis carinii* in patients with AIDS. Also imprints are very useful when familiarizing oneself with the cytology of a new site. Freshly biopsied unfixed material is preferable, although autopsy material can be used. Care must be taken in the microscopic interpretation of the latter material due to degenerative (autolytic) changes. Three methods of preparing these smears are described.

Dry slide method

(1) Press or smear the cut tissue surface against the slide
(2) Rapidly fix in alcohol or PEG fixative.
(3) Prepare an additional air-dried smear for Romanowsky staining.

Albuminized slide method

(1) Place a small drop of albumin on the slide and smear over the whole surface.
(2) Prepare imprints by pressing the slide against the cut tissue surface.
(3) Rapidly fix in alcohol or PEG fixative.
(4) Prepare an additional air-dried smear for Romanowsky staining.

Frozen slide method

Useful for dry consolidated tissue.

(1) Coat the slide with albumin.
(2) Wearing protective gloves, immerse the slide in liquid nitrogen for 10 s.
(3) Immediately press the frozen coated slide surface against the tissue for a second or two.
(4) Lift the slide from the tissue and fix rapidly.
(5) Prepare air-dried imprints for Romanowsky staining.

Electron microscopy of cytological specimens

Scanning electron microscopy (SEM) and transmission electron microscopy (TEM) have been used extensively to study the morphology of cells in cytological samples in an attempt to improve diagnostic accuracy. There are many reports of the value of TEM for the diagnosis of tumours of non-epithelial origin such as mesothelioma (Kobzik, Anthan and Warhol,

1985) and carcinoid (Miles et al., 1985) and for the diagnosis of virus infections (Coleman et al., 1977; Hills and Laverty 1979; Smith and Coleman, 1983). Electron microscopy (EM) has been used to discriminate benign from malignant cells in cervical smears (Stanbridge, Butler and Langley, 1980) and in urine sediment.

Techniques have been developed for studying the same cell or cell cluster at both light and electron microscope level. Thus a cell identified initially in a Papanicolaou smear can be reprocessed for SEM (Becker et al., 1981; Domagala, Kahan and Koss, 1979; Mikel and Johnson, 1980; Takenaga et al., 1980), or TEM (Mather, Stanbridge and Butler, 1981; Ruiter, Mauw and Beyer-Boon, 1979). This section describes the techniques available for preparing fresh cells for EM and also for reprocessing Papanicolaou-stained smears for ultrastructural studies.

Preparation of cell deposit, tissue fragment or clot for TEM

Disruption of cell ultrastructure must be kept to a minimum during fixation. Alcoholic fixatives cause disruption of organelles although cell membranes and intracellular virions are well preserved (Becker et al., 1981; Coleman et al., 1977). Buffered formalin is acceptable for demonstration of neurosecretory granules (Carson, Martin and Lynn, 1973) but the use of glutaraldehyde is mandatory if cellular ultrastructure is to be demonstrated to best advantage.

Solutions

4% Glutaraldehyde in 0.1 M cacodylate buffer

Stock solutions

Solution a: 0.2 M sodium cacodylate (42.8 g $Na(CH_3)_2 AsO_2 \cdot 3H_2O$ in 1000 ml distilled water)
Solution b: 0.1 N hydrochloric acid (HCl)
Solution c: 0.5 M calcium chloride ($CaCl_2$ anhydrous)
Solution d: 25% aqueous glutaraldehyde

Mix 25 ml Solution a and 1.35 ml Solution b. Check the pH, adjust to pH 7.4 if necessary, and make up to 50 ml with distilled water. Add 0.5 ml Solution c agitating all the time. To 42 ml of this solution add 8 ml of Solution d.

0.1 M cacodylate buffer

Stock solutions

Solution a: 0.2 M sodium cacodylate
Solution b: 0.1 N hydrochloric acid
Solution c: 0.5 M calcium chloride

Mix 50 ml Solution a and 2.7 ml Solution b. Check the pH, adjust to pH 7.4 if necessary and make up to 100 ml with distilled water. Add 1.0 ml Solution c.
To 4 g of sucrose add the above solution until the volume is 100 ml.

Procedure

(1) Rinse the unfixed cell deposit three times in isotonic saline.
(2) Resuspend the deposit in glutaraldehyde fixative for 4 h.
(3) Wash the deposit three times in cacodylate buffer.
(4) Resuspend in fresh cacodylate buffer.
(5) Store at 4 °C until transfer to the EM laboratory for embedding.

Reprocessing of alcohol or glutaraldehyde-fixed Papanicolaou-stained smears for TEM

A technique is described whereby inclusion-bearing cells identified by light microscopy in stained smears can be reprocessed for examination in the electron microscope. It is particularly valuable for retrospective studies of mounted cytological or histological material when suitable specimens are no longer available for virological investigation. The method can be applied to any routine smear but is particularly useful for further investigation of urine, CSF and bronchial aspirates. It can be used to detect any viral, or chlamydial agents present in the cell, i.e. herpes simplex, cytomegalovirus, human polyomavirus, adenovirus, *Chlamydia trachomatis*.

1% Osmium tetroxide in 0.12 M phosphate buffer
Stock solutions

Solution a: 2.26% sodium dihydrogen orthophosphate ($NaH_2PO_4 \cdot 2H_2O$)
Solution b: 2.52% sodium hydroxide (NaOH)
Solution c: 5.4% glucose

Mix 41.5 ml Solution a with 8.5 ml Solution b. Check the pH and adjust to pH 7.4. Add 45 ml of this mixture to 5 ml Solution c. Add to 0.5 g of osmium tetroxide.

Uranyl acetate

Saturated aqueous solution of uranyl acetate.

Harris haematoxylin (*see* p. 79).

Resin mixture

Epon resin	8.5 ml
DDSA (hardener)	4.0 ml
MNA (hardener)	5.25 ml
DMP 30 (accelerator)	0.27 ml

Mix well before use.

Propylene oxide

Procedure

(1) Screen the alcohol-fixed Papanicolaou-stained smear for cells showing possible cytopathic changes due to virus infection. Photograph or make a map of the cells to be investigated further. Mark the position of the cells selected for EM investigation with an ink ring on the coverslip. Remove the slide from the microscope stage and trace the ink ring on the under surface of the slide using a diamond scriber.
(2) Immerse in xylene to remove coverslip. Rehydrate through graded alcohols prior to rinsing in distilled water.
(3) Post-fix in 1% phosphate-buffered osmium tetroxide (pH 7.4) at room temperature for 15 min.
(4) Rinse in distilled water.
(5) Immerse in saturated aqueous solution of uranyl acetate for 10 min at room temperature.
(6) Rinse in distilled water.
(7) Restain in haematoxylin for 2 min. Blue in tap water (the osmium tetroxide and uranyl acetate will have reduced nuclear staining).
(8) Examine the marked area to check that the cells under investigation are obvious at low magnification.
(9) Dehydrate as follows:
75% ethanol, 3 min, room temperature
85% ethanol, 3 min, room temperature
95% ethanol, 3 min, room temperature
absolute ethanol, 5 min, room temperature
(10) Immerse in propylene oxide for 5 min at room temperature.
(11) Flood slide with propylene oxide/resin mixture 1:1 for 5 min at room temperature.
(12) Replace with resin mixture 30–60 min at 37 °C.

Embedding procedure:

(13) Cut the end off the polythene embedding capsule (TAAB) leaving a hole 2–3 mm in diameter at the end. Heat polish the cut surface on a hot microscope slide. This should leave the cut surface flat and perpendicular to the sides of the capsule, so that leakage is prevented.
(14) Place the capsule in position on the microscope slide with the hole surrounding the selected cells and half fill it with resin mixture.
(15) Incubate at 70 °C for 18 h or until polymerization is complete.
(16) Crack the capsule and resin block off the microscope slide using liquid nitrogen.
(17) Examine face of resin block under low power magnification to identify cells under investigation using photomicrograph to aid identification.
(18) Trim face of resin block.
(19) Cut ultra-thin sections, prepare for EM and examine under transmission EM.

Technical note

Step 14/15: A special frame has been designed by the Department of Pathology, St Mary's Hospital, London W2, to hold the polythene embedding capsule in position during incubation.

Special note

This procedure can be used for glutaraldehyde-fixed smears. Smears should be prepared in the normal way and fixed in glutaraldehyde for 1 h. Wash three times in cacodylate buffer. Stain by Papanicolaou technique and mark cell under investigation on reverse of slide and proceed as from Step 2.

Technical aspects of smear preparation

One of the most challenging techniques for the cytotechnologist is the preparation of smears from unfixed samples. An ideal smear should be evenly spread and well stained with cell morphology clearly displayed. Successful smear preparation comes with practice but success can be achieved more quickly if a few main rules are observed.

(1) The optimal method of preparing the specimen should be selected. For example, cell concentration techniques are needed to prepare large volumes of fluid, whereas mucinous samples are best prepared as direct smears. CSF should be prepared by cytocentrifugation or filtration techniques, never by centrifugation and direct smearing. Haemolysis of red blood cells in a blood-stained effusion should be carried out before smears are prepared. The decision to use centrifugation or membrane filtration for the preparation of urine depends largely on personal choice. Both have their advantages and disadvantages and both give consistently good results in expert hands. Barratt and King (1976) have shown that tumour cells can be detected more frequently after membrane filtration than after centrifugation but the morphology of the cells on the filter is often poorly displayed.

(2) Cells should be evenly spread across the slide to form a monolayer under the area of the coverslip. There is no advantage in preparing numerous slides containing scanty cells. Studies of smear preparations of sputum have shown that the chances of detecting tumour cells improve very little after the first two or three smears have been examined and the time and effort involved in screening the extra slides may be counterproductive. However, it may be useful to prepare extra smears from each sample if enough material is available and set them aside for 'special staining' if this is indicated after examination of the Papanicolaou-stained smears.

(3) Care must be taken to preserve the morphology of the cells in the sample. Morphology can be impaired by delay in transit, inappropriate or inadequate fixation or air drying. For optimal cytopreparations, unfixed specimens must be processed immediately they are received by the laboratory. Rapid transit to the laboratory is essential. Hot, humid conditions encourage cell degeneration, enzyme activity and the growth of bacterial contaminants, therefore the specimen container should be packed in ice or stored at 4 °C if delay is anticipated. If this is not possible the specimen should be fixed before dispatch. Specimens sent between hospital centres should be transported in a cold box. However, specimens should not be frozen as ice crystals destroy cell morphology.

As the Papanicolaou stain is the preferred method of staining in cytology, smears must be fixed rapidly while the cells are wet, in order to preserve the morphological detail of the epithelial cells. Air drying results in a poor tinctorial quality and blurred nuclear detail. Care must be taken to ensure slides are fully immersed in the fixative solution. After fixation, smears can be removed prior to staining, but must not be allowed to dry unless a coating fixative was used. In any event, slides should not be left for more than a week in an alcoholic solution as subsequent staining will be impaired.

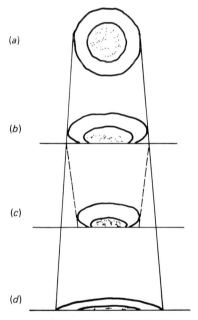

Figure 4.13 The effects of wet fixation and air drying on a spherical cell. (a) An unfixed cell in suspension; (b) the unfixed cell settles onto the slide and partially flattens down; (c) a wet-fixed cell for comparison – after initial swelling the cell shrinks back, probably to a diameter equivalent to that of the suspended cell; (d) the air-dried cell – as the suspension fluid evaporates from the unfixed cell (b), it flattens and spreads out due to the influence of atmospheric pressure and surface tension. When water leaves the cell itself, the membrane collapses and shrinks downwards. A spherical cell can dry down to as much as twice the original diameter whereas a mature squamoid cell shows little increase in size

Particular care is needed in the preparation of air-dried smears for Romanowsky stains. Rapid drying of the preparation is important as slow evaporation of water from the cells causes structural damage together with morphological aberrations (Bahr, 1976). This is why smears stained by the Romanowsky method often stain poorly at the centre of the smear where drying may be slow and satisfactorily at the periphery where rapid drying occurs. Thus air-dried smears should be evenly and thinly spread. Thin smears dry quite quickly in an exhaust cabinet where air is drawn past them. Thick smears can be dried quickly with a hair dryer. Alternatively, they can be placed on a hot plate at 40 °C. Pure methanol must be used for fixation which should be complete within 5 min. Once fixed, the morphological appearance is not appreciably altered if the slides are allowed to dry again. They will remain unimpaired over many weeks of storage.

Cells in Romanowsky-stained smears often appear larger than wet fixed cells. This is because air drying tends to cause flattening of the cells. The probable sequence of events leading to a larger cell area during air drying is shown diagrammatically in *Figure 4.13*.

(4) Cell loss from the surface of smears during staining must be minimized. Rubio (1981) has drawn attention to the problems of tumour cell loss during fixation and staining of cytological smears.

Factors affecting cell loss

The factors affecting cell loss are complex and ill-understood and further research is needed in this field. However, it is probable that a combination of factors are involved in determining the strength of bonding of cells to glass when smears are prepared from cytological samples. These include:

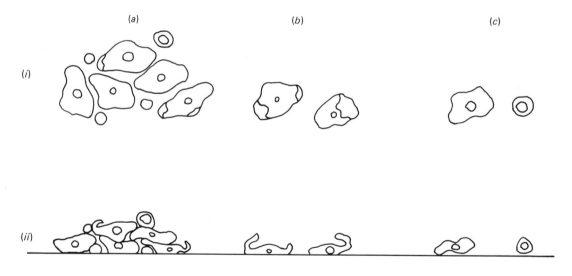

Figure 4.14 A diagrammatic view of (i) cells in suspension, (ii) cells at wet fixation, (iii) air-dried cells. (*a*) Multi-layered cells, (*b*) folded cells, (*c*) pre-fixed cells

(a) Sedimentation rate of the cells.
(b) Adhesiveness of cells.
(c) Viscosity of the specimen.
(d) Atmospheric pressure.
(e) Surface tension.
(f) Shape of cells.
(g) Type of fixative.
(h) Electrostatic charge.

By taking these factors into account in smear preparation it is possible to minimize cell loss.

Sedimentation rate

Any cells in suspension will spontaneously begin to sediment immediately they are smeared on a glass slide, before fixation occurs. They jostle for position and slide over one another under the influence of gravity to form layers of cells. The sedimentation rate depends largely on cell size; large cells tend to overlie the smaller ones, although occasionally they may be trapped beneath them (*Figure 4.14*). Whatever the situation, the surface layers will be more readily detached than the deeper layers.

Adhesiveness of cells

Cells exfoliating from mucous membranes (e.g. endocervix and respiratory tract) are coated with mucin and protein while cells from other sites (e.g. serous fluids) are bathed in protein alone. Some cells, including tumour cells, are also bathed in hyaluronic acid (Bard, McBride and Ross, 1983). When such cells are in contact with the slide and allowed to air dry, the mucin and protein act as a 'glue' causing the cells to adhere firmly to the slide. If the cells are fixed with alcohol before they are allowed to dry the mucoprotein coat coagulates into strands forming a mesh-like structure around and between the cells which can be seen on scanning electron microscopy (*Figure 4.15a*) and which secures them to the slide.

Cells from non-serous fluids and salinated washings have little or no mucoprotein available to cement them to the slide. Even if protein is present on the cells it will be diluted in the aqueous environment. Cells are likely to be lost from smears prepared from such samples unless the slide is pretreated with an adhesive substance. If cells are fixed in solution before being spread onto a slide, their mucoprotein is coagulated and their adhesive properties diminished.

Their fixed solid state prevents them flattening down onto the slide surface and they are washed away during subsequent staining (*Figure 4.14c(ii)*).

Viscosity of specimen

Many cytological samples are composed of cells suspended in a saline, serous or mucinous fluid. The more dense and viscous the suspension fluid, the more cell sedimentation is retarded. Smears prepared from very viscous fluid will, when wet, contain cells that are not closely adherent to the slide so that they may be lost during processing. On microscopy they give the appearance of multilayering and appear as loose groups of cells (*Figure 4.14a(ii)*). Cells lost from such smears are a common cause of 'floaters' and are a particular hazard of any multiple slide fixing and staining system (Husain, Grainger and Sims, 1978) as they may float from one slide to another leading to misdiagnosis (Rubio, Kock and Berglund, 1980).

Atmospheric pressure

Air pressure (760 mmHg) causes flattened cells to resist being lifted off the slide. Air-dried cells have more resistance to being disturbed during fixation and staining than cells fixed whilst wet as they are flatter and the effect of atmospheric pressure is more pronounced. This is one reason why Romanowsky-stained smears are usually more cellular than Papanicolaou-stained smears.

Surface tension

Alcoholic fixation causes rapid coagulation of protein which fixes the cells in position on the slide. However, the low surface tension of alcohol (approximately three times less than water) counteracts the fixative effect by causing turbulence on the slide which lifts up the cells and detaches them.

As an experiment, place a drop of water on a clean slide and next to it add a drop of alcohol. Tilt the slide to mix them together and notice how the water is repelled at the interface with the alcohol. If larger amounts of water and alcohol are added to the slide 'lines of miscibility' are seen. They are caused by the difference in surface tension, a change in the dielectric constant and generation of heat. The 'lines of

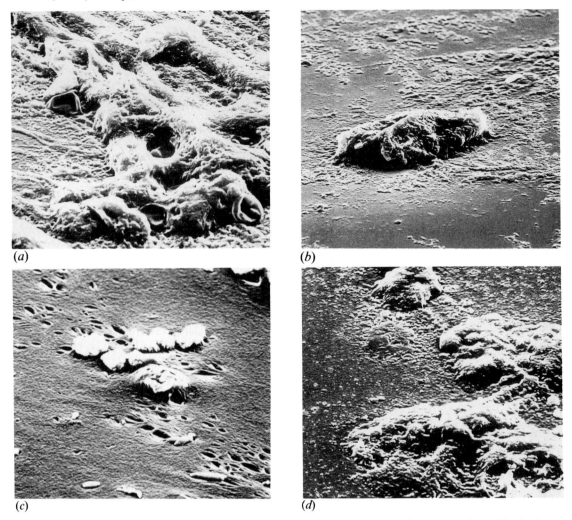

Figure 4.15 Comparison of surface morphologies of wet-fixed and air-dried smears. (*a*) Sputum – wet fixed. Clearly shows mucus strands on and around groups of cells; (*b*) Sputum – air dried. A group of squames – surrounding mucus has retracted down to slide surface; (*c*) Urine – wet fixed. Cells are rounded up and sitting on the slide surface – bacilli are present and holes suggest loss of material; (*d*) Urine – air dried. Appreciably more material is present which has dried down to the slide surface. Courtesy of EM Unit, Department of Botany and Zoology, University of Manchester

miscibility' exhibit great turbulence and it is this 'tidal wave' effect which can wrench cells away from the slide surface (and each other) before mucoprotein coagulation. Cells which are retained by a few fine threads of coagulum only are in danger of being lost during the staining sequence (*Figure 4.10b*).

Shape of cells

The shape of unfixed cells also presents problems. Those cells having a smaller cell/slide contact area relative to the diameter of the cell, are at risk of being wrenched away (*Figure 4.14b(iii)*). Cells which are folded, curled or creased probably have less resistance to the turbulence caused by alcoholic fixation (*Figure 4.14b(ii)*).

Type of fixative

Non-alcoholic fixatives are notoriously slow in coagulating mucoprotein, thereby diminishing the adhesive effect of the fibrinous mesh (*Figure 4.15a*) leading to increased cell loss. When non-alcoholic fixatives, e.g. formal saline are used on smears that are not allowed to air dry, the high surface tension of water in the fixative will tend to lift the cells off the slide surface. The fixative may also seep under the cells reducing the effect of atmospheric pressure.

Electrostatic charge

All cells in solution have a minute negative electrical charge. When a glass slide is in contact with plastic (e.g. Cytospin sample chamber) it becomes positively charged with static electricity. Thus the cells will be attracted to the slide surface. This, in essence, is the principle of the poly-L-lysine coating technique.

Methods of minimizing cell loss

Factors which affect cell adhesiveness were reviewed in the previous section. It is important for the cytotechnologist to bear them in mind when preparing smears from cytological samples. This is particularly important when processing non-serous fluids and salinated washings which have a low protein content. The following precautions are recommended:

(1) Use graded alcohols during dehydration and rehydration. This will diminish the effect of differences in surface tension. It is advised, wherever practicable, to use 30%, 50%, 70% and 90% alcohol steps.
(2) Prepare air-dried smears in preference to wet-fixed smears if possible. If wet-fixed smears are preferred, ensure rapid fixation while the surface of the preparation is still moist *at the edges*. Spray fixation has been recommended (Beyer-Boon and Voorn-den-Hollander, 1978).
(3) Avoid vigorous agitation of smears in reagents.
(4) Add coating adhesives to cell deposit. After the specimen has been centrifuged, a drop of bovine albumin, fetal calf serum (Nagasawa and Nagasawa, 1983) or 2% carbowax should be added to the deposit before preparing the smear.
(5) Coat slide surface with glycerine/egg albumin, chrome alum/gelatin mixture or poly-L-lysine (Husain, Page-Roberts and Millet, 1978; Mazia, Schatten and Seele, 1975). Cells will adhere more strongly to this surface than to a plain glass slide. The former two, however, take up dyes giving a standard background of variable intensity.
(6) Coat the slide with celloidin after fixation of the smear to minimize cell loss.

Celloidin procedure

(1) Remove the slide from alcohol and stand upright to drain almost dry.
(2) Introduce 1% celloidin (dissolved in ether/alcohol) dropwise to the top of the slide and let it run down over the whole area of the preparation. The celloidin should be stored in a tightly stoppered bottle to prevent evaporation.
(3) Leave for a few seconds to evaporate (a celloidin film is now covering the cells).
(4) Before the celloidin dries, place the slide in 70% alcohol to harden it for 5–10 min.
(5) Rehydrate and proceed with staining. (At no time during this technique should the preparation be allowed to dry off. Irreparable damage is caused to the cells.)
(6) Immerse in equal parts isopropanol/ether for 10 min to remove celloidin during the dehydration process since celloidin retains acid dyes. (This process may cause over-differentiation of the counterstains OG6 and EA 50 used in the Papanicolaou technique.)

A combination of these methods should appreciably minimize cell loss from slides.

References

BAHR, G. F. (1976). Some considerations of basic cell chemistry. In *Compendium on Cytopreparatory Technique*, 3rd Edition, (Keebler, C. M., Reagan, J. W. and Wied, G. L., eds), p. 1. Chicago: Tutorials of Cytology

BARD, J. B. L., MCBRIDE, W. H. and ROSS, A. R. (1983). Morphology of hyaluronidase-sensitive cell coats as seen in the SEM after freeze-drying. *J. Cell Sci.*, 62, 371

BARRETT, D. L. and KING, E. B. (1976). Comparison of cellular recovery rates and morphologic detail obtained using membrane filter and cytocentrifuge techniques. *Acta Cytol.*, 20, 174

BECKER, S. N. *et al.* (1981). Scanning electron microscopy of alcohol-fixed cytopathology specimens. *Acta Cytol.*, 25, 578

BEYER-BOON, M. E. and VOORN-DEN-HOLLANDER, M. S. A. (1978). Cell yield obtained with various cytopreparatory techniques for urinary cytology. *Acta Cytol.*, 22, 589

BEYER-BOON, M. E., WICKEL, A. F. and DAVOREN, R. A. M. (1983). Role of the air bubble in increasing cell recovery using Cytospin I and II. *Acta Cytol.*, 27, 699

CARSON, F. L., MARTIN, J. H. and LYNN, J. A. (1973). Formalin fixation for electron microscopy: a re-evaluation. *Am. J. Clin. Pathol.*, 59, 365.

COLEMAN, D. V. *et al.* (1977). Human papovavirus in Papanicolaou smears of urinary sediment detected by transmission electron microscopy. *J. Clin. Pathol.*, **30**, 1015

DOMAGALA, W., KAHAN, A. V. and KOSS, L. G. (1979). A simple method of preparation and identification of cells for scanning electron microscopy. *Acta Cytol.*, **23**, 140

ELEQUIN, F. T. *et al.* (1977). A quick method for concentrating and processing cancer cells from smears, fluids and fine-needle nodule aspirates. *Acta Cytol.*, **21**, 596

EVANS, D. I. K., O'ROURKE, C. and MORRIS JONES, P. (1974). The cerebrospinal fluid in acute leukaemia of childhood: studies with the cytocentrifuge. *J. Clin. Pathol.*, **27**, 226

GANDOLFI, A., TEDESCHI, F. and BRIZZI, R. (1983). The squash-smear technique in the diagnosis of spinal cord neurinomas: report of three cases. *Acta Cytol.*, **27**, 273

GREEN, G. H. and WAGSTAFFE, J. A. (1973). Nuclepore membrane filter techniques for diagnostic cytology of urine and other body fluids. *Med. Lab. Technol.*, **30**, 265.

HIGGINS, A. A. and SMITH, J. P. (1963). Fixation and transport in a postal cytodiagnostic service. *J. Clin. Pathol.*, **16**, 489

HILLS, E. and LAVERTY, C. R. (1979). Electron microscopic detection of papilloma virus particles in selected koilocytotic cells in a routine cervical smear. *Acta Cytol.*, **23**, 53

HUSAIN, O. A. N., PAGE-ROBERTS, B. A. and MILLET, J. A. (1978). A sample preparation for automated cervical cancer screening. *Acta Cytol.*, **22**, 15

HUSAIN, O. A. N., GRAINGER, J. M. and SIMS, J. (1978). Cross contamination of cytological smears with automated staining machines and bulk manual staining procedures. *J. Clin. Pathol.*, **31**, 63

KOBZIK, L. ANTHAN, K. H. and WARHOL, M. J. (1985). The distinction of mesothelioma from adenocarcinoma in malignant effusions by electron microscopy. *Acta Cytol.*, **29**, 219

KUPER, S. W. A. *et al.* (1966). The use of soluble swabs in exfoliative cytology of the bronchus and hollow viscera. *Lancet*, **ii**, 680

LINEHAN, J. J., MELCHER, D. H. and STRACHAN, C. L. (1983). Rapid outpatient detection of rectal cancer by gloved digital scrape cytology. *Acta Cytol.*, **27**, 146

MASIN, F. and MASIN, M. (1974). Formaldehyde-distilled water as haemolysing medium for cell suspensions. *Acta Cytol.*, **18**, 13

MATHER, J., STANBRIDGE, C. M. and BUTLER, E. B. (1981). Method for the removal of selected cells from cytological smear preparations for transmission electron microscopy. *J. Clin. Pathol.*, **34**, 1355

MAZIA, K. W., SCHATTEN, G. and SEELE, W. (1975). Adhesion of cells to surfaces coated with poly-L-lysine. *J. Cell Biol.* **66**, 198

MCGREW, E. A. and NANOS, S. (1975). Serous fluids: flotation method. In *A Manual of Cytotechnology*, 4th Edition, (Keebler, C. M. and Reagan, J. W., eds), p. 305. Chicago: The American Society of Clinical Pathologists

MIKEL, U. V. and JOHNSON, F. B. (1980). A simple method for study of the same cells by light and scanning electron microscopy. *Acta Cytol.*, **24**, 252

MILES, P. A., HERRERA, G. A., MENA, H. and TRUJILLO, I. (1985). Cytologic finding in primary malignant carcinoid tumour of the cervix: including immunohistochemistry, and electron microscopy performed on cervical smears. *Acta Cytol.*, **29**, 1003–1008.

NAGASAWA, T. and NAGASAWA, S. (1983). Enrichment of malignant cells from pleural effusions by percoll density gradients. *Acta Cytol.*, **27**, 119

NEILSON, M. L., FISCHER, S. HOGSBORG, E. and THERKELSEN, K. (1983). Adhesives for retaining prefixed urothelial cells on slides after imprinting from cellulosic filters. *Acta Cytol.*, **27**, 371

O'SULLIVAN, J. L. and PROCTOR, D. T. (1982). Cytospinning made easy. *The CMIAC Forum*, **VIII**, 3

OUD, P. S. (1984). Pressure-fixation method of transferring cells from polycarbonate filters to glass slides. *Anal. Quant. Cytol.*, **6**, 131

PELC, S. (1982) Cytocentrifugation of cerebrospinal fluid with dextran. *Acta Cytol.*, **26**, 721

PIESLOR, P. C., OERTEL, Y. C. and MENDOZA, M. (1979). The use of 2-molar urea as a haemolysing solution for cytologic smears. *Acta Cytol.*, **23**, 137

RUBIO, C. A. (1981). False negatives in cervical cytology: Can they be avoided? *Acta Cytol.*, **25**, 199

RUBIO, C. A., KOCK, Y. and BERGLUND, K. (1980). Studies of the distribution of abnormal cells in cytologic preparations. *Acta Cytol.*, **24**, 49

RUITER, D. J., MAUW, B. J. and BEYER-BOON, M. E. (1979). Ultrastructure of normal epithelial cells in Papanicolaou-stained cervical smears: an application of a modified open-face embedding technique for transmission electron microscopy. *Acta Cytol.*, **23**, 507

SACCOMANNO, G. *et al* (1963). Concentrations of carcinoma or atypical cells in sputum. *Acta Cytol.*, **7**, 305

SEAL, S. H. (1956). A method of concentrating cancer cells suspended in large quantities of fluid. *Cancer*, **9**, 866

SMITH, J. and COLEMAN, D. V. (1983). Electron microscopy of cells showing viral cytopathic effect in Papanicolaou smears. *Acta Cytol.*, **27**, 605

SMITH, M. J., KINI, S. R. and WATSON, E. (1980). Fine needle aspiration and endoscopic brush cytology. *Acta Cytol.*, **24**, 456

SPRIGGS, A. I. (1975). A simple density gradient method for removing red cells from haemorrhagic serous fluids. *Acta Cytol.*, **19**, 470

STANBRIDGE, C. M., BUTLER, E. B. and LANGLEY, F. A. (1980). Problems in cervicovaginal cytology: fine structure as an aid to diagnosis. *Acta Cytol.*, **24**, 335

TAKENAGA, A. *et al.* (1980). Giant cell carcinoma of the lung: comparative studies of the same cancer cells by light microscopy and scanning electron microscopy. *Acta Cytol.*, **24**, 190

TO, A. *et al.* (1983). Indirect immunoalkaline phosphatase staining of cytologic smears of serous effusions for tumor marker studies. *Acta Cytol.*, **27**, 109

WATSON, P. (1966). A slide centrifuge: an apparatus for concentrating cells in suspension on to a microscope slide. *J. Lab. Clin. Med.*, **68**, 494

YAM, L. T. and JANCKILA, A. J. (1983). A simple method of preparing smears from bloody effusions for cytodiagnosis. *Acta Cytol.*, **27**, 114

5

Staining techniques

David T. Proctor

Staining procedure

The staining sequence is both an art and a science. Each slide requires individual attention to realize the optimal staining qualities of the dyes; in practice, however, automated batch staining machines are used in busy routine laboratories and these need careful monitoring to give consistently reliable results. Many of the procedures in use today are empirical and numerous modifications abound so that methods differ between laboratories. Because no single dye can demonstrate all cellular elements selectively, combination staining sequences were developed. It is only recently that a clear understanding of dye chemistry has evolved and the reader is referred to recent works by Horobin (1982a) and Pearce (1980) for further details.

A stain must colour the preparation, resist decoloration in solvents and show minimal fading after long periods of storage. The main objectives of staining are:

(1) Microscopic visualization of the preparation.
(2) Differentiation of cellular constituents.
(3) Optical clarity.
(4) Reproducibility.
(5) Standardization.

In cytology, high quality staining of the cell nucleus is of paramount importance as the distinction between benign and malignant cells is based on nuclear morphology. Cytoplasmic staining supplies additional information with regard to cell maturity and origin.

Haematoxylin stains

There are many formulae for haematoxylin stains but alum haematoxylins are widely used in cytology as they produce clear nuclear staining, are simple to prepare and give consistent reliable results. A further advantage of alum haematoxylin is that a large through-put of slides can be accommodated before renewal. Erlich's haematoxylin will stain mucin so is not recommended for cytology.

Harris's haematoxylin (Harris, 1900)

Haematoxylin	5 g
Ethanol	50 ml
Potassium alum	100 g
Distilled water (50 °C)	1000 ml
Mercuric oxide	2.5 g
Glacial acetic acid (optional)	40 ml

Dissolve the alum in the water and dissolve the haematoxylin in the alcohol. Combine the two solutions in a 2 litre flask and bring rapidly to the boil. Carefully add the mercuric oxide while mixing, then cool quickly in a cold water bath. Add glacial acetic acid at this stage (if required) and filter. (The inclusion of glacial acetic acid improves the quality of nuclear staining in cytological material.)

Developed principally for use as a regressive stain in histology, Harris's haematoxylin is incorporated into the Papanicolaou stain but does have some limitations. One of these is uneven staining which is most likely to occur during differentiation of the nuclear stain in hydrochloric acid. In batch staining, hyperchromatic and hypochromatic staining of the nuclei may occur due to variability in the thickness of the smear. Harris's haematoxylin needs frequent filtering to remove the metallic scum which shortens the life of the stain. It should be stored in the dark to prevent over-oxidation and will gradually deteriorate after 3–4 months.

Mayer's haematoxylin (Mayer, 1903)

Haematoxylin	1 g
Potassium or ammonium alum	50 g
Citric acid	1 g
Chloral hydrate	50 g
Sodium iodate	0.2 g
Distilled water (50 °C)	1000 ml

Dissolve the haematoxylin, alum and sodium iodate in the water. This can be hastened by using a magnetic stirrer. Add the citric acid and chloral hydrate and boil for 5 min. After cooling and filtering the stain is ready for immediate use. If stored in the dark it should remain stable for about six months.

Mayer's haematoxylin can be used as both a regressive and progressive stain. As a progressive stain with short immersion time and low haematoxylin concentration, it selectively stains chromatin allowing for more control over nuclear density without any appreciable coloration of the cytoplasm.

Gill's haematoxylin (Gill, Frost and Miller, 1974)

Haematoxylin	2 g
Sodium iodate	0.2 g
Aluminium sulphate	17 g
Ethylene glycol	250 ml
Glacial acetic acid	40 ml
Distilled water	710 ml

Mix all the ingredients using a magnetic stirrer until dissolved. After filtering, the stain is ready for use although it may perform better after ripening at 40 °C for one week.

After carefully evaluating the qualities of other haematoxylins, Gill developed this half-oxidized derivative of Mayer's recipe. It retains the qualities of alum haematoxylin and stains chromatin at a controllable rate without the risk of over-staining. Being highly selective, differentiation is unnecessary and subsequent cytoplasmic staining is enhanced. If stored in the dark it should remain stable for up to one year.

Carazzi's haematoxylin (Carazzi, 1911)

Haematoxylin	1 g
Potassium alum	50 g
Potassium iodate	0.2 g
Glycerol	200 ml
Distilled water	800 ml

Dissolve the haematoxylin in the glycerol and the alum in 600 ml of water overnight. Combine the two solutions slowly with continual mixing. Dissolve the potassium iodate in the rest of the water using gentle heat if necessary, and add to the combined haematoxylin–alum–glycerol solution, mixing well. After filtration the haematoxylin is ready for use and should remain stable for up to one year.

This haematoxylin may be used progressively or regressively in a similar fashion to Mayer's recipe. It gives selective nuclear staining with minimal cytoplasmic coloration.

Routine stains used in cytology

The Papanicolaou stain

After many years of experimentation, Papanicolaou developed his trichrome stain in 1942 (Papanicolaou, 1942) which has survived with little modification and is the internationally accepted staining sequence in cytology. The smears must not be allowed to dry at any time before fixation or during processing.

Solutions

Haematoxylin of choice (see pp. 79–80)

Scott's tap water substitute

Magnesium sulphate (anhydrous)	10 g
Sodium bicarbonate	2 g
Tap or deionized water	1000 ml

Orange G6 (OG6)

Orange G (10% aqueous)	50 ml
Alcohol	950 ml
Phosphotungstic acid	0.15 g

Filter before use

Gill's working EA solution (Gill and Miller, 1973)

0.04 M light green SF	10 ml
0.3 M eosin Y	20 ml
Phosphotungstic acid	2 g
Alcohol	700 ml

Methanol	250 ml
Glacial acetic acid	20 ml
Filter before use	

Basic method

(1)	Remove polyethylene glycol fixative (PEG) in 50% alcohol	2 min
(2)	Rinse in water	1 min
(3)	Stain in Harris's haematoxylin	5 min
(4)	Rinse in water	2 min
(5)	0.5% aqueous hydrochloric acid	10–20 s
(6)	Rinse in water	2 min
(7)	Scott's tap water substitute	1 min
(8)	Rinse in water	2 min
(9)	Commence dehydration in 30% alcohol	1 min
(10)	Rinse in 50% alcohol	1 min
(11)	Rinse in 70% alcohol	1 min
(12)	Rinse in 90% alcohol	1 min
(13)	Rinse in 96% alcohol	1 min
(14)	Stain in Orange G6 (OG6)	1 min
(15)	Rinse in 96% alcohol	0.5 min
(16)	Rinse in 96% alcohol	0.5 min
(17)	Stain in Gill's EA (eosin-azure) solution	2.5 min
(18)	Rinse in 96% alcohol	1 min
(19)	Rinse in alcohol	1 min
(20)	Rinse in alcohol	1 min
(21)	Xylene	1 min
(22)	Xylene	1 min
(23)	Xylene until ready to mount or	1 min

Results

Nuclei	blue/black
Cytoplasm (non-keratinized) and effete red cells	green
Cytoplasm (keratinized)	pink
Red cells	orange

Technical notes

Step 1: Failure to remove PEG adequately will result in 'carbowax artefact' which obscures the cells and is very difficult to remove retrospectively. Smears should be washed in large volumes of fresh 50% alcohol before commencing the staining sequence.

Step 2: Rehydration is completed.

Step 3: Staining time depends on the haematoxylin formula used and personal preference. Coloration of some nuclei by OG6 or EA represents actual staining and does not necessarily indicate an impairment of the haematoxylin dye (Galbraith and Marshall, 1984). Haematoxylin is used progressively in some centres.

Step 4: Unbound haematoxylin is removed.

Step 5: Aqueous acid is preferable to the traditional acid/alcohol mixture as it lessens the likelihood of cells being detached by the 'tidal wave' phenomenon (see p. 76).

Step 7: The colour of brown acid haematein is changed to blue/black by using a weak alkaline solution. Scott's tap water substitute is recommended but 0.1% ammoniated water or weak aqueous solution of lithium carbonate may be used.

Step 8: Wash thoroughly to remove alkaline salts.

Steps 9–13: Use graded alcohols to minimize the loss of cells.

Step 14: Orange G6 is also available commercially. It has been shown by Drijver and Boon (1983) and Wittekind et al. (1982) that omission of OG6 causes little or no alteration in diagnostic accuracy. This step could be sacrificed in rapid staining techniques.

Steps 15–16: Excess stain is removed.

Step 17: Commercially available EA 50 is consistently reliable on all samples. EA is a combination of bismark brown, eosin Y and light green SF. However, bismark brown is rarely included in modern formulae as it contributes little to the cytoplasmic stain (Gill, Frost and Miller, 1974; Drijver and Beyer-Boon, 1983).

Step 18: Excess stain is removed.

Steps 19–20: Dehydration is completed.

Steps 21–23: The preparation is cleared and ready for coverslipping.

It now seems certain that the presence of the colourless macromolecular dye phosphotungstic acid (PTA) in both OG6 and EA is responsible for producing the selective staining attributed to the technique (Horobin, 1982b). Drijver and Beyer-Boon (1983) ascribe the function of PTA as both an accentuator and mordant.

Villanueva and Levine (1982) combined the four acid dyes of OG6 and EA in one staining bath thus reducing the number of sequential steps. Barr bodies are reported to stain well with this technique.

The Romanowsky stains (Romanowsky, 1891)

Often regarded as a special stain in histology the Romanowsky (ROM) technique must be regarded routine in any cytology department engaged in non-gynaecological specimen preparation.

Developed as a haematological stain for blood smears, these dyes have become a most useful diagnostic aid in cytology. Many ROM formulae are commercially available either as a dry powder or reagent ready for use. They are neutral dyes in that they consist of a combination of basic thiazine dyes (methylene blue and/or azure B) and the acid stain eosin which should be dissolved in acetone-free methanol for use (Wittekind, 1979).

Solutions

ROM stain of choice—available commercially ready for use.

Preparation of powdered dye:

Jenner	3 g/l in methanol
Leishman	1.5 g/l in methanol
May-Grünwald	3 g/l in methanol

Place the crystalline powder in a mortar and add a small aliquot of methanol. With a pestle grind the dye until a near saturated solution is apparent. Decant the stain into a dark bottle. Repeat the grinding and decanting process with more methanol until no more dye is extracted from the crystals. Store at room temperature and allow a few days for ripening. Filter fresh aliquots each day before use.

Stock Giemsa stain (Lillie modification, 1943)—also available commercially as a reagent ready for use.

Azure A-eosinate	0.5 g
Azure B-eosinate	2.5 g
Methylene blue-eosinate	2 g
Methylene blue chloride	1 g
Glycerol	375 ml
Methanol	375 ml

Mix the methanol and glycerol, and dissolve the dyes in the mixture overnight. Complete dissolution by shaking the stain for 10 min. Store in a dark bottle at room temperature.

Buffered Giemsa stain

To 10 ml of stock Giemsa stain add 90 ml of phosphate buffer (pH 6.8). Mix and filter before use. Prepare fresh each day.

Phosphate buffer (pH 6.8)

Available commercially in concentrated form in glass vials. Dilute with deionized water according to instructions.

Basic method

Use Coplin jars throughout.
(1) Fix air-dried smears in methanol   2 min
(2) ROM stain of choice   2–5 min
(3) Buffered Giemsa stain   6 min
(4) Rinse in phosphate buffer (pH 6.8)   2 min
(5) Drain, air dry, clear and mount

Results

Nuclei	purple
Degenerate nuclei	pink
Cytoplasm	pink or blue
Red cells and eosinophils	pink/red

This stain is particularly recommended for body cavity fluids and fine needle aspirations where haematological disorders may be expected. It can be helpful in distinguishing between lymphomas and small cell undifferentiated carcinoma. It is also useful for detecting fungi and parasites in smears.

Technical notes

The use of Coplin jars is advised to minimize the introduction of stain deposit to the slide surface. If this does happen, decolorize the preparation in methanol and proceed from Step 2.

Step 1: The slide and cell preparation must be absolutely dry to ensure an optimal result. Some methods combine Steps 1 and 2 (fix and stain together) but initial fixation in methanol is recommended.

Step 2: ROM stains are used undiluted, must be filtered fresh each day, and must not be allowed to dry on the surface of the slide as precipitation occurs readily.

Step 3: Buffered Giemsa stain, when used secondarily to Jenner, Leishman or

May-Grünwald stains, considerably enhances cytoplasmic staining.

Step 4: This step differentiates the stain. Blue dye is removed slowly to accentuate the eosin coloration. Take into account the drying of the slide in the overall buffer rinse time.

Step 5: Drain the slide and set aside vertically in a rack to dry. Drying may be hastened by placing the slide in a current of air or on a hot plate at 40 °C. The slide must be completely dry before clearing in fresh xylene otherwise water droplets will be seen microscopically. ROM stains are soluble in alcohol, therefore slides stained by this method should not be allowed to come into contact with this chemical.

The haematoxylin and eosin stain (H & E)

Originally introduced by Wissowzky (1876), the H & E has remained the basic histological tissue-staining sequence throughout the world. Thus it can be termed a 'reference stain' which means that histology sections can be studied retrospectively or by referral between histopathologists in different centres without the need for restaining. The staining sequence can be varied according to personal requirements and numerous modifications abound by the use of different haematoxylin (used progressively or regressively) and eosin formulae. Alterations in immersion time result in darker or lighter tinctorial characteristics.

Applied to cytology the H & E is most useful in the staining of cell blocks, clots and imprints when a direct comparison can be made to the histology section. Some cytopathologists prefer H & E-stained smears on non-gynaecological specimens, but ideally Papanicolaou-stained smears should be made at the same time.

Smears must not be allowed to dry at any time before fixation and during processing. Histology blocks should have been fixed in formal saline (*see* pp. 53–54).

Solutions

Haematoxylin of choice—usually Harris's or Mayer's (*see* pp. 79–80).

Scott's tap water substitute (*see* p. 80).

Eosin Y—1% aqueous.

Basic method

Smears

(1)	Remove PEG in 50% alcohol	2 min
(2)	Rinse in water	1 min
(3)	Stain in haematoxylin	5 min
(4)	Rinse in water	2 min
(5)	0.5% aqueous hydrochloric acid	10–20 s
(6)	Rinse in water	2 min
(7)	Scott's tap water substitute	1 min
(8)	Rinse in water	2 min
(9)	Eosin	1–2 min
(10)	Rinse in water	2 min
(11)	50% alcohol	0.5–1 min
(12)	70% alcohol	0.5 min
(13)	Rinse in 90% alcohol	0.5 min
(14)	Rinse in 96% alcohol	1 min
(15)	Rinse in alcohol	1 min
(16)	Rinse in alcohol	1 min
(17)	Xylene	1 min
(18)	Xylene	1 min
(19)	Xylene	until ready to mount or 1 min

Paraffin wax sections

Dewax in xylene	5 min
Xylene	5 min
Alcohol	2 min
Alcohol	2 min
96% alcohol	2 min
70% alcohol	2 min
Rinse in water	1 min
Stain in haematoxylin	5 min
Rinse in water	2 min
1% hydrochloric acid in 70% alcohol	10–20 s
Rinse in water	2 min
Scott's tap water substitute	1 min
Rinse in water	2 min
Eosin	1–2 min
Rinse in water	2 min
70% alcohol	0.5 min
70% alcohol	0.5 min
Rinse in 90% alcohol	0.5 min
Rinse in 96% alcohol	1 min
Rinse in alcohol	1 min
Rinse in alcohol	1 min
Xylene	1 min
Xylene	1 min
Xylene	until ready to mount or 1 min

Results

Nuclei	blue/black
Cytoplasm	shades of pink
Tissue elements	shades of pink
Red cells	red

Technical notes

Step 1: PEG in smears and paraffin wax in sections must be completely removed before commencing the stain.

Step 2: Rehydration is completed.

Step 3: Staining time depends on the haematoxylin formula used and personal preference. Haematoxylin may be used progressively if preferred. Stain in haematoxylin 1–2 min and omit Steps 5 and 6

Step 4: Unbound haematoxylin is removed.

Step 5: Aqueous acid is preferable to the traditional acid/alcohol mixture for smears as it prevents cells being detached by the 'tidal wave' phenomenon.

Step 7: Brown acid haematin is converted to a blue/black colour by using a weak alkaline solution. Alternatives to Scott's tap water substitute are 0.1% ammoniated water or a weak aqueous solution of lithium carbonate.

Step 8: Wash thoroughly to remove alkaline salts.

Step 9: Personal preference governs the eosin formula and immersion time.

Step 10: This step removes excess stain and can also be used to differentiate the eosin coloration by rinsing for longer periods.

Steps 11–12: Differentiation of eosin and partial dehydration are accomplished in this step.

Steps 13–16: Dehydration is completed.

Steps 17–19: The preparation is cleared and ready for coverslipping.

Rapid staining techniques

As all experienced laboratory workers are aware, any technique can be performed rapidly if necessary. Personal preference and trial and error governs the methodology used under these circumstances. So long as it is realized that artefacts may be introduced and the stain may fade more quickly, rapid staining is a useful adjunct in routine laboratory practice.

Special stains commonly used in cytology

Wherever possible unstained smears should be used for best results. Some techniques require alternative fixation and this should be borne in mind during the preparatory stage. All too frequently, a Papanicolaou or Romanowsky-stained smear contains the only material requiring a selective staining procedure. Decoverslipping, destaining and masking techniques are described on pp. 99–100. Staining techniques to demonstrate specific cellular components are described below. This is not a complete list and the reader is referred to more specialized texts and journals for additional information.

Wherever practicable, when a special stain is undertaken, control slides must be used to guard against false positive and negative results. The following selective staining techniques are included in this chapter:

Eosinophils	Carbol chromotrope 2R
Glycogen	PAS
	Diastase digest
	Best's carmine
Lipid	Nile blue sulphate
	Oil red O
	Sudan black B
Microorganisms	Gram
	Gram–Weigert
	Modified Dieterle stain
	Modified hexamine silver
	Ziehl–Neelsen (ZN)
Mucin	Alcian blue/PAS
	Metachromatic stain
	Southgate's mucicarmine
Mucin, prekeratin and keratin	Modified Kreyberg's stain
Non-specific esterase	Hexazonium salt

Nuclear sex chromatin	Acetic–orcein Cresyl fast violet
Nucleic acids	Feulgen reaction Gallocyanin–chrome alum Methyl green–pyronin
Pigments	Bleaching techniques Diazo reaction DOPA reaction Masson–Fontana Perl's stain Schmorl's reaction Sudan black B
Reticulin	Gordon and Sweets
Spermatozoa	Bryan's stain

Eosinophils

In air-dried preparations eosinophils are demonstrated adequately by using Romanowsky stains (see p. 82).

Lendrum's carbol chromotrope 2R technique (Lendrum, 1944)

Solutions

Mayer's haematoxylin (see p. 80)

Lithium carbonate solution

Saturated aqueous lithium carbonate	1 part
Deionized water	9 parts

Carbol chromotrope 2R

Phenol	1 g
Chromotrope 2R	0.5 g
Distilled water	100 ml

Dissolve the phenol in a flask under a hot water tap, add the chromotrope 2R and mix to dissolve. Ripening of the phenol/stain mixture at 37 °C for 2–4 days enhances the result (Lees, 1966, personal communication). Add the water, mix well and filter. Shelf life is at least 3 months.

Method

(1) Rehydrate the preparation	
(2) Haematoxylin	1 min
(3) Rinse in water	1 min
(4) Lithium carbonate solution to 'blue'	1 min
(5) Rinse in water	2 min
(6) Carbol chromotrope 2R	15–30 min
(7) Rinse in water	
(8) Rapidly dehydrate in alcohol, clear and mount	

Results

Eosinophil granules	bright red
Nuclei	blue/black
Red cells	rust colour

Glycogen

PAS technique

Solutions

Periodic acid—1% aqueous

Schiff's reagent—commercially available

Mayer's haematoxylin (see p. 80)

Ammoniated water—0.1% aqueous

Tartrazine—1% in 2-ethoxyethanol

Method

(1) Rehydrate the preparation	
(2) Periodic acid	5 min
(3) Wash well in tap water	5 min
(4) Rinse in deionized water	1 min
(5) Schiff's reagent	10–30 min
(6) Wash in three changes of deionized water	5 min
(7) Haematoxylin	0.5–1 min
(8) Rinse in water	
(9) Ammoniated water	0.5 min
(10) Rinse in water	
(11) Tartrazine (optional)	0.5 min
(12) Dehydrate in alcohol, clear and mount	

Results

Nuclei	pale blue/black
Background	pale yellow (optional)
Glycogen	bright red

(Other PAS positive material includes: adrenal lipofuscin; agar; albumin; cellulose; colloid; compound lipids; hyaline casts; mucins of intestinal, gastric, endocervical, bronchial and conjunctival epithelium; mucins of salivary glands, ovarian cysts and corpora amylacea; pancreatic zymogen granules; phospholipids; Russell bodies and starch grains. N.B. this list is incomplete.)

Diastase digestion technique to remove glycogen

Solutions

Diastase—1% aqueous

Method

A positive and negative control and the test slide/s are treated as follows:

(1) Rehydrate the preparation
(2) Diastase at 37 °C   30 min
(3) Wash in water   5 min
(4) Continue from Step 2 in technique 2 above

Results

Nuclei   pale blue/black
Background   pale yellow (optional)

Loss of red coloration from the diastase-treated test slide when compared to an untreated test slide (technique 2) confirms the presence of glycogen. Persistent PAS positive staining is *not* glycogen (*see* list after technique 2).

Best's carmine technique (Best, 1906)

Solutions

Harris's haematoxylin (*see* p. 79)

Hydrochloric acid—1% aqueous

Carmine stain
Stock solution:
    Carmine   2 g
    Potassium chloride   5 g
    Potassium carbonate   1 g
    Ammonia (sp. gr. 0.88)   20 ml
    Distilled water   60 ml

Add the carmine, potassium chloride and potassium carbonate to the water and boil gently for 5 min. After cooling, add the ammonia, mix well and store at 4 °C, tightly stoppered. Shelf life is 3–4 months.
Working solution:
    Stock solution   12 ml
    Ammonia (sp. gr. 0.88)   18 ml
    Methanol   18 ml

Prepare fresh and filter before use.

Method

(1) Rehydrate the preparation
(2) Haematoxylin   5–10 min
(3) Rinse in water   1 min
(4) Differentiate the nuclei in hydrochloric acid   5–10 s
(5) Rinse in water (nuclei 'blued' in next step)   30 s
(6) Working carmine stain (in a Coplin jar)   10 min
(7) Without letting the slide dry, wash in alcohol until excess stain ceases to flow out of the smear
(8) Clear and mount

Results

Glycogen   red
Nuclei   blue/black
(Some mucins and fibrin may stain pink.)

Lipid

The techniques described here are for simple fats (neutral lipids). They are particularly useful in assessing fetal maturity (lipid is present in fetal squames after the 34th week of gestation), and for identifying lipid-laden macrophages and extracellular and intracellular fat globules. Lipid stains are more soluble in lipid than the reagent in which they are dissolved. Solvents must not be used in the following techniques.

Fixation prior to lipid stains

10% Formal saline
Formal-calcium fixative (*see* p. 54)
Formalin vapour (*see* p. 54)
Air-dried preparations may be used but nuclear staining will be impaired.

Nile blue sulphate technique (Cain, 1947)

Solutions

Nile blue sulphate—1% aqueous
Acetic acid—1% aqueous

Method

(1) Rinse in distilled water
(2) Nile blue sulphate at 40 °C   5 min

(3) Rinse in distilled water
(4) Acetic acid   5 s
(5) Rinse in water
(6) Apply aqueous mountant and coverslip
   (Use glycerol for temporary preparations; Farrant's or Apathy's for permanancy.)

Results

Neutral lipids	orange/pink
Phospholipids	blue
Basophilic elements	blue

Oil red O technique (Casselman, 1959)

Solutions

Triethyl phosphate—60% aqueous

Oil red O
   Oil red O   1 g
   60% triethyl phosphate   100 ml
   Heat to almost boiling point and mix to dissolve (5 min). Filter, cool and filter again before use.

Mayer's haematoxylin (see p. 80)

Method

(1) Rinse well in the triethyl phosphate
(2) Oil red O   10–15 min
(3) Rinse off excess stain with triethyl phosphate
(4) Rinse in water
(5) Haematoxylin   1 min
(6) 'Blue' in tap water
(7) Apply aqueous mountant and coverslip

Results

Neutral lipids	bright red
Phospholipids	pink
Mineral oil	bright red
Nuclei	blue

Sudan black B technique (Chiffelle and Putt, 1951)

Solutions

Propylene glycol

Sudan black
   Sudan black B   1 g
   Propylene glycol   100 ml
   Boil for 3 min and filter while hot through coarse filter paper. When cool, refilter before use.

85% propylene glycol—aqueous

50% propylene glycol—aqueous

Neutral red—1% aqueous

Method

(1) Rinse in propylene glycol
(2) Sudan black   5 min
(3) Differentiate in warm 85% propylene glycol   2 min
(4) Rinse in 50% propylene glycol
(5) Rinse in water
(6) Neutral red   10 s
(7) Rinse in water
(8) Apply aqueous mountant and coverslip

Results

Neutral lipids	black
Phospholipids	grey
Nuclei	red

Microorganisms

Bacteria are recognized in routine smears as rods or cocci. Culture and the Gram stain are necessary for positive identification of bacterial species. Identification of fungi, parasites and viral inclusions from their characteristic morphology is described in *Table 5.1*.

Gram's stain

Solutions

Cresyl violet or *methyl violet 6B*—0.5% aqueous

Lugol's iodine
   Iodine   1 g
   Potassium iodide   2 g
   Distilled water   100 ml

Gram's differentiater
   Acetone   1 part
   96% alcohol   1 part

Neutral red or *safranin O*—1% aqueous

88 *Staining techniques*

Table 5.1 The staining reaction of microorganisms

Stain	Bacteria	Fungi	Parasites	Viral inclusions
PAP	Grey	*Candida* red	Blue/green	Pink/red
ROM	Violet	Violet	Variable but characteristic	Blue
PAS		Red	Variable	
Gram	Positive blue Negative red	Hyphae blue Spores red		Red
Silver		Black		

Method

(1) Rehydrate the preparation
(2) Cresyl or methyl violet — 1–3 min
(3) Rinse briefly in water
(4) Lugol's iodine — 1–3 min
(5) Rinse in water
(6) Gram's differentiater — 2–3 s
(7) Rinse in water
(8) Neutral red or safranin O — 15–30 s
(9) Rinse in water and drain almost dry
(10) Dehydrate rapidly in alcohol, clear and mount

Results

Gram positive bacteria	blue/black
Gram negative bacteria	red
Background	shades of red

Gram–Weigert technique

Solutions

Aniline crystal violet
Crystal violet	5 g
Aniline	2 ml
Alcohol	10 ml
Distilled water	88 ml

Mayer's haematoxylin (see p. 80)

Lugol's iodine (see above)

Eosin or *phloxine*—2.5% aqueous
Aniline–xylene mixture
Aniline	1 part
Xylene	1 part

Method

(1) Rehydrate the preparation
(2) Haematoxylin — 1 min
(3) 'Blue' in tap water
(4) Eosin or phloxine at 56 °C — 10 min
(5) Rinse in water
(6) Aniline crystal violet — 0.5–1 h
(7) Rinse in water
(8) Lugol's iodine — 1 min
(9) Rinse in water and drain almost dry
(10) Aniline–xylene mixture — 2–3 min
(Differentiation is complete when only bacteria and fibrin are stained blue/black.)
(11) Rinse in xylene and mount

Results

Gram positive bacteria	blue/black
Fibrin	blue/black
Nuclei	blue
Background	shades of red

Modified Dieterle spirochaete stain for Legionnaires' disease (Van Orden and Greer, 1977)

Solutions

Uranyl nitrate—5% in 70% ethanol

Alcoholic gum mastic
Gum mastic	10 g
Ethanol	100 ml

Mix and leave for 2 or 3 days to dissolve. Filter and store at 4 °C in an airtight container.

Silver nitrate—1% aqueous

Developer
Hydroquinone	1.5 g
Sodium sulphite	0.25 g
Distilled water	60 ml
Acetone	10 ml
Formalin (concentrated)	10 ml

| Pyridine | 10 ml |
| Alcoholic gum mastic | 10 ml |

Add the hydroquinone and sodium sulphite to the water and mix. Swirl slowly as each of the other fluids is added. The solution turns a milky orange/brown as the mastic is added, and a medium brown on standing. Ready for use after 6 h. Discard after use.

Method

(1) Preheat the uranyl nitrate in a Coplin jar in the oven at 56 °C. Preheat the silver nitrate in a Coplin jar in the oven at 56 °C.
(2) Rehydrate the preparation
(3) Rinse in distilled water
(4) Uranyl nitrate at 56 °C — 1 h
(5) Rinse in distilled water
(6) 95% ethanol — 5 s
(7) Alcoholic gum mastic — 3 min
(8) 95% ethanol — 5 s
(9) Rinse in distilled water — 1 min
(10) Drain almost dry — 10–15 min
(11) Wipe off excess water from around the smear/section
(12) Silver nitrate at 56 °C — 4 h (or at 40 °C overnight)
(13) Rinse in distilled water — 10 s
(14) Developer (until pale yellow to light tan) — 2–5 min
(15) Rinse in distilled water
(16) 95% alcohol — 10 s
(17) Acetone — 10 s
(18) Clear in xylene and mount

Results

Spirochaetes and
Legionnaires' bacilli — brown/black
(Other structures which stain similarly include *Pneumocystis carinii*, most other bacteria, fungi, melanin and foreign material in macrophages.)
Background — yellow/light tan

Modified hexamine silver technique
(Gomori, 1946; Grocott, 1955; Pintozzi, 1978)

Grocott's method has the disadvantage of being rather a slow technique requiring some 3 h to complete. Pintozzi (1978) introduced a modification which reduced the staining time to around 10 min. This method has been found to give consistent results. Greater control over silver deposition can be achieved by reducing the temperature of the silver solution to 75 °C and including dimethyl sulphoxide in the formula (Campbell and McCorriston, 1987).

Solutions

Chromic acid—5% aqueous

Sodium bisulphate—1% aqueous

Hexamine silver solution
Stock solution:
| 5% aqueous silver nitrate | 5 ml |
| 3% aqueous hexamine | 100 ml |

Store at 4 °C.
Working solution:
5% aqueous powdered borax	2 ml
Distilled water	25 ml
Stock solution	25 ml

Discard after use.

Gold chloride—0.2% aqueous

Sodium thiosulphate—2% aqueous

Light green SF—0.03% in 0.03% acetic acid

Method

Five minutes before commencing, pour the chromic acid and silver solution into clean Coplin jars and place them in a heated water bath at 80 °C.

(1) Rehydrate the preparation
(2) Chromic acid at 80 °C — 2 min
(3) Rinse in three changes of distilled water
(4) Sodium bisulphate — 30 s
(5) Rinse in three changes of distilled water
(6) Hexamine silver solution at 80 °C — 5 min
(7) Rinse in three changes of distilled water
(8) Gold chloride — 10 s
(9) Rinse in three changes of distilled water
(10) Sodium thiosulphate — 10 s
(11) Rinse in water
(12) Light green — 30 s
(13) Rinse in water
(14) Dehydrate, clear and mount

Results

Pneumocystis carinii cysts — black

Other fungi — black
Background — pale green

Ziehl–Neelsen technique (ZN)

Solutions

Carbol fuchsin—available commercially

Acid alcohol—1% hydrochloric acid in 70% alcohol

Sulphuric acid—25% aqueous

Loeffler's methylene blue
0.8% methylene blue in 100% alcohol	30 ml
Distilled water	99 ml
1% aqueous potassium hydrate	1 ml

Method

(1) Rehydrate the preparation
(2) Carbol fuchsin at 56 °C — 30 min
(Alternatively, place the slide on a slide rack over the sink and flood with stain. Apply heat until steam rises and leave for 10 min. Flood with more stain and repeat heat process twice more. Do not allow the stain to precipitate on the slide.)
(3) Wash off excess stain and differentiate in acid alcohol — 5 min
(4) Sulphuric acid — 5 min
(5) Rinse in water until a pale pink colour — 1–2 min
(6) Loeffler's methylene blue — 5–10 s
(7) Rinse in water
(8) Dehydrate rapidly, clear and mount

Results

Acid-fast bacilli	red
Nuclei	blue
Background	pale blue

Mucins

Most epithelial mucins are demonstrated by the PAS technique (*see* p. 85).

Alcian blue/PAS technique
(Mowry, 1956)

Solution

Alcian blue 8 GX—1% in 3% aqueous acetic acid

Method

(1) Rehydrate the preparation
(2) Alcian blue — 10 min
(3) Rinse in water
(4) Continue from Step 2 of diastase digest technique on p. 86. (If PAS stain is not required, counterstain lightly with haematoxylin.)

Results

Acid mucin	sky blue
Neutral mucin	red (optional)
Nuclei	dark blue/black

Metachromatic stain (Vassar and Culling, 1959)

Solutions

Toluidine blue—0.25% in veronal acetate–hydrochloric acid buffer at pH 4.5

Veronal acetate buffer at pH 4.5
Stock solution A:
Sodium acetate trihydrate	1.94 g
Sodium barbitone	2.94 g
Distilled water	to 100 ml

Stock solution B [0.1 M HCl]:
Hydrochloric acid	0.85 ml
Distilled water	to 100 ml

Working solution:
Stock solution A	5.0 ml
Stock solution B	10.5 ml
Distilled water	7.5 ml

Method

(1) Rehydrate the preparation
(2) Toluidine blue — 10 s
(3) Rinse in distilled water
(4) Drain dry (or hot plate at 40 °C)
(5) Clear and mount

Results

Mucin	pink
Other constituents	blue

Southgate's mucicarmine technique
(Southgate, 1927)

Solutions

Mayer's haematoxylin (*see* p. 80)

Scott's tap water substitute (see p. 80)

Carmalum stain

Carmine	1 g
Aluminium hydroxide	1 g
Aluminium chloride (anhydrous)	0.5 g
50% alcohol	100 ml

Dissolve the carmine and aluminium hydroxide in the alcohol then add the aluminium chloride. Boil gently for 3 min, cool, reconstitute to original volume with 50% alcohol and filter. Stored at 4 °C the stain will keep for about 6 months.

Method

(1) Rehydrate the preparation
(2) Haematoxylin — 1–2 min
(3) Rinse in water — 1 min
(4) Scott's tap water substitute — 1 min
(5) Rinse in water — 2 min
(6) Carmalum stain — 20 min
(7) Rinse in water
(8) Drain almost dry, rinse in 100% alcohol, clear and mount

Results

| Mucin | red |
| Nuclei | blue/black |

Mucin, prekeratin and keratin

Modified Kreyberg's stain

Kreyberg (1962) developed a technique for identification of tumour cell types in cytological material, particularly sputa, which was later modified by Dane and Herman (1963). The stain has been found useful in cervical cytology for distinguishing between cells from CIN III and adenocarcinoma, and for its concise staining of warty (HPV) cells (Davies, 1986, personal communication).

Solutions

Celestine blue

Celestine blue B	2.5 g
Ferric ammonium sulphate	25 g
Glycerol	70 ml
Distilled water	500 ml

Dissolve the ferric ammonium sulphate in the water and add celestine blue. Boil for 3 min, cool, filter and add the glycerol. Shelf life about 6 months.

Mayer's haematoxylin (see p. 80)

Hydrochloric acid—1% aqueous

Phloxine—1% aqueous

Alcian blue

| 1% aqueous alcian blue 8 GX | 1 part |
| 1% aqueous acetic acid | 1 part |

Orange G—0.5% in 2% alcoholic phosphotungstic acid

Method

(1) Rehydrate the preparation
(2) Celestine blue — 5 min
(3) Rinse in water
(4) Haematoxylin — 5 min
(5) Rinse in water
(6) Differentiate in hydrochloric acid if necessary — 10–30 s
(7) 'Blue' in warm water
(8) Rinse in distilled water
(9) Phloxine — 3 min
(10) Wash in water until excess stain removed
(11) Rinse in distilled water
(12) Alcian blue — 8 min
(13) Rinse in distilled water
(14) Orange G — 13 min
(15) Rinse in 95% alcohol
(16) Dehydrate, clear and mount

Results

Nuclei	brown
Acid mucin	blue/green
Prekeratin	orange
Keratin	red/orange
Cytoplasm	grey/brown

Non-specific esterase

Hexazonium salt technique

Solutions

Formal–calcium fixative, pH 7.0 (see p. 54)

Phosphate buffer (pH 7.0–7.1)

| Potassium dihydrogen phosphate | 9.53 g |

92 Staining techniques

Di-sodium hydrogen phosphate
(dihydrate)   23.14 g
Distilled water   to 100 ml

Pararosaniline hydrochloride (Sigma P3750)
Pararosaniline hydrochloride   2 g
2 M Hydrochloric acid   50 ml

Mix both reagents and heat gently to dissolve. Allow to cool, filter and store at 4 °C.

Sodium nitrate
Sodium nitrate   400 mg
Distilled water   10 ml

Prepare fresh before use.

Carazzi's haematoxylin (see p. 80)

Hexazonium salt
Pararosaniline hydrochloride
solution   2 drops
Sodium nitrate solution 2 drops
Phosphate buffer   10 ml

Prepare just prior to use but allow 1–2 min before the addition of the phosphate buffer.

Alpha-naphthyl acetate (Sigma N6750)
Alpha-naphthyl acetate 20 mg
Methoxyethanol   1 ml
Phosphate buffer   40 ml

Dissolve the alpha-naphthyl acetate in the methoxyethanol with vigorous shaking then slowly add the phosphate buffer. (The alpha-naphthyl acetate should be seen as small globules in the buffer; a white precipitate will not work.)

Incubation mixture
Hexazonium salt   10 ml
Alpha-naphthyl acetate 40 ml

Method

For batch staining, smears may be stored frozen. Fresh or frozen smears are fixed in formal–calcium for 10 min then rinsed in distilled water and allowed to dry. Cell blocks are cryostat sectioned and fixed as for the smears.

(1) Incubation mixture   20 min–1 h
    (Each laboratory must determine its own optimum time. The mixture forms a very fine orange-brown suspension which is normal.)
(2) Wash in distilled water   2–3 min
(3) Haematoxylin   3–5 min
(4) 'Blue' in tap water
(5) Allow slide to dry
(6) Clear and mount

Results

Non-specific esterase   reddish-brown
                        granulation
Nuclei   blue/black

Technical notes

(a) Always use chemicals and reagents of fine quality.
(b) Take care in handling the chemicals and reagents from a health and safety point of view.
(c) Allow the slides to dry after fixation.

Nuclear sex chromatin

Seen as a chromatin body or 'Barr body' on the membrane of the interphase nucleus in normal female cells, they are fragments of heterochromatin derived from inactivated X chromosomes (*Figure 5.1*). X chromosome material is also seen as 'drumsticks' on the nuclear lobes of polymorphs in Romanowsky-stained blood smears. Their distribution in sex-linked disorders is shown in *Table 5.2*.

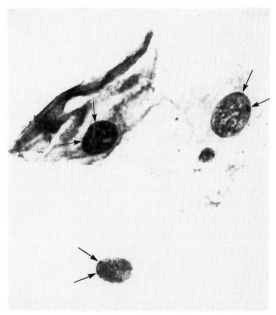

Figure 5.1 Buccal cells from XXX syndrome showing two Barr bodies on the nuclear membrane (arrowed). Stain: acetic–orcein

Buccal smear preparation

Instruct the patient to rinse the mouth with

water. Using the rounded end of an Ayre's spatula, or preferably a metal spatula or blade, scrape the inside of the cheek against the pressure of the fingers placed on the outside surface. The resultant creamy specimen is spread onto pre-labelled slides and fixed with PEG fixative (*see* p. 53).

Table 5.2 Distribution of Barr bodies in sex-linked disorders

Condition	Sex	Phenotype	Barr bodies
Normal male (M)		46XY	0 (<1%)
Normal female (F)		46XX	1 (>16%)
Turner's syndrome	F	45X	0
Pseudo-male	M	46XX	1
Pseudo-female	F	46XY	0
Kleinfelter's syndrome	M	47XXY	1
XXX syndrome	F	47XXX	2
Kleinfelter's XXX variant	M	48XXXY	2
XXXX syndrome	F	48XXXX	3
Kleinfelter's XXXX variant	M	49XXXXY	3
XXXXX syndrome	F	49XXXXX	4

Acetic–orcein technique (after Lillie, 1965a)

Solutions

Acetic–orcein stain
   Synthetic orcein   1 g
   Glacial acetic acid   45 ml
   Distilled water   55 ml

Heat the glacial acetic acid to 80°C and add the orcein. Mix to dissolve and pour into the distilled water at room temperature, mixing constantly. Cool under a cold water tap, filter and store in the dark. Shelf life about 1 year.

Fast green FCF—0.03% in 96% alcohol

Method

(1) Rehydrate the preparation
(2) Acetic–orcein   5 min
(3) Rinse in distilled water
(4) Dehydrate through graded alcohols to 96% alcohol
(5) Fast green FCF   5 min
(6) Rinse in 96% alcohol
(7) Complete dehydration, clear and mount

Results

Nuclei	pink/red
Barr body (nuclear sex chromatin)	intense red
Background	pale green

Cresyl fast violet technique
(Moore and Barr, 1955)

Solution

Cresyl fast violet—1% aqueous

Method

(1) Rehydrate the preparation
(2) Cresyl fast violet   5 min
(3) Rinse in water and drain almost dry
(4) Differentiate in 96% alcohol   2–4 min
(Differentiation is complete when the cytoplasm is colourless.)
(5) Dehydrate rapidly, clear and mount

Results

Nuclei	blue/mauve
Barr body (nuclear sex chromatin)	intense mauve
Background	shades of blue

Technical note

Barr bodies are seen in Papanicolaou-stained smears but their demonstration is not selective. Villanueva and Levine (1982), however, reported excellent results with their modified Papanicolaou stain.

Nucleic acids

Occasionally the demonstration of deoxyribonucleic acid (DNA) and ribonucleic acid (RNA) is requested for the identification of nucleoli, myeloma cells, plasma cells and for Sternberg-Reed cells in Hodgkin's lymphoma.

Feulgen reaction (Feulgen and Rossenbeck, 1924)

In this technique, mild acid hydrolysis exposes reactive aldehyde groups after cleavage of the purine–deoxyribose bond. The aldehyde groups are then stained red with Schiff's reagent.

Solutions

N/1 hydrochloric acid

Schiff's reagent—commercially available

94 *Staining techniques*

Sulphite rinses 1, 2 and 3
    10% potassium metabisulphite   7.5 ml
    N/1 hydrochloric acid   7.5 ml
    Distilled water   135 ml
Prepare fresh before use and dispense into three Coplin jars.

Tartrazine—1% in 2-ethoxyethanol (optional)

Method

(1) Rehydrate the preparation
(2) Rinse in cold N/1 hydrochloric acid
(3) N/1 hydrochloric acid at 60 °C   10 min
(4) Rinse in cold N/1 hydrochloric acid
(5) Rinse in distilled water
(6) Schiff's reagent   30–90 min
(7) Transfer direct to sulphite rinse 1   1 min
(8) Sulphite rinse 2   2 min
(9) Sulphite rinse 3   2 min
(10) Rinse well in distilled water
(11) Tartrazine (optional)   0.5 min
(12) Dehydrate in alcohol, clear and mount

Results

DNA   red
Background   shades of yellow (optional)

Gallocyanin–chrome alum technique
(Einarson, 1951)

Solution

Gallocyanin–chrome alum
    5% aqueous chrome alum   100 ml
    Gallocyanin   0.15 g
Mix and shake well. Bring slowly to the boil and simmer for 5 min. Allow to cool and filter. Make up to 100 ml with distilled water.

Method

(1) Rehydrate the preparation
(2) Gallocyanin–chrome alum   48 h
(3) Rinse in water
(4) Dehydrate, clear and mount

Result

DNA and RNA   deep blue

Methyl green–pyronin technique
(d'Ablaing et al., 1970)

Solutions

Methyl green–pyronin solution
    2% aqueous methyl green
    (chloroform washed)   40 ml
    2% aqueous pyronin Y   40 ml
    Glycerol   80 ml
    Distilled water   240 ml
    Liquid phenol   1 ml
    Tris buffer (pH 9.1)   8.1 ml
    Citric acid/phosphate buffer
    (pH 5.5)   3.2 ml
Add the methyl green to the glycerol and water, mixing well. Upon adding the pyronin mix well again and add the liquid phenol. The mixture is vigorously shaken to ensure dissolution. Add Tris buffer, mix well and allow to stand for 2 h. The citric acid/phosphate buffer is added and after mixing the solution is allowed to stand for a further 1 h. The solution is ready for use and has an indefinite shelf life.

Tris buffer (pH 9.1)
    0.2 M Tris(hydroxymethyl)-
      aminomethane   25 ml
    0.1 N Hydrochloric acid   5 ml
    Distilled water   70 ml

Citric acid/phosphate buffer (pH 5.5)
    0.04 M citric acid   9 parts
    0.08 M disodium hydrogen phosphate   11 parts

Butyl alcohol

Method

(1) Rehydrate the preparation
(2) Methyl green–pyronin solution   15 min
(3) Rinse in distilled water
(4) Allow to drain almost dry then dehydrate in butyl alcohol. If at this stage the nuclei are stained red, differentiate in 80% ethanol for a few seconds followed by a brief rinse in water. Drain the slide almost dry and dehydrate again in butyl alcohol. Repeat differentiation if necessary. When the nuclei are blue/green:
(5) Clear in xylene and mount

Results

DNA   green to blue/green
RNA   pink to red

Pigments

A variety of pigments may be encountered in cytological preparations. They are classified as endogenous or exogenous, the former being products of biosynthesis and the latter having entered the body from the environment. Their characteristic results with the more common staining techniques are set out in *Table 5.3*. Artefacts and contaminants are considered later.

Argentaffin is produced by endocrine cells of the APUD system and is diagnostic of carcinoid tumours seen in the gastrointestinal and respiratory tracts.

Chromaffin is only seen when the preparation has been fixed in dichromate-containing fluids. The classical 'chromaffin reaction' is a chrome salt–adrenaline complex in cells of the APUD system which normally originate from the adrenal medulla. Pheochromocytoma is a tumour derived from these cells.

Glycogen is included here since yellow pigmentation is commonly seen.

Haemosiderin is a breakdown product of haemoglobin bound to protein; seen in macrophages from haemorrhagic areas, 'heart failure' cells of the respiratory tract and as a ferruginous coat around asbestos fibres.

Lipofuscin is produced by the oxidation of lipids and often known as 'brown atrophy' pigment. Widely distributed among animal tissues.

Melanin is normally present in skin, the eye and hair and is produced from tyrosine via dehydroxyphenylalanine (DOPA). Mature melanocytes will contain pigment but immature melanocytes are very often amelanotic and will require the DOPA reaction for a definitive diagnosis. Malignant melanocytes produce a 'black cancer' and melanin is often seen in active macrophages from the site.

Asbestos fibres are invisible in routine light microscopy except by birefringent techniques. Visible asbestos 'bodies' are formed when the fibre becomes coated with protein and haemosiderin. They may be observed as beaded rods or 'dumb-bells' in sputa and pleural cavity aspirations. (Asbestos fibres are exogenous elements; the asbestos 'body' is an endogenous reaction.)

Carbon is introduced to the respiratory tract by inhalation and is found in respiratory macrophages and the bronchial lymphatics.

Pseudomelanin e.g. *Pseudomelanosis coli* in which there is pigmentation of the intestinal epithelium. Usually caused by laxative therapy when cascara is used over a prolonged period.

Bleaching techniques

Melanin and argentaffin pigments are bleached (removed) from preparations by any of the three following methods.

Table 5.3 Pigments: their characteristic results with the more common techniques

	Technique used							
	Papanicolaou	Romanowsky	Perls'	Masson-Fontana	Bleach	Schmorl's	PAS	Other techniques
Endogenous:								
Argentaffin	Light brown	Greenish	–	+	+	+	–	Diazo +
Chromaffin (after chromation)	Brown	Yellow/green	–	+	–	+ Greenish	+	
Glycogen	Yellow	Greenish	–	–	–	–	+(A) −(B)	
Haemosiderin	Brown	Green/blue	+	–	–	–	±	
Lipofuscin	Brown	Green/blue	–	+	–	+	+	Sudan black +
Melanin	Dark brown	Blue/black	–	+	+	+	–	DOPA +
Exogenous:								
Asbestos 'bodies'	Gold/brown	Greenish blue	+	–	–	–	±	
Carbon	Black	Black	–	–	–	–	–	
Pseudomelanin	Dark brown	Blue/black	±	+	±	+	±	DOPA –

+, good reaction; –, no reaction; ±, variable reaction; (A), PAS technique; (B), Diastase digest technique.

96 Staining techniques

(a) Lillie's performic acid method (1965b)

Solutions

Performic acid reagent

90% aqueous formic acid	8 ml
30% hydrogen peroxide	31 ml
Sulphuric acid (concentrated)	0.22 ml

Mix the ingredients immediately before use. Approximately 4.7% performic acid is formed in 2 h after which the mixture deteriorates.

Neutral red—1% aqueous

Method

(1) Rehydrate the preparation
(2) Performic acid reagent — 2 h
(3) Rinse carefully in water
(4) Neutral red — 30 s
(5) Dehydrate, clear and mount

(b) Chlorate method

Solutions

Chlorate reagent

Potassium chlorate	1 g
50% alcohol	about 30 ml
Hydrochloric acid (concentrated)	0.2 ml

Place the potassium chlorate in a Coplin jar and add 50% alcohol (enough to cover the material on the slide). With a pipette add the hydrochloric acid to the bottom of the jar and without mixing introduce the slide.

Neutral red—1% aqueous

Method

(1) Rehydrate the preparation to 50% alcohol
(2) Chlorate reagent — 24 h
(3) Rinse carefully in water
(4) Neutral red — 30 s
(5) Dehydrate, clear and mount

(c) Permanganate method

Solutions

Potassium permanganate—0.5% aqueous

Oxalic acid—1% aqueous

Neutral red—1% aqueous

Method

(1) Rehydrate the preparation
(2) Potassium permanganate — 12–18 h
(3) Rinse in water
(4) Oxalic acid — 3–5 min
(5) Rinse in water
(6) Neutral red — 30 s
(7) Dehydrate, clear and mount

Results of the bleaching methods

Nuclei	red
Background	shades of red

Melanin and argentaffin pigment are no longer visible in the control and test slides that have been bleached.

Diazo reaction (after Gomori, 1952)

Solutions

Diazo reagent

Fast red B salt (1% aqueous) at 4 °C	5 ml
Lithium carbonate (saturated aqueous) at 4 °C	2 ml

Mix and use immediately

Mayer's haematoxylin (see p. 80)

Method

(1) Rehydrate the preparation
(2) Diazo reagent at 4 °C — 2 min
(3) Carefully rinse in water — 3 min
(4) Haematoxylin — 30 s
(5) Rinse in tap water to 'blue'
(6) Dehydrate, clear and mount

Results

Argentaffin granules	orange/red
Nuclei	blue
Background	yellow

Technical note

This technique will only perform well on fresh cells and tissue fixed in aldehyde solutions.

DOPA oxidase technique (after Bloch, 1917)

Solutions

Formal alcohol (*see* p. 54)

DOPA incubating solution
DL3:4-dihydroxyphenylalanine	100 mg
Phosphate buffer (pH 7.4)	100 ml

Prepare immediately before use.

Neutral red—1% aqueous

Method

(1) Fix preparations in formal alcohol   1 h
(2) Rinse in distilled water
(3) DOPA incubating solution   2–3 h
(4) Rinse in water
(5) Neutral red   30 s
(6) Dehydrate, clear and mount

Results

Melanin (DOPA oxidase)	dark brown-black
Nuclei	red

Masson–Fontana silver reduction technique (Masson, 1914; Fontana, 1912)

Solutions

Fontana silver solution
10% silver nitrate	20 ml
Ammonia (sp. gr. 0.88)	a few drops
Distilled water	20 ml

Add the ammonia dropwise from a pipette to the silver nitrate while agitating the mixture until the precipitate initially formed is just about dissolved. Add the distilled water and the solution is ready for use. *Caution*: Ammoniacal silver compounds are liable to form an explosive mixture on storage so they should be freshly prepared in clean glassware. After use, inactivate by adding 5 ml of sodium chloride (saturated solution) before discarding.

Gold chloride—0.2% aqueous

Sodium thiosulphate—5% aqueous

Neutral red—1% aqueous

Method

(1) Rehydrate the preparation
(2) Rinse in distilled water
(3) Fontana silver solution   12–18 h
    (Silver reduction may be hastened in a 60 °C oven for up to 1 h but there is a risk of over blackening.)
(4) Rinse in several changes of distilled water
(5) Gold chloride   3 min
(6) Rinse in distilled water
(7) Sodium thiosulphate   2 min
(8) Wash in water
(9) Neutral red   30 s
(10) Dehydrate, clear and mount

Results

Melanin and argentaffin	black
Chromaffin and lipofuscin	black (but variable)
Background	shades of red

Perls' prussian blue reaction (Perls, 1867)

Solutions

Ferrocyanide reagent
2% potassium ferrocyanide	1 part
2% hydrochloric acid	1 part

The potassium ferrocyanide should be freshly prepared. Discard after use.

Neutral red—1% aqueous

Method

(1) Rehydrate the preparation
(2) Rinse in distilled water
(3) Ferrocyanide reagent   20–30 min
(4) Rinse in distilled water
(5) Neutral red   15–30 s
(6) Dehydrate, clear and mount

Results

Haemosiderin and ferric salts	deep blue
Asbestos 'bodies'	deep blue
Background	shades of red

Schmorl's reaction (after Lillie, 1965c)

Solutions

Ferric ferricyanide reagent
1% potassium ferricyanide	4 ml
1% ferric chloride	30 ml
Distilled water	6 ml

The stock solutions and working reagent should be freshly prepared.

Glacial acetic acid—1% aqueous

Neutral red—1% aqueous

Method

(1) Rehydrate the preparation
(2) Rinse in distilled water
(3) Ferric ferricyanide reagent   2–5 min
(4) Rinse in glacial acetic acid
(5) Rinse in water
(6) Neutral red   1 min
(7) Dehydrate, clear and mount

Results

Lipofuscin	dark blue
Melanin and argentaffin	shades of blue
Chromaffin	greenish blue
Background	shades of red

Sudan black B technique

Solution

Sudan black B—saturated solution in 70% alcohol

Method

(1) Rinse in 70% alcohol
(2) Sudan black B (filter before use)   12–16 h
(3) Rinse in 70% alcohol until pale grey in colour
(4) Rinse well in water
(5) Apply aqueous mountant and coverslip

Results

Lipofuscin, lipid and red cells	black
Background	pale grey

Reticulin

It is useful to demonstrate reticulin in imprints of lymph nodes.

Gordon and Sweets' technique (1936)

Solutions

Acidified potassium permanganate
    0.5% aqueous potassium permanganate   95 ml
    3% sulphuric acid   5 ml

Oxalic acid—1% aqueous

Iron alum (ferric ammonium sulphate)—2.5% aqueous

Silver solution
10.2% silver nitrate	5 ml
Ammonia (sp. gr. 0.88)	a few drops
3.1% sodium hydroxide	5 ml
Distilled water	up to 50 ml

Add the ammonia dropwise from a pipette to the silver nitrate, mixing well between each addition, until the precipitate first formed is just dissolved. Add the sodium hydroxide and repeat the dropwise addition of ammonia until the resulting precipitate is almost dissolved. (The solution should be very slightly opaque.) Make up to 50 ml with distilled water. N.B. See the cautionary note in Masson–Fontana technique on p. 97.

Formalin—10% aqueous

Gold chloride—0.2% aqueous

Sodium thiosulphate—5% aqueous

Neutral red—1% aqueous

Method

(1) Rehydrate the preparation
(2) Oxidize in acidified potassium permanganate   1–5 min
(3) Rinse in water
(4) Bleach in oxalic acid   3–5 min
(5) Rinse well in water followed by distilled water
(6) Sensitize in iron alum   10–30 min
(7) Rinse in distilled water   2 min
(8) Impregnate with silver solution   30 s
(Flood the slide and agitate it. The material will become transparent when impregnation is complete.)
(9) Rinse well in several changes of distilled water
(10) Reduce in formalin (with agitation)   10–60 s
(11) Rinse well in distilled water   2 min
(12) Tone in gold chloride   1 min
(13) Rinse in distilled water
(14) Fix in sodium thiosulphate   5 min
(15) Wash well in water   2 min
(16) Counterstain in neutral red   30 s
(17) Dehydrate, clear and mount

Results

Reticulin fibres — black
Collagen fibres — purplish
Background — shades of red

Spermatozoa

Modified Bryan's sperm stain

Solutions

Alpha-naphthol solution
Alpha-naphthol — 1 g
40% ethanol — 100 ml

Analine–pyronin
Pyronin Y — 0.1 g
Analine — 4 ml
40% ethanol — 96 ml

Citrate buffer
Sodium citrate — 7 g
0.9% sodium chloride — 1000 ml
Adjust the pH to 7.5

Bryan's stain
Eosin Y — 0.5 g
Fast green FCF — 0.5 g
Naphthol yellow S — 0.5 g
1% glacial acetic acid — 1500 ml
Mix thoroughly. Stored in a tightly stoppered bottle it has an indefinite shelf life. Filter before use.

Buffered Leishman's stain
Leishman's stain (*see* p. 82) — 5 ml
pH 6.8 buffer (*see* p. 82) — 15 ml
Filter before use.

Method

Freshly prepared air-dried smears should be used.
(1) Fix in 10% formal alcohol (*see* p. 54) — 1 min
(2) 80% alcohol — 5 min
(3) 70% alcohol — 5 min
(4) 50% alcohol — 5 min
(5) Alpha-naphthol solution — 4 min
(6) Wash in running water — 15 min
(7) Analine–pyronin — 4 min
(8) Wash in three changes of water — 1 s each
(9) Citrate buffer — 3 min
(10) Distilled water — 1 min
(11) Bryan's stain — 15 min
(12) Rinse in two changes of 1% glacial acetic acid — 1 s each
(13) Wash in running water — 1 min
(14) Leishman's stain — 20 min
(15) Rinse in phosphate buffer (pH 6.8)
(16) Drain completely dry, clear in xylene and mount

Results

Mature spermatozoa—blue nucleus with a magenta crescent-shaped acrosomal cap. The mid and end portions stain violet and the tail a grey/violet.
Immature germ cells—in general their nuclei stain violet/purple and the cytoplasm grey. The magenta-staining acrosomal cap may be visualized at an early stage of development.
Leucocytes—polymorphs have a violet/purple nucleus with pale blue cytoplasm. The peroxidase positive cytoplasmic granules will stain blue/black enabling them to be easily distinguished from immature germ cells. Lymphocyte nuclei stain a dark violet and the cytoplasm a pale grey.

Secondary staining

More often than not, pre-fixed smears are routinely stained by the Papanicolaou or the Romanowsky method on receipt. Having been screened, it is sometimes discovered that additional specific stains are needed. The usual practice is to request further smears, which can then be fixed and stained according to requirements.

Destaining a smear

Occasionally, however, additional material may not be available and the smears may need to be destained and restained. Obviously lipids cannot be demonstrated due to their prior removal by solvents.

Method

(1) Transfer cell reference marks to the back of the slide with a diamond writer.
(2) Remove the coverslip by immersing the

slide in xylene. A recently mounted coverslip can be detached in about an hour whereas others will require much longer.

(3) Rinse the slide for 30 min in three changes of xylene to ensure removal of all mounting media.
(4) Rehydrate the preparation.
(5) Destain Papanicolaou-stained slides by immersion in 1% hydrochloric acid in 70% alcohol for 1–12 h. Romanowsky-stained smears can be decolorized by alcohol alone in under 1 min.

Technical note

Mallory's bleach can be used for rapid decolorization if it is compatible with the subsequent staining method: flood the slide with 0.25% aqueous potassium permanganate for 2 min, rinse in water for 5 min, flood with 5% oxalic acid for 5 min and wash in running tap water for 10 min.

Proceed with the staining technique of choice.

Paraffin wax masking technique

This technique is useful in cases where more than one stain technique is required on the same smear.

Method

(1) Transfer reference marks to the back of the slide.
(2) Remove the coverslip in xylene (*see* Step 2 above).
(3) Rinse in xylene to ensure removal of mounting media (*see* Step 3 above).
(4) Immerse the smear in hot paraffin wax up to the area required for secondary staining. Hold rigidly in position until the slide has reached the temperature of the wax then slowly withdraw it. The original stain is now protected by a wax coat (*Figure 5.2a*).
(5) Allow the slide to cool before rehydrating through graded alcohols.
(6) Destain the unmasked portion of the smear (*see* Step 5 above).
(7) Proceed with the staining technique of choice avoiding those which involve the use of wax solvents or heated solutions.
(8) Dehydrate and immerse the whole slide in xylene to clear and dewax.
(9) When the wax has dissolved, rinse in two changes of clean xylene over 5 min.
(10) Coverslip, label and replace reference marks.

Technical notes

Multiple staining techniques are possible using the above method (*Figure 5.2b*). Satisfactory results are usually obtained although silver techniques may prove troublesome.

Mounting medium (with or without a coverslip) or Vaseline jelly (Pieslor, Oertel and Mendoza, 1979) are alternative methods for masking part of the slide, but are more tedious than the paraffin wax method.

Mounting and coverslipping

This heralds the completion of cytopreparation and while some regard it as a chore, it is no less important than the preceding sections. With this final step in the preparatory process, the cytotechnologist is expressing satisfaction at having produced a slide of optimal quality for cytoscreening and reporting, given a satisfactory sample initially.

The objectives in applying mounting media and coverslips are to:

(1) Render the preparation visible with dry objectives.

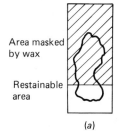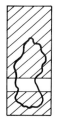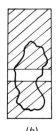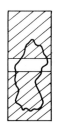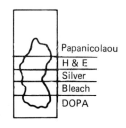

(a)　　　　　　　　　　　　(b)

Figure 5.2 Masking technique for secondary staining; (*a*) one additional stain, (*b*) multiple stains

(2) Preserve the various entities and their staining characteristics.
(3) Protect the preparation from dust and damage.

Mounting media

Mountants must all fulfil the following criteria. They must:

(1) be chemically inert;
(2) have a refractive index similar to glass (1.515) to avoid light scatter;
(3) show optical clarity;
(4) not crystallize on long storage;
(5) harden sufficiently within 16 h.

Canada Balsam (a natural resin) was used for many years for permanent preparations in microscopy; in the last few years it has been replaced by synthetic styrene mountants.

Coverslipping

The coverslip must be optically flat, clean and free from glass dust and no more than 0.17 mm in thickness to permit the use of high power objectives which have short working distances. Popular sizes are 22×22 mm for Cytospin preparations; 32×22 mm, 40×22 mm and 50×22 mm for smears, spreads and filters. The coverslip must completely cover the preparation and the whole slide screened otherwise tumour cells may be missed. This fact was demonstrated effectively by Rubio, Kock and Berglund (1980) in a study of tumour cell distribution in smears.

Automated coverslipping machines are now available. However, these machines are probably less useful in cytology than in the histology laboratory, because smears vary in their thickness and position on the slide.

Method

Routine smears

(1) Remove the slide from the xylene reservoir and allow excess solvent to drain.
(2) Select a suitably sized coverslip and lay it on a clean flat surface. Apply free-flowing mountant along the middle of the coverslip and invert the preparation (still moist with xylene) onto it. Turn the slide face up, position the coverslip and set aside.

An alternative method may also be used in which free-flowing mountant is applied along the middle of the preparation. Drop the coverslip on top, position it and set aside.
(3) Allow the mountant to spread under the coverslip as it settles into position.
(4) Air bubbles may be removed at this stage.
(5) A clearly numbered self-adhesive label is affixed after checking the number and/or name on the slide.
(6) The cytopreparations are delivered with their appropriate forms to the screening laboratory.
(7) File the slides after screening.

Technical notes

Step 1: Never allow the xylene to dry as xylene artefact occurs progressively the drier the material becomes.

Step 2: Mountants may be dropped onto the slide from a glass rod or squeezed from a tube with a narrow nozzle.

Step 3: Although it is not recommended in histology, mountant should be allowed to extrude beyond the edges of the coverslip in cytology for the following reasons:

(a) more cells may be available for screening along the edges of the slide;
(b) if there is a tendency to 'drying-back', the excess mountant will be drawn under the coverslip to fill the void;
(c) When the clearing agent has evaporated the mountant should be clear. If opalescence is noted, this is a sure sign of contamination by water or oil droplets.

Step 4: Gentle pressure should be applied with a pointed instrument behind the bubble in the direction of an adjacent edge. If the mountant retracts from the edges of the coverslip, more may be added as the coverslip is held down. As the pressure is gradually released the mountant will be drawn into the space. Excess mountant can be removed with a linen rag moistened with clearing agent.

Step 5: Labels will not adhere to glass if polyethylene glycol has not been removed or if the xylene on the slide has not evaporated.

Step 6: Keep the preparations flat on open slide trays until the mountant has hardened.

Step 7: The slides should be allowed to stand for a few days before filing otherwise they will adhere to each other. Should this occur, soak the slides in xylene for a few hours to soften the mountant before attempting to separate them.

Membrane filters

Always use flat-nosed forceps to handle filters and never let them dry at any stage.

Method for Gelman and Millipore filters

(1) Cut large filters in half with scissors after removal from xylene.
(2) Lay the filter (cell side up) on clean filter paper to remove excess xylene.
(3) Immerse in mountant in a flat enclosed dish for up to 10 min to allow perfusion through the pores.
(4) Remove from the mountant and hold vertically for a few seconds to drain.
(5) Remove excess mountant from the back of the filter by drawing it over the edge of a slide.
(6) Place the filter (face uppermost) on a xylene-wet slide, lowering it progressively from one end. Position it and place a coverslip over the whole filter area. There should be enough mountant present to spread to the edges of the coverslip.
(7) Remove air bubbles and flatten the filter by first laying filter paper over the slide. Place a roller (a length of 0.75 or 1 inch dowel rod is ideal) toward the centre of the preparation and roll with gentle pressure toward one end then back to the other.
(8) Remove the filter paper cover and set aside to dry.

Method for Nuclepore filters

The refractile pores on these filters are visible on screening unless removed with chloroform. The procedure must be carried out in an extraction cabinet.

(1) The filters should not remain in the xylene reservoir for too long as they will curl up and be difficult to flatten. Divide large filters with scissors.
(2) Place the filter cell side down, on a labelled slide; quickly blot and flatten with filter paper. Pour a few drops of chloroform on to the centre of the filter and allow them to spread over the whole surface. Before evaporation is complete, dip the slide in xylene and coverslip the preparation. Alternatively, run a little chloroform round the edges of the filter which will adhere to the slide as the chloroform evaporates. Flood the slide with chloroform for 2 or 3 s, drain the excess and before evaporation is complete, dip in xylene, apply mountant and coverslip the preparation.

Artefacts and contaminants

An artefact is an artificial product or morphological change in a cytological smear which is visible by light microscopy and is induced during sampling, transportation or smear preparation. It may also be the result of physical degradation of the specimen.

Contaminants are 'foreign bodies' introduced during sampling, transportation or processing.

There are numerous varieties of artefacts and contaminants which may cause diagnostic error; the more commonly seen types are discussed below. Iatrogenic changes (due to therapy) are not considered here.

Common artefacts

(1) Air drying before fixation causes flattening and spreading of the cells. The nuclear/cytoplasmic ratio is altered and the staining lacks crispness; cells appear 'washed-out' and there is a pervading eosinophilia. These changes are commonly seen in badly prepared Papanicolaou-stained smears.
(2) Trauma artefact. Aggressive smearing technique causes 'streaming' of the cells in the direction of spread; the nuclear membrane is ruptured resulting in the smearing of the contents. This can cause confusion in differentiation between lymphocytes and oat cells on light microscopy.
(3) Degenerative changes may occur due to delayed delivery of unfixed material. If a delay is anticipated, specimens should be stored at 4 °C.

(4) Xylene artefact is seen when clearing agent evaporates before coverslipping. It results in a black or brown 'deposit' on the cells and is believed to be due to air bubbles trapped within the cell. Remove the coverslip and mountant, rinse in alcohol, clear and remount.
(5) Air bubble artefact may be due to large air bubbles forming as a result of retraction of the mounting medium. Small air bubbles throughout the smear are due to aerated mountant, and both can be removed by taking the smear back to xylene and remounting.
(6) Carbowax artefact causes golden brown, granular 'pigmentation' on cells. It is thought to be due to incomplete removal of PEG before staining. In practice, this artefact is difficult to remove from stained preparations, indicating the probability that PEG is modified by some ingredient in the staining technique.
(7) Detergent artefact is due to troughs contaminated with soap after cleaning and results in pale staining of the cells in the smear.
(8) Refractile crystals are often seen in older smears and believed to be due to impure resinous mounting media.
(9) Bleaching. Continuous exposure of a stained slide to intense light, e.g. during microscopy, will cause bleaching of the stain. Bright sunlight will have a similar effect. Never leave stained slides uncovered by a window.

Common contaminants

(1) Contaminants due to delay in delivery. Specimens left in a warm damp atmosphere may provide a culture medium for non-pathogenic fungi of which *Alternaria sporidium* is the most common. This fungus can be identified by its typical racquet-shaped conidium which colours a deep brown/red after Papanicolaou staining. Other non-pathologic fungi are usually larger than *Candida*; some have double-walled hyphae, others a bizarre appearance.
(2) Dust and carbon particles from a dirty environment. Always ensure the specimen containers are tightly sealed.
(3) Fibres. Cotton wool and cellulose fibres are commonly seen, the latter being birefringent. Human hair is readily identifiable. Suture material is usually a hyaline structure which has an avidity for acid dyes. Fibres from filter paper may also contaminate a specimen.
(4) Lubricating fluid. Medical cream and jelly used during examination in clinics produces large empty areas in the smear. Cells tend to collect around the edge of the globules. (Use the oil red O technique for demonstration—*see* p. 87.)
(5) Talc crystals introduced via surgical gloves are recognized as translucent, refractile bodies which exhibit a Maltese cross appearance in polarized light (birefringence).
(6) Airborne contamination includes pollen, fungi and stellate structures (Medak and Burlakow, 1980; Jackson, 1981). These may contaminate specimens left uncovered near an open window.
(7) Vegetable cell and meat fibres may contaminate specimens, especially sputa or gastric washings (Weaver, Novak and Naylor, 1981).
(8) Water-borne contaminants from fish tanks and stagnant or dirty water include algae, diatoms, peritrichae, varicellae and vorticellae. These contaminants can be avoided by changing water troughs daily.
(9) Cross-contamination can occur when several smears are transported in the same container. Smears should be transported in slotted slide boxes with smear surfaces facing the same way if this problem is to be avoided. Cells may float from one slide to the other during the staining sequence and this can be avoided if stains are filtered daily, and membrane filters and effusion smears are fixed and stained separately. The 'floaters' will appear in a different plane of focus on microscopy. Transport fluids must be discarded after use.
(10) Mineral oil droplets in the mountant may appear as refractile globules on the smear. When the solvent has evaporated the mountant becomes macroscopically opaque. This is due to manufacturing error and all mountant of the same batch number should be discarded.
(11) Stain deposit. The presence of crystals or lakes of dye indicates inadequate filtration or removal of excess stain. It is also caused by precipitation of stains during the stain-

ing sequence, e.g. Romanowsky and ZN techniques.

(12) Water droplets in the smear may be due to inadequate dehydration of the slide or contaminated xylene and mounting media. The slides will become opaque and can be treated by removing the coverslip and dehydrating the slide before remounting. If the mountant is contaminated with water it should be discarded.

References

BEST, F. (1906). Über Karminfärbung des Glykogens und der Kerne. *Z. wiss. Mikrosk.*, **23**, 319

BLOCH, B. (1917). Des problem der pigmentbildung in der Haut. *Archives of Dermato-syphiligraphiques*, **124**, 129

CAIN, A. J. (1947). Use of nile blue in the examination of lipoids. *Quart. J. Microsc. Sci.*, **88**, 383

CAMPBELL, G. M. and MCCORRISTON, G. (1987). *Pneumocystis carinii* in bronchioalveolar lavage. *J. Clin. Pathol.*, **40**, 354

CARAZZI, D. (1911). Eine neue Hämatoxylinlösung. *Z. wiss. Mikrosk.*, **28**, 272

CASSELMAN, W. G. B. (1959). Lipids. In *Histochemical Techniques*, p. 71. London: Methuen

CHIFFELLE, T. L. and PUTT, F. A. (1951). Propylene and ethylene glycol as solvents for Sudan IV and Sudan black B. *Stain Technol.*, **26**, 51

D'ABLAING, G. et al. (1970). Laboratory suggestions: a simplified and modified methyl green pyronin stain. *Am. J. Clin. Pathol.*, **54**, 667

DANE, E. T. and HERMAN, D. (1963). Haematoxylin–phloxine–alcian blue–orange G differential staining of prekeratin, keratin and mucin. *Stain Technol.*, **38**, 97

DRIJVER, J. S. and BEYER-BOON, M. B. (1983). Manipulating the Papanicolaou staining method—role of acidity in the EA counterstain. *Acta Cytol.*, **27**, 693

EINARSON, L. (1951). On the theory of gallocyanin–chromalum staining and its application for quantitative estimation of basophilia. Selective staining of exquisite progressivity. *Acta Pathol. Microbiol. Scand.*, **28**, 82

FEULGEN, R. and ROSSENBECK, H. (1924). Mikroskopisch-chemischer Nachweis einer Nucleinsäure von Typus der Thymonucleinsäure und die darauf beruhende elektive Färbung von Zellkernen in mikroskopischen Präparaten. *Z. Phys. Chem.*, **135**, 203

FONTANA, A. (1912). Verfahren zur intensiven und raschen Färbung des treponema pallidum und anderer Spirochäten. *Derm. Wschr.*, **55**, 1003

GALGRAITH, W. and MARSHALL, P. N. (1984). Studies on Papanicolaou staining: III. Quantitative investigations of orangeophilia and cyanophilia. *Stain Technol.*, **59**, 133

GILL, G. W. and MILLER, K. A. (1973). In *Laboratory Cytopreparatory Techniques for Specimen Preparation*, p. 8. Baltimore: Johns Hopkins University School of Medicine

GILL, G. W., FROST, J. K. and MILLER, K. A. (1974). A new formula for a half-oxidised haematoxylin solution that neither overstains nor requires differentiation. *Acta Cytol.*, **18**, 300

GOMORI, G. (1946). A new histochemical test for glycogen and mucin. *Am. J. Clin. Pathol.*, **10**, 177

GOMORI, G. (1952). Organic substances. In *Microscopic Histochemistry*, p. 127. Chicago: University Press

GORDON, H. and SWEETS, H. H. (1936). A simple method for the silver impregnation of reticulin. *Am. J. Pathol.*, **12**, 545

GROCOTT, R. G. (1955). A stain for fungi in tissue sections and smears using Gomori's methenamine–silver nitrate technic. *Am. J. Clin. Pathol.*, **25**, 975

HARRIS, H. F. (1900). On the rapid conversion of haematoxylin into haematein in staining reactions. *J. Appl. Lab. Methods*, **3**, 777

HOROBIN, R. W. (1982a). *Histochemistry*. London: Butterworths

HOROBIN, R. W. (1982b). Staining methods involving dyeing. In *Histochemistry*, (Horobin, R. W. ed.), p. 109. London: Butterworths

JACKSON, R. N. (1981). Stellate structures in cytologic preparations. *Acta Cytol.*, **25**, 430

KREYBERG, L. (1962). Histological lung cancer types. In *Norwegian Monographs on Medical Science*, p. 88

LENDRUM, A. C. (1944). The staining of eosinophil polymorphs and enterochromaffin cells in histological sections. *J. Pathol. Bacteriol.*, **56**, 441

LILLIE, R. D. (1943). A Giemsa stain of quite constant composition and performance made in the laboratory from eosin and methylene blue. *Public Health Report*, **55**, 440

LILLIE, R. D. (1965a). Nuclei, nucleic acids, general oversight stains. In *Histopathologic Technic and Practical Histochemistry*, 3rd Edition, p. 158. New York: McGraw-Hill

LILLIE, R. D. (1965b). Chemical end groups. In *Histopathologic Technic and Practical Histochemistry*, 3rd Edition, p. 183. New York: McGraw-Hill

LILLIE, R. D. (1965c). Chemical end groups. In *Histopathologic Technic and Practical Histochemistry*, 3rd Edition, p. 211. New York: McGraw-Hill

MASSON, P. (1914). La glande endocrine de l'intestine chez l'homme. *C.R. Hebd. Séanc. Acad. Sci.*, **158**, 59

MAYER, P. (1903). Notiz über Hamatein und Hamalaun. *Z. wiss. Mikrosk.*, **20**, 409

MEDAK, H. and BURLAKOW, P. (1980). Stellate structures in oral and vaginal smears. *Acta Cytol.*, **24**, 269

MOORE, K. L. and BARR, M. L. (1955). Smears from the oral mucosa in the detection of chromosomal sex. *Lancet*, **ii**, 57

MOWRY, R. W. (1956). Observations on the use of sulphuric ether for the sulphation of hydroxyl groups in tissue sections. *J. Histochem. Cytochem.*, **4**, 407

PAPANICOLAOU, G. N. (1942). A new procedure for staining vaginal smears. *Science*, **95**, 438

PEARCE, A. G. E. (1980). *Histochemistry—Theoretical and Applied*, 4th Edition. Edinburgh: Churchill Livingstone

PERLS, M. (1867). Nachweis von Eisenoxyd in gewissen Pigmentation. *Virchows Arch. Pathol. Anat. Physiol.*, **39**, 42

PIESLOR, P. C., OERTEL, Y. C. and MENDOZA, M. (1979). The use of 2-molar urea as a haemolyzing solution for cytologic smears. *Acta Cytol.*, **23**, 137

PINTOZZI, R. L. (1978). Modified Grocott's methenamine silver nitrate method for quick staining of *Pneumocystis carinii*. *J. Clin. Pathol.*, **31**, 803

ROMANOWSKY, D. (1891). Zur frage der Parasitologie und Therapie der Malaria. *St. Petersb. Med. Wschr.*, **viii**, 297, 307

RUBIO, C. A., KOCK, Y. and BERGLUND, K. (1980). Studies of the distribution of abnormal cells in cytologic preparations. *Acta Cytol.*, **24**, 49

SOUTHGATE, H. W. (1927). Notes on preparing mucicarmine. *J. Pathol. Bacteriol.*, **30**, 729

VAN ORDEN A. E. and GREER P. W. (1977). Modified Dieterle spirochaete stain for Legionnaire's disease. *J. Histotechnol.*, Vol. 1, 2, 51–53

VASSAR, P. S. and CULLING, C. F. A. (1959). Fibrosis of the breast. *Arch. Pathol.*, **67**, 128

VILLANUEVA, A. R. and LEVINE, A. J. (1982). A modified Papanicolaou stain for cytological structures. *J. Histotechnol.*, **5**, 74

WEAVER, K. M., NOVAK, P. M. and NAYLOR, B. (1981). Vegetable cell contaminants in cytologic specimens: their resemblance to cells associated with various normal and pathologic states. *Acta Cytol.*, **25**, 210

WISSOWZKY, A. (1876). Ueber das Eosin als Reagenz auf Hämoglobin und die Bildung von Blugefässen und Blutkörperchen bei Säugethier und Hühnerembryonen. *Arch. Mikr. Anat.*, **13**, 479

WITTEKIND, D. (1979). On the nature of Romanowsky dyes and the Romanowsky–Giemsa effect. *Clin. Lab. Haematol.*, **1**, 247

WITTEKIND, D. *et al.* (1982). A new and reproducible Papanicolaou stain. *Anal. Quant. Cytol.*, **4**, 286

6

Immunocytochemical methods

Alan R. Potter

Immunocytochemical techniques for the demonstration of cell antigens have been available for diagnostic use since 1941 when Coons described a method of protein labelling using fluorescent dyes conjugated to an appropriate antibody (Coons, Creech and Jones, 1941). However, the fluorescent system was not ideally suited to the type of retrospective study usually required in diagnostic cell pathology as the fluorescent labels soon faded. A system in which the protein under investigation is permanently labelled was needed. In 1966 a technique for labelling antibodies permanently was described by Nakane and Pierce. This technique which involved the use of the enzyme horseradish peroxidase has proved to be the single most important advance in diagnostic cell pathology for several decades.

The value of immunocytochemical staining of histological specimens in the diagnosis, prognosis and classification of disease is now widely acknowledged. The use of these techniques in clinical cytology is currently being explored. Immunocytochemical techniques have been applied to serous fluids, fine needle aspiration samples and cervical smears in an attempt to improve discrimination between benign and malignant cells, thereby improving diagnostic accuracy. Immunological methods have also been used to identify tumour types in serous fluids, fine needle aspiration biopsy samples and touch preparations. They have also been directed towards the use of immunoalkaline phosphatase labelling as a method of sorting cells by flow cytometry with a view to using immunocytochemical methods for automated screening of cervical smears.

In this chapter the techniques of immunocytochemical staining as applied to cytological specimens are described and the potential value of this new approach to the analysis of cytological specimens is discussed.

Immunocytochemical techniques

Several methods have been developed for demonstrating cellular antigens. In the simplest system, specific antiserum raised by animal inoculation with the antigen to be localized can be conjugated to a variety of fluorescent or enzyme markers without impairment to its function. The conjugated antiserum is applied to tissue sections or smears and becomes fixed to the appropriate antigen forming a stable immune complex. Non-antibody proteins are removed by washing, and the result observed by the appropriate system for the marker. Fluorescent conjugated antibody can be visualized by exposure to ultraviolet light. Enzyme markers are visualized by reaction with an appropriate substrate to produce a colour compound recognizable with light microscopy.

Irrespective of the label used, an antigen may be visualized by the direct application of a labelled antibody or, more commonly, in a multiple layer indirect technique. This type of system has been used extensively for immunofluoresence work and is illustrated by the model shown in *Figure 6.1*. The unlabelled middle layer of the 'sandwich' acts both as an antibody to the antigen to be localized and also as an antigen to the second labelled antibody raised in a different species. This double layer technique

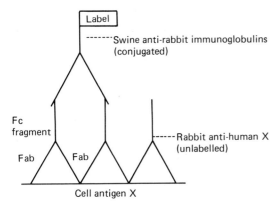

Figure 6.1 Indirect antibody technique

has distinct advantages compared with the direct labelling of the antibody:

(1) Numerous specific antisera may be localized by using only one labelled antibody specific for the serum protein of the species used.
(2) Inherent in the technique is the possibility of a negative control, usually the replacement of the first antibody with a non-immune serum in a parallel preparation.

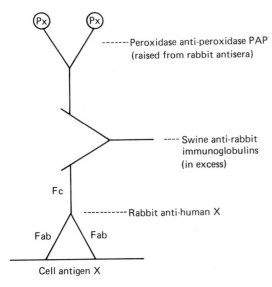

Figure 6.2 Diagrammatic representation of the peroxidase/anti-peroxidase (PAP) technique

(3) Sensitivity of the technique is increased. This effect is produced by several antigenic sites being available on the first antibody for each site of the original cell antigen. The use of a further layer or triple 'sandwich' would produce a further increase in sensitivity and is frequently used in immunoenzyme systems, as is demonstrated by the peroxidase/anti-peroxidase (PAP) system shown in *Figure 6.2*. The general methodology for the indirect and peroxidase/anti-peroxidase systems is given on (p. 117). The factors which influence each stage of the techniques are discussed below.

Antisera

The commercial availability of an ever increasing number of specific antisera to human antigens makes only a brief outline of the methods used for their production sufficient for this text. The antisera in current use are of two types—polyclonal and monoclonal.

Polyclonal antibodies

The chosen species, usually rabbit or goat, is given three inoculations of purified antigen, each at fortnightly intervals. For a rabbit, approximately 1 mg of antigen solubilized in phosphate buffered saline and homogenized with 1 ml of complete Freund's adjuvant is injected intradermally at multiple sites in the neck region. Up to 60 ml of blood can be taken from the central ear artery without damage to the animal, and the globulin fraction of the serum separated by precipitation with ammonium sulphate.

Monoclonal antibodies

The immune response of even inbred animals to a given antigen is unpredictable, often producing antisera with differing characteristics. Since each antigenic determinant of the inoculated antigen may stimulate development of a clone of antibody-producing cells, the resultant antiserum will be of polyclonal origin. Exact replication of such antibodies is not possible. A single B cell lymphocyte produces one type of antibody with a specificity to a single antigenic determinant. In theory, a large population of identical cells would produce large amounts of identical antibody. The practical problems of culturing B lymphocytes were overcome by Kohler and Milstein (1976). These workers fused the sensitized antibody-producing B lymphocyte with a myeloma cell, producing a hybrid with antigenic specificity and immortality.

The resultant hybridoma produces monoclonal antibodies specific for the one antigenic determinant chosen. The principles and practice of such technology are well described (Blann, 1979; Milstein *et al.*, 1979; Kennett, 1980; Scharff, Roberts and Thammana, 1981).

Hybridomas producing monoclonal antibody are commonly prepared using mice or rats as the host species. Sensitized spleen cells from an immunized mouse are disrupted mechanically in sterile saline and mixed with mouse myeloma cells. Fusions between the two cell lines are encouraged with polyethylene glycol. Spleen B lymphocyte/myeloma hybrids are isolated from the culture. These may be grown in large scale culture, or reintroduced into the peritoneum of a mouse to generate an ascites-forming tumour. The fundamental advantage of monoclonal antibody is that it represents a limitless source of large quantities of a monospecific antibody which will react in a uniform and reproducible way with its particular antigen. In addition, it is not essential to purify the immunizing antigen, since specificity depends on the choice of final hybrid used for cloning.

Multiple layer systems use the same initial antibodies irrespective of the marker. Routine diagnostic departments must depend upon the commercial availability of specific antisera for their purposes. A growing number are available, Dako* now offering a sufficient range for most current needs.

The antisera are supplied either desiccated ready for reconstitution, or in 0.5, 1 or 2 ml aliquots. It may be convenient to divide antisera into microlitre amounts, sufficient for one slide or case preparation. Otherwise antisera should be stored at 4 °C. Flocculation may occur with age but can be removed by centrifugation. For use, antisera are diluted either in Coons' phosphate buffered saline (PBS) or Tris buffered saline at pH 7.5. The optimum working dilution for each antibody should be assessed by serial dilution against standardized antigen positive material, the greatest dilution to give strong selective results being the optimum. In general the more dilute the antiserum, the less the risk of a non-specific protein interaction producing false positive results. Similarly, the use of only the gamma globulin fraction of an immune serum as opposed to whole serum will reduce the amount of non-antibody protein present,

* Dako Ltd, 22 The Arcade, The Octagon, High Wycombe, Bucks, UK.

and therefore the risk of non-specificity.

Some manufacturers recommend further purification of particular antisera before use, for example, polyclonal anti-carcinoembryonic antigen commonly requires absorption with tissue powders to prevent cross reaction with non-specific cross reacting antigen (Isaacson and Judd, 1977). Two methods of purification are in common use: (i) affinity purification and (ii) tissue powder absorption.

Affinity purification

The antibody to be purified is mixed with its pure antigen which has been absorbed on to Sepharose beads. Unbound proteins are washed off and the specific antibody eluted from the Sepharose beads by washing with either a low or high pH buffer.

Tissue powder absorption

Absorption with dried liver powder may be quite effective in some cases, e.g. for the purification of commercial polyclonal carcinoembryonic antigen, human placental lactogen, or human chorionic gonadotropin. The tissue powder is prepared as follows:

(1) Finely dice approximately 5 g of fresh liver tissue.
(2) Mechanically homogenize in acetone.
(3) Filter, and dry filtrate at 37 °C.
(4) Grind into a fine powder with a mortar and pestle. The powder may be stored in a closed container at room temperature.
(5) For use: add 100 mg of powder to 1 ml of antiserum
(6) Shake well and agitate continuously at room temperature for 2 h.
(7) Centrifuge at 5000 rev/min for 15 min.
(8) Decant the supernatant and use in the standard technique.

Antibody markers

Antibodies can be labelled with fluorescent dyes or stable enzymes such as horseradish peroxidase or alkaline phosphatase. Fluorescent dyes are widely used as labels in immunology, whereas enzyme markers are more commonly used by the histopathologist and cytopathologist. The advantages and disadvantages of the two labelling systems are discussed in this section.

Fluorescent markers

Historically, the first label successfully attached to antibodies was the fluorescent dye, fluorescein isocyanate (Coons, Creech and Jones, 1941). This was eventually replaced by fluorescein isothiocyanate (FITC) which was more readily conjugated to antibody protein (Riggs et al., 1958). These fluorescein compounds emit a bright apple green fluorescence. An alternative fluorescent dye, rhodamine isothiocyanate, is also in common use. This dye produces a red fluorescence.

Immunofluorescence does have inherent disadvantages when applied to any diagnostic procedure based on morphology.

(1) Specialized microscopy is needed, an HBO 200 watt mercury vapour lamp or quartz iodine light source with an FITC excitation filter is required.
(2) The visualization of the labelled immune complexes by fluorescence against a dark background allows little assessment of morphology.
(3) Preparations are impermanent, making critical comparisons difficult.
(4) The technique is not retrospective. This is a particular disadvantage to the histologist who is usually in receipt of formalin-fixed paraffin wax embedded material. The cytologist has more opportunity to duplicate smears for other techniques.
(5) Granulocytes and histiocytes can be a source of confusion to the cytologist, due to particularly bright non-specific staining which may easily be mistaken for specific fluorescence with the poor morphology afforded by the system.
(6) Immunofluorescence is not applicable to electron microscopy.

However, it should be remembered that the parameters and use of immunofluorescence are well understood, and that tried and tested reagents are readily available (*Plate 1*). Indeed fluorescence may well be the system of choice for the demonstration of discrete structures such as fungal elements, where the lack of clear morphology would not make interpretation difficult. When considering intracellular antigens, however, the indistinct morphology associated with fluorescence microscopy must remain the most notable disadvantage of this technique to the cytologist.

Enzyme labels

The requirement of improved light microscopy has resulted in the widespread use of enzymes. Horseradish peroxidase (Nakane and Pierce, 1966) and alkaline phosphatase (Mason and Sammons, 1978) have been widely advocated. The more recently introduced β-galactosidase (Bondi et al., 1982) and glucose oxidase (Rathlev et al., 1981) have useful practical advantages.

Enzyme labels are visualized by conventional histochemistry at the end of the antigen/antibody interaction, producing coloured insoluble reaction products which may be viewed with conventional light microscopy.

Immunoperoxidase

Immunoperoxidase methods use the stable enzyme marker horseradish peroxidase, which is visualized with a chromogenic substrate. The most widely employed is diaminobenzidine which, in the presence of hydrogen peroxide, deposits at the site of the antigen/antibody reaction an insoluble brown polymer, resistant to alcohol and xylene. Preparations are counterstained with haematoxylin, mounted in a resinous medium and examined by conventional microscopy. The result is a permanent preparation affording essentially the same morphological detail as a Papanicolaou or haematoxylin and eosin stained smear. Peroxidase conjugated antibodies are directly comparable with fluorescent labelled antibodies, and may be used in an indirect method as already described (Nakane and Kawaoi, 1974). However, the actual conjugation procedure does have well documented disadvantages:

(1) Partial denaturation of the antibody.
(2) Some inactivation of the enzyme.
(3) Residual free unlabelled antibody or enzyme. This can result in non-specific staining and reduced sensitivity.

Procedures that do not use directly labelled

antibodies (non-conjugated) have therefore been developed to reduce this risk. The two most commonly used are the enzyme bridge method (Mason, Phifer and Spicer, 1969; Ford and Stoward, 1974), and the peroxidase/anti-peroxidase (PAP) method (Sternberger et al., 1970). The PAP method is an example of such a system, and has now become the most widely used for immunoperoxidase staining.

The first and second antibodies are unlabelled, and the third is a complex of peroxidase/anti-peroxidase (PAP) (*Figure 6.2*). In the model shown, the first stage antibody (anti-human X) raised in rabbit is applied to the smear. The second stage is an excess of swine anti-rabbit immunoglobulin, which will couple to the first stage antibody using one Fab valency. The second Fab valency is left free to couple with the third stage PAP complex. To further elucidate the nature of the third stage PAP complex the preparation of PAP will be outlined (for exact times and quantities refer to the original paper by Sternberger et al., 1970).

(1) A rabbit is immunized with horseradish peroxidase, the animal is duly bled, and the resultant rabbit antiserum to peroxidase obtained.
(2) A solution of horseradish peroxidase is mixed with the antiserum, and the immune peroxidase/anti-peroxidase complex is allowed to develop.
(3) The precipitate is removed by centrifugation, washed and redissolved in a solution of excess horseradish peroxidase by adjusting the pH to 2.3.
(4) The solution is immediately neutralized to pH 7.4, producing re-equilibrium of peroxidase/anti-peroxidase into soluble complexes of homogenous composition.
(5) These soluble complexes are harvested from the excess of peroxidase by precipitation with ammonium sulphate. The result is an immune complex of peroxidase/anti-peroxidase which is antigenically recognizable as rabbit.

The most important step in this technique is to obtain the correct dilution of the second stage swine anti-rabbit serum. If this is not present in excess, both valencies of the Fab portion of the antibody could become coupled to the first stage antibody, leaving no available link to the rabbit PAP complex.

The PAP method provides high sensitivity, estimated at 100–1000 times that of a conventional indirect fluorescent technique (Sternberger, 1974), and avoids the need for the preparation of a specifically purified labelled antibody. Commercially prepared PAP complexes of high concentration, stability and homogeneity are available. If the first stage antibody is raised in a species other than rabbit, for example goat or mouse, then the anti-horseradish peroxidase used for the preparation of the PAP must also be produced in that species.

The PAP unlabelled antibody method satisfies both criteria for a sensitive cytochemical method, i.e. high sensitivity of the cytochemical detector, and absence from the serum of interfering proteins and antibodies.

The enzyme bridge method

The first two stages are as for the PAP method. The swine anti-rabbit serum again has to be used in excess, in relation to the first stage antibody. The third stage is rabbit anti-horseradish peroxidase. Localization is achieved by the addition of free horseradish peroxidase and subsequent reaction with diaminobenzidine. This system is particularly time-consuming having numerous steps and washes, and has generally been superseded by the use of PAP, which has combined the last two sequential steps.

Development of peroxidase activity

All the chromogenic substrates in present use are best regarded as potential carcinogens or toxins (Heyderman, 1979). Alternative non-carcinogens have been suggested, but their non-carcinogenicity is in most cases presumptive. The more common substrates are listed below.

3,3-Diaminobenzidine (DAB)

This is still probably the best substrate for demonstrating peroxidase activity. The final reaction product is osmophilic, making immunoperoxidase directly applicable to electron microscopy.

Formula:

3,3-Diaminobenzidine	50 mg
Tris buffer pH 7.6	100 ml
Hydrogen peroxide 30 vol	0.15 ml

Incubate for 10 min at room temperature. Coplin jars are convenient for this purpose. The reaction is stopped by washing in tap water. The haematoxylin counterstain should be relatively light, or overstaining of the final brown reaction product may result. Harris' haematoxylin diluted 50:50 with alum, and used for 30 s gives good results (Graham and Karnovsky, 1966). If required the reaction product of DAB may be enhanced or changed in colour by the use of metallic ions (Hsu and Soban, 1982).

3-Amino,9-ethylcarbazole

Peroxidase catalyses the oxidation of 3-amino,9-ethylcarbazole by hydrogen peroxide, producing an insoluble red product (Burstone, 1960).

Formula:
  (a) Dissolve 10 mg 3-amino,9-ethylcarbazole in 2.3 ml N,N-dimethylformamide.
  (b) Add 0.05 M acetate buffer pH 5 to produce a final volume of 50 ml.
  (c) Add 0.10 ml 30 vol hydrogen peroxide.

Incubate for 5 min at room temperature. The red reaction product is alcohol soluble, requiring preparations to be mounted in an aqueous medium.

Tetramethyl benzidine

Tetramethyl benzidine has been put forward as a possible non-carcinogenic substrate, producing a blue insoluble reaction product (Mesulam, 1978). According to the author, the substrate shows superior sensitivity to the DAB reaction. Its main disadvantage is that the technique has three stages lasting for a total of 1 h, which when added to the time for the PAP method is prohibitive. The recommended counterstain is neutral red.

4-Chloro-1-naphthol

This may be substituted for the diaminobenzidine in the DAB technique. However, the blue reaction product is not as stable, which precludes its satisfactory routine use.

Endogenous peroxidase

Peroxidase enzymes are present in several normal tissues, the levels being particularly high in red blood cells and granulocytes. Peroxidase activity can also be found in some neoplastic tissues, and has been reported as markedly high in some oestrogen-sensitive tissues (Desombre, Anderson and Kang, 1974). The activity of these peroxidases will survive even formalin fixation and paraffin wax processing with avidity. In cytological preparations this activity would be well shown by any of the substrates discussed, and could cause considerable confusion. A blocking procedure is therefore an essential step in any immunoperoxidase technique. This possibility represents a major disadvantage of the immunoperoxidase system, and the techniques for the blocking of endogenous peroxidase are numerous (Burns, Hambridge and Taylor, 1974). Blocking techniques include:

(1) *Methanol/hydrogen peroxide.* After fixation, smears are transferred directly to absolute methanol containing 0.3% hydrogen peroxide (30 vol) for 15 min. This achieves a considerable reduction (but not complete abolition) of peroxidase activity without adversely affecting the subsequent technique (Steefkerk, 1972).
(2) *Periodic acid-borohydride.* Preparations are treated with 2.28% periodic acid in distilled water for 5 min then washed off with tap water and treated with 0.02% potassium borohydride in distilled water for 2 min. This method blocks even the 'stubborn' activity present in eosinophils, and is probably now regarded as the most effective endogenous peroxidase blocking system (Heyderman, 1979).
(3) *HCl/ethanol.* Endogenous peroxidase is 'abolished' by treating with 0.2 ml of concentrated hydrochloric acid in 100 ml of absolute ethanol for 15 min at room temperature. It is worthy of note that this blocking procedure fails to prevent the endogenous positivity of eosinophils (Weir et al., 1974).

Alternative enzyme markers

Alkaline phosphatase

Alkaline phosphatase is frequently used for both single and double labelling, and has been recommended by some workers as the label of choice for cytology preparations because of the difficulties with the complete blocking of endogenous peroxidase (To et al., 1983). Although endogenous alkaline phosphatase is blocked in tissue sections by paraffin processing itself, it is little affected by procedures used for the preparation of cytological samples. This presents difficulties as isoenzymes of alkaline phosphatase are found in many cells present in normal pathologic tissues (Fishman, 1974). Endogenous alkaline phosphatase may be blocked either by the pre-treatment of preparations with 20% acetic acid (Ponder and Wilkinson, 1981) or, most effectively, by the addition of levamisole hydrochloride to the chromogenic substrate (Borgers, 1973). The levamisole hydrochloride used at a final concentration of 1 mM suppresses the endogenous alkaline phosphatase of most mammalian tissues with the exception of resistant calf intestine alkaline phosphatase, which is the isoenzyme most commonly used for antibody conjugation.

It is now recognized that the conventional methods of blocking endogenous peroxidase are not acceptable to the antigenic integrity of cells when using monoclonal antibodies (Straus, 1979). Alkaline phosphatase is therefore a valuable alternative for cytological preparations which are invariably rich in endogenous peroxidase. For equivalent sensitivity alkaline phosphatase/anti-alkaline phosphatase (APAAP) should be used in the same way as PAP (Cordell et al., 1984). The enzyme can then be demonstrated with a conventional histochemical procedure.

Preparation of chromogenic substrate for alkaline phosphatase technique

Solution A: Dissolve 5 mg Brentamine Fast Red TR in 10 ml of veronal acetate buffer (pH 9.2).
Solution B: Dissolve 5 mg of naphthol AS:Bl phosphoric acid sodium salt in 1 drop of dimethyl formamide.

Mix solutions A and B immediately before use and readjust the pH to 9.2. Filter onto smears, and leave at room temperature for 15 min. Rinse in distilled water followed by tap water. Counterstain lightly with Harris' haematoxylin. Coverslip using an aqueous mountant.

This substrate yields a bright red reaction product which is particularly photogenic. The main disadvantages are that the preparations are not permanent and the substrate must be prepared fresh each time.

Glucose oxidase

Glucose oxidase conjugated antibodies and glucose oxidase/anti-glucose oxidase (GAG) complex are potentially valuable markers of cell and tissue antigens in human tissue, as they are microbial enzymes not normally present in mammalian tissue (Clark, Downs and Primus, 1982). Glucose oxidase labelling has been shown to be most effective for staining tissue which is rich in endogenous peroxidase that cannot readily be eliminated (Falini and Taylor, 1983). The use of a tetrazolium salt in the chromogenic substrate gives a blue formazan reaction product which is stable to dehydration and clearing (Suffin et al., 1979).

Other markers

Variations on the basic procedures already outlined are numerous; three in particular have some potential application in cytology.

Protein A

Protein A is derived from the cell wall of *Staphylococcus aureus* and has a specificity for the Fc portion of immunoglobulin molecules from several mammalian species (Goding, 1978). Protein A can therefore be directly attached to any of the enzyme markers and used to localize primary antibodies via their Fc fragments (Falini et al., 1980).

Biotin–avidin

Avidin is a large glycoprotein found in egg white with a high affinity for the small molecular weight protein, biotin, which is a vitamin found in egg yolk. Each avidin molecule has the potential to bind biotin at four sites. The Fc portion of immunoglobulins has the ability to bind several biotin molecules, and both avidin and biotin may be labelled with fluorescent dyes (Berman and Basch, 1980) or enzyme markers (Guesdon, Ternynck and Avrameas, 1979; Warnke and Levy, 1980) without inhibiting the formation of complexes.

This affinity has the potential for technology which can be both flexible and engineered to provide great sensitivity. As a basic technique preparations are incubated with a biotinylated primary antibody followed by an avidin–peroxidase conjugate. To increase the sensitivity of the system, numerous combinations of avidin, biotin and antibody may be built up. The use of a preformed avidin–biotin–peroxidase (ABC) complex (Hsu, Raine and Fanger, 1981) is claimed to enhance sensitivity in excess of the PAP system allowing the use of primary antisera at far higher dilutions. This ABC technique has been shown to be more effective than direct immunofluorescence for the enumeration of subpopulations of lymphocytes (Paradis et al., 1984).

For practical purposes a blocking procedure is required to prevent avidin used in the technique from binding to biotin which is widespread in human tissues. This can be achieved by applying unconjugated avidin to occupy endogenous biotin sites. The binding sites of the avidin are then saturated with unlabelled biotin and are therefore no longer available to take up any introduced avidin-labelled biotin complex.

Colloidal gold labels

Electron-dense colloidal gold particles form an excellent immunocytochemical marker for transmission electron microscopy (Feldherr and Marshall, 1962; Faulk and Taylor, 1971). Colloidal gold has the ability to absorb proteins without affecting their immunological capability. The potential use of gold as a marker has been increased by the realization that deposits of colloidal gold on antigens can be seen with light microscopy as a characteristic red colour (Gu et al., 1981; Roth, 1982). In addition, with transmission electron microscopy colloidal gold gives a more easily recognized artefact than the osmophilic reaction product of immunoperoxidase (Geuze et al., 1981).

The labelling of immunoglobulins with gold probes has produced many technical difficulties (Horisberger, 1981). The improvement of manufacturing methods and the homogeneity of monoclonal antibodies has resolved the situation. Reliable and stable immunoglobulin–gold probes for indirect techniques are now commercially available. The size of the individual gold particles can be made to vary between 3 nm and 150 nm depending on the method of production (Horisberger, 1979). It has been shown that the smaller size particles are able to penetrate cell membranes and localize intracellular antigens (De Mey et al., 1981).

It would seem that colloidal gold probes are most useful for transmission electron microscopy because of ease of recognition. They also represent a bridge which allows correlation of the electron microscopic findings with light microscopy. Established methods now exist for immunocytochemistry using immunogold on cell suspensions and cytospin preparations (Polak and Van Noorden, 1983).

Enzyme digestion

Much interest has been shown in the enhancement of immunoenzyme staining by the use of proteolytic enzymes. Unexpected negative results have been reported with the PAP method when demonstrating gamma globulins and other intracellular antigens (Curran and Gregory, 1977). The controlled use of proteolytic enzymes is advocated to obtain consistent results (Mepham, Frater and Mitchell, 1979). This effect could be due to the unmasking of antigens (Huang, Minassian and More, 1976), to an increase in cell permeability, or to the freeing of cross-linked antigenic molecules. Trypsin appears to be the enzyme of choice but its use must be carefully controlled with regard to concentration, time and temperature. Paraffin wax sections, given the optimum treatment, exhibit an apparent increase in immunoreactivity, allowing the use of antisera at high titre with a reduction in background staining.

Caution has been advised in applying the use of tissue digestion (Heyderman, 1979). A theoretical objection is the possibility that the proteolytic action may reveal fragments common to many antigenic determinants with resultant loss of specificity. Alternative fixation is recommended.

In studies on impression smears the author has been able to show no increase in the number of positive cells by the use of trypsinization. Exposure to enzyme for more than a few minutes has resulted in loss of cell nuclei, and severe loss of specificity. If it is to be assumed that any inconsistencies in results on wax sections are a product of formalin fixation and wax processing, then it is unlikely that trypsinization will be necessary for cytological preparations.

Preparation of cytological specimens for immunological staining

Special care has to be taken with the preparation of cytological material for immunocytochemical staining. Unlike tissue sections, smears are not uniform and vary in thickness with cells commonly heaped on one another. Blood and mucus often interfere with results, and special techniques need to be applied to render some of the specimens suitable for staining.

Impression smears

Impression smears make good material providing they are thin and very rapidly air-dried. In general terms, smears that show good morphology also show effective immunocytochemical localization and vice versa. Staining must be immediate or the preparations should be stored in aluminium foil at −20 °C. Smears that are thick and poorly prepared will often show ill-defined positivity involving large groups of cells, which cannot safely be interpreted. Smears left at room temperature for any length of time will commonly become negative for the more labile antigens.

Cervical smears

The routine preparation of cervical smears is not ideal for the majority of polyclonal antisera as the cells overlap and the specimen contains much blood and mucus. Moreover, since morphological examination is the first consideration, and usually one smear only is taken, material is not available for routine immunocytochemistry. It is possible to reprocess fixed and stained smears for immunocytochemical staining by removing the coverslip with xylene and bleaching with 0.25% acid alcohol for 30 min. The smear is then stained by a standard PAP technique. However, in the experience of the author a degree of artefact staining will often be present and the majority of results are unacceptable. In addition it has to be accepted that one destained Papanicolaou smear is insufficient material for adequate negative controls. This approach has been used to demonstrate the presence of herpes simplex 2 viral antigens in smears first collected and stained eight years previously. Providing a degree of interpretation is permissible, and it is remembered that results should be used with discretion as an adjunct to morphology, this approach can be useful. Indeed, subtle diagnosis has been reported using this technique (O'Hara, Gardener and Bennett, 1980; Craig et al., 1983).

Far more satisfactory immunocytochemical staining can be obtained if a concentration of cells is prepared from the cervical scrape and multiple smears made using this approach.

A cell washing and concentration technique has also been applied to material scraped from cervical epithelium at colposcopy (Lloyd et al., 1984). Scrapings are suspended in 20 ml PBS, washed by repeated centrifugation, resuspended in buffer to give a cell density of 100 000 cells/ml and cytocentrifuged.

Sputum

Specimens of sputum do not make good preparations. Untreated samples provide smears that are generally too thick and usually show heavy extracellular background staining, due probably to physical entrapment of the antisera. Digestion of the substance of the sputum may adversely affect intracellular and surface antigens. Immunocytochemistry can satisfactorily be carried out on cell block sections (Boon et al., 1982). Material is fixed in 4% buffered formalin and routinely processed to paraffin wax as with histological tissue. 4 µm sections are then cut and stained.

Serous effusions

Sources of body fluids are very suitable for immunocytochemical staining providing they are free of blood. Multiple identical smears can be prepared from one sample. Aspirates should be collected in sterile containers with either EDTA or sodium citrate as anticoagulant. Specimens received containing fibrin clots may be treated with 10 ml of reconstituted Varidase Topical for 5 min to loosen trapped cells (To et al., 1983).

The fluid should be centrifuged at approximately 3000 rev/min for 5 min, a smear made from the deposit, and either wet fixed or air dried (Ghosh, Spriggs and Taylor-Papadimitriou, 1983). More control over the quality of preparations can be achieved by following the cell washing and concentration regime proposed by To et al. (1983). This technique advocates washing the cells free of protein before smear preparation. The procedure is as follows:

(1) The specimen is centrifuged at 2500 rev/min for 5 min.

(2) The supernatant is discarded, and the cell pellet resuspended in a suitable volume (10–20 ml) of phosphate buffered saline pH 7.4 and recentrifuged.
(3) The process is repeated.

This washing of cells has two benefits; the viscosity of the fluid is reduced which allows the cells to be more effectively flattened, and non-specific proteins coating the cells are removed, thereby eliminating heavy background staining which can cause unreliable interpretation (To et al., 1983). However, there is the risk of removing the protein under investigation during the washing.

Preparations also benefit from the removal of red blood cells, which when present in any numbers obscure the morphological detail of the nucleated cells, and when using immunoperoxidase represent a large reservoir of endogenous peroxidase. Either the microhaematocrit technique (To et al., 1983; Yam and Janckila, 1983) or a density gradient (Nagasawa and Nagasawa, 1983) may be used for this purpose.

Fine needle aspiration biopsy specimens

Like impression smears, fine needle aspiration smears are very suitable for immunocytochemical staining. They can be air-dried or fixed immediately as required. The morphology of the cells in wet-fixed preparations is usually easier to interpet.

Fixation

The choice of fixative is dependent not only upon the antigen to be localized, but also the material available, the marker system and the type of antibody used. It is becoming obvious that fixation does not destroy the immunological reactions of as many antigens as previously thought, but as the subject has received relatively little systematic consideration, a degree of trial and error has to be adopted to find the optimum treatment for each system to be used.

Irrespective of the type of preparation or marker, the use of most monoclonal antisera requires a differing fixation procedure than polyclonal antisera. Polyclonal antibodies are in essence a mixture of antibodies with different affinities against several antigenic determinants on a particular molecule. By definition, monoclonal antisera contain a single 'species' of antibody with a sole affinity, which may be high or low, together with monospecificity. It is now understood that many monoclonal antibodies that successfully demonstrate antigens in cell suspension or in frozen sections, may fail in tissue or smears that have been fixed with the more standard reagents (Naritoku and Taylor, 1982). This is presumed to be the result of denaturation of the critical antigenic determinant recognized by the antibody.

Brief fixation of cell preparations in water-free 'analar' grade acetone appears to precipitate most cellular protein *in situ* and gives acceptable localization of antigens with most monoclonal antibodies. The recommended fixation times vary from 30 s to 30 min (Warnke and Levy, 1980; Poppema et al., 1981; Erber, Pinching and Mason, 1984). A few monoclonal antibodies are known to be effective after formalin or alcohol fixation, notably common leucocyte antigen (Pizzolo et al., 1980), cytokeratin (Lane, 1982) and human milk fat globulin (Taylor-Papadimitriou et al., 1981) but in the absence of exact information concerning the fixation requirements of a given antigenic site, the following procedure is recommended.

(1) Air dry preparations at room temperature for 30 min.
(2) Fix for 30 min in water-free 'analar' grade acetone at room temperature.
(3) Dry and immunostain within 6 h, or
(4) Wrap the slides individually in aluminium foil and store at $-20\,°C$ until required. Allow to return to room temperature before unwrapping.

Alternative methods of fixing antigens for localization have been described. Absolute methanol is recommended for immunofluorescence (Singh, Whiteside and Dekker, 1979) and 95% ethanol for immunoenzyme techniques (Nadjii, 1980; To et al., 1981). For the latter, immediate 'wet' fixation for 30 min is preferred. Smears are then sprayed with a polyethylene glycol containing fixative and ideally stored in closed containers at $-20\,°C$. It is worthy of note that good results have been obtained with cytocentrifuge preparations stored at ambient room temperature for four years.

Control of specificity

Any immunocytochemical staining system can only be as good as the antibodies it uses. It therefore follows that trace antibodies of unwanted specificity, which may be present in acceptable amounts for immunodiffusion techniques, could cause considerable problems in a PAP type procedure. Truly monospecific commercial conjugates are rare in the extreme (Preud'homme and Labaune, 1975); as a consequence carefully planned controls and some degree of interpretation is mandatory.

Staining of the background proteins of smears, particularly from body effusions, is evident to some degree if the cells are not washed before centrifugation. This can be due to either specific or non-specific reactions. Specific background staining may result from the presence of the antigen in some concentration in the fluid to be examined. For example, carcinoembryonic antigen may not only be present in malignant cells found in pleural and ascitic fluids, but high concentrations have been estimated in the fluid itself (Booth et al., 1977). When precipitated by wet fixation, this will show as a degree of specific background staining in the smear unless the cells are washed. Diffusion of antigen from cells prior to fixation or poor air drying technique will produce a similar effect. Non-specific background staining caused by contaminant antibodies of unknown specificity is a greater problem. The interference of non-specific staining can be controlled to some extent by diluting the primary antisera until only the wanted specificity is apparent. Initially chequerboard dilutions of each layer antibody should be compared to assess the optimum for each stage. In reality it is the primary antiserum that may vary considerably with each type of antigen to be localized. The inclusion of negative and positive controls with each test run is essential.

Negative controls

Absorption

If an antibody is absorbed with an excess of the antigen under test, then all staining due to the presence of that antibody in the primary antiserum should cease. Any subsequent staining by such an absorbed antiserum must be due to contaminating antibody. This is the basis of the most useful form of specificity control. There are two methods of absorption.

(1) An excess of antigen is added to the diluted antiserum and incubated overnight at 4 °C. Stored absorbed antisera may show free antibody due to disassociation; the addition of a 0.1% solution of bovine serum albumin is recommended (Polak and Van Noorden, 1983).
(2) The antiserum is incubated with antigen insolubilized with glutaraldehyde (Avrameas and Ternyck, 1969). This should avoid the production of soluble immune complexes, and is the more effective of the two methods. Antigen-absorbed Sepharose beads are also a useful way of removing antibody from solution.

Non-immune serum

The traditional negative control of a multiple layer technique is the use on a parallel smear of serum from an animal of the same species which did not receive an inoculation with the antigen under test. This type of control is inadequate and serves only to prove that the first two sera are different (Heyderman, 1979) and not that the immunized animal recognized and reacted to only the test antigen and not an impurity. Even worse is the complete omission of the first stage antibody from the technique; this proves virtually nothing.

Blocking with antisera from another species

The smear may be blocked by a specific antiserum raised to the antigen under test in a different species, e.g. goat instead of rabbit. This will not afford a link to the second stage swine anti-rabbit serum, and will also block the subsequent coupling of the same antibody from the correct species (rabbit) to the antigen. Specific staining should be markedly reduced by this procedure.

Comparison with other antisera

If batches of antisera show variations in staining distribution, this may be taken as evidence of differing specificity. However, it should be remembered that different cell antigens can

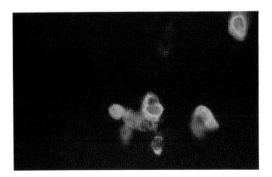

Plate 1 Cervical smears stained for herpes associated transforming protein using a direct fluorescent antibody technique. Note positive (yellow/green) staining of cytoplasm of dyskaryotic cells. Nuclei of dyskaryotic cells and normal squames in the smear are negative (red). FITC and rhodamine counterstain. Magnification × 800. Courtesy of Dr S. Tyms, Department of Virology, St Mary's Hospital, London

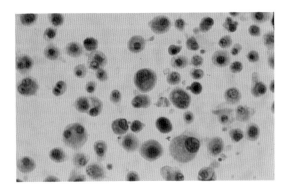

Plate 2 Pleural fluid from patient with multiple myelomatosis. Note malignant plasma cells. Papanicolaou. Magnification × 800. Courtesy of Dr D. V. Coleman, St Mary's Hospital, London

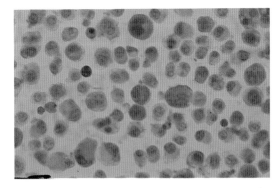

Plate 3 Same specimen as *Plate 2* stained for Kappa and Lambda light chains. The plasma cells stained with the anti-Kappa antibody only. Immunoperoxidase. Magnification × 800. Courtesy of Dr D. V. Coleman, St Mary's Hospital, London

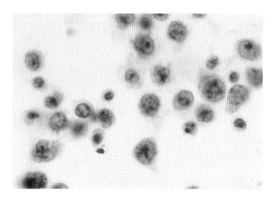

Plate 4 Effusion containing numerous discrete malignant cells suggestive of lymphoma. Papanicolaou. Magnification × 800. Courtesy of Dr D. V. Coleman, St Mary's Hospital, London

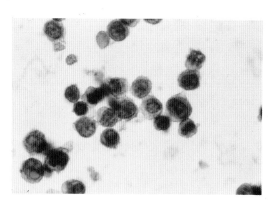

Plate 5 Same specimen as *Plate 4* stained with leucocyte common antigen confirming the non-epithelial origin of the tumour. CAM 5.2 was negative. Immunoperoxidase. Magnification × 800. Courtesy of Dr D. V. Coleman, St Mary's Hospital, London

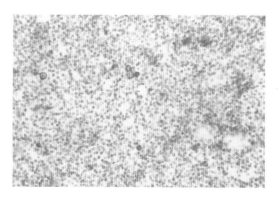

Plate 6 FNA lymph node stained with CAM 5.2, a low molecular weight keratin marker. Note discrete positively stained tumour cells against background of non-staining lymphocytes. Immunoperoxidase. Magnification × 200. Courtesy of Dr D. V. Coleman, St Mary's Hospital, London

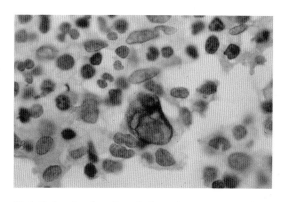

Plate 7 Another lymph node from the same case as in *Plate 6* but at higher magnification showing CAM 5.2 positive cancer cell. Immunoperoxidase. Magnification × 2000. Courtesy of Dr D. V. Coleman, St Mary's Hospital, London

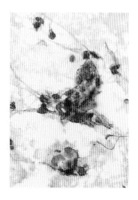

Plate 8 (*a & b*) FNA of tumour of calf from male aged 54 years thought clinically to be lipoma. Abnormal cells present stained positively for vimentin, an intermediate filament protein present in connective tissue. Biopsy confirmed malignant fibrous histiocytoma. Immunoperoxidase. Magnification × 800. Courtesy of Dr D. V. Coleman, St Mary's Hospital, London

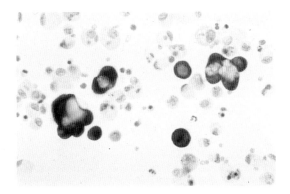

Plate 9 Pleural effusion stained for epithelial membrane antigen (EMA). Note positively stained malignant cells consistent with a carcinoma. The lymphocytes and mesothelial cells are not staining. Alkaline phosphatase. Magnification ×800. Courtesy of Dr D. V. Coleman, St Mary's Hospital, London

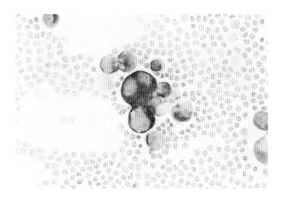

Plate 10 Pleural effusion stained for EMA using an immunoperoxidase technique. Note positive staining of tumour cells only. Immunoperoxidase. Magnification ×800. Courtesy of Dr D. V. Coleman, St Mary's Hospital, London

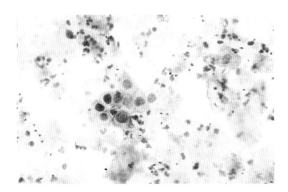

Plate 11 Dyskaryotic cells in cervical smear stained for carcinoembryonic antigen. Normal squames are negative. Immunoperoxidase. Magnification ×800. Courtesy of Dr D. V. Coleman, St Mary's Hospital, London

occupy the same morphologic site, so that even a non-specific contaminant could show a similar localization (Heyderman and Monaghan, 1979).

Chromogenic substrate alone

This ensures that there has been efficient blocking of any endogenous label. It is also useful to test the avidity of the substrate with an unblocked smear, to show that any unexpected negative result is not due to an unreactive substrate.

Positive controls

It is essential to have a bank of suitably stored known positive smears. Where these are not available histology sections can be used.

Staining schedules

Generally, routine immunoperoxidase schedules involve either the peroxidase/anti-peroxidase method or the indirect method, therefore only procedures for these techniques are given. The rationale for each step is given in the chapter.

Peroxidase/anti-peroxidase (PAP) method

This is suitable for polyclonal antisera.

Tris Buffered Saline (TBS)

Sodium chloride	34 g
Tris	24.2 g
2 M Hydrochloric acid	75 ml
Distilled water	4 l

Adjust pH to 7.4 if necessary.

All other solutions as applicable.

Procedure
(1) 'Wet fix' preparations in alcohol for 30 min.
(2) Wash in TBS pH 7.4 for 5 min.
(3) Block endogenous enzyme.
(4) Wash in TBS pH 7.4 for 5 min.
(5) Apply rabbit anti-human primary antiserum diluted in TBS for 30 min at room temperature.
(6) Wash with TBS for 15 min., three changes.
(7) Apply swine anti-rabbit immunoglobulins diluted in TBS for 30 min at room temperature.
(8) Wash for 15 min. with TBS, three changes.
(9) Apply rabbit peroxidase/anti-peroxidase (APAAP or GAG) diluted in TBS for 30 min.
(10) Wash for 15 min. with TBS, three changes.
(11) Develop colour of reaction products.
(12) Counterstain with Harris' haematoxylin diluted 50 : 50 with 2% potassium alum for 20 s.
(13) Select mountant to suit label reaction product.

Indirect method

Suitable for monoclonal antisera.

All solutions as applicable.

Procedure
(1) Thoroughly air dry preparations for 30 min at room temperature.
(2) Fix in acetone for 30 min at room temperature.
(3) Rinse in Tris buffered saline (TBS) pH 7.4.
(4) Wipe excess buffer from around preparation and place in a moist chamber.
(5) Apply monoclonal antiserum diluted in TBS for 30 min.
(6) Rinse gently with TBS, three changes over 2 min.
(7) Apply suitably conjugated rabbit anti-mouse serum (absorbed against insolubilized human immunoglobulins) diluted in TBS for 30 min.
(8) Rinse gently with TBS, three changes over 2 min.
(9) Wash in tap water.
(10) Develop coloured reaction product.
(11) Counterstain lightly with Harris' haematoxylin diluted 50 : 50 with 2% potassium alum for 10 s.
(12) Mount in accordance with the label reaction product.

Technical note

Traditionally smears are incubated at room temperature in a moist chamber to avoid drying of the anti-sera, which can cause gross false staining. However, providing adequate volumes are used, and the slides are periodically rocked, the use of a staining rack over a sink will suffice for the relatively short reaction time. Between each application of polyclonal antibody the smears should be thoroughly washed using a large volume of buffer, under continuous agitation by a magnetic stirrer. At least three changes of buffer are needed over 15 min. When monoclonal antibodies are used only a relatively gentle passive wash should be carried out; three changes over 5 min. Vigorous washing should be avoided when antibodies with low antigen affinities are used. For immunofluorescence, Coons' Phosphate Buffered Saline (PBS) is recommended throughout. Since phosphate ions are said to inhibit peroxidase demonstration, it is advisable to use tris-buffered saline pH 7.4 in immunoperoxidase systems, either buffer being acceptable for other markers.

Applications

The exact areas and the extent of useful application of immunocytochemistry in clinical cytology has yet to be clearly defined. Some examples of how immunocytochemical staining can usefully be applied in cytology are shown in *Plates 1–11*, but new applications are constantly being identified – especially with the expansion of fine needle aspiration cytology. In this section, the use of immunocytochemical markers on touch preparations, cervical smears, serous effusions and fine needle aspiration biopsy is described. In some cases the antibodies have been applied in an attempt to distinguish benign from malignant cells; in other cases the antibodies have been used to identify tumour type or establish the presence of viral antigen.

There is now increasing evidence that immunocytochemistry can contribute to the solution of diagnostic problems in cytology (*Plates 1–8*) (Chess and Hajdu, 1986). It is also clear that relatively minor variations in technique can produce significant changes in staining reaction. The general assumption that the specificity of an antigen, as expressed in a tissue section, will be retained in cytological smears needs more careful assessment.

Touch preparations

Despite the simplicity of the technique and its increasing popularity for the study of solid tumours, published works concentrating on impression smears or touch preparations are few.

Halliday (1977) reports in full on a clinical case of multiple myeloma showing alteration in morphology of the neoplastic cells during the course of the disease. The atypia was such that touch preparations normally used by that author for the identification of lymphoreticular cells were of no use morphologically in confirming the diagnosis of plasma cell neoplasm. However, immunoperoxidase studies showed the cells to be producing gamma globulin of one type only (IgG, Kappa). Similar results have been reported (Nadjii, 1980) on cytology preparations from myelomas and lymphomas of B cell origin.

Fine needle aspirates (FNA)

Several surveys of fine needle aspirates have been undertaken (Craig *et al.*, 1983; Ramaekers *et al.*, 1984; Domagala, Weber and Osborn, 1986; Chess and Hajdu, 1986). In the light of the current revaluation of the role of immunocytochemistry in surgical pathology, a degree of contention is inevitable. The particular problems associated with obtaining a differential diagnosis have been studied on aspirates from 21 lymph nodes using monoclonal antisera to cytokeratin subsets, desmin and vimentin (Domagala, Weber and Osborn, 1986). Results indicate that these antibodies can improve the accuracy of cytological diagnosis and specifically assist in distinguishing (i) malignant melanoma from squamous carcinoma; (ii) malignant lymphoma from small cell anaplastic carcinoma; and (iii) carcinoma from sarcoma. In some instances small numbers of tumour cells have been identified using immunocytochemistry in smear aspirates apparently negative by light microscopy.

Much potential is claimed for antibodies to intermediate filaments (Gabbiani *et al.*, 1981). The fact that tumour cells retain the intermediate filament characteristics of their cell of origin, and in general do not acquire any new types in metastases, accounts for the increasing significance of intermediate filaments in surgical pathology. Several recent studies support the diagnostic application of intermediate filament antibodies in clinical cytology (Altmannsberger *et al.*, 1981; Ramaekers *et al.*, 1984).

The significance of keratin staining in cells from lymph node aspirates is established as a direct indication of metastatic carcinoma. A study of lymph node aspirates (Ramaekers *et al.*, 1984) showed that in all cases lymphocytes were negative for keratin, but weakly positive for vimentin. Examination of frozen sections of total lymph nodes revealed no keratin positive cells in normal lymph follicles (Ramaekers *et al.*, 1983). As a result the detection of keratin positive cells in aspirates from a lymph node is a very strong indication of the presence of metastatic carcinoma cells. Experience has shown that a keratin positive reaction using immunofluorescence can be detected quickly, even in cases containing only a few cells in the preparation.

A definitive diagnosis of choriocarcinoma has been reported from a lung aspiration specimen (Craig *et al.*, 1983). A successful PAP reaction for anti-human chorionic beta subunit was achieved in alcohol-fixed, Papanicolaou-stained slides that were destained in xylene.

Sputum

Due to the nature of the material, untreated sputum does not generally make good preparations for immunocytochemistry. Cell blocks are preferred for this type of specimen. In a study of sputum cell blocks, carcinoembryonic antigen (CEA) was not only demonstrated in a proportion of malignant cells, but also in some examples of benign squamous cells (Boon *et al.*, 1982).

Effusions

For the clinical cytologist, the ability to identify positively a cell type can be potentially as useful as a specific tumour marker. For example, demonstration of epithelial cells in pleural or ascitic specimens is in itself diagnostic in that epithelial cells are not normal constituents of these fluids. There is a growing number of commercially available antisera that are particularly useful in indicating cell lineage.

Serous effusions present one of the problem areas in diagnostic cytology. Difficulties in routine preparations can be caused by small numbers of malignant cells being overwhelmed by numerous reactive mesothelial cells and active macrophages. The use of a single antiserum has serious limitations; the best approach is effected by the use of a panel of antisera. CEA, human milk fat globulin (HMFG), epithelial membrane antigen (EMA), cytokeratin, alpha-1 antitrypsin, alpha-1 antichymotrypsin constitute a useful grouping. It should be remembered that the antigenic sites recognized by a number of polyclonal antisera remain intact for some time after phagocytosis. This can result in positivity for epithelial markers within macrophages, but such cells should also show alpha-1 antitrypsin and chymotrypsin activity.

In extensive studies of EMA staining in pleural and peritoneal effusions (To *et al.*, 1981, 1983), positive staining of malignant cells was found in 86% of preparations from proven cases of carcinoma (*Plates 9* and *10*) and mesothelioma. Benign mesothelial cells were either negative or showed pale 'rim' staining of the cytoplasm. A similar type of reaction was also seen in some malignant cells, although the majority gave dense central cytoplasmic staining, which was not seen in benign mesothelial cells. This type of dense cytoplasmic staining was argued to be useful for indicating epithelial lineage of cells in suspicious smears. Using a monoclonal antibody to HMFG, similar results have been obtained by Taylor-Papadimitriou *et al.*, 1981 and Ghosh, Spriggs and Taylor-Papadimitriou, 1983.

High levels of expressed protein can be present in the fluid of pleural and ascitic effusions associated with malignancy (in particular breast and gastrointestinal primaries). In a study of 640 cases the authors found that the accuracy of diagnosis using a morphological approach was poor, resulting in a high rate of false negatives (37%). A large number of these false negative results showed a remarkably high titre of CEA in the fluid. The suggestion was made, therefore, that actual CEA measurements on fluids could improve the sensitivity of this type of cytological examination (Whiteside and Dekker, 1979). It has also to be recognized that high levels of such proteins within the fluid itself may give rise to phagocytosis by macrophages, thus feigning the staining results associated with epithelial cells. In this situation, the demonstration of indicators of lysosomal activity and phagocytic capability are useful. Such cells can be stained with antisera to the antiproteases, alpha-1 antitrypsin and alpha-1 antichymotrypsin (Isaacson *et al.*, 1981).

The cytological diagnosis of serous fluids containing high levels of cells from the lymphoid series may present difficulty. Large numbers of lymphocytes may be the result of inflammation, non-specific reactions to an epithelial tumour, or to lymphocytic leukaemia/lymphoma. In these cases, a differential study of T and B cell numbers may augment morphological criteria. A retrospective immuno-alkaline phosphatase technique applied to air-dried smears from serous fluids (Ghosh, Spriggs and Mason, 1985) has shown the enumeration of such populations with monoclonal anti-B, anti-T and anti-HLA-DR sera to be of value. In this study the majority of lymphoid cells in reactive effusions were T cells which lacked HLA-DR, and showed a marked excess of helper/inducer cells. By contrast, in most cases of chronic lymphocytic leukaemia and lymphoma studied, the populations shown on immunocytochemical staining were clearly different from those in reactive effusions. In some cases of B cell chronic lymphocytic leukaemia 99% of lymphoid cells were labelled with anti-B antibody. This was in marked contrast to the staining patterns provided by a control group of patients with no demonstrable lymphoid disease. Preparations from these subjects gave a mean of 85.2% of cells reacting with anti-T cell antibody, with the numbers of lymphocytes reacting with anti-B and anti-HLA-DR representing less than 20% of the lymphoid cells. These results agree well with those of other workers using a different technology (Domagala, Emeson and Koss, 1981; Krajewski, Dewar and Ramage, 1982). The number of HLA-DR positive lymphocytes was often shown to be slightly higher than the total of anti-B positive cells, and suggested to the authors that a minority of T lymphocytes in serous effusions had undergone 'activation'. Similarly, the ratio of helper/inducer to suppressor cytotoxic T cells in reactive effusions was found to be higher than that of peripheral blood (Erber, Pinching and Mason, 1984) and much closer to the proportions demonstrated in lymphoid tissues (Ralfkiaer et al., 1984).

Blood/bone marrow smears

The use of immunocytochemistry on haematological material has application not only in the classification of neoplastic blood disorders, but also the detection of metastatic carcinoma cells.

The use of an alkaline phosphatase/anti-alkaline phosphatase (APAAP) technique is to be preferred to avoid the problems caused by endogenous enzyme activity of the immunoperoxidase techniques. Labelling can be performed on smears of whole blood or bone marrow so that the trauma of cell separation is avoided. Enumeration of lymphocyte sub-populations and the classification of leukaemias on such material is now in use involving T and B cell markers, common acute lymphoblastic leukaemia antigen, terminal transferase and common leucocyte antigen.

Immunocytochemistry has much potential for detecting metastatic carcinoma cells in bone marrow smears. Very small numbers of malignant cells which could not be detected by routine haematological examination alone have been demonstrated (Ghosh et al., 1985). Detection of isolated mammary carcinoma cells in the marrow of patients with primary breast carcinoma by using epithelial cell markers has been reported (Dearnaley et al., 1983).

Cervical smears

The evidence that human papillomavirus (HPV) is implicated in a high percentage of pre-malignant cervical lesions has been substantiated by immunocytochemistry (Dyson, Walker and Singer, 1984). Immunofluorescent staining of cervical smears with monoclonal antibody to detect chlamydial infection has also been advocated (Forster et al., 1983). Herpes simplex virus infection of the genital tract can also be detected by immunofluorescence of cervical smears (*Figure 6.3*).

(a) (b)

Figure 6.3 Cells in tissue culture stained for herpes virus antigen using a direct immunofluorescence technique. Stain: FITC. Magnification × 1000. (a) Multinucleate giant cell infected with herpes simplex-2, showing cytoplasmic staining of plasma membrane and dense central Golgi apparatus; (b) herpes simplex-1 infected culture showing cytoplasmic staining of single cells. Courtesy of Dr S. Tyms, Department of Virology, St Mary's Hospital, London

Considerable attention is now being focused on the potential of immunocytochemical labelling of the cells produced from cervical scrapings, not only as an aid to the light microscopy diagnosis of malignancy, but also as a suitable signal for automated cytometry. As already discussed, the use of conventional cervical smears has distinct disadvantages for good immunocytochemistry. In two series of studies (Moncrieff, Ormerod and Coleman, 1984; Valkova et al., 1984) alternative preparation techniques were used, which gave good quality immunocytochemical staining. Essentially the cervical smears were washed free of blood and mucus before smearing. The material obtained by Ayre spatula scrape was placed directly into a solution of Cellfix. This is a mucus-dissolving fixative with the ability to lyse red blood cells, whilst preserving the morphology of the epithelial cells. The spatulae were soaked overnight in Cellfix, and the material eluted by washing in phosphate buffered saline and centrifugation. The final cell suspension was spread in small aliquots on slides and air-dried. These preparations were then stained with a routine immunocytochemical technique, using a polyclonal antibody to EMA and monoclonal equivalents. Conventional Papanicolaou stained smears prepared at the time of cervical scrapings were used as a reference to assess staining pattern.

Squamous epithelial cells from normal cervices showed no staining for EMA and CEA but dyskaryotic cells from cervices with cervical intraepithelial neoplasia (CIN) did express positivity for both. Metaplastic cells from both normal and abnormal cervices also frequently stained. The authors concluded that over 60% of negative cases could be eliminated by semi-automated screening of cervical smears using such a marker as the signal, and were encouraged by the evidence that false negative results should not be occasioned. This conclusion is supported by a subsequent study using a Leitz Tas (Textur Analyse System) video camera system (Valkova and Laurence, 1985). Material was again prepared using the Cellfix technique, immunostained for EMA and counterstained with haematoxylin. In the series of 87 cases the authors obtained a false positive error rate of 3% and a false negative error rate of 2%. This compares very favourably with reported manual pre-screening error rates of 4% false positive and 6% false negative respectively (Collins and Kaufman, 1974), and is clearly worthy of further study.

The extent to which any of these recent advances can, or will, be used in routine diagnostic cytology remains to be seen. Although some applications of the technique are open to criticism, and caution should be exercised when applying it to diagnostic material, the considerable potential of immunocytochemistry is now obvious.

References

ALTMANNSBERGER, M., OSBORN, M., SCHAUER, A. and WEBER, K. (1981). Antibodies to different intermediate filament proteins: cell type-specific markers on paraffin-embedded human tissues. *Lab. Invest.*, **45**, 427–434

AVRAMEAS, S. and TERNYCK, R. (1969). The cross linking of proteins with glutaraldehyde and its use for the preparation of immunoabsorbents. *Immunochemistry*, **6**, 53–66

BERMAN, J. W. and BASCH, R. S. (1980). Amplification of the biotin–avidin immunofluorescence technique. *J. Immunonol. Methods*, **36**, 335–338

BLANN, A. D. (1979). Cell hybrids: an important new source of antibody production. *Med. Lab. Sci.*, **36**, 329–338

BONDI, A., CHIEREGATTI, G., EUSEBI, V., FULCHERI, E. and BUSSOLATI, G. (1982). The use of beta-galactosidase as a tracer in immunocytochemistry. *Histochemistry*, **76**, 153–158

BOON, M. E., LINDEMAN, J., MEEUWISSEN, A. L. J. and OTTO, A. J. (1982). Carcinoembryonic antigen in sputum cytology. *Acta Cytol.*, **26**, 389–394

BOOTH, S. N., LAKIN, G., DYKES, P. W., BURNETT, D. and BRADWELL, A. R. (1977). Cancer associated proteins in effusion fluids. *J. Clin. Pathol.*, **30**, 537–540

BORGERS, M. (1973). The cytochemical application of new potent inhibitors of alkaline phosphatases. *J. Histochem. Cytochem.*, **21**, 812–824

BURNS, J., HAMBRIDGE, M. and TAYLOR, C. R. (1974). Intracellular immunoglobulins. A comparative study on three standard tissue processing methods using horseradish peroxidase and fluorochrome conjugates. *J. Clin. Pathol.*, **27**, 548–557

BURSTONE, M. S. (1960). The demonstration of peroxidase activity by 3-amino,9-ethylcarbazole. *J. Histochem. Cytochem.*, **8**, 63

CHESS, Q. and HAJDU, S. (1986). The role of immunoperoxidase staining in diagnostic cytology. *Acta Cytol.*, **39**, 1–7

CLARK, C. A., DOWNS, E. C. and PRIMUS, E. J. (1982). An unlabelled antibody method using glucose oxidase anti-glucose oxidase complexes (GAG): a sensitive alternative to immunoperoxidase for the detection of tissue antigens. *J. Histochem. Cytochem.*, **30**, 27–34

COLLINS, D. N. and KAUFMAN, W. (1974). Quality evaluation of cytology laboratories in New York state: expanded program 1971–73. *Acta Cytol.*, **18**, 404–413

COONS, A. H., CREECH, H. J. and JONES, R. N. (1941). Immunological properties of an antibody containing a fluorescent group. *Proc. Soc. Exp. Biol. Med.*, **47**, 200–202

CORDELL, J. L., FALINI, B., ERBER, W. N., GHOSH, A. K., ABDULAZIZZ, MACDONALD, S. and PULFORD, K. A. F. (1984). Immunoenzymatic labelling of monoclonal antibodies using immune complexes of alkaline phosphatase and monoclonal anti-alkaline phosphatase (APAAP complexes). *J. Histochem. Cytochem.*, **32**, 219–229

CRAIG, I. D., SHUM, D. T., DESROSIERS, P., MCLEOD, C., LEFCOE, M. S., PATERSON, N. A. M., FINLEY, R. J., WOODS, B. and ANDERSON, R. J. (1983). Choriocarcinoma metastatic to the lung; a cytologic study with identification of human choriogonadotropin with an immunoperoxidase technique. *Acta Cytol.*, **27**, 647–650

CURRAN, R. C. and GREGORY, J. (1977). The unmasking of antigens in paraffin sections of tissues by trypsin. *Experientia*, **33**, 1400–1401

DEARNALEY, D. P., ORMEROD, M. G., SLOANE, J. P. and LUMLEY, H. (1983). Detection of isolated mammary carcinoma cells in marrow of patients with primary breast cancer. *J. Roy. Soc. Med.*, **76**, 359–364

DE MEY, J., MOEREMANS, M., DE WAELE, M., GEUENS, G. and DE BRABANDER, M. (1981). The IGS (Immunogold staining) method used with monoclonal antibodies. In *Proceedings of Colloquium on the Protides of the Biological Fluids*, (Peters, M., ed.), pp. 946–947. Oxford: Pergamon Press

DESOMBRE, E. R., ANDERSON, W. A. and KANG, Y. H. (1974). Identification, subcellular localisation and oestrogen regulation of peroxidase in 7,12-dimethyl benzanthracene-induced rat mammary tumours. *Cancer Res.*, **35**, 172–179

DOMAGALA, W., WEBER, K. and OSBORN, M. (1986). Differential diagnosis of lymph node aspirates by intermediate filament typing of tumour cells. *Acta Cytol.*, **39**, 225–234

DYSON, J. L., WALKER, P. G. and SINGER, A. (1984). Human papillomavirus infection of the uterine cervix: histological appearances in 28 cases identified by immunohistochemical techniques. *J. Clin. Pathol.*, **37**, 126–130

ERBER, W. N., PINCHING, A. J. and MASON, D. Y. (1984). Immunocytochemical detection of T and B cell populations in routine blood smears. *Lancet*, **i**, 1042–1046

FALINI, B. and TAYLOR, C. R. (1983). New development in immunoperoxidase techniques and their application. *Arch. Pathol. Lab. Med.*, **107**, 105–117

FALINI, B., TABILIO, A., ZUCCACCIA, M. and MARTELLI, M. F. (1980). Protein A-peroxidase conjugates for two-stage immunoenzyme staining of intracellular antigens in paraffin-embedded tissues. *J. Immunol. Methods*, **39**, 111–120

FAULK, W. P. and TAYLOR, G. M. (1971). An immunocolloid method for the electron microscope. *Immunocytochemistry*, **8**, 1081–1083

FELDHERR, C. M. and MARSHALL, J. M. (1962). The use of colloidal gold for studies of intracellular exchanges in the amoeba *Chaos chaos*. *J. Cell Biol.*, **12**, 640–645

FISHMAN, W. H. (1974). Perspectives on alkaline phosphatase isoenzymes. *Am. J. Med.*, **56**, 617–650

FORD, P. M. and STOWARD, P. J. (1974). The detection of autoantibodies by an enzyme bridge method. *J. Clin. Pathol.*, **27**, 118–121

FORSTER, G., JHA, R., CHEETHAM, D., MUNDAY, P., COLEMAN, D. and TAYLOR-ROBINSON, D. (1983). Cytological diagnosis of chlamydial infection of female genital tract. *Lancet*, **ii**, 578

GABBIANI, G., KAPANCI, Y., BARAZZONE, P. and FRANKE, W. W. (1981). Immunochemical identification of intermediate sized filaments in human neoplastic cells: a diagnostic aid for the surgical pathologist. *Am. J. Pathol.*, **104**, 206–216

GEUZE, H., SLOT, J., VAN DER LEY, P., SCHUFFER, R. and GRIFFITH, J. (1981). Use of colloidal gold particles in double labelling immuno-electron microscopy of ultrathin frozen tissue sections. *J. Cell Biol.*, **89**, 657–665

GHOSH, A. K., SPRIGGS, A. I. and TAYLOR-PAPADIMITRIOU, J. (1983). Immunocytochemical staining of cells in pleural and peritoneal effusions with a panel of monoclonal antibodies. *J. Clin. Pathol.*, **36**, 1154–1164

GHOSH, A. K., SPRIGGS, A. T. and MASON, D. Y. (1985). Immunocytochemical staining of T and B lymphocytes in serous effusions. *J. Clin. Pathol.*, **38**, 608–612

GHOSH, A. K., ERBER, W. N., HATTON, C., FALINI, B., O'CONNOR, N. J., OSBORN, M. and MASON, D. Y. (1985). Detection of metastatic tumour cells in routine bone marrow smears by immuno-alkaline phosphatase labelling with monoclonal antibodies. *Br. J. Haematol.*, **61**, 21–30

GODING, J. W. (1978). Use of staphylococcal protein A as an immunological reagent. *J. Immunol. Methods*, **20**, 241–253

GRAHAM, R. C. and KARNOVSKY, H. J. (1966). The early stages of the absorption of injected horseradish peroxidase in the proximal tubules of mouse kidney; ultrastructural cytochemistry by a new technique. *J. Histochem. Cytochem.*, **14**, 291–302

GU, J., DE MEY, J., MOEREMANS, M. and POLAK, J. (1981). Sequential use of the PAP and immunogold staining methods for the light microscopical double staining of tissue antigens. Its application to the study of regulatory peptides in the gut. *Regul. Pept.*, **1**, 365–374

GUESDON, J. L., TERNYNCK, T. and AVRAMEAS, S. (1979). The use of avidin–biotin interaction in immunoenzymatic techniques. *J. Histochem. Cytochem.*, **27**, 1131–1139

HALLIDAY, D. (1977). Identification of intracellular immunoglobulin in extramedullary myeloma. *Arch. Pathol. Lab. Med.*, **101**, 522–525

HEYDERMAN, E. (1979). Immunoperoxidase techniques in histopathology: applications, methods and controls. *J. Clin. Pathol.*, **32**, 971–978

HEYDERMAN, E. and MONAGHAN, P. (1979). Immunoperoxidase reactions in resin embedded sections. *Invest. Cell Pathol.*, **2**, 119–122

HORISBERGER, M. (1979). Evaluation of colloidal gold as a cytochemical marker for transmission and scanning electron microscopy. *Biologie Cellulaire*, **36**, 253–258

HORISBERGER, M. (1981). Colloidal gold: a cytochemical marker for light and fluorescent microscopy and for transmission and scanning electron microscopy. In *Scanning Electron Microscopy II*, (Johari, O., ed.), pp. 9–31. AMF O'Hare, Illinois: SEM Inc.

HSU, S. and SOBAN, E. (1982). Color modification of diaminobenzidine (DAB) precipitation by metallic ions and its application for double immunohistochemistry. *J. Histochem. Cytochem.*, **30**, 1079–1082

HSU, S. M., RAINE, L. and FANGER, H. (1981). Use of avidin–biotin–peroxidase (ABC) in immunoperoxidase techniques; a comparison between ABC and unlabelled antibody (PAP) procedures. *J. Histochem. Cytochem.*, **29**, 577–580

HUANG, S. N., MINASSIAN, H. and MORE, J. D. (1976). Application of immunofluorescent staining on paraffin sections improved by trypsin digestion. *Lab. Invest.*, **35**, 383–390

ISAACSON, P. and JUDD, M. A. (1977). Immunohistochemistry of carcinoembryonic antigen: characterisation of cross reactions with other glycoproteins. *Gut*, **18**, 779–785

ISAACSON, P., JONES, D. B., MILLWARD-SADLER, G. H., JUDD, M. A. and PAYNE, J. (1981). Alpha-1-antitrypsin in human macrophages. *J. Clin. Pathol.*, **34**, 982–990

KENNETT, R. H. (1980). *Monoclonal Antibodies. Hybridomas: A New Dimension in Biological Analysis.* London: Plenum Press

KOHLER, G. and MILSTEIN, C. (1976). Derivation of specific antibody-producing tissue culture and tumour lines by cell fusion. *Eur. J. Immunol.*, **6**, 511–519

KRAJEWSKI, A. S., DEWAR, A. E. and RAMAGE, E. F. (1982). T and B lymphocyte markers in effusions of patients with non-Hodgkin's lymphoma. *J. Clin. Pathol.*, **35**, 1216–1219

LANE, E. B. (1982). Monoclonal antibodies provide specific intramolecular markers for the study of epithelial monofilament organisation. *J. Cell Biol.*, **92**, 655–673

LLOYD, J. M., O'DOWD, T., DRIVER, M. and TEE, D. E. H. (1984). Immunohistochemical detection of Ca antigenic normal, dysplastic and neoplastic squamous epithelia of the human uterine cervix. *J. Clin. Pathol.*, **37**, 14–19

MASON, D. Y. and SAMMONS, R. E. (1978). Alkaline phosphatase and peroxidase for double immunoenzymatic labelling of cellular constituents. *J. Clin. Pathol.*, **31**, 454–462

MASON, F. E., PHIFER, R. F. and SPICER, S. S. (1969). An immunoperoxidase enzyme bridge method for localising tissue antigens. *J. Histochem. Cytochem.*, **17**, 563–569

MEPHAM, B. L., FRATER, W. and MITCHELL, B. S. (1979). The use of proteolytic enzymes to improve immunoglobulin staining by the PAP technique. *J. Histochem.*, **11**, 345–357

MESULAM, M. (1978). Tetramethyl benzidine for horseradish peroxidase. *J. Histochem. Cytochem.*, **26**, 106–117

MILSTEIN, C., GALFRE, G., SECHER, D. S. and SPRINGER, T. (1979). Monoclonal antibodies and cell surface antigens. *Cell Biology International Reports (London)*, **3**, 1–16

MONCRIEFF, D., ORMEROD, M. and COLEMAN, D. V. (1984). Immunocytochemical staining of cervical smears for diagnosis of cervical intraepithelial neoplasia. *Analyt. Quant. Cytol.*, **6**, 201–205

NADJII, M. (1980). The potential value of immunoperoxidase techniques in diagnostic cytology. *Acta Cytol.*, **24**, 422–447

NAGASAWA, T. and NAGASAWA, S. (1983). Enrichment of malignant cells from pleural effusions by Percoll density gradients. *Acta Cytol.*, **27**, 119–123

NAKANE, P. K. and KAWAOI, A. (1974). Peroxidase labelled antibody. A new method of conjugation. *J. Histochem. Cytochem.*, **22**, 1084–1091

NAKANE, P. K. and PIERCE, G. B. (1966). Enzyme labelled antibodies: preparation and application for the localisation of antigens. *J. Histochem. Cytochem.*, **21**, 855–894

NARITOKU, W. Y. and TAYLOR, C. R. (1982). A comparative study of the use of monoclonal antibodies using three different immunohistochemical methods. An evaluation of monoclonal and polyclonal antibodies against human prostatic acid phosphatase. *J. Histochem. Cytochem.*, **30**, 253–260

PARADIS, I. L., MERRALL, E. J., KRELL, J. M., DAUBER, J. H., ROGERS, R. M. and RABIN, B. S. (1984). Lymphocyte enumeration: a comparison between a modified avidin–biotin–immunoperoxidase system and flow cytometry. *J. Histochem. Cytochem.*, **32**, 358–362

PIZZOLO, G., SLOANE, J., BEVERLEY, P. C. L., THOMAS, J. A., BRADSTOCK, K. F., MATTINGLY, S. and JANOSSY, G. (1980). Differential diagnosis of malignant lymphoma and non-lymphoid tumours using monoclonal anti leukocyte antibody. *Cancer*, **6**, 2640–2647

POLAK, J. M. and VAN NOORDEN, S. (1983). *Immunocytochemistry. Practical Applications in Pathology and Biology.* Bristol, London, Boston: Wright

PONDER, A. and WILKINSON, M. (1981). Inhibition of endogenous tissue alkaline phosphatase with the use of alkaline phosphatase conjugates in immunohistochemistry. *J. Histochem. Cytochem.*, **29**, 981–984

POPPEMA, S., BHAN, A. K., REINHERZ, E. L., MCCLUSKEY, R. T. and SCHLOSSMAN, S. F. (1981). Distribution of T cell subsets in human lymph nodes. *J. Exp. Med.*, **153**, 30–41

PREUD'HOMME, J. L. and LABAUNE, S. (1975). Immunofluorescent staining of human lymphocytes for the detection of surface immunoglobulins. *Ann. N.Y. Acad. Sci.*, **254**, 254–261

RALFKIAER, E., PLESNER, T., LANGE, W., THOMSEN, K., NISSEN, N. I. and HOUE-JENSEN, K. (1984). Immunohistological identification of lymphocyte subsets and accessory cells in human hyperplastic lymph nodes. *Scand. J. Haematol.*, **32**, 536–543

RAMAEKERS, F. C. S., PUTS, J., MOESKER, O., KANT, A., JAP, P. and VOOIJS, P. (1983). Demonstration of keratin in human adenocarcinomas. *Am. J. Pathol.*, **111**, 213–223

RAMAEKERS, F. C. S., HAAG, D., JAP, P. and VOOIJS, P. G. (1984). Immunochemical demonstration of keratin and vimentin in cytological aspirates. *Acta Cytol.* **28**, 385–392

RATHLEV, T., HOCKO, J. M., FRANKS, G. F., SUFFIN, S. C., O'DONNELL, C. M. and PORTER, D. D. (1981). Glucose oxidase immunoenzyme methodology as a substitute for fluorescence microscopy in the clinical laboratory. *Clin. Chem.*, **27**, 1513–1515

RIGGS, J. L., SEIWALD, R. J., BURCKHALTER, J. H., DOWNS, C. M. and METCALF, T. G. (1958). Isothiocyanate compounds as fluorescent labelling agents for immune serum. *Am. J. Pathol.*, **34**, 1081–1097

ROTH, J. (1982). Applications of immunocolloids in light microscopy. *J. Histochem. Cytochem.*, **30**, 691–696

SCHARFF, M. D., ROBERTS, S. and THAMMANA, P. (1981). Monoclonal antibodies. *J. Infect. Dis.*, **143**, 346–351

SINGH, G., WHITESIDE, T. L. and DEKKER, A. (1979). Immunodiagnosis of mesothelioma: use of antimesothelial cell serum in an indirect immunofluoresence assay. *Cancer*, **43**, 2288–2296

SLEMMON, J. R., SALVATERRA, P. M. and SAITO, K. (1980). Preparation and characterisation of peroxidase antiperoxidase Fab complex. *J. Histochem. Cytochem.*, **28**, 10–15

STEEFKERK, J. G. (1972). Inhibition of erythrocyte pseudoperoxidase activity by treatment with hydrogen peroxide following methanol. *J. Histochem. Cytochem.*, **20**, 829–831

STERNBERGER, L. A. (1974). *Immunocytochemistry*. Englewood Cliffs, New Jersey: Prentice-Hall

STERNBERGER, L. A., HARDY, P. H., CUCCLIS, J. J. and MEYER, H. G. (1970). The unlabelled antibody enzyme method of immunohistochemistry. Preparation and properties of soluble antigen–antibody complex and its use in the identification of spirochaetes. *J. Histochem. Cytochem.*, **18**, 315-333

STRAUS, W. (1979). Peroxidase procedures: technical problems encountered during the application. *J. Histochem. Cytochem.*, **27**, 1349–1351

SUFFIN, S. C., MUCK, K. B., YOUNG, J. C., LEWIN, K. and PORTER, D. D. (1979). Improvement of the glucose oxidase immunoenzyme technic: use of a tetrazolium whose formation is stable without heavy metal chelation. *Am. J. Clin. Pathol.*, **71**, 492–496

TAYLOR-PAPADIMITRIOU, J., PETERSON, J. A., ARKLIE, J., BURCHELL, J., CERIANI, R. L. and BODMER, W. F. (1981). Monoclonal antibodies to epithelium-specific components of the human milk fat globule membrane: production and reaction with cells in culture. *Int. J. Cancer*, **28**, 17–21

TO, A., COLEMAN, D. V., DEARNALEY, D. P., ORMEROD, M. G., STELLE, K. and NEVILLE, A. M. (1981). Use of antisera to epithelial membrane antigen for the cytodiagnosis of malignancy in serous effusions. *J. Clin. Pathol.*, **34**, 1326–1332

TO, A., DEARNALEY, D. P., ORMEROD, M. G., CANTI, G. and COLEMAN, D. V. (1983). Indirect immunoalkaline phosphatase staining of cytologic smears of serous effusions for tumour marker studies. *Acta Cytol.*, **27**, 109–113

VALKOVA, B. and LAURENCE, D. (1985). Automated screening of cervical smears using immunocytochemical staining: a possible approach. *J. Clin. Pathol.*, **38**, 886–892

VALKOVA, B., ORMEROD, M. G., MONCRIEFF, D. and COLEMAN, D. V. (1984). Epithelial membrane antigen in cells from the uterine cervix: immunocytological staining of cervical smears. *J. Clin. Pathol.*, **37**, 984–989

WARNKE, R. and LEVY, R. (1980). Detection of T and B cell antigens with hybridoma monoclonal antibodies: a biotin-avidin–horseradish peroxidase method. *J. Histochem. Cytochem.*, **28**, 771–776

WEIR, E. E., PRETLOW, T. E., PITTS, A. and WILLIAMS, E. E. (1974). Destruction of endogenous peroxidase activity in order to locate cellular antigens by peroxidase labelled antibodies. *J. Histochem. Cytochem.*, **22**, 51–54

WHITESIDE, T. L. and DEKKER, A. (1979). Diagnostic significance of carcinoembryonic antigen levels in serous effusions. *Acta Cytol.*, **23**, 443–448

YAM, L. T. and JANCKILA, A. J. (1983). A simple method of preparing smears from bloody effusions for cytodiagnosis. *Acta Cytol.*, **27**, 114–118

7

Microscopy

Rod Setterington

The microscope is one of the most important tools of the clinical laboratory and every reader of this book will be more or less conversant with it. Those who decide they belong to the 'less' rather than the 'more' familiar class will find in this chapter a summary of the facts most vital for practical work. For the beginner in microscopy, familiarization with the components of the microscope is recommended as a starting point. *Figure 7.1* illustrates the basic parts of a standard laboratory binocular microscope. The user should be familiar with the components identified on the microscope. Various adaptations to the illuminating and imaging light beams in the microscope enable the image to be inspected in a number of specialized ways, each providing different interpretations of the specimen. These techniques are summarized in the following paragraphs while the protocol necessary to ensure the optimum performance of the microscope appears in the latter part of the chapter.

Bright field illumination

This is the basic illumination technique comprising most of routine microscopy. With this method either the brightness or the colour of the light penetrating different areas of the specimen is differentially affected resulting in bright–dark contrast or, with coloured specimens, in colour contrast. Most microscope specimens, however, are not coloured and since staining is not always possible, methods offering contrast without staining are sometimes desirable. The simplest

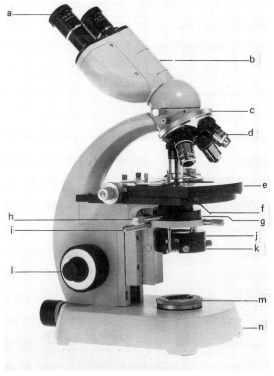

Figure 7.1 A standard binocular microscope. a, Eyepiece; b, Binocular tube; c, Revolving nosepiece for changing objectives; d, Objectives; e, Specimen stage; f, Lever for aperature iris diaphragm; g, Substage condenser; h, Swing-out front lens lever; i, Condenser centering knob; j, Swing-out holder for filters; k, Swing-out auxiliary lens holder; l, Coarse and fine adjustment knobs; m, Field iris diaphragm; n, Base with integral low-voltage illuminator.

way to increase contrast in such cases is to reduce the aperture (condenser) even though a resulting loss of sharpness of detail cannot be avoided. Similar contrast enhancement can be achieved, again at the cost of sharpness, by fine-focusing slightly off the plane of maximum sharpness (so-called extrafocal adjustment).

Dark field illumination

Dark field illumination is an excellent means of increasing the contrast of small light-reflecting particles in tissue. A special purpose condenser blocks out the central portion of the illuminating light. The specimen is thus illuminated by a hollow cone of light striking it at an angle. Only the light diffracted by the specimen enters the objective to form an image. The tiniest particles inside a specimen diffract light into the objective and become visible. On the other hand light rays undeviated by the transparent portions of the specimen do not enter the objective with the result that the areas in the field of view that contain no object structures remain dark (*Figure 7.2a*). Dark field illumination requires a powerful light source since only a small portion of the light reaching the specimen actually enters the microscope's lens system.

Phase contrast illumination

The best solution to the contrast problem with transparent objects was discovered by the Dutch physicist Frits Zernike, an achievement for which he was awarded the Nobel Prize for Physics in 1953.

The solution is based on the observation that when light passes through transparent material, its amplitude and wavelength are not appreciably altered, but its phase is altered by the tissue. For example, as light passes through a group of unstained cells, the cell membranes will accelerate or retard light to a different extent than will the cytoplasm or the nuclei which results in a phase difference. With appropriate optics, these differences in phase are transformed into differences in brightness of different parts of the specimen (*see Figure 7.2c*).

To facilitate the separation of the unchanged light and the phase-shifted light, a phase condenser has an annular diaphragm rather than an iris so that the specimen is illuminated with a hollow cone of light. Phase objectives have the means both to intensify the phase differences of light transmitted through the specimen and to dim the unchanged light. When adjusting a phase set-up, the bright annulus of the condenser is covered by a dark ring in the objectives; this is possible only if the annulus and the ring have the same aperture (angular size). Since various powers of phase objectives have various apertures, a phase condenser contains a rotating disc with annular diaphragms of different sizes.

Polarized light illumination

Various crystalline structures in biological material can be revealed by the use of polarized light. In its simplest form the technique requires that a polarizing filter is inserted in the illuminating path, often just below the substage condenser, and a second filter with its vibration direction set at right angles to the first is inserted in the imaging path just above the objective.

The effect of these opposed filters means that unless the character of the polarized light is influenced by the specimen no light will reach the observer. If, however, parts of the specimen are birefringent or dichroic the character of the polarized light will be affected. This provides a simple but valuable method of identifying uric acid crystals, banded muscle tissue, etc. by allowing the relevant areas of the specimen to be seen and identified.

Differential interference contrast (after Nomarski)

The problem of increasing contrast in low contrast specimens can also be solved with the differential interference contrast (DIC) method. With the DIC method, as in phase contrast, the light path is modified but in a different manner. Polarized light is produced and is split into two separate beams by a special prism in such a way that the planes of polarization of the two beams are perpendicular to each other. These two closely adjacent beams (the separation is below the resolving power of the optics) are subject to different phase shifts because of the different properties of neighbouring object areas. To produce contrast by interference, the two beams are first reunited by another prism and then brought into the same plane of polarization by an 'analyser'. In addition, the second prism is adjustable so that the user can add or subtract

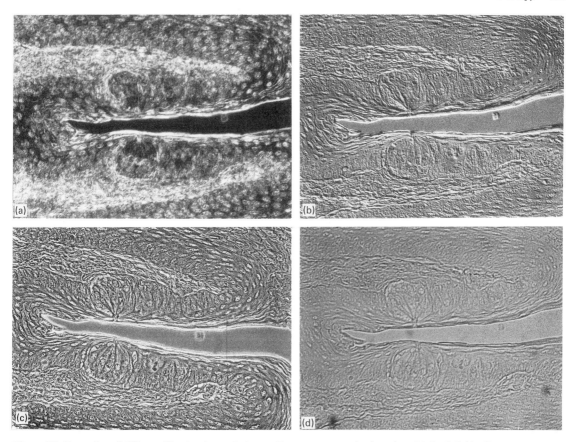

Figure 7.2 Examples of different illumination techniques. Cat tongue, unstained section. (*a*) Dark field; (*b*) differential interference contrast; (*c*) phase contrast; (*d*) bright field

phase differences to achieve any desired contrast up to a dark field image or to colour contrast. DIC produces the final image in relief contrast with shadow effects (*Figure 7.2b*).

Fluorescence microscopy

Fluorescence can be defined as the emission of visible light by a substance as a result of its irradiation by energy of a shorter wavelength. 'Fluorochromes' are dyes applied in a manner quite similar to other staining methods but can generally be used in much lower concentrations. Depending on their chemical composition, they can be selected to adhere to specific areas or components in the specimen while leaving others unstained. Most microscopical disciplines including immunology, histology, bacteriology, haematology, as well as all branches of cytology have evolved techniques which take advantage of fluorescence and it is normally possible to add the necessary accessories to any existing microscope.

The first requirement is for a light source which supplies enough exciting radiation for the fluorochrome being used and generally this will be a high pressure mercury vapour lamp although a 100 W quartz halogen illuminator is perfectly adequate if only blue light excitation is required. The optimum wavelength is then selected by inserting an 'exciter' filter in the illumination path which has the function of limiting the energy which reaches the specimen to that range which excites the maximum emission of fluorescence. Since excessive excitation light which is not absorbed by the specimen would interfere with the image it must be suppressed by a 'barrier' filter placed in the imaging path. Various sets of these complementary filters have been evolved to provide optimum results with different fluorochromes.

For many years now the technique has been further refined by developing 'incident' fluor-

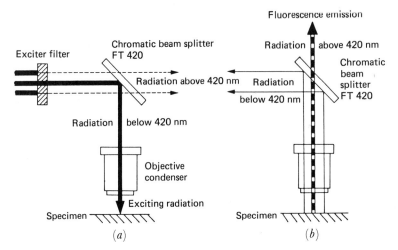

Figure 7.3 Principle of fluorescence microscopy with reflected light excitation (a) and dichromatic beam splitter (b)

escence which provides much brighter images by arranging for the excitation light to reach the specimen via the objective in use. A 'chromatic' beam splitter is set at 45° to the optical axis above the objective and has the effect of preferentially reflecting the shorter excitation wavelengths and also preferentially transmitting the longer wavelengths emitted by the fluorescing parts of the specimen. Particular combinations of exciter and barrier filters along with the appropriate chromatic beam splitter are available for all conventional fluorochromes and most manufacturers provide these mounted together in an interchangeable block which can be inserted into the optical path of an incident fluorescence illuminator mounted between the objective and the eyepiece of the microscope (*Figure 7.3*).

Using the microscope

With relatively few exceptions the material of interest to cytologists consists of smears mounted on glass slides and covered by a thin cover glass. The slide is placed on the microscope stage with the cover glass on top. The condenser should be raised to its maximum height and the condenser front lens swung into place. The specimen and the objective should be cautiously moved closer together using the coarse adjustment until they are only a few millimetres apart (observe from the side!). In modern microscopes, a stop and the spring mount of the objectives prevent ordinary specimens (1 mm slide, 0.17 mm cover glass) from being damaged by the objective. Only a few special purpose objectives are exceptions.

At the beginning of an examination a low power objective of $\times 2.5$ to $\times 10$ should be used because it permits coverage of a wider field and is thus best suited for scanning. In addition, its depth of field is greater than that offered by objectives of higher power, and thus it is easier to find the correct plane of focus. The power of an objective is its initial magnification indicated by the first value engraved on each objective. The total magnification of the microscope is determined by multiplying the magnification of the objective by that of the eyepiece. A $\times 10$ objective used in conjunction with a $\times 12.5$ eyepiece yields a total magnification of $\times 125$.

The condenser iris should be closed approximately halfway. Now the microscopist can look into the eyepiece and increase the distance between the objective and the specimen by the coarse adjustment until the details of the specimen can be recognized even if they are still blurred. Exact focusing is achieved with the aid of the fine adjustment. When the area to be examined is in the centre of the field of view and focused, the objective of the next higher magnification may be moved into position, and focused with the fine control.

The adjustment of the condenser is a frequent source of error. The condenser is correctly adjusted when it is just below its highest position, and when its iris has been closed only far enough to obtain sufficient contrast. The beginner usually closes (stops down) the condenser far too much. The condenser iris should never be used to control the image brightness! Other means (e.g. neutral density filters) must be used for this purpose. Lowering the condenser to increase contrast, popular with many microscopists, has the same effect as closing the condenser iris and is inappropriate. If the illuminated field should be too small with the objectives of very low power, the front lens of the condenser should be swung out and its diaphragm fully opened.

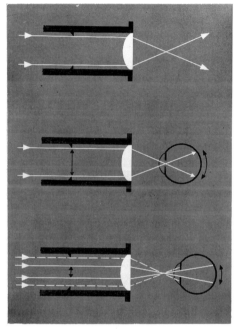

Figure 7.4 Correct distance of eye from eyepiece. Top: location of microscope's exit pupil. Centre: exit and eye pupils in the same place. Bottom: eye too far away. Eye pupil reduces the field

The depth of field in the specimen, corresponding to a certain setting of the fine adjustment, is extraordinarily shallow. This shallow depth of field is more pronounced with high power objectives than with lower powers, and more with high quality optics than with lower quality optics. In microscopic work, the fine adjustment is continuously moved up and down in order to cover the entire depth of the specimen. Beginners often make the mistake of focusing the microscope as if the magnified image they want to view was located very close to the eye—as one would view an object closely with optical aids. This near setting of the eye (accommodation) is extremely tiring if maintained for a long time and, therefore, it is best to work with perfectly relaxed eyes. At the beginning this is done more easily if it is imagined that the image being viewed is at infinity. This encourages looking 'through the microscope' instead of 'into the microscope'. Another important point is to maintain the eye at the correct distance from the eyepiece. The best procedure is to move in from a few inches distance until the field of view in the eyepiece is widest and sharply defined. At this distance all the light rays emerging from the microscope find their way into the eye. As is evident from *Figure 7.4*, this is possible only if the pupil of the eye is brought very exactly to the point of constringence of the microscope's light rays, which has been named, therefore, the exit pupil of the microscope. When you have acquired a certain amount of practice in the use of a microscope you will automatically find and keep the right distance from the eyepiece.

It is usually advantageous for those who wear prescription eyeglasses to keep their glasses on when using the microscope. Special rubber guards on the eyepieces prevent scratching.

Optical components of the microscope

The objective provides the initial, and usually the major, magnification of the specimen. Of the three optical components in the microscope (condenser, objective and eyepiece) the objective is the part which has the greatest effect on image quality. By using a cheap magnifying lens as an example, it becomes immediately apparent that a simple lens exhibits quite a number of image errors: black figures show coloured fringes along their edges, sharpness rapidly decreases towards the edges of the field, and intolerable distortions are produced if the object is viewed at only a slight angle through the lens. If such a simple magnifier exhibits aberrations that are so obvious, how much more evident must these errors be at the far higher magnification of a microscope objective when, to make things worse, the magnified image is viewed through an additional magnifier—the eyepiece. The ingenious selection and combination of lenses, lens curvatures, different glass types, etc. allows the microscope manufacturer to eliminate many of the image errors to varying degrees. It is obvious that the elimination of more errors—that is, higher correction—requires greater sophistication which results in higher costs. Thus, the differences in microscopist's requirements and available funds have prompted microscope manufacturers to offer several 'quality categories' of objectives.

The greatest number of microscopes are supplied with so-called achromats. Up-to-date versions of this simplest type of objective have a remarkably high performance and are entirely satisfactory for many routine purposes in teaching and research. The fact that certain colour fringes will still be noticed in certain critical work is hardly an obstacle to observation, especially since this can easily be remedied by using a filter, normally a green one. However, even the beginner will notice that while using a ×40 achromat for example, the centre of the field and the outer zones cannot be seen in sharp focus at the same time. The surface of best definition is 'dished'. This aberration, called 'curvature of field', is not a great handicap in microscopic examination as movement of the stage or adjustment of the fine focus can allow the viewer to see all areas clearly. Field curvature may be a nuisance in photomicrography, however, where a sharp image of the entire field of view is required. The extent to which curvature of field becomes visible also depends considerably on the type of specimen observed. It will be less noticeable in a thick section than a smear slide.

On the other hand, there are some very positive characteristics that, in general, can only be found in achromats. Because their numerical aperture is low, achromats offer greater depth of field and have a longer working distance than more highly corrected objectives. This allows greater freedom in the selection of cover glasses, which, in this case will not degrade image quality as quickly should they deviate from the optimum thickness of 0.17 mm (see p. 134). These factors favour the use of such objectives by the beginner and routine worker.

Specifying the next higher 'quality category' after the achromat is somewhat difficult because opinions vary as to which of the aberrations inherent in achromats should be eliminated first. While flat field achromats have been corrected for curvature of field, fluorite objectives and apochromats will produce images distinguished from those of achromats by greater colour fidelity, better contrast and higher resolving power. The differences between fluorite objectives and apochromats is less of a basic than of a transitional nature; the fluorite objectives lie roughly between achromats and apochromats. Investigators frequently using photomicrography will give preference to planachromats. Workers demanding optimum definition, such as those in critical research work, will choose fluorite objectives or apochromats. The uncompromising microscopist who demands both a completely flat field and optimum colour fidelity combined with extremely high definition, may choose from the range of planapochromats—microscope objectives of absolutely top quality.

Although the layman generally believes that results will improve with increasing magnification, the experienced microscopist will use high powers only if absolutely necessary. For technical reasons resolution does not increase in proportion with magnification. High power objectives, therefore, give relatively weak images, have a very low depth of field, and offer comparatively less brilliance and sharpness than do low power objectives. The magnifications to be chosen for a set of objectives will always depend on the type of specimen to be examined.

Although a few years ago it was necessary to use different types of eyepieces (oculars) for low and high objectives, microscope objectives now are used generally with compensating oculars. Consequently eyepieces no longer need to be exchanged when changing objectives. In spite of this, two classes of eyepiece are still being manufactured. Although simple eyepieces are used primarily in conjunction with achromatic objectives, compensating flat field eyepieces are intended for the other objectives of higher correction but may, of course, also be combined with achromats. The most widely used eyepiece magnifications are ×10 and ×12.5.

Particular mention should be made of the eyepieces specially designed for spectacle wearers. The exit pupil of the microscope and the eye pupil must coincide if the entire field of view is to be observed. With normal eyepieces the distance between the top edge of the eyepiece and the exit pupil (eye point) is so short that glasses cannot be kept on if the two points are to coincide. The high eye point of the special eyepieces for spectacle wearers allows glasses to be worn: the rubber guards supplied with these eyepieces prevent scratching of the spectacle lenses. If these special types of eyepieces are used by observers not wearing glasses, care must be taken to keep the eyes at an adequate distance from the eyepiece. If necessary, finding the correct distance can be facilitated by slipping eyecups onto the eyepieces.

Apart from its magnification, an eyepiece is characterized by its field of view. With the aid of this number it is easy to calculate the diameter of the field covered in the specimen plane. Field diameter is equal (in millimetres) to the field of view number of the eyepiece divided by the initial magnification of the objective. For example, a ×10 objective in combination with an eyepiece of the field of view number 12.5 permits a field of 1.25 mm diameter to be covered in the specimen plane.

The primary purpose of the condenser is to provide appropriate illumination of the specimen. To meet the high standards of most microscopists the condenser performs two functions. It provides homogeneous illumination of the object plane and, more importantly, it delivers light rays with a particular angle of incidence. This latter function results in complete filling of the rear lens of the objective (the back focal plane or exit pupil of the objective) with light, which is a prerequisite for obtaining an image of high resolution.

Low power objectives take a large object field with a small angle of incidence (low aperture), whereas high power objectives require a small object field and large angle of incidence. No single condenser can meet these contradictory demands, so most condenser units are combinations of a regular condensing lens and an auxilliary front lens which can easily be removed from the optical path for use with low power objectives.

The demands generally made on condensers are partly inadequate and partly exaggerated. A standard condenser has a numerical aperture of 0.9 and a front lens that can be removed to illuminate large object fields. Achromatic–aplanatic condensers are recommended for colour photomicrography and microphotometry. Condenser apertures above 0.9 are useful only if the condenser is immersed in oil (*see* p. 133).

If phase contrast work should be contemplated—perhaps at a later date—it is advisable to buy a phase contrast condenser right at the beginning because this type of condenser is useful for conventional bright field microscopy as well.

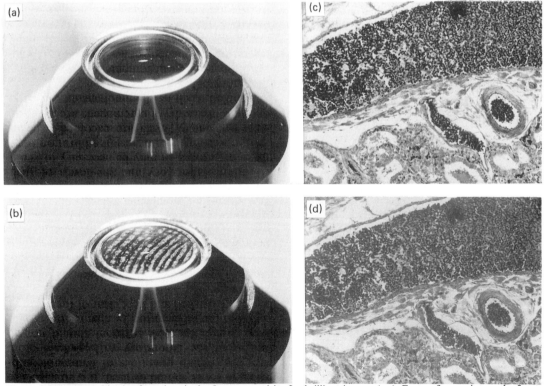

Figure 7.5 A clean objective front lens is the first prerequisite for brilliant images (*a,c*). Even a fingerprint on the front lens will ruin the contrast (*b,d*)

Care and cleaning of the microscope

An instrument that has to satisfy the most exacting requirements regarding mechanical and optical precision naturally demands a certain amount of care. Dust on optical elements will degrade the image quality. Although all surfaces exposed to dust are also easily accessible for cleaning, no glass surface is improved by 'cleaning'. The best advice is always to avoid exposure of the microscope to dust by covering it with a hood when it is not in use or by keeping it in a cabinet. Special care should be taken to ensure that the tubes of a microscope are always closed either by an eyepiece or a dust plug.

While dust particles on the eyepiece will only give rise to patches in the image, a dirty objective front lens may hopelessly reduce the sharpness of the image, or at least its contrast. Because it is close to the specimen and possibly to the immersion oil on the cover glass, as well as being particularly close to the hand operating the nosepiece, the front lens of the objective is especially prone to getting soiled. Even the lightest fingerprint may have grave consequences (*Figure 7.5*). Before starting important work it is advisable to unscrew every objective and check it carefully with the aid of a magnifier. Should structures be found in the image which are suspected of being extraneous to the specimen, the fault may be traced as follows. If the trouble can be eliminated by slight adjustment of the condenser, the cause may be in the bulb of the lamp, the lamp condenser, or the filter in front of it. However, if adjustment of the condenser does not produce any result, the next step is to turn to the focusing adjustment which should identify all faults due to soiling of the condenser front lens or the specimen. If this does not identify the source of the problem, slightly rotate first the objective and then the eyepiece, and you will immediately notice in which case the foreign body follows the rotation. Dust particles are most clearly seen when the aperture diaphragm has been fully closed, because in this case the depth of focus is at its greatest.

In almost all cases it will be sufficient to clean the outer lens faces with the aid of a grease-free brush (if necessary, wash in ether first) or with a frequently washed, absolutely dust-free linen rag and distilled water. If any organic solvent must be used it is advisable to use very little ether or benzene rather than water. Alcohol might destroy the cement between the lens elements if older optical systems are being used. Ether is usually preferred to alcohol for cleaning because it evaporates more quickly, and any harmful effect is thus less likely. Finally, residues are always removed with water as described above. Should compressed air be available for cleaning, be sure to use a filter of cotton wool.

Topics for the advanced microscopist

To understand this section better it is helpful to think of a projector and a projected slide. At first the image of the slide is observed on the screen from far behind the projector. In this case magnification is still low because the angle of view is quite small. Magnification improves upon moving closer to the screen and previously unrecognized details become evident. Finally, however, a point is reached where higher magnification (moving closer) does not give better results. Instead of single leaves on a tree, only the structure of the colour film is seen—an accumulation of dye particles (the grain). Further reducing the viewing distance could, of course, further increase magnification, but this would be 'empty magnification' i.e. magnification that does not reveal any new detail.

In the microscope the situation is similar. The projected slide image corresponds to the aerial image formed in the tube by the microscope objective. As with the projected slide, there is also a limit for the projected aerial image where 'useful magnification' ends and 'empty magnification' begins. It is true that the microscope image will not exhibit any actual grain structure, but one may speak of 'grain' in this case because there is no optical image in which an object point is really reproduced as a point (i.e. with no measurable diameter). Owing to the phenomenon of diffraction, an object point is always transformed into a small disc in the image. It is these so-called 'airy discs' that form the image 'grain'. This type of grain can be found in any optical image if the magnification is high enough. It is obvious that image details will no longer be clearly recognized as soon as they reach the size of the airy disc, just as it is impossible to recognize the leaves in the slide when they are the same size as the dye grain.

If a photographic slide is viewed with higher

magnification than is useful, no harm is actually done because it is very well known that the above-mentioned grain structure does not belong to the object. However, in microscopy the situation is different. Empty magnification may show structures in the image that do not exist in the specimen, and with unfamiliar specimens there is a danger of erroneous conclusions.

It follows from the above remarks that the performance limit of the microscope objective in use should be known, i.e. the point at which its useful magnification ends. This is indicated by a value engraved on every objective called 'numerical aperture'. The numerical aperture is the factor that determines the size of the airy discs in the microscopic image i.e. the size of the 'grain'.

The following rule is helpful at this point. The total magnification of the microscope (i.e. the magnification of the objective multiplied by that of the eyepiece multiplied by any tube factor) should not exceed 1000 times the numerical aperture. Microscopists who have particularly good visual acuity should even use 500 times the numerical aperture as a limit. Exceptions to this rule may be allowed in order to facilitate measuring and counting work.

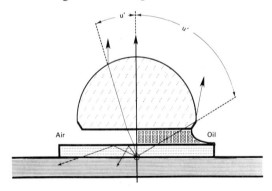

Figure 7.6 Diagrammatic representation of numerical aperture. Since air has a different refractive index from glass, dry objectives are limited by the angle of light they capture. The use of immersion oil with a refractive index similar to glass renders considerable improvement

An aid to understanding numerical aperture is given in *Figure 7.6*, which gives a diagrammatic view of an objective front lens and specimen. The angle u' subtended by the optical axis and the outermost ray accepted by the objectives is a measure of the aperture of the objective: it is half the aperture angle. However, the magnitude of this angle is not indicated in degrees but in the form of a sine value, i.e. a numerical value. This explains the origin of the term 'numerical aperture'. It is the sine of one-half the aperture angle multiplied by the refractive index of the medium filling the space between the cover glass and the front lens.

Numerical aperture = $n \sin u'$

Since air has a refractive index of 1, n may be neglected when dry objectives are used. Immersion oil, which in the case of an oil immersion objective fills the space between the cover glass and the front lens, typically has a refractive index of about 1.5. It is obvious that such an objective makes it possible to achieve a gain of 50% in numerical aperture.

With the aid of the numerical aperture we can compute the limit of resolution or maximum resolving power. This is the smallest distance, designated d, by which two structural elements may be separated in order to be seen as two distinct elements instead of one. If NA_{obj} and NA_{cond} are the numerical apertures of objective and condenser respectively, and λ is the wavelength of the light used for observation, then the smallest resolvable separation between two object points is:

$$d = \frac{\lambda}{NA_{obj} + NA_{cond}}$$

Example: Using a green filter, the wavelength is 0.55 µm; assuming the objective used has an aperture of 1.25 and the condenser one of 0.9 then $d = 0.25$ µm. However, structural elements which are just resolved are not reproduced with full fidelity.

Oil immersion objectives are used with immersion oil instead of air between the objective front lens and cover glass. The refractive index of the oil is similar to that of glass, resulting in a gain in numerical aperture (*see Figure 7.6*) and less diffraction. In regard to correction, for the same magnification an oil immersion objective is always superior to a comparable dry objective. The user should, if possible, employ only the type of immersion oil supplied by the manufacturer of the objective because the refractive index and the dispersion of the oil must have certain values.

If the condenser in use has a numerical aperture higher than 0.9, it is advisable also to immerse the condenser front lens for use in combination with an oil immersion objective. For this purpose a drop of oil is applied to the underside of the specimen slide and another one to the condenser front lens. This will effectively reduce the danger of bubbles forming when the condenser is immersed. Even if the immersion oil is of the non-drying type, it should be removed at the end of the day by means of a clean rag of the kind used for cleaning optical elements.

Knowing the high precision required to produce microscope objectives that are to give first-rate images, it should not be surprising that the cover glass, which is located between the specimen and the objective front lens, has optical effects and must have certain specifications. The characteristics of the cover glass must be taken into account in the computation and design of the objective. With normal microscope objectives, the cover glass is part of the image-forming system. It is a 'lens element' (though with infinite radii of curvature) which is located outside the objective mount but still forming a part of the optical system. The optimum thickness of a cover glass is 0.17 mm.

Strictly speaking this applies only to objectives whose numerical apertures exceed a certain value. With objectives having an aperture not exceeding 0.2, specimens may be examined both uncovered or with cover glasses as thick as 1 mm without the spherical aberration being noticeable. With apertures between 0.3 and 0.7 the cover glass thickness should not deviate from the nominal value by more than 0.03 mm. With higher apertures even a deviation of 0.01 mm will considerably impair the quality of the image.

In order not to restrict the cover glass to specified thicknesses when using dry objectives of particularly high apertures, some objectives have been designed in which the optimum cover glass thickness is not a fixed 0.17 mm, but can be varied within a range of 0.12–0.22 mm. These are objectives with correction collars. If the thickness of the cover glass on the specimen to be examined is unknown, first set the correction collar at 0.17 mm and find a high contrast area in the specimen. Then test whether the contrast is improved if the correction collar is set for a greater or lesser cover glass thickness. This also requires refocusing. It is obvious that in the hands of inexperienced users who usually set the collar incorrectly, these objectives may give very disappointing results (*Figure 7.7*).

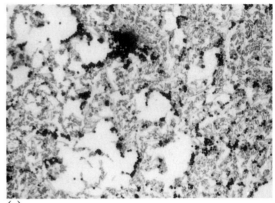

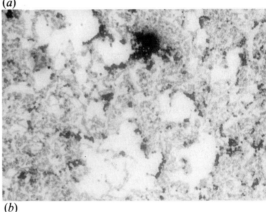

Figure 7.7 Image deterioration that occurs when the thickness of the cover glass is incorrectly matched to the objective. A planachromat 40/0.95 objective with correction collar and 0.17 mm cover glass are used. In (*a*) the collar is adjusted to 0.17 mm and in (*b*) to 0.19 mm

The situation is somewhat different with oil immersion objectives because a medium of high refractive index, namely oil, is in front of the first lens. In this case the thickness of the cover glass is not very critical. Still, it should be remembered that immersions are not 'homogeneous' enough (i.e. glass and oil do not have identical refractive indexes) to allow the issue of cover glass to be dispensed with as a matter of course. For practical work with specimens that are to give optimum images, it is necessary always to use cover glasses that are exactly 0.17 mm thick. These may be purchased from most scientific supply houses.

It should be mentioned that in those cases in which the cover glass is not in direct contact with the specimen, the intermediate layer of mounting medium has the same effect as additional thickness of cover. It is therefore advisable to weigh down the covers of specimens during drying.

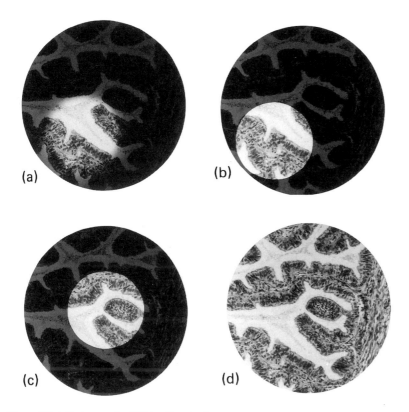

Figure 7.8 Adjustments for Köhler illumination; (*a*) close the field iris diaphragm; (*b*) raise the condenser to sharpen the iris image; (*c*) centre the light beam; (*d*) open the iris until the field is just totally illuminated

Köhler illumination

Microscope illumination has two important requirements. First, the field of view should be evenly illuminated, and when the condenser iris is open and the condenser is in the uppermost position the back lens of the objective should be filled with light as completely as possible. For microscopic research work, photomicrography, microprojection, etc. precise control of the light path should start before the light reaches the specimen, i.e. at the light source. Professor August Köhler first used this exact control of the light path in the illuminating beam of the microscope and the method is today called 'Köhler illumination'. The procedure for applying the method is as follows (*see Figure 7.8*):

(1) Raise the condenser to its maximum and take the eyepiece out of the tube. The back lens of the objective will be filled with light. Dimming the light by a neutral density filter would be advisable. Insert the eyepiece and focus the specimen with a ×10 or ×16 objective.

(2) Close the iris diaphragm of the field stop (*Figure 7.1*, m) almost completely (*Figure 7.8a*) and focus the edge of this diaphragm image by slightly lowering the condenser (*Figure 7.8b*).

(3) Now move the sharp image of the field stop into the centre of the field (*Figure 7.8c*) by operating the centering knobs (*Figure 7.1*, i) of the condenser. Both the specimen and the lamp field stop are now sharply defined. Then open the field stop until the entire field of view is just clear (*Figure 7.8d*).

(4) Slightly readjust the lamp socket or the lamp condenser, until the field of view is evenly illuminated.

(5) At first open the condenser iris (aperture stop) completely (image very bright) and close it only far enough to eliminate glare in the most important image elements and to

make them appear with satisfactory contrast. With stained specimens this is the case even at a relatively wide aperture, while fresh specimens require slightly more stopping down. Stopping down further than is absolutely necessary (for example, to reduce the brightness of the image) is one of the biggest mistakes a microscopist can make since it entails a loss of resolving power. If the intensity of the light is too high, reduce the lamp voltage or insert grey or green filters into the filter holder.

(6) If with low power objectives only part of the field is illuminated, swing out the condenser front lens (*Figure 7.1*, h), and open the condenser fully. Contrast is then controlled with the aid of the lamp diaphragm.

When adjusting the equipment for Köhler illumination, it is possible that a certain granularity may be seen over the entire field of view. This rises from the texture of a ground-glass screen or an etched collector lens, which is used to reduce the irregularity of the light source. If this occurs it can be eliminated by slightly altering the height of the condenser.

Further reading

BRADBURY, S. A. (1984). *An Introduction to the Optical Microscope*. Royal Microscopical Society Microscopy Handbook 01

HARTLEY, W. G. (1979). *Hartley's Microscopy*. Charlbury, Oxon: Senecio Publishing Co.

JAMES, J. (1976). *Light Microscopic Techniques in Biology and Medicine*. The Hague: Martinus Nijhoff

MOLLRING, F. K. (1981). *Microscopy from the Very Beginning*. W. Germany: Carl Zeiss

8

Female genital tract: normal cytology

Dulcie V. Coleman

The cervical scrape is a safe, simple and non-invasive method of sampling the uterine cervix. The technique was developed by Dr George Papanicolaou (Carmichael, 1973) whose name it bears and who first realized its potential as a method for the diagnosis and prevention of cervical cancer. The accessibility of the cervix and the propensity of neoplastic cells to shed from its surface has enabled us to study the process of malignant transformation in the cervix in a way that has not been possible at any other site.

The microscope analysis of stained and mounted smears prepared from cervical scrapes has become a major component of the workload of most cytology laboratories. The task of analysing these smears is complex and the pitfalls of diagnosis are many. A proper awareness of the wide range of cytological patterns which may be encountered in health and disease is a prerequisite for accurate diagnosis.

In this chapter many aspects of the cervical smear test are discussed. These include collection of specimens, methods of analysis and reporting, and applications of the test. The many cytological patterns found in smears from the normal cervix are also described and their physiological basis discussed. The appearance of the smear in the presence of cervical inflammatory disease, and neoplasia are described in subsequent chapters. In order to fully understand the range of patterns seen in the smears, the cytologist must have a sound working knowledge of the normal anatomy, physiology and histology of the cervix and its relation to other organs in the pelvic cavity and this information is also provided in this chapter.

Anatomy and histology

The female genital tract consists of the vulva (external genitalia), vagina, uterus (cervix and body), and paired fallopian tubes and ovaries (*Figure 8.1*). The upper vagina, uterus, fallopian tubes and ovaries are situated in the pelvis

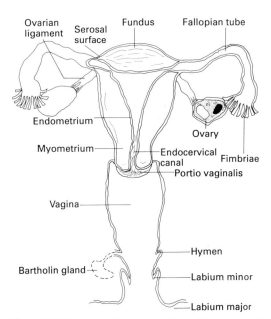

Figure 8.1 Diagrammatic representation of the organs of the female genital tract

138 *Female genital tract: normal cytology*

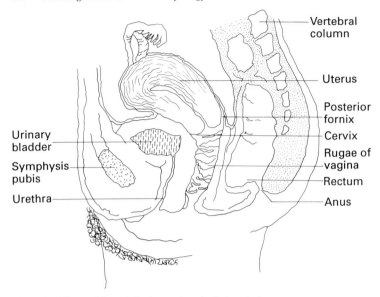

Figure 8.2 The position of the internal genitalia in relation to the bladder and rectum in the female

where they lie behind the bladder and in front of the rectum (*Figure 8.2*). The uterus is covered by peritoneum which dips down behind the uterus and in front of the rectum to form the pouch of Douglas.

Vulva

The vulva consists of two longitudinal folds of skin (labia majora and minora) on either side of the midline, with a central opening (the vestibule) bounded in front by the clitoris and behind by perineal skin surrounding the anus (*Figure 8.3*). These tissues are covered by stratified squamous epithelium, which is keratinizing over the labia and non-keratinizing in the vestibule. Sweat and sebaceous glands are present in the labia.

The labia majora are the larger, outer folds; the lateral aspect is covered by hair, whereas the medial aspect and the smaller inner folds (labia minora) are not.

The vestibule lies between the labia minora and contains the orifices of the vagina, urethra and the two mucus-secreting glands of Bartholin—one on either side of the vaginal orifice. The glands of Bartholin consist of acini of pale staining cells with basal nuclei, with ducts lined by stratified columnar or transitional epithelium.

Vagina

The vagina is a fibromuscular tube extending from the vestibule to the uterus. The uterine cervix projects into the vagina at an angle

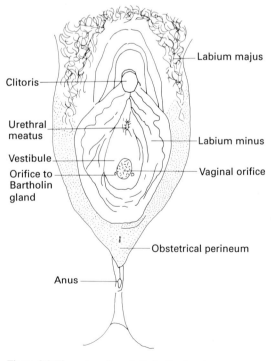

Figure 8.3 The external genitalia in the female

(*Figure 8.2*) and creates recesses called fornices at the upper end of the vagina. The posterior fornix is the deepest, and cellular debris accumulates there.

The wall is lined by non-keratinizing stratified squamous epithelium (*Figure 8.4*). There are normally no mucous glands in the vagina, and the mucus which is found there comes from the glands of the cervix.

The hymen is a thin fold of the same epithelium over a loose connective tissue core, present at the lower end of the vagina. It has a central opening, which in the virgin varies in size from very small up to the diameter of the vagina.

The vaginal epithelium is under the influence of ovarian hormones and the number of its cell layers varies at different periods of life. During childhood the epithelium is thin, but at menarche the epithelium responds to oestrogen stimulation by becoming highly stratified. After the menopause, when menstruation ceases, the epithelium becomes atrophic because of low oestrogen levels.

Cervix

The cervix is the lower portion of the uterus (about one-third of its total length) which protrudes into the vagina. It has a thick muscular wall. Running through the cervix is a narrow channel 1.5–2.0 cm long, the endocervical canal. The opening into the vagina is called the external os and is well-defined, but at its upper end—the internal os—it merges gradually into the cavity of the body of the uterus. The endocervical canal is lined by columnar mucus-secreting cells and ciliated cells (*Figure 8.5*). Small rounded cells may be seen occasionally in the basal layers of the epithelium or in the superficial layers of the connective tissue on which it stands. These are the reserve cells which become active when the columnar epithelium is stimulated to undergo metaplastic change. The underlying connective tissue stroma contains large branched tubular glands lined by similar epithelium. The mucus secreted by these glands varies: at the time of ovulation it becomes

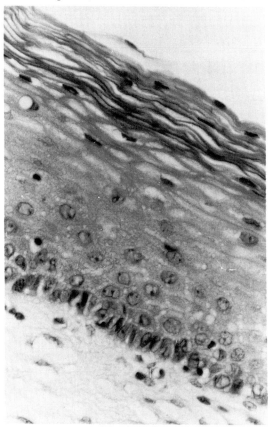

Figure 8.4 Non-keratinizing squamous epithelium of the vagina. Stain: haematoxylin and eosin. Magnification × 330

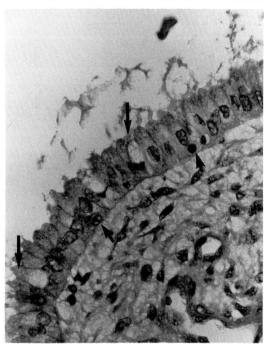

Figure 8.5 Endocervical epithelium showing occasional mucus-secreting cells (long arrow) interspersed among ciliated cells. Also note the reserve cells between columnar cells (short arrow). Stain: haematoxylin and eosin. Magnification × 770

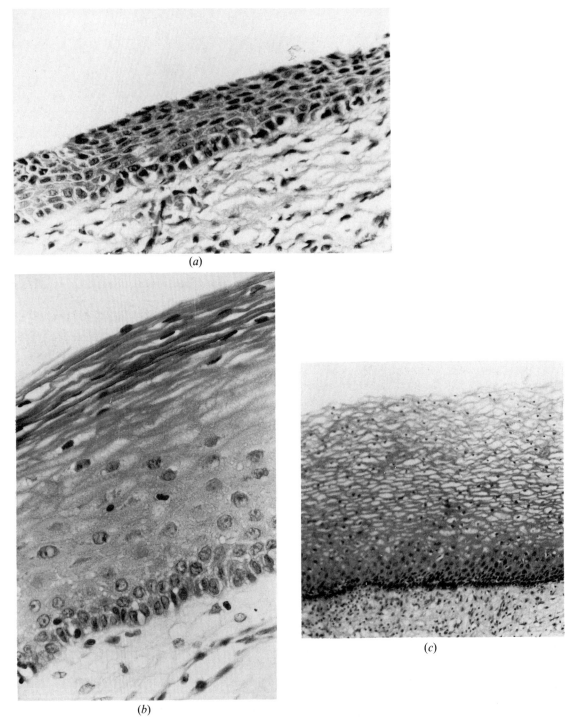

Figure 8.6 (*a*) Atrophic epithelium lining the entire cervix. The epithelium is composed of a few layers of delicate rounded cells. (*b*) Fully mature squamous epithelium showing clearly a basal, suprabasal, mid and superficial zone. The surface layers are composed of large cells with abundant cytoplasm and pyknotic nuclei. Stain: haematoxylin and eosin. Magnification ×330. (*c*) Cervical epithelium showing incomplete maturation. The mid zone is expanded and the cells show a lesser degree of stratification and nuclear degeneration compared with (*b*). Stain: haematoxylin and eosin. Magnification ×120

thinner in response to oestrogen, while under the influence of progesterone it is thick and mucoid. Sometimes the glands become obstructed and dilate to form cysts called Nabothian follicles.

At or about the external os, the columnar epithelium gives way to non-keratinized stratified squamous, without glands, and this covers the portion of cervix exposed to the vagina, i.e. the ectocervix. This epithelium is similar to that which lines the vagina (*Figure 8.4*) and is continuous with it. It differs from the non-keratinized squamous epithelium of the vagina in two ways: the cervical epithelial cells are less sensitive to ovarian hormones than the vagina and the superficial layers contain less glycogen.

Although the squamous epithelium of the cervix is less susceptible to the influence of the ovarian hormones than that of the vagina, changes in the thickness of the epithelium can be recognized during the menstrual cycle and in the postmenopausal period. Thus during childhood and at the menopause the epithelium is thin and atrophic (*Figure 8.6(a)*) and consequently susceptible to infection.

Under oestrogen stimulation the epithelium proliferates to its full thickness (*Figure 8.6(b)*). The epithelium can then be seen to consist of four zones—a basal zone, a suprabasal zone, a mid zone and a superficial zone. The cells in the basal layer are undifferentiated and have the potential for mitotic division. The cells in the suprabasal layers show features of differentiation such as intercellular bridge formation, and contain glycogen granules. As they mature the amount of cytoplasm increases and nuclei degenerate. The superficial zone in a fully mature epithelium is composed of cells with abundant cytoplasm and small pyknotic nuclei. Under the influence of oestrogen, the cytoplasm of the cells in these superficial layers often contains a keratin precursor.

When the influence of oestrogen is only slight, or there are high levels of circulating progesterone, the squamous epithelium of the cervix appears less than fully mature, i.e. the nuclei of the surface cells are vesicular and the cytoplasm is devoid of prekeratin (*Figure 8.6(c)*). Occasionally intense irritation of the cervical epithelium, such as that caused by a ring pessary, may lead to keratinization of the surface epithelium, a condition described clinically as leukoplakia (*see* Chapter 9).

The transition between the squamous epi-

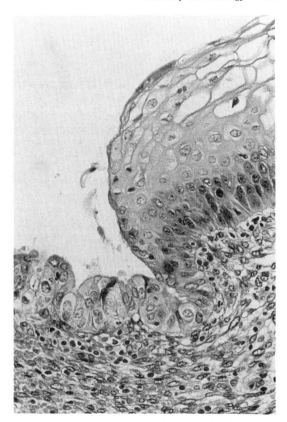

Figure 8.7 Squamocolumnar junction. Stain: haematoxylin and eosin.

thelium of the ectocervix and columnar epithelium lining the endocervical canal is known as the *squamocolumnar junction* (*Figure 8.7*). The exact location of the squamocolumnar junction varies throughout life (*Figure 8.8*). Before puberty, it is usually located within the endocervical canal. With the onset of menstruation or at the time of the first pregnancy, increase in volume of the cervix leads to eversion of the columnar epithelium (ectopy) so that the squamocolumnar junction is located distal to the external os (*Figure 8.8*). In the second and third decade of life the ectopic columnar epithelium undergoes *metaplastic change* to a non-keratinizing stratified squamous epithelium so that the position of the squamocolumnar junction changes again. The process of metaplastic change is a patchy one occurring at several different sites in the ectopic columnar epithelium at different times. Eventually a confluent layer of mature metaplastic squamous epithelium is formed creating a new squamocolumnar junction at or near the external os. As the menopause approaches the cervix atrophies

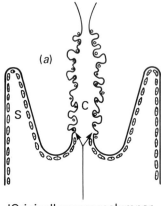
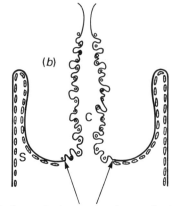

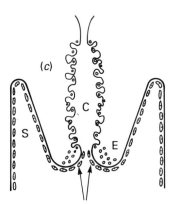
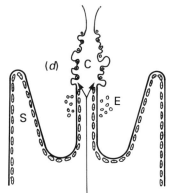

(a) 'Original' squamocolumnar junction

(b) Endocervical eversion (ectropion) with 'original' squamocolumnar junction

(c) Transformation zone with 'new' squamocolumnar junction

(d) Squamocolumnar junction within endocervical canal after menopause

Figure 8.8 Changing positions of squamocolumnar function throughout life. (*a*) Before puberty; (*b*) puberty and first pregnancy; (*c*) during reproductive life; (*d*) after the menopause

and the location of the squamocolumnar junction changes yet again as it recedes into the endocervical canal.

Three stages to the metaplastic process have been described, namely:

(1) Reserve cell hyperplasia (*Figure 8.9*)
(2) Immature squamous metaplasia (*Figure 8.10*)
(3) Mature squamous metaplasia (*Figure 8.11*)

The pathophysiological basis and the histological and cytological presentation of these stages are discussed in detail in Chapter 10. Current concepts of the dynamics of the cervical epithelium are the result of a synthesis of colposcopical, histological, and cytological observations recorded over the last 30 years. Pioneers in this field include Meyer (1910), Fluhmann (1961) and Reid (1964). Metaplastic change is currently conceived of as a protective mechanism whereby the fragile monolayer of columnar cells on the ectocervix is replaced by a protective multilayered squamous epithelium. It should be considered a normal physiological process. The stimulus for metaplastic change is believed to be the acid pH of the vagina although chronic infection or chronic irritation can evoke a similar response.

The epithelium in the region of the external os which is susceptible to metaplastic change is frequently described as the *transformation zone*.

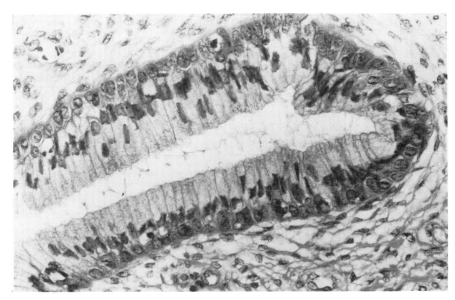

Figure 8.9. Reserve cell hyperplasia in a cervical gland. Note proliferation of reserve cells to produce a double layer of cells. The reserve cells lie deep to the original columnar cells. Stain: haematoxylin and eosin. Magnification ×770

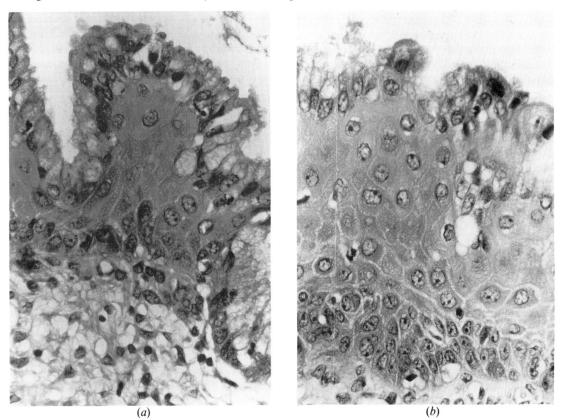

(a) (b)

Figure 8.10 Immature metaplasia. The continued proliferation of the reserve cells results in a multilayered epithelium. In (a) the original glandular epithelial cells lining the cervix can still be identified. In (b) the metaplastic change is slightly more advanced and there is a tendency for the cells to differentiate and form intercellular bridges. Stain: haematoxylin and eosin. Magnification ×770

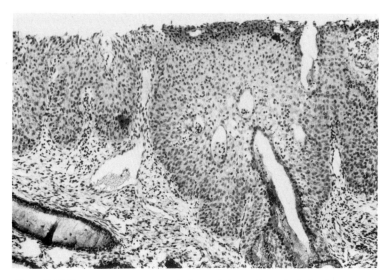

Figure 8.11 Mature squamous metaplasia overlying endocervical glands

It is a region which has a special significance for the pathologist as it is in this area that neoplastic change occurs and precancerous and cancerous lesions of the cervix arise. For this reason a cervical smear taken from a woman of childbearing age should always contain a sample of cells from this area.

In its earliest stages, metaplastic change can be recognized in histological section by proliferation of the reserve cells to produce two layers of cells in the ectopic columnar epithelium (*Figure 8.9*). Further division of the reserve cells results in replacement of the columnar cell monolayer by a multilayered epithelium composed of undifferentiated epithelial cells (*Figure 8.10(a)*). At this stage the epithelium is described as showing 'immature metaplastic changes' and the cells from the surface of this type of epithelium are readily recognizable in cervical smears. As the metaplastic epithelium matures it assumes some of the characteristics of stratified squamous epithelium, e.g. intracellular bridges are formed (*Figure 8.10(b)*). As maturation proceeds, the epithelium resembles the original squamous epithelium of the cervix (*Figure 8.6b* or *8.6c*). Thus mature metaplastic cells cannot be distinguished from native cells from the original squamous epithelium in a cervical smear. However, in histological section mature metaplastic epithelium can be recognized by the fact that the underlying connective tissue may contain endocervical glands (*Figure 8.11*).

Corpus uterus

The uterus is a pear-shaped organ approximately 7 cm long, of which the body forms the upper two-thirds. The wall is 1–2 cm thick and is composed of three layers: an outer layer of peritoneum (one cell thick), a thick coat of smooth muscle (the myometrium), and an inner secretory layer (the endometrium) 1–5 mm thick.

The uterine walls are in contact except for a small triangular cavity which is in continuity with the endocervical canal below and the fallopian tubes above.

The endometrium has a surface layer of epithelium from which tubular glands extend into the underlying connective tissue (endometrial stroma) (*Figure 8.12*). It consists of two layers: a basal layer next to the myometrium and a superficial functional layer. The functional layer undergoes cyclic changes in response to the ovarian hormones and is shed at menstruation. The basal layer does not show cyclic changes and remains as the regenerative layer at the end of each cycle.

The menstrual cycle is normally taken as 28 days but it may be shorter or longer than this. The day when the menstrual flow begins is counted as the first day of the cycle and bleeding usually lasts four or five days. The proliferative phase lasts from menstruation until ovulation occurs, about the fourteenth day of the cycle: under the influence of oestrogen secreted by a

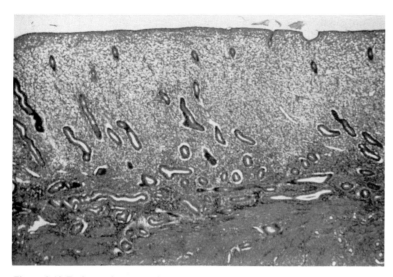

Figure 8.12 Endometrium: proliferative phase showing tubular glands
Stain: haematoxylin and eosin. Magnification ×330, reduced to 90% in reproduction

maturing follicle in the ovary the endometrium grows from 1 mm to about 3 mm thick. At ovulation the mature follicle releases an ovum and becomes a corpus luteum, which secretes both oestrogen and progesterone. The progesterone produces secretory changes in the endometrium, with a further increase in thickness up to about 5 mm, and this secretory phase lasts about 14 days. At the end of this time there is an abrupt fall in hormone production by the corpus luteum, and vasoconstriction occurs causing ischaemic degeneration of the functional layer of endometrium, which is shed in small fragments during menstruation.

The histological appearance of the endometrium varies with the reproductive cycle. Before the menarche and after the menopause the endometrium is thin (about 1 mm) and lacks a functional layer (*Figure 8.13*); it shows neither mitotic nor secretory activity.

In the proliferative phase mitoses can be readily identified in the cells of the stroma and glands (which are lined by columnar cells); in the early part of this phase the glands are straight tubules (*Figure 8.12*) which develop increasing convolution as the phase progresses to late proliferative.

In the secretory phase secretory vacuoles first appear between the nucleus and the base of the cell (*Figure 8.14*) and this subnuclear vacuolation is characteristic of the early secretory phase. In the mid-secretory phase the secretory vacuoles accumulate between the nucleus and

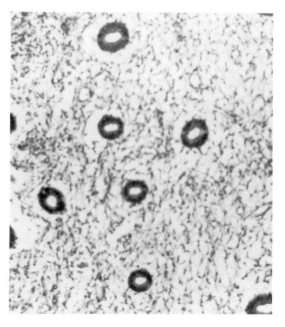

Figure 8.13 Atrophic endometrium in a postmenopausal woman showing scanty glands and stroma. Stain: haematoxylin and eosin. Magnification ×330

the free border of the cell, and increasingly secretion enters the lumen of the gland. Throughout the secretory phase the glands become increasingly tortuous.

In the late secretory phase (*Figure 8.15*) the distended glands develop a characteristic saw-toothed outline, and the epithelial cells develop ragged luminal borders. One or two days before menstruation the nuclei of the epithelial cells begin to fragment and the stroma shows enlargement and swelling (decidual change) of the cells around the arteries and prominence of stromal granulocytes. The stromal granulocytes have an irregular nucleus and large eosinophilic cytoplasmic granules, thus resembling neutrophils: they produce lytic enzymes and a polypeptide 'relaxin' which are thought to assist in the fragmentation of the endometrium at menstruation.

In the menstrual phase the functional layer of endometrium (*Figure 8.16*) shows degenerative changes with clumping and swelling of the stromal cells and a variable level of residual secretory changes in the glands. The basal endometrium proliferates in response to oestrogen in the proliferative phase, but develops little or no secretory response to progesterone in the secretory phase. The endometrial response to hormones is greatest in the upper part (fundus) of the uterus and diminishes towards the endocervical canal.

If pregnancy occurs the secretory phase continues with increasing thickness of the endometrium, elongation and convolution of the glands, and decidual changes in the stromal cells (*Figure 8.17*). These decidual changes also occur in stroma elsewhere in the genital tract, notably in the cervix and ovary. The epithelium of the glands commonly develops areas of 'hypersecretory' (Arias-Stella) change, which less often occurs in some late secretory phases in the non-pregnant uterus.

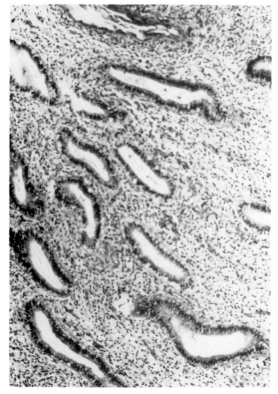

Figure 8.14 Endometrium: early secretory phase showing slightly convoluted glands. Note subnuclear vacuolation. Stain: haematoxylin and eosin. Magnification × 330

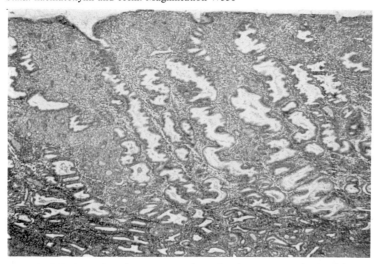

Figure 8.15 Endometrium: late secretory phase showing tortuous glands with sawtoothed appearance. Stain: haematoxylin and eosin. Magnification × 330, reduced to 90% in reproduction

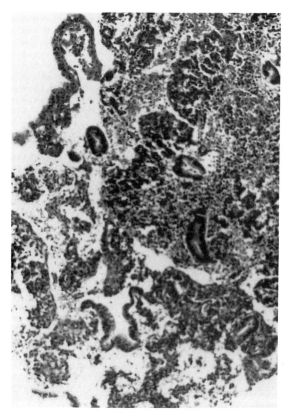

Figure 8.16 Degenerative changes in endometrium in menstrual phase of cycle. Note swelling of stromal cells and fragmentation of epithelium. Stain: haematoxylin and eosin. Magnification ×330

Fallopian tube (oviduct)

The fallopian tubes are each about 10 cm long and 0.6 cm in diameter. The wall has three layers: peritoneum on the outer surface, a thin layer of smooth muscle, and an inner layer of mucosa. The epithelium of the mucosa consists of a single layer of columnar cells, some of which are secretory and others ciliated, which rests on a thin connective tissue stroma.

At the medial end of the fallopian tube its lumen is in continuity with the uterine cavity. At the lateral end the wall forms multiple folds (fimbriae) which extend towards the ovary: the lateral end of the lumen opens into the peritoneal space close to the ovary.

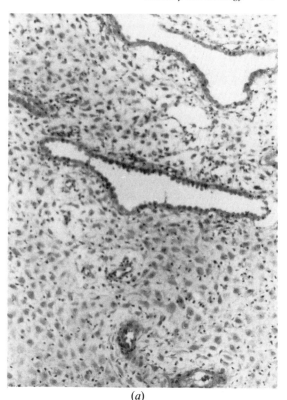

(a)

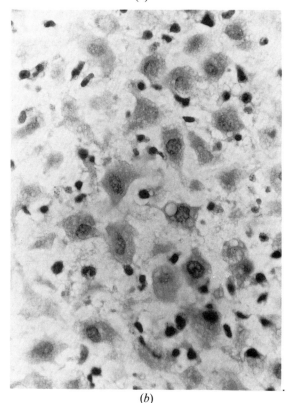

(b)

Figure 8.17 (a) Uterus in pregnancy showing decidual changes in stromal cells. Stain: haematoxylin and eosin. Magnification ×330; (b) stromal cells showing decidual changes. Stain: haematoxylin and eosin. Magnification ×770

Ovary

Situated on either side of the uterus, each ovary is about 3 × 1.5 × 1 cm. On the cut surface there is an outer cortex containing follicles (*Figure 8.18(a)*) and a central medulla with many blood vessels. The cortex is covered by a layer of flattened cuboidal cells which lies on a specialized connective tissue stroma. The entire complement of primordial follicles for an individual is formed before birth and many thousands are present in each ovary. Each primordial follicle contains an oocyte surrounded by a single layer of granulosa stromal cells (*Figure 8.18(b)*). From puberty one follicle develops to maturity each month and at ovulation liberates an ovum. The mature (Graafian) follicle is up to 1.5 cm in diameter and consists of a central ovum lying within a cavity of follicular fluid and surrounded by layers of specialized stromal cells (theca interna and theca externa). The ovum lies within a mound of granulosal cells which form the innermost layer. When the follicle ruptures it releases the ovum into the peritoneal space near the fimbrial (lateral) end of the fallopian tube. The follicle then collapses, but the cells in its wall become considerably enlarged by accumulation of lipid in their cytoplasm, forming the corpus luteum (*Figure 8.19*) which is 1.5–2.5 cm in diameter. If fertilization occurs the corpus luteum increases to a diameter of about 3 cm and persists until the twelfth week of gestation, when its function of hormone production is taken over by the placenta. If pregnancy does not occur the corpus luteum continues to function for about 12 days and then begins to involute and is gradually replaced by fibrosis to form the corpus albicans.

Specimen collection

The collection of a cell sample and the preparation of the cervical smear is not usually in the hands of the cytologist. However, it is essential that he or she is aware of the advantages and limitations of the different methods of specimen collection so that advice can be given in the event of the sample being improperly taken. The most efficient method of sampling the cervix is the cervical scrape, which is taken with a wooden (*Figure 8.20*) or plastic (*Figure 8.21*) spatula which has been specially designed for the purpose. Other methods include aspiration

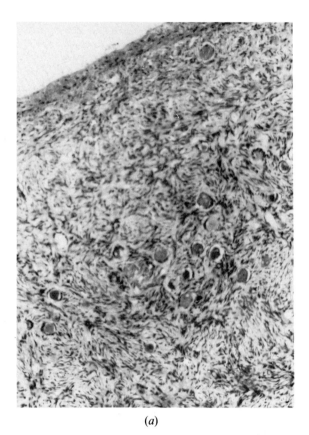

(*a*)

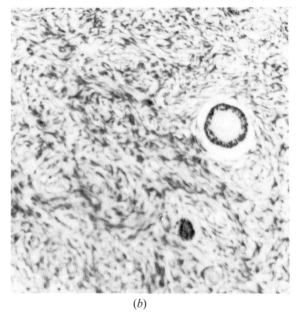

(*b*)

Figure 8.18 (*a*) Cortical layer of ovary containing numerous follicles. Stain: haematoxylin and eosin. Magnification ×330; (*b*) a maturing follicle. Stain: haematoxylin and eosin. Magnification ×770

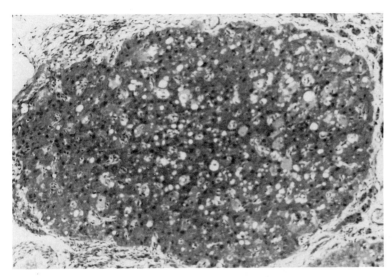

Figure 8.19 Corpus luteum composed of lipid-containing cells. Stain: haematoxylin and eosin. Magnification ×330, reduced to 90% in reproduction

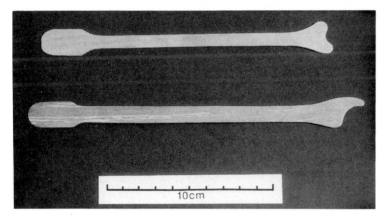

Figure 8.20 Cervical spatulas: Ayre spatula, Aylesbury spatula. These are probably the most commonly used spatulas in the UK

from the posterior fornix or endocervical brush specimens. A vault smear or a vaginal smear may be taken under special circumstances. An excellent illustrated account of cervical sampling has been prepared by Macgregor (1981) for distribution to doctors and nurses involved in smear taking.

Cervical scrape

These specimens are taken using an Ayre spatula (Ayre, 1944) or a modification of it. The cervix must be clearly visualized and the pointed end of the spatula inserted into the cervical os and rotated through 360°. The cervical mucus and cells adhering to the spatula are spread evenly across a glass slide which has been previously labelled in pencil or diamond pen with the patient's name (*Figure 8.22*). The slide must be fixed immediately by immersion in 95% ethanol. It must be left in the fixative for a minimum of 30 min after which it may be safely removed and stored dry at room temperature. Fixatives such as spray fixative or carbo wax fixatives described in Chapter 4 are satisfactory alternatives to immersion in alcohol. On no account should the smear be allowed to dry before fixation.

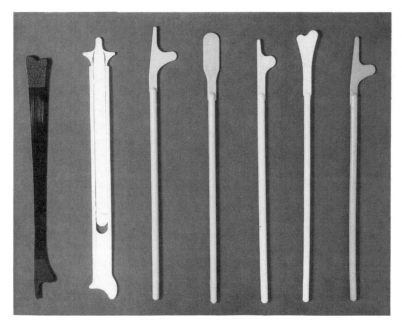

Figure 8.21 A variety of spatulas used for taking cervical smears. The large number of designs suggests that none is ideal for this purpose

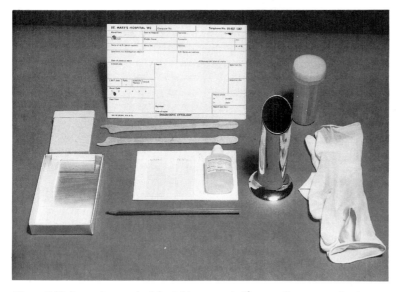

Figure 8.22 Apparatus required for taking a cervical smear. Smear tray should contain spatulas, fixative, glass slides, pencil, slide transport boxes and request form. A vaginal speculum and disposable gloves can also be included in the tray

Posterior fornix aspirate

Papanicolaou and Traut (1943) used posterior fornix aspirates as a source of material for cytological study in their classic report on the diagnosis of uterine cancer by the vaginal smear. The aspirates which were taken with a pipette, contained cells exfoliated from the vagina, cervix and body of the uterus and were considered to be an effective method of detecting tumours at all three sites. Comparative studies (Macgregor, Fraser and Mann, 1966) showed that a posterior fornix aspirate is less efficient than a cervical scrape for the detection

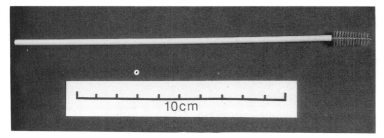

Figure 8.23 Endocervical brush used for sampling the endocervix (Medscand, Sweden)

of cervical neoplasia and less efficient than direct endometrial aspirations for the detection of endometrial cancer. This method of sampling is now largely abandoned for the purposes of cervical cancer screening and should be reserved for those cases where the cervix cannot be visualized. Thus it has a place in the investigation of women with a very narrow introitus, e.g. due to scarring or atrophy. The vaginal aspirate has in the past been advocated as a method of 'self testing' but women should be discouraged from using it in this way due to the low sensitivity of the test.

Endocervical brush specimen

The squamocolumnar junction is frequently located within the endocervical canal after the menopause and after cone biopsy and a routine cervical scrape taken from the external os may fail to detect a small focus of neoplastic change in these cases. Thus it may be appropriate to routinely sample the canal in older women and in women who have been treated for cervical intraepithelial neoplasia (CIN). A nylon brush (*Figure 8.23*) (Medscand, Malmo, Sweden) has been developed for this purpose although a spatula with a long pointed end (*Figures 8.20, 8.21*) such as that designed by Wolfendale *et al.* (1987) or Pistofidis *et al.* (1984) or even a cervical swab can be used for this purpose. Endocervical samples may also be helpful whenever the endocervical canal cannot be clearly visualized at colposcopy, particularly to exclude the skip lesions of endocervical adenocarcinoma. When endocervical and ectocervical samples are combined, 98% of all neoplastic lesions of the cervix can be detected (Wilbanks *et al.*, 1968).

Vault smears

One of the key uses of cytology is to monitor the progress of the patient who has been treated for carcinoma of the cervix by amputation of the cervix or hysterectomy. In these cases a smear from the vaginal vault may detect residual carcinoma or provide evidence of recurrence. The rounded end of an Ayre spatula can be used for taking the sample.

Scrape from upper third of vaginal wall

One of the earliest uses of exfoliative cytology was for the investigation of the effect of oestrogen and progesterone on the epithelium of the vagina. Much of the early work in this field was carried out on posterior fornix aspirates but later studies showed that a scrape from the lateral wall of the upper third of the vagina was more reliable. The rounded end of the spatula can be used for taking the smear and a rapid staining method for demonstrating the proportion of superficial and intermediate squamous cells in the smear has been developed (Schorr, 1941).

Several methods of estimating the effect of oestrogen have been described. These include the karyopyknotic index, maturation index, etc. These tests are of limited value as the response of the vaginal epithelium to oestrogen stimulation is slow and the full effect of the hormone on the epithelium may only be evident after 36 h. The tests have largely been superseded by biochemical assays of urine or blood. However, where facilities for these more sophisticated tests are not available, cytological studies can be used to investigate primary or secondary amenorrhoea and infertility. In these circumstances it is advisable to obtain sequential smears rather than report on a single smear. A report on the hormonal status of the patient should not be given in the presence of vaginal infection (*see* also p. 164).

The normal cervical smear

It is often said that cervical cytology is easy and that a technologist or a pathologist can become competent to screen in a matter of weeks. Nothing could be further from the truth. Although the basic principles of cervical screening may be acquired in a few weeks, the smear patterns reflecting the cytological presentations of inflammatory, regenerative, reactive and neoplastic change are so many and varied that a screener needs to examine several thousand smears in order to acquire the basic knowledge that would make him or her a safe screener.

Moreover, pattern recognition is only one of the skills that a cytologist needs. He or she needs to be capable of intense concentration as the slide is scanned frame by frame, otherwise a single abnormal cell present in the smear may be missed. He/she also needs enough clinical knowledge to be able to interpret the cytological findings in the light of the clinical data provided. For example, an atrophic smear pattern in a woman of reproductive age may indicate a failure of ovulation, a postpartum smear or even an error of labelling! Finally, he or she has to have a sound knowledge of the histology of the cervix in health and disease to be able to build up a picture of the cervical epithelium from the appearance of a few surface cells in the smear, so that a meaningful report can be given to the clinician.

The content of a normal smear

A cervical smear taken with an Ayre spatula contains exfoliated cells which have been trapped in the cervical mucus together with cells that have been forcibly detached from the cervical mucosa. The content of the normal smear can be categorized as follows:

(1) Surface epithelial cells from the original squamous epithelium of the cervix and vagina.
(2) Columnar epithelial cells from the endocervical canal.
(3) Mature and immature metaplastic cells from the transformation zone.
(4) Cells from the endometrial lining and stroma.
(5) Leucocytes and erythrocytes.
(6) Commensal organisms.
(7) Mucous strands.
(8) Contaminants, e.g. spermatozoa.

Cells shed from original squamous epithelium

Three types of cells may be shed from the surface of the original squamous epithelium of the cervix and vagina. These have been designated superficial cells, intermediate cells and parabasal cells according to a convention recommended by the International Academy of Cytology in 1958 and endorsed by the World Health Organization in their atlas on the *Cytology of The Female Genital Tract* (Riotton and Christopherson, 1973). The different cell types reflect the response of the epithelium to ovarian hormones at the time the smear was taken.

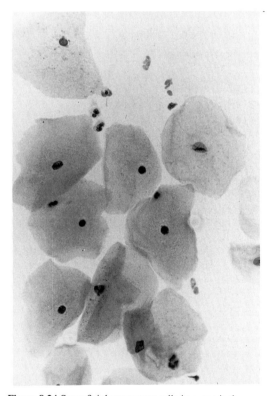

Figure 8.24 Superficial squamous cells in a cervical smear. Note abundant translucent cytoplasm and pyknotic nuclei. Stain: Papanicolaou. Magnification × 330

Superficial cells

These are shed from the surface of fully mature squamous epithelium which has developed to its full thickness under the influence of oestrogen. They appear as large polygonal squames with transparent cytoplasm and angular borders measuring 45–50 µm in diameter. They are usually found in large sheets and stain a delicate pink with the Orange G in the Papanicolaou stain. Scanning electron microscopy (Ferenczy and Richart, 1973) reveals that the cells form a mosaic pattern with the surface of each cell being covered by a network of microridges which have a protective function. Transmission electron microscopy reveals that occasional membrane-bound keratinosomes can be found in the cytoplasm on transmission electron microscopy but extensive development of keratin is not a feature of the normal cervical epithelium (Hafez, 1982). Superficial cells have small structureless pyknotic nuclei 5 μm or less in diameter. Karyorrhexis is sometimes seen.

distinguished from them by the structure of the nucleus which is vesicular and by the less angular appearance of the cytoplasm (*Figures 8.25 and 8.26*). They usually assume an azure hue with the Papanicolaou stain but it must be remembered that the staining pattern is only a rough guide to cell type and cell morphology is of paramount importance when classifying cells in cervical smears. In pregnancy, intermediate cells appear elongated and boat shaped and are termed *navicular cells* (*Figure 8.27*). The cytoplasmic borders are thickened and the vesicular nucleus is eccentrically placed. Electron microscopy reveals that these cells contain an abundance of glycogen granules in the cytoplasm which may account for the change in shape.

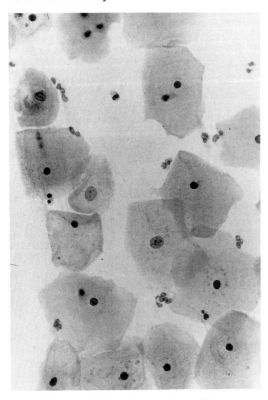

Figure 8.25 Superficial and intermediate cells in smear. Note vesicular nucleus of intermediate cell in centre of field. Stain: Papanicolaou. Magnification × 330

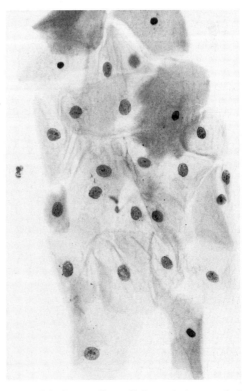

Figure 8.26 Intermediate cells in cervical smear. Stain: Papanicolaou. Magnification × 330

Intermediate cells

These are shed from the surface of semimature squamous epithelium which is showing a diminished response to oestrogen or the effect of progesterone. These flat squames are only slightly smaller than superficial cells but can be

Parabasal cells

These are much smaller than superficial and intermediate cells, measuring 15–30 μm in diameter, and have a granular nucleus which may contain an occasional chromocentre or small nucleolus and occupies no more than a third of

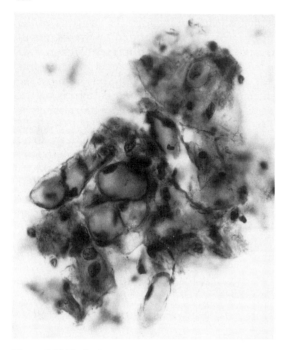

Figure 8.27 Navicular cells in cervical smear in pregnancy. Note boat-shaped cells with thickened borders and eccentric nuclei. Stain: Papanicolaou. Magnification × 330

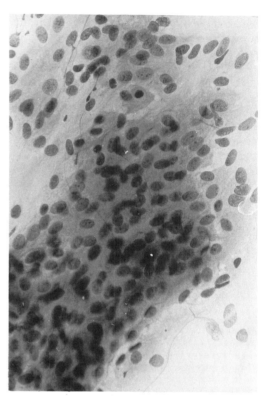

Figure 8.29 Sheets of parabasal cells in atrophic smear. The delicate cytoplasm of some cells has degenerated leaving some free nuclei in the smear. Stain: Papanicolaou. Magnification × 330

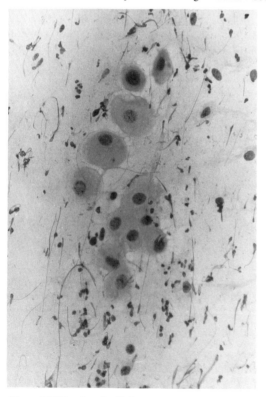

Figure 8.28 Parabasal cells in an atrophic smear. Note mucin strands. Stain: Papanicolaou. Magnification × 330

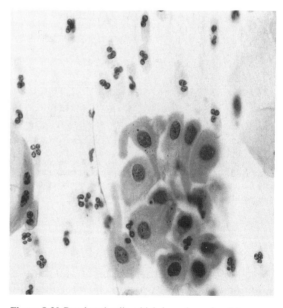

Figure 8.30 Parabasal cells which have been forcibly detached. Note long cellular processes. Stain: Papanicolaou. Magnification × 330

the cell volume (*Figure 8.28*). The cytoplasm is usually cyanophilic, although amphophilic or strong eosinophilic staining is not uncommon. The nucleus occupies one third of the area of the cell. The cells may be found singly or in sheets. Single cells appear rounded when they have been shed spontaneously (*Figure 8.28*). The cytoplasm of parabasal cells is so fragile that it may fragment, leaving only free nuclei in the smear (*Figure 8.29*). When parabasal cells have been forcibly detached they may have long cytoplasmic processing (*Figure 8.30*).

Parabasal cells are found in smears from prepubertal females and postmenopausal women where the cervical epithelium is atrophic (*Figure 8.6(a)*). They are also found in smears taken from an area of immature squamous metaplasia (*Figure 8.10(b)*). Although the morphology of the parabasal cells in the smear is very similar in both these situations the overall smear patterns are quite distinct and are discussed in a subsequent section.

Cells shed from the endocervical canal

These cells may appear singly or in sheets (*Figures 8.31* and *8.32*). Individual cells are readily recognized by their columnar shape, vacuolated cytoplasm and basal nucleus (*Figure 8.31*). Sheets of cells form a palisade (*Figure 8.31*) or honeycomb pattern (*Figure 8.32*).

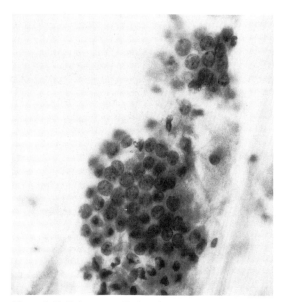

Figure 8.32 Columnar cells forming honeycomb pattern. Stain: Papanicolaou. Magnification × 330

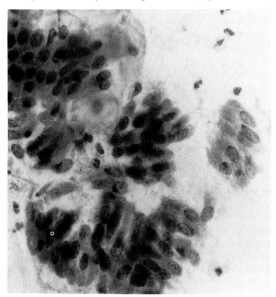

Figure 8.31 Sheets of columnar cells derived from the endocervical epithelium. Stain: Papanicolaou. Magnification × 330

Endocervical cells are very delicate and their morphology is often poorly preserved so that the outline of the nucleus and cytoplasm may be indistinct. Cilia are rarely seen. Occasionally, free nuclei or fragments of cytoplasm are all that remain of these cells. An unusual finding is ciliated fragments of cytoplasm in the smears which may reflect apoptosis of the endocervical epithelium.

Distinction between an atrophic smear pattern and a smear taken from an area of immature metaplasia

A smear taken from an atrophic cervix will present a uniform appearance and be composed entirely of parabasal cells (*Figure 8.33(a)*). Because of the delicate nature of the atrophic cervical epithelium the parabasal cells show degenerative changes and may be devoid of cytoplasm or exhibit nuclear pyknosis or karyorrhexis. Leucocytes and mucus strands are also commonly present in the smear due to a low grade cervicitis.

In contrast, a smear from the transformation zone will contain several different types of squamous cells. Parabasal cells from an area of immature metaplasia as well as superficial and intermediate cells from an area of mature metaplastic change will be present, reflecting the patchy process of metaplastic change (*Figure 8.33(b)*). Superficial and intermediate squames

156 Female genital tract: normal cytology

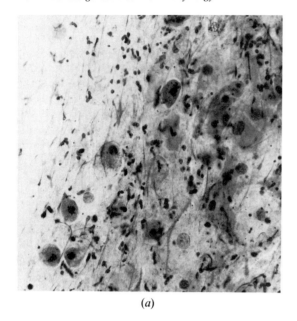

(a)

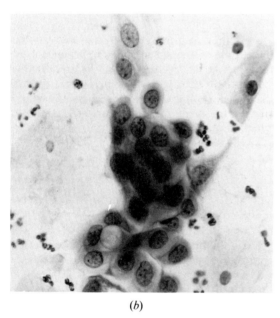

(b)

Figure 8.33 (a) Smear from an atrophic cervix composed entirely of parabasal cells showing degenerative changes (e.g. pyknosis, karyorrhexis). There are numerous leucocytes and mucous strands indicating the presence of atrophic cervicitis. No superficial squamous cells are seen. (b) Parabasal cells and superficial cells in same smear indicating that the smear has been taken from an area of squamous metaplasia. Contrast this smear pattern with those shown in *Figures 8.28, 8.29,* and *8.33* (a). These smears were taken from atrophic cervices and are composed entirely of parabasal cells. Stain: Papanicolaou. Magnification × 330

shed from an area of mature squamous epithelium cannot be distinguished from superficial and intermediate cells shed from original squamous epithelium. A full description of the cells shed from the transformation zone is given in Chapter 10.

Cells shed from other parts of the genital tract

Cells of endometrial origin (*Figure 8.34*) are frequently found in smears taken at or near the time of menstruation. They are also found in smears from women fitted with an IUCD and in smears from women with intermenstrual or postmenstrual bleeding. A full description of exfoliated endometrial cells is given in Chapter 11. They can be distinguished from endocervical cells by their small size (5–20 nm) and their coarse chromatin structure. They may be shed singly (*Figure 8.34*) or in tightly packed berry-like clusters, some of which are composed of a dense core of stromal cells and an outer pale rim of glandular cells. It must be remembered that

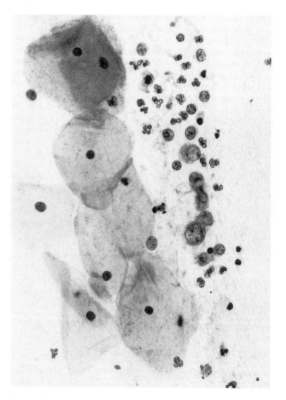

Figure 8.34 Discrete endometrial cells in cervical smear. Stain: Papanicolaou. Magnification × 330

the presentation of endometrial cells found in cervical smears differs from that of endometrial cells in endometrial aspirates. The former have exfoliated spontaneously and are often found in rounded clumps of poorly preserved cells, whereas the latter appear as sheets of well preserved glandular epithelial cells (*see* Chapter 11).

Commensal organisms

A variety of organisms have been isolated from the vagina in the absence of disease. The saprophytes most commonly found are lactobacilli (Doderlein bacilli) and *Gardnerella vaginalis* but streptococci, staphylococci, micrococci, non-pathogenic *Neisseria, Pseudomonas, Mycoplasma, Leptothrix* and *Candida* species have also been isolated.

Most of these organisms are also capable of causing local tissue damage and are discussed in Chapter 9 on pages 174–175. Lactobacilli are the exception in that they are always commensal. They appear in Papanicolaou smears as pale blue staining rods 1–2 µm in length (*Figure 8.35*). They are particularly abundant in smears taken during the secretory phase of the menstrual cycle or during pregnancy. They metabolize the glycogen in the intermediate cells which predominate in these smears, destroying the cytoplasm (cytolysis) and giving a ragged appearance to the cell. Free nuclei may abound in these smears.

Other components of the smear

Some leucocytes and erythrocytes are present in almost every smear. The leucocytes are present in cervical mucus and are deemed to reflect the immune response of the cervix to local factors. Only when their presence interferes with the interpretation of the smear do they render a smear unsuitable for cytology. A heavy exudate of polymorphs reflecting an acute inflammatory response (*Figure 8.36*) or a heavily bloodstained smear taken during menstruation may fall into this category (*Figure 8.37*). Postcoital smears may contain many spermatozoa (*Figure 8.38*) or cells from seminal fluid which are occasionally mistaken for dyskaryotic cells. Mucin strands are transparent with the Papanicolaou stain and rarely interfere with the interpretation of the smear. However, in atrophic smears dense inspissated masses of mucus may be found (blue blobs) (*Figure 8.39*) which may be misdiagnosed as malignant cells. Their amorphous structure should permit a correct diagnosis. Some authors consider these artefacts are derived from degenerating parabasal cells (Boon and Tabbers-Boumester, 1981).

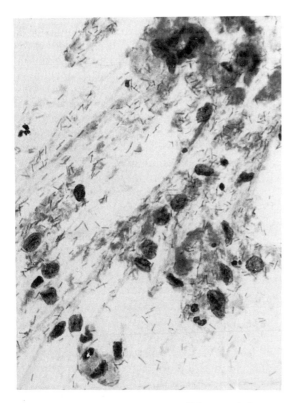

Figure 8.35 Numerous Doderlein bacilli in a cytolytic smear taken from a woman attending the antenatal clinic. Note ragged appearance of the intermediate cells and free nuclei. Such a smear may be unsuitable for reliable cytological assessment. Stain: Papanicolaou. Magnification ×330

Some common smear patterns

The epithelial cells found in smears taken from the external os in the absence of inflammatory or neoplastic disease are determined by two factors:

(1) The hormonal status of the woman at the time the smear is taken.
(2) The position of the squamocolumnar junction.

The smear pattern can generally be predicted provided these two factors are known.

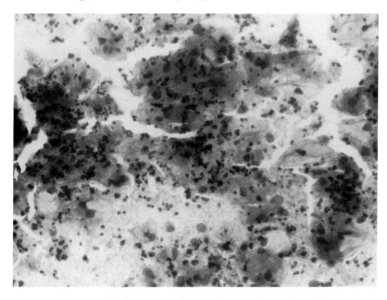

Figure 8.36 An unsatisfactory smear due to a heavy polymorph exudate obscuring the epithelial cells. Stain: Papanicolaou. Magnification × 330

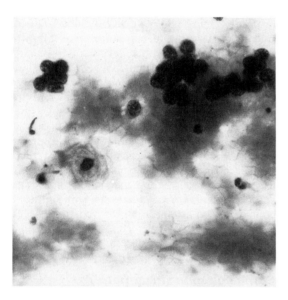

Figure 8.37 Heavily bloodstained menstrual smear. Unsuitable for reliable cytological assessment. Stain: Papanicolaou. Magnification × 330

Effect of hormonal status

As mentioned in the previous section, the squamous epithelium of the cervix and vagina is hormone-sensitive and responds to oestrogen by becoming thicker and by cell maturation. Thus a smear taken mid-cycle will be composed almost entirely of fully mature squamous cells (superficial cells) (*Figure 8.24*). In the early proliferative stage of the menstrual cycle and in the secretory phase the squamous epithelium is slightly less mature and intermediate cells dominate the smear (*Figure 8.26*). In the absence of oestrogen, e.g. in the prepubertal female, postpartum and after the menopause, the squamous epithelium of the cervix is thin and fragile and parabasal cells predominate (*Figures 8.28, 8.29*). These smears often contain numerous polymorphs and the parabasal cells show degenerative changes (*Figure 8.33(a)*). When interpreting a smear it should be remembered that the endogenous hormonal status of the woman is not the only factor to affect the smear pattern. Exogenous hormones in the form of oral contraceptives may also influence the appearance of the smear. Thus many smears from women of childbearing age reflect an intermediate degree of epithelial maturation. Local infection, particularly *Trichomonas vaginalis*, or chronic irritation such as that induced by a pessary, may induce maturation of the cervical epithelium which is reflected in the smear.

A smear taken at or near the time of menstruation will contain endometrial cells and histiocytes (*Figure 8.40*). Towards the end of menstrual flow a large streak of histiocytes with round oval or bean-shaped nuclei and foamy cytoplasm may be seen (*Figure 8.40*). This pattern

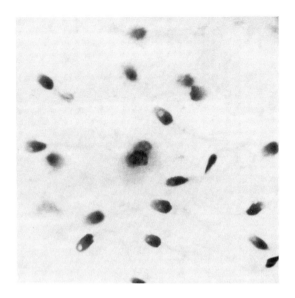

Figure 8.38 Spermatozoa in a cervical smear. Stain: Papanicolaou. Magnification ×770

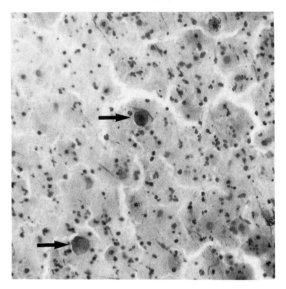

Figure 8.39 Inspissated masses of mucus (arrowed) in atrophic smear. They are sometimes mistaken for malignant cells. Stain: Papanicolaou. Magnification ×330

has been termed the 'exodus'. If the woman is menstruating at the time the smear is taken the smear will be so bloodstained that cytological analysis will not be possible (*Figure 8.37*). For this reason it is recommended that smears should be taken mid-cycle whenever possible.

The smear in pregnancy also has a characteristic pattern which reflects the effect of the gestational hormones on the cervical and vaginal squamous epithelium. The smear is composed of glycogen-filled intermediate cells which are usually present in large sheets or clusters. Navicular cells are commonly seen (*Figure 8.27*). Doderlein bacilli are numerous and cause extensive cytolysis, releasing the nuclei and disrupting the cytoplasm of the epithelial cells.

The free nuclei and cytoplasmic fragments may be mistaken for trichomonads, particularly karyorrhexis and pyknosis, reflecting the vulnerability of the atrophic epithelial lining to infection and trauma unless care is taken. However, it should be remembered that *Trichomonas vaginalis* rarely thrives in the acid environment which favours the growth of the Doderlein bacilli. Cytolysis can be so extensive that in some cases it may be impossible to determine cell morphology with the high degree of accuracy required for screening and abnormal cells may be missed. Such smears should be reported as unsatisfactory and a repeat smear requested mid-cycle.

Location of squamocolumnar junction

Fundamental research has shown that preinvasive and invasive squamous cancers of the uterine cervix arise within the transformation zone so that it is important that this area is sampled whenever a smear is taken. The cytologist can sometimes recognize smears taken from this area as they may contain immature metaplastic cells. The transformation zone is most readily sampled when it is located at or near the external os. If it is located high in the endocervical canal, e.g. in postmenopausal women, an endocervical brush specimen may be advisable.

It has been observed that columnar cells or immature metaplastic cells are rarely found in the cytolytical smears in pregnancy despite the fact that the squamocolumnar junction is readily accessible to the smear taker and a large ectopy may be present. It has been postulated that this may be due to the fact that the cervix is covered with a thick layer of viscid mucus which prevents the cells from being detached. This explanation is not entirely convincing.

Reporting of cervical smears

The preparation of a cytology report requires just as much care as the preparation of a surgical pathology report. It should be accurate and concise and contain all the information that

160 *Female genital tract: normal cytology*

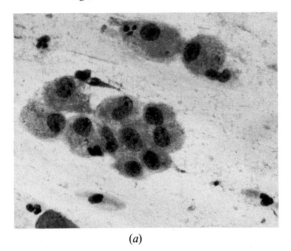

(a)

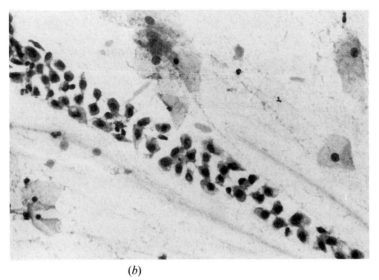

(b)

Figure 8.40 (a) A cluster of histiocytes showing round, oval or bean-shaped nuclei and foamy cytoplasm. Stain: Papanicolaou. Magnification ×770; (b) a streak of histiocytes in postmenstrual smear. Stain: Papanicolaou. Magnification ×330

the clinician needs for the correct management of the patient. The report should be written in narrative form and shorthand forms or classification systems such as that devised by Papanicolaou (1954) should be avoided. Guidelines for the construction of a report and recommended terminology have been published by the British Society for Clinical Cytology (Evans *et al.*, 1986).

Several important principles should be followed when composing a cytology report.

(1) The cytological findings must be described and the underlying histological changes predicted. Thus a normal smear predicts a normal cervix. A mild dyskaryosis in the smear is suggestive of CIN-1 (mild dysplasia) in the cervix. A description of the cytological changes in cervical smears and the underlying histological findings is given in *Table 8.1*.

(2) Every attempt should be made to predict the underlying histological pattern as accurately as possible. Where this cannot be achieved the fact should be clearly stated in the report. The use of the term 'atypia' is discouraged as this has come to mean anything from inflammatory change to

frank malignancy. The British Society for Clinical Cytology (Evans et al., 1986) has suggested the phrase 'bordering on mild dyskaryosis' be used to describe those smears where it is impossible to discriminate between inflammatory and neoplastic changes. Unfortunately, there will always be a small number of cases where the cytologist is unable to make a diagnosis from the smear. This may be due to inadequate sampling or an unusual cytological pattern or equivocal cytological changes. In these cases the cytologist should describe the appearance of the smear. A repeat sample should be requested only if it is thought this will further the diagnosis.

(3) The nomenclature used to describe the cytological findings must accurately reflect the smear pattern and must be distinct from that used to describe the tissue changes seen in histological section. The presence of abnormal cells in the smear which are suggestive of preinvasive or invasive carcinoma of the cervix are described as 'dyskaryotic' cells (Greek *dys* = abnormal; *karion* = nut or nucleus). The terms 'dysplastic cells' or 'malignant cells' should be avoided as they do not accurately describe the cytological findings. Dysplasia (abnormal growth of tissue) and malignancy (tissue invasion by tumour cells) can only be diagnosed from histological sections.

(4) The smear should be assessed at the outset for its cell content, staining properties and cytomorphology and in those cases where these are found to be inadequate, the smear should be rejected as unsuitable for reliable assessment.

(5) The terminology used and method of reporting should be consistent from laboratory to laboratory to permit comparative studies and quality control.

(6) The smear must always be interpreted in the light of the clinical data provided, otherwise errors of judgement may occur. For example, a well oestrogenized smear from a postmenopausal woman could reflect the presence of an oestrogen-producing tumour or hormone replacement therapy. In either case the proliferative pattern should be reported and a note made of the presence or absence of endometrial cells.

In our laboratory we strongly discourage the reporting of 'inflammatory changes' as this information is of no value to the clinician. However, findings of pathological significance which may affect patient management, e.g. specific infection, are included in the report. We also include advice on follow-up and referral of patients with abnormal smears although we recognize that the decision-making processes regarding patient management are the responsibility of the clinician who has taken the smear. Nevertheless our clinicians have indicated that they find our guidelines quite helpful. Consequently we have outlined our current recommendations for management in *Table 8.2*. Management schemes are discussed in more detail in Chapter 10 and readers are also referred to the document prepared by the British Society for Cytology on the subject (Evans et al., 1987). All schemes must be sufficiently flexible to allow for the special requirements of individual patients. They may need updating as new information about the pathobiology of cervical cancer becomes available.

Table 8.1 A description of the cytological changes in cervical smears and predicted histological findings.

Cytology report	Predicted histological findings
Smear unsuitable for cytological assessment	
Findings essentially normal	Normal cervix. No evidence of malignancy
Borderline nuclear changes	Probably inflammatory changes not amounting to neoplasia
Mild, moderate dyskaryosis, suggestive of CIN I or II	CIN I or II
Severe dyskaryosis suggestive of CIN III or invasive squamous carcinoma	CIN III or microinvasive or invasive squamous carcinoma
Abnormal glandular cells present	Endocervical adenocarcinoma, endometrial hyperplasia or adenocarcinoma
Abnormal cells suggestive of other tumour types	Other tumour types

N.B. Other findings of pathological significance e.g. specific vaginal infection, endometrial cells in smears from postmenopausal women, or hormonal effect inconsistent with clinical condition of patient should also be described.

Table 8.2 Recommendations for management and follow-up of women with abnormal cervical smears

Cytology	Management
Smear unsuitable for reliable cytological assessment	Repeat smear soon
Findings normal	Routine recall according to local well women screening policy
Mild dyskaryosis; borderline nuclear changes	Repeat in six months then manage according to results. Persisting abnormalities are an indication for colposcopy
Moderate dyskaryosis	Repeat in three months. Persisting abnormalities are an indication for colposcopy
Severe dyskaryosis	Refer for colposcopy

N.B. (1) All dyskaryotic smears require careful follow-up, even after treatment or apparent regression of cervical lesion.
(2) If there is a clinical suspicion of malignancy the patient should be referred for biopsy even though the smear is negative and there is cytological evidence of malignancy.
(3) The presence of wart virus changes should not influence the management of the patient.
(4) Colposcopy should not be delayed where a mild or moderate dyskaryosis is detected in women aged 35 or over.

Validity of smear report

The accuracy of a cervical smear can be expressed in terms of the sensitivity and specificity of the test and its predictive value. The sensitivity of the test can be calculated from the formula:

$$\frac{\text{Number of cases screened positive correctly}}{\text{Total with cervical neoplasia (true positives)}}$$

The specificity of the test can be calculated as follows:

$$\frac{\text{Number of cases screened negative correctly}}{\text{Total without the disease (true negatives)}}$$

The false negative and false positive values can be calculated from these data.

The predictive value expresses the frequency with which the result of a test is confirmed by an acceptable diagnostic procedure. It can be calculated as follows:

$$\frac{\text{Number of cases screened positive correctly}}{\text{Number of cases reported positive by cytology}}$$

In defining the accuracy it is often difficult to establish the baselines as the interpretation of a smear is a very subjective exercise and even the 'gold standard' provided by histology is not completely free from error. Moreover, there is no clear definition of what constitutes a positive test. Should a positive test include borderline changes, and all grades of CIN, or only CIN III? Small wonder then that the degree of confidence that can be placed in a smear report has been a subject for debate for many years.

Despite these problems of definition there are many studies which show that it is relatively rare for an experienced cytologist to issue a false positive report indicative of CIN III or invasive squamous carcinoma and the specificity of cervical cytology is very high. This has been demonstrated by careful correlation of cytological and histological findings. Reagan and Hamonic (1958) were able to demonstrate a focus of neoplasia in 100% of cone biopsy specimens from women who were reported as having an abnormal cervical smear. In order to prove the point, however, it was necessary in a number of cases to cut additional sections from the blocks. Unfortunately, correlation between colposcopic punch biopsies and cytology is much less satisfactory, largely because the biopsies are small and the lesions difficult to target. It is the practice at our hospital to proceed to cone biopsy in cases where the punch biopsy is negative and the smear is abnormal. The recent trend toward laser excision biopsy (Wright, Davies and Riopelle, 1984) as a method of diagnosing and removing a focus of CIN in a single treatment is welcome, as it permits much better cytological/colposcopic correlation than can be achieved by colposcopic punch biopsy.

Although the specificity of cervical cytology is very high, the sensitivity of the technique has been estimated to lie between 75 and 95%. These estimates are based on reviews of many thousands of smears from women who were found to have an abnormal smear within 12 months of a previous negative smear (Coppleson and Brown, 1974; Gay, Donaldson and Goellner, 1985; Van der Graaf and Vooijs, 1987). There are many reasons for false negative reporting:

(1) The smear was not taken from the squamocolumnar junction.
(2) Tumour cells occasionally fail to exfoliate.
(3) The smear may have been unsuitable for diagnosis.
(4) There may have been an error of screening.

Of particular concern is the number of false negative reports that are issued in cases of invasive cervical cancer (Figge, Bennington and Schweid, 1970; Rylander, 1977; Walker, Hare and Cooper, 1983; Paterson, Peel and Vosun, 1984). Several studies have shown that a smear from a fungating or ulcerating cancerous lesion may contain only blood, leucocytes and necrotic debris which obscure many of the epithelial cells in the smear. Thus cytology is not a satisfactory method for diagnosing frank malignant disease, and whenever malignancy is suspected clinically the patient should be referred for colposcopy regardless of the cervical smear report.

The unsatisfactory smear

One of the limitations of cervical cytology is that a proportion of smears submitted for analysis in the light microscope are unsuitable for reliable cytological assessment. Generally, the number of cases that fall into this category are low. If more than 10% of the laboratory workload are rejected as unsuitable there is cause for concern. The decision as to whether a smear is suitable for analysis is made by the cytologist and is largely a subjective one. However, some guidelines can be given.

Smears are considered inadequate if they contain too few epithelial cells. Cytology can be regarded as a cell sampling technique so that the more epithelial cells present in the smear the greater the chances of detecting an abnormality if it is present, always assuming the sample is correctly taken from the transformation zone. There is often controversy about the significance of endocervical cells in smears. It is assumed that their presence in a smear is an indication that the smear has been taken from the squamocolumnar junction. This is not necessarily the case as the endocervical cells may have been shed from high in the endocervical canal and transferred to the smear in the cervical mucus.

Vooijs et al. (1985) studied endocervical cells as a parameter of smear quality and showed that a significantly higher number of epithelial abnormalities can be detected in smears containing endocervical cells than in smears without them. In contrast, Beilby et al. (1983) found no difference between the two groups. In view of the conflicting data, it would be unwise at present to rely on endocervical cells as a parameter of smear quality. A more useful approach is to ensure that smear taking techniques are optimized so that the transformation zone is sampled in every case. This can be achieved by using the pointed spatulae or endocervical brush shown in *Figures 8.20, 8.21* and *8.23*. There are those who would advocate two smears—an Ayre scrape and an endocervical brush specimen—in every case but this is not practical in a national screening programme where cost effectiveness is important.

The presence of excesssive blood or pus in the smear which obscures the epithelial cells renders it unsuitable for analysis as does the marked cytolysis associated with pregnancy (*Figures 8.35–8.37*). Air drying impairs the quality of the stain so that cell morphology is poorly preserved and such smears should also be rejected. It is advisable when reporting a smear as unsuitable for cytology to give the reason in the report so that the clinician can improve his performance next time.

Uses of the cervical smear test

The acceptability of the Papanicolaou smear test as a method of detecting preinvasive and invasive lesions of the uterine cervix has given it an important role in modern gynaecological practice. It can be used as a diagnostic test for cervical cancer in those women who have clinical signs or symptoms such as intermenstrual, postcoital or postmenopausal bleeding which are suggestive of malignant disease. However, it is important to remember that a negative smear test should be ignored if the suspicion of cervical cancer is strong, as a false negative cytology report may be issued in cases of fungating or necrotic ulcerating lesions (*see* also p. 214).

The test also has an important role to play in the follow-up of women who have been treated for invasive cervical cancer by hysterectomy or radiotherapy and for those treated for pre-invasive cancer by local ablative therapy or cone biopsy. Regular screening after treatement can provide the first evidence of recurrent disease or signs of residual tumour. Problems may arise in the interpretation of smears after therapy and these are discussed in Chapter 24. Occasionally cone biopsy may result in a stenosed cervical os and recurrence of tumour may not be detected in a routine smear taken with an Ayre spatula. An endocervical brush specimen (*Figure 8.23*) may be advisable in these cases.

In the majority of cytology laboratories, however, the major source of smears submitted for microscope analysis is from well women who are participating in regional or national screening programmes for the prevention of cervical cancer. The aim of these programmes is to reduce morbidity and mortality from invasive squamous carcinoma of the cervix by detecting and treating the preinvasive forms of the disease which are collectively described as cervical intraepithelial neoplasia (CIN). The pathogenesis, epidemiology, histology and cytological presentation of CIN are discussed in Chapter 10, together with management protocols for this disease.

Although many cervical screening programmes have been implemented without proper trials or careful consideration of the costs and benefits of such an enormous exercise, the results of several well run programmes have more than justified the expense. In British Columbia, where a provincial screening programme has been in operation for over 20 years, there has been a fall in morbidity and mortality from cervical cancer of over 80% (Walton Report, 1976; Miller, 1981). Other countries including Finland, Sweden and Iceland have shown similar improvement in morbidity (Laara, Day and Hakama, 1987). These results bear testimony to the fact that cervical screening for CIN can prevent invasive disease and represents a major advance in primary health care.

Although the majority of cervical smears are taken with the aim of detecting premalignant lesions of the cervix, the cervical smear may coincidentally provide information about other disease in the genital tract. For example, specific infections such as herpes simplex and *Trichomonas vaginalis* can be diagnosed on a routine smear. This makes the test a particularly valuable investigation in the context of a sexual diseases clinic. However, the smear should not be used as a substitute for conventional microbiological tests, but rather as an adjunct to them, as the sensitivity of the cervical smear for this purpose is not high.

Hormone cytology

The techniques of exfoliative cytology were first used to study the effect of hormones on the vaginal epithelium in the early part of this century. Papanicolaou was a pioneer in this field as he was in so many other aspects of cytology (Stockard and Papanicolaou, 1917). He recorded the changes in vaginal smears taken at intervals throughout the menstrual cycle from women admitted to the Women's Hospital, New York, and showed that the hormonal status of these women correlated with the smear patterns in many instances. His observations started a vogue for using serial vaginal smears to monitor ovulation, hormone therapy, ovarian dysfunction and placentation. As experience grew it became clear that the test was unreliable and that blood and urine estimations of endocrine function were far more efficient than cytology for estimating hormone levels in clinical practice. The technique of taking vaginal smears for hormonal studies has largely been abandoned but is mentioned here for completeness.

Smears intended for hormonal studies should be taken from the upper third of the lateral vaginal wall. A posterior fornix aspirate is also acceptable. As the cervical epithelium is less sensitive than the vaginal epithelium to the effects of steroid hormones, a cervical scrape is not recommended for hormonal studies. Serial smears are of more value than a single smear. The presence of infection or a pessary can confound the results. A rapid method of staining smears for hormonal studies has been developed by Schorr (1941). Several sets of cellular indices have been developed for the purposes of hormone evaluation. The most widely used are the *karyopyknotic index*, the *eosinophilic cell index* and the *maturation index*. Other indices have been described including the *folded cell index* and the *crowded cell index*. The multiplicity of indices is a clear indication of the limited value of these measurements.

Karyopyknotic index (KPI)

This is determined by the ratio of superficial

cells to intermediate cells in a smear. Three hundred cells are counted. Superficial cells are determined by the presence of pyknotic nuclei which must be 6 µm or less. To carry out this test with any degree of accuracy, nuclear diameter should be measured using a micrometer screw gauge. However, the ratio of these two cell types could equally well reflect the vigour with which the smear was taken.

Eosinophilic index (EI)

This is determined by the ratio of eosinophilic squamous cells to basophilic squamous cells regardless of nuclear size. As the colouration of a Papanicolaou-stained smear depends on fixation, vaginal pH, etc., this is a most unreliable test.

Maturation index (MI)

This is determined by counting 100 squamous cells and determining the ratio of parabasal, intermediate and superficial cells in the smear. If all three types are found it is valueless for hormone estimation as this could indicate local infection and ulceration of the epithelium. A value can be assigned to each cell type so that a maturation value can be quoted.

Interpretation of the smears must take into account the patient's age and clinical history. The date of the last menstrual period should be recorded and contraceptive methods indicated. Although many different smear patterns have been described including a progesterone effect and an androgen effect, Weid and Bibbo (1971) are prepared to acknowledge only two cell patterns, namely a proliferative pattern consistent with oestrogen stimulation and an atrophic pattern which is found in the absence of any hormones. Cytological findings in various endocrinopathies have been described by Rakoff and Daley (1971), Wachtel (1964) and Weid and Bibbo (1975), but should be interpreted with caution.

References

AYRE, J. E. (1944). A simple test for uterine cancer diagnosis. *Can. Med. Assoc. J.*, **51**, 17

BOON, M. E. and TABBERS-BOURMEESTER, M. L. (1981). Benign changes, infection and inflammation. In *Gynaecological Cytology*, pp. 57–71. London: Macmillan

CARMICHAEL, D. E. (1973). Diagnosis by the vaginal smear. In *The Papanicolaou Smear: Life of George N. Papanicolaou.* pp. 54–61. Springfield, Illinois: Charles C. Thomas

COPPLESON, L. W. and BROWN, B. (1974). Estimation of the screening error rate from the observed detection rates in repeated cervical cytology. *Am. J. Obstet. Gynecol.*, **119**, 953–958

EVANS, D. M. D., HUDSON, E. A., BROWN, C. L., BODDINGTON, M. M., HUGHES, H. C., MACKENZIE, E. F. D. and MARSHALL, T. (1986). Terminology in gynaecological cytopathology: report of the working party of the British Society for Clinical Cytology. *J. Clin. Pathol.*, **39**, 933–934

EVANS, D. M. D., HUDSON, E. A., BROWN, C. L., BODDINGTON, M. M., HUGHES, H. C. and MACKENZIE, E. (1987). Management of women with abnormal cervical smears: Supplement to Terminology in gynaecological cytopathology. *J. Clin. Pathol.*, **40**, 530–531

FERENCZY, A. and RICHART, R. M. (1973). Scanning electron microscopy of the cervical transformation zone. *Am. J. Obstet. Gynecol.*, **115**, 151–157

FIGGE, D. C., BENNINGTON, J. L. and SCHWEID, A. L. (1970). Cervical cancer after initial negative and atypical vaginal cytology. *Am. J. Obstet. Gynecol.*, **108**, 422–428

FLUHMANN, C. F. (1961). *The Cervix Uteri and its Diseases.* Philadelphia: Saunders

GAY, D., DONALDSON, L. M. and GOELLNER, J. R. (1985). False negative cervical cytologic studies. *Acta Cytol.*, **29**, 1043–1046

HAFEZ, E. S. (1982). Structural and ultrastructural parameters of the uterine cervix. *Obstet. Gynaecol. Survey*, **37**, 507–516

LAARA, E., DAY, N. E. and HAKAMA, M. (1987). Trends in mortality from cervical cancer in the Nordic countries: association with organised screening programmes. *Lancet*, **i**, 1247–1249

MACGREGOR, J. E. (1981). *Taking Uterine Cervical Smears.* A British Society for Clinical Cytology Publication, Aberdeen University Press

MACGREGOR, J. E., FRASER, M. E. and MANN, E. M. F. (1966). The cytopipette in the diagnosis of early cervical carcinoma. *Lancet*, **i**, 252–256

MEYER, R. (1910). Die Epithelenwicklung der Cervix und Porto Vaginalis Uteri und die Pseudoerosio Congenita. *Arch. Gynäk.*, **91**, 579–598.

MILLER, A. B. (1981). An evaluation of population screening for cervical cancer. In *Advances in Clinical Cytology*, (Koss, L. G. and Coleman, D.V., eds), pp. 64–89. London: Butterworth

PAPANICOLAOU, G. N. and TRAUT, A. F. (1943). *Diagnosis of Uterine Cancer by the Vaginal Smear.* New York: Commonwealth Fund

PAPANICOLAOU, G. N. (1954). *Atlas of Exfoliative Cytology.* Cambridge, Massachusetts: Harvard University Press

PATERSON, M., PEEL, K. R. and VOSUN, C. A. F. (1984). Cervical smear histories of 500 women with invasive cervical cancer in Yorkshire. *Br. Med. J.*, **289**, 896–898

PISTOFIDIS, G. A., HOUSE, F. R., MOIR SHEPHERD, J. and VALE, J. C. (1984). The multispatula: a new dimension in sampling the cervix. *Lancet*, **i**, 1214–1215

RAKOFF, A. E. and DALEY, J. G. (1971). Vaginal hormonal cytology. In *Laboratory Diagnosis of Endocrine Diseases*, (Sunderman, F. W. and Sunderman, F. W. Jr., eds), St Louis: W. H. Green

REAGAN, J. W. AND HAMONIC, M. J. (1958). The cellular pathology of carcinoma *in situ*. *Cancer*, **9**, 385–402

REID, B. L. (1964). Autoradiographic analysis of uptake of tritiated thymidine and ^{35}S-cystine by cultured human cervical explants undergoing metaplasia. *J. Natl Cancer Inst.*, **32**, 1059–1073

RIOTTON, G. and CHRISTOPHERSON, W. M. (1973). *Cytology of the Female Genital Tract*, International Histological Classification of Tumours, No 8. Geneva: World Heath Organization

RYLANDER, E. (1977). Negative smears in women developing invasive cervical cancer. *Acta Obstet. Gynaecol. Scand.*, **56**, 115–118

SCHORR, E. (1941). New techniques for staining vaginal smears. III A simple differential stain. *Science*, **94**, 545–546

STOCKARD, C. R. and PAPANICOLAOU, G. N. (1917). The existence of a typical oestrus cycle in the guinea pig with a study of its histological and physiological changes. *Am. J. Anat.*, **22**, 2205

VAN DER GRAAF, Y. and VOOIJS, G. P. (1987). False negative rate in cervical cytology. *J. Clin. Pathol.*, **40**, 438–442

VOOIJS, P., ELIAS, A., VAN DER GRAAF, Y. and VELING, S. (1985). Relationship between the diagnosis of epithelial abnormalities and the composition of cervical smears. *Acta Cytol.*, **29**, 323–328

WACHTEL, E. (1964). *Exfoliative Cytology in Gynaecologic Practice*. Washington DC: Butterworth

WALKER, E. M., HARE, M. J. and COOPER, P. (1983). A retrospective review of cervical cytology in women developing invasive squamous cell carcinoma. *Br. J. Obstet. Gynaecol.*, **90**, 1087–1091

WALTON REPORT (1976). Cervical cancer screening programmes. *Can. Med. Assoc. J.*, **114**, 1003–1033

WEID, G. L. and BIBBO, M. (1971). Hormonal cytology of the female genital tract. In *Pathways to Conception*, Springfield, Illinois: Charles C. Thomas

WEID, G. L. and BIBBO, M. (1975). Evaluation of endocrinologic conditions in exfoliative cytology. In *Gynaecology Endocrinology*, 2nd edn, (Gold, J. J., ed.), pp. 117–155. Harper and Row

WILBANKS, G. D., IKOMIE, E., PRADO, R. B. and RICHART, R. M. (1968). An evaluation of a one-slide cervical cytology method for the detection of cervical intraepithelial neoplasia. *Acta Cytol.*, **12**, 157–158

WOLFENDALE, M. R., HOWE-GUEST, R., USHERWOOD, M. and DRAPER, G. J. (1987). Controlled trial of a new cervical spatula. *Br. Med. J.*, **294**, 33–35

WRIGHT, V. C., DAVIES, E. and RIOPELLE, M. A. (1984). Laser cylindrical excision to replace conisation. *Am. J. Obstet. Gynecol.*, **150**, 704–709

9

Female genital tract: infection, inflammation and repair

Patricia A. Chapman

Inflammation is a local tissue response to injury. Infection usually, though not invariably, causes inflammation. Other factors associated with inflammation in the female genital tract include physical trauma, indwelling pessaries and other foreign bodies and obstruction to menstrual flow. An inflammatory response is also induced by therapeutic procedures such as cautery, laser treatment and irradiation. Three main patterns of inflammatory response have been described: acute, chronic and granulomatous. The pathological basis of inflammatory change is discussed in Chapter 7. This chapter is concerned with the cytological manifestation of the inflammatory reaction and its histological basis. It is important to note that some of the inflammatory changes seen in cervical smears are very subtle and are not apparent in histological sections, probably due to the superior nuclear and cytoplasmic preservation that can be achieved with alcohol fixatives compared with formalin fixation. Moreover, it should be remembered that inflammation is usually followed by regeneration and repair and all three changes may occur simultaneously in the cervix and may be reflected in the same smear. The cytological pattern of regeneration and repair is described on p. 190.

Acute inflammation

The first stage in the acute inflammatory response is dilation of small blood vessels. This results in congestion producing warmth and redness of the involved area. The walls of these blood vessels become more permeable so that plasma (fluid and protein) exudes through the wall into the tissue, producing swelling. At the same time circulating leucocytes, mainly polymorphs, migrate through the vessel walls into the tissue. Swelling and a variety of chemical changes produce the fourth striking feature of acute inflammation—pain.

The outcome of acute inflammation is variable. Lesser degrees of inflammation will usually resolve completely over a few days. Greater degrees are associated with tissue destruction and ulceration. This is usually followed by healing and restoration of the normal tissue architecture, but specialized tissue such as muscle may not be restored; in this event, healing is by fibrosis rather than resolution, producing a scar.

Severe inflammation is associated with massive infiltrates of neutrophils, many of which die; these are pus cells and collections of pus cells may form an abscess. The neutrophils are phagocytic, i.e. they ingest particles, particularly bacteria. They have a short life (3–4 days) and after the first day or two their phagocytic role is taken over by macrophages, which are blood monocytes that have migrated into the tissue. These are the most important cells in the demolition phase which follows the acute inflammatory phase: they ingest and break up particles and debris, and carry them away from the damaged site in the lymphatic channels.

Cytology of acute cervicitis and vaginitis

Acute non-specific inflammation of the cervix

and vagina produce a wide variety of nuclear and cytoplasmic changes in cervical smears. The changes seen in the smears fall into three groups:

(1) Presence of an inflammatory exudate.
(2) Degenerative changes in the nuclei of the epithelial cells in the smear.
(3) Degenerative changes in the cytoplasm of the epithelial cells in the smear.

The smear pattern is dominated by the inflammatory exudate (*Figure 9.1*). Large numbers of polymorphonuclear leucocytes are usually present, together with histiocytes and sometimes fibrin and red blood cells. There is much cellular debris as a result of cytolysis and cell necrosis and often many bacteria in the background.

ties. Anisonucleosis is frequently seen and many nuclei show irregularity in the nuclear outline (*Figure 9.1*). The chromatin may be clumped so that the nuclei appear hyperchromatic. This can usually be differentiated from the coarser chromatin found in dyskaryotic cells because the granules are evenly spaced and equal in size. Other degenerative nuclear changes associated with inflammation include margination of chromatin, karyopyknosis, karyorrhexis and vacuolation (*Figure 9.2*).

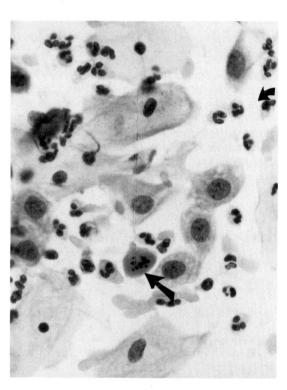

Figure 9.2 Smear pattern in non-specific cervicitis. The epithelial cells show a slight coarsening of the chromatin with one cell exhibiting karyorrhexis (arrowed). Note fine vacuolation of cytoplasm and leucocytes in the background. Stain: Papanicolaou. Magnification ×770

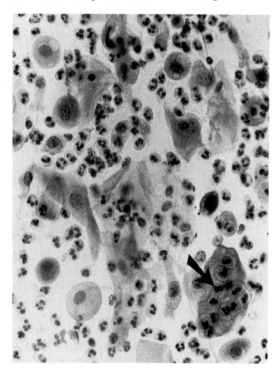

Figure 9.1 Smear pattern in non-specific cervicitis. Note numerous polymorphs and phagocytosis at arrow. The epithelial cells show some degree of anisonucleosis and cytoplasmic vacuolation. Stain: Papanicolaou. Magnification ×770

Another striking feature of the smear is the degenerative changes in the nucleus and cytoplasm which reflect the effect of bacterial toxins on the epithelial cells. Inflammatory nuclear changes encompass a wide variety of irregulari-

The cytoplasm of epithelial cells also shows degenerative changes in inflammatory conditions. Small and large vacuoles are a common finding (*Figures 9.2* and *9.3*) and phagocytosed polymorphs can be found in them (*Figure 9.1*). Perinuclear vacuolation, which gives the nucleus a 'haloed' appearance, is frequently seen (*Figure 9.24*). Variation in the staining reaction of the cytoplasm is another common feature of inflammation. Cells which would normally be basophilic stain pink or red, and in trichomonal infestations the squamous epithelial cells show 'pseudo-eosinophilia' and stain a brassy-red. Some cells exhibit an amphophilic staining. The reason for this is not understood.

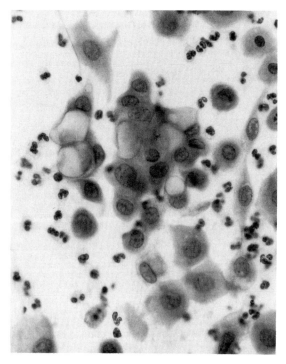

Figure 9.3 Smear pattern in non-specific cervicitis showing cytoplasmic vacuolation of epithelial cells. Stain: Papanicolaou. Magnification × 770

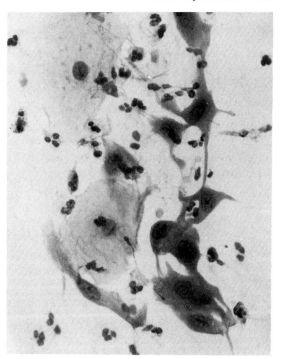

Figure 9.4 Cervical smear from patient with non-specific cervicitis. Note slight hyperchromasia of nuclei and ragged cytoplasm of the epithelial cells. Stain: Papanicolaou. Magnification × 770

Acute inflammation due to local infection is common in the endocervical crypts due to blockage of the mucous glands. Immature metaplastic epithelium is also susceptible to infection. It is most likely to occur postpartum or in the presence of an indwelling pessary or following obstruction due to a cervical polyp. For this reason inflammatory changes are most prominent in endocervical cells and metaplastic cells in cervical smears (*Figures 9.2, 9.3* and *9.4*).

When inflammation affects the endocervical epithelium the cytoplasm of the endocervical cells is often poorly defined or absent and many bare nuclei are observed. Rarely, ciliated tufts comprising cilia on a terminal plate are found in the smear, often alongside vacuolated columnar cells. This phenomenon—ciliocytophthoria (CCP)—is probably a degenerative change (Muller Kobald-Wolterbeek and Beyer-Boon, 1975).

Chronic inflammation

If the agent causing an acute inflammatory response persists, the congestion due to the increased vascularity and exudation of fluid and protein does not resolve completely. The infiltrate of polymorphs and macrophages is expanded by the migration of plasma cells and lymphocytes. This process—chronic inflammation—usually follows a definite episode of acute inflammation, but in some cases it appears to develop at the same time as the acute inflammation. The combination of acute and chronic inflammatory infiltrate is sometimes called 'active' chronic inflammation. In a few situations, notably in the liver, chronic inflammation can occur without any acute component, and this also applies to the sub-type of chronic inflammation which is described as granulomatous change.

A heavy lymphocytic infiltrate consistent with chronic inflammation is frequently seen in histological section of the cervix and reported as evidence of chronic cervicitis. It is often found in association with reactive changes in the cervical epithelium. The fragile columnar epithelium may react to persistent infection or irritation by undergoing squamous metaplasia. The squamous epithelium may react by becoming thicker either as a result of the production of a keratin layer (hyperkeratosis) or as the result of acanthosis and parakeratosis. The different tissue changes which characterize chronic cervicitis are often reflected in the smear.

Cytology of chronic cervicitis

The presence of chronic cervicitis can be suspected when the predominant inflammatory cells in the smear are lymphocytes, plasma cells and macrophages which are often multinucleated. Cellular changes reflecting hyperkeratosis and parakeratosis may also be seen in cases of long-standing chronic cervicitis. However, in those cases where chronic inflammation is superimposed on acute inflammation and reactive changes in the epithelial cells are not seen, a distinction between acute and chronic cervicitis cannot be made on cytological grounds.

Hyperkeratosis and parakeratosis

The normal non-keratinizing stratified squamous epithelium of the cervix, whether original or metaplastic, will undergo changes in response to continuous irritation. Two changes that are commonly found in the most superficial layers of the squamous epithelium of the cervix and vagina are hyperkeratosis and parakeratosis. These may exist separately or in combination and reflect a protective surface reaction overlying normal epithelium. Clinically they may confer a whitish appearance on the cervix, often described as leukoplakia. In some circumstances the clinical or cytological presentation of these conditions may present diagnostic difficulty and colposcopy with tissue biopsy may be necessary to exclude neoplastic change.

Hyperkeratosis is a condition in which there is an increased number of otherwise normal keratin plates resulting in greater overall thickness of the superficial layers of the squamous epithelium. When the outer layers of the cervical and vaginal epithelium undergo complete keratinization the epithelium may eventually become histologically indistinguishable from epidermis. Hyperkeratosis presents in cervical smears as anucleate polygonal squames, either singly or in sheets (*Figure 9.5*). A small clear

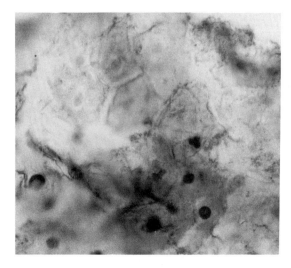

Figure 9.5 Highly keratinized anucleate squames in a cervical smear reflecting hyperkeratosis of cervical epithelium. The site of the nucleus can be seen as a 'ghost' in a few cells. Stain: Papanicolaou. Magnification ×512. (With permission of E. B. Butler)

area may be seen at the centre of each cell indicating the site of the lysed nucleus. All stages of transition between mature nucleated squamous cells and anucleate cells may be seen. The cytoplasmic staining is characteristically yellow, orange or red rather than pink as in normal superficial cells stained by the Papanicolaou method. Occasionally keratin granules can be found in the cytoplasm.

Parakeratosis in the female genital tract is a condition represented by the presence of very small nucleated squamous cells in the surface layers of the squamous epithelium. Because there is a diffuse attempt at keratinization of these cells, this process is known as parakeratosis. In cytological specimens the cells shed from an area of parakeratosis are commonly seen in sheets (*Figure 9.6*) but can occur singly. The cells are much smaller than normal superficial cells with denser cytoplasm which usually stains red or orange; elongated or oval forms may be seen. The nuclei are usually pyknotic but somewhat larger than the pyknotic nuclei of normal superficial cells and they may vary slightly in shape and size.

Hyperkeratosis and parakeratosis can be associated with many conditions including chronic cervicitis, atrophic cervicitis and cervical polyp. Hyperkeratosis is particularly common in procidentia whereas both hyperkeratosis

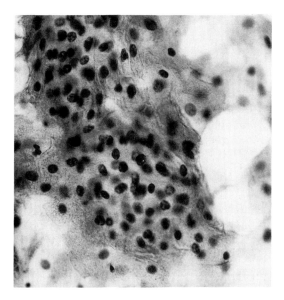

Figure 9.6 A sheet of highly keratinized small squamous cells with pyknotic nuclei. Note the multilayering and closely packed nuclei. Stain: Papanicolaou. Magnification × 330

and parakeratosis are a common manifestation of human papillomavirus infection (Meisels and Fortin, 1976).

Variants of chronic cervicitis

Several histological variants of the chronic inflammatory reaction described above can be seen in the cervix. These include granulomatous change, follicular cervicitis, plasma cell cervicitis, chronic papillary cervicitis and atrophic (senile) cervicitis. A characteristic cellular pattern may be seen in cervical smears when these conditions are present.

Granulomatous inflammation

This type of chronic inflammation is characterized by the presence of aggregates of macrophages and small lymphocytes. The best known example of granulomatous change is tuberculosis. Granulomatous inflammation is also seen in other infections such as syphilis, schistosomiasis, amoebiasis, Donovania granulomatis, lymphogranuloma venereum, and as a response to the inorganic substances silica and starch. It is a feature of sarcoidosis and Crohn's disease.

The macrophages become large with pale nuclei and abundant cytoplasm and are designated 'epithelioid' cells due to a supposed resemblance to the epithelial cells of the epidermis. Some of the macrophages fuse to form giant cells which often have a distinct morphology. Giant cells with nuclei disposed around the periphery of the cell are commonly found in granulomata associated with tuberculosis (*Figure 9.14*), whereas giant cells with nuclei scattered throughout the cytoplasm are found in granulomata associated with foreign material (*Figure 9.7*). As the granulomata increase in size they may undergo central necrosis, which in the case of tuberculous infection is called 'caseation'.

Granulomatous change is characterized in cervical smears by a mixed inflammatory cell population including polymorphs and lymphocytes. Giant multinucleated cells are also seen (*Figure 9.14*). Epithelioid cells are abundant (*Figure 9.13*) and present as narrow elongated cells with round or oval nuclei with soft frayed cytoplasm outlining the nucleus.

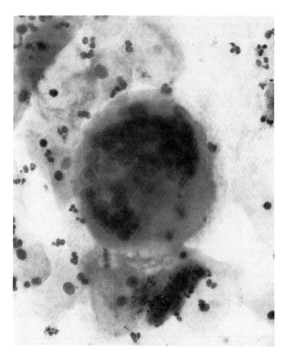

Figure 9.7 Multinucleate giant cell in cervical smear. This type of cell may reflect a foreign body reaction. Stain: Papanicolaou. Magnification × 770

Follicular cervicitis

This is a form of chronic cervicitis in which there is a dense lymphocytic infiltration of the subepithelial connective tissue associated with the formation of lymphoid follicles. In tissue sections these follicles can be seen to have a pale germinal centre containing reticular cells, immature lymphocytes and macrophages which is surrounded by a peripheral ring of mature lymphocytes. The surrounding stroma is usually diffusely infiltrated by lymphocytes and plasma cells.

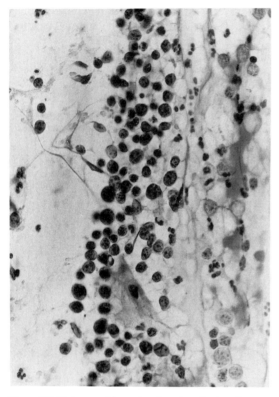

Figure 9.8 Mature and immature lymphocytes together with histiocytes and reticular cells in a cervical smear from a case of follicular cervicitis. Stain: Papanicolaou. Magnification ×770

During smear taking the fragile epithelial surface may be scraped off and a mixture of mature and immature lymphocytes released together with histiocytes, plasma and reticular cells. These lymphoid cells characteristically appear in the smear in long dense streaks (*Figure 9.8*). They must be distinguished from cells shed from lymphomas involving the cervix, benign and malignant endometrial cells, and small cell squamous carcinoma. The distinction between endometrial cells, carcinoma cells and cells shed from follicular cervicitis can be made fairly easily as the lymphoid cells in the smear are always discrete. Distinction between the cells shed from follicular cervicitis and metastatic lymphoma is more difficult but fortunately the latter condition is very rare. Follicular cervicitis may be found in association with *Chlamydia trachomatis* (Hare *et al.*, 1981) but is also seen in asymptomatic postmenopausal patients with normal appearing cervices (Roberts and Ng, 1975).

Plasma cell cervicitis

Plasma cell cervicitis is a rare condition in which a non-specific dense plasma cell infiltrate is found in both the ectocervical and endocervical stroma. Bizarre, sometimes multinucleate, plasma cells which may be confused with malignant cells may be found in cervical smears (Qizilbach, 1974).

Atrophic cervicitis

Atrophic cervicitis is a condition which results from a combination of factors: lack of hormonal stimulation, change of vaginal pH and a low grade infection in a locally altered environment. It can occur as a result of ovarian dysfunction, castration and menopause.

Single round parabasal cells dominate the smear showing remarkable variation in size and staining reaction, some having abundant pale-staining cytoplasm, others having scanty dense, red-staining cytoplasm. The nuclei of the parabasal cells may be enlarged, pale and contain prominent chromocentres and small single nucleoli. Alternatively the nuclei may appear hyperchromatic and irregular, suggesting squamous cell carinoma (*Figures 9.9* and *8.33a*). Numerous polymorphs may be seen. Sometimes giant macrophages are found which must be distinguished from multinucleate cancer cells; Nasiell (1961) based the differential diagnosis on the uniform nuclear configuration of the giant macrophages. Atrophic smears may also contain basophilic bodies ('blue blobs', which are similar in size and shape to degenerating parabasal cells *Figure 9.10*). In the middle of these 'bodies' the remains of a nucleus can nearly always be seen. It is important to differentiate 'blue blobs' from stripped nuclei of cancer cells.

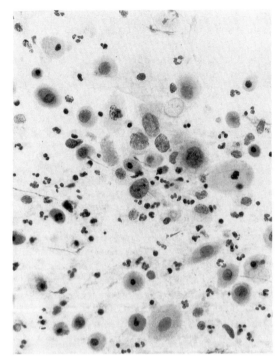

Figure 9.9 Cells from a case of chronic atrophic cervicitis showing variation in nuclear size and staining appearance. Some cells have condensed hyperchromatic nuclei (pyknosis), others show enlarged, pale staining, finely granular nuclei. Stain: Papanicolaou. Magnification × 770, reduced to 80% in reproduction

In some instances the differential diagnosis between atrophic changes and malignancy can be extremely difficult. It has been suggested by Keebler and Weid (1974) that in problematic cases the clinician should give the patient oral or local oestrogens and take a repeat smear 2–3 days later. The diagnosis should then be easier to make.

Psammoma bodies

Psammoma bodies or calcospherites are laminated, homogeneous spherules measuring 30–100 μm in diameter. They can be seen in cervical smears as isolated deeply basophilic staining bodies to which may be attached fragments of hyalinized stroma. Psammoma bodies give a positive staining reaction with Von Kossa, PAS, Alizarin red and Perle's prussian blue. They are found in cervical smears in many conditions, e.g. chronic cervicitis, carcinoma of the ovary and endometrium, tuberculosis and microglandular endocervical hyperplasia. Psammoma bodies can also develop in the hyalinized endometrial stroma surrounding an IUCD and may be found in Papanicolaou smears from women fitted with these devices (Highman, 1971).

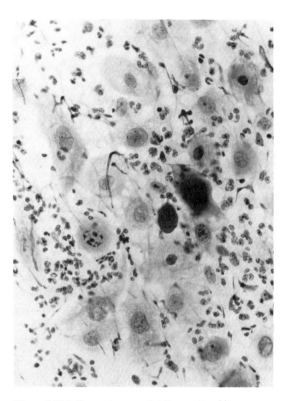

Figure 9.10 Inflammatory exudate in an atrophic smear. Note also pleomorphism of the parabasal cells and 'blue blobs'. Stain: Papanicolaou. Magnification × 770

Specific infections

Numerous microorganisms live and multiply within the human body (host). When harmful these microorganisms are termed pathogens; when harmless they are designated commensals. Under certain conditions a pathogen may become a commensal and a commensal a pathogen which may, in part, explain why all infections do not invariably cause an inflammatory reaction.

Pathogenicity of microorganisms

Pathogenic microorganisms are commonly found in the female genital tract. The disease they produce is the result of the interaction between pathogen and host. The outcome of this exchange depends, not only on the virulence of the organism, but on the susceptibility of the host and the growth inhibiting effect of other microorganisms present.

Before genital infection can occur pathogenic microorganisms must first gain access to the host tissue, either by direct contact or via the circulation. Once established the pathogens bring about damage to the host tissue by the production of toxins, but the host reaction is responsible for many of the destructive features associated with infection. Some microorganisms occur at the surface of the epithelium but other organisms tend to proliferate in the subepithelial connective tissue. The multi-layered squamous epithelial surfaces of the cervix and vagina normally act as barriers to infectious agents having the mechanism for ridding themselves of microorganisms and resisting their penetration. However, endocervical and metaplastic epithelium is vulnerable to penetration and is most susceptible to attack. Many pathogens are helped in the early stages by attaching themselves to specific receptors on the cell membrane where they remain and multiply, producing a spreading infection in the epithelium. Others invade the subepithelial tissue where they are susceptible to assault by scavenging cells or the neutralizing action of host antibodies.

Phagocytic cells (polymorphonuclear leucocytes and macrophages) normally ingest pathogens and kill them. Some microorganisms, however, have evolved methods of overcoming this system; for example, by producing surface antigens which prevent phagocytosis, by secreting substances which repel phagocytes, or by producing leucocidins which destroy white cells. Other microorganisms have the ability to live inside phagocytic cells by producing substances in their cell walls enabling them to resist the killing action of the phagocytes. Microorganisms have also developed the means to overcome the host immune response by releasing antigens which neutralize the antibody, or by altering their own cell surface in order that the antibody does not recognize them.

In this section we discuss those infectious agents which provide changes in a cervical smear which are pathognomonic. These changes are superimposed on the changes of acute or chronic cervicitis described in the preceding section. The organisms themselves may be recognizable with the Papanicolaou stain or they may induce specific cellular changes in the epithelial cells, e.g. inclusions. In some cases both features are present. For the purposes of this chapter microorganisms will be classified as bacterial, mycotic, chlamydial, viral and parasitic.

Bacterial flora of the cervix and vagina

Bacteria are microscopic, unicellular organisms which can be classified according to their shape, their affinity for Gram stain, their dependence on oxygen and their nutritional requirements. Cocci are round or oval-shaped bacteria, bacilli rod-shaped or cylindrical, vibrios are typically comma-shaped and spirochaetes are coiled. All of the groups may be Gram-positive or Gram-negative and may be aerobic or anaerobic.

The lower female genital tract is known to harbour a wide range of aerobic and anaerobic bacteria. The bacteria most commonly found in this area are shown in *Table 9.1* but are not in strict taxonomic order, nor in order of frequency. Although in some cases an attempt at grouping by morphology can be made on Papanicolaou and Gram-stained smears, culture on selective media is mandatory for the identification of specific bacteria.

Table 9.1 Bacteria commonly found in the female genital tract

Gram positive	*Gram negative*
Cocci	
Staphylococcus sp.	*Neisseria* sp.
Streptococcus sp.	
Rods	
Lactobacilli sp.	Coliform sp.
Listeria sp.	*Haemophilus* sp.
Mycobacterium sp.	*Proteus* sp.
Gardnerella vaginale	*Gardnerella vaginale*

As explained on p. 157, many of the bacteria shown in *Table 9.1* can be isolated from the genital tract in the absence of disease. They can, however, assume pathogenic properties and cause an acute cervicitis or vaginitis. The morphological and biological characteristics of the most common organisms are discussed in this chapter.

Lactobacillus sp. (Doderlein bacillus)

The *Lactobacillus* sp. are the most common non-pathogenic inhabitants of the vagina and it is debatable whether they ever cause infections. These bacteria are aerobic Gram-positive rods, varying greatly in length. In Papanicolaou smears they stain pale grey and are found on cell surfaces and in the smear background. *Lactobacillus* sp. produce enzymes which cause cytolysis of glycogen-producing cells. As glycogen is present mainly in intermediate squamous cells, the bacteria predominate during the luteal phase of the menstrual cycle, during pregnancy, premenarche and early menopause.

At these times the smear is distinguished by the presence of numerous 'free' nuclei and a ragged appearance to the cytoplasm of the squamous cells. The lactobacilli transform glycogen into lactic acid which lowers the acidity of the vagina (pH 3.9–4.5), providing an optimum environment for reproduction of the organism. It is this acidity which appears to offer a degree of protection to the human vagina against pathogens which usually require a more neutral environment. Superficial squamous cells and parabasal cells appear to resist cytolysis presumably because of their low glycogen content. For this reason *Lactobacillus* sp. are absent in atrophic smears.

Leptothrix

Leptothrix are long, filamentous, often curving, bacilli (frequently described as looking like a child's scribble) which stain pale grey in Papanicolaou smears (*Figure 9.11*). Controversy remains as to which species of bacteria they belong to or whether they represent a form of lactobacillus. Certainly the distinction can be difficult to make from cervical smears. Bibbo and Harris (1972) consider that two types of *Leptothrix* are found in the vagina, one which is easily identifiable and the other which is indistinguishable from Doderlein bacilli. *Leptothrix* is frequently associated with *Trichomonas vaginalis* infection. In a study of 1000 consecutive cases of *Leptothrix* at the University of Chicago, 75% showed an association with trichomonads (Bibbo and Weid, 1988). Experience suggests that *Leptothrix* may also occur as a contaminant.

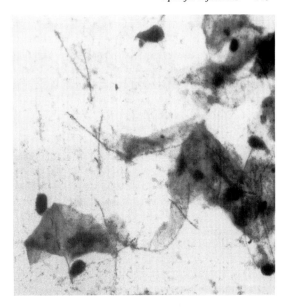

Figure 9.11 *Leptothrix* appearing as long, filamentous curving organisms. The bacilli are often associated with *Trichomonas vaginalis* infection. Stain: Papanicolaou. Magnification × 770

Gardnerella vaginalis

The bacterium *Gardnerella vaginalis* was identified as a cause of vaginitis in 1955 by Gardner and Dukes. The species was previously known under the names of *Haemophilus vaginalis* and *Corynebacterium vaginalis* but the careful work of Gardner showed that these classifications were erroneous. The infection is common and usually mild, producing discomfort, an offensive smell resembling rotten fish and a thin grey discharge which is often frothy. Pruritus is present in 12% of patients (Dawson and Harris, 1983). Leucocytes are rarely seen in the discharge unless other organisms are also present. This has been borne out by microscopical, histological and colposcopic findings (Pheifer *et al.*, 1978). Recent work suggests that coinfection with anaerobic bacteria such as *Bacteroides* sp. may account for the symptoms and signs associated with *G. vaginalis* infection (Dattani *et al.*, 1982; Taylor *et al.*, 1982).

Gardnerella vaginalis is a small pleomorphic Gram-negative or Gram-variable bacillus or coccobacillus which stains dark blue in Papanicolaou-stained smears. Infection may be suspected when large numbers of these bacteria are seen adhering to the surface of squamous cells giving them a uniquely stippled appearance; these cells are described as 'clue' cells (*Figure 9.12*). Not all cells show this phenomenon. Some are only partially covered; the same bacteria are frequently seen in the background of the smear. *Gardnerella vaginalis* prefers an environment at pH 5.0–5.5 which may explain why *Lactobacillus* sp. are rarely found together with this organism. Errors of diagnosis may occur if too much reliance is placed on 'clue' cells alone because other bacteria also adhere to epithelial cells. Confirmatory tests should therefore be carried out

If untreated the organism can persist for several years but treatment with metronidazole is very effective. Transmission of the organism is mainly sexual, although male consorts are symptomless. *Gardnerella vaginalis* has been found in 25% of sexually transmitted disease clinic patients.

Neisseria gonorrhoeae

Neisseria gonorrhoeae infection is a common venereal disease which produces an acute or chronic inflammatory reaction in the endocervix. If left untreated the infection may spread via the fallopian tubes to the peritoneum causing infertility or acute peritonitis. Occasionally it can produce an acute necrotizing cervicitis which may mimic ulcerating carcinoma. Rarely infection is found in children and is manifest as a vaginitis. *Neisseria gonorrhoeae* is caused by a Gram-negative diplococcus which may be observed in Papanicolaou-stained cervical smears when screening with a high power objective. These small bean-shaped cocci are arranged in pairs or mosaic clusters. Opinion as to their exact location differs. Traditionally described as adhering to the surface of intermediate and parabasal cells, Heller (1974) observed the organisms mainly in squamous metaplastic cells. Koss (1979), on the other hand, described the organisms in mature squamous epithelial cells and Bibbo and Weid (1988) found them in the cytoplasm of polymorphonuclear leucocytes. Clearly the organisms may be found in a wide variety of cells. However, it must be remembered that *Neisseria gonorrhoeae* cannot be differentiated from other species of *Neisseria* on morphological grounds alone and further tests are needed for a definitive diagnosis.

Treponema pallidum (syphilis)

Syphilis is caused by the spirochaete *Treponema pallidum*. The primary lesion (chancre) of syphilis occurs most commonly on the labia or fourchette, but is found on the cervix in 40% of cases. Smears taken from the surface of the primary chancre contain spirochaetes in variable numbers. Fresh unstained material may be examined using dark ground illumination or phase contrast microscopy. Dried, fixed smears may be stained by a dilute Giemsa or the silver method of Hage-Fontana. The spirochaetes cannot be identified in Papanicolaou-stained smears and the cytological findings are those of chronic granulomatous change (*see* p. 40).

Donovania granulomatis

Donovania granulomatis is a venereal disease also known as granuloma inguinale which may cause ulcerative lesions on the uterine cervix and vagina. It is caused by small Gram-negative rods which appear as bipolar encapsulated bodies which look like tiny safety pins. These are called 'Donovan bodies'. Donovan bodies

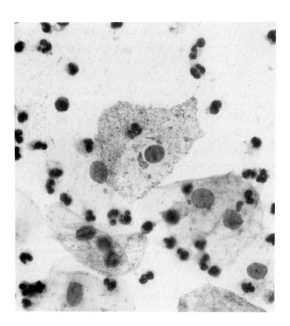

Figure 9.12 *Gardnerella vaginalis* bacteria adhering to a squamous epithelial cell – 'clue' cell. Stain: Papanicolaou. Magnification × 770

can be demonstrated by the Warthin-Starry Silver technique or by Giemsa stain, and are found in giant macrophages which may be 100 μm or more in diameter.

Mycobacterium tuberculosis

Tuberculosis of the female genital tract is rare in developed countries and is almost always a secondary disease from the lung or intestine. The fallopian tubes and endometrium are also often infected, although tuberculosis of the uterine cervix and vagina is rare.

Mycobacterium tuberculosis is an acid-fast bacillus which cannot be recognized in Papanicolaou-stained smears but which can be identified using Ziehl-Neelson stain or by fluorescence microscopy using an auramine and rhodamine stain. However, the presence in smears of elements from tuberculous granulomata such as epithelioid cells, Langhan's giant cells and fragments of amorphous acidophilic material originating from the caseous necrosis surrounding the tubercle may be seen, and suggest the diagnosis (Coleman, 1969; Mischi *et al.*, 1976).

Epithelioid cells are large, slender cells with a finely vacuolated often ill-defined cytoplasm (*Figure 9.13*). The nuclei are round or oval, have a fine chromatin structure and often contain a small nucleolus. Langhan's cells are large multinucleate cells with characteristically placed nuclei which form a horseshoe shape around the periphery of the cell (*Figure 9.14*).

The difficulty of finding epithelioid cells and differentiating Langhan's cells from other types of multinucleate giant cells cannot be overemphasized and the final diagnosis must be made on microbiological or histological studies.

It is interesting to note that commensal *Mycobacterium* sp. are a common finding in the female genital tract.

Other bacteria

A large variety of Gram-positive and Gram-negative cocci and rods may be isolated from the lower genital tract in the female. Most are harmless but may become pathogenic at any time. In Papanicolaou-stained smears all bacteria stain grey or dark blue and are usually unobtrusive. Occasionally, however, they obscure the cells and give a 'dirty' appearance to the smear. The most common types of cocci

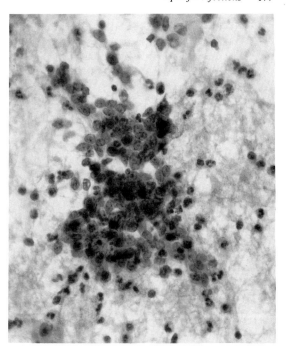

Figure 9.13 Cervical smear from a patient with tuberculosis of the cervix showing a cluster of epithelioid cells. Stain: Papanicolaou. Magnification × 770. (Published by permission of Chapman and Hall)

found are either *Streptococcus* sp. or *Staphylococcus* sp. which appear in chains or clusters resembling bunches of grapes. Recently *Staphylococcus aureus* received particular attention as it was implicated in 'toxic shock syndrome' in women using a particular type of internal tampon (McKenna, Meadows and Brewer, 1980).

Other bacteria often isolated in vaginal cultures include *Escherichia coli*, an aerobic Gram-negative bacillus which causes vaginitis in teenagers and postmenopausal women. *Bacteroides* sp. is also a common finding and *Bacteroides fragilis* is now thought to work with other bacteria to produce a discharge. *Listeria monocytogenes*, a Gram-positive rod which causes abortions and perinatal death, is occasionally found.

In general one must be very cautious about reading too much into the identification of bacteria in smears as their presence may be transient. For example Boon and Tabbers-Boumeester (1980) have shown that coitus influences the bacterial content of the female genital tract by altering the pH of the vagina. The neutral protein-rich ejaculate affects bacterial growth causing the disappearance of *Lacto-*

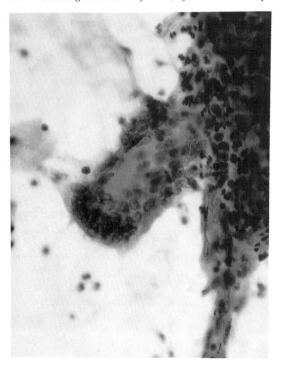

Figure 9.14 Langhans giant cell found in a cervical smear from the same case as *Figure 9.13*. Stain: Papanicolaou. Magnification × 770

bacillus sp. which are replaced by cocci from the perineum. This phenomenon reaches a maximum 16 h after intercourse and is known as the 'coitus effect'. At this time the smear may be covered by a cloud of coccoid bacteria; 32 h after coitus the normal flora of the vagina—the lactobacilli—reappear.

Higher bacteria: *Actinomyces*

Actinomyces are a heterogeneous genus of filamentous microorganisms which are related to 'true bacteria'. Many are free-living in the soil; however, the genus also includes the obligate anaerobic pathogens which are the causative organisms of actinomycosis in animals and man.

Gupta, Hollander and Frost (1976) were the first to detect the presence of *Actinomyces*-like organisms in cervical smears from women using an IUCD. Subsequent reports corroborated this finding (Luff *et al.*, 1978; Jones *et al.*, 1979; Duguid, Parratt and Traynor, 1980; Fry, Linder and Bull, 1980). Special stains, immunofluorescence and anaerobic cultures indicate that these organisms are *Actinomyces israelii* (Bhagavan and Gupta, 1978; Christ and Haja, 1978; Spence *et al.*, 1978; Pine *et al.*, 1981).

In Papanicolaou-stained smears, *Actinomyces* can be recognized at low magnification as isolated aggregates. The organisms usually appear as dark, dense balls with an indistinct central core (*Figure 9.15*). At high magnification the organisms appear as branching filamentous structures which stain strongly with haematoxylin. Sometimes these filaments appear to possess radially arranged 'clubs' but it has been shown that these clubs are host protein deposits and not part of the organism. This probably explains why the filaments are positive with Gram, Giemsa and Ziehl-Neelson stain and PAS-negative, and the clubs exhibit the reverse staining reaction.

Morphologically *Actinomyces* resembles several other filamentous agents, e.g. *Candida*, *Aspergillus*, *Nocardia*, *Alternaria*, *Trichophyton* and *Leptothrix*. Fibrin threads, cotton and synthetic fibre may also be misdiagnosed as *Actinomyces* (Gupta, 1982).

The route of entry of *Actinomyces* into the female genital tract is uncertain and various theories have been suggested, including self-

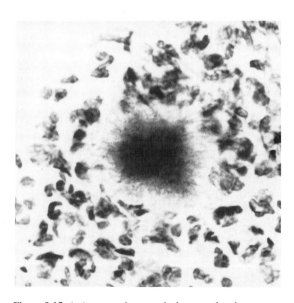

Figure 9.15 *Actinomyces* in a cervical smear showing a dense central core of organisms and branching filamentous threads. Stain: Papanicolaou. Magnification × 700. (With permission of T. Kobayashi)

contamination from rectum to vagina (Bhagavan and Gupta, 1978). The practice of orogenital and anal sex (Spence et al., 1978) or haematogenous spread from a focus of infection in the ileocaecal region, appendix or jaw (Sykes and Shelley, 1981) may be contributory factors. Infection with *Actinomyces israelii* is almost always associated with the presence of an IUCD, although a few cases occur following surgical instrumentation or criminal abortion. Overall about 7% of women with an IUCD have *Actinomyces* in their cervical smear; the frequency appears to be higher in women with a plastic IUCD than in women with a copper device (Duguid, Parratt and Traynor, 1980). The frequency also increases with duration of use (Blenkinsopp and Chapman, 1982). In the vast majority of these women there is no evidence that the colonization is harmful and it is now believed from the studies of Christ and Haja (1978) who demonstrated *Actinomyces* sp. in smears of women using a vaginal pessary, and from the work of Pine et al. (1981) who carried out immunofluorescence of cervical smears from IUCD and non-IUCD users, that *Actinomyces* sp. is a common commensal in the vagina and the organism grows freely on the string of the IUCD. There is a low risk of progression to pelvic inflammatory disease.

Ruehsen et al. (1980) reported the presence of non-pathogenic amoebae in 1% of cervical smears from IUCD users. The amoebae resemble those of *Entamoeba gingivalis* and can be distinguished from pathogenic amoebae such as *E. histolytica* by their blunt pseudopodia and failure to ingest erythrocytes. Care must be taken when making this distinction so as not to overlook the presence of the parasite *E. histolytica* in areas where this parasite is endemic.

Mycotic infections

Fungi are dimorphic, non-photosynthesizing organisms which are classified according to the method of spore formation and the presence or absence of mycelium. Because of the carbohydrate content of fungal walls, most are PAS-positive. Specific classification of fungal organisms is not possible from cytological examination alone although a provisional identification can usually be made from the morphological appearance in a Papanicolaou-stained smear. It has been suggested (Kearns and Gray, 1963) that 80% of cases of fungal infections in the female genital tract are caused by *Candida* species. These include *Candida albicans* (79%), *C. tropicalis* (13%), *C. parakrusei* (6%), *C. krusei* (1%) and *C. guilliermondi* (1%).

Candida sp.

Candida albicans infection is common in reproductive life but is an infrequent cause of clinical vaginitis. Infection with this Gram-positive yeast occurs most frequently in the 16–35 year age group. It is found in smears from pregnant women, in women using high progesterone contraceptive pills, in patients taking broad spectrum antibiotics or immunosuppresive drugs and in patients with diabetes melitus.

Candida albicans stains pinkish red in Papanicolaou smears. The long filamentous pseudomycelium, which is a conglomerate of pseudohyphae, resembles a bamboo cane (*Figure 9.16*). Budding spores which are small, oval, encapsulated bodies approximately 1.5–8 µm in diameter may also be seen. Squamous cells often lie along or cluster across the pseudo-hyphae in a distinctive way. Non-specific inflammatory changes may be found in the smear.

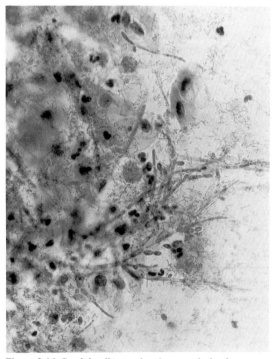

Figure 9.16 *Candida albicans* showing pseudo-hyphae and spores in a cervical smear. Stain: Papanicolaou. Magnification ×770

Comparative studies involving culture of high vaginal swabs (Cassie and Stevenson, 1973; Thin et al., 1975) have shown that only 25% of cases of vaginal candidiasis can be detected by a cervical smear.

Candida species can penetrate, invade, develop and reproduce within the deep layers of the vaginal and cervical epithelium. The organism's glycoprotein coat is an important factor in the development of human vaginal candida infection because it produces a cytolytic enzyme. This, together with a capacity for intracellular growth (Schnell and Voigt, 1976) helps protect the organism from host defences and from antimycotic therapy (Schmidt, 1987).

Other fungal infections causing cervicovaginitis

A variety of other fungi may rarely cause vaginitis. Less than 12% of mycotic infections of the vagina are caused by *Torulopsis glabrata* or *Geotrichum candidum*. *Torulopsis* sp. which has encapsulated double walled spores and long branching hyphae reproduces by budding. *Geotrichum* reproduces by forming arthrospores. Other fungal infections that may be found in the vagina include *Cryptococcus* sp., *Blastomyces* sp., *Aspergillus fumigatus*, *Trichosporon* sp. and *Saccharomyces cerevisiae*.

Chlamydia trachomatis

Chlamydiae are obligate intracellular organisms which contain both DNA and RNA. They are Gram-negative and stain with aniline dyes including Giemsa stain. Chlamydiae are divided into two sub-groups, A and B (*Table 9.2*). Sub-group A or *Chlamydia trachomatis* is exclusively a human pathogen which is further sub-divided by serotyping; serotypes A, B, Ba and C cause hyperendemic trachoma; serotypes D–K cause genital infections in both sexes and eye infections such as inclusion conjunctivitis; serotypes L1–L3 cause lymphogranuloma venereum. Sub-group B or *Chlamydia psitacci* is composed of many serotypically distinct chlamydiae some of which can cause pneumonia in man. The organisms that comprise sub-group A form compact inclusions that are glycogen positive. They are also sensitive to sulphadiazines. The chlamydiae which comprise sub-group B form diffuse inclusions that are glycogen negative, resistant to sulphadiazines, and mainly infect birds and mammals.

Table 9.2 Human diseases caused by *Chlamydia*

Species	Serotype	Disease
Sub-group A		
C. trachomatis	A,B,Ba,C	Hyperendemic trachoma
C. trachomatis	D,E,F,G,H,I, J,K and L	Inclusion conjunctivitis (adult and neonate); non-gonococcal urethritis, cervicitis, salpingitis, proctitis, epididymitis; pneumonia of newborn
C. trachomatis	L1,L2,L3	Lymphogranuloma venereum
Sub-group B		
C. psittaci	Many	Psittacosis

Developmental cycle

The developmental cycle of *Chlamydia trachomatis* is complex and occurs over a period of 36–40 h. Infectious elementary bodies 250 μm in diameter attach to the surface of a host cell and are engulfed by a process of pinocytosis. A vacuole consisting of glycogen matrix develops around the particle and during the next few hours the elementary body transforms into a larger initial body which then multiplies by binary fission. At this stage replication stops and the initial bodies reorganize into elementary bodies which are then released from the host cell to infect other cells.

Laboratory diagnosis

Laboratory diagnosis of *C. trachomatis* may be achieved by several methods. These include isolation in cell culture (Schachter, 1978), diagnosis by immunofluorescence using monoclonal antibody to detect chlamydial elementary bodies (Foster et al., 1985) and enzyme-linked immunoassays (Pugh et al., 1985). Because these methods are expensive and time-consuming, cervical smears have been studied in an effort to detect cellular changes pathognomonic for infection.

The cytological changes attributed to *Chlamydia trachomatis* infection of the cervix were originally described by Naib (1970) and later modified by Gupta et al. (1979). These authors described granular, uniformly sized coccoid

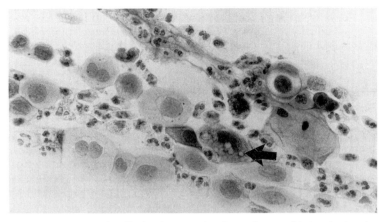

Figure 9.17 Metaplastic cells showing vacuolated cytoplasm and intracytoplasmic inclusions. These changes might reflect *Chlamydia* infection, but must be interpreted with caution. Stain: Papanicolaou. Magnification ×770

bodies in the cytoplasm of cells in cervical smears which they assumed were chlamydial inclusions. Frequently these inclusions appeared to mould to or overlap each other (*Figure 9.17*). Columnar and metaplastic cells are said to be more frequently involved than squamous epithelial cells. Elementary bodies in the background of the smear were also described.

The value of cytology as a method of diagnosing *Chlamydia trachomatis* in the cervix has been challenged by Dorman, Danos and Wilson (1983) and Forster *et al.* (1985). Both groups compared cytological diagnosis of this infection in Papanicolaou-stained cervical smears with culture and immuofluorescence of cervical scrapes; each investigation was carried out independently. Correlation was noted in less than 40% of cases, no greater than would be expected by chance. These studies indicate that the cytodiagnosis of chlamydiae in cervical smears is unreliable. Even with experience the distinction between chlamydial inclusions, ingested red cell fragments and degenerated polymorphs is extremely difficult to make.

On the other hand cytology is an effective method of diagnosing inclusion conjunctivitis in both adults and neonates (Coleman, 1979). *Chlamydia trachomatis* can be identified in 75% of patients when conjunctival smears are taken in the acute stages of the disease.

Chlamydia trachomatis is now recognized as an extremely common sexually transmitted disease with an incubation period of 6–14 days. It can infect both sexual partners. In the pregnant female the fetus may be infected during parturition resulting in chlamydial conjunctivitis or pneumonia in the neonate. In men it is the commonest cause of non-specific urethritis. In women cervicitis caused by *C. trachomatis* may be complicated by endometritis and salpingitis (Oriel, 1983), although half the women who are infected are symptom-free. Genital infection in the female has been associated with follicular cervicitis (Hare *et al.*, 1981), and the presence of an atypical transformation zone with punctation and/or mosaicism (Paavonen, 1979; Paavonen *et al.*, 1982). However, a causal relationship between *C. trachomatis* and cervical cancer has not been established.

Lymphogranuloma venereum

Lymphogranuloma venereum is a genital infection which is almost always acquired in the tropics and is rarely seen in temperate climates. It is due to infection with chlamydial serotypes L1–L3 and is a much more severe infection than the chlamydial infections due to serotypes D–K. It is characterized by the development of a painless primary sore (like syphilis) followed days or even weeks later by tender, enlarged lymph nodes in the groin; abscesses form within the lymph nodes which discharge and heal with fibrosis. As the disease progresses the inflammatory process may result in fistulae and scarring, commonly involving the rectum and less commonly the vagina. Diagnosis is by culture and serology. No specific cytological changes have been described.

Viruses

Pathogenic viruses are obligate intracellular parasites which produce a variety of diseases in their host. Their cytotoxic effect is due in part to the complex biochemical disturbances that accompany viral replication. They can be classified according to their nucleic acid content (which may be composed of RNA or DNA) and according to the symmetry of the capsomers (which may be helical or cubical). Infection may affect the cell in several ways. For example, the virus may replicate within the cell destroying it completely (lytic infection). This type of infection results in the formation of inclusions which can be readily identified in cervical smears or may cause cellular degeneration and cell death. Lytic infections usually resolve completely with the virus being totally eliminated from the body. However, virus particles may persist in the host cells in very small numbers resulting in latent infection. Reactivation may occur at any time to produce a lytic infection particularly if the patient's immune system is impaired.

Some viruses have the potential to transform cells and induce neoplastic change. This phenomenon has been demonstrated in laboratory animals although the evidence for malignant transformation in man is largely circumstantial. Herpes viruses have been implicated in the pathogenesis of Burkitt's lymphoma and nasopharyngeal carcinoma. The evidence for the involvement of human papillomavirus in cervical carcinogenesis is discussed in Chapter 10.

Several viruses are known to infect the anogenital region in both male and female. These include herpes simplex, cytomegalovirus, herpes zoster, human papillomavirus and pox viruses. These cause cytopathic changes which can be detected in cervical smears.

Herpes viruses

Several species of the genus *Herpes* are known to infect man (*Table 9.3*). These include herpes simplex virus types 1 and 2, varicella/zoster virus, Epstein-Barr virus and cytomegalovirus. All are DNA viruses with a characteristic capacity to cause lytic infection or to remain latent in their hosts. The viruses are composed of a double-stranded DNA core surrounded by capsid of hollow hexagonal and pentagonal capsomers arranged in icosahedral symmetry. The mature virus is 120 nm in diameter and has an outer lipid envelope.

Herpes simplex virus (HSV)

Herpes simplex virus (HSV) is a common human pathogen. Two main antigenic types have been identified. Type 1 is associated with vesicular and ulcerating lesions of mouth, lips, eyes and skin on the upper limbs, chest and back. It may also cause an encephalitis. Type 2 virus is associated with genital and anal infections and skin lesions of the thighs. It has also been identified as a cause of meningitis (Oriel, 1983). The viruses may be transmitted by sexual intercourse or after physical contact and according

Table 9.3 Human viral diseases which can be detected by cytology

Family	Virus	Disease	Cytological specimen
Herpes			
	Herpes simplex 1	Face and other sites	Skin scrape, sputum
	2	Genital infections	Cervical smears
	Herpes varicella/zoster	Chickenpox/shingles	Scrape from vesicle
	Cytomegalovirus	Cytomegalic inclusion disease	
		Pneumonia in immunosuppressed patients	Sputum, bronchoalveolar lavage
		Opportunistic infection	Cervical smear, urine sediment
Papova			
	Papilloma	Condylomatous lesions	Cervical smear
	Polyoma	Ureteric obstruction in immunosuppressed patients, haemorrhagic cystitis, progressive multifocal leucoencephalopathy (PML)	Urine sediment

to Barton *et al.* (1982) up to 60% of genital lesions may be due to type 1 virus possibly through anogenital contact.

The diagnosis of herpes simplex can be made in several ways. Tissue culture and isolation of the organism is successful in about 60% of cases. Serological tests may be used as an aid in the diagnosis of primary infection or to type the viral strains; and electron microscopy can be used to demonstrate HSV particles in infected cells. Cytology is a reliable method of detecting active infection but is not as sensitive as culture (Naib, 1981). Immunoperoxidase and immunofluorescence techniques using labelled anti-HSV antiserum have been applied to cervical smears with varying degrees of success (Gardner *et al.*, 1968).

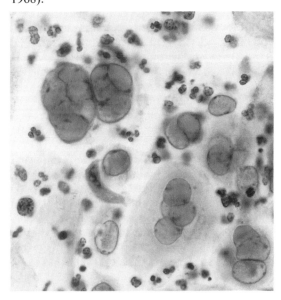

Figure 9.18 Herpes simplex infected multinucleate cells showing nuclear moulding, margination of chromatin and a 'ground glass' appearance to the nuclei. Stain: Papanicolaou. Magnification ×770

The cytological diagnosis of genital herpes infection in Papanicolaou smears depends upon the recognition of large multinucleate cells (*Figure 9.18*). Infection with HSV initially causes hypertrophy of both the nucleus and the cytoplasm of the affected cells followed by enlargement of the nucleoli and displacement of the nuclear chromatin to the periphery of the nucleus. Multinucleate giant cells between 20 and 30 µm in diameter develop as a result of cell fusion. They contain up to 30 nuclei which show a characteristic moulding without overlapping which helps to differentiate the cells from multinucleate histiocytes. The marginated chromatin confers an empty ballooned 'ground glass' appearance to the nucleus. Sometimes intranuclear inclusions may be seen which appear as large round centrally placed acidophilic bodies surrounded by a clear halo. The absence of inclusion bodies does not invalidate the diagnosis. It merely represents a different stage of the disease and is not as some authors suggest evidence of recurrent infection. Single squamous and columnar cells may also have a 'ground glass' nucleus but these cells alone without the multinucleate cells are not pathognomonic of herpes simplex.

Primary genital herpes is a sexually transmitted disease with an incubation period of about seven days. The first attack usually presents with painful genital vesicles which ulcerate very rapidly. They are found in the female on the vulva, vagina, cervix and perineum and in the male on the penis and scrotum. In homosexuals the vesicles are found around the anus. Inguinal lymphadenopathy occurs in approximately 50% of primary infections and sacral radiculitis and meningitis are occasional complications. Recurrent infections are usually less severe and shorter than the initial attack, the virus lying dormant in the sacral ganglia between attacks. Initial and recurrent infection by herpes may be asymptomatic which raises considerable problems with regard to transmission of this highly infectious disease. Transmission of HSV 2 to the fetus via the placenta during pregnancy causes fetal malformation and miscarriages but fortunately is rare. Infection can occur at birth as a result of delivery of the fetus through an infected cervix (Nahmias, 1978; Kobayashi *et al.*, 1982). For many years there was no effective treatment for herpes infection. Recently, acyclovir has been used to control the disease. This drug inhibits the replication of the virus and reduces the level and duration of the infection.

It has been suggested that there is an association between herpes infection of the lower female genital tract and cervical neoplasia. This is based on the fact that herpes antibodies are more frequently found in patients with cervical cancer than in normal women. However, failure to demonstrate the viral DNA or viral antigens in the tumour has cast doubt on the validity of this hypothesis. If herpes does play a role in cervical carcinogenesis it is probably as an initiator of neoplastic change (Galloway and McDougall, 1983).

Herpes zoster

Herpes zoster can cause infection of the vulva. Scrapings taken from the edges of the vesicles show cells with the same morphological characteristics as those from herpes simplex so that the two types of infection cannot be differentiated by cytology. However, the Tzanck test (Tzanck, 1948) can be used to distinguish between the vesicles of varicella zoster infection and those of pemphigus. Scrapings from the floor of the vesicles are spread on a glass slide and stained by the May-Grünwald–Giemsa method. The presence of rounded acantholytic cells is characteristic of pemphigus. These can readily be distinguished from the multinucleate giant cells of herpes in Papanicolaou or Giemsa-stained smears (Canti, 1984).

Cytomegalovirus

Cytomegalovirus (CMV) may be sexually transmitted and can affect the cervix without any evidence of systemic disease. In women of child-bearing age infection is usually mild or sympmay threaten the health of the baby. Transproduce characteristic morphological changes placental transmission of virus may cause disease of the fetus *in utero* affecting the central nervous system (Harris, 1975). Fortunately less than 2% of babies harbouring CMV will have impairment of their CNS. Although 10% (or more) of pregnant women are found to have the virus in the cervix, cytopathic changes are rarely seen in cervical smears. A large basophilic inclusion body separated from the nuclear membrane by a clearly defined halo may be seen in some cells. This gives the cells an owl's eye appearance. Occasionally intracytoplasmic inclusion bodies may be found. Endocervical cells are more commonly affected by the virus than other cell types. Verification that the inclusion-bearing cells are the result of cytomegalovirus can be made by immunofluorescence. However, serological tests or virus isolation techniques are to be preferred to cytology for the diagnosis of this infection.

Papovaviruses

Papovaviruses are a group of species-specific small DNA viruses which infect a wide range of animals and which replicate exclusively in the nucleus of the host cell. Two sub-groups are recognized: the *papillomaviruses* which are the causal agent of a wide range of papillomas and warts in animals and man, and the *polyomaviruses* which have been isolated from the tissue of a number of mammals (including man) and which are highly tumorigenic under laboratory conditions.

Human papillomavirus

Human papillomaviruses (HPV) are a heterogeneous group of viruses which cause a variety of proliferative lesions in squamous epithelium (Gissmann, 1984). The virions, which measure approximately 50 nm in diameter, are unenveloped particles with an icosahedral symmetry. The genome is a single molecule of double-stranded circular DNA (Andrewes, Pereira and Wildy, 1978). Recent advances in DNA technology using restriction enzyme analysis and molecular hybridization have shown that there are many distinct types of HPV. Over 50 have now been characterized and each HPV type appears to be associated with a specific pathological condition (Singer, Campion and McCance, 1985). For example genotype 1 is associated with plantar warts, types 5 and 8 are associated with the rare autosomal recessive disorder epidermodysplasia verruciformis and type 11 is associated with laryngeal papillomas.

To date 15 types have been isolated in the female genital tract of which types, 6, 11, 16 and 18 have been studied most intensively. HPV 6 and 11 have been shown to have homology to the order of 85% and may be put into the same typing group. It is this group which has been associated with benign genital warts and cervical dysplasia (CIN I). On the other hand, types 16 and 18 have been implicated in CIN III and invasive carcinomas. This association of human papillomaviruses with malignant disease is discussed further in Chapter 10.

All genotypes of HPV appear to infect the basal cells of the epithelium (Beckmann et al., 1985). The initial activity of the viral genome has a stimulatory effect on growth of the host cell and induces proliferation of the epithelium. Many HPV infections allow subsequent differentiation of the infected cells but this differentiation is often modified towards the surface. The complete virus is only found in keratinizing cells but it should be noted that many of the virus types that have been identified exist only in an incomplete form and can be identified only as DNA sequences.

Infection of the genital tract

Genital human papillomavirus infection is a sexually transmitted disease with an incubation period of 6–8 weeks (Oriel, 1983). In men the commonest site of infection is the penis and in women the vulva and perineum. It is also a common disease in anoreceptive homosexuals. Infection with human papillomavirus produces specific morphological alterations in cells. These changes were first noticed in smears by Papanicolaou (1933), and later Koss and Durfee (1956) coined the term koilocytotic atypia to describe these cellular anomalies. Some years on, Meisels and Fortin (1976), and Meisels, Fortin and Roy (1977) and Purola and Savia (1977) showed by cytological, histological and colposcopical observations that two different types of wart virus infection are found in the cervix and vagina, namely flat warts and exophytic warts. Meisels, Fortin and Roy (1977) and Purola and Savia (1977) suggested that because of similar cytomorphology flat warts and exophytic wart lesions shared the same viral aetiology.

Today it is accepted that in the female genital tract HPV is associated with three types of infection:

(1) Subclinical papillomavirus infection (SPI) which has also been called non-condylomatous wart virus infection or flat warts;
(2) Exophytic warts or condylomata acuminatum; and
(3) Endophytic warts.

Subclinical papillomavirus infection (SPI)

It has now been shown that most cervical infections caused by HPV are sub-clinical, i.e. the lesions are flat and invisible macroscopically but produce characteristic morphological changes in the cervical squamous epithelium in histological section (Singer et al., 1985). These lesions may develop into a condylomatous lesion but usually regress spontaneously after a few months. Three histological types of subclinical lesions have been described (*Figure 9.19*).

(1) A discrete focus of HPV infection in an otherwise normal squamous epithelium. This is characterized histologically by basal cell hyperplasia, acanthosis, parakeratosis, hyperkeratosis, multinucleation and an unusual cellular pattern that has been designated koilocytotis atypia. This is characterized by cells with an empty or hollow cytoplasm, and slight atypia of the nuclei with minimal nuclear enlargement.
(2) A discrete focus of HPV infection but adjacent to an era of cervical intraepithelial neoplasia (CIN).
(3) HPV infection arising in an area of cervical intraepithelial neoplasia of any grade. Histologically this is characterized by an area of CIN in which the surface dyskaryotic cells show koilocytotic changes. These koilocytes have the dyskaryotic nuclei characteristic of the CIN lesion in which the infection is located.

HPV 6, 11, 16 and 18 may be found in any of these lesions although types 16 and 18 are more commonly associated with those infections which occur adjacent to an area of CIN III.

Cytology

The cells in the cervical smear obviously reflect the underlying histological changes, therefore in subclinical HPV infections the smear pattern differs depending on the type of lesion present. However, the cytological hallmark of all subclinical papillomavirus infections is the koilocyte. Koilocytes are large rounded cells in which the nucleus is surrounded by a large clear or transparent zone bordered by a dense condensation of cytoplasm (*Figure 9.20*). Binucleation and multinucleation may be present (*Figure 9.21*). The degree of nuclear abnormality varies from slight to moderate anisonucleosis (*Figure 9.22*) to enlarged, hyperchromatic nuclei with irregular chromatin. Keratinized intensely eosinophilic parakeratotic cells (dyskeratocytes) are also frequently found.

Exophytic warts and condyloma acuminatum

Condylomatous lesions of the female tract are frequently caused by HPV types 6 and 11. They present as papillary masses which may be discrete or multiple. Condylomata are common on the vulva and in the vagina but are rare in the cervix, being found in less that 6% of women with genital warts (Oriel, 1971; Walker et al., 1983). During pregnancy HPV infection can produce large masses which mimic fungating squamous carcinoma. Condylomata have also been recorded following laser therapy. Histologically the epithelium shows changes similar

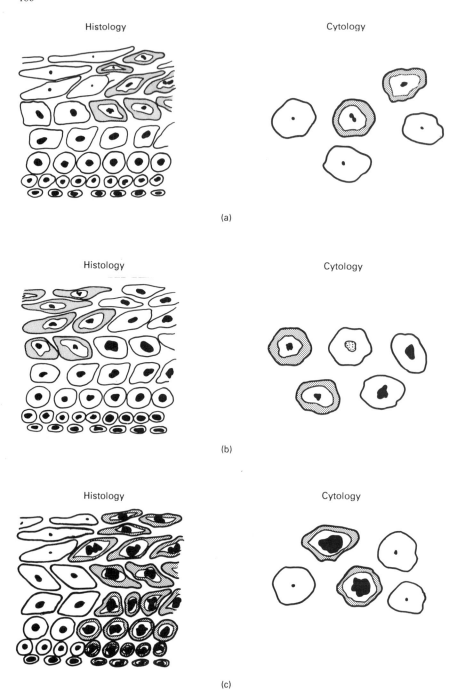

Figure 9.19 Schematic drawing showing the histological and cytological presentation of koilocytotic atypia in subclinical papilloma virus infection of the cervix. (*a*) Wart virus infection involving squamous epithelium; (*b*) wart virus infection adjacent to an area of CIN; (*c*) wart virus infection involving an area of CIN

Specific infections 187

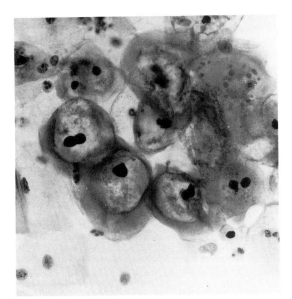

Figure 9.20 Koilocytes in a cervical smear showing a large clear zone around the cell nuclei. Stain: Papanicolaou. Magnification × 770

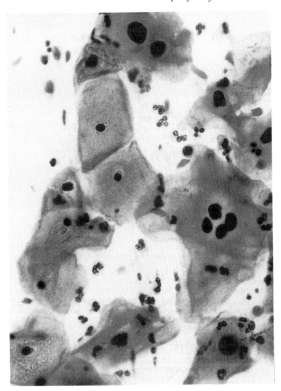

Figure 9.22 Cervical smear exhibiting changes associated with wart virus infection including nuclear hyperchromasia and multinucleation. Stain: Papanicolaou. Magnification × 770

to those seen in sub-clinical lesions i.e. acanthosis, multinucleation, parakeratosis, hyperkeratosis, koilocytosis. In addition papillomatosis is a prominent feature. In cervical smears, koilocytes showing minor alterations of the nucleus are found together with parakeratotic cells. Sometimes bizarre cell changes may also be noticed.

Endophytic warts

Endophytic or inverted warts are rare and may be mistaken for carcinoma, particularly verrucous carcinoma. They do not produce koilocytes.

Laboratory diagnosis

To date HPV cannot be propagated by *in vitro* culture methods. The diagnosis, therefore, other than the morphological changes seen in histological and cytological preparation which are not definitive, depends on the demonstration of viral particles, viral antigen and viral DNA.

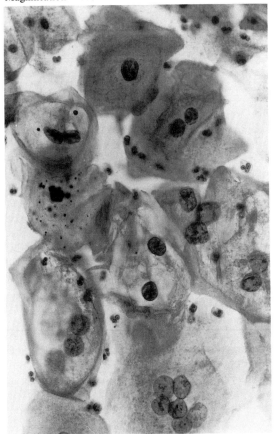

Figure 9.21 Koilocytosis in a cervical smear. Note binucleation and multinucleation in cells. Stain: Papanicolaou. Magnification × 770

Viral particles can be demonstrated by electron microscopy, ideally using fresh material but a diagnosis can be made retrospectively using alcohol-fixed smears or paraffin-embedded tissue (see Chapter 4). Alternatively HPV can be demonstrated by immunocytochemistry using an antibody raised against a common protein of the internal capsid of the virion. Detection rates by this method are not high as the antigen is only expressed in 30–50% of cells. The most sensitive method of detecting the virus is by DNA hybridization of unfixed cervical cells or fresh biopsy tissue using radioactive probes (Byrne et al., 1986). By this method up to 90% of cases can be detected. The specific viral type associated with the lesion can also be defined. More recently Beckmann et al. (1985) have used non-radioactive probes to study the distribution of viral DNA by in situ hybridization of Papanicolaou smears and paraffin-embedded cervical tissue with encouraging results.

Human papillomaviruses have been implicated in the pathogenesis of cervical cancer (Walker et al., 1983; McCance et al., 1985; Mitchell, Drake and Medley, 1986) and yet studies in normal women have shown that at least 30% harbour the virus with no ill effect. Whether these viral genotypes are simply common infectious agents that run a chronic course or are truly oncogenic agents remains to be seen. At the present time large-scale epidemiological studies are underway to determine whether the presence of HPV DNA in the genome increases the risk of cervical cancer.

Adenoviruses

Adenoviruses vary in size from 60–90 nm, the capsid has a cubical symmetry and there is no outer envelope. They cause upper respiratory tract infection and keratoconjunctivitis. Cellular changes consistent with adenovirus infection have been described in cervical smears. The epithelial cells are reported to show single or multiple eosinophilic amorphous intranuclear inclusions (Laverty et al., 1977; Vesterinen et al., 1978). However, the virus has never been isolated from the cervix and the cytological changes attributed to these viruses need to be interpreted with caution.

Poxviruses

Poxviruses vary in size from 50–200 nm, the capsid has a complex symmetry and there is an outer envelope. They cause smallpox vaccinia and molluscum contagiosum in man. Molluscum contagiosum is a sexually transmitted disease with an incubation period of between two and six weeks. The pearly, raised, umbilicated lesions mainly affect the penis, vulva and perigenital tissues. The diagnosis may be made from the cellular sample by scraping the lesion and finding large inclusions within the cytoplasm of parabasal cells (Canti, 1984; Bibbo and Weid, 1988).

Parasites

Parasitic infections of the cervix and vagina are common but there is marked geographical variation in the frequency with which they can be found. The more common parasites found in the lower female genital tract are mentioned briefly in this chapter for completeness (for a more detailed description see Chapter 23).

Trichomonas vaginalis

Trichomonas infection of the human genitalia occurs in both sexes. Although normally regarded as a sexual transmitted disease, the accidental spread of *Trichomonas vaginalis* by inanimate objects, e.g. toilet seats, and by cross infection resulting from gynaecological examination cannot be ruled out. Trichomonas infection can be found in 20% of women and is associated with a frothy yellow-green discharge with an unpleasant odour. The parasite may, however, also be demonstrated in the absence of clinical symptoms or signs.

In cervical smears the protozoa are usually present in large numbers. They appear as pyriform, faintly greyish-blue organisms with an ill-defined elongated nucleus (*Figure 9.23*). Pink-staining cytoplasmic granules are usually seen. The epithelial cells frequently show pseudo-eosinophilia and perinuclear vacuolation (halo effect) (*Figure 9.24*) as well as non-specific inflammatory changes. The diagnosis of trichomoniasis from Papanicolaou-stained smears is as reliable as either wet films or culture but for the best results all three methods should be used in conjunction (Cassie and Stevenson, 1973; Thin et al., 1975). Morton (1975) has shown that *Trichomonas vaginalis* frequently accompanies other venereal diseases and Berggren (1969)

Specific infections

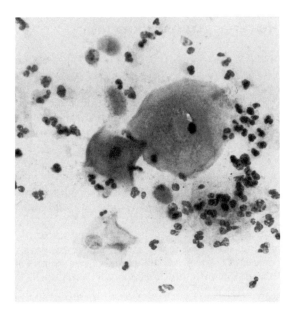

Figure 9.23 *Trichomonas vaginalis* infection in a cervical smear. The pyriform parasites vary in size and show an ill-defined elongated nucleus. Stain: Papanicolaou. Magnification × 770

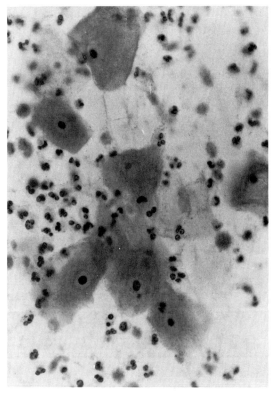

Figure 9.24 Squamous epithelial cells showing perinuclear haloes. These changes are frequently seen in *Trichomonas vaginalis* infection but may be found in association with other infections causing severe inflammatory cellular reactions. Stain: Papanicolaou. Magnification × 770

has shown that, contrary to some early reports, there is no association between *Trichomonas* infection and cervical cancer.

Entamoeba histolytica

Infections caused by *Entamoeba* are widespread in tropical and sub-tropical areas of the world. *Entamoeba histolytica* is usually associated with intestinal and hepatic disease but Cohen (1973) has shown that spread from anus to perineum and thence to vagina may occur. Mylius and Ten Seldam (1962) have suggested that direct venereal infection is also a possibility. Cervical smears taken from women with *Entamoeba histolytica* of the genital tract contain numerous trophozoite forms of the parasite. These, however, must be differentiated from trophozoites of *E. gingivalis* (Chapter 23), another member of the genus which is a common commensal of the mouth and is sometimes found in the genital tract in patients with an IUCD (Gupta, 1982). Ulceration of the cervix caused by *E. histolytica* can simulate cancer of the cervix, but as the parasite is known to coexist with cervical carcinoma (Carter, Jones and Thomas, 1954), a biopsy may be necessary to establish the diagnosis.

Balantidium coli

Balantidium coli is a protozoan which usually inhabits the intestinal tract but may rarely be found in the lower female genital tract. In cervical smears the precystic form may be seen.

Schistosomiasis

Schistosomiasis is a widespread disease in the tropics and is caused by *S. haematobium*, *S. mansoni* or *S. japonicum*. The life cycle is complex but it has been shown that in the female genital tract the preferential sites for the deposition of schistosome ova are the cervix and vagina (Williams, 1967).

The reaction of the cervical and vaginal tissue to this invasion varies. There may be no inflammatory reaction when the ova are freshly deposited; however, a severe granulomatous reaction may be seen in large-scale infection. Giant histiocytes may also be found. The cytological diagnosis of schistosomiasis depends on the identification of the ova and miracidia (Berry, 1971). Visualization of the spine is essential but may be difficult particularly if the ova are degenerate. *S. mansoni* ova are strongly acid-fast and give a positive staining reaction with Ziehl-Neelsen stain which may prove helpful.

Other parasites found in smears

Numerous other parasites may be found in cervical smears (Koss, 1979), the most common being *Enterobius vermicularis* (pin worm) (Bhambhani *et al.*, 1985) which is frequently found in children. Other nematodes include *Trichuris trichura* (Garud, Saratya and Paraskar, 1980), *Ascaris lumbricoides* (Bhambhani, 1984) and *Strongyloides stercoralis*. Rare findings such as vorticella (Hermann and Deininger, 1963) and filariae (De Borges, 1971) have also been recorded. Occasionally small insects such as mites or water fleas may be observed.

Regeneration and repair

When tissue is destroyed by inflammation the surviving epithelial cells endeavour to cover the defective surface. The defect is made good by proliferation of the cells from adjacent intact epithelium. This process of repair can be seen in both histological and cytological preparations (Bibbo, Keebler and Weid, 1971; Geirsson *et al.*, 1977). In Papanicolaou smears undifferentiated regenerating epithelial cells occur predominantly in sheets with indistinct cell boundaries. Nuclei are larger than normal and variation of nuclear size in the sheet can be marked (*Figure 9.25*). The chromatin pattern is regular, usually finely granular and evenly dispersed (*Figure 9.26*). Chromocentres may be found. Eosinophilic nucleoli are often prominent and multiple, showing some variation of size from one nucleus to another in the same sheet. Most nucleoli are regular in shape but occasionally irregular configurations are seen. Cytoplasmic staining is usually basophilic but

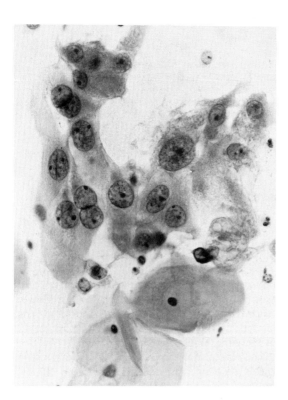

Figure 9.25 Cells from tissue repair showing pleomorphism and prominent nucleoli. Stain: Papanicolaou. Magnification × 770

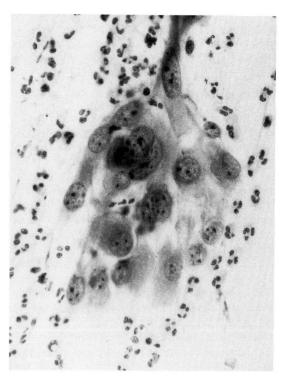

Figure 9.26 Regenerating epithelial cells showing marked variation in nuclear size but a fine chromatin pattern. Chromocentres can also be seen. Stain: Papanicolaou. Magnification × 770

can be amphophilic. Normal mitotic figures may be found (*see* also Chapter 3).

Sheets of cells from regenerating epithelium can cause problems of differential diagnosis (Geirsson *et al.*, 1977), especially if too much weight is given to pleomorphism and prominent nucleoli. It is important to recognize the benign appearance of the finely granular, dispersed chromatin pattern and to note that cellular cohesiveness is also maintained (*Figure 9.27*).

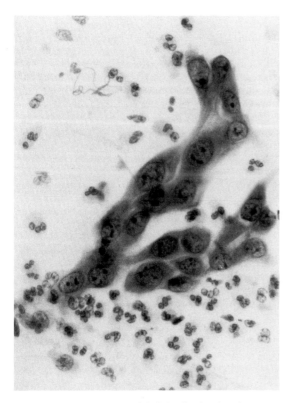

Figure 9.27 Regenerating epithelial cells showing that cellular cohesiveness is maintained during the repair process. Stain: Papanicolaou. Magnification ×512. (With permission of E. B. Butler)

These features serve to distinguish regenerating epithelium from neoplastic epithelium. It should also be remembered that repair may occur simultaneously with acute inflammation and destruction of the epithelium so that degenerative and regenerative changes may be found in the same smear.

Fibrosis

If tissue destruction is so extensive that reconstruction of the epithelium by regeneration cannot be effected, repair is achieved by fibrosis.

Fibrosis is associated with the production of collagen fibres by fibroblasts which are supported in the early stages by the growth of numerous capillaries (granulation tissue) and by a mixture of inflammatory cells. As the area of fibrosis matures the collagen fibres shrink and the vascularity and cellularity of the tissue diminish over about six months, producing a scar. In such cases fibroblasts may be noted in the smear.

References

ANDREWES, C., PEREIRA, H. G. and WILDY, P. (1978). *Papovaviridal Viruses of Vertebrates*, 4th edn, pp. 273–292. London: Baillière Tindall

BARTON, I. G., KINGHORN, G. R., NAJEM, S., AL-OMAR, L. S. and POTTER, C. W. (1982). Incidence of Herpes simplex virus types 1 and 2 isolated in patients with Herpes genitalis in Sheffield. *Br. J. Vener. Dis.*, **58**, 44–47.

BECKMANN, A. M., MYERSON, D., DALING, J., KIVIAT, N. B., FENOGLIO, C. M. and MCDOUGALL, J. K. (1985). Detection and localisation in human genital condylomas by *in situ* hybridisation with biotinylated probes. *J. Med. Virol.*, **16**, 265–273

BERGGREN, O. (1969). Association of carcinoma of the uterine cervix and *Trichomonas vaginalis* infestations. *Am. J. Obstet. Gynecol.*, **105**, 166–168

BERRY, A. (1971). Evidence of gynaecologic bilharziasis in cytologic material. *Acta Cytol.*, **15**, 482–498

BHAGAVAN, B. S. and GUPTA, P. K. (1978). Genital actinomycosis and intrauterine contraceptive devices: cytopathologic diagnosis and clinical significance. *Human Pathol.*, **9**, 567–578

BHAMBHANI, S. (1984). Eggs of *Ascaris lumbricoides* in cervical vaginal smears. *Acta Cytol.*, **28**, 92

BHAMBHANI, S., MILNER, A., PANT, J. and LUTHRA, U. (1985). Ova of *Taenia* and *Enterobius vermicularis* in cervicovaginal smears. *Acta Cytol.*, **29**, 913–914

BIBBO, M. and HARRIS, M. J. (1972). *Leptothrix. Acta Cytol.*, **16**, 66–74

BIBBO, M. and WEID, G. L. (1988). Microbiology and inflammation of the female genital tract. In *Compendium on Diagnostic Cytology* (Weid, G. L., Keebler, C. M., Koss, L. G. and Reagan, J. W., eds). Chicago: Tutorials of Cytology

BIBBO, M., KEEBLER, C. M. and WEID, G. L. (1971). The cytologic diagnosis of tissue repair in the female genital tract. *Acta Cytol.*, **15**, 133–137

BLENKINSOPP, W. K. and CHAPMAN, P. A. (1982). Prevalence of cervical neoplasia and infection in woman using intrauterine contraceptive devices. *J. Reprod. Med.*, **27**, 709–713

BOON, M. E. and TABBERS-BOUMEESTER, M. L. (1980). Benign changes, infection and inflammation. *Gynaecological Cytology: A Textbook and Atlas*, pp. 57–70. London: Macmillan

BYRNE, M. A., MØLLER, B. R., TAYLOR-ROBINSON, D. et al. (1986). The effect of interferon on human papilloma virus associated with CIN monitored by cytology, colposcopy and DNA hybridisation. Br. J. Obstet. Gynaecol., 93, 1136–1144.

CANTI, G. (1984). Rapid cytological diagnosis of skin lesions. In Advances in Clinical Cytology, Vol. 2, (Koss, L. G. and Coleman, D. V., eds), pp. 243–266. New York: Masson

CARTER, B., JONES, C. P. and THOMAS, W. L. (1954). Invasion of squamous cell carcinoma of the cervix uteri by Entamoeba histolytica. Am. J. Obstet. Gynecol., 68, 1607–1610

CASSIE, R. and STEVENSON, A. (1973). Screening for gonorrhoea, trichomoniasis, moniliasis and syphilis in pregnancy. J. Obstet. Gynaecol. Br. Commonw., 80, 48–51

CHRIST, M. L. and HAJA, J. (1978). Cytologic changes associated with vaginal pessary use: with special reference to the presence of Actinomyces. Acta Cytol., 22, 146–149

COHEN, C. (1973). Three cases of amoebiasis of the cervix uteri. J. Obstet. Gynaecol. Br. Commonw., 80, 476–479

COLEMAN, D. V. (1969). A case of tuberculosis of the cervix. Acta Cytol., 13, 104–107

COLEMAN, D. V. (1979). Cytological diagnosis of virus infected cells in Papanicolaou smears and its application in clinical practice. J. Clin. Pathol., 32, 1075–1089

COLEMAN, D. V. and EVANS, D. M. (1988). Infection, inflammation and repair. In Biopsy, Pathology and Cytology of the Cervix, pp. 77–130. London: Chapman and Hall

DATTANI, I. M., GERKEN, A. and EVANS, B. A. (1982). Aetiology and management of non-specific vaginitis. Br. J. Vener. Dis., 58, 32–35

DAWSON, S. G. and HARRIS, J. R. (1983). Venereal disease, Gardnerella vaginalis and non-specific vaginitis. Br. J. Hosp. Med., 1, 28–37

DE BORGES, R. (1971). Findings of microfilarial larval stages in gynecologic smears. Acta Cytol., 15, 476–478

DORMAN, S. A., DANOS, L. M. and WILSON, D. J. (1983). Detection of chlamydial cervicitis by Papanicolaou stained smears and cultures. Am. J. Clin. Pathol., 79, 421–425

DUGUID, H. L. D., PARRATT, D. and TRAYNOR, R. (1980). Actinomyces-like organism in cervical smears from women using intrauterine contraceptive devices. Br. Med. J., 281, 534–535

FORSTER, G. E., COOKEY, I., MANDRAY, P. E. et al. (1985). Investigation into the value of Papanicolaou-stained cervical smears for the diagnosis of chlamydial cervical infection. J. Clin. Pathol., 38, 399–402

FRY, R., LINDER, A. M. and BULL, M. M. (1980). Actinomyces-like organisms in cervicovaginal smears. S. Afr. Med. J., 57, 1041–1043

GALLOWAY, D. A. and MCDOUGALL, J. K. (1983). The oncogenic potential of HSV: evidence for a hit and run mechanism. Nature, 302, 21–24

GARDNER, H. and DUKES, C. D. (1955). Haemophilus vaginalis: a newly defined specific infection previously classified 'non-specific vaginitis'. Am. J. Obstet. Gynecol., 69, 962–976

GARDNER, P. S., MCQUILLIN, J., BLACK, M. M. and RICHARDSON, J. (1968). Rapid diagnosis of Herpesvirus hominis infections in superficial lesions by immunofluorescent antibody techniques. Br. Med. J., iv, 89–90

GARUD, M. A., SARAIYA, U. and PARASKAR, M. (1980). Vaginal parasitosis. Acta Cytol. 24, 34–35

GEIRSSON, G., WOODWORTH, F. E., PATTEN, S. F. and BONFIGLIO T. A. (1977). Epithelial repair and regeneration in the uterine cervix. I. An analysis of the cells. Acta Cytol., 21, 371–378

GISSMANN, L. (1984). Papillomaviruses and their association with cancer in animals and in man. Cancer Surveillance, 3, 161–181

GUPTA, P. K. (1982). Intrauterine contraceptive devices: vaginal cytology, pathological changes and clinical implications. Acta Cytol., 26, 571–613

GUPTA, P. K., HOLLANDER, D. H. and FROST, J. K. (1976). Actinomyces in cervico-vaginal smears: an association with IUD usage. Acta Cytol., 20, 295–297

GUPTA, P. K., LEE, E. F., EROZAN, Y. S., FROST, J. K., GEDDES, S. T. and DONOVAN, P. A. (1979). Cytologic investigations in Chlamydia infection. Acta Cytol., 23, 315–320

HARE, M. J., TOONE, E., TAYLOR-ROBINSON D. et al. (1981). Follicular cervicitis—colposcopic appearances and association with Chlamydia trachomatis. Br. J. Obstet. Gynaecol., 88, 174–180

HARRIS, J. R. W. (1975). Cytomegalovirus infection. In Recent Advances in Sexually Transmitted Diseases, (Morton, R. S. and Harris, J. R. W., eds), pp. 361–364. Edinburgh: Churchill Livingstone

HELLER, C. J. (1974). Neisseria gonorrhoeae in Papanicolaou smears. Acta Cytol., 18, 338–340

HERMANN, G. III and DEININGER, J. T. (1963). Vorticella, an unusual protozoan found on endocervical smears (letter). Acta Cytol., 7, 129–130

HIGHMAN, W. J. (1971). Calcified bodies and the intrauterine device. Acta Cytol., 15, 473–475

JONES, M. C., BUSCHMANN, B. O., DOWLING, E. A. and POLLOCK, H. M. (1979). The prevalence of Actinomyces-like organisms found in cervicovaginal smears of 310 IUD wearers. Acta Cytol., 23, 282–286

KEARNS, P. R. and GRAY, J. E. (1963). Mycotic vulvovaginitis. J. Obstet. Gynaecol., 22, 621–625

KEEBLER, C. M. and WEID, G. L. (1974). The estrogen test: an aid in differential cytodiagnosis. Acta Cytol., 18, 482–493

KOBAYASHI, T. K., UMEZAWA, Y., UEMURA, M. et al. (1982) Cytodiagnosis of herpes simplex virus infection in the newborn infant. Acta Cytol., 26, 65–68

KOSS, L. (1979). Diagnostic Cytology and its Histopathologic Basis, 3rd edn, pp. 223–269. Philadelphia: J. B. Lippincott

KOSS, L. and DURFEE, G. R. (1956). Unusual patterns of squamous epithelium of the uterine cervix. Cytological and pathologic study of koilocytotic atypia. Ann. NY. Acad. Sci., 63, 1245–1261

LAVERTY, C. R., RUSSELL, P., BLACK, J., KAPPAGODA, N., BENN, R. A. V. and BOOTH, N. (1977). Adenovirus infection of the cervix. Acta Cytol., 21, 114–117

LUFF, R. D., GUPTA, P. K., SPENCE, M. R. and FROST, J. K. (1978). Pelvic actinomycosis and the intrauterine contraceptive device: a cyto-histomorphologic study. Am. J. Clin. Pathol., 69, 581–586

MCCANCE, D. J., CAMPION, M. J., CLARKSON, P. K., CHESTERS, P. M., JENKINS, D. and SINGER, A. (1985a). Prevalence of human papillomavirus type 16 DNA sequences in cervical intraepithelial neoplasia and invasive carcinoma of the cervix. *Br. J. Obstet. Gynaecol.*, **92**, 1101–1105

MCCANCE, D. J., CLARKSON, P. K., DYSON, J. L., WALKER, P. G. and SINGER, A. (1985b). Human papillomavirus types 6 and 16 in multifocal intraepithelial neoplasia of the female genital tract. *Br. J. Obstet. Gynaecol.*, **92**, 1093–1100

MCKENNA, U. G., MEADOWS, J. A. III and BREWER, N. S. (1980). Toxic shock syndrome, a newly recognised disease entity. *Mayo Clinic Proc.*, **55**, 663–672

MEISELS, A. and FORTIN, R. (1976). Condylomatous lesions of the cervix and vagina. I. Cytologic patterns. *Acta Cytol.*, **20**, 505–509

MEISELS, A., FORTIN, R. and ROY, M. (1977). Condylomatous lesions of the cervix, II. *Acta Cytol.*, **21**, 379–390

MISCH, K. A., SMITHIES, A., TWOMEY, D., O'SULLIVAN, J. C. and ONUIGOBO, W. (1976). Tuberculosis of the cervix: cytology as an aid to diagnosis. *J. Clin. Pathol.*, **29**, 313–316

MITCHELL, H., DRAKE, M. and MEDLEY, G. (1986). Prospective evaluation of risk of cervical cancer after cytological evidence of human papillomavirus infection. *Lancet*, **ii**, 573–575

MORTON, R. S. (1975). Trichomoniasis. In *Recent Advances in Sexually Transmitted Diseases*, No. 1, (Morton, R. S. and Harris, J. R. W., eds), pp. 201–228. Edinburgh: Churchill Livingstone

MULLER KOBOLD-WOLTERBEEK, A. C. and BEYER-BOON, M. E. (1975). Ciliacytophthoria in cervical cytology. *Acta Cytol.*, **19**, 89–91

MYLIUS, R. E. and TEN SELDAM, R. E. J. (1962). Venereal infection by *Entamoeba histolytica* in a New Guinea native couple. *Trop. Geog. Med.*, **14**, 20–26

NAHMIAS, A. J. (1978) Herpes simplex infection of the fetus and newborn. In *Viral Diseases of the Fetus and Newborn*, (Hanshaw, J. B. and Dudgeon, J. A., eds), pp. 153–181. Philadelphia: W. B. Saunders

NAIB, Z. M. (1970). Cytology of TRIC agent infection of the eye of newborn infants and their mothers' genital tracts. *Acta Cytol.*, **14**, 391–395

NAIB, Z. M. (1981). In *The Human Herpesvirus*, (Nahmias, A. J., Dowdle, W. R. and Schinazi, R. F. eds), pp. 381–382. New York: Elsevier

NASIELL, M. (1961). Histiocytes and histiocytic reaction in vaginal cytology. A survey of the risks of false positive diagnosis. *Cancer*, **14**, 1223–1225

ORIEL, J. D. (1971). Natural history of genital warts. *Br. J. Vener. Dis.*, **47**, 1–13

ORIEL, J. D. (1983). Chlamydiae and viruses causing human sexually transmitted diseases—analogies with infections in animals: a review. *J. R. Soc. Med.*, **76**, 602–607

PAAVONEN, J. (1979). Genital *Chlamydia trachomatis* infections in patients with cervical atypia. *Obstet. Gynecol.*, **54**, 289–291

PAAVONEN, J., VESTERINEN, E., MEYER, B. and SAKSELA, E. (1982). Colposcopic and histologic findings in cervical chlamydial infection. *Obstet. Gynecol.*, **59**, 712–714

PAPANICOLAOU, G. N. (1933). The sexual cycle in the human female as revealed by vaginal smears. *Am. J. Anat.*, **52**, 599–611

PHEIFER, T. A., FORSYTH, P. S., DURFEE, M. A., POLLACK, H. M. and HOLMES, K. K. (1978). Non-specific vaginitis. Role of *Haemophilus vaginalis* and treatment with metronidazole. *N. Engl. J. Med.*, **298**, 1429–1434

PINE, L., BRADLEY, G., MALCOLM, B., CURTIS, E. M. and BROWN, J. M. (1981). Demonstration of *Actinomyces* and *Arachnia* species in cervicovaginal smears by direct staining with species-specific fluorescent-antibody conjugates. *J. Clin. Microbiol.*, **13**, 15–21

PUGH, S. F., SLACK, R. C., CAUL, E. O., PAUL, I. D., APPLETON, P. N. and GATLEY, S. (1985). Enzyme-amplified immunoassay. A novel technique applied for direct detection of *Chlamydia trachomatis* in clinical specimens. *J. Clin. Pathol.*, **38**, 1139–1141

PUROLA, E. and SAVIA, E. (1977). Cytology of gynecologic condyloma acuminatum. *Acta Cytol.*, **21**, 26–31

QIZILBACH, A. H. (1974). Chronic plasma cell cervicitis. A rare pitfall in gynecological cytology. *Acta Cytol.*, **18**, 198–200

ROBERTS, T. H. and NG, A. B. P. (1975). Chronic lymphocytic cervicitis: cytologic and histopathologic manifestations. *Acta Cytol.*, **19**, 235–243

RUEHSEN, M. DE M., MCNEILL, R. E., FROST, J. K. et al. (1980). Amoeba resembling *Entamoeba gingivalis* in the genital tract of IUD users. *Acta Cytol.*, **24**, 413–420

SCHACHTER, J. (1978). Chlamydial infections. *N. Engl. J. Med.*, **298**, 428–435; 490–495; 540–549

SCHIFFER, M. A., ELGUEZABAL, A., SULTANA, M. AND ALLEN, A. C. (1975). Actinomycosis infection associated with intrauterine contraceptive devices. *Obstet. Gynecol.*, **45**, 67–72

SCHMIDT, W. A. (1987). Pathology of the vagina. In *Obstetrical and Gynaecological Pathology*, 3rd edn, Vol. 1, (Haines and Taylor), (Fox, H., ed.) pp. 146–216. Edinburgh: Churchill Livingstone

SCHNELL, J. D. and VOIGT, W. H. (1976). Are yeasts in vaginal smears intracellular or extracellular? *Acta Cytol.*, **20**, 343–346

SINGER, A., CAMPION, M. J. and MCCANCE, D. J. (1985). Human papillomavirus. *Br. J. Hosp. Med.*, **34**, 104–108

SPENCE, M. R., GUPTA, P. K., FROST, J. K. and KING, T. M. (1978). Cytological detection and clinical significance of *Actinomyces israelii* in women employing intrauterine contraceptive devices. *Am. J. Obstet. Gynecol.*, **131**, 295–300

SYKES, G. S. and SHELLEY, G. (1981). *Actinomyces*-like structures and their association with intrauterine contraceptive devices, pelvic infection and abnormal cervical cytology. *Br. J. Obstet. Gynaecol.*, **88**, 934–937

TAYLOR, E., BLACKWELL, H. L., BARLOW, D. and PHILLIPS, I. (1982). *Gardnerella vaginalis*, anaerobes and vaginal discharge. *Lancet*, **i**, 1376

THIN, R. N. T., ATIA, W., PARKER, J. D. J., NICOL, C. S. and CANTI, G. (1975). Value of Papanicolaou-stained smears in the diagnosis of trichomoniasis, candidiasis and cervical herpes simplex virus infection in women. *Br. J. Vener. Dis.*, **51**, 116–118

TZANCK, A. (1948). Le cytodiagnostic immediat en dermatologie. *Annales de Dermatologie et Syphiligraphie* **8**, 205–218

VALICENTI, J. F., PAPPAS, A. A., GRABER, C. D., WILLIAMSON, H. O. and WILLIS, N. F. (1982). Detection and prevalence of IUD-associated *Actinomyces* colonization and related morbidity: a prospective study of 69,925 cervical smears. *J. Am. Med. Assoc.*, **247**, 1149–1152

VESTERINEN, E., VAHERI, A., PAAVONEN, J. and SAKSELA, E. (1978). Adenovirus infection and cytopathic alterations of human cervical epithelial cells *in vitro*. *Acta Cytol.*, **22**, 566–569

WALKER, P. G., SINGER, A., DYSON, J. L., SHAH, D. V., TO, A. and COLEMAN, D. V. (1983). The prevalence of human papillomavirus antigen in patients with cervical intraepithelial neoplasia. *Br. J. Cancer*, **48**, 99–101

WILLIAMS, A. O. (1967). Pathology of schistosomiasis of the uterine cervix due to *S. haematobium*. *Am. J. Obstet. Gynecol.*, **98**, 784–791

10

Intraepithelial and invasive carcinoma of the cervix

E. Blanche Butler

Previous chapters have dealt with the normal histology and cytology of the uterine cervix, but for a proper understanding of the development of the commonest forms of intraepithelial and invasive carcinoma it is necessary to consider the physiological changes in the cervix at various stages of a woman's life. This depends mainly on variation of hormonal levels which are associated with an increased volume of cervical stroma causing eversion of the columnar epithelium lining the endocervical canal to produce an *ectopy*. Exposure of the everted columnar epithelium to the acid pH of vaginal secretions stimulates replacement of columnar epithelium by squamous epithelium and this process is known as *squamous metaplasia*.

Ectopy

An ectopy is still called a 'cervical erosion' by many gynaecologists, although with increasing use of the colposcope this is becoming less common. In the older literature a 'cervical erosion' was considered to be a pathological condition as cases in which the lesion was seen usually presented with mucopurulent discharge, groin pain and backache. It was not realized that the symptoms reflected infection (cervicitis) superimposed upon the physiological changes of an ectopy. In recent years routine examination of many women in family planning clinics has demonstrated that ectopy is a frequent finding in the normal cervix, and serial colposcopic examinations have shown that the morphology of the cervix varies throughout life (Singer, 1975).

An ectopy is usually present in newborn female infants due to the influence of the hormone levels of the mother (*Figure 10.1*) and this has to be distinguished from cases where columnar epithelium extends onto the vagina as a form of congenital abnormality. This latter appearance can occur spontaneously but it has been associated, in particular, with exposure of the fetus to diethylstilboestrol (DES) in the first trimester of pregnancy (see p. 435). The normal physiological ectopy of the neonate resolves due to inversion of the cervix following loss of maternal hormones from the circulation of the child after delivery. However, some metaplasia can occur *in utero* (*Figure 10.2*).

With the establishment of a normal menstrual cycle at puberty, eversion of columnar epithelium occurs again with formation of an ectopy and this is exaggerated during pregnancy and in women on oral contraceptives, particularly those with a high oestrogen content (*Figure 10.3*). Replacement of columnar epithelium by squamous epithelium (squamous metaplasia) is a repetitive process, and so by the time the woman reaches the menopause and hormonal levels fall, the consequent inversion of the cervix results in a squamocolumnar junction which is well inside the endocervical canal (*Figure 10.4*). It will be noted that it is possible to distinguish mature squamous epithelium which is metaplastic from original squamous epithelium because of the presence of endocervical crypts in the stroma (*Figure 10.5*).

Cytology

As would be expected, columnar cells which are

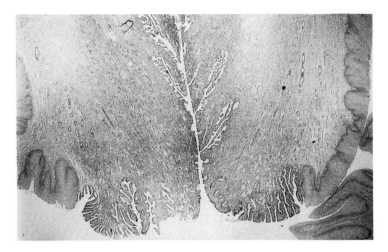

Figure 10.1 Neonatal cervix. In this section of cervix from a stillborn, full-term infant, columnar epithelium is seen on the portio vaginalis of the cervix. The neonatal ectopy so formed is due to the effect of maternal hormones on the child. Haematoxylin and eosin. Magnification ×16, reduced to 75% in reproduction

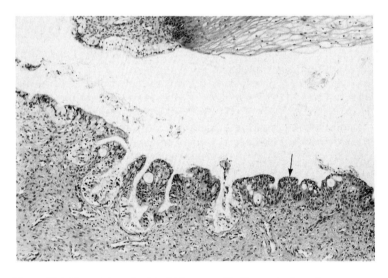

Figure 10.2 Neonatal cervix. At a higher magnification early, immature squamous metaplasia can be seen (arrow). Haematoxylin and eosin. Magnification ×80, reduced to 75% in reproduction

scraped from a simple ectopy show no abnormality and are indistinguishable from endocervical columnar cells collected from the canal. Marked reactive changes can be seen when cervicitis is present and also in women with hyperplasia of endocervical glands (adenomatous hyperplasia). The latter can occur in pregnancy and in women on high oestrogen oral contraceptives.

The columnar epithelium is usually shed in sheets and is characterized by large nuclei, dispersion of finely granular chromatin and prominent, but uniformly shaped, nucleoli. Normal mitoses can be seen and cytoplasmic vacuoles may be present; with infection these can contain polymorphs (*Figure 10.6*).

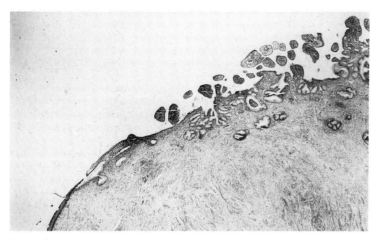

Figure 10.3 Ectopy of the cervix, (compare with *Figure 10.1*). This section shows part of the cervix with the squamocolumnar junction at the lateral angle. Haematoxylin and eosin. Magnification × 16, reduced to 75% in reproduction

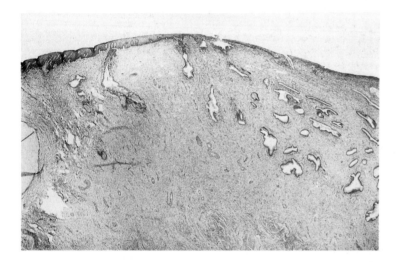

Figure 10.4 Post-menopausal cervix. In this section of the endocervical canal, the surface is covered by metaplastic squamous epithelium. Note the columnar cell-lined crypts in the stroma. Haematoxylin and eosin. Magnification × 16, reduced to 75% in reproduction

Metaplasia (*see also* Chapter 8)

The tall, mucus-secreting columnar epithelium which lines the endocervical canal is thrown into folds which may be coarse and longitudinal or more finely grouped as papillary villi which gives a 'bunch of grapes' appearance. As a result, on histological section crypts extending into the stroma are seen (*Figures 10.1–10.5*) and the appearance in sections imitates that of discrete glandular structures. Consequently these crypts are usually described as endocervical glands; however, it is important to remember that these are not true glands and that the columnar cell lining of the crypts is continuous with the columnar epithelium of the surface.

It is generally accepted that the initial stimulus which begins the metaplastic process is the

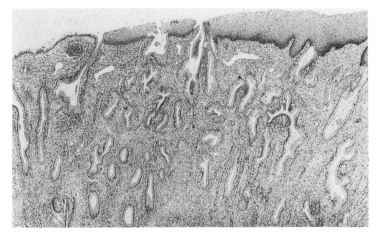

Figure 10.5 Immature and mature squamous metaplasia. The new squamocolumnar junction is seen to the left of the field. Moving to the right there is a zone of immature squamous epithelium followed by mature squamous epithelium. Both are known to be metaplastic because of the presence of crypts in the stroma. Haematoxylin and eosin. Magnification × 32, reduced to 75% in reproduction

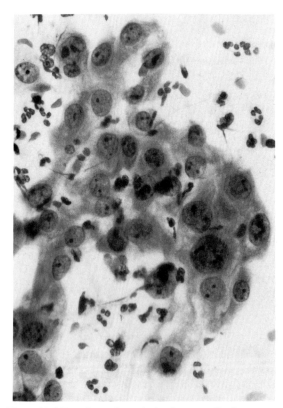

Figure 10.6 Reactive endocervical columnar cells. A loose cluster of columnar cells is seen. These show pleomorphism and prominent nucleoli but the nuclear chromatin pattern is regular and finely granular. In addition, nucleoli are uniformly round to oval in shape. Papanicolaou stain. Magnification × 512

effect of the low pH of vaginal secretions on columnar epithelium exposed as an ectopy (Singer and Jordan, 1976). When the metaplastic process begins, the area is referred to as the transformation zone. The process begins at the tips of the villi which later fuse; the metaplastic process can also extend into the crypts. In the earliest stages columnar epithelium is seen to overlie marked cellular activity in the stroma. In reporting histological sections this is described as *reserve cell hyperplasia* (*Figure 10.7*). The term derives from earlier theories of the origin of metaplastic cells. Primitive cells called reserve cells are present in most cervices between columnar cells and the basement membrane and it was thought that proliferation of these cells reflected the first stage of the metaplastic process (Howard, Erickson and Stoddart, 1951). However, more recent work has demonstrated a similar process of development of new squamous epithelium following destruction with cautery or laser, and this supports the theory that metaplastic cells originate from mononuclear-like cells which migrate to the stimulated area.

The next stage of the metaplastic process is the transformation of the reserve cells to immature squamous epithelium which in due course becomes mature, glycogenated squamous epithelium (*Figure 10.5*). As noted above, this epithelium can be recognized as metaplastic because of residual crypts in the underlying stroma.

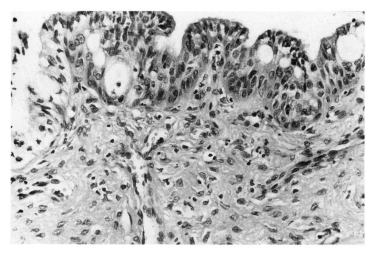

Figure 10.7 Reserve cell hyperplasia. Note the proliferation of multilayered undifferentiated cells deep to the columnar epithelium. Haematoxylin and eosin. Magnification × 256, reduced to 75% in reproduction

Cytology of reserve cell hyperplasia

Reserve cells (*Figure 10.8*) are usually closely associated with columnar cells as they are collected by forcible removal when delicate columnar epithelium is scraped. The nuclear chromatin pattern is similar to that of endocervical columnar cells and is finely granular with small chromocentres. The nucleus is centrally placed and round or oval in shape with a uniform nuclear membrane. The cytoplasm is poorly differentiated and scanty with a cyanophilic staining reaction. These cells are found in syncytial sheets or clusters and can be distinguished from columnar cells seen en fosse by focusing through the cell with the fine adjustment when it is found that the depth of cytoplasm present in columnar cells, either above or below the nucleus, cannot be demonstrated. These cells may be present in tight clusters in which it is difficult to evaluate the chromatin pattern; in addition they have a high nucleocytoplasmic ratio and there can be confusion with dyskaryotic cells (*see* below). Problems of differential diagnosis are increased in the presence of infection when nuclei are larger and dispersion of chromatin exaggerates the granular pattern. It is fortunate that although the cellular manifestation of reserve cell hyperplasia is as described above, it is a relatively uncommon finding in routine cervical smears. Single reserve cells cannot be distinguished from small, histiocyte-like cells.

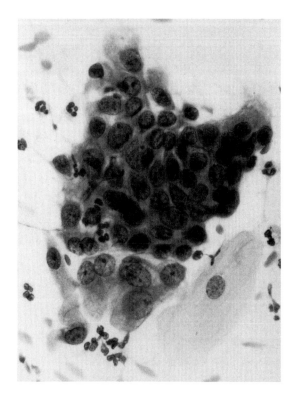

Figure 10.8 Reserve cells. A tight cluster of poorly differentiated cells is seen in the centre of the field with a group of endocervical columnar cells to the left. In this picture it is possible to recognize the benign nuclear chromatin pattern of the reserve cells. Papanicolaou stain. Magnification × 512

Cytology of immature squamous metaplasia

Stages of the metaplastic process are reflected in the cervical smear by the presence of immature squamous cells of the parabasal type. It is important to remember that 'parabasal' cells do not appear in smears taken from mature squamous epithelium; their presence indicates immature epithelium which can be the immature metaplasia of an incomplete transformation zone. When parabasal cells are seen in clusters surrounded by a population of otherwise mature squamous cells, such cells can be identified and described as 'metaplastic cells'. Smears in which parabasal cells predominate usually indicate an atrophic epithelium or, in younger women, the puerperal state. However, there can be confusion when a very large transformation zone is present and where there is doubt, examination of a lateral vaginal wall smear will establish the hormonal status of the patient.

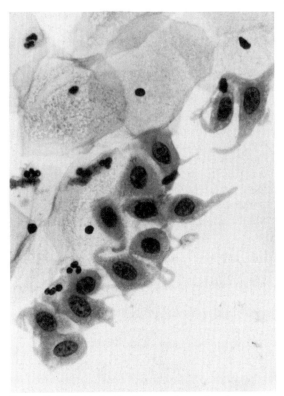

Figure 10.10 'Spider' cells. Elongated cytoplasmic processes are seen. These are seen in metaplastic cells which have been removed forcibly by scraping the cervix rather than exfoliated cells which have been collected from the surface. Papanicolaou stain. Magnification × 512

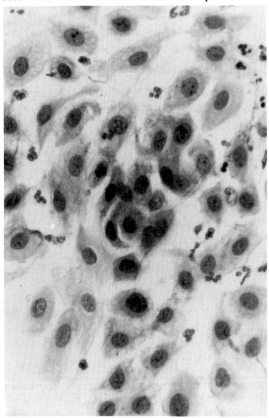

Figure 10.9 Immature metaplastic cells. This field shows a wide range of differentiation of metaplastic cells. Many have poorly defined cytoplasm while others are clearly squamous cells of the parabasal type. Papanicolaou stain. Magnification × 512, reduced to 86% in reproduction

The morphological appearance of metaplastic cells in the cervical smear varies widely and depends on the degree of maturation of the metaplastic epithelium. Metaplastic cells are a common finding in routine cervical smears taken between puberty and the menopause, and their presence in a smear indicates that it has been collected from the transformation zone and that the specimen is suitable for reporting. Metaplastic cells show a wide range of maturation and this reflects the spectrum of histological patterns which can be present.

Metaplastic cells usually present singly, but the most immature forms can be seen as loose clusters, similar to reserve cells but with more distinct cytoplasm (*Figure 10.9*). With increasing maturity the cell shape changes from round to oval to a polygonal form. In some smears, metaplastic cells showing cytoplasmic processes between cells are seen. These have been called 'spider cells' (*Figure 10.10*) but they have no significance except to indicate that they have been forcibly removed instead of exfoliating spontaneously (Patten, 1978). The cytoplasm of immature metaplastic cells is dense and homogeneous with a cyanophilic staining reaction. Cytoplasmic vacuoles are sometimes seen but this usually indicates the presence of infection or degeneration. The more mature polygonal forms may show the same cytoplasmic appearance, but in some the cytoplasm seems to be divided into a dense outer zone (ectoplasm) and a pale perinuclear zone (endoplasm) (*Figure 10.11*). With complete maturation the cells are indistinguishable from normal intermediate and superficial squamous cells.

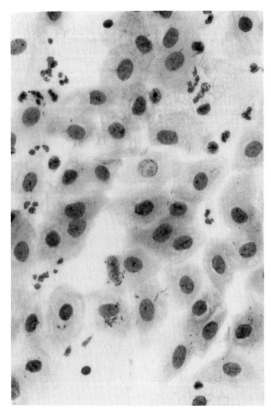

Figure 10.11 Metaplastic cells. This field shows metaplastic cells which are more mature than those seen in *Figure 10.9*. Cytoplasm is well defined and while some cells have a rounded outline, others are becoming polygonal in shape. Papanicolaou stain. Magnification × 512, reduced to 85% in reproduction

The nuclei of immature metaplastic cells are centrally placed and are round or oval in shape; the nucleocytoplasmic ratio decreases with increased maturity of the cells. The nuclear chromatin pattern is uniform and finely granular with occasional small chromocentres. Nucleoli are not seen in normal metaplastic cells.

Metaplastic cells are immature cells and must be distinguished from dyskaryotic cells. The absence of hyperchromasia is usually sufficient to prevent error but marked reactive changes and overstaining in unsatisfactory preparations can cause difficulties which bring these cells into a borderline category necessitating repeat smears.

Neoplasia of the cervix

It is probable that most cancers pass through a stage in which transformed cells are confined to the epithelium and that this precedes invasion through the basement membrane into the stroma. Because of the accessibility of the uterine cervix, this sequence of events has been studied in more detail than is possible elsewhere. The concept of a preinvasive stage in the development of cervical cancer was first proposed in the early years of this century (Schauenstein, 1908) but it was the introduction of cytological screening for cervical cancer which showed the importance of this phase of the disease and its implications in the field of early diagnosis and treatment. The preinvasive lesions were initially described as 'incipient cancer' or 'carcinoma *in situ*' but are now designated cervical intraepithelial neoplasia (CIN).

Epidemiology

It has become accepted that most types of cancer of the cervix are the result of sexually transmitted disease. The evidence for this is reviewed in detail in standard texts (Anderson, 1976; Reid and Coppleson, 1976; Koss, 1979; Singer, 1985). The disease is more common in women who have an early age of first coitus and a multiplicity of partners, and the implication has been that the known association with multiparity was a logical sequence of an early age of first coitus. However, Miller, Barclay and Choi (1980) have shown that parity increases the risk of carcinoma *in situ* at all ages and of invasive carcinoma in women over the age of 50. Although the effect of the pregnant state seems to influence progression of the disease, to recognize carcinoma of the cervix as a sexually transmitted disease implies that it is caused by transmissible agents.

Transmissible agents which have been implicated as carcinogens or co-carcinogens include herpes simplex virus type 2, human papillomavirus and sperm basic protein. In addition, there is evidence of secondary 'trigger' mechanisms. These include parity (Miller, Barclay and Choi, 1980), treatment with immunosuppressants (Stanbridge and Butler, 1983), radiotherapy (Choo and Anderson, 1982) and smoking (Trevathan, Layde and Webster, 1983).

The association of carcinoma of the cervix, both intraepithelial and invasive, with female promiscuity has caused distress to non-promiscuous women who are found to have the disease. It may even inhibit some women from presenting themselves for cervical smear tests. Recent work has emphasized the male factor and more should be made of this in writing for lay readers. Gynaecologists and others who work in colposcopy clinics should also have this in mind when counselling patients.

It is suggested that the 'male factor' has its effect in two ways:

(1) The level of sperm basic protein in the semen. This is due to its ability to bind nuclear DNA which is present on cell surfaces of both spermatozoa and viruses (Reid, 1981).
(2) Male promiscuity. Skegg et al. (1982) refer to the incidence of carcinoma of cervix in countries where wives are secluded but where there is a tradition of their husbands resorting to prostitutes.

The concept of a 'male factor' is supported by a number of workers. Martines (1969) found a relationship between carcinoma of cervix and carcinoma of penis in marital partners and Kessler (1977) identified 29 marital clusters of cervical cancer in which two women married to the same man developed the disease. Reid, French and Singer (1978) calculated sperm basic protein ratios according to social class and found that the slope of the curve approximated to the incidence of positive cervical cytology and death rate from cervical cancer amongst the classes. An interesting case-controlled study of the behaviour of husbands of women with dysplasia and carcinoma *in situ* has been reported by Buckley et al. (1981).

Pathogenesis

The majority of cervical squamous carcinomas develop in the transformation zone between original squamous epithelium and columnar epithelium. This has been demonstrated histologically, particularly at the intraepithelial stage, as it can be shown that peripheral spread beyond the last crypt is unusual (Burghardt, 1976). The physiological development of squamous metaplasia in this area has been described above and Coppleson and Reid (1967) suggest that at the early, active phase of metaplasia, cells are vulnerable to carcinogens. At coitus these agents are transmitted to active metaplastic cells with alteration of the cell genome to produce a line which has neoplastic potential. This may remain dormant for months or years before other factors trigger progression to histologically recognizable intraepithelial neoplasia and invasive carcinoma.

The value of screening for cervical cancer has depended on the prolonged duration of the intraepithelial phase which has been estimated at between 10 and 15 years. However, there has been concern in recent years about the increase in the number of cases of invasive cancer in women under 30 years of age. This may be related to changing sexual mores with earlier initiation of the epithelium and more rapid progression in younger women. An alternative theory is that a new form of cervical cancer is being seen with shorter natural history. These are usually poorly differentiated small cell tumours and it can be difficult to distinguish between squamous and adenocarcinoma without the use of mucin stains. In practice, it is important for the cytotechnologist to be alert for the presence of clusters of poorly differentiated cells in the smears of young women and not to underestimate them as reactive columnar cells or cells from reserve cell hyperplasia.

Intraepithelial neoplasia (histology)

Terminology (histology)

The nomenclature used to distinguish between different degrees of abnormality at the intraepithelial stage is in a state of transition. The older terms of 'dysplasia' and 'carcinoma *in situ*' are still in use but there is increasing acceptance of reporting grades of CIN as this recognizes continuity of the disease process.

In 1962 an International Committee on Histological Terminology defined carcinoma *in situ* as 'a lesion of the epithelium in which, throughout its thickness, no differentiation takes place' and in 1975, in a WHO publication, Poulsen, Taylor and Sobin said 'dysplasia is a lesion in which part of the thickness of the epithelium is replaced by cells showing varying degrees of atypia'. The chief problem with this dual nomenclature is the implication that there are two disease processes, with dysplasia having a less malign significance. Dysplasia is graded as mild, moderate or severe and so serious problems of management are possible depending on whether the pathologist calls a given lesion 'severe dysplasia' or 'carcinoma *in situ*'. Richart (1967) proposed a single diagnostic category of Cervical Intraepithelial Neoplasia (CIN) and this can be defined as 'a spectrum of intraepithelial change which begins as a generally well differentiated neoplasm which has traditionally been classified as mild dysplasia and ends with invasive carcinoma' (Ferenczy, 1977a).

For descriptive purposes it is convenient to subdivide CIN into three categories:

CIN I which equates to mild dysplasia;
CIN II which equates to moderate dysplasia;
CIN III which includes severe dysplasia and carcinoma *in situ*.

However, these subdivisions have no prognostic value on an individual basis, although collectively a lesion graded as CIN III (severe dysplasia/carcinoma *in situ*) is more likely to progress to invasion than a lesion graded as CIN I (mild dysplasia). Indeed, a patient with a small area of CIN III can have a better prognosis than a patient with a large area of CIN I or II (Ferenczy, 1977b). In addition, it is now generally recognized that moderate or mild dysplasia (CIN I or II) can progress to invasion without transition through a carcinoma *in situ* (CIN III) stage (Burghardt, 1976).

Those who object to the new terminology point out that not all lesions diagnosed as CIN are neoplastic, but result from the non-specific reaction to infection or trauma. This emphasizes the need for strict diagnostic criteria particularly in cases of CIN I. It has also been suggested that use of the word 'neoplasia' can lead to overtreatment but as it has been shown that only 6% of cases of 'mild dysplasia' regress spontaneously (Richart and Barron, 1969), it is probably sensible to treat, even at this early stage (Buckley, Butler and Fox, 1982).

Histological features of CIN

CIN is characterized by loss of polarity of cells and disruption of the orderly stratification found in normal squamous epithelium. The division of CIN into three grades on histological criteria is, to a large extent, subjective but as the grading is descriptive rather than prognostic this is of less importance than when the distinction was attempted between dysplasia and carcinoma *in situ*.

Grading of CIN depends on:

(1) The proportion of epithelial thickness occupied by undifferentiated neoplastic cells.
(2) The level within the epithelium at which mitotic figures are found.
(3) The presence of abnormal mitoses.

CIN I

Undifferentiated neoplastic cells occupy the lower third of the epithelium while cytoplasmic differentiation is seen in the middle and upper thirds. However, the nuclei of these differentiated cells are also abnormal showing pleomorphism with a hyperchromatic granular chromatin pattern and prominent nucleoli. Single cell keratinization can be seen deep to the surface while mitoses are infrequent and usually normal (*Figure 10.12*).

CIN II

Undifferentiated, pleomorphic, non-stratified neoplastic cells reach the middle third of the epithelium but do not extend into the upper third. Abnormal cells showing varying degrees of cytoplasmic differentiation and stratification are seen in the upper third of the epithelium and mitoses are confined to the lower two-thirds; these may be abnormal (*Figure 10.13*).

CIN III

Three main histological patterns are recognized:
(1) Small cell undifferentiated (*Figure 10.14*). The full thickness of the epithelium is occupied by undifferentiated non-stratified cells of the basaloid type. Nuclear crowding and pleomorphism are usually more marked

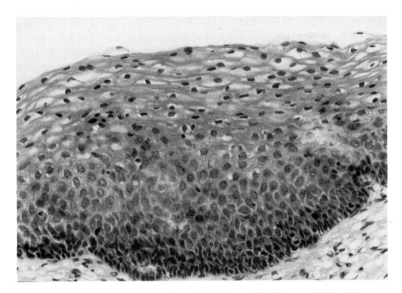

Figure 10.12 CIN I. Failure of differentiation is seen in the lower third of the epithelium with cytoplasmic differentiation in the upper two-thirds. However, in the upper two-thirds nuclei are large and irregular. Haematoxylin and eosin. Magnification × 256, reduced to 65% in reproduction

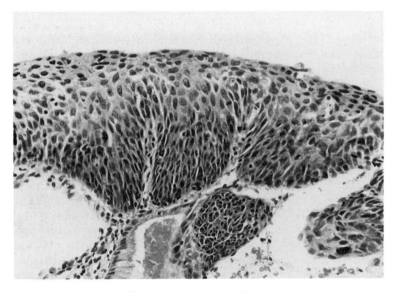

Figure 10.13 CIN II. Undifferentiated neoplastic cells occupy less than two-thirds of the thickness of the epithelium. Haematoxylin and eosin. Magnification × 256, reduced to 65% in reproduction

than in CIN I or CIN II. Mitoses are frequent and are present in the upper third of the epithelium; they are often abnormal.
(2) Large cell undifferentiated (*Figure 10.15*). Although the cells are larger with more cytoplasm, cell boundaries are often indistinct and there is no evidence of stratification. These cells are neoplastic in type and extend into the upper third of the epithelium and often occupy the full thickness. Mitoses which can be abnormal are also seen in the upper third of the epithelium. In this type of

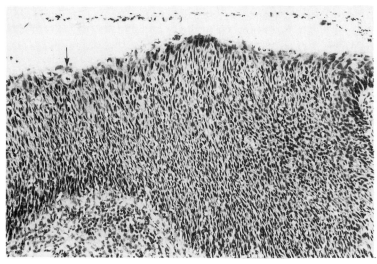

Figure 10.14 CIN III (small cell undifferentiated). Undifferentiated cells of the basaloid type occupy the whole thickness of the epithelium. A mitosis is seen near the surface (arrow). Haematoxylin and eosin. Magnification ×256, reduced to 65% in reproduction

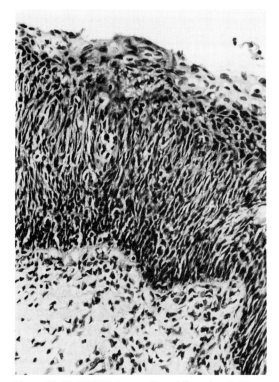

Figure 10.15 CIN III (large cell undifferentiated). There is failure of differentiation throughout almost the full thickness of the epithelium. The constituent cells have more cytoplasm and are of the parabasal type. Haematoxylin and eosin. Magnification ×256, reduced to 55% in reproduction

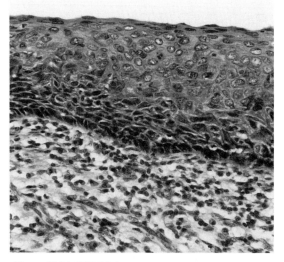

Figure 10.16 CIN III (keratinizing). In this example undifferentiated neoplastic cells reach the outer third of the epithelium and the surface shows differentiated keratinized cells with variable loss of stratification. In other cases abnormal differentiated cells with loss of stratification may occupy the full thickness of the epithelium. Haematoxylin and eosin. Magnification ×256, reduced to 55% in reproduction

lesion the cells are thought to be parabasaloid in type.
(3) Differentiated (large cell keratinizing) (*Figure 10.16*). Cells with well-defined cell boundaries and with a relatively low nucleo-

cytoplasmic ratio extend into the upper third of the epithelium. There is loss of polarity and stratification with complete disorganization of the growth pattern and there may be surface keratinization. Nuclear pleomorphism is marked and abnormal mitoses are frequent.

Cytological diagnosis of CIN

The cytological diagnosis of cervical cancer by means of the cervical smear was first described by Aurel Babes, a Rumanian pathologist in 1927 (Daniel and Babes, 1927). The following year George Papanicolaou, a Greek zoologist working in the Cornell Medical Center in New York, whose main interest was hormonal cytology reported that tumour cells could be detected in vaginal smears from women with cervical cancer (Papanicolaou, 1928). Neither paper attracted much interest at the time.

Fifteen years later, Papanicolaou and Traut published their now famous monograph (Papanicolaou and Traut, 1943) in which they described the range of cells seen in vaginal smears. Their observation was so complete and their work so thorough that to this day there has been little new to add to their original report. The potential of this technique as a method of screening for cervical cancer and detecting it at its earliest and most treatable stage was realized and the first cervical cancer screening clinics were established in many countries in the Western hemisphere in the next 10 years.

Terminology (cytology)

In recent years attempts have been made to standardize the terms used to describe the range of cells shed from preinvasive and invasive cancer of the cervix in cervical smears. Papanicolaou (1954) introduced the word 'dyskaryotic' to describe all cells with abnormal nuclei ranging from mild degrees of abnormality to cells showing features suspicious of an invasive lesion. Confusion followed with the practice of calling cells from carcinoma *in situ* 'malignant cells' and restricting the term 'dyskaryotic' to cells thought to be shed from dysplasia. By using a dual terminology the terms ceased to be descriptive and became an interpretation of cell potential which may vary from one observer to another.

The tendency today is to revert to Papanicolaou's terminology and use the term 'dyskaryo-tic' to describe cells shed from an area of CIN and to reserve the term 'malignant' for cells which the observer is confident are shed from an invasive lesion. This is the terminology currently recommended by the British Society for Clinical Cytology (Evans *et al*, 1986, 1987) and is the one used in this book. We strongly recommend that this terminology be adhered to, to ensure uniformity of reporting in different laboratories for the purposes of quality control and comparative analysis of results.

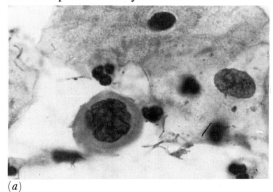

(a)

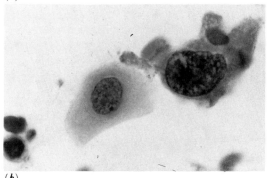

(b)

Figure 10.17 (*a*) Normal superficial and intermediate squamous cells compared with a cell showing a moderate dyskaryosis; (*b*) normal small intermediate squamous cell compared with a cell showing criteria of malignancy (shed from invasive carcinoma). Papanicolaou stain.
Magnification × 1280, reduced to 70% in reproduction

Dyskaryotic cells shed from an area of squamous intraepithelial neoplasia are remarkable for their pleomorphism. They can be identified mainly by the appearance of the nucleus which is enlarged (at least twice the size of a normal intermediate cell nucleus), hyperchromatic and irregular in outline. (*Figure 10.17*) The chromatin content is increased and granular and anisonucleosis is marked. Abnormal nucleoli and mitotic figures are rarely seen although multinucleate cells are a common finding. The cytoplasm may be abundant (*Figures 10.18 and 10.19*) resembling that of a superficial or intermediate cell or very scanty indeed (*Figures 10.20 and 10.21*).

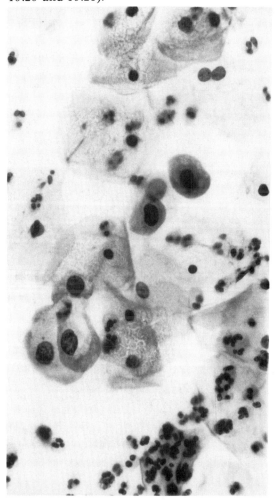

Figure 10.18 Dyskaryotic cells. In addition to normal squamous cells a group of cells showing a mild or moderate dyskaryosis can be seen. The appearances suggest CIN I to II. Papanicolaou stain. Magnification ×512.

Dyskaryotic cells have been further classified into subgroups to enable the cytologist to predict the underlying histological changes in the cervix with a greater degree of accuracy. One of the first attempts at classification subdivided dyskaryotic cells according to the degree of differentiation of the cytoplasm, and described several different types of dyskaryosis including superficial cell dyskaryosis, parabasal cell dyskaryosis and undifferentiated dyskaryotic cells (British Society of Clinical Cytology, 1978;

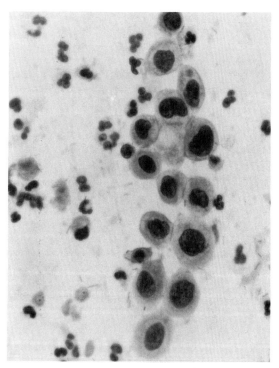

Figure 10.19 Dyskaryotic cells. Cells showing moderate dyskaryosis are seen in this field. The appearances suggest CIN II. Papanicolaou stain. Magnification ×512.

Spriggs and Boddington, 1980). This terminology has been superseded by one which places more emphasis on the degree of nuclear abnormality which is the key feature of dyskaryosis (Evans *et al.*, 1987). Three grades of dyskaryosis are recognized—mild, moderate and severe. In general, the presence of mild or moderate dyskaryosis in a cervical smear reflects a low grade of intraepithelial neoplasia (CIN I or II) whereas severe dyskaryosis indicates an underlying area of CIN III or invasive cancer.

Interpretation of the abnormal smear

A clinician receiving a cytology report is most interested in the interpretation of the whole smear in terms of the anticipated histological findings. Thus it is usual to describe the types of dyskaryotic cell present in the smear (mild, moderate or severe dyskaryosis) and predict the underlying lesion (CIN I, II or III). This may be followed by a recommendation for action (*see* p. 159).

Cells showing *mild dyskaryosis* usually have abundant cytoplasm with angular borders resembling superficial or intermediate cells. The cytoplasm is often keratinized. The nucleus

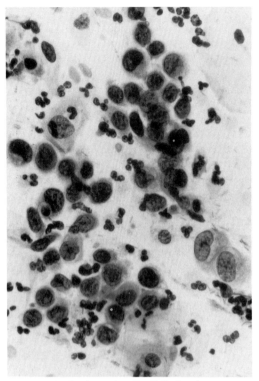

Figure 10.20 Dyskaryotic cells. Cells showing a severe dyskaryosis present singly and in loose cluster. An occasional moderately dyskaryotic cell is also present. The appearances suggest CIN III. Papanicolaou stain. Magnification ×512, reduced to 90% in reproduction

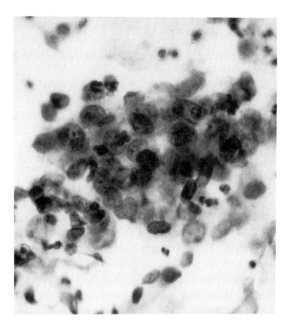

Figure 10.21 Dyskaryotic cells. A tight cluster of severely dyskaryotic cells is seen in this field. The patient had CIN III but the difficulties in evaluating the nuclear chromatin pattern in such a case are apparent. Compare with *Figure 10.8*. Papanicolaou stain. Magnification ×512, reduced to 70% in reproduction

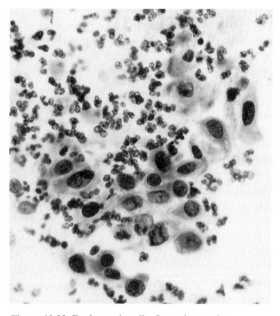

Figure 10.22 Dyskaryotic cells. Several severely dyskaryotic cells are seen in this field. Some of the latter were keratinized and some elongated cell forms are present. This smear was taken from a patient with CIN III of the keratinizing type. Papanicolaou stain. Magnification ×512, reduced to 70% in reproduction

occupies less than half the total area of the cytoplasm. The cells may occur singly or in sheets or highly keratinized plaques of cells. *Moderate dyskaryosis* is characterized by a greater degree of nuclear enlargement so that the nucleus occupies one-half to two-thirds of the total area of the cytoplasm. The nuclei of cells showing a moderate dyskaryosis tend to exhibit a greater degree of hyperchromasia and a greater irregularity of nuclear outline.

Severely dyskaryotic cells typically have a narrow rim of cytoplasm and an abnormal nucleus which practically fills the cell or at least two-thirds of it (*Figure 10.22*). Occasionally cells showing a severe dyskaryosis have abundant cytoplasm. In this respect they resemble cells showing a mild dyskaryosis but can be distinguished from them by the much greater size of the nucleus. Occasionally severely dyskaryotic cells are shed in sheets or dense clusters with indistinct cell borders. They can be recognized by nuclear size, increased chromatin content, abnormal chromatin structure and irregularity of nuclear outline.

The number of dyskaryotic cells in a smear is very variable—there may be one or two or many hundreds. It should be remembered that the number of dyskaryotic cells does not reflect the extent of the lesion. The smears may be bloodstained or contain numerous polymorphs if there is coexisting cervicitis, although this is a more common finding in smears from patients with invasive carcinoma.

Serial sections of a cervical cone biopsy often show a range of changes in the cervical epithelium. Thus the biopsy may contain areas of immature metaplasia, CIN I, II and III. This will be reflected in the smear which will contain mild, moderate and severely dyskaryotic cells as well as normal metaplastic cells. It is conventional to phrase the report with reference to the most severe histological abnormality anticipated.

Human papillomavirus (HPV) infection and cervical neoplasia

One of the most frequent findings in biopsies and smears from patients with CIN is cells showing changes suggestive of human papillomavirus infection of the cervix. Koilocytosis can be demonstrated in at least 70% of biopsies from patients with CIN. Some authorities report koilocytosis in every case of CIN. It is more readily found in low grade lesions (CIN I or II) than in CIN III and is often accompanied by other evidence of wart virus infection e.g. individual cell keratinization, hyperkeratosis and multinucleation. The focus of infection in the biopsy may be adjacent to an area of CIN or superimposed on it. In either case, koilocytes will be found in the smear.

The smear pattern will vary slightly according to the location of the koilocytes in the cervical epithelium. In those cases where the virus infection is associated with an area of CIN, the nuclei of the koilocytes are usually within normal limits or only slightly enlarged and hyperchromatic. Other evidence of HPV infection (dyskeratocytes, anucleate keratinized squames and multinucleate cells) may also be found in the smear (see also Chapter 9).

The close association of HPV infection and cervical cancer has led to speculation that the viruses are involved in the oncogenic process. The evidence to support this hypothesis is drawn from DNA analysis of cervical biopsy material and cervical smears. These studies have shown that HPV DNA is present in 70% of biopsies showing the changes of intraepithelial or invasive cancer. The prevalence of the different HPV types in the biopsies varies with the severity of the cervical lesion. Thus HPV 6 and 11 are found most commonly in low grade CIN lesions (CIN I and II) whereas HPV 16 DNA can be detected more frequently in CIN III and invasive cancers. The affinity of HPV 16 for invasive cancer has led to the suggestion that the various types of HPV vary in their pathological potential and women who harbour HPV 16 may have a greater risk of developing invasive cancer of the cervix than women who harbour HPV 6 or 11. This theory is strengthened by the fact that HPV 16 is integrated into the nucleus of the invasive cancer cells in the invasive lesion, whereas in CIN the viral DNA is not integrated. Moreover, DNA sequences have been detected in the human virus which are analogous to DNA sequences in bovine papillomaviruses which have been shown to have a transforming function. Although the evidence linking the viruses with cervical cancer is intriguing, it is important to remember that the full significance of the association is not known. HPV infection of the genital tract is exceedingly common, and it has been shown that HPV DNA can be detected in 21% of women with cytologically and colposcopically normal cervices (Coleman et al., 1985). It has been suggested that cofactors may be needed to promote neoplastic change in such cases. Alternatively, the presence of the virus in the cervix in neoplasia may represent opportunistic infection.

Until the role of the viruses can be clarified, evidence of HPV infection in a smear or biopsy should not influence the clinical approach and women with abnormal smears should be managed according to the degree of dyskaryosis (Kaufman et al., 1983).

Causes of error in cytological diagnosis

The range of cytological patterns encountered in cervical smears is very variable. When errors are made in evaluating the degree of abnormality present in cytological material, these are usually due to infection causing cell degeneration, or cell reaction due to trauma or healing. Difficulties of differential diagnosis are increased when either of these is superimposed on a population of normal, immature metaplastic cells. Special problems of diagnosis may occur when cellular changes due to human papillomavirus or herpes simplex virus are present in the smear, or when smears are taken from women who have received cytotoxic therapy or radiography. In addition, it can be difficult to identify dyskaryotic cells in a cytolytic smear or an atrophic smear. These problems of smear interpretation are discussed below.

Infection

The cell changes are non-specific and include a degree of nuclear enlargement comparable to that seen in severely dyskaryotic cells. The nuclear chromatin is dispersed which gives a granular effect, but the nuclear staining is hypochromatic rather than hyperchromatic. Moreover, cells showing inflammatory change rarely exhibit the pleomorphism which is characteristic of dyskaryotic cells. Infection with the human papillomavirus (HPV) presents a special problem which is discussed elsewhere (*Figure 10.23*). However, as already stated in the section on histological differential diagnosis, it is important not to allow cell markers associated with HPV infection to lead to an underestimation of any other abnormality present (Kaufman *et al.*, 1983).

Immature metaplasia and reserve cells

Immature metaplastic cells and reserve cells have a granular chromatin pattern and when superimposed infection causes dispersion of chromatin, the result can mimic dyskaryosis. The distinction is even more difficult when the cells are present as coherent sheets or tight clusters (Spriggs and Boddington, 1980). In these cases, repeat smears should be examined and it may be necessary to proceed to colposcopy and biopsy to ascertain the diagnosis (compare *Figures 10.8* and *10.21*).

Viral changes

For many years the multinucleate giant cells which characterize herpes simplex infection were misinterpreted as malignant and women with a herpes cervicitis were subjected to unnecessary cone biopsy. Naib (1966) was one of the first to recognize the true nature of these giant cells which reflect the cell fusion induced by these genital viruses. It must be remembered

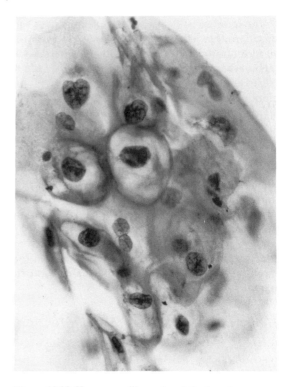

Figure 10.23 Human papillomavirus infection. Squamous cells with abnormal nuclei are seen associated with koilocytes. Papanicolaou stain. Magnification × 512.

that herpes infection and CIN can occur in the same cervix, and both virus-infected cells and dyskaryotic cells may be found in the same smear. Cells scraped from an exophytic wart on the cervix frequently have abnormal nuclei, leading to misdiagnosis of CIN. In fact, the cell morphology may be so bizarre and the cytoplasm so heavily keratinized that invasive cancer may be suspected. Clinical and colposcopic inspection of the cervix will reveal the true nature of these cells.

Occasionally, discrete cells shed from a focus of HPV infection in the flat squamous epithelium of the cervix may contain enlarged, irregular, hyperchromatic nuclei which cannot be distinguished from cells from CIN I showing a mild dyskaryosis. Cytological surveillance is advised.

Iatrogenic changes

Smears taken after radiotherapy or after insertion of an intrauterine device often contain abnormal cells. These can usually be distinguished from dyskaryotic cells by the presence of macrocytosis without any change in the nucleocytoplasmic ratio. Nuclei are often pale staining. A careful history is essential for the correct diagnosis of these cells.

Cytolytic smears and atrophic smears

Dyskaryotic cells may be particularly difficult to detect in cytolytic smears. If there is any difficulty in verifying nuclear morphology, a repeat smear mid-cycle or post partum should be requested. Atrophic smears are often air-dried and pale staining and dyskaryotic nuclei may be only slightly darker staining than the nuclei found in the normal parabasal cells in the smear. It is not uncommon to find that the cytoplasm of the cells has disintegrated and the smears contain many free nuclei. It may be difficult to identify dyskaryotic cells under such conditions. The irregular nuclear outline and presence of anisonucleosis are useful markers of CIN. Mucin deposits in atrophic smears may be mistaken for hyperchromatic nuclei if care is not taken.

Hyperkeratosis

The presence of keratinized squames in a smear may be due to several causes—reaction to trauma or chronic infection, local infection with HPV, differentiated CIN or keratinizing squamous carcinoma. Unfortunately, the keratinized plaques in the smear may obscure the severity of the underlying lesion and a high level of suspicion should be maintained when such smears are being screened. Biopsy is advisable in those cases where keratinized plaques studded with small, irregular pyknotic nuclei are found.

Invasive carcinoma

Squamous cell carcinoma is the commonest form of cancer to develop in the uterine cervix. In recent years, increasing numbers of adenocarcinomas have been described (Ng, 1983); in some series these have accounted for 34% of cases (Davies and Moon, 1975). This may be a real increase but could also be due to better diagnostic techniques such as the use of mucin stains when evaluating poorly differentiated tumours. It has also been suggested that the increase is more apparent than real, due to decrease in the numbers of squamous cell carcinomas. Other primary tumours such as lymphoma and melanoma are very rare.

Squamous cell carcinoma (*Figure 10.24*)

Three main histological patterns occur: keratinizing, large cell non-keratinizing and small cell non-keratinizing. These categories have been adopted by the World Health Organization (Riotton and Christopherson, 1973).

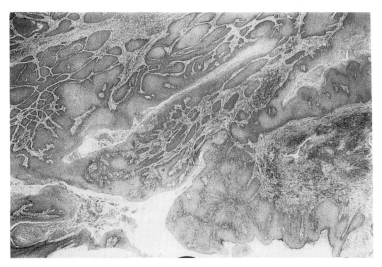

Figure 10.24 Squamous cell carcinoma. This low power view shows masses of tumour cells invading the stroma. Haematoxylin and eosin. Magnification ×16, reduced to 65% in reproduction

Keratinizing

These tumours are well differentiated and, in addition to irregular masses of malignant epidermoid cells, circular whorls of cells with central nests of keratin (epithelial pearls) are also present. Intercellular bridges are seen together with keratohyaline granules and cytoplasmic keratinization. Nuclei are large and hyperchromatic with an irregular coarse chromatin pattern. When present, mitoses are usually found in the less well differentiated parts of the tumour.

Tumour masses invade the stroma and the infiltrating margins are usually sharp and irregular, although rounded or blunt advancing margins can occur.

Large cell non-keratinizing

Masses of polygonal cells are seen with buds or cords at the advancing margins. Individual cell keratinization is seen but epithelial pearls are not a feature. Mitoses are present and central necrosis can be seen in the centre of the large cell masses.

Small cell non-keratinizing

The tissue pattern is diffuse with syncytial masses of small cells which have poorly defined cytoplasmic borders. Nuclei are hyperchromatic and nucleoli are often present. Mitoses are frequent. Differentiation is poor and the tumour may not show features which are typical of squamous cell origin. In these cases, special stains may be needed to establish the diagnosis.

Adenocarcinoma

Several patterns of adenocarcinoma can be primary to the cervix and this reflects the developmental potential of the Müllerian duct from which it derives. Endocervical, endometrial, clear cell and mixed cell types have been described. In addition, in the same way as with squamous tumours, an *in situ* phase can be recognized.

Adenocarcinoma in situ

This is confined to the surface epithelium and epithelium lining the crypts. Nuclei are hyperchromatic and often centrally placed with loss of mucus secretion. Pseudostratification is present and the general pattern is clearly different from normal endocervical columnar epithelium. Adenocarcinoma *in situ* can be associated with squamous carcinoma *in situ* and both may appear as discrete lesions or they may be intermingled.

Adenocarcinoma of the endocervical type (*Figure 10.25*)

These are usually well differentiated tumours which resemble normal endocervical columnar epithelium and diagnosis may depend on irregularity of the glands and evidence of invasion. In less well differentiated tumours, glands are more closely packed and the cells show a greater degree of pleomorphism. Papillary forms are seen and in some cases there is excessive production of mucus resulting in a mucoid or colloid carcinoma.

Endocervical adenocarcinoma of endometrial type

These tumours mimic the histological pattern seen in adenocarcinoma of the endometrium and, in the same way, squamous change can occur to give the appearance of an adenoacanthoma.

Clear cell adenocarcinoma

These tumours consist of sheets of cells with clear cytoplasm giving a signet ring effect. In most cases tubular structures are absent or poorly developed. A few clear cell tumours are true mesonephric tumours developing in remnants of the Wolffian duct. These show papillary areas, tubules, microcysts and 'hobnail' epithelium.

Mixed carcinomas of cervix

A combined adenosquamous carcinoma can occur and these have to be distinguished from adenoacanthoma of the endometrial type. In adenosquamous carcinoma both the glandular component and the squamous component of the tumour are malignant, whereas in adenoacanthoma the glandular component of the tumour is malignant but the squamous component is benign. Various combinations of invasive and intraepithelial carcinoma are also seen, i.e. both

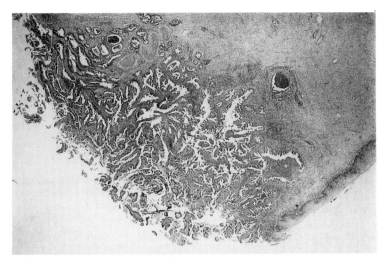

Figure 10.25 Adenocarcinoma. A low power view of a well-differentiated adenocarcinoma of the endocervical type. This patient had a history of previous cone biopsy from squamous cell carcinoma *in situ*. Haematoxylin and eosin. Magnification ×16, reduced to 65% in reproduction

elements may be invasive or both may be intraepithelial or one may be intraepithelial and one invasive.

Other primary tumours of cervix

These are all rare but have been described. They include adenoid cystic carcinoma, sarcomas, mixed mesodermal tumours, lymphomas and malignant melanoma.

Secondary tumours of cervix

Metastases from primary tumours elsewhere occur in the cervix but are not often diagnosed. This may reflect the failure to take blocks from the cervix at necropsy. Metastases from breast are the most often reported but a secondary deposit from a pancreatic carcinoma has been seen by the author, and Anderson (1976) reports metastatic spread from primary gastric carcinoma. Direct spread or metastatic spread of primary adenocarcinoma from the endometrial cavity to the cervix must always be borne in mind when malignant glandular cells are found in the cervical smears.

Cytology of invasive cancer

The presence of invasive cancer can be predicted when cells showing the features of malignancy described below are seen in the smear. Unfortunately, the distinction between invasive cancer and intraepithelial carcinoma is not always possible from the appearance of the epithelial cells in the smear. However, when a confident cytodiagnosis of malignancy can be made it is usually possible to identify the type of tumour present in the cervix from the smear pattern.

Criteria of malignancy

Many of the features of the malignant cell are similar to those found in a dyskaryotic cell. There are, however, a number of additional changes which indicate invasive cancer rather than CIN.

One of the most useful diagnostic features of malignancy is the coarse irregular appearance of the nuclear chromatin. This may take the form of sharp-edged, angular clumps with irregular clear areas between them. This pattern is commoner in malignant cells shed from a squamous cell carcinoma than in adenocarcinoma. In addition, the outline of the nucleus is extremely irregular with angular indentations and sharp peaks, and in cells shed from poorly differentiated carcinoma the nucleocytoplasmic ratio is extremely high. A second feature of malignancy is the presence of large, prominent nucleoli in the tumour cells. These are rarely seen in dyskaryotic cells shed from an area of CIN III. The nucleoli in malignant cells are usually eosinophilic and have an irregular outline with sharp irregularities. Moreover, nucleoli in the same nucleus vary in shape and size. In this respect the nucleoli differ from those seen in cells undergoing reactive change which are usually small and even. A third feature of smears from an invasive cancer is that the malignant cells are shed in tissue fragments whereas in smears from an area of CIN, the dyskaryotic cells are often discrete. Sometimes the nuclei are very densely packed and moulding can be seen. These closely packed cells must be distinguished from clusters of degenerating endocervical or metaplastic cells which are sometimes found in an inflammatory smear. Finally, smears from patients with an ulcerating malignant lesion of the cervix frequently contain blood and necrotic debris—often termed the 'malignant diathesis'. This is an unreliable sign of malignancy as it reflects an associated cervicitis rather than the presence of tumour.

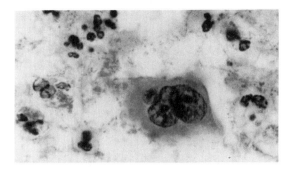

Figure 10.26 Malignant squamous cell. The chromatin pattern is irregular and a sharp indentation is seen in one nucleus. The cytoplasm is clearly defined and squamoid in type. Papanicolaou stain. Magnification × 1280, reduced to 70% in reproduction

Cytological diagnosis of tumour type

Squamous cell carcinoma (Figures 10.17(b) and 10.26)

Diagnosis depends on recognizing squamous differentiation of the cytoplasm. In well differentiated tumours keratinization of cytoplasm is seen and discrete bizarre highly keratinized cells may be present in the smear. Tadpole or spindle cells may be seen. Plaques of keratinized squames or anucleate fragments may also be seen. In less well differentiated squamous carcinomas it is usually possible to recognize single cells with a rim of sharply outlined, deeply stained cyanophilic or eosinophilic cytoplasm. The presence of flat sheets of malignant cells giving a pavement effect also suggest squamous differentiation. Smears from a non-keratinizing small cell squamous carcinoma may contain only dense clusters of undifferentiated malignant cells and it may be impossible to distinguish the tumour from a poorly differentiated adenocarcinoma (Figure 10.27).

Adenocarcinoma of endocervix

The glandular origin of the tumour cells is readily recognized in smears from well differentiated adenocarcinoma. Papillary clusters of

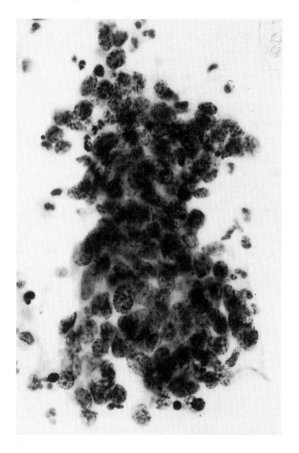

Figure 10.27 Undifferentiated malignant cells. This field shows a tight cluster of malignant cells with poorly defined cytoplasm. The general form of the group is similar to that seen in Figures 10.8 and 10.21 but in this example irregularities of chromatin pattern are seen in nuclei at the periphery. Papanicolaou stain. Magnification × 512.

glandular cells with irregular nuclei, vacuolated cytoplasm and large nucleoli are seen. However, in the poorly differentiated forms distinction between adeno and squamous carcinoma is difficult.

When the tumour is of the *endocervical type* (*Figure 10.28*), it is possible to recognize the similarities of cytoplasmic differentiation to those seen in normal endocervical columnar cells. The columnar shape of the cell with a basal nucleus is apparent in the sheets and clusters of malignant cells present.

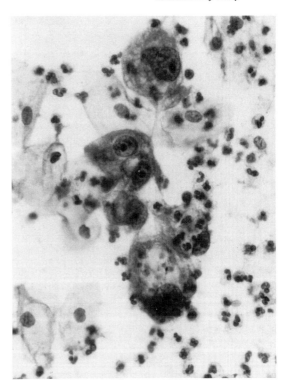

Figure 10.29 Malignant columnar cells. These cells are shed from a clear cell adenocarcinoma. Cytoplasm is plentiful and soft and foamy in type. Note the irregularities of nuclear outline and the irregular shape of the nucleolus in the isolated cell. Papanicolaou stain. Magnification × 512

Adenocarcinoma of the endometrial type exfoliates loose clusters, papillary and acinar fragments with scanty cytoplasm but sometimes vacuoles may be present and these can contain polymorphs. This cytological picture is the same as that seen with adenocarcinoma of the endometrium.

Metastatic carcinoma should be suspected when the form of acinar fragments and the appearance of the constituent cells is unlike that usually seen from genital cancers. An example is shown in *Figure 10.30* in which the primary tumour was in the breast and the similarities to a breast lump aspirate will be noted. Whenever malignant glandular cells are found in smears, the possibility of them being derived from a primary site other than the cervix must be borne in mind.

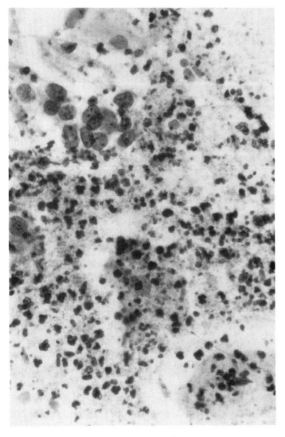

Figure 10.28 Malignant columnar cells. This is a field from a smear taken from the case illustrated in *Figure 10.29*. In addition to malignant columnar cells of the endocervical type this field shows the cell degeneration and serous background known as the cancer diathesis. Papanicolaou stain. Magnification × 512.

It is also possible to identify *clear cell adenocarcinoma* by the presence of clusters, and sometimes acinar forms of cells with abundant clear cytoplasm. In these cases, the malignant cells present often resemble those seen in renal adenocarcinoma (*Figure 10.29*).

Carcinoma of ovary

Malignant cells from ovarian tumours can

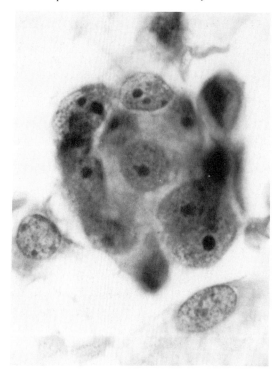

Figure 10.30 Malignant columnar cells. This acinar fragment was present in the smear of a patient with metastatic carcinoma of cervix. The primary tumour was in the breast. Similarities with smears made from fine needle aspiration of breast can be recognized. Papanicolaou stain. Magnification × 512

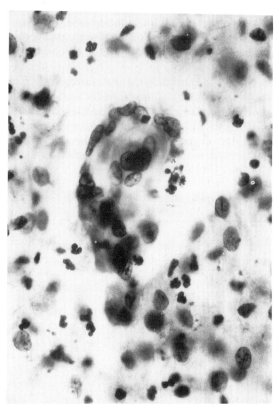

Figure 10.31 Malignant columnar cells. This acinar fragment of malignant cells is moulded into a 'cast'-like form by passage through the fallopian tube. Note the clean background of the smear. The patient had a carcinoma of ovary with no deposits in endometrium or cervix. Papanicolaou stain. Magnification × 512

travel through the tubes and uterus and present in the cervical smear without secondary deposits in the lower genital tract. Unlike cancers of the cervix and endometrium, the 'cancer diathesis' which reflects tissue necrosis is not seen (*Figure 10.28*). In addition, the tumour fragments have been moulded by passage through the fallopian tube and this results in a cast-like packet of tumour cells (*Figure 10.31*).

Carcinoma of vulva

Most women who present with this disease are elderly and complain of postmenopausal bleeding. The lesions are usually very advanced at this stage and present as cauliflower-like growths but may present as ulcerated or indurated areas of the vulval or vaginal mucosa. The tumours are usually squamous in type and smears prepared from vulval and vaginal scrapes from well differentiated tumours contain bizarre highly keratinized squames. Intraepithelial forms of the tumours are recognized.

The distinction between squamous carcinoma of the vulva and vagina and leukoplakia due to hyperkeratosis of the vulval or vaginal epithelium is difficult both clinically and cytologically. Indeed, it has been suggested that half the cases of squamous carcinoma of the vulva arise in an area of leukoplakia. If there is clinical doubt, biopsy is advised.

Recommendations for management

Cervical smears taken as part of a screening programme are usually taken by General Practitioners or Community Health doctors. These people are often dependent on advice from the laboratory on appropriate action following any report which is not completely negative. Even when smears are received from gynaecological clinics dialogue with the clinician is helpful both to the laboratory and to the gynaecologist.

Further action is straightforward in the case of the completely negative smear when the woman remains in the screening programme or in the case of the severely abnormal smear when immediate biopsy is mandatory. Recommended action with unsatisfactory smears, inflammatory smears and smears containing mildly dyskaryotic cells has been less precise. Recent reports have shown that there can be more serious underlying abnormality in such cases (Campion, Singer and Mitchell, 1987) and the British Society for Clinical Cytology (BSCC) have made recommendations on the management of women with abnormal cervical cytology (Evans et al., 1987).

Ideally women with cervical smears in these borderline categories should be referred for colposcopy and biopsy but in many areas this is not practical because of the lack of trained colposcopists. In consequence, until there are adequate colposcopy services throughout the country compromises are necessary. The recommendations presented here are based on the BSCC report. In every case, recommendations must take into account not only the cytological findings but also the clinical condition and age of the woman.

Negative smear

With a normal cervix on clinical examination, a repeat smear at the usual screening interval is sufficient. When there is strong clinical suspicion of malignancy immediate biopsy is essential regardless of the cervical smear report.

Inadequate or unsuitable smear

Repeat smear soon. If the smear is still unsatisfactory without identification of a cause for this e.g. specific infection which can be treated or drying artefact which can occur with atrophic smears, colposcopy is advisable.

Severe inflammatory changes with or without recognizable cause of infection

Repeat smear in three months after treatment of infection. If severe inflammatory changes persist, colposcopy is advisable.

Keratinized cells or koilocytes

Repeat in six months and if the changes persist colposcopy is advisable. It is important to note that keratinization of the cervix as seen with wart virus infection is a common cause of failure of exfoliation of severely dyskaryotic cells from an underlying CIN III.

Borderline changes and mild dyskaryosis

Repeat in three to six months. If the changes persist or become worse, colposcopy is advisable. If the smear has become negative it should be repeated in one year before returning the woman to the usual screening programme. Some workers would advise annual smears for at least three years before return to the usual screening interval. This also applies if the original abnormal smear showed warty changes.

Moderate dyskaryosis

Depending on the availability of colposcopy, referral may be immediate or after a confirmatory second cervical smear.

Severe dyskaryosis

Immediate referral for colposcopy and biopsy.

Monitoring after treatment

Most authors advise an annual cervical smear after treatment of CIN or wart virus infection.

Acknowledgement

It is a pleasure to thank Mrs Linda Chawner for producing the photographic prints.

References

ANDERSON, M. C. (1976). The aetiology and pathology of cancer of the cervix. In *Clinics in Obstetrics and Gynaecology. Cancer of the Vulva, Vagina and Uterus*, Volume 3/No. 2, (Langley, F. A., ed.), pp. 317–337. Philadelphia: W. B. Saunders

BRITISH SOCIETY FOR CLINICAL CYTOLOGY (1978). Problems of cell nomenclature in cervical cytology. *J. Clin. Pathol.*, **31**, 1226–1227

BUCKLEY, C. H., BUTLER, E. B. and FOX, H. (1982). Cervical intraepithelial neoplasia. *J. Clin. Pathol.*, **35**, 1–13

BUCKLEY, J. D., HARRIS, K. W., DOLL, R., VESSEY, M. P. and WILLIAMS, P. T. (1981). A case controlled study of the husbands of women with dysplasia and carcinoma *in situ*. *Lancet*, **2**, 1010–1015

BURGHARDT, E. (1976). Pre-malignant conditions of the cervix. In *Clinics in Obstetrics and Gynaecology. Cancer of the Vulva, Vagina and Uterus*, Volume 3/No. 2, (Langley, F. A., ed.), pp. 257–294. Philadelphia: W. B. Saunders

CAMPION, M. J., SINGER, A. and MITCHELL, H. S. (1987). Complacency in the diagnosis of cervical cancer. *Br. Med. J.*, **294**, 1337–1339

CHOO, Y -C. and ANDERSON, D. G. (1982). Neoplasms of the vagina following cervical carcinoma. *Gynaecol. Oncol.*, **14**, 125–132

COLEMAN, D. V. (1985). Genital warts, human papillomavirus and cervical cancer. In *Clinical Problems in Sexually Transmitted Diseases*, (Taylor-Robinson, D., ed.). Lancaster: Martinus Nijhoff

COPPLESON, M. and REID, B.L. (1967). *Preclinical Carcinoma of the Cervix Uteri*. Oxford: Pergamon Press

DANIEL, C. and BABES, A. (1927). Posibilitatea diagnosticului cancerului uterin cu ajutorul frotiului. *Proc. Bucharest Gynaecol. Soc.* January

DAVIES, J. R. and MOON, L. B. (1975). Increased incidence of adenocarcinoma of the uterine cervix. *Obstet. Gynaecol.* **45**, 79

EVANS, D. M. D., HUDSON, E. A., BROWN, C. L., BODDINGTON, M. M., HUGHES, H. E., MACKENZIE, E. F. D. and MARSHALL, T. (1986). Terminology in Gynaecological Cytopathology: Report of the Working Party of the British Society for Clinical Cytology. *J. Clin. Pathol.*, **39**, 933–944

EVANS, D. M. D., HUDSON, E. A., BROWN, C. L., BODDINGTON, M. M., HUGHES, H. E. and MACKENZIE, E. F. D. (1987). Management of women with abnormal cervical smears: supplement to Terminology in Gynaecological Cytopathology. *J. Clin. Pathol.*, **40**, 530–531

FERENCZY, A. (1977a). Cervical intraepithelial neoplasia. In *Pathology of the Female Genital Tract*, (Blaustein, A., ed.), pp. 143–165. Berlin: Springer-Verlag

FERENCZY, A. (1977b). Carcinoma and other malignant tumours of the cervix. In *Pathology of the Female Genital Tract*, (Blaustein, A., ed.), pp. 171–205. Berlin: Springer-Verlag

HOWARD, L., ERICKSON, C. C. and STODDART, L. D. (1951). A study of the incidence and histogenesis of endocervical metaplasia and intraepithelial carcinoma. *Cancer*, **4**, 1210–1233

INTERNATIONAL COMMITTEE ON HISTOLOGICAL TERMINOLOGY (1962). An international agreement on histological terminology for lesions of the uterine cervix. *Acta Cytol.*, **6**, 235–236

KAUFMAN, R., KOSS, L. G., KURMAN, R. J. *et al.* (1983). Statement of caution in the interpretation of papilloma virus associated lesions of the epithelium of the uterine cervix. *Acta Cytol.*, **27**, 107–108

KESSLER, I. I. (1977). Venereal factors in human cervical cancer: evidence from marital clusters. *Cancer*, **39**, 1912–1919

KOSS, L. G. (1979). *Diagnostic Cytology and its Histopathologic Bases*, Volume 1, 3rd Edition, pp. 287–289. Philadelphia: J. B. Lippincott

MARTINES, I. (1969). Relationship of squamous cell carcinoma of the cervix uteri to squamous cell carcinoma of the penis among Puerto Rican women married to men with penile carcinoma. *Cancer*, **24**, 777–780

MILLER, A. B., BARCLAY, T. H., CHOI, M. G. (1980). A study of cancer, parity and age at first pregnancy. *J. Chronic Dis.*, **33**, 595–605

NAIB, Z. M. (1966). Exfoliative cytology of viral cervicovaginitis. *Acta Cytol.*, **10**, 126–129

NG, A. B. P. (1983). Microinvasive adenocarcinoma and precursors of adenocarcinoma of the uterine cervix. In *Compendium of Diagnostic Cytology*, 5th Edition, (Wied, G. L., Koss, L. G. and Reagan, J. W., eds), pp. 148–154. Chicago: Tutorials of Cytology

PAPANICOLAOU, G. N. (1928). New cancer diagnosis. In *Proc. 3rd Race Betterment Conference*, p. 528. Battle Creek: Race Betterment Federation

PAPANICOLAOU, G. N. (1954). *Atlas of Exfoliative Cytology*. Commonwealth Fund, Cambridge, Massachussetts: Havard University Press

PAPANICOLAOU, G. N. and TRAUT, H. F. (1943). *Diagnosis of Uterine Cancer by the Vaginal Smear*. New York: Commonwealth Fund

PATTEN, S. F. (1978). *Diagnostic Cytopathology of the Uterine Cervix*, 2nd Edition. Monographs in Clinical Cytology No. 3, p. 65. Basel: Karger

POULSEN, H. E., TAYLOR, C. W. and SOBIN, L. H. (1975). *Histological Types of Female Genital Tract Tumours*. International Histological Classification of Tumours, No. 13. Geneva: World Health Organization

REID, B. L. (1981). Carcinogenesis. In *Gynaecological Oncology*, (Coppleson, M., ed.), p. 36. London: Churchill Livingstone

REID, B. L. and COPPLESON, M. (1976). Natural history; recent advances. In *The Cervix*, (Jordan, J. A. and Singer, A., eds), pp. 317–331. Philadelphia: W. B. Saunders

REID, B. L., FRENCH, P. W. and SINGER, A. (1978). Sperm basic proteins in cervical carcinogenesis. *Lancet*, **2**, 60–64

RICHART, R. M. (1967). Natural history of cervical intraepithelial neoplasia. *Clin. Obstet. Gynaecol.*, **10**, 748–784

RICHART, R. M. and BARRON, B. A. (1969). A follow-up study of patients with cervical dysplasia. *Am. J. Obstet. Gynecol.*, **105**, 386–393

RIOTTON, G. and CHRISTOPHERSON, W. M. (1973). *Cytology of the Female Genital Tract*. International Histological Classification of Tumours, No. 8. Geneva: World Health Organization

SCHAUENSTEIN, W. (1908). Histologische Untersuchungen uber atypisches platten-epithel an der portio und an der innenflache der cervix uteri. *Arch. Gynak.*, **85**, 576–616

SINGER, A. (1975). The uterine cervix from adolescence to the menopause. *Br. J. Obstet. Gynaecol.*, **82**, 81–99

SINGER, A. (1985). High risk groups for cervical cancer development. In *Cancer of the Uterine Cervix*, Volume 8 of Cancer Campaign, (Bender, H. G. and Beck, L., eds), pp. 49–58. Stuttgart: Fischer Verlag

SINGER, A. and JORDAN, J. A. (1976). The anatomy of the cervix. In *The Cervix*, (Jordan, J. A. and Singer, A., eds), pp. 13–36. Philadelphia: W. B. Saunders

SKEGG, D. C., CORWEN, P. A., PAUL, C and DOLL, R. (1982). Importance of the male factor in cancer of the cervix. *Lancet*, **2**, 581–583

SPRIGGS, A. I. and BODDINGTON, M. M. (1980). Progression and regression of cervical lesions: review of smears from women followed without initial biopsy or treatment. *J. Clin. Pathol.*, **33**, 517–525

STANBRIDGE, C. M. and BUTLER, E. B. (1983). Human papillomavirus infection of the lower female genital tract: association with multicentric neoplasia. *Int. J. Gynaecol. Pathol.*, **2**, 264–274

TREVATHAN, E., LAYDE, P. and WEBSTER, L. (1983). Cigarette smoking and dysplasia and carcinoma *in situ* of the uterine cervix. *J. Am. Med. Ass.*, **250**, 499–502

11

The corpus uteri

Elizabeth Hudson and W. K. Blenkinsopp

The success of the cervical smear in screening for preclinical cervical cancer has led to expectations that early diagnosis of endometrial cancer could be achieved by a comparable method. Progress has been slow because of the technical problem of obtaining suitable specimens from the uterine cavity and because of the difficulty in distinguishing between benign and malignant endometrial cells in the smears. The distinction between hyperplastic endometrial cells and adenocarcinoma can be particularly difficult when the tumour is well differentiated and skill and experience are needed for reliable interpretation of these specimens.

It has been known for many years that benign and malignant endometrial cells may be present as a coincidental finding in cervical smears taken primarily for the detection of precancerous lesions of the cervix, and a diagnosis of endometrial cancer can be made from the smear. However, endometrial exfoliation is inconsistent and cervical smears are not a reliable method of diagnosing this tumour. In consequence, alternative methods of diagnosing endometrial cancer have been developed and direct endometrial sampling is now an acceptable method for the investigation of women with dysfunctional uterine bleeding.

Specimen collection

Indirect endometrial sampling by examination of cervical and vaginal smears

Cells exfoliated from the endometrial cavity may be present in cervical and vaginal smears, thereby presenting an opportunity to report on pathological changes in the endometrium as part of the cervical cancer screening programme. However, as only 50% of cervical smears from women with endometrial cancer will contain tumour cells, this method of detecting endometrial cancer is insensitive and cannot be used as a reliable test for these lesions (Jafari, 1978). Vaginal sampling from the posterior fornix is considered by some to give a better diagnostic yield for endometrial malignancy. A posterior fornix sample contains an accumulation of cells exfoliated from all parts of the genital tract and is more likely to contain endometrial cells than a cervical scrape (Koss, 1979). On the other hand, exfoliated endometrial cells become trapped in the cervical mucus as they pass through the endocervical canal and this accumulation will be sampled by a cervical scrape. A marginally higher rate of identification of endometrial cells will be achieved if both cervical and posterior fornix specimens are taken rather than just one or the other. These two smears may be recommended for patients over 40 years, especially if endometrial pathology is suspected.

Direct endometrial sampling

The poor results from cervical and vaginal smears for the detection of endometrial cancer have led to the development of a range of instruments for collecting cell and tissue samples directly from the endometrium which can be used in an outpatient setting, as dilatation of the cervix and general anaesthesia is not required.

Direct endometrial sampling without dilatation of the cervix was first carried out by Ayre in 1955 who showed that diagnosis of endometrial carcinoma could be made from brush specimens of the endometrium. A decade later, Torres and coworkers showed that washing the endometrial cavity with saline under positive pressure provided suitable samples for light microscope analysis (Torres, Holmquist and Danos, 1969; Anderson *et al.*, 1976). Concern about the possible complications of this procedure such as uterine infection, dissemination of tumour through the fallopian tubes into the peritoneal cavity and perforation of the uterus, caused gynaecologists to resist this approach for several years. Subsequently, instruments were designed for washing the endometrial cavity with physiological saline under negative pressure (Dowling, Gravlee and Hutchins, 1969) so that the risk of dissemination of tumour into the peritoneal cavity was eliminated. The saline washings were centrifuged in the laboratory and smears made from the cell deposit and stained by the Papanicolaou method. Fragments of tissue, if present, were prepared for histological section. The success of the technique depended on maintaining an airtight seal between the instrument and the cervix throughout the procedure. Unfortunately, endometrial washings were difficult to interpret (Afonso, 1975; Lewis and Chapman, 1977; Richart, Marchbein and Sherman, 1979) and the laboratory processing was time-consuming, so the method was not widely accepted.

In the 1970s a new generation of endometrial sampling instruments was introduced. These depended on abrasion of the endometrium by a plastic helix, the *Mi-Mark* (Milan and Markley, 1973) or curette, the *Accurette* (Goldberg, Tsalacopoulos and Davey, 1982) and immediate transfer of the cytological material to a glass slide. The specimen was smeared over one or more slides and fixed and stained by the Papanicolaou method. A third instrument, the *Isaacs Endometrial Sampler* (Isaacs and Wilhoite, 1974) depended on aspiration of the endometrial cavity. The three instruments are illustrated in *Figure 11.1*.

The Mi-Mark endometrial sampler consists of two plastic instruments: a flexible helix which is rotated in the endometrial cavity, to which the abraded endometrium sticks in a coagulum of

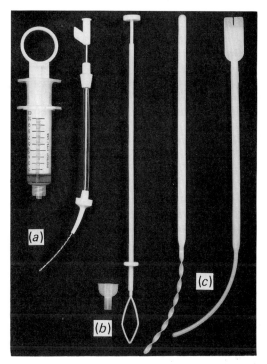

Figure 11.1 Endometrial sampling devices. (*a*) The Isaacs aspiration curette consists of a perforated semi-rigid stainless steel instrument in a plastic cannula with an endocervical stop. The syringe is used to create suction whereby endometrial material is drawn in through the perforations. After withdrawal of the instrument, positive pressure from air in the syringe expels the endometrial material onto one or more slides. (*b*) The Accurette instrument consists of an abrasive diamond-shaped head which is extruded from the cannula when the instrument has passed through the cervical canal. The adaptor and a syringe are attached to the cannula after the curette has been withdrawn and loose residual material is aspirated. (*c*) The Mi-Mark consists of a flexible plastic helix and a paddle-shaped instrument with a slit for removal of the material from the helix. The curved end is a uterine sound

blood and mucus, and a paddle-shaped instrument with a slot which is used to remove the specimen from the helix. Smears are made and any remaining material is fixed in formalin for histological section.

The Accurette endometrial sampler also abrades the endometrium with its diamond-shaped flexible plastic head. This is introduced through the cervical canal inside a cannula. After sampling and withdrawal of the curette, residual endometrial material can be aspirated through the cannula using the adaptor and a disposable plastic syringe.

The Isaacs endometrial cell sampler consists of a semi-rigid stainless steel cannula with 40 small perforations. The cannula can be bent to the angle of the uterus. A cervical stop lessens the risk of perforation of the uterus and makes an airtight seal so that negative pressure is produced within the uterus when the plunger on the attached syringe is withdrawn, thus aspirating cellular material from the surface of the endometrium. Laboratory processing of specimens from these three instruments is simple and quick in comparison with the saline washings produced by lavage techniques. Some reports recommend immersion of the slides in Carnoy's fixative to lyse the blood which is always present, but others do not find this necessary. The Mi-Mark, Accurette and Isaacs instruments do not usually cause more than minimal discomfort to the patient and there are no reports of complications of their use.

Cytology of normal endometrial cells

Endometrial cells vary in their appearance according to whether they have been exfoliated or whether they have been forcibly detached by direct sampling of the uterine cavity. Exfoliated endometrial cells in cervical and vaginal smears are usually present in smaller numbers, and are often degenerated and compacted into dense balls or tight clusters of cells. In contrast, smears prepared as a result of direct sampling contain sheets of endometrial cells with cytoplasm intact and nuclear detail perfectly preserved.

Exfoliated endometrial cells

Menstrual shedding of the endometrium may result in the presence of endometrial cells in cervical scrapes and in smears prepared from cells aspirated from the posterior fornix. This occurs most frequently in cervical smears taken during the first 2–3 days of the menstrual cycle but there is considerable normal variation. It is accepted that endometrial cells may be found in smears taken at any time from the first to the tenth day of the menstrual cycle. In a few individuals endometrial cells may be found up to the twelfth day of the cycle. Nevertheless it should be remembered that abnormal conditions of the endometrium may also cause exfoliation at this time.

A normal endometrial cell is slightly larger than a polymorph and comparable in size to the nucleus of an intermediate squamous cell. It has a hyperchromatic nucleus with granular chromatin which is evenly distributed apart from one or two slightly larger aggregations. The cytoplasm is confined to a narrow rim around the nucleus. Degenerative nuclear and cytoplasmic changes are common and result in poor definition of the chromatin pattern and cytoplasmic vacuolation. Prominent nucleoli which stain pink with Papanicolaou stain are occasionally seen if the cell preservation is unusually good. The endometrial cells appear in clusters in which they are most easily recognized (*Figure 11.2*) and as single cells in which case their identification is more difficult.

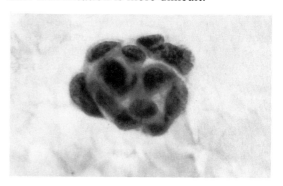

Figure 11.2 A cluster of endometrial cells in a cervical smear taken on the fourth day of the menstrual cycle. The nuclei are small and hyperchromatic and the cytoplasm is scanty. The concentric arrangement of the cells is characteristic. Papanicolaou stain. Original magnification × 250 (enlarged × 5), reduced to 80% in reproduction

One-third to one-half of the thickness of the endometrium is shed at menstruation, therefore stromal cells may be exfoliated as well as glandular epithelium. The distinction between exfoliated stromal and epithelial endometrial cells is evident in the endometrial plaque. This consists of a microscopic fragment of endometrium which assumes a spherical shape on exfoliation with a single layer of glandular endometrial cells on the surface and stromal endometrium in the centre (*Figure 11.3*). The different morphological characteristics of stromal and epithelial endometrial cells are described in detail by Reagan and Ng (1973). The distinction is difficult to make in cervical and vaginal smears and is not of diagnostic significance in routine laboratory work.

Histiocytes are seen in smears in large numbers around the fourth day of menstruation. This was referred to as the 'exodus' by

Cytology of normal endometrial cells

Endometrial cells in smears have to be distinguished from endocervical cells, histiocytes and severely dyskaryotic squamous cells. Careful attention to the nuclear and cytoplasmic characteristics and to their arrangement in clusters usually confirms their identity. The tendency for exfoliated endometrial cells to assume a rounded shape is helpful for the identification of endometrial material in smears. It contrasts with the angular shape of the endocervical epithelial cells which appear in flat sheets as a result of direct abrasion with the cervical spatula. Minor variations in their small size and degenerative changes sometimes cause diagnostic problems. Familiarity can be established by studying smears taken during menstruation.

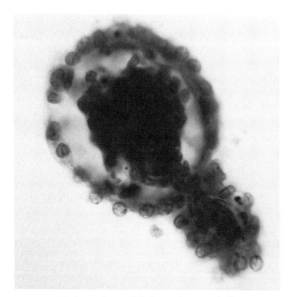

Figure 11.3 An endometrial plaque in a cervical smear taken during menstruation. The circumference of the spherical structure consists of glandular endometrial cells. Their nuclei are on the periphery external to the clear zone of cytoplasm. The centre consists of a dense mass of mainly stromal endometrial cells. Papanicolaou stain. Original magnification ×160 (enlarged ×5), reduced to 70% in reproduction

Direct sampling

Interpretation of the cytological specimen obtained by direct endometrial sampling requires experience. The sheets of fresh endometrial epithelial and stromal cells present different diagnostic criteria from exfoliated endometrial cells. There is usually an abundance of material which may take many minutes to examine thoroughly. The appearance of normal endometrium varies according to the stage of the cycle but in all normal specimens the endometrial cell nuclei are round or oval and of uniform size and regularly arranged in sheets without overlapping. The nuclei have finely granular chromatin which is evenly distributed apart from a few chromocentres. Nucleoli and occasional mitotic

Papanicolaou (1953) who suggested that these histiocytes which were closely associated with endometrial exfoliation might be derived from the stroma. The histiocytes are recognized by their reniform, oval or rounded nuclei which are eccentrically placed and by their thin, foamy cytoplasm which may contain phagocytosed cells or debris (*Figure 11.4*).

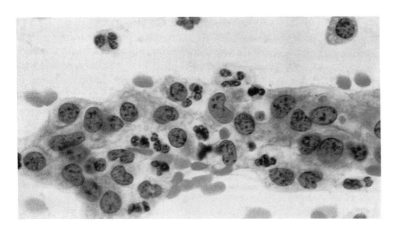

Figure 11.4 Sheets of histiocytes in a cervical smear taken near the end of menstruation. This is known as the 'exodus'. The reniform shape of the nucleus is seen in some cells. The cytoplasm is thin and foamy. Papanicolaou stain. Original magnification ×250 (enlarged ×5), reduced to 60% in reproduction

figures are seen in the proliferative phase and in the secretory phase the chromatin becomes more finely granular, nucleoli are less apparent and the nucleocytoplasmic ratio decreases. Subnuclear vacuoles may also be seen in smears taken in the secretory phase. Samples taken after the menopause from an atrophic endometrium often contain scanty pale staining endometrial cells (*Figure 11.5*).

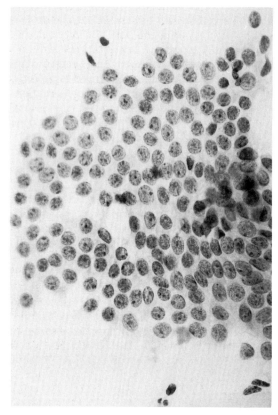

Figure 11.5 Inactive postmenopause endometrium (obtained with Accurette). A sheet of uniform cells; the nuclei show granular chromatin and one or two nucleoli. Papanicolaou stain. Original magnification × 160 (enlarged × 5), reduced to 55% in reproduction

Cytology of benign and malignant lesions of the endometrium

The distinction between benign and malignant endometrial cells is often difficult in cervical and vaginal smears and in smears prepared from endometrial aspirates. A high level of expertise is needed for their interpretation, especially for the distinction between hyperplastic lesions of the endometrium and adenocarcinoma. An awareness of the pathological changes which may lead to endometrial exfoliation is of value when attempting the analysis of this material.

Conditions which cause abnormal endometrial exfoliation are listed in *Table 11.1*.

Table 11.1 Causes of abnormal endometrial exfoliation

Benign	Menstrual irregularities
	Hormone therapy
	Acute endometritis
	Abortion
	Recent curettage
	Intrauterine contraceptive device
	Endometrial polyp
	Endometrial hyperplasia
Malignant	Endometrial carcinoma
	Other primary or metastatic tumours involving the endometrium

Hormone imbalance

Disturbances of the menstrual cycle are the commonest reason for the presence of endometrial cells in smears after the tenth day of the cycle. These irregularities may be due to imbalance of hormone secretion (endogenous) but they may also be the result of hormone therapy (exogenous). The breakthrough bleeding associated with oral contraceptives is a common example of the latter. Current low-dose oral contraceptives are commonly associated with a reduced level of cyclical changes—the proliferative phase may be shorter, with little convolution of the glands, and the secretory phase may be poorly developed; focal decidual change may occur in the stroma from mid-cycle. In some patients neither proliferative nor secretory activity is seen, and after prolonged use atrophy may occur. Focal hyperplasia has been recorded as a possible effect, but in general oral contraceptives do not produce endometrial hyperplasia and they do not cause endometrial carcinoma. Indeed, there is evidence that combined oral contraceptives halve the risk of both endometrial carcinoma and ovarian carcinoma (Drife and Guillebaud, 1986).

Oestrogen may be administered alone as postmenopausal hormone replacement therapy and in the treatment of breast cancer. It produces glandular hyperplasia which is usually cystic and occasionally atypical. Prolonged therapy is associated with an increased incidence of endometrial carcinoma.

Progesterone given alone, as in the treatment

of menstrual irregularities and endometriosis, produces marked gland atrophy, with prominent decidual change, stromal granulocytes, and dilated venules in the stroma (*Figure 11.6*). The progesterone component of oral contraceptives often produces similar but less marked changes.

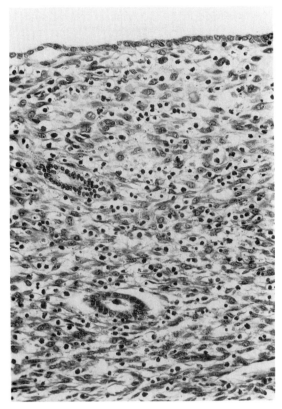

Figure 11.6 Endometrium with exogenous progesterone effect—gland atrophy, decidual change, and stromal granulocytes. Haematoxylin and eosin. Magnification × 350, reduced to 60% in reproduction

Acute endometritis

Lymphocytes and lymphoid foci are occasionally found in small numbers in apparently normal endometrium. During the late secretory phase stromal granulocytes (K cells) become prominent and are easily mistaken for polymorphs. Neutrophils are usually present as menstruation begins. Collections of neutrophils otherwise indicate an acute endometritis, and chronic endometritis is characterized by an infiltrate of plasma cells, usually with lymphocytes and histiocytes.

Inflammation is commonly due to nonspecific infection associated with indwelling intrauterine contraceptive devices (IUCD), retained products of conception or recent curettage. Specific infections are rare; tuberculosis produces granulomatous inflammation but is now rarely seen in Western countries, other than in immigrants from developing countries. Actinomycetes are sometimes present with an IUCD (*see* below), but usually they appear to be commensals and cause no harm.

Acute endometritis is an uncommon cause of endometrial cells in smears. When they do occur, the cells appear enlarged with prominent nucleoli and may suggest carcinoma to the unwary. In endometritis, they will be accompanied by an inflammatory exudate and the patient is usually considerably below the age for endometrial carcinoma. It is unusual for a smear to be taken within a week of curettage but it can cause confusion unless the history is known.

The intrauterine contraceptive device (IUCD)

Insertion commonly produces a mild endometritis which may be bacterial, and a chronic inflammatory reaction is usually present in the endometrium for at least some months. This may persist indefinitely as a mild focal reaction of neutrophils and plasma cells, and in a few women this persistent reaction is severe and accompanied by loss of cyclical changes, which may be focal or diffuse. An IUCD may cause exfoliation of endometrial cells at any time of the menstrual cycle. Changes occur in the endometrium at points of contact with the device resulting in abnormal appearances of the endometrial epithelium and sometimes squamous metaplasia. When the abnormal endometrial cells appear in the cervical smear, they are frequently enlarged with either hyperchromatic or pale nuclei. Multinucleation and prominent nucleoli are common. The cytoplasm may be vacuolated and in a cluster of cells this looks alarmingly like adenocarcinoma (*Figure 11.7(a)*). (Fornari, 1974).

Although patients with IUCDs have an increased incidence of pelvic inflammatory disease they do not have an increased risk of endometrial carcinoma. Nevertheless, it is always possible that the patient has a coexistent, more serious endometrial lesion such as carcinoma, so that abnormal endometrial cells in patients with IUCDs should always be viewed with suspicion. It may be appropriate (especi-

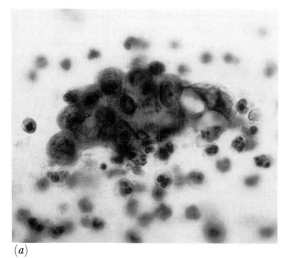

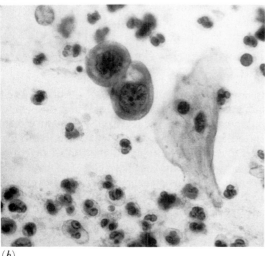

Figure 11.7 Abnormal endometrial cells in the cervical smear at mid-cycle in the presence of an IUCD. (*a*) A cluster of enlarged endometrial cells. These are indistinguishable from endometrial exfoliation due to hyperplasia or malignancy. (*b*) Two enlarged endometrial cells with irregular chromatin clumping and a narrow rim of dense cytoplasm. One cell has a cytoplasmic vacuole, the other has an irregular nuclear membrane. Papanicolaou stain. Original magnification × 250 (enlarged × 5), reduced to 60% in reproduction

ally in women aged over 40) to recommend the removal of the IUCD. Another smear taken after the next menstrual period will be normal unless there is other endometrial pathology. This usually provides a rapid solution to the problem but a curettage is a more reliable way of excluding serious pathology. Abnormal cells associated with IUCDs may also be difficult to distinguish from the severe dyskaryosis of cervi-cal intraepithelial neoplasia (CIN III) (*Figure 11.7(b)*). If this problem cannot be resolved from the morphology of the cells or by colposcopic examination of the cervix, the IUCD should be removed to determine whether the supposedly dyskaryotic cells disappear from the smear.

A description of the appearances of smears from patients with IUCDs would be incomplete without mention of actinomyces-like organisms. These may be present in 1–31% of smears from patients with IUCDs (Duguid, Parratt and Traynor, 1980; Kaebler, Chatwani and Schwartz, 1983). They are seen more frequently with inert plastic devices than with copper IUCDs and their occurrence increases with the length of time the device has been in the uterus. The difficulty of culturing the organism delayed the confirmation of its presence although the characteristic appearance in smears which are shown in *Figure 11.8* had been associated with the IUCD for some years. Gupta, Erozan and Frost (1978) demonstrated *Actinomyces israelii* by immunofluorescence and Duguid, Parratt and Traynor (1980) were able to grow the organism by a special method from endocervical samples. There are case reports of serious pelvic inflammatory disease due to actinomyces and even death in patients with IUCDs. This risk is considered remote compared with the large numbers of women with IUCDs. The presence of actinomyces-like organisms in smears is not an indication for treatment of asymptomatic women. However, the association of the organism with the length of time that IUCDs have been in the uterus or the presence of other symptoms frequently stimulates the medical attendant to change the device, with subsequent disappearance of the actinomyces from the smear.

Microscopic identification in the routine smear depends on recognition of aggregations of fine, branching filaments, some of which appear beaded. The filaments tend to run parallel to form a central core to the vegetations from which they branch out at acute angles. High magnification is necessary to distinguish these microorganisms from aggregations of more common bacilli. The cytologist can only report a morphological appearance. Culture is necessary for proof if this is required. Actinomyces-like organisms are rarely seen in the absence of an IUCD. It is probably the presence of a foreign body and the resultant trauma which stimulates growth of this organism which may have spread to the lower genital tract from the intestinal tract. Other anaerobic organisms may also flourish and produce a foul-smelling vaginal discharge.

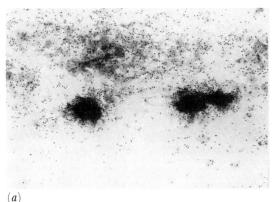

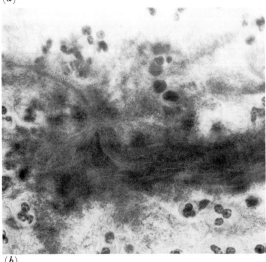

Figure 11.8 (*a*) Actinomyces-like organisms in cervical smear at low magnification. Note dense colonies of filamentous material. Papanicolaou stain. Magnification ×250. (*b*) Parallel filaments of the organism are seen in the centre of the colony with numerous fine radiating branches giving a fluffy appearance. Papanicolaou stain. Magnification ×1000, reduced to 60% in reproduction

Endometrial hyperplasia

This is the result of prolonged oestrogen stimulation in the absence of progesterone. It occurs in anovulatory cycles or in the presence of excess endogenous oestrogen secretion or exogenous oestrogen therapy. It can occur at any age but anovulatory cycles are commonest soon after the menarche and in the years before the menopause. Irregular menstruation, infertility or low parity are frequent associations. Oestrogen stimulation in the absence of progesterone also produces a mature squamous epithelium of the upper vagina and ectocervix which may be noticed in the smears of patients with endometrial hyperplasia. Although a mature squamous epithelium is normal during the first half of the menstrual cycle (proliferative phase), and up to 10% of postmenopausal women without endometrial disease maintain some superficial squamous cells in their smears (Meisels, 1966), a smear consisting almost entirely of superficial squamous cells and anucleate keratinized squames in a postmenopausal woman should raise a suspicion of coexisting endometrial hyperplasia.

Two types of hyperplasia are recognized— cystic, and atypical. The glands in cystic hyperplasia (*Figure 11.9*) vary considerably in size, including some which are abnormally dilated, but the lining cells are not abnormal; squamous metaplasia is common, but secretory activity is rare unless treatment with progesterone has been given. There is a high ratio of stroma to glands. This change is no longer considered to indicate an increased risk of subsequent endometrial carcinoma.

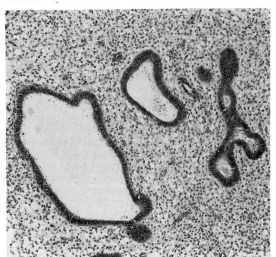

Figure 11.9 Endometrium in cystic hyperplasia—variation in gland size with plentiful stroma, without cytological dysplasia. Haematoxylin and eosin. Magnification ×280, reduced to 65% in reproduction

In atypical hyperplasia there is irregular epithelial proliferation with a reduction in the ratio of stroma to glands which are therefore

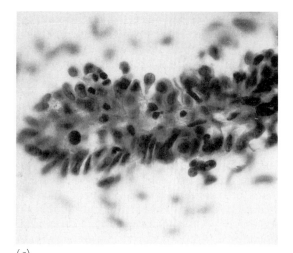

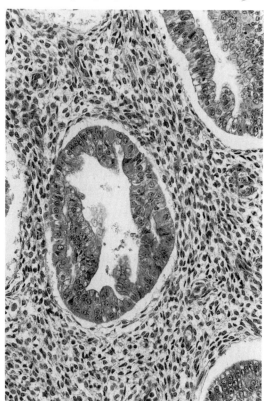

glands (*Figure 11.11*) will develop invasive carcinoma. A minority of cases with atypical hyperplasia will revert to a normal endometrium after treatment with progesterone.

Figure 11.11 Endometrium in atypical hyperplasia—severe cytological dysplasia. Haematoxylin and eosin. Magnification × 680, reduced to 70% in reproduction

Figure 11.10 (*a*) A fragment from endometrium (obtained with Accurette) which had cystic hyperplasia and some atypical hyperplasia. Most of the cells are hyperchromatic and crowded. A few nuclei contain a prominent nucleolus. Papanicolaou stain. Original magnification × 160 (enlarged × 5), reduced to 65% in reproduction. (*b*) Histological section of endometrium in atypical hyperplasia—gland crowding with mild cytological dysplasia. Haematoxylin and eosin. Magnification × 280, reduced to 65% in reproduction

crowded together. The lesion is characterized by atypia of the gland architecture and of the cells lining the glands. It is probable that architectural atypia alone, manifested by crowding and irregular distribution of the glands (*Figure 11.10(a)* and (*b*)) carries a very low risk of subsequent endometrial carcinoma, whereas perhaps 30–40% of patients with moderate or severe cytological atypia of the cells lining the

Accurate diagnosis of endometrial hyperplasia is not always possible from the appearances of endometrial cells in smears. Cells from cystic glandular hyperplasia are similar to menstrual endometrial cells, although prominent nucleoli may be seen. On the other hand, the cells from atypical hyperplasia (*Figure 11.12*) may be indistinguishable from a well differentiated adenocarcinoma (*Figure 11.13*). The cytology report should note the presence of abnormal endometrial cells and curettage will provide the precise histological diagnosis.

Endometrial carcinoma

Endometrial carcinoma is the commonest malignant tumour of the corpus uteri. It is responsible for approximately the same number of deaths as occur from cancer of the uterine cervix in England and Wales. The disease is increasing in wealthy societies but this may be partly attributable to longevity (Weiss, Szekely and Austin, 1976). It usually presents at a later age than cervical cancer.

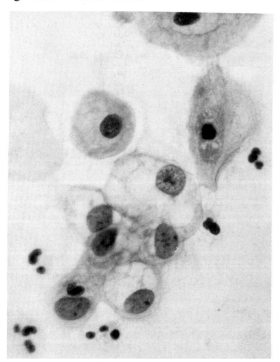

Figure 11.12 Endometrial hyperplasia. A cluster of six enlarged endometrial cells in a cervical smear from a patient with atypical hyperplasia. The nuclei have prominent nucleoli and the cytoplasm is abundant and vacuolated. A parabasal squamous cell is nearby for comparison. Papanicolaou stain. Original magnification × 250 (enlarged × 5), reduced to 55% in reproduction

Most cancers occur in women aged between 50 and 60 years who present with postmenopausal bleeding. Careful attention to this symptom should lead to early diagnosis. However, 5–10% of cases occur before the menopause, therefore the presence of abnormal endometrial cells in smears must be reported at all ages.

Endometrial cancer is particularly likely to affect obese women of low parity and it is also associated with diabetes, hypertension and previous endometrial hyperplasia. Strong links between endogenous oestrogen production and endometrial cancer have been demonstrated. Obese women have higher levels of endogenous oestrone as a result of conversion of androstenedione in peripheral fat than their slimmer counterparts (MacDonald and Siiteri, 1974). A similar link has been demonstrated between exogenous oestrogen and endometrial cancer. Antunes *et al.* (1979) and Jick *et al.* (1979) showed that long term users of oestrogens have a considerably higher risk of endometrial cancer than non-users. The risk increases with the dosage and duration of treatment and declines when treatment is stopped. Thus both these groups of women represent a high risk group in whom a satisfactory method of screening for premalignant or early disease could be effective.

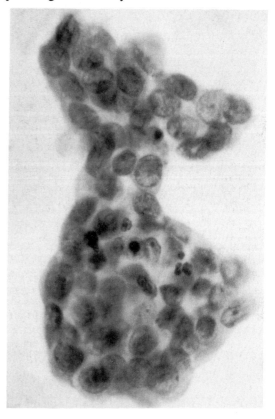

Figure 11.13 A group of exfoliated endometrial cells from a well differentiated adenocarcinoma. The nuclear chromatin clumping is fine and nucleoli are prominent in some cells but firm diagnosis of malignancy depends on the histology of the curettings. Papanicolaou stain. Original magnification × 250 (enlarged × 5), reduced to 55% in reproduction

Almost all endometrial cancers are adenocarcinoma (*Figure 11.14*), although areas of squamous metaplasia are common. A variety of morphological types can be seen; some are endocervical in type. Adenosquamous carcinoma, composed of squamous carcinoma and adenocarcinoma cells also occurs but is rare. The prognosis depends mainly on the extent of spread: the five year survival for tumours limited to superficial myometrial invasion is 90%; for tumours involving more than the inner third of the myometrium this figure falls considerably (to between 50% and 80%). Prognosis is also related to the degree of differentiation. The adenosquamous type (malignant squamous and glandular elements) has the poor prognosis of the poorly differentiated end of the spectrum.

tumour. The cells may be present in clusters or they may be discrete. The squamous metaplasia which is commonly seen in histological sections of endometrial carcinoma may be represented in the smears by atypical, but benign, squamous cells. This possibility must be borne in mind when deciding whether abnormal cells in the smear originate from the endometrium or the cervix.

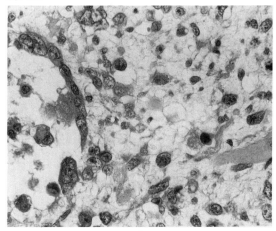

Figure 11.15 Heterologous mixed Müllerian tumour of endometrium—malignant gland and malignant mesenchymal stroma, including voluntary muscle. Haematoxylin and eosin. Magnification × 680, reduced to 60% in reproduction

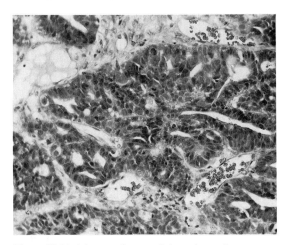

Figure 11.14 Adenocarcinoma of the endometrium. Haematoxylin and eosin. Magnification × 280, reduced to 60% in reproduction

A small proportion of endometrial malignancies are mixed Müllerian tumours which contain both carcinomatous and sarcomatous elements. If the sarcomatous component is composed of tissue homologous for the uterus e.g. undifferentiated spindle cells and/or smooth muscle and/or stroma the tumour is called a carcinosarcoma. If the sarcomatous component includes tissue types not normally found in the uterus, such as cartilage or voluntary muscle, it is designated heterologous (*Figure 11.15*). These tumours behave much more aggressively than the simple adenocarcinomas.

The cytological appearances of adenocarcinoma cells in Papanicolaou-stained smears show the pleomorphism and nuclear characteristics of malignancy (*Figures 11.16* and *11.17*). Cellular enlargement, macronucleoli and mucus-filled cytoplasmic vacuoles are strongly suggestive of a well differentiated adenocarcinoma. Dense clusters of small hyperchromatic tumour cells suggest a poorly differentiated

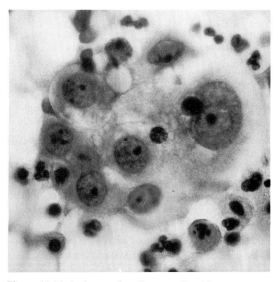

Figure 11.16 A cluster of malignant cells with characteristics of adenocarcinoma. The nuclei vary in size, macronucleoli are present and the cytoplasm is abundant and vacuolated. From a moderately well differentiated endometrial adenocarcinoma. Papanicolaou stain. Original magnification × 250 (enlarged × 5), reduced to 65% in reproduction

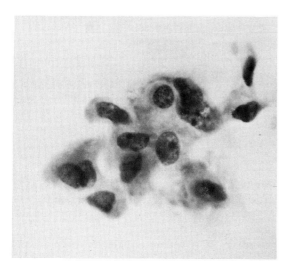

Figure 11.17 Endometrial adenocarcinoma (obtained with Accurette). The nuclei vary in size, the chromatin distribution is abnormal and macronucleoli are present. In this suction specimen the cellular arrangement is disorderly. Papanicolaou stain. Original magnification ×160 (enlarged ×5), reduced to 60% in reproduction

Principles of reporting

The first principle of reporting endometrial cells in smears must be a statement of their presence. They may be described as malignant if the morphology is diagnostic but the possibility of benign abnormalities of the endometrium must be considered. Curettage will provide material for precise histological diagnosis so it is not necessary or desirable to hazard a guess. It may be difficult to identify the tumour type when the tumour cells are poorly differentiated (*Figure 11.18*). It is unwise to assume that undifferentiated malignant cells in the cervical smear of a postmenopausal patient are from the endometrium. The tumour may turn out to be a poorly differentiated squamous carcinoma of the cervix and the gynaecologist will be ungrateful if he has been misled by the cytology report.

Adenocarcinoma in cervical smears is most frequently from the endometrium but other possible origins are the endocervix, the ovary and the fallopian tubes. Endocervical adenocarcinoma is characterized by the abundance of abnormal glandular cells which appear in flat sheets on the slide in contrast to endometrial adenocarcinoma which, unless it has extended through the cervix, will have exfoliated and presents as rather scanty dense clusters of tumour cells. When both glandular and squamous malignant cells are found in the smear, adenosquamous carcinoma of the endometrium should be borne in mind. An alternative diagnosis of adenocarcinoma of the endocervix associated with CIN must also be considered. The two lesions occur together in 43% of cases of adenocarcinoma (Maier and Norris, 1980). Secondary infection of endometrial carcinoma results in the presence of inflammatory exudate and necrotic cell debris in the smear. A relatively 'clean' smear with abnormal endometrial cells and a mature squamous epithelium in a postmenopausal woman is more likely (although not inevitably) to represent endometrial hyperplasia.

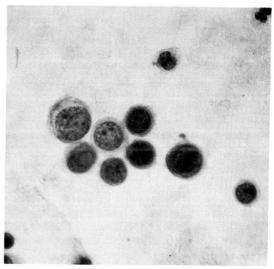

Figure 11.18 Poorly differentiated tumour cells from an endometrial adenocarcinoma. Abnormal nuclear chromatin distribution and the high nucleocytoplasmic ratio indicate malignancy but the tumour type is not apparent from these cells. Papanicolaou stain. Original magnification ×250 (enlarged ×5), reduced to 60% in reproduction

After the tenth day of the menstrual cycle or in a postmenopausal patient, endometrial cells in the smear are an indication of endometrial pathology. Unless the clinician is convinced that the presence of the endometrial cells is due to benign endometrial pathology, curettage is indicated. After the menopause, curettage should be mandatory if the endometrial cells are present in a smear. It should always be remembered that the positive identification of endometrial cells in a smear may be of diagnostic importance but the absence of endometrial cells does not exclude endometrial pathology. Thus any clinical suspicion of endometrial malignancy, for example because of postmenopausal bleeding,

Table 11.2 Diagnosis of endometrial carcinoma by endometrial sampling device compared with curettage

Authors	No. of patients	Tumours detected	
		Sampling device	Curettage
Swingler, Cave and Mitchard (1979)	101	Mi-Mark	8[a]	9
Crow, Gordon and Hudson (1980)	115	Mi-Mark	3	4
Ginsberg, Padleckas and Javaheri (1983)	138	Mi-Mark	8	8
Hutton et al. (1978)	121	Isaacs	6	6
Isaacs and Ross (1978)	69	Isaacs	8[a]	9
Segadal and Iverson (1980)	150	Isaacs	17	17
Goldberg, Tsalacopoulos and Davey (1982)	40	Accurette	3	5
Sonnendecker et al. (1982)	42	Accurette	1	1

[a] Malignant and suspicious reports

should override a negative cytology report and curettage should not be delayed while a repeat smear is examined. The cytologist must be aware of these limitations of cytodiagnosis and must make them known to the clinician when necessary.

Role of endometrial cytology in clinical practice

The introduction of direct endometrial sampling techniques has greatly increased the value of cytology in gynaecological practice. The various sampling techniques have been carefully evaluated for their accuracy, their safety and their acceptability to both patient, gynaecologist and pathologist for the diagnosis of endometrial cancer and its precursor, endometrial hyperplasia.

Several studies have been carried out to compare the accuracy of endometrial sampling with curettage as a method of detecting endometrial cancer and a high degree of correlation was noted (*Table 11.2*). Indeed, instances are recorded of tumours missed by the curette which were detected by cytology. As an outpatient test for women with abnormal bleeding, the results also compared favourably with Vabra suction curettage (Haack-Sorensen et al., 1979; Ellice, Morse and Anderson, 1981). Failure when it did occur was attributed to incomplete sampling of the endometrium due to faulty technique or because of abnormalities of the uterus, such as a fibroid or polyp, which distort the endometrial cavity. At least one centre has established enough experience with the Isaacs instrument to select patients for curettage (Morse et al., 1982). Others with less experience do not have so much confidence in the technique because of an occasional missed tumour or benign abnormality when compared with the results of curettage (Studd et al., 1979).

One of the objects of direct sampling of the endometrial cavity is to detect endometrial hyperplasia in view of its supposed role as a precursor of endometrial cancer. There is some controversy about the value of cytology in diagnosing this condition. Although the features of endometrial hyperplasia in endometrial aspirates have been described by Morse (1981), other groups report that the lesions are underdiagnosed (Studd et al., 1979; Crow, Gordon and Hudson, 1980; Ginsberg, Padleckas and Javaheri, 1983). This is probably because cytology can never provide the architectural detail needed for an accurate diagnosis of this disease. However, some authorities prepare cell blocks from the aspirated material and base their diagnosis on what can be regarded as mini-biopsies.

Most authorities who advocate the use of direct endometrial sampling in gynaecological practice recommend it as a diagnostic test for patients with dysfunctional uterine bleeding. Because anaesthesia is unnecessary, the specimens can be taken as an outpatient procedure in appropriate cases (Lewis, Melcher and Chapman, 1978; Morse et al., 1982). There is little doubt that direct endometrial sampling is particularly suitable for screening women at high risk of developing endometrial cancer. Women who are obese, nulliparous or on oestrogen therapy would provide a target population. Koss et al. (1984) investigated 2586 asymptomatic women aged 45 and over with a view to evaluating the technique as a screening test for endometrial cancer in healthy individuals; 16 occult cancers and 21 cases of endometrial hyperplasia were detected in this study. The low number of endometrial hyperplasias detected led Koss to question the value of this approach as a method of preventing endometrial cancer, although there is little doubt that the diagnosis of occult cancer in asymptomatic women is beneficial in terms of primary health care.

Acknowledgement

We are grateful to the Department of Medical Illustration, Clinical Research Centre, for assistance with the photographs.

References

AFONSO, J. F. (1975). Value of the Gravlee jet washer in the diagnosis of endometrial cancer. *Obstet. Gynecol.*, **46**, 141–146

ANDERSON, D. G., EATON, C. J., GALINKIN, L. J., NEWTON, C. W., HAINES, J. P. and MILLER, N. F. (1976). The cytologic diagnosis of endometrial carcinoma. *Am. J. Obstet. Gynecol.*, **125**, 376–383

ANTUNES, C. M. F., STOLLEY, P. D., ROSENSHEIN, N. B., DAVIES, J. L., TONASCIA, J. A., BROWN, C., BURNETT, L., RUTLEDGE, A., POKEMPNER, M. and GARCIA, R. (1979). Endometrial cancer and estrogen use. *N. Engl. J. Med.*, **300**, 9–13

AYRE, J. E. (1955). Rotating endometrial brush: new technique for the diagnosis of fundal carcinoma. *Obstet. Gynecol.*, **5**, 137–141

CROW, J., GORDON, H. and HUDSON, E. (1980). An assessment of the Mi-Mark endometrial sampling technique. *J. Clin. Pathol.*, **33**, 72–80

DOWLING, E. A., GRAVLEE, L. C. and HUTCHINS, K. E. (1969). A new technique for the detection of adenocarcinoma of the endometrium. *Acta Cytol.*, **13**, 496–501

DRIFE, J. and GUILLEBAUD, J. (1986). Cancer in women: hormonal contraception and cancer. *Br. J. Hosp. Med.*, **35**, 25–29

DUGUID, H. L. D., PARRATT, D. and TRAYNOR, R. (1980). Actinomyces-like organisms in cervical smears from women using intrauterine contraceptive devices. *Br. Med. J.*, **2**, 534–537

ELLICE, R. M., MORSE, A. R. and ANDERSON, M. C. (1981). Aspiration cytology versus histology in the assessment of the endometrium of women attending a menopause clinic. *Br. J. Obstet. Gynaecol.*, **88**, 421–425

FORNARI, M.L. (1974). Cellular changes in the glandular epithelium of patients using IUCDs—a source of cytologic error. *Acta Cytol.*, **18**, 341–343

GINSBERG, N. A., PADLECKAS, R. and JAVAHERI, G. (1983). Diagnostic reliability of Mi-Mark helix technique in endometrial neoplasia. *Obstet. Gynecol.*, **62**, 225–230

GOLDBERG, G. L., TSALACOPOULOS, G. and DAVEY, D. A. (1982). A comparison of endometrial sampling with the Accurette and Vabra aspirator and uterine curettage. *S. Afr. Med. J.*, **61**, 114–116

GUPTA, P. K., EROZAN, Y. S. and FROST, J. K. (1978). Actinomyces and the IUD: an update. *Acta Cytol.*, **20**, 295–297

HAACK-SORENSEN, P. E., STARKLINT, H., ARONSEN, A., KERN HANSEN, M. and KRISTOFFERSEN, K. (1979). Diagnostic Vabra (aspiration) curettage. *Dan. Med. Bull.*, **26**, 1–6

HUTTON, J. D., MORSE, A. R., ANDERSON. M. C. and BEARD, R. W. (1978). Endometrial assessment with Isaacs cell sampler. *Br. Med. J.*, **1**, 947–949

ISAACS, J. H. and ROSS, F. H. (1978). Cytologic evaluation of the endometrium in women with postmenopausal bleeding. *Am. J. Obstet. Gynecol.*, **131**, 410–415

ISAACS, J. H. and WILHOITE, R. W. (1974). Aspiration cytology of the endometrium: office and hospital sampling procedures. *Am. J. Obstet. Gynecol.*, **118**, 679–684

JAFARI, K. (1978). False-negative Pap smear in uterine malignancy. *Gynecol. Oncol.*, **6**, 76–82

JICK, H., WATKINS, R. N., HUNTER, J. R., DINAN, B. J., MADSEN, S. J., ROTHMAN, K. J. and WALKER, A. M. (1979). Replacement estrogens and endometrial cancer. *N. Engl. J. Med.*, **300**, 218–222

KAEBLER, C., CHATWANI, A. and SCHWARTZ, R. (1983). Actinomyces infection associated with intrauterine contraceptive devices. *Am. J. Obstet. Gynecol.*, **145**, 596–599

KOSS, L. G. (1979). Proliferative disorders and carcinoma of the endometrium. In *Diagnostic Cytology and its Histopathologic Basis*, 3rd edition, Volume 1, pp. 412–455. Philadelphia: J. B. Lippincott

KOSS. L. G., SCHREIBER, K., OBERLANDER, S. G., MOUSSOURIS, H. F. and LESSER, M. (1984). Detection of endometrial carcinoma and hyperplasia in asymptomatic women. *Obstet. Gynecol.*, **64**, 1–11

LEWIS, B. V. and CHAPMAN, P. A. (1977). Screening for endometrial carcinoma using a negative pressure intrauterine 'Jet Wash'. *Ach. Gesghwulstforschung*, **47/4**, 314–319

LEWIS, B. V., MELCHER, D. H. and CHAPMAN, P. A. (1978). Outpatient diagnosis of endometrial carcinoma. In *Endoemetrial Cancer*, Brush, M., Taylor, R. and King, R., eds, pp. 124–133. London: Baillière Tindall

MACDONALD, P. C. and SIITERI, P. K. (1974). The relationship between the extraglandular production of oestrone and the occurrence of endometrial neoplasia. *Gynecol. Oncol.*, **2**, 259–263

MAIER, R. C. and NORRIS, H. J. (1980). Coexistence of cervical intraepithelial neoplasia with primary adenocarcinoma of the endocervix. *Obstet. Gynecol.*, **56**, 361–364

MEISELS. A. (1966). The menopause: a cytohormonal study. *Acta Cytol.*, **10**, 49–55

MILAN. A. R. and MARKLEY, R. L. (1973). Endometrial cytology by a new technique. *Obstet. Gynecol.*, **42**, 469–475

MORSE, A. R. (1981). The value of endometrial aspiration in gynaecological practice. In *Advances in Clinical Cytology*, (Koss L. G. and Coleman, D. V., eds), pp. 44–63. London: Butterworths

MORSE, A. R., ELLICE, R. M., ANDERSON, M. C. and BEARD, R. W. (1982). Reliability of endometrial aspiration cytology in the assessment of endometrial status. *Obstet. Gynecol.*, **59**, 513–518

PAPANICOLAOU. G. N. (1953). Observations on the origin and specific functions of the histiocytes in the female genital tract. *Fertil. Steril.*, **4**, 472–478

REAGAN, J. W. and NG, A. B. P. (1973). *The Cells of Uterine Adenocarcinoma*, 2nd Edition, pp. 12–24. Basel: Karger

RICHART, R. M., MARCHBEIN, H. and SHERMAN, A. (1979). Studies of the Gravlee jet washer in the detection of endometrial neoplasia. *Gynecol. Oncol.*, **8**, 49–59

SEGADAL, E. and IVERSON, O. E. (1980). The Isaacs cell sampler: an alternative to curettage. *Br. Med. J.*, **281**, 364–366

SONNENDECKER, E. W. W., SIMON, G. B., SEVITZ, H. and HOFMEYR, G. J. (1982). Diagnostic accuracy of the Accurette endometrial sampler. *S. Afr. Med. J.*, **61**, 109–113

STUDD, J. W. W., THOM, M., DISCHE, F., DRIVER, M., WADE-EVANS, T. and WILLIAMS, D. (1979). Value of cytology for detecting endometrial abnormalities in climacteric women receiving hormone replacement therapy. *Br. Med. J.*, **1**, 846–848

SWINGLER, G. R., CAVE, D. G., and MITCHARD, P. (1979). Diagnostic accuracy of the Mi-Mark endometrial cell sampler in 101 patients with postmenopausal bleeding. *Br. J. Obstet. Gynaecol.*, **86**, 816–818

TORRES, J. E., HOLMQUIST, N. D. and DANOS, M. L. (1969). The endometrial irrigation smear in the detection of adenocarcinoma of the endometrium. *Acta Cytol.*, **13**, 163–168

WEISS, N. S., SZEKELY, D. R. and AUSTIN, D. F. (1976). Increasing incidence of endometrial cancer in the United States. *N. Engl. J. Med.*, **294**, 1259–1262

12

The respiratory tract: normal structure, function and methods of investigation

Gordon M. Campbell and Jocelyn E. A. Imrie

The value of cytology as a method of diagnosing disorders of the respiratory tract has been appreciated by pathologists and clinicians for many years. The classical description of tumour cells in sputum by Dudgeon and Wrigley in 1935 did much to promote the use of sputum cytology for the investigation of patients suspected of having bronchial cancer. With increasing experience it became clear that cytology could also be used to diagnose a number of benign conditions. The introduction of fibreoptic bronchoscopy and fine needle aspiration has further expanded the range of diagnoses that can be made from cytological material by providing new methods of sampling. Respiratory tract cytology now contributes a major part of the workload of many cytology laboratories (Clee, Duguid and Sinclair, 1982; MacMahon, Courtney and Little, 1983).

In order to provide a reliable cytology service it is essential for the cytologist to be familiar with the full range of cytological patterns which can be seen in smears from the respiratory tract. In this chapter we provide the background information necessary for him or her to achieve this goal. Thus we include a brief review of the anatomy, physiology and histology of the respiratory tract together with an account of the specimen collection and processing. We conclude with a description of the types of cells seen in Papanicolaou-stained smears of sputum and other specimens in the absence of malignant disease.

Anatomy, histology and physiology

The respiratory system comprises the upper respiratory tract (which includes the nasal passages, nasal sinuses, oral cavity and pharynx), trachea and bronchi (*Figure 12.1*) and the lungs.

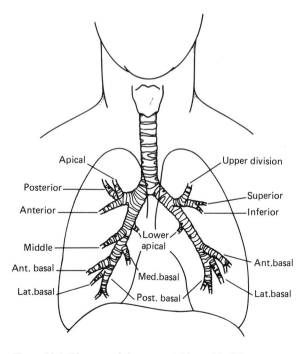

Figure 12.1 Diagram of the segmental bronchi of the lung. All lie within the reach of the fibreoptic bronchoscope

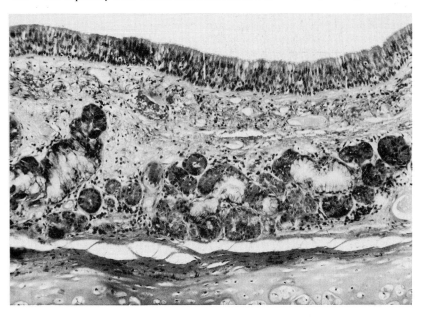

Figure 12.2 Normal bronchial wall showing ciliated pseudostratified columnar epithelium with underlying mucous and serous glands, then cartilage. Stain: haematoxylin and eosin. Magnification × 160, reduced to 40% in reproduction

Each lung weighs 300–400 g but they differ anatomically in that the right lung has three lobes and the left only two. The purpose of the system is to oxygenate the blood and remove excess carbon dioxide from the body. To allow for gaseous exchange, inspired air must be brought into close contact with the blood in the capillaries in the alveolar walls. The upper respiratory tract serves as a pathway for the gases to and from the lungs; the main bronchi dividing again and again ending in small channels (terminal bronchioles) which deliver the air to the alveolar sacs in the respiratory lobules.

The upper respiratory tract also provides some anatomical and physiological defences against foreign materials in the air e.g. bacteria, atmospheric pollutants. By virtue of its close contact with these external agents, the upper respiratory tract is susceptible to virus infection; such infections are usually mild and self-limiting. Despite this susceptibility, the lungs themselves normally remain sterile due to the efficiency of a variety of defence mechanisms sited at different levels in the respiratory tract. In the nose, particles deposited on the non-ciliated epithelium of the vestibule are removed by sneezing or blowing the nose. Those deposited further back are transported by ciliated epithelium to the nasopharynx from where they are swallowed. Within the respiratory passages themselves, cilia beat to move a film of mucus upwards from the lungs and particles deposited are taken up in this film and transported with it. Within the alveoli, macrophages ingest any particles which reach this level of the respiratory tract. They either digest the foreign material or remove them to bronchioles from where ciliary movement clears them. Other factors involved in the defence against infection are the cough reflex which helps remove mucus from the respiratory tract, and IgA antibodies secreted within the mucus.

The mucous membranes lining the upper and lower respiratory tract are continuous. The epithelial lining of the mouth and pharynx is squamous in type whereas the posterior nasal space, trachea and bronchi are lined by ciliated pseudostratified columnar epithelium (*Figure 12.2*). Scattered between the columnar cells are mucus secreting cells and very occasional argyrophil (Feyrter) cells. Underlying this epithelium in the trachea and bronchi there are mucous and serous glands. The smaller bronchioles are lined by a single layer of low columnar epithelium. Within the bronchioles the epithelium includes apocrine secretory cells called Clara cells which produce a protein thought to be involved in surfactant production.

Bronchioles subdivide to form preterminal then terminal bronchioles. Each terminal bronchiole supplies one acinus, which contains several alveoli and 3–5 acini form one respiratory lobule. Respiratory lobules are separated from each other by fibrous septa which may be seen on the cut surface of the lung. The multiple branching of the airways and the 'bunch of grapes' structure of alveoli provides a large area of alveolar wall (estimated at over 100 square metres) across which gas exchange occurs.

Each alveolus is lined by different cell types, which comprise:

(1) Type 1 pneumocytes (membranous pneumocytes): cells which line approximately 95% of the area of alveolar wall.
(2) Type 2 pneumocytes (granular pneumocytes): rounded cells with microvilli which act as reserve cells in cases of Type 1 pneumocyte loss. They contain osmiophilic lamellar bodies which are thought to be precursors of surfactant, a substance involved in lowering surface tension within alveolar spaces.
(3) Type 3 pneumocytes (alveolar macrophages): cells which are loosely attached to the epithelial cells or lie free within the alveoli.

Between the epithelial surface of the alveolus and the endothelium of the capillaries is a narrow interstitial space which may contain ground substance, fine elastic fibres, collagen and occasional smooth muscle cells or mast cells. If this interstitium thickens the diffusion of gas is lessened and respiration is impaired. This also happens if alveolar walls are obscured or destroyed, thus giving less surface for gas exchange.

The diffusion of gases from alveolus to capillary is only part of respiration because movement of the air in and out of the lungs (ventilation) is essential to renew the gases within the alveoli. In a normal adult 7 litres of air are inspired per minute but because the respiratory passages constitute a non-functioning dead space, only approximately 5 litres/min reach alveolar walls. This volume may be diminished by conditions causing restrictions of chest movement (which involves both active and passive movement) or obstruction of the respiratory passages. The other factor affecting gas exchange is the provision of adequate blood flow to the alveolar capillaries. This may be diminished by occlusion of large vessels by emboli, slowing of the circulation in venous congestion or by destruction of the capillary bed in diffuse lung disease. Thus in the broad sense of the term, respiration is dependent on three factors:

(1) An adequate air supply to the alveoli.
(2) Normal diffusion of gases across alveolar membranes.
(3) An adequate rate of blood flow to the pulmonary capillary bed.

Specimen collection

The range of specimens suitable for cytological examination is shown in *Table 12.1*. The upper respiratory tract can be investigated cytologically by direct swabbing. The most readily available samples from the lower respiratory tract are sputum samples but specimens can also be obtained at bronchoscopy. Transthoracic fine needle aspiration (FNA) under fluoroscopic control is assuming an increasing importance in the investigation of lung disease. The place of each of the different specimen types in clinical practice is considered in this section.

Table 12.1 Types of specimen suitable for cytological examination of respiratory tract

Nasal, pharyngeal swabs
Sputum
Bronchial aspirates { Bronchial secretions / Bronchial washings / Bronchoalveolar lavage specimens
Bronchial brushings
Transbronchial FNA
Touch preparations
Transthoracic percutaneous FNA

Specimens from upper respiratory tract

Nasal secretions from patients with allergic disease are occasionally submitted for examination for eosinophils. These samples are collected by means of a moistened cotton swab. Smears should be prepared before the mucus on the swab has time to dry. Specimens from the pharynx and larynx are collected directly from abnormal areas during clinical examination using a wooden or metal spatula or a swab. It should be remembered that samples of saliva may contain malignant cells arising from lesions in the upper respiratory tract as, for example, in patients presenting with hoarseness. Specimens from the larynx or trachea may be obtained by blind aspiration or at bronchoscopy.

Sputum

Sputum production in significant amounts is itself evidence of lung pathology. In the majority of cases this is benign but sputum cytology remains the simplest test for obtaining a diagnosis of carcinoma of the lung. The procedure involves no risk to the patient and can be repeated as often as necessary (Payne *et al.*, 1981; Clee, Duguid and Sinclair, 1982). Three separate samples should be examined in every patient. This gives an optimum diagnostic rate of 60–70% (Ng and Horak, 1983) although a higher rate can be achieved with an increased number of samples. Post-bronchoscopy sputum samples also yield positive results in a higher percentage of patients.

It is of paramount importance that sputum samples contain cells from the lower respiratory tract and it is advisable to instruct the patient how this can best be achieved. Ideally a sample of the first sputum produced in the morning should be sent to the laboratory on each of three successive days. The specimen should be collected before the patient has breakfasted or cleaned his teeth in order to avoid contamination with food or toothpaste. Some patients find it difficult to produce sputum spontaneously and in these cases postural drainage or physiotherapy might be of value. Methods are available to provide forced cough specimens: these include inhalation of polyethylene glycol or percutaneous injection into the trachea of 25 ml of sterile saline. These methods are particularly valuable for the detection of *Pneumocystis carinii* in AIDS patients.

It is an unfortunate fact that it is not possible, in the vast majority of cases, to assess accurately the quality of specimens macroscopically and they must all, therefore, be examined microscopically. If possible, samples should be submitted whilst the patient is awaiting admission to hospital. If this is done, a significant proportion of patients will have a diagnosis available when they are admitted (Buirski *et al.*, 1981).

Bronchoscopic sampling

A major drawback of sputum cytology is that it cannot be used to localize a lesion within the lung. For this bronchoscopy may be required if the patient's chest X-ray is negative. Today most bronchoscopy specimens are obtained using the fibreoptic bronchoscope which was developed in Japan in 1964 and introduced to Britain in the early 1970s. This instrument allows direct vision and sampling of the respiratory tract to the level of most segmental and some subsegmental bronchi (*Figure 12.3*). It is not necessary to administer a general anaesthetic for any of these techniques. Local anaesthetic and mild sedation are all that is required in the majority of cases.

The principal advantages of bronchoscopic sampling are that the procedure allows localization of lesions to one lung or part of the lung and samples are free from contamination by cells from other sites in the respiratory tract. The main disadvantages of the method are that it is unpleasant for the patient and bronchial examination requires a high level of clinical and cytological expertise if interpretation of bronchoscopic findings is to be of clinical value (Koss, 1979). Although the fibreoptic bronchoscope is now widely used, the rigid instrument is still preferred in some circumstances, e.g. in investigation of patients who have had recent massive haemoptysis. Unsatisfactory bronchoscopy samples are usually due to an inadequate amount of material or to drying artefact. Several different types of cytological specimen can be obtained through the bronchoscope, including bronchial secretions, bronchial washings, bronchoalveolar lavage specimens and bronchial brushings. Touch preparations and fine needle aspiration biopsies can also be obtained at bronchoscopy.

Bronchial secretions

These are obtained by applying suction and collecting aspirated material into a trap.

Bronchial washings

This involves the introduction of a small amount of physiological saline through the instrument to wash the surface of the mucosa. Usually about 10 ml is instilled in 2–3 ml portions and then aspirated.

Bronchoalveolar lavage

This was developed to facilitate investigation of

interstitial lung disease. It is a safe procedure (Strumpf et al., 1981) which yields plentiful material and allows repeated sampling at all stages of the disease. Repeated sampling, however, if carried out within four days, may cause a dramatic increase in the cell content of the fluid (Krombach et al., 1985). Sampling is carried out by impacting the fibreoptic bronchoscope in segmental bronchi of the middle lobe or lingula in patients with diffuse lung disease or in the appropriate segmental bronchus of those with localized abnormalities. Approximately 100 ml of sterile saline is introduced through the bronchoscope in 20 ml aliquots and is aspirated into a trap. Roughly half of the instilled saline is normally recovered. Quantification of the cells in the fluid is of value in assessing the type and severity of interstitial lung disease. The method can also be used for the diagnosis of peripherally sited tumours outside the field of bronchoscopic inspection and unusual conditions such as alveolar proteinosis and *Pneumocystis carinii* pneumonia.

Bronchial brushings

These are carried out using a nylon brush, which may be either disposable or reusable, which abrades the surface of the suspect area. The surface cells adhere to the brush and can be transferred directly on to a microscope slide.

Touch preparations

These can be prepared from transbronchial biopsy specimens when rapid diagnosis is required. Smears prepared in this way are particularly valuable for the diagnosis of *Pneumocystis carinii* in AIDS patients.

Transbronchial fine needle aspiration (FNA)

This was introduced in 1978 to overcome the problem of diagnosing respiratory disease in patients whose bronchoscopies showed either normal mucosa or extrinsic compression of the lumen (Wang et al., 1983). The technique is also of value in obtaining material from necrotic endobronchial lesions (Rosenthal and Wallace, 1984). Direct biopsy of the bronchial wall in these patients is usually unrewarding. The method is highly successful in the diagnosis of patients with central bronchial carcinoma (Buirski et al., 1981) and parabronchial lesions (Muers et al., 1982) and for sampling the carinal nodes for accurate staging of lung tumours. It is advisable to perform the aspiration after other samples have been obtained in order to avoid spreading blood or cellular material within the airway. The instrument used in this technique is a flexible metal needle 0.6 mm in diameter attached to a plastic catheter surrounded by a flexible metal sheath into which the needle can be retracted. The device is 111 cm in length and the needle can be extended 8 cm beyond the external sheath. The proximal end of the instrument is machined to accept a syringe. The flexible needle is passed through the instrumental channel of the bronchoscope and aspiration is performed under direct vision (Lundgren, 1980).

Transthoracic fine needle aspiration

Percutaneous transthoracic fine needle biopsy was introduced in the 1930s for the diagnosis of neoplasms of the lung (Dahlgren, 1974). It is of particular value in the investigation of patients who are poor surgical risks or who have lesions sited at the periphery of the lung, beyond the reach of the bronchoscope. The combination of the fine needle (19–22 G) and sophisticated radiological imaging has made the method safe and reliable (Bonfiglio, 1982). Use of fluoroscopic imaging allows lesions as small as 5 mm in diameter to be aspirated. This means, in fact, that virtually all radiologically visible lesions can be sampled. As the needle tip can be seen during the entire procedure, it is possible to target different areas of a lesion for sampling. This is particularly important where the tumour is necrotic or cystic. The technique has also recently been applied to hilar and mediastinal lesions (MacMahon, Courtney and Little, 1983).

The aspiration procedure is relatively simple and does not require general anaesthesia. The skin, chest wall and pleura are infiltrated with local anaesthetic and the aspiration needle is introduced until the point is within the lesion. The needles used are long, 9–16 cm, and have solid stylets inserted to prevent material entering the lumen from the thoracic wall. When the needle tip is within the lesion, the needle is rotated once and the stylet removed. A syringe is fitted and negative pressure applied. The patient must be encouraged to hold his breath at this stage of the procedure. After equalizing pressure, the needle is withdrawn and the aspirated material expelled on to a glass slide. Major vessels and airways must be avoided during this procedure.

The principal disadvantages of the method are the risk of complications and failure to obtain an adequate and representative sample. Pneumothorax is the commonest complication of the technique and may occur in up to 40% of patients (Dahlgren, 1974) although only a small percentage of these will require insertion of a chest tube and drainage (Bonfiglio, 1982; MacMahon, Courtney and Little, 1983). Significant haemorrhage can occur and the procedure is contraindicated in patients with uncorrected coagulation defects. Air embolism and implantation metastases are very rare complications if a fine needle is used. Patients must be kept under observation for a period following aspiration and a post-aspiration chest X-ray is advisable (see also Chapter 19).

Laboratory procedures

A high standard of specimen preparation is essential for accurate cytological diagnosis. Material improperly prepared in the initial stages may be unsuitable for diagnosis and may not be easily replaced. Preparations should, therefore, be made by an experienced member of staff who is thoroughly versed in the techniques involved and not delegated to the most junior and least experienced person in the laboratory.

Specimens from the respiratory tract are potentially infectious and strict adherence to safe working practice is essential. A proportion of patients with suspected lung cancer will, in fact, be suffering from tuberculosis or some other infectious disease. For example, in Scotland there are several hundred new cases of respiratory tract tuberculosis notified each year indicating that the risk of infection is a very real problem for the cytotechnologist. All specimens must be prepared in a Class 1 microbiological safety cabinet.

Sputum

Specimens should be collected into wide-mouthed, screw-capped, 60 ml containers which are translucent to facilitate macroscopic examination of the specimen. It is not usually necessary to add any preservative to the specimen, but if undue delay is anticipated, an equal volume of 70% alcohol should be added. There are three methods available for preparing sputum samples: direct smear preparation, liquefaction and cell block techniques.

Direct smear preparation

This is the simplest, most rapid and probably the most widely used method. The number of smears made will influence the likelihood of obtaining an accurate diagnosis. Three smears per sample are optimal. Smears should be prepared from any bloody, purulent or discoloured areas of the specimen. Selecting these areas is made easier by holding the translucent specimen pot against a black background. Smears can be made with a loop, applicator stick or, very conveniently, with an Ayre's spatula. The selected material is placed on to the surface of two slides and a third is used to 'crush-smear' it evenly over their surfaces (Hughes and Dodds, 1968). The three smears thus produced are immediately fixed in alcohol containing polyethylene glycol, applied using a laboratory spray or dropper bottle. Alcohol may be used alone if preferred. Sputum smears can be air-dried if special staining methods such as Romanowsky or fat stains are required. The practice of examining unfixed smears stained with methylene blue can no longer be recommended because of the risk of infection inherent in the method.

Sputum samples which appear to the naked eye to be watery can be concentrated either by centrifugation or membrane filtration. Slides should be pretreated with poly-L-lysine (Husain, Millett and Grainger, 1980) to minimize cell loss during fixation and staining.

Liquefaction

Liquefaction and subsequent concentration of sputum samples offers some possible advantages over the direct smear technique. The yield of abnormal cells is high and large numbers of smears can be made from the concentrated material. This is useful if teaching material is being prepared. Concentration techniques are time consuming and the smears prepared from the concentrated material often contain large numbers of cells of upper respiratory tract origin. As a result, they require considerable time to screen. There is also a danger that diagnostically important features may be disrupted by digestion and concentration.

Liquefaction of sputum samples can be achieved using dilute mineral acids (Taplin, 1966) followed by neutralization in alkaline solution. To obtain satisfactory nuclear staining after this treatment, an extended immersion in haematoxylin is necessary.

Enzymatic digestion of mucus may also be carried out using enzymes such as pancreatin, papain or trypsin. Sputum is incubated at 37 °C in a solution of the chosen enzyme and when digestion appears complete, usually after 30–60 min, the suspension is centrifuged. Alternatively, amino acids such as L-cysteine can be used. These are more rapid in action than the enzymes and break down mucus more completely.

The third alternative is to use a mechanical blender to liquefy the sputum. Probably the best known of this type of method is that of Saccomanno (Saccomanno et al., 1963) which involves blending a mixture of sputum and fixative for 5–10 seconds, centrifuging the resultant blend and preparing smears from the deposit. As mucus is destroyed by these methods, slides should be treated with albumin or poly-L-lysine before use in order to minimize cell loss during processing. Alternatively, cells can be concentrated on to a cellulose or polycarbonate filter.

Preparation of cell blocks

This is a particularly time consuming approach which, in the authors' opinion, offers no advantage over direct smears for routine diagnosis. The method involves fixation of the whole specimen in a solution such as Dubosq Brazil, overnight processing on an automatic tissue processor to paraffin wax and then cutting sections at three levels. Sections are stained by the haematoxylin and eosin method.

Bronchial secretions and bronchial washings

These are often submitted in the bronchoscope traps used to collect them. The two types of trap used in the authors' hospitals are illustrated in *Figure 12.3(a)* and *(b)*. These traps are attached to the proximal end of the bronchoscope. Samples of both bronchial secretions and washings can be collected in the larger trap whereas the smaller one is suitable only for the collection of secretions. Both traps can be sealed for transportation to the laboratory, the larger by uniting the male and female connections of the trap, the smaller by the application of a small cap and bung. The specimens can be processed in much the same way as sputum samples. Material collected in the smaller trap is best removed by using an orange stick or applicator stick which fits easily into the trap. Smears can be prepared directly with the stick or the material can be transferred to the slides and spread by the 'crush smear' technique.

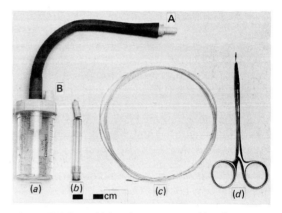

Figure 12.3 Bronchial aspirator traps and brush. (*a*) Large, external, bronchial trap. Placing the connection A into the socket B effectively seals the trap for transportation; (*b*) The small trap is shown with cap and bung in position; (*c*) Disposable bronchial brush with plastic catheter; (*d*) Disposable brush clamped in Halstead forceps

An additional source of diagnostic material is to examine the mucus which may adhere to the bronchoscope biopsy channel. Viscous secretions may adhere to the walls of the channel and are normally lost when the instrument is cleaned. By recovering this adherent material for cytological examination, a 10% improvement in diagnostic yield can be achieved (Clee and Duguid, 1983). It may be necessary to concentrate bronchial washings and secretions if they appear watery; for example, if the bronchoscope channel has been rinsed through with saline after the sample was collected. If these samples are centrifuged at 3000 rev/min for 10 min, they separate into a mucus layer which floats at the top of the centrifuge tube, and a denser layer at the bottom of the tube.

Larger cells, e.g. squamous cells, tend to congregate in the top layer whereas small, denser cells will aggregate at the bottom. If both these layers are sampled, more representative smears are obtained. Two smears should be prepared from each layer and immediately fixed or air-dried.

Brush biopsy

Bronchoscopy samples may be obtained with a reusable or a disposable brush. If a reusable brush is used, undoubtedly the best procedure is to have a member of the laboratory staff in attendance in theatre to prepare and fix smears from the brush immediately it is removed from the bronchoscope. This ensures that the risk of drying artefact is reduced to a minimum. It is not always possible, however, to arrange this and in these circumstances it is advisable to have the bronchoscopist or his assistant agitate the brush vigorously in a small volume of sterile saline (5–10 ml) after the smears have been made. The saline suspension is a valuable source of additional material and should be sent to the laboratory together with the smears. The saline suspension can be prepared for examination using the cytocentrifuge or by membrane filtration. Cytocentrifuge preparations are preferred in the authors' department as they provide a small area of cells which can be rapidly screened if an urgent diagnosis is required. They also facilitate diagnosis as the cells tend to be rather less compacted than they are in the direct smears.

The disposable brush (*Figure 12.3(c)*) consists of a long thin wire (150 cm long) with a nylon brush incorporated at one end. The device is enclosed in a plastic sheath into which the brush can be drawn for removal from the bronchoscope. The use of a disposable brush successfully overcomes the need for a member of the laboratory staff to be present at bronchoscopy as the brush can be sent to the laboratory for processing. This is best done by cutting off the distal 5–10 cm of the wire and placing it in a suitable container with 5–10 ml of saline. This procedure precludes drying artefact.

Handling these tiny brushes does present some problems in the laboratory. Due to the thinness of the wire, it is difficult to make smears from the brush unless it is held very firmly. Halstead artery forceps (*Figure 12.3(d)*) provide a convenient and completely successful means of achieving this. If the wire is gripped immediately above the brush (*Figure 12.4*), it can easily be manipulated and rotated to prepare the smears. After the direct smears are made, the brush is returned to the saline and is vigorously agitated to dislodge any adherent cells. The cell suspension thus produced is processed as already described.

Figure 12.4 Disposable brush and forceps. Magnification ×4, reduced to 60% in reproduction

Bronchoalveolar lavage fluid

This has a relatively low cell count of between 5 and 10 million cells in a recovered aspirate of approximately 50 ml. In healthy non-smokers, the differential count yields 93% macrophages and 7% lymphocytes. Neutrophils are not normally present, except in smokers, and if they are found they are indicative of lung disease (Dunnill, 1982). Eosinophils and mast cells may be encountered in some diseases. An absolute count of cells should be carried out using a haemocytometer.

Smears are prepared from the centrifuged deposit or after cytocentrifugation, air-dried and stained by a Romanowsky method for differential cell counting. Investigations into the sub-populations of lymphocytes obtained by bronchoalveolar lavage have been used in some centres to differentiate the various types of interstitial lung disease. Immunogold staining has been shown to be of value in these investigations (Duffy, Stevens and McLennan, 1986). Recently bronchoalveolar lavage has been used in conjunction with immunochemical staining to diagnose and assess pulmonary rejection after heart-lung transplantation (Schafers *et al.*, 1987; Zeevi *et al.*, 1987). If specific infection is suspected (as in immunocompromised patients), Papanicolaou staining should be carried out routinely as this allows recognition of most organisms although special stains should be used for more precise identification.

Transbronchial and transthoracic FNA specimens

These tend to be of very small volume and, because of this, are prone to drying artefact. Direct smears can be prepared from the material aspirated into the syringe by removing the needle, charging the syringe with air, refitting the needle and gently expelling the contents on to slides where it can be spread (*see* Chapter 19). Immediate fixation in alcohol is essential for Papanicolaou staining. Air-dried smears should be prepared for Romanowsky staining. The syringe should then be washed out by drawing 5–10 ml of saline through the needle into the barrel. After removal of the needle, this cell suspension can be transferred to a suitable container for transport to the laboratory for processing. Occasionally, aspiration will yield several millilitres of fluid and these specimens should be sent for laboratory processing. Membrane filtration or cytocentrifugation are suitable for these samples and for cell suspensions. The samples must be handled gently as rough usage may damage or destroy up to 20% of the aspirated cells. Rinsing the syringe and needle through with sterile heparin solution is sometimes recommended to minimize damage to cells and prevent clotting. In many cases referred for FNA, there is a clinical uncertainty as to whether the lung lesion is neoplastic or inflammatory and in these cases part of the aspirate should be expelled into sterile saline for microbiological examination.

Normal cytology

The cellular content of respiratory tract samples depends upon the nature of the sample and the site from which it was obtained. It is essential that this information is available to the cytologist as it can exert considerable bearing on the diagnosis. This is particularly important in assessing the suitability of a sample for analysis.

Upper respiratory tract

In samples obtained from the upper respiratory tract, which is lined very largely by non-keratinizing squamous epithelium, the normal cell content will include large numbers of squamous cells of superficial and intermediate types. In Papanicolaou-stained preparations these cells

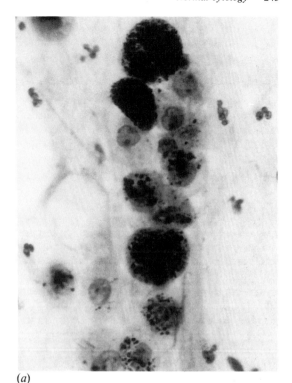
(*a*)

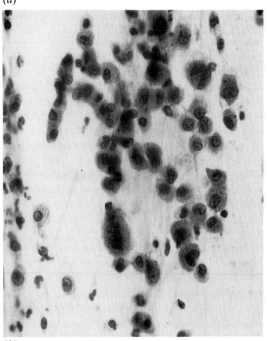
(*b*)

Figure 12.5 (*a,b*) Carbon-containing pulmonary macrophages in sputum seen against a background of mucus and polymorphonuclear leucocytes. In all of the macrophages in (*a*) carbon particles completely obscure the nuclei of the cells. Papanicolaou stain. Magnification ×400, reduced to 60% in reproduction

have either eosinophilic or cyanophilic abundant cytoplasm with pyknotic or vesicular nuclei. Keratinization is not normally seen. Polymorphonuclear leucocytes are usually present in moderate numbers. Columnar cells may also occur in upper respiratory tract specimens, and these tend to be shorter and stubbier than bronchial epithelial cells. They can, however, closely resemble bronchial epithelial cells on occasions. Both ciliated and mucus-secreting types occur, the former more frequently than the latter.

Sputum

Sputum is characterized by the presence of carbon-laden histiocytes (*Figure 12.5*). These are large, rather pleomorphic cells with round or oval nuclei and abundant, foamy cytoplasm which contains variable amounts of carbon particles. The particles range in size from fine granules to large angular masses. Pulmonary histiocytes do not always contain carbon, but the cells can be recognized on morphological grounds and they constitute evidence of the lower respiratory tract origin of the sample. Multinucleated macrophages are frequently seen.

Bronchial epithelial cells may also be present and can be ciliated or mucus-secreting (*Figure 12.6*). These cells may occur singly, or in loose groups or in clumps. Ciliated cells are tall with basal vesicular nuclei with granular chromatin often containing several nucleoli. The luminal end of the cells may carry an eosinophilic terminal plate and cilia. Mucus-secreting cells appear barrel-shaped and have vacuolated, pale staining cytoplasm. Nuclei are basal and vesicular and may be compressed by the mucin within the cell. The cytoplasmic staining reaction of both cell types is cyanophilic with Papanicolaou stain. Squamous cells from the mouth and pharynx are usually present in samples of sputum, often in considerable numbers. Polymorphonuclear leucocytes and lymphocytes are invariably present.

A proportion of specimens submitted to the laboratory as sputum will, in fact, consist only of saliva. This is recognized by the presence of large numbers of squamous cells and the absence of carbon-laden histiocytes. The squamous cells are often degenerate and contaminated with large numbers of bacteria and/or yeasts. Some mucus and columnar cells of upper respiratory tract origin may also be present.

Samples showing these features must not be regarded as adequate for the investigation of lower respiratory tract disease.

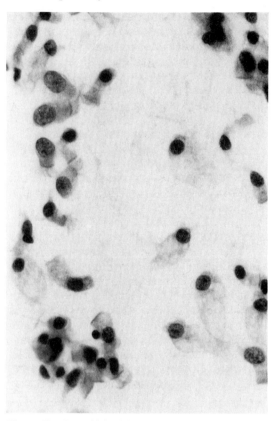

Figure 12.6 Bronchial epithelial cells. Large numbers of ciliated bronchial epithelial cells are seen. Mucus-secreting bronchial cells are also present, particularly in the lower right quadrant. Bronchial aspirate. Papanicolaou stain. Magnification × 400, reduced to 60% in reproduction

Bronchial secretions, bronchial washings and bronchial brushings

These contain large numbers of bronchial epithelial cells. Palisade and honeycomb arrangements are encountered quite often and tightly packed groups of 'deep' bronchial cells are frequently seen in brush specimens. These deep cells appear as small, round or oval nuclei with granular chromatin. Nucleoli are often seen, although these are not prominent. Nuclei often appear distorted as a result of crowding and the cells can resemble small anaplastic tumour cells (oat cells) as they have very scanty cyanophilic cytoplasm in Papanicolaou-stained preparations. Careful examination reveals that the nuclei are not moulded around each other, rather they are pressed together. Carbon-laden histiocytes may be few in number or absent altogether; this is particularly true of brush specimens. Occasional squamous cells may be present as may a small number of polymorphonuclear leucocytes.

Bronchoalveolar lavage spcimens

These are characterized by the large number of macrophages they contain. The proportion of leucocytes is very variable.

Transbronchial and transthoracic FNA specimens

These are often very cellular and may contain ciliated and mucus-secreting bronchial epithelial cells, bronchial reserve cells and alveolar macrophages. Reserve cells must be differentiated from small anaplastic tumour cells. They lack the nuclear features of malignancy and are often seen attached to normal columnar cells. Mesothelial cells may occasionally be seen in transthoracic specimens and they may occur either singly or in small sheets.

Non cellular elements

Mucus is often present in sputum, bronchial aspirates and wash specimens, less noticeably in brush specimens and bronchoalveolar lavage specimens. It generally forms an uneven blue background to the smears but may also occur as inspissated masses of very variable size. These tend to be round or oval in shape and stain very intensely blue with haematoxylin. Occasionally, if a small mass of mucus overlies a squamous cell, the appearance may suggest malignancy. Careful examination will reveal the featureless 'nucleus' and cytoplasm to be on different focal planes. Mucus may also be seen in the form of Curschmann's spirals (*Figure 12.7*), which are inspissated mucus casts of small bronchioles and are seen in chronic respiratory diseases, particularly asthma. Other non-cellular elements of diagnostic significance which may be encountered include asbestos bodies (ferruginous bodies) recognized by their characteristic dumb-bell appearance (*Figure 12.8*), and Charcot-Leyden crystals which are large, needle-shaped, brightly orangeophilic crystals seen occasionally in asthmatics (*Figure 12.9*). They are derived from the eosinophil granules

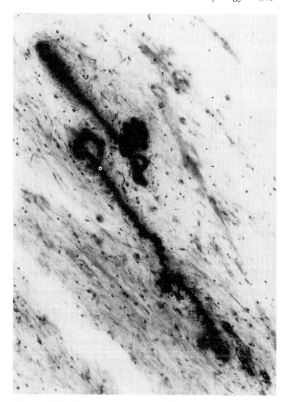

Figure 12.7 Curschmann's spirals. These are casts of small bronchioles and composed of inspissated mucus. Papanicolaou stain. Magnification × 540

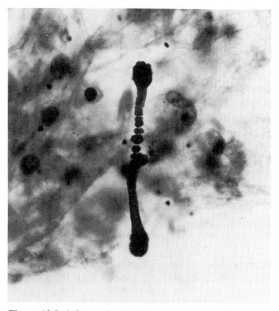

Figure 12.8 Asbestos body. These structures can be recognized by their dumb-bell shape. They are golden brown in Papanicolaou stained smears. Papanicolaou stain. Magnification × 750

of leucocytes. Pollen grains, deposited on smears from the atmosphere especially in spring and summer, may occasionally be encountered. These can often be recognized by their thick double wall.

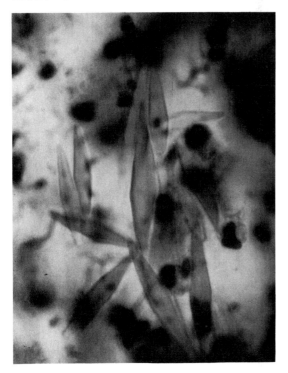

Figure 12.9 Charcot-Leyden crystals. These are needle shaped crystals found in sputum samples from asthmatic patients. Papanicolaou stain. Magnification × 750

Samples of sputum are rendered unsuitable for diagnosis by the presence of food particles which can mask large areas of the smears (*Figure 12.10*). In the case of vegetable fibres, these can be recognized by the presence of a cellulose membrane. The cytoplasm is usually cyanophilic and may appear quite angular. Nuclei are deeply stained and usually featureless. Meat fibres have an elongated, usually eosinophilic or orangeophilic cytoplasm in which striations can often be seen. Dense acute inflammatory exudate also renders samples unsuitable as it is impossible to assess the quantity and nature of any underlying epithelial cells. Drying distortion may be present to such an extent as to render specimens, particularly smears made in theatre from bronchial brushes of fine needle aspirates, unsuitable for diagnosis. In addition to these specific causes, samples may be unsatisfactory because they do not, in the cytologist's opinion, contain sufficient diagnostic material to constitute a representative sample. This can only be assessed with the benefit of experience and, therefore, will vary from one cytologist to another.

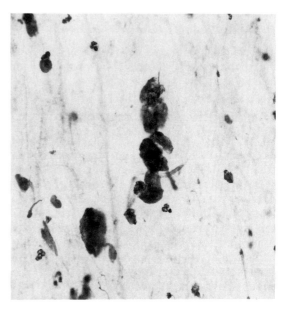

Figure 12.10 Amorphous debris in sputum which could be mistaken for tumour cells. Papanicolaou stain. Magnification × 540

References

BONFIGLIO, T. A. (1982). Fine needle aspiration biopsy of the lung. *Pathol. Ann.*, **16**, 159–180

BUIRSKI, G., CALVERLEY, P. M., DOUGLAS, N. J., LAMB, D., MCINTYRE, M., SUDLOW, M. F. and WHYTE, H. (1981). Bronchial needle aspiration in the diagnosis of bronchial carcinoma. *Thorax*, **36**, 508–511

CLEE, M. D. and DUGUID, H. L. D. (1983). Retained secretions in the bronchoscope: a source of material for cytology. *Br. J. Dis. Chest*, **77**, 91–93

CLEE, M. D., DUGUID, H. L. D. and SINCLAIR, D. J. M. (1982). Accuracy of morphological diagnosis of lung cancer in a department of respiratory medicine. *J. Clin. Pathol.*, **35**, 414–419

DAHLGREN, S. (1974). Lungs. In *Aspiration Biopsy Cytology. Part 1. Cytology of Supradiaphragmatic Organs*, (Zajicek, J., ed.), pp. 195–208. Basel: S. Karger

DUDGEON, L. S. and WRIGLEY, C. H. (1935). On the demonstration of particles of malignant growth in sputum by means of the wet film method. *J. Laryngol. Otol.*, **50**, 752–753

DUFFY, A. M., STEVENS, M. W. and MCLENNAN, G. (1986). The immunogold staining technique for the measurement of lymphocyte subpopulations in bronchoalveolar lavage fluid. *Acta Cytol.*, **30**, 152–156

DUNNILL, M. S. (1982). Pulmonary fibrosis. In *Pulmonary Pathology*, pp. 216–244. Edinburgh: Churchill Livingstone

HUGHES, H. E. and DODDS, T. C. (1968). *Handbook of Diagnostic Cytology*, Chapters 6, 7, 8 and 23. Edinburgh: E & S Livingstone

HUSAIN, O. A. N., MILLETT, J. A. and GRAINGER, J. M. (1980). Use of poly-L-lysine coated slides in the preparation of cell samples for diagnostic cytology. *J. Clin. Pathol.*, **33**, 309–311

KOSS, L. G. (1979). *Diagnostic Cytology and its Histopathologic Basis*, 3rd Edition, Volume 2, Chapters 19 and 21. Philadelphia: J. B. Lippincott

KROMBACH, F., KONIG, G., NANDERS, A. *et al.* (1985). Effect of repeated bronchoalveolar lavage on free lung cells and peripheral leucocytes. *Transplantation Proc.*, **XVII**(5), 2134–2136

LUNDGREN, R. (1980). A flexible thin needle for transbronchial aspiration biopsy through the flexible fibreoptic bronchoscope. *Endoscopy*, **12**, 180–182

MACMAHON, H., COURTNEY, J. V. and LITTLE, A. G. (1983). Diagnostic methods in lung cancer. *Semin. Oncol.*, **10**, 20–33

MUERS, M. F., BODDINGTON, M. M., COLE, M., MURPHY, D. and SPRIGGS, A. I. (1982). Cytological sampling at fibreoptic bronchoscopy: comparison of catheter aspirations and brush biopsies. *Thorax*, **37**, 457–461

NG, A. B. P. and HORAK, G. C. (1983). Factors significant in the diagnostic accuracy of lung cytology in bronchial washings and sputum samples. II. Sputum samples. *Acta Cytol.*, **27**, 397–402

PAYNE, C. R., HADFIELD, J. W., STOVIN, P. G., BARKER, V., HEARD, B. E. and STARK, J. E. (1981). Diagnostic accuracy of cytology and biopsy in primary bronchial carcinoma. *J. Pathol.*, **34**, 773–778

ROSENTHAL, D. L. and WALLACE, J. M. (1984). Fine needle aspiration of pulmonary lesions via fibreoptic bronchoscopy. *Acta Cytol.*, **28**, 203–210

SACCOMANNO, G., SAUNDERS, R. P., ELLIS, H., ARCHER, V. E. and WOOD, B. G. (1963). Concentration of carcinoma of atypical cells in sputum. *Acta Cytol.*, **7**, 305–310

SCHAFERS, H.-J., HAVERICH, A., DAMMENHAYN, L., TAKAYAMA, T., WAHLERS, T. L., WORCH, K. and KEMNITZ, J. (1987). The role of bronchoalveolar lavage in diagnosing pulmonary rejection after heart-lung transplantation. *Transplantation Proc.*, **XIX**, 2551

STRUMPF, I. J., FELD, M. K., CORNELIUS, M. J., KEOGH, B. A. and CRYSTAL, R. G. (1981). Safety of fibreoptic bronchoalveolar lavage in evaluation of interstitial lung disease. *Chest*, **80**, 268–271

TAPLIN, D. J. (1966). Malignant cells in sputum: a simple method of liquefying sputum. *J. Med. Lab. Technol.*, **23**, 252–255

WANG, K. P., BROWER, R., HAPONIK, E. F. and SIEGELMAN, S. (1983). Flexible transbronchial needle aspiration for staging of bronchial carcinoma. *Chest*, **84**, 571–576

ZEEVI, A., FUNG, J. J., PARADIS, I. L., GRYZAN, S., DAUBER, J. H., HARDESTY, R. L. GRIFFITH, B., TRENTO, A., SAIDMAN, S. and DUQUESNOY, R. J. (1987). Bronchoalveolar macrophage–lymphocyte reactivity in heart-lung recipients. *Transplantation Proc.*, **XIX**, 2537–2540

13

The respiratory tract: pathology and cytology of respiratory tract diseases

Jocelyn E. A. Imrie and Gordon M. Campbell

Although the respiratory tract is subject to a wide variety of benign diseases, relatively few produce cellular changes in sputum or other cytological samples which permit their recognition with the light microscope. In contrast, cytology has an important role to play in the diagnosis of malignant lesions of the respiratory tract as most bronchial carcinomas can be detected readily in smears of sputum and more peripheral lesions can be detected by bronchoscopic sampling or fine needle aspiration. In this chapter we review the pathobiology and cytology of benign and malignant lesions of the respiratory tract in order to acquaint the cytologist with the range of cytological patterns he may expect to encounter. We also present a critical evaluation of the role of cytology in the diagnosis and management of the patient with lung cancer as it is in this area that respiratory tract cytology can make a major contribution to clinical practice.

Pathology of upper respiratory tract disease

The most common pathology of the upper respiratory tract is infection, notably acute coryza, i.e. the common cold, which may be caused by several different viruses. Allergic rhinitis may also cause a runny nose due to hypersensitivity to agents such as pollen. Infections of the larynx may be viral or bacterial (e.g. *Streptococcus pyogenes*). Most are minor, although rarer conditions such as acute epiglottitis in children or diphtheria may be fatal. Unusual inflammatory conditions also occur such as Wegener's granuloma which is a necrotizing granulomatous disorder of unknown aetiology affecting the nose, lung and kidney.

Tumours of the upper respiratory tract are not uncommon and these may be benign or malignant. In the nose, the tumour most frequently seen is a nasal polyp which consists of loose gelatinous connective tissue containing mucous glands and covered by respiratory epithelium. Such polyps may be of infective or allergic aetiology; in the latter case eosinophils are present in significant numbers within the stroma. Papillomas also occur in the nose and these may be composed of squamous or transitional cells. These may sometimes recur and some progress to squamous carcinoma. Other malignancies seen in the nose include anaplastic carcinoma or, less commonly, adenocarcinoma. In the nasopharynx squamous carcinoma is seen and is known to be associated with the Epstein–Barr virus. Benign tumours of the larynx include 'singer's nodes', which are inflammatory polyps occurring on the vocal cords in adults. These are usually single and usually simple squamous papillomas on histological examination. Pharyngeal carcinoma occurs mainly in males over 50 years. This is almost always squamous in type and in many cases carcinoma *in situ* precedes the invasive lesion.

Pathology of lower respiratory tract disease

A wide range of diseases affect the lower respiratory tract. Those most likely to come to the attention of the cytologist are described in the following section and are broadly grouped as:

(1) Inflammatory disease,
(2) Obstructive airways disease,
(3) Diseases of vascular origin,
(4) Diffuse interstitial fibrosis, and
(5) Tumours.

Inflammatory disease

Many of the sputum samples referred to a cytologist are from patients with lower respiratory tract infection. This is because the symptoms and signs of inflammatory lung diseases are often very similar to those present in patients with lung cancer. Thus patients who complain of cough and haemoptysis or who are found to have a slowly resolving pneumonia present a problem of differential diagnosis. Indeed the problem is often complicated by the fact that inflammatory lung disease and malignant disease frequently coexist. The pathological bases for a number of common inflammatory diseases of the bronchi and lungs are described below.

Acute bronchitis and pneumonia

Acute tracheitis and bronchitis are common self-limiting entities which are seldom seen in pathological practice. Acute infection of the lung parenchyma is termed pneumonia and has various causes. The commonest form is bronchopneumonia. In this condition bacteria colonize bronchioles and bronchi and extend into surrounding alveoli from the air passages giving rise to multiple foci of inflammation with consolidation. This disease is most often seen in the very young, the old and debilitated and is caused by different kinds of bacteria notably streptococci, staphylococci, *Klebsiella* and *Haemophilus influenzae*. It is found in a large proportion of hospital post mortems where it may either be a direct or an indirect cause of death.

Lobar pneumonia is a less common form and, as its name suggests, it remains restricted to an anatomical lobe of the lung. The causative organism is *Streptococcus pneumoniae* in 90–95% of cases and the characteristic clinical picture is of fever, pleuritic pain, rusty sputum and herpetic lesions of mouth and nose. Another less common form of pneumonia is Legionnaire's disease which was first recognized as an epidemic in the United States in 1976. Since then, several smaller epidemics and sporadic cases have been recognized in other countries including the UK. The causative organism is a very small coccobacillus, *Legionella pneumophila*, which is found in water coolant (e.g. in air conditioning) systems and also in natural running water. The organism is difficult to culture but may be recognized in the alveolar exudate of patients with the disease using Dieterle's silver stain which shows up the coccobacilli as small black or dark brown granules.

Aspiration pneumonia is a disease seen in the elderly or debilitated. In this form, the inflammation of the lung is due to an irritant, i.e. acid gastric contents or food particles, but bacterial infection often supervenes. Aspiration of oils, e.g. in children or in adults using nasal sprays or oil-based laxatives, gives rise to so called 'lipid pneumonia' in which the oils may be seen in alveolar macrophages if fat stains are used.

Lung abscess

The most common cause of a lung abscess is aspiration of infected material but other causes include infection secondary to obstruction by a bronchial tumour. Lung abscesses may also form following bacterial pneumonia, bronchiectasis or tuberculosis and rarely are due to septic emboli from distant sites. Initially an abscess forms a poorly defined cavity containing pus and debris but as it becomes chronic, surrounding fibroblastic proliferation produces a fibrous wall and the lining of an abscess may form squamous or, less commonly, columnar epithelium. If close to a bronchus, an abscess may rupture into it and the patient may suddenly produce a large volume of purulent sputum. Some abscesses heal spontaneously but others persist causing chronic debility and respiratory symptoms. In a proportion of cases infective emboli from a lung abscess may give rise to secondary infection in the brain causing cerebral abscess or meningitis.

Tuberculosis

It is only in the last 30 years that effective treatment has been available for this disease and the number of cases encountered in clinical practice has been significantly reduced. Lack of awareness of the disease may result in underdiagnosis and unnecessary death from the disease. Nowadays the lungs are more often affected by tuberculosis than any other organ largely because inhalation is the commonest route of infection. The causative organism, *Mycobacterium tuberculosis*, is a slender slightly curved or straight rod measuring 1–4 µm, which stains red by the Ziehl-Neelsen method because it resists decolourizing by acid and alcohol, thus earning it the alternative name, Acid and Alcohol Fast Bacillus (AAFB).

Two main forms of pulmonary infection are recognized. Primary tuberculosis may occur in childhood or adulthood and may resolve or prove fatal. Post primary infection may occur later and may heal or progress to chronic pulmonary disease. The initial lesion seen in primary tuberculosis is a small focus of infection, called the Ghon focus, which forms in the mid zone just below the pleura. Infection spreads from here via the lymphatics to hilar lymph nodes. The Ghon focus and the enlarged hilar lymph nodes constitute the primary (Ghon) complex. In both the Ghon focus and the lymph nodes, caseating tubercules form. Most cases of primary tuberculosis heal with fibrosis and calcification of the complex but in some there may be extension of infection from the primary complex, this extension being local or disseminated to distant organs. Blood-borne spread may result in the formation of large numbers of small uniform tubercules in many organs (generalized or miliary tuberculosis) or in only one or two distant foci. Involved hilar lymph nodes may obstruct bronchi or disseminate bacilli through the bronchial tree causing acute bronchopneumonia.

Post primary infection may be due to reinfection or reactivation. The evolution of the disease is generally much slower with cavitation and fibrosis. Local spread may give rise to a pleural effusion (tuberculous or reactive) or even tuberculous empyema.

Opportunistic infections

This group of infections, caused by organisms of low infectivity, occurs in patients whose body defences are lowered. Suppression of the body's defence mechanisms may be induced by agents used in the treatment of malignancy or to allow the patient to accept an organ transplant, but currently the most widely recognized risk group is patients with acquired immune deficiency syndrome (AIDS). This is a disease recently recognized in the Western world, occurring predominantly in homosexuals but also in drug users and haemophiliacs. It is caused by human immune deficiency virus (HIV) previously known as human T lymphotropic virus type III (HTLV III) or lymphadenopathy associated virus (LAV). The virus is transmitted by sexual intercourse or via body fluids. This syndrome is characterized by opportunistic infections, often life threatening, and unusual forms of malignancy, notably Kaposi sarcoma or cerebral lymphoma.

The most usual infection in AIDS patients is *Pneumocystis carinii* pneumonia which is the presenting disease in approximately 50% of cases and accounts for 85% of opportunistic infections in such patients. Young, Hopkin and Cuthbertson (1984) diagnosed *P. carinii* pneumonia in alveolar lavage fluid from 27% of patients on immunosuppressive regimes or on treatment for leukaemia, lymphoma or allied conditions. The organism is a protozoan which occurs in trophozoite and cyst form, these forms being recognizable histologically and cytologically. Early diagnosis is important as the infection can be treated effectively.

The lungs are a common site for other opportunistic infections and the usual picture is of a diffuse pulmonary infiltrate. Many organisms including fungi, notably *Candida albicans* and *Aspergillus fumigatus*, and viruses such as cytomegalovirus and herpes simplex virus may cause such infections.

Airways disease

Chronic bronchitis

Chronic bronchitis is the commonest form of respiratory disability seen in the west of Scotland. It is caused by irritation of bronchial mucosa, usually by cigarette smoke which may be the only factor in heavy smokers, or in conjunction with atmospheric pollution in industrialized areas. It is seen most often in the middle-aged and elderly and is commonest in males, presumably because of the higher incidence of smoking in males of this age group. Irritation by these agents leads to hypertrophy and hyperplasia of the bronchial mucous glands and an increase in the number of goblet cells within the epithelium itself, with associated loss of ciliated cells. Thus, mucus is produced in excess by the larger number of goblet cells present and, because of loss of cilia, this is not so readily eliminated. There are often foci of squamous metaplasia within this bronchial mucosa as a result of irritation.

Chronic bronchitis is a clinical rather than a pathological diagnosis and is defined as 'the production of sputum on most days over a three month period or more for at least two years'. It is misnamed as it is not an inflammatory process but a combination of metaplastic and hyperplastic changes. Exacerbations are common, often occurring serveral times in any one year. They are caused by bacterial colonization of the stagnant mucus secretions, notably *Haemophilus influenzae* or *Streptococcus pneumoniae*, leading to acute infection. With repeated exacerbations, changes tend to spread down the respiratory tree and may develop in the terminal bronchioles. In many cases emphysema develops. Respiratory disability is progressive and chronic bronchitis may lead to cor pulmonale and then cardiac failure. Because of their common causal factor, cigarette smoking, chronic bronchitis and lung cancer are often associated.

Emphysema

Emphysema is defined as a permanent increase in the size of the air spaces distal to the terminal bronchioles, with destruction of their walls, and is thus a pathological diagnosis. Two main forms occur of which centrilobular is more common, being associated with chronic bronchitis and the pneumoconioses. It affects the respiratory bronchioles and so shows abnormal enlargement of the spaces in the centre of the respiratory lobules. The other form is panacinar emphysema in which the dilatations appear at the periphery of the respiratory lobules. This is not associated with other respiratory diseases but may occur in alpha-1 antitrypsin deficiency.

Asthma

This is a form of acute dyspnoea brought about by spasmodic narrowing of bronchi and plugging of small airways by viscous mucus causing difficulty in breathing which is more marked on exhaling. There are two main forms of the disease, an extrinsic and an intrinsic form. Extrinsic asthma is commonly seen in children and young adults and is usually an atopic reaction to a range of exogenous allergens, e.g. the house dust mite, animal fur. Often the patient gives a history of preceding infantile eczema or food allergy and there is often also a family history of asthma. This form tends to become less severe with age and often ceases during adolescence.

Intrinsic asthma starts in adult life and is not associated with atopy or family history. The prognosis is less good than in extrinsic asthma because the attacks may become progressively more severe and drug sensitivities may develop especially to aspirin and penicillin. This form is sometimes associated with nasal polyps and may progress to chronic bronchitis.

The most severe form of asthmatic attack is status asthmaticus in which attacks occur incessantly. This may prove fatal.

Diseases of vascular origin
Pulmonary thromboembolus

This is commonly seen in hospital pathology practice. Thrombi which embolize to the lung usually form in the veins of the calf but also in veins in the pelvis. Classically thrombi develop if the patient is immobilized or bedbound and they may become detached, either in part or completely, and pass through the venous circulation to the right side of the heart to reach the pulmonary vasculature. Small emboli are found in most hospital post mortem cases if thorough examination is undertaken. These small emboli are dissolved away and larger ones resolved by recanalization. Symptoms caused by emboli vary with the size, site and rapidity of occlusion. Large emboli causing occlusion of the pulmonary tree or its main branches constitute one of the most acute medical emergencies and are often rapidly fatal.

Pulmonary emboli are not all composed of blood. Fat emboli may occur in bony fractures as may bone marrow emboli; a rare complication of childbirth is amniotic fluid embolism. Air emboli may also occur but these are virtually always iatrogenic. Occlusion of a pulmonary artery leads to infarction of an area of lung and the characteristic X-ray appearance of this is of a wedge-shaped lesion at the periphery of the lung fields, usually in the lower lobes. In more than 50% of cases lesions are multiple.

Pulmonary oedema

Fluid accumulates in the lungs in various circumstances, either because of increased back pressure in the venous system which lessens the resorption of fluid into the vessels, e.g. in congestive cardiac failure or left ventricular failure, or because of increased amounts of fluid in the vascular system (volume overload). Pulmonary oedema also occurs in raised intracranial pressure and at high altitude but the mechanism in these conditions is less well understood.

Fluid initially collects in the interstitium and only in the late stages do the alveoli become filled with it. Prolonged oedema may lead to small intra-alveolar haemorrhages and the formation of haemosiderin-laden macrophages ('heart failure cells') and in some cases diffuse fibrosis may occur.

Pulmonary hypertension

This has many different causes and is always present in cor pulmonale as it leads to right ventricular hypertrophy. In a minority of cases it is idiopathic. Chronic lung disease of any kind may lead to pulmonary hypertension by increasing pulmonary vascular resistance or by formation of vascular shunts. Thus, chronic bronchitis, emphysema, tuberculosis and fibrotic conditions of the lung all may give rise to it. Pulmonary hypertension may also result from cardiac pathology and is seen in some forms of congenital heart disease e.g. ventricular septal defect, as well as in mitral stenosis. Atheroma is only seen within the pulmonary vasculature if pulmonary hypertension is present.

Diffuse interstitial lung disease

Several disease entities are classified together to form this heterogeneous group. The appearance is of diffuse, usually chronic, involvement of lung tissue involving approximately 25% of the lung. The changes give rise to restrictive rather than obstructive lung disease and result in reduced oxygen diffusion capacity. They may lead to right-sided heart failure. The typical chest X-ray appearance is of 'ground glass' due to the presence of small nodules throughout the lung fields.

Although it is possible to differentiate the disease forms in this group at an early stage, all may progress to what is termed 'honeycomb lung' at which stage it may not be possible to recognize aetiology. Classification is difficult and various forms have been attempted. It is further complicated by the variety of terms used within the group. One possible form of classification is given in *Table 13.1*.

Table 13.1 Classification of diffuse interstitial lung diseases

Forms with recognized aetiology:
  Pneumoconiosis
  Post-infection e.g. tuberculosis
  Ingestion of toxin e.g. paraquat
  Physical agents e.g. radiation, oxygen
  Drugs e.g. busulphan
  Immunologic mechanism—extrinsic allergic alveolitis
  e.g. farmer's lung, bird fancier's lung

Forms with unknown aetiology:
  Diffuse fibrosing alveolitis
  Sarcoidosis
  Pulmonary alveolar proteinosis
  Diffuse malignant neoplasms e.g. lymphoma

Pneumoconiosis

The pneumoconioses are lung diseases caused by inhalation of dust encountered in industry. They have been recognized in the UK since the eighteenth century and various minerals may give rise to them. To do so, the mineral must exist in a very small particle form (less than 5 μm diameter) to reach the alveoli, but larger particles are removed higher in the respiratory system. The form of lung disease induced varies with the nature of the mineral and the size of its dust particles but all show a degree of fibrosis as the main feature. The commonest forms seen in this country are related to coal, silica and asbestos.

In simple coal worker's pneumoconiosis there are few symptoms, although the whole lung fields may become impregnated with coal dust particles. However, progressive massive fibrosis, i.e. irreversible damage with fibrous obliteration of large areas of lung, may develop from this. Silicosis is seen in patients working with glass or sand, quarrying certain stones and in some forms of mining. The silica dust particles which give rise to the fibrosis are 2–3 μm in size. Silicosis is known to predispose to tuberculosis.

Asbestos is a widely used insulating material which occurs in three main forms of which the most commonly used in the UK is chrysotile, and the most dangerous is crocidolite. The fibres which are inhaled may be up to 50 μm long but are narrow. They become impacted in the alveoli where they are attacked by macrophages. Iron and protein become encrusted on the fibres to form the characteristic segmented asbestos bodies which may be seen most readily on Prussian blue-stained sections. The fibrosis in asbestosis is diffuse but may be most marked subpleurally and is often associated with pleural thickening. There is no increase in the incidence of tuberculosis in asbestosis but it is known to predispose to carcinoma of the bronchus which arises in approximately 50% of males with asbestosis. It also is associated with mesothelioma.

Other occupational lung diseases are caused by organic dusts which give rise to extrinsic allergic alveolitis. The various forms of this are named after the associated occupation, e.g. farmer's lung caused by inhalation of the thermophilic actinomycete *Micropolyspora faeni*, and bird fancier's lung caused by the inhalation of an agent (probably avian proteins) in the droppings of birds such as parrots or pigeons.

Diffuse fibrosing alveolitis

This entity has a variety of names such as idiopathic diffuse pulmonary fibrosis, interstitial pneumonia or Hamman-Rich syndrome. It consists of a spectrum of appearances varying with the stage of the disease although some authorities consider that these appearances may, in fact, represent different disease processes. These are described as interstitial pneumonitis in which there is alveolar cell loss with resulting hyaline membrane formation, metaplastic change, interstitial cell infiltration and fibrosis; or desquamative pneumonitis in which intraalveolar macrophages are numerous and there is little interstitial fibrosis.

Sarcoidosis

This is an uncommon generalized disease which usually occurs in young adults. Its cause is unknown but it may take various forms, being acute or chronic and affecting different organs notably lymph nodes, lungs, skin and eyes. The characteristic lesion is a non-caseating granuloma but sarcoidosis is usually diagnosed by performing a Kveim test, a skin test which is positive in most cases.

Tumours

The pathology occurring in the lungs which is most relevant to cytologists is undoubtedly neoplasia and most tumours occurring within the lung are malignant. The incidence of lung cancer has increased dramatically in this century and its epidemiology has been a source of great interest. It is the commonest malignancy in man and accounts for 18.6% of all cancers and 27% of cancer deaths in the UK. In the 1920s and 1930s it was almost exclusively a disease of males, but in the post-war years the incidence in females has increased rapidly. Numerous epidemiological studies have shown that cigarette smoking has a causal role in lung cancer. (This probably accounts for the increasing incidence in women as their smoking habits are changing.) Other factors also involved in the genesis of lung cancer include exposure to asbestos and other industrial materials as well as atmospheric pollutants such as polycyclic hydrocarbons. The involvement of these factors is supported by the higher incidence of the disease in urban rather than rural communities.

The term lung cancer is inaccurate as the majority of malignant tumours of the lung arise in the bronchi. The tumours may arise in different cells in the bronchial epithelium giving rise to different types of malignancy (*Table 13.2*). There are various classifications in use, the importance of this being that the different forms of tumour respond to different therapy. However, it is recognized that the degree of differentiation may vary within a tumour and that two types may co-exist within one tumour, and it is worth remembering that classification of a malignant tumour from biopsy material is not completely reliable.

Table 13.2 Classification of epithelial tumours of the lung

Squamous (epidermoid) carcinoma—keratinizing type
                                        non-keratinizing type
Adenocarcinoma—usual (bronchial) type
                     bronchoalveolar or alveolar cell
Small cell anaplastic ('oat cell')
Large cell anaplastic
Bronchial carcinoid
Others e.g. giant cell carcinoma, mucoepidermoid carcinoma, lymphoma sarcoma and carcinosarcoma

Based on the classification used by the Working Party for Therapy for Lung Cancer.

Squamous carcinoma

This is the most common type of bronchial carcinoma accounting for 70% of all lung cancers. Its development is thought to be an example of the 'multistep' theory of carcinogenesis and the progression through squamous metaplasia, atypical squamous metaplasia and carcinoma *in situ* to invasive squamous carcinoma is widely accepted. Various trigger factors act at different times to cause this progression and although these have not been accurately identified, the time sequence from atypical metaplasia to malignancy appears to be in the order of 3–9 years.

Metaplasia is a mechanism in which a mucosal lining is altered to another form which usually is more protective. Thus, in the bronchi, in response to irritation or inflammation (notably caused by cigarette smoking) columnar epithelium is replaced by squamous epithelium. (This may be reversible if the precipitating factor is removed.) Progression may occur through increasing degrees of atypia and the point at which changes amount to squamous carcinoma *in situ* is difficult to define.

Squamous carcinoma usually presents with signs of consolidation and infection of the lung distal to the blockage. Thus a persistent cough and bloodstained sputum are common presenting symptoms. Its association with cigarette smoking is well established.

Adenocarcinoma

The more usual form of this tumour originates in bronchial epithelium and is composed of columnar cells which may secrete mucus. Such tumours are seen more often in the peripheral lung field and they often arise in an old scar. Because of its peripheral site, adenocarcinoma is often clinically silent and may present only when it metastasizes. It is also thought to be slower growing than squamous carcinoma. The incidence of adenocarcinoma increases with age, occurring almost equally in men and women, and it is not considered to be closely related to cigarette smoking. It accounts for about 4% of lung cancer.

A less common form of adenocarcinoma recognized in the lung is bronchoalveolar carcinoma, also called alveolar cell carcinoma. A characteristic finding in this condition is the presence of discrete tumour cells in the bronchiolar and alveolar walls, thereby causing diffuse lung involvement without forming a definite tumour mass and without disrupting the interstitial framework of the lung. These tumours are also more often seen in the peripheral lung fields. It is thought that not all bronchoalveolar carcinomas arise from the same cell type but in some cases electron microscopy has shown the presence of lamellar bodies within tumour cells such as are seen in type 2 pneumocytes suggesting that at least some of these tumours arise from such cells.

Large cell anaplastic carcinoma

Many tumours classified as large cell anaplastic carcinoma are undifferentiated squamous cell carcinomas and some may be undifferentiated adenocarcinomas, but the degree of differentiation within the tumours in this group is such that they can no longer be classified under a specific cell type.

Small cell anaplastic ('oat cell') carcinoma

These tumours tend to rise near the hilum and account for about 18% of lung cancers. They used to be called lymphosarcomas and were thought to arise in mediastinal lymph nodes. They are now considered to arise from Feyrter cells, i.e. they are neuroendocrine in origin and are often associated with inappropriate hormone secretion and tumour-associated endocrine abnormalities. At presentation these tumours are often already inoperable and the treatment of choice is chemotherapy.

Carcinoid tumour

This uncommon slow growing tumour is also sometimes known as bronchial adenoma but since it can metastasize it is not benign (as implied by the term adenoma). It is not possible to determine which tumours will metastasize from their histological appearance and therefore all carcinoids must be considered to have malignant potential and be treated accordingly. They tend to occur centrally, often growing into the lumen of the bronchus and therefore present with signs of obstruction of the main bronchus. They originate from cells of the APUD system (i.e. neuroendocrine cells) and show the neurosecretory granules characteristic of these cells electron microscopically. Carcinoid tumours tend to lie underneath the epithelium and therefore do not exfoliate (although occasionally they may ulcerate). The origin of oat cell carcinomas and carcinoid tumours appears to be very similar and some people consider them to represent different ends of the spectrum of neuroendocrine tumours of the lung.

Others

In some classifications this group would be further subdivided into such categories as giant cell tumours, mucoepidermoid carcinomas, lymphoma and carcinosarcomas. It must also be remembered that within this group is metastatic carcinoma as the lung may be the site of lymphatic or blood-borne metastases from various sites.

Cytological patterns in benign disease of the respiratory tract

Cytology is of limited value in the diagnosis of benign disorders of the respiratory tract as very few of them produce specific cellular changes in Papanicolaou smears. Characteristic smear patterns have been described for the following disorders but even these must be interpreted with caution as there is considerable overlap in the cellular changes that can be seen.

Benign lesions of the mouth and pharynx

The squamous epithelium lining the mouth and pharynx frequently contaminates sputum samples and causes problems of diagnosis. Buccal and pharyngeal epithelium may appear necrotic in the presence of acute infection of the mouth or fauces, and ulceration can give rise to the presence of deep squamous cells with intact intercellular bridges in saliva or sputum. Numerous leucocytes are usually present in the smears. Squamous epithelium subjected to chronic irritation, for example from ill-fitting dentures, tends to keratinize resulting in the appearance in smears of anucleated cells with quite dense eosinophilic or orangeophilic cytoplasm. These must be differentiated from the anucleated squamous cells seen in epidermoid carcinoma which have a denser, more brilliantly staining cytoplasm. Keratinized and often atypical cells with abnormally-shaped, irregular pyknotic and karyorrhectic nuclei may arise from abnormal squamous metaplasia in the trachea of patients fitted with a tracheostomy tube.

Allergic rhinitis (hay fever)

Nasopharyngeal smears from patients with allergic rhinitis may contain a large number of eosinophil leucocytes which are best demonstrated by Romanowsky or haematoxylin and eosin staining. There may also be many goblet or mucus-secreting cells present. Metaplastic changes also occur in the mucosa of the nasopharynx and there is often a large number of free nuclei which have been denuded of cytoplasm, in the background of the smears. These nuclei can be hyperchromatic and may contain prominent nucleoli. The cells must be carefully assessed if a false diagnosis of malignancy, especially of small cell anaplastic carcinoma, is to be avoided. The denuded nuclei do not show the characteristic moulding of oat cells.

Asthma

In histological section, asthma shows bronchi filled with viscous mucus containing eosinophils and shed bronchial epithelial cells. There is a characteristic hyaline appearance to the basement membrane of the epithelium below which is an eosinophil infiltrate in the submucosa which may be oedematous. The underlying muscle is hypertrophied as a result of prolonged spasm.

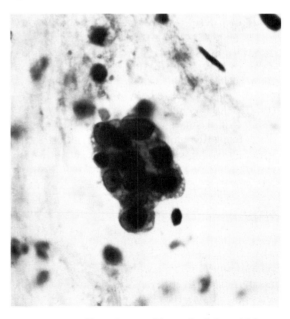

Figure 13.1 Papillary clusters of hyperplastic bronchial epithelial cells (Creola bodies) seen in asthmatics. These may be confused with adenocarcinoma. Papanicolaou stain. Magnification × 540

Sputum samples from patients with asthma reflect the underlying histology. Hyperplasia of the epithelium results in shedding of papillary clusters of bronchial epithelial cells (*Figure 13.1*). These clusters, known as 'Creola bodies' can be found in 42% of samples from asthmatics (Naylor, 1962). They may also be seen occasionally in bronchitis and bronchiectasis. These clusters can easily be mistaken for adenocarcinoma due, at least in part, to the presence of mucus-secreting cells within the fragment. Close examination of these groups reveals normal nuclear chromatin patterns and regular, uniform nucleoli. Cilia can be recognized on the surface of the groups. Eosinophils and Charcot-Leyden crystals (*Figure 12.9*) are also found in sputum samples from patients with asthma.

Acute inflammatory conditions of the airways and lung parenchyma

In acute bronchitis and bronchiolitis the sputum may be purulent and blood streaked. Smears are composed almost entirely of polymorphonuclear leucocytes and mucus strands. Occasional bronchial epithelial cells may be seen with ragged cytoplasm and pyknotic nuclei. Degenerate epithelial cells may be swollen with enlarged nuclei and nucleoli. Nuclear chromatin may become coarsely granular. Often these cells retain their terminal plates although the cilia are fragile and easily lost. Multinucleated cells may be seen. Macrophages may be scanty and may show phagocytosis. Lymphocytes are seen in smears when the surface epithelium is damaged or lost (*Figure 13.2*). These present as clusters of small, densely staining cells with scanty cyanophilic cytoplasm. The cells are about 10 µm in diameter with uniformly sized, rather dark, round or oval nuclei. These cell clusters must be distinguished from the small anaplastic tumour cell groups of oat cell carcinoma.

Pneumonia

The histological features common to the various forms of pneumonia are the presence of an inflammatory infiltrate which may contain polymorphonuclear leucocytes, lymphocytes, histiocytes and plasma cells. The proportions of these vary in different forms of pneumonia and different stages of the disease. The sputum is usually scanty and watery and in the acute congestive stages may be bloodstained. Smears contain few

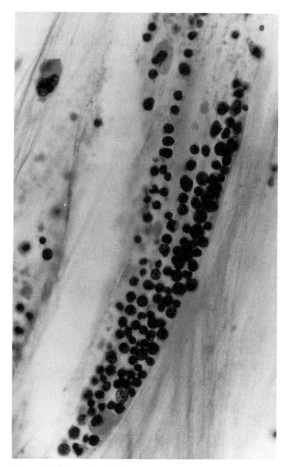

Figure 13.2 Streaks of lymphocytes may be confused with oat cells in sputum. Oat cells can be recognized by their characteristic pattern of nuclear moulding. Papanicolaou stain. Magnification × 540

polymorphonuclear leucocytes but macrophages (some containing blood pigments) may be present as the pneumonia resolves.

Tuberculosis

The histological appearance of tuberculous lesions is characterized by caseous (resembling cheese) necrosis. This usually forms the centre of the tubercle and is surrounded by lymphocytes and histiocytes, some with an epithelioid appearance and others forming Langhan's giant cells which are multinucleated and characteristically have peripheral nuclei. Fibroblastic proliferation may also be a feature as the body attempts to wall off the lesion (*Figure 13.3*).

The cytological appearances of tuberculosis have been described in sputum and bronchial secretions (Nasiell et al., 1972), in bronchial lavage fluid (Verma, Sandhyamani and Pande, 1983) and in material obtained by fine needle aspiration of the lung (Robicheux, Moinuddin and Lee, 1985). In the latter report, the importance of demonstrating acid-alcohol fast bacilli is stressed. Cytological diagnosis depends on identifying cells arising in the tubercle, i.e. epithelioid cells and Langhan's giant cells. Epithelioid cells are seen as loose clusters of elongated, slender cells which are rather shorter than bronchial epithelial cells. The cytoplasm is usually eosinophilic and has ill-defined borders. The nucleus is elongated, central and pale staining. The multinucleated Langhan's cells have a characteristic peripheral, horseshoe arrangement of nuclei.

Viral infections

Among the viral infections most commonly seen are herpes simplex, herpes zoster and cytomegalovirus infection.

Herpes simplex infection is characterized by the presence of large multinucleated giant cells with nuclear moulding (*Figure 13.4*). Nuclei may contain large eosinophilic inclusions or they may have a featureless ground glass appearance. In sputum the cells may have an eosinophilic cytoplasm and the nuclei may be overlapping and crowded to give the appearance of a single large irregular hyperchromatic nucleus. Consequently they are readily mistaken for malignant squamous cells. It should be remembered that the presence of herpes giant cells in sputum may reflect infection anywhere in the upper and lower respiratory tract. A careful inspection of the mouth and pharynx is advised before concluding that the trachea or bronchi are involved in the herpetic process.

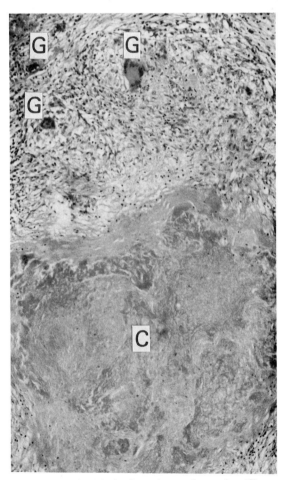

Figure 13.3 Tuberculosis. Central area of caseation (C) with surrounding lymphocytes and histiocytes and occasional Langhan's giant cells (G). Stain: haematoxylin and eosin. Magnification × 160, reduced to 65% in reproduction

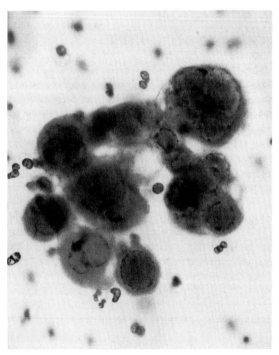

Figure 13.4 Herpes infected cells in sputum. Note presence of large multinucleate cells. The individual nuclei have a 'glassy' appearance and there is margination of the chromatin. Papanicolaou stain. Magnification × 540

Cytomegalovirus infection is characterized by the presence of large cells bearing large intranuclear and intracytoplasmic inclusions. The intranuclear inclusions are usually basophilic, but may be eosinophilic, and are surrounded by a clear area giving rise to an 'owl's eye' appearance. Unlike herpes simplex, multinucleation is not a feature of cytomegalovirus infection.

One of the most striking changes seen in patients with viral infection is ciliocytophthoria (CCP). This has been described in sputum of patients with influenza virus, parainfluenza virus, adenovirus, respiratory syncytial virus and measles (Naib, 1970). This change is characterized in the smear by the presence of fragments of ciliated cytoplasm which are frequently mistaken for flagellate protozoa. Ciliocytophthoria reflects degeneration and exfoliation of the bronchial epithelial cells as a result of virus-induced necrosis of the epithelium (Johnston and Frable, 1976). Ciliocytophthoria is also seen in sputum samples taken immediately after bronchoscopy and in association with bronchial carcinoma and is not specific for respiratory viral infection.

Fungal infections

Fungi are frequently encountered in sputum as a result of contamination from the mouth or from the air if the specimen is not fresh. However, pathogenic infection must always be borne in mind, particularly in immunocompromised patients. Many fungi have a recognizable morphology and can, therefore, be identified specifically. Pathogenic fungi which can be diagnosed with confidence include *Cryptococcus neoformans* and *Blastomyces dermatidis*. Recognizable opportunistic fungi include *Candida albicans*, *Aspergillus fumigatus*, *Histoplasma capsulatum* and zygomycetes. The organisms are easily detected in Papanicolaou-stained smears and special staining is rarely necessary. It should be remembered that some fungi are so thick that they may not be present at the focal plane of the cells on the slide but can be detected above this level. Mycotic infections provoke non-specific inflammatory changes in respiratory cells.

Pneumocystis carinii and other opportunistic infections

Diagnosis of *P. carinii* depends on the microscopic examination of biopsies of lung parenchyma and aspirates from the respiratory lobules and alveolar sacs. Histology shows a pneumonia in which the distinguishing feature is the presence within the alveolar sacs of characteristic granular foamy material which appears eosinophilic in haematoxylin and eosin stained sections and within which cysts may be identified. Cytology offers a range of methods suitable for obtaining diagnostic samples and these have the advantage of being safer and more rapid than either bronchial biopsy or open lung biopsy. The most suitable samples are bronchoalveolar lavage fluid, touch preparations (Ognibene *et al.*, 1984; Fleury *et al.*, 1985; Orenstein *et al.*, 1985) and material obtained by transbronchial fine needle aspiration (Wallace *et al.*, 1985). Most workers do not find sputum samples of value for the diagnosis of *P. carinii*. However, Pitchenik and his colleagues are an exception (Pitchenik *et al.*, 1986).

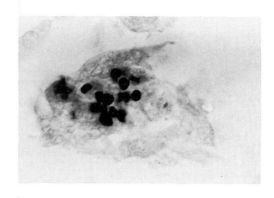

Figure 13.5 *Pneumocystis carinii*. The protozoon, in its encysted form, is seen against a background of precipitated protein. Section of lung. Stain: Grocott's methenamine silver nitrate (Pintozzi's modification). Magnification × 1000, reduced to 60% in reproduction

Several techniques are useful for demonstrating *Pneumocystis* in cytological specimens. These include Toluidine blue, Gram-Weigert, PAS, Cresyl Violet and Grocott's modification of Gomori's methenamine silver method (*Figure 13.5*). Of these, the latter gives the best contrast between the organism and its background and should always be included as the tinctorial methods are open to misinterpretation (Kim and Hughes, 1973; Young, 1985). In methenamine silver-stained preparations the organism appears as a spherical cyst (*Figure 13.6*) 6–8 μm in diameter, approximately the size of an erythrocyte. Some forms appear cup-shaped or crinkled. A small globoid structure can sometimes be seen attached to the inner wall of the cyst. The cysts are often found in clusters of 40–50, although an accurate diagnosis can be made when only a few cysts are found if care is taken to ascertain the morphological features. Histoplasmosis, *Candida* spores and red blood cells must be considered in the differential diagnosis. A trophozoite form of *P. carinii* has been described which can sometimes be seen in Giemsa-stained preparations as a shadowy elliptical body, 8–10 μm in length. However, diagnosis of *Pneumocystis* based on this pattern is not reliable, and a search should always be made for the cysts.

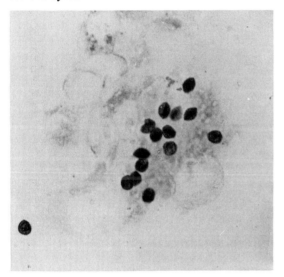

Figure 13.6 *Pneumocystis carcinii* in bronchoalveolar lavage specimen. Note spherical shape of cyst each of which contains a darkly staining trophozoite in the centre. Grocott stain. Magnification × 1000

Grocott's method has the disadvantage of being rather slow, requiring 3 h to complete. Pintozzi (1978) introduced a modification which reduces the staining time to 10 min. This is the method routinely used in this department where it has been found to give consistent results. Greater control over silver deposition can be achieved by reducing the temperature of the silver solution to 75 °C and including dimethyl sulphoxide in the silver solution (Campbell and McCorriston, 1987). It has been reported that *P. carinii* can be identified by direct fluorescence of Papanicolaou-stained smears. The organism gives a distinct yellow-green fluorescence when viewed under ultraviolet light (Ghali, Garcia and Skolom, 1984). This method has the advantage of being rapid (Gaurner, Robey and Gupta, 1986). However, non-specific fluorescence of other material in the sample is often compromising. Recently, *P. carinii* has been accurately identified in pulmonary material by immunofluorescence using a monoclonal antibody directed against the organism (Kovacs *et al.*, 1986). In Papanicolaou-stained smears, *P. carinii* infection can be suspected by the occurrence of characteristic clumps of fine frothy mucus in which the outline of numerous cysts can be detected (*Figure 13.7*). The clumps usually stain pale grey-green in colour, although eosinophilic staining has been described (Young *et al.*, 1986), and are often associated with alveolar macrophages. They must be distinguished from clumps of erythrocytes and from coarser clumps of frothy mucus. Erythrocytes are slightly larger than the cysts and stain green or greenish red. With experience, the appearances are easily recognized and should be detected on routine screening, particularly if a relevant clinical history is supplied. However, confirmation by examination of a Grocott-stained preparation must be undertaken. It should be noted that a negative result does not guarantee that the patient is free from *P. carinii* infection.

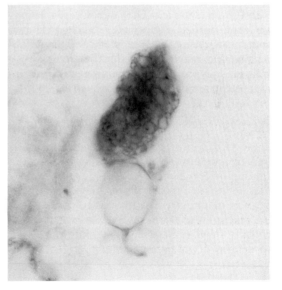

Figure 13.7 *Pneumocystis carinii*. The individual trophozoites can be clearly seen in this characteristic foamy cluster. Bronchoalveolar lavage fluid. Papanicolaou stain. Magnification × 1000, reduced to 80% in reproduction

The widespread use of cytotoxic therapy has resulted in reactivation of other latent infections which are rarely seen in individuals where the immune response is intact. The cytologist must be alert to rarer manifestations such as *Strongyloides stercoralis* and filarial worms which present a striking appearance in the smears (*see* Chapter 23).

Chronic bronchitis

Histological sections of bronchi show hypertrophy and hyperplasia of the submucosal mucous glands. The overlying epithelium contains increased numbers of goblet cells and there is loss of cilia on the epithelial cells. Squamous metaplasia may also be present.

The sputum from patients with chronic bronchitis is characteristically copious. If acute infection supervenes, the sputum may be purulent and bloodstained. Polymorphonuclear leucocytes and mucus strands predominate in the smears and Curshmann's spirals are commonly seen (*Figure 12.7*). Evidence of squamous metaplasia may be seen as discrete cells or clusters of cells which are smaller than mature squamous cells and have relatively large nuclei (*Figure 13.8*). Cells at various stages of maturity can be seen. In Papanicolaou-stained smears the cytoplasm may be cyanophilic, eosinophilic or even orangeophilic. The nuclei are usually uniform in size and shape. Nuclear atypia, sometimes severe, may be seen, and single abnormal cells may detach from the groups. Such cells may be precursors of malignancy and epidermoid carcinoma must be excluded.

Pulmonary oedema and hypertension

Some of the most striking cytological changes may be found in the pulmonary macrophages in sputum from patients with cardiac failure. Haemosiderin-laden macrophages may be found in cases of pulmonary oedema due to cardiac failure and volume overload. They reflect the presence of small intra-alveolar haemorrhages associated with pulmonary oedema.

Iatrogenic changes

Radiation therapy damages both squamous and bronchial epithelium and the effects may be very long lasting. The epithelial cells may show marked enlargement. The nuclei tend to appear structureless and are often wrinkled or pyknotic. Alternatively, enlarged nucleoli or a coarse chromatin pattern may be seen. Multinucleation and cytoplasmic vacuolation are often pres-

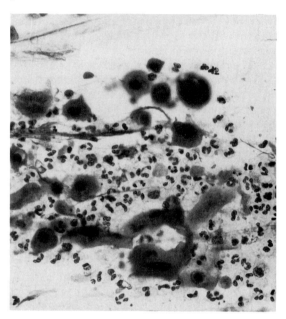

Figure 13.8 Small keratinized squamous cells with rounded pyknotic nuclei. These may reflect metaplastic change. There is some variation of nuclear size and shape and additional sputum samples should be examined to exclude squamous carcinoma. Papanicolaou stain. Magnification × 500

ent. Sometimes the cells may be so large, bizarre and intensely eosinophilic that they resemble malignant cells from an epidermoid carcinoma. A careful search will often reveal the remnants of terminal plates or cilia. The most striking effect of radiation is seen in squamous metaplastic cells. A history of radiation therapy should discourage a diagnosis of malignancy unless viable malignant cells unaffected by radiation changes are identified (Koss, 1979).

Patients undergoing treatment for chronic myeloid leukaemia with the alkylating agent busulphan may develop severe alterations of both the lung parenchyma and bronchial epithelium. Exfoliated cells show great cytoplasmic enlargement with corresponding enlargement of nuclei, which are hyperchromatic. Only an occasional affected cell is seen in the sputum of these patients but their extremely abnormal appearance may lead to a mistaken diagnosis of malignancy (*see also* Chapter 24).

Diffuse interstitial lung disease

Cytology has been particularly useful for the differential diagnosis of cases of diffuse interstitial lung disease. There are several disease entities classified together in this heterogeneous group. In a proportion of patients with interstitial lung disease, the cause of the condition is unknown. Bronchoalveolar lavage has been used to distinguish between those patients with idiopathic pulmonary fibrosis and those with diffuse interstitial lung disease of known cause. Both groups of patients have been found to have an inflammatory eosinophilic response which responds to steroid therapy. In contrast to the idiopathic lesions, the lavage fluid from cases classified as having a known aetiology has been found to contain a striking increase in the number of T lymphocytes suggesting an independent immune mechanism for the two conditions. This concept has been reinforced by the observation that bronchoalveolar lung lavage from patients with interstitial lung disease of known aetiology have high levels of IgM compared with cases of idiopathic pulmonary fibrosis.

Asbestos bodies

These may be found in sputum although it must be remembered that their presence is not pathognomonic of asbestosis (*Figure 12.8*). Haemosiderin and protein become encrusted on the fibre to form the characteristic dumbbell shaped body which in Papanicolaou-stained smears appears a golden brown colour. The diagnosis can be confirmed with a Perls' Prussian Blue stain.

Other lesions

In recent years there have been numerous reports of cytological investigations of less common benign diseases of the respiratory tract. The cytology of such diverse diseases as, for example, respiratory distress syndrome in neonates (Doshi *et al.*, 1982), pulmonary embolism (Bewtra, Dewar and O'Donahue, 1983) and squamous cell papilloma (Rubel and Reynolds, 1979) have been described.

D'Ablaing *et al.* (1975) have described the cytological changes seen in bronchopulmonary dysplasia of the newborn. This condition results from the use of positive pressure respirators and a high oxygen level in the treatment of respiratory distress syndrome. The progress of the condition can be charted by the changing cytological appearances of pulmonary secretions from initially intact and well preserved columnar cells, through degenerate columnar cells to metaplasia and dysplasia. The cytological changes are accompanied by altered X-ray appearances. The authors suggest that with careful monitoring by X-ray and cytology, high risk infants can be identified.

Cytological patterns in malignant lesions of the respiratory tract

The primary aim of respiratory tract cytology remains the diagnosis of malignant disease. Increasing use of non-invasive methods of investigation in cases of suspected lung cancer (Clee, Duguid and Sinclair, 1982) and the use of fine needle aspiration biopsy ensure that cytology continues to play an important role in both the diagnosis and typing of lung tumours. Accurate tumour typing is essential whenever possible as the different forms of tumour respond to different therapies.

Squamous carcinoma
Histology

The histological features of squamous carcinoma are keratinization, with formation of characteristic cell nests and intercellular bridges. At least one of these features should be present if the tumour is to be classified as squamous. Necrosis is a common feature within the cell masses formed in this tumour. Keratinizing and non-keratinizing forms of squamous carcinoma are recognized (*Figure 13.9*).

Cytology

Keratinizing epidermoid carcinoma (*Figures 13.10 and 13.11*) is recognized in Papanicolaou-stained smears by the presence of irregularly shaped squamous cells with dense brightly orangeophilic cytoplasm. This cytoplasm has a refractile quality which is very characteristic. The nuclei are usually pyknotic or contain dense irregular chromatin clumps and show striking pleomorphism. The shapes of these highly keratinized cells may be very bizarre and tadpole or spindle-shaped forms may be seen. It is not uncommon to find only two or three bizarre keratinized cells in a smear. In such cases, a tentative diagnosis of squamous carcinoma can be made. Anucleated keratinized cells also occur and these are often termed 'ghost' cells. They cannot be regarded as diagnostic of squamous carcinoma. When they are found, however, a careful search for similar cells with malignant nuclei should be undertaken. Often a few less well differentiated malignant cells will be found (Hughes and Dodds, 1968). Malignant cells from non-keratinizing tumours have dense, cyanophilic cytoplasm and dark staining, large nuclei with irregular chromatin and an irregular or notched nuclear outline. They tend to occur singly or in small loosely cohesive groups. In bronchial aspirates and brushings, the tumour cells tend to be less well differentiated and more uniform in pattern than in sputum. Nucleoli are also seen more frequently in the bronchial samples.

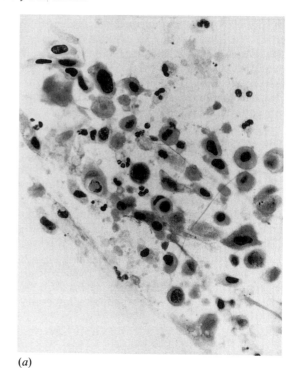

(a)

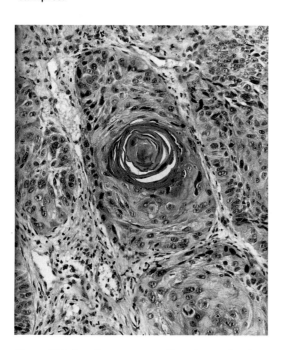

Figure 13.9 Invasive squamous cell carcinoma showing keratinization in the centre of the cell nest. Stain: haematoxylin and eosin. Magnification × 180, reduced to 65% in reproduction

As epidermoid carcinomas can attain large size, they are often necrotic and the background of smears may contain considerable numbers of degenerate cells mixed with an inflammatory exudate of polymorphonuclear leucocytes, lymphocytes and histiocytes. Whilst the predominant cell type is squamous, poorly differentiated and undifferentiated malignant cells may also occur with this tumour.

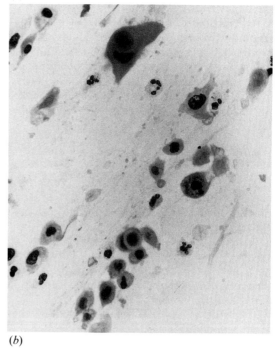

(b)

Figure 13.10 (a, b) A streak of abnormal highly keratinized squamous exfoliated into sputum from case of squamous carcinoma of bronchus. Note nuclear hyperchromasia, pleomorphism and bizarre cell shapes. There is also necrotic debris in the field composed of degenerating keratinized cells. Papanicolaou stain. Magnification × 500

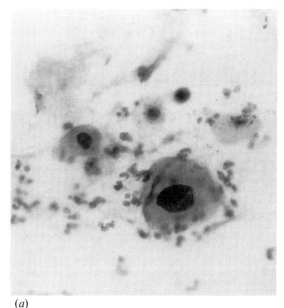

carcinoma *in situ*. The atypical metaplastic cells exhibit nuclear hyperchromasia and anisonucleosis and show a greater tendency to keratinization. The distinction of these atypical cells from those of carcinoma *in situ* is one of degree. The cells of a carcinoma *in situ* are larger, tend to lose their cohesiveness and are often frankly keratinized. In any case where such cells are found, a careful examination of the respiratory tract as described on p. 235 is essential to establish their source.

Adenocarcinoma

Histology

Histologically, the tumour consists of acinar structures composed of cuboidal or columnar cells (*Figure 13.12*) in which mucin production, if present, may be demonstrated by special stains. Bronchoalveolar carcinoma shows a more diffuse pattern of infiltration with the malignant cells growing round alveolar walls.

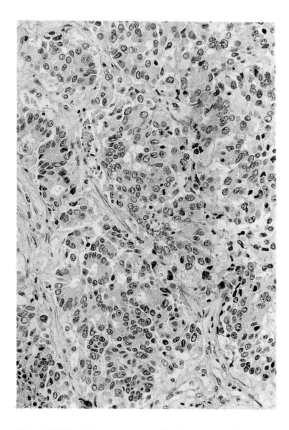

Figure 13.11 (*a, b*) Abnormal squamous cells with large irregular nuclei and coarse chromatin structure. There is abundant opaque cytoplasm. The findings are consistent with squamous carcinoma. Sputum. Papanicolaou stain. Magnification × 500

Carcinoma *in situ*

The morphology of the premalignant cells in smears has been described by Saccomanno *et al.* (1974). They range from atypical metaplastic cells to the abnormal cells from an area of

Figure 13.12 Adenocarcinoma showing some acinar formation. Stain: haematoxylin and eosin. Magnification × 200, reduced to 65% in reproduction

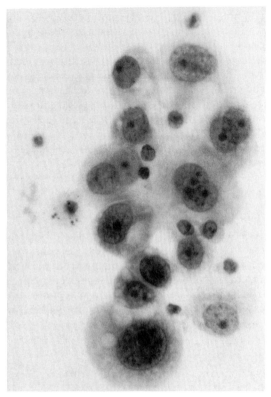

Figure 13.13 Differentiated malignant cells from an adenocarcinoma. This group of malignant cells shows the characteristics of adenocarcinoma with eccentric nuclei, prominent nucleoli and vacuolated cytoplasm. Sputum. Papanicolaou stain. Magnification × 400

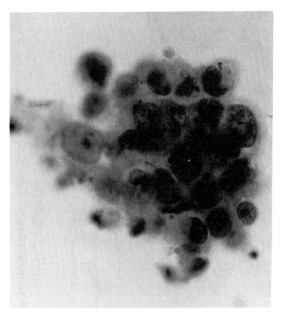

Figure 13.14 A cluster of poorly differentiated tumour cells in a sputum sample. Note abnormal chromatin structure. One cell contains a prominent nucleolus but it is impossible to determine on the basis of the cytological presentation of this tumour, whether these cells are derived from an adenocarcinoma or a squamous carcinoma. Papanicolaou stain. Magnification × 500

Cytology

Malignant cells from an adenocarcinoma (*Figure 13.13*) are usually present in considerable numbers, either as single cells or in groups. The groups often have a knobbly outline giving them a raspberry-like appearance. Individual malignant cells have enlarged, eccentric nuclei in which the chromatin pattern is strikingly abnormal. Nucleoli are invariably present and these are large, prominent and angular and may be multiple. The cytoplasm is delicate and faintly cyanophilic and foamy. The cytoplasm may be almost completely occupied by several large vacuoles which contain mucin. The amount of cytoplasm present depends upon the degree of differentiation of the tumour cells. It may be either abundant or relatively scanty (*Figure 13.14*).

Alveolar cell carcinoma (bronchoalveolar carcinoma) can occasionally be diagnosed from the cytological appearances of the neoplastic glandular cells in Papanicolaou-stained smears (*Figure 13.15*). Although these tumours are frequently peripheral, the diagnosis is most often made on sputum samples. The malignant cells are present in well demarcated, round or elongated groups which give the impression of considerable depth or thickness. There is often a characteristic mucus background. Individual malignant cells are round or oval, with round, uniform nuclei and prominent nucleoli. Chromatin is finely granular and of rather bland appearance. The cytoplasm, which is cyanophilic, is either clear or finely vacuolated. The cells are smaller than those of adenocarcinoma. The diagnosis of alveolar cell carcinoma is not often made as many of these tumours exfoliate cells which are indistinguishable from those of adenocarcinoma. A danger of misdiagnosing groups of reactive bronchial epithelial cells as alveolar cell carcinoma exists and must be borne in mind.

Large cell anaplastic carcinoma

Histology

By definition these tumours consist of large cells showing little differentiation but having an

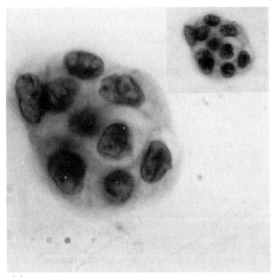

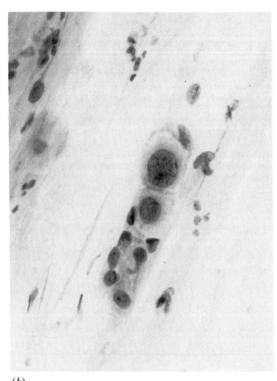

Figure 13.15 (*a*) Malignant cells from an alveolar cell carcinoma. Note the 'three-dimensional' appearance of the group and the relatively bland appearance of the nuclear chromatin. Sputum. Papanicolaou stain. Magnification ×1000 (inset ×400). (*b*) Bronchial cast in sputum from a case of bronchoalveolar cell carcinoma. Note anisonucleosis of the tumour cells which make up the cast. Papanicolaou stain. Magnification ×400

enlarged nucleus and prominent nucleolus. Most are considered to be dedifferentiated squamous cell carcinomas.

Cytology

Large cell anaplastic carcinomas generally exfoliate readily and sputum contains many malignant cells which may occur singly or in groups. The single cell pattern tends to occur more often in sputum samples whereas clumps are more usual in bronchial aspirate and brush biopsy samples. The scanty cytoplasm is often vacuolated with indistinct borders. This is seen in both the single cells and the cells in groups. The amount of cytoplasm present is very variable and may be completely absent. Nuclei are pleomorphic and often angular and adjacent nuclei may show moulding. Their chromatin pattern is very abnormal with irregular condensation of chromatin at the nuclear border. Nuclei are usually hyperchromatic but occasionally pale staining, abnormal nuclei can be seen.

Small cell anaplastic carcinoma (oat cell carcinoma)

Histology

The histology shows small uniform oval dark cells with very scanty cytoplasm which occur in solid masses and columns. In biopsy material the cells often appear squashed and traumatized and it can be very difficult to distinguish them from lymphocytes (*Figure 13.16*).

Cytology

Small cell anaplastic carcinoma cells (oat cells) tend to occur in clusters or in an elongated sheaf of cells which gives them their name (*Figure 13.17*). When the cells are arranged in tight clusters they show striking moulding of adjacent nuclei which is a diagnostic feature. These cells are slightly larger than the diameter of a normal lymphocyte. Nuclei are usually hyperchromatic, round or oval with abnormal chromatin patterns. In poorly preserved cells the chromatin structure may be difficult to discern and there may be condensation of chromatin at the nuclear border. The malignant cells have very scanty cyanophilic cytoplasm. Oat cells seen in brush biopsy specimens and fine needle aspirates show nuclear moulding to a greater extent than in other types of sample and the cells also appear considerably larger than they do in sputum. If concentration methods are used, the groups of oat cells tend to lose their cohesiveness and present a rather 'blown apart' appearance (Johnston and Frable, 1976). Small cell anaplastic tumours are often necrotic and the background of the smears may reflect this by the presence of cellular debris and inflammatory cells. Oat cells, however, may be seen in smears which are essentially 'clean' and free from inflammatory cells. Occasionally, these tumour cells are seen in very poor samples of sputum which are composed mainly of saliva.

Carcinoid

Histology

Carcinoid tumours show a characteristic packeted appearance of uniform small clear cells occurring in nests and islands in histological sections.

Cytology

These tumours are rarely diagnosed cytologically as they generally lie under the epithelium and, therefore, do not tend to exfoliate unless ulceration into the bronchus occurs. When this happens, the sputum may contain small cells, 10–20 µm in diameter, which resemble histiocytes or epithelial cells rather than cancer cells. Diagnosis of this lesion is much more likely after bronchial brushing or biopsy when the mucosa is penetrated. The uniform appearance of the tumour cells, their delicate cytoplasm and eccentric nuclei should raise the suspicion of carcinoid tumour. A retrospective study by Gephardt and Belovich (1982) stressed the simi-

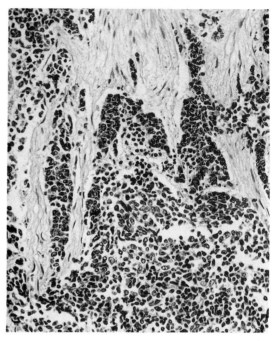

Figure 13.16 Small cell anaplastic carcinoma composed of small darkly staining cells with scanty cytoplasm. Stain: haematoxylin and eosin. Magnification ×270, reduced to 65% in reproduction

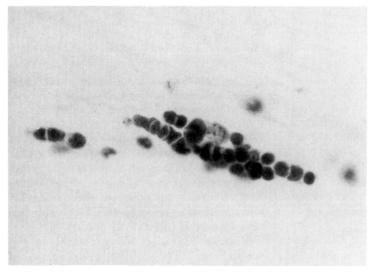

Figure 13.17 Small, undifferentiated malignant cells. This elongated group of anaplastic malignant cells shows the characteristics of oat cells. Note moulding between some of the adjacent nuclei. Sputum. Papanicolaou stain. Magnification ×400, reduced to 60% in reproduction

larity of carcinoid cells to normal respiratory epithelium. Atypical carcinoid tumours such as spindle cell types may occur (Craig and Finley, 1982).

Other neoplastic lesions of the respiratory tract

Cytology

Giant cell carcinoma usually resembles adenocarcinoma but occasionally can be recognized by the presence of large multinucleated cells roughly 2–3 times the size of other malignant cells. The dominant feature of mucoepidermoid carcinoma is that of the glandular component. The epidermoid cells tend to be single, isolated and poorly differentiated with rather dense, cyanophilic cytoplasm. These squamous cells may be attached to groups of adenocarcinoma cells. It is the juxtaposition of the two cell types which leads one to suspect a diagnosis of mucoepidermoid carcinoma. Much more rarely encountered primary lung tumours are lymphomas, both Hodgkin and non-Hodgkin, sarcomas and carcinosarcomas.

Metastatic tumours to the lung may be diagnosed in approximately 50% of cases (Koss, 1979). Sputum, bronchial brushings and fine needle aspiration biopsies tend to give the best results; bronchial washings and aspirates are less successful. If surgical tissue is available for comparison, it may be possible at least to state that the tumour cells are compatible with the primary tumour. (Johnston and Frable, 1976).

Evaluation of techniques

Accuracy of sputum cytology

Examination of sputum is a relatively simple test that can yield a positive diagnosis in approximately 50–80% of patients with lung cancer. The method is most successful in detecting central tumours of more than 2 cm diameter, but is much less so in detecting small peripheral tumours. It has the considerable disadvantage of not providing any information as to the location of the tumour and is dogged by a high percentage of unsatisfactory samples. These usually contain very little lower respiratory tract material, and consist mainly of saliva and may account for 20–25% of all 'sputum' samples submitted to the laboratory unless clinical staff receive proper instruction in specimen collection. Nevertheless, repeated sampling improves the diagnostic yield without distressing the patient.

At first glance, sputum cytology, together with chest radiography, appears to offer a means of screening the population at high risk of developing lung cancer such as heavy smokers with a history of chronic cough, workers in nickel, arsenic and asbestos industries and in uranium mining, and patients with chronic lung disease (Tao et al., 1982). Both techniques are capable of detecting presymptomatic lesions. Radiologically occult tumours comprise carcinoma *in situ*, microinvasive carcinoma and early invasive carcinoma less than 2 cm in diameter. Although radiology can detect approximately four times as many lesions as cytology, 12–15% of cytologically positive tumours are radiologically negative.

An excellent account of the problems encountered in a pulmonary cytology screening programme is given by Woolner who also reviews methods for the diagnosis, localization and treatment of occult lung tumours (Woolner, 1981). When a radiologically occult carcinoma is suspected as a result of cytological screening, very careful investigation is needed to establish the site of the lesion. Careful examination of the head and neck should be carried out to exclude malignant cells emanating from the tongue, tonsils, nasopharynx, larynx, epiglottis or upper oesophagus (Loke, Matthay and Ikeda, 1982). Bronchoscopy is essential and segmental and subsegmental bronchi must be systematically sampled by brushing, biopsy or differential segmental lavage to identify the source of the malignant cells. In particular, areas of abnormal mucosa, narrowing or blunting of spurs should be sampled. Tantalum bronchography and fluorescent bronchoscopy may facilitate localization.

Improved prognosis will result in some cases from early diagnosis particularly in patients with resectable lesions, but the extent of the protection given is hard to quantify. The introduction of bronchial cancer screening programmes cannot be justified on medical or economic grounds at present (Grant, 1982; Loke, Matthay and Ikeda, 1982; Feld, 1983). The predictive value of sputum cytology has been reviewed by Benbassat, Regev and Slater (1987) who state that sputum cytology should only be undertaken if the estimated positive predictive value is considered high enough for a positive diagnosis to bring about a change in the patient's management.

Accuracy of bronchial aspirate and bronchial brush cytology

Examination of specimens obtained at bronchoscopy is also a very efficient method of diagnosing malignancy. Bronchial aspiration and brush biopsy cytology used in conjunction will give positive results in up to 96% of cases (Muers et al., 1982). The factors which influence the success of cytological investigation of bronchial aspirates are similar to those affecting sputum cytology, i.e. the method is very successful in tumours over 2 cm diameter and located centrally. If the tumour is small or peripherally located, then the results are poor (Ng and Horak, 1983). Bronchial brush biopsies are less successful than bronchial aspirations in detecting peripheral lesions.

Accuracy of fine needle aspiration of lung

Fine needle aspiration biopsy is a very accurate method of diagnosing malignant tumours, most series reporting accuracy rates of between 85 and 90% (Dahlgren, 1974; Payne et al., 1981; Bonfiglio, 1982; Clee, Duguid and Sinclair, 1982; Pilotti et al., 1982; Michel, Lushpihan and Ahmed, 1983; MacMahon, Courtney and Little, 1983; Mitchell et al., 1984). The incidence of false positive diagnoses is low (1–2%). Previous radiotherapy, chemotherapy and diffuse lymphocytic infiltrations of the lung may result in errors of diagnosis.

The results of staging carcinoma of the bronchus by transbronchial fine needle aspiration have been shown to compare well with those of X-ray techniques for determining the presence of mediastinal metastases (Wang et al., 1983). Patients with abnormal mediastinal findings on X-ray or computerized tomography (CT scan) can benefit from firm diagnosis by this method as can those with a clinical history suggestive of mediastinal involvement. It can also be used to diagnose bronchogenic cysts (Schwartz et al., 1985) (see also Chapter 19).

Comparison of cytology and bronchial biopsy

Cytological investigation of neoplastic lesions of the respiratory tract offers several advantages over bronchial biopsy. Most notable among these is the ability to sample all areas of the respiratory tract including those which are inaccessible to the bronchoscope. Cytology provides a better means of sampling mucosa which shows either minimal abnormality or no abnormality or shows suspicious mucosal change at more than one site. Specimens can also be obtained from submucosal lesions. The tiny size of the biopsy often makes histological processing difficult, and the frequent occurrence of crush artefact often precludes accurate diagnosis (Clee, Duguid and Sinclair, 1982). Cytology is less likely to cause complications than biopsy. Haemorrhage is the major complication of biopsy and may occur in 9% of the patients. This figure rises to 29% in immunocompromised patients and to 45% in uraemic patients. Although bleeding associated with biopsy is usually slight, occasional deaths have occurred. Pneumothorax is a possible complication if the biopsy forceps puncture the visceral pleura (MacMahon, Courtney and Little, 1983).

Tumour typing by cytological methods

Respiratory tract cytology is a very successful method of typing both squamous carcinomas and adenocarcinomas, providing the tumours are well differentiated. Predictive typing of tumours is accurate in approximately 70% of squamous carcinomas, 95–100% of small cell anaplastic carcinomas and 33–86% of adenocarcinomas. The distinction between large cell undifferentiated carcinomas and squamous and adenocarcinomas gives rise to the greatest difficulties in tumour typing.

Where tumours are less well differentiated the correlation is much less accurate and cytology and histology may attribute these tumours to quite separate classifications. There are three possible reasons for this discrepancy. Firstly, a tumour may show different degrees of differentiation within its substance. Moreover, cytological samples are more likely to be derived from necrotic areas of tumour which exfoliate readily, while histological samples tend to be taken from better preserved areas. A second factor is that a tumour may be composed of more than one type of cell, i.e. both squamous carcinoma and adenocarcinoma may be present within the same tumour. Thirdly, carcinoma *in situ* cannot be accurately diagnosed by cytology and may be interpreted as an invasive carcinoma.

Accurate typing of lung tumours is essential whenever possible as the treatment and/or prognosis differs for the various tumour types. Squamous carcinoma tends, for example, to be detected relatively early as it causes early symptoms. It tends to be a rapidly growing and aggressive tumour in which surgery is the treatment of choice. Adenocarcinoma, on the other hand, is often a peripheral lesion and is not usually detected until relatively late as it does not give rise to early symptoms. It is generally less aggressive and more slowly growing than squamous carcinoma. The treatment is by surgery in suitable cases. It is very important that small cell anaplastic carcinoma (oat cell carcinoma) is distinguished from other poorly differentiated or anaplastic tumours as chemotherapy (often combined chemotherapy) is the treatment of choice. Surgery and radiotherapy are contraindicated. Recent work involving immunostaining of sputum samples for neuron-specific enolase and keratin has shown that these markers are a valuable aid to diagnosis in cases where definite diagnosis of small cell carcinoma cannot be made on the basis of morphology alone (Kyrkou et al., 1986).

Hess, McDowell and Trump (1981) in their review article on the status of cytological typing of lung tumours stress the importance of detail in differentiating tumour cells. Poorly differentiated epidermoid carcinomas and adenocarcinomas can be accurately differentiated if such features as tonofilament bundles, which represent epidermoid differentiation, or secretory granules or well developed Golgi apparatus, which suggests adenocarcinoma, are taken into account. These and other features can be recognized in the electron microscope. However, their presence may be suspected from the Papanicolaou-stained smears. Epidermoid differentiation is suggested by features such as a glassy appearance and/or dense concentric rings or fragments which represent the tonofilament bundles. Adenocarcinomatous differentiation is indicated by foamy cytoplasm, which corresponds with the secretory granules and well developed Golgi apparatus, and cytoplasmic basophilia which represents a well developed endoplasmic reticulum. If close attention is paid to these and other features, more accurate classification of poorly differentiated carcinomas can be achieved.

References

BENBASSAT, B., REGEV, A. and SLATER, P. E. (1987). Predictive value of sputum cytology. *Thorax*, **42**, 165–172

BEWTRA, C., DEWAR, N. and O'DONAHUE, W. (1983). Exfoliative sputum cytology in pulmonary embolism. *Acta Cytol.*, **27**, 489–496

BONFIGLIO, T. A. (1982). Fine needle aspiration biopsy of the lung. *Pathol. Ann.*, **16**, 159–180

CAMPBELL, G. K. and MCCORRISTON, J. (1987). *Pneumocystis carinii* in bronchoalveolar lavage (letter). *J. Clin. Pathol.*, **40**, 354

CLEE, M. D., DUGUID, H. L. D. and SINCLAIR, D. J. M. (1982). Accuracy of morphological diagnosis of lung cancer in a department of respiratory medicine. *J. Clin. Pathol.*, **35**, 414–419

CORRIN, B. and SPENCER, H. (1981). Some aspects of pulmonary pathology. In *Recent Advances in Histopathology, II*, (Anthony, P. P. and MacSween, R. N. M., eds), pp. 83–98. Edinburgh: Churchill Livingstone

CRAIG, L. D. and FINLEY, R. J. (1982). Spindle-cell carcinoid tumour of the lung. *Acta Cytol.*, **26**, 495–498

D'ABLAING, G., BERNARD, B., ZAHAROV, I., BARTON, L., KAPLAN, B. and SCHWINN, C. P. (1975). Neonatal pulmonary cytology and bronchopulmonary dysplasia. *Acta Cytol.*, **19**, 21–27

DAHLGREN, S. (1974). Lungs. In *Aspiration Biopsy Cytology. Part 1. Cytology of Supradiaphragmatic Organs*, (Zajicek, J., ed.), Chapter 8. Basel: S. Karger

DOSHI, J., KANBOUR, A., FUJIKURA, T. and KLOINSKY, B. (1982). Tracheal aspiration cytology in neonates with respiratory distress. *Acta Cytol.*, **26**, 15–21

DUNNILL, M. S. (1982). *Pulmonary Pathology*. Edinburgh: Churchill Livingstone

EDITORIAL (1985). *Pneumocystis carinii* pneumonia. *Thorax*, **40**, 561–570

FELD, R. (1983). Symposium on the diagnosis and treatment of common cancers. 1. Lung cancer—1983. *Can. J. Surg.*, **26**, 266–268

FLEURY, J., ESCUDIER, E., POCHOLLE, M. J., CARRE, C. and BERNAUDIN, J. F. (1985). Cell population obtained by bronchoalveolar lavage in *Pneumocystis carinii* pneumonitis. *Acta Cytol.*, **29**, 721–726

GAURNER, J., ROBEY, S. S. and GUPTA, P. K. (1986). *Pneumocystis carinii*: a comparison of Papanicolaou and other histochemical stains. *Diag. Cytopathol.*, **2**, 133–137

GEPHARDT, G. N. and BELOVICH, D. M. (1982). Cytology of pulmonary carcinoid tumours. *Acta Cytol.*, **26**, 434–438

GHALI, V. S., GARCIA, R. L. and SKOLOM, J. (1984). Fluorescence of *Pneumocystis carinii* in Papanicolaou stained smears. *Human Pathol.*, **15**, 907–909

GRANT, I. M. B. (1982). Screening for lung cancer. *Br. Med. J.*, **284**, 1209–1210

HEATH, D. and KAY, J. M. (1980). Respiratory system. In *Muir's Textbook of Pathology*, (Anderson, J. R., ed.), pp. 432–503. London: Edward Arnold

HESS, F. G., MCDOWELL, E. M. and TRUMP, B. F. (1981). Pulmonary cytology; current status of cytological typing of respiratory tract tumours. *Am. J. Pathol.*, **103**, 323–333

HUGHES, H. E. and DODDS, T. C. (1968). *Handbook of Diagnostic Cytology*, Chapters 6, 7, 8 and 23. Edinburgh: E & S Livingstone

JOHNSTON, W. W. and FRABLE, W. J. (1976). The cytopathology of the respiratory tract. *Am. J. Pathol.*, **84**, 372–413

KIM, H. and HUGHES, W. T. (1973). Comparison of methods for identification of *Pnuemocystis carinii* in pulmonary aspirates. *Am. J. Clin. Pathol.*, **60**, 462–466

KOSS, L. G. (1979). *Diagnostic Cytology and its Histopathologic Bases*, 3rd Edition, Volume 2, Chapters 19 and 21. Philadelphia: J. B. Lippincott

KOVACS, J. A., GILL, V., SWAN, J. C. *et al.* (1986). Prospective evaluation of a monoclonal antibody in diagnosis of *Pneumocystis carinii* pneumonia. *Lancet*, **2**, 1–3

KYRKOU, K. A., IATRIDIS, S. G., ATHANASSIADOU, P. P., LAMBROPOULOU, S. and LIOSSI, A. (1986). Immunodetection of neuron-specific enolase and keratin in cytological preparations as an aid to the differential diagnosis of lung cancer. *Diag. Cytopathol.*, **2**, 217 220

LOKE, J., MATTHAY, R. A. and IKEDA, S. (1982). Technique for diagnosing lung cancer. *Clin. Chest Med.*, **21**, 321–329

MACMAHON, H., COURTNEY, J. V. and LITTLE, A. G. (1983). Diagnostic methods in lung cancer. *Semin. Oncol.*, **10**, 20–33

MICHEL, R. P., LUSHPIHAN, A. and AHMED, M. N. (1983). Pathological findings of transthoracic needle aspiration in the diagnosis of localized pulmonary lesions. *Cancer*, **51**, 1663–1672

MITCHELL, M. L., KING, D. E., BONFIGLIO, T. A. and PATTEN, S. F. (1984). Pulmonary fine needle aspiration cytopathology. A five year correlation study. *Acta Cytol.*, **28**, 72–76

MUERS, M. F., BODDINGTON, M. M., COLE, M., MURPHY, D. and SPRIGGS, A. I. (1982). Cytological sampling at fibreoptic bronchoscopy: comparison of catheter aspirations and brush biopsies. *Thorax*, **37**, 457–461

NAIB, Z. (1970). *Exfoliative Cytology*, Chapter 11. Boston: Little, Brown and Co.

NASIELL, M., ROGER, V., NASIELL, K., ENSTAD, I., VOGEL, B. and BISTHER, A. (1972). Cytological findings indicating pulmonary tuberculosis. 1. The diagnostic significance of epitheliod cells and Langhan's giant cells found in sputum or bronchial secretions. *Acta Cytol.*, **16**, 146–151

NAYLOR, B. (1962). The shedding of the mucosa of the bronchial tree in asthma. *Thorax*, **17**, 69–72

NG, A. B. P. and HORAK, G. C. (1983). Factors significant in the diagnostic accuracy of lung cytology in bronchial washings and sputum samples. II. Sputum samples. *Acta Cytol.*, **27**, 397–402

OGNIBENE, F. P., SHELHAMER, J., GILL, V. *et al.* (1984). The diagnosis of *Pneumocystis carinii* pneumonia in patients with the acquired immunodeficiency syndrome using subsegmental bronchoalveolar lavage. *Am. Rev. Resp. Dis.*, **129**, 929–932

ORENSTEIN, M., WEBBER, C. and HEURICH, A. E. (1985). Cytologic diagnosis of *Pneumocystis carinii* infection by bronchoalveolar lavage in acquired immune deficiency syndrome. *Acta Cytol.*, **29**, 727–731

PAYNE, C. R. HADFIELD, J. W., STOVIN, P. G., BARKER, V., HEARD, B. E. and STARK, J. E. (1981). Diagnostic accuracy of cytology and biopsy in primary bronchial carcinoma. *J. Pathol.*, **34**, 773–778

PILOTTI, S., RILKE, F., GRIBAUDI, G. and DAMASCELLI, B. (1982). Fine needle aspiration biopsy cytology of primary and metastatic pulmonary tumours. *Acta Cytol.*, **26**, 661–666

PINTOZZI, R. S. (1978). Modified Grocott's methenamine silver nitrate method for quick staining of *Pneumocystis carinii*. *J. Clin. Pathol.*, **3**, 803–805

PITCHENIK, A. E., GANJEI, P., TORRES, A. *et al.* (1986). Sputum examination for the diagnosis of *P. carinii* pneumonia in the acquired immune deficiency syndrome. *Am. Rev. Resp. Dis.*, **133**, 226–229

ROBBINS, S. L. and COTRAM, R. S. (1979). The respiratory system. In *Pathologic Basis of Disease*, pp. 814–885. Philadelphia: Saunders

ROBICHEUX, G., MOINUDDIN, S. M. and LEE, L. H. (1985). The role of aspiration biopsy cytology in the diagnosis of pulmonary tuberculosis. *Am. J. Clin. Pathol.*, **83**, 719–722

RUBEL, L. R. and REYNOLDS, R. E. (1979). Cytological description of squamous papilloma of the respiratory tract. *Acta Cytol.*, **23**, 227–230

SACCOMANNO, G., ARCHER, V. E., AUERBACH, O., SAUNDERS, P. P. and BRENNAN, L. M. (1974). Development of carcinoma of the lung as reflected in exfoliated cells. *Cancer*, **33**, 256–270

SCHWARTZ, D. B., BEALS, T. F., WIMBISH, K. *et al.* (1985). Transbronchial fine needle aspiration of bronchogenic cysts. *Chest*, **88**, 573–575

TAO, L. C., CHAMBERLAIN, D. W., DELARUE, N. C., PEARSON, F. G. and DONAT, E. E. (1982). Cytological diagnosis of radiologically occult squamous cell carcinoma of the lung. *Cancer*, **50**, 1580–1586

VERMA, K., SANDHYAMANI, S. and PANDE, J. N. (1983). Cytological diagnosis of pulmonary tuberculosis by bronchoalveolar lavage. *Acta Cytol.*, **27**, 211–212

WALLACE, J. M., BATRA, P., GONG, H. *et al.* (1985). Percutaneous needle lung aspiration for diagnosing pneumonitis in the patient with the acquired immunodeficiency syndrome (AIDS). *Am. Rev. Resp. Dis.*, **131**, 389–392

WANG, K. P., BROWER, R., HAPONIK, E. F. and SIEGLEMAN, S. (1983). Flexible transbronchial needle aspiration for staging of bronchial carcinoma. *Chest*, **84**, 571–576

WOOLNER, L. B. (1981). Recent advances in pulmonary cytology: early detection and localization of occult lung tumours. In *Advances in Clinical Cytology*, (Koss, L. and Coleman, D. V., eds), Chapter 4. London: Butterworths

YOUNG, J. A. (1985). Infections and infestations. In *Colour Atlas of Pulmonary Cytopathology*, Chapter 3. London and Oxford: Harvey Miller and Oxford University Press

YOUNG, J. A., HOPKIN, J. M. and CUTHBERTSON, W. P. (1984). Pulmonary infiltrates in immunocompromised patients: diagnosis by cytological examination of bronchoalveolar lavage fluid. *J. Clin. Pathol.*, **37**, 390–397

YOUNG, J. A., STONE, J. W., MCGONIGLE, R. J. S., ADU, D and MICHAEL, J. (1986). Diagnosing pneumonia by cytological examination of bronchoalveolar lavage fluid: report of 15 cases. *J. Clin. Pathol.*, **39**, 945–949

14

Serous effusions

Michael M. Boddington

Accumulation of fluid in the serous cavities is a reflection of local or systemic disease, and examination of the cells in the fluid can be of value in establishing the underlying disease process. Light microscopic analysis of smears or cell blocks prepared from the accumulated fluid can reveal whether the disease process is benign or malignant, inflammatory or reactive and is one of the most useful tests available for the investigation of the patient with a serous effusion. It is a test that has been in use for many years (Grunze and Spriggs, 1983). A review of the literature in 1903 indicates that most of the early pathologists prepared smears from the deposit rather than cell blocks and used the haematological stains which had been developed by Ehrlich and Romanowsky very much as they do today. A method of staining wet-fixed smears with eosin–orange–haematein was also described which preceded the Papanicolaou stain used today.

Anatomy, physiology and histology of the serous membranes

The serous membranes (the peritoneum, the pleura and the pericardium) are thin layers of loose connective tissue covered on their free surface by a simple layer of flat or low cuboidal cells designated the mesothelium. The membranes line the walls of the body cavities (the parietal layer) and the viscera (the visceral layer). The parietal and visceral layers are separated by a potential space which under normal conditions is filled with a small amount of fluid so that the layers glide smoothly on each other. In the healthy individual, the amount of fluid present is too small to be aspirated, although a small amount of free fluid may collect in the pelvic cavity in the absence of disease.

The movement of fluid across the pleural space is determined by several factors, the most important of which are the colloid osmotic pressure exerted by the plasma proteins, and the hydrostatic pressure within the capillary lumen. Other factors affecting fluid movement are capillary permeability and lymphatic drainage. Normally these are in equilibrium, but a change in one or more factors due to disease may result in accumulation of fluid until a new equilibrium is reached. There is normally a continuous removal of fluid from the cavity at the rate of 500 ml/h. The visceral pleural capillary bed has a large capacity to absorb protein-free fluid; protein removal is by the lymphatic system.

The majority of cells present in normal fluid are a degenerative form of mesothelial cell accompanied by a few lymphocytes although some aspirates may be contaminated with more active mesothelial cells and fragments of mesothelium mechanically detached during the procedure.

Specimen collection

Specimens should be collected into a sterile container and transferred to the laboratory as soon as possible. If delay is anticipated, the specimen can safely be stored at 4 °C for up to four days. A satisfactory deposit can be obtained with as little as 10 ml of fluid.

Anticoagulants commonly employed for a 20 ml sample are: 2 ml of 3.8% sodium citrate, 20 mg of EDTA (ethylenediaminetetraacetic acid) or 2 mg (200 units) of heparin. Sodium citrate has the disadvantage of being in solution which can become contaminated or leak, whereas EDTA can be sterilized in solid form and is more conveniently stored. Certain other anticoagulants such as oxalate destroy white cells and must not be used.

A note should be made of the appearance of the fluid as bloodstained samples require special preparatory techniques.

Preparation

Samples which are not bloodstained should be transferred to a glass or plastic conical tube and centrifuged at 1500 rev/min for 5 min. In the majority of cases a deposit will be formed. The button of cells will consist largely of nucleated cells with relatively few red cells. It is then a matter of choice how these cells are prepared for examination. Some laboratories use the cell block technique and fix this button of cells in a perspex tube which can be readily dissolved from the embedded deposit. This method is particularly appropriate if the cells are to be studied in the electron microscope. Other laboratories use the cytocentrifuge, although cytocentrifugation of the deposit without dilution often produces smears which are too concentrated. Yet another approach is the preparation of fresh unfixed smears for phase contrast examination or for staining by polychrome methylene blue as quick methods of examination.

In this laboratory we prepare permanent direct smears from the deposit; both wet-fixed and air-dried slides are made. Usually, no more than two smears of each type of preparation are required for diagnostic purposes but it is recommended that spare smears be made for special stains.

Air-dried smears

These are prepared for staining by a Romanowsky method where the Giemsa stain is the most important and widely used. In order that the cells take up the stains differentially, they must be spread thinly so that drying is instant. The deposit therefore must be as concentrated as possible. This is achieved by decanting the supernatant and inverting the tube so that excess fluid drains away from the cell button. A small sample of the cells is removed by inserting a wire loop or Pasteur pipette into the deposit.

Only a pinhead sized sample is necessary for the preparation of satisfactory smears. This is placed on a glass slide, and spread quickly and evenly to obtain thin, quick-drying areas. The sample can be spread as for blood and marrow films using another glass slide to sweep the cells across the slide. If, as is often the case, the concentrate is too sticky for this approach, the cells can be drawn out into a single streak with the pipette and the thick portions of the streak drawn out in single strokes to produce short thin areas of quick-drying cells. Care must be taken not to go over areas which have already dried. If small clots are present these can be swept to and fro across the slide to remove excess fluid and deposit cells contained in the clot. Suitable thin smears will be formed in the tail of the sweep.

Wet-fixed smears

After making dry films, a small amount of supernatant fluid should be added to the deposit to dilute the cells for wet-fixed smears. The cells can be spread in a similar fashion as for dry smears but must be fixed instantly in alcohol, whilst still wet. Most cytology laboratories stain these smears by the Papanicolaou method, although histopathologists may prefer a haematoxylin and eosin stain.

Heavily bloodstained fluids and clotted samples

Too many red cells will dilute the white cell population and interfere with the search for abnormal cells in smears. Therefore it is important to avoid excessive red blood cells in the preparation of bloodstained samples. Similarly, tumour cells may be trapped in the small clots and these, too, should be processed separately.

Methodology for bloodstained samples

Bloodstained samples which give a heavy deposit of red blood cells after centrifugation require special care. The supernatant fluid must be decanted by pipette. If many white cells are also present, these may be seen as a creamy 'buffy coat' on the surface of the deposit. Gently remove this layer by pipette and centrifuge the aspirate again in a clean tube so that smears can be made in the usual way. Search among the red cells remaining from the first centrifugation with a platinum loop for small pieces of fibrin which are almost always present. These clots can then be swept across the slide in a close zigzag pattern to prepare thin, quick-drying smears. Several slides may be required before well spread smears are obtained.

These methods involve less manipulation and will give better preparations than flotation techniques. For details of methods of preparing bloodstained fluids see Chapter 5.

Advantages and disadvantages of different methods

Cell block technique

Although histological sections are excellent for displaying organized structures such as papillary clusters of mesothelial or adenocarcinoma cells, it is often difficult to distinguish between benign and malignant structures and almost impossible to recognize discrete free-cell varieties of tumour.

Wet-fixed smears

This method is widely used in many laboratories but is associated with a risk of contamination between samples. It is well known that tumour cells readily float from one smear to another. Moreover, the presence of a large number of red cells masks the staining of other cells and the detection of small free tumour cells such as oat cells and some lymphomas in bloodstained smears can be very difficult.

The Papanicolaou counterstains were designed to demonstrate the maturation of squamous cells which are rarely seen in serous fluids and are, therefore, largely irrelevant to these preparations. This method does, however, serve to confirm the presence of keratinized squamous cells on those rare occasions when these are present and can sometimes display the papillary collections of adenocarcinoma beautifully. The wet preparations often deteriorate with age and the staining effect can fade considerably on storage. Alcohol fixation preserves the nuclear detail so that the interpretation of cell morphology is facilitated.

Air-dried smears

When prepared properly, the staining is permanent and never fades on storing. The presence of red cells in large numbers does not interfere with the staining reaction of the nucleated cells. Since most of the benign cells present are of haemic origin and the mesothelial cells not of epithelial type, a haematological method is more appropriate. These cells are also small and the flattening effect of air drying displays the difference in size between benign and malignant forms more readily. The cell content, particularly the various granules of polymorphs and cytoplasmic basophilia, are also well shown.

The criteria for the interpretation of malignancy in air-dried smears are completely different from those applied to Papanicolaou-stained smears, so that additional training in the interpretation of the Giemsa-stained smears is necessary. Although nuclear hyperchromasia is not a feature of Giemsa smears, chromatin structure is well shown and subtle degrees of cytoplasmic basophilic staining may be seen which are not present in Papanicolaou smears. The Romanowsky stains permit greater diagnostic accuracy in many of the free cell types of tumour, e.g. lymphoma, myeloma, as well as allowing an easier distinction between some of the well differentiated benign and malignant papillary structures.

Types of effusion

Effusions are usually classified clinically into transudates or exudates according to the specific gravity or protein content of the fluid. Classically, these are distinguished by a protein value greater or less than 30 g/l, the higher values being found in exudates whilst low values constitute a transudate. Transudates imply a systemic cause such as high venous pressure (as in cardiac failure) or hypoproteinaemia (due to renal failure), whilst exudates suggest a local inflammatory cause.

The value of using protein estimation to distinguish between the two types of effusion has been questioned and various investigations have shown that there is a 30% overlap in the diagnostic correlation with the disease process using a cut off point of 30 g/l. This finding is common to many quantitative estimations where experienced observers may expect to find only an 80% diagnostic correlation with the mean value.

Transudates

Cytologically, the classic feature of transudates with a low protein value is the very small deposit because of the low cell content consisting mainly of degenerating mesothelial cells and macrophages. This picture is commonly seen in ascites due to renal failure or cirrhosis of the liver, but is rare in pleural fluids. Congestive cardiac failure often is associated with a pleural transudate where the protein value is high possibly due to a local complication such as pulmonary congestion or infarction. The cell picture in these cases is more like that of an exudate.

Exudates

For practical purposes, a discussion on the cytology of effusions can be confined to the appearances seen in exudates of which two main causes can be defined: those due to inflammatory processes and those due to malignant cell infiltration. Approximately one-half of the fluids seen in a routine district hospital laboratory are associated with malignant disease whilst the remainder may be due to acute or chronic inflammation of the organs contained within the cavities. Bacterial infection causing empyema or lymphocytic effusions is not commonly seen, so that the great majority of all fluids are sterile.

The cells of effusions

Benign cells

Leucocytes

The predominant cells in most of the benign and many of the malignant fluids are leucocytes. These have exactly the same morphology as those in blood films but are present in different proportions. Neutrophil, eosinophil and basophil polymorphonuclear granulocytes occur commonly. Very occasionally, primitive cells of the myeloid series can be found and the unexpected appearances of these may be confused with single tumour cells of small cell type. Lymphocytes are equally common and plasma cells may also be found.

Plasma cells are classically described as having a 'cartwheel' nucleus because of the regular chromatin clumping which forms blocks like spokes of a wheel. In practice, this is rarely seen in Giemsa-stained films and these cells are clearly recognized by the deep blue cytoplasm which contains a clear zone adjacent to the nucleus which represents the site of a large Golgi apparatus (*Plate 12*). In Papanicolaou stains, this zone is a pale apple green in colour. Mitoses can be frequent and care must be taken not to confuse dividing plasma cells with malignant cells.

Macrophages

Sometimes cells identical to the monocytes seen in blood are present, but it is much more usual to see a vacuolated form of a histiocyte which is much larger than the monocyte. It is very likely that some of these are derived from blood, but others no doubt migrate from the subpleural connective tissue whilst others are almost certainly altered mesothelial cells which have been shed into the fluid.

A whole range of cells may be seen including a degenerative type of mesothelial cell with frayed cytoplasm and cells filled with vacuoles of varying size with or without phagocytosed cell debris or blood pigment (*Plate 13*). Other cells distended by a large single hydropic vacuole confer a 'signet-ring' appearance on the macrophages. The nucleus in these cells is flattened by the vacuole to the shape of a new moon (*Plate 14*). The phagocytic forms are common in bloodstained effusions due to direct trauma, whilst the signet-ring forms are common in transudates.

Macrophages are readily distinguished in Romanowsky-stained preparations where the cytoplasmic staining is very pale, almost colourless. In contrast, the cytoplasm of cells of mesothelial type stains various shades of blue. In Papanicolaou preparations the counterstains are not so useful and it is not possible to distinguish macrophages from mesothelial cells. Electron microscopy reveals a clear difference between the two types of cell. Macrophages have complex surface folds and ridges forming ruffles, whereas mesothelial cells have only irregular short microvilli.

Mesothelial cells

These cells are usually found in smears as discrete cells or in small clusters. They vary in size from 15–30 μm and binucleate or multinucleate forms may be seen. When the mesothelium becomes thickened and multilayered, the number of mesothelial cells in the fluid may be very large indeed. The cells may undergo proliferation in the fluid so that mitotic figures may be seen.

It is this benign but excessive proliferation of the mesothelial cell in serous fluid which has made the cytology of effusions notoriously difficult. Once the appearances of these benign changes are mastered, the distinction between these and most of the malignant appearances they can resemble becomes relatively simple. The distinction between reactive mesothelial cells, macrophages and malignant cells is difficult in Papanicolaou smears. Romanowsky-stained preparations make this distinction much easier. Active mesothelial cells with basophilic cytoplasm stain bright blue and as the cells age or change to histiocytic forms, so the staining reaction of the cytoplasm becomes paler. A complete range of cell types from the most active with bright blue cytoplasm, to the 'degenerative' form with dull pale blue staining, to the histiocytic form which is colourless may be present (*Plate 15*).

Mesothelial cells are usually uniform in size and shape with a slightly eccentric nucleus occupying about half of the cell diameter. Their uniformity is apparent even when they are present in sheets or clusters and it is this uniformity which distinguishes them from most malignant cells. The presence of benign mesothelial cells showing a marked cellular pleomorphism and nuclear irregularity is generally quite rare, and in benign conditions comprises only a small proportion of the total mesothelial cell population. Sometimes papillary clusters or rosettes of mesothelial cells are found. These contain a central core of pink staining collagen which may be the connective tissue element of the tip of a papillary frond. This collagen core may calcify and form a psammoma body which, although usually associated with malignancy, is occasionally seen in benign effusions.

Clustering and rounded collections of mesothelial cells are much more apparent in Papanicolaou-stained smears than in Giemsa preparations where these cells form loose sheets. Occasionally brush borders of microvilli can be seen in Giemsa-stained films (*Plate 16*); cannibalistic activity where one cell ingests another is sometimes present, and mitoses are not infrequent when active cells are present. Such cells also have prominent nucleoli.

Sometimes extremely degenerate forms of mesothelial cells are present, somewhat resembling plasma cells with bizarre or fragmented nuclei and cytoplasmic vacuolation or droplets. These are more readily identified in Giemsa-stained smears and are known as Mott cells.

In certain conditions, mesothelial cells are remarkable by their absence from the fluid. In effusions due to tuberculosis or rheumatoid disease, the mesothelium is often covered by a fibrinous exudate and mesothelial cells can no longer be shed into the fluid.

Other benign cells

It is quite common in Giemsa-stained smears to see discrete tissue mast cells, which can be recognized by their content of intensely basophilic granules which are not readily displayed in Papanicolaou-stained smears.

Malignant cells

Malignant cells in effusions are most commonly metastatic from primary sites elsewhere in the body. Usually the morphological changes are so obviously grossly abnormal that a diagnosis of malignancy can be made with absolute certainty.

Because no two tumours are exactly the same and the tumour cells are unique to each case, there can be no standard descriptive appearance to cover all the cellular patterns seen. The one rewarding feature of fluid cytology is that unusual appearances, unique even to the expert, are constantly being seen. The main characterisitics of malignant cells in effusions are summarized below.

Adenocarcinoma: papillary and acinar forms

The majority of tumour cell types seen in effusions fall into this category. The aggregates may be irregular or rounded and papillary in shape or gland-like hollow balls of cells (*Plate 17*). They can be composed of relatively few cells of uniform size and shape or reach enormous size containing thousands of cells. Such large aggregates of cells are visible to the naked eye as cloudy shimmering particles in the fluid when it is held up to the light. They may be recognized during smear preparation as an exceedingly granular deposit and produce a mottled staining effect at the edges of the smear when the smear is stained. Individually, the cells may resemble mesothelial cells but usually the cytoplasmic basophilia is much less marked and the cells are generally larger in size. The nucleocytoplasmic ratio is greater than that found in normal mesothelial cells. These features are more readily displayed in Romanowsky-stained preparations than in Papanicolaou-stained smears.

Other characteristics of papillary clusters of adenocarcinoma cells are nuclear pleomorphism, marked nuclear enlargement, the presence of giant nucleoli, and in many instances the presence of vacuoles in the cytoplasm of the cell. These large single vacuoles give an extremely swollen appearance to the cell and in this respect they are more like macrophages than mesothelial cells. Unlike macrophages, however, the giant vacuoles rarely compress the nucleus to a crescent shape as in signet-ring cells but usually overlie a portion of the nucleus which largely retains its spherical shape.

A useful diagnostic feature is the position of the nuclei in the tumour cells which make up the papillary clusters seen in malignant effusions. They are usually peripherally placed around the sphere unlike similar aggregations of mesothelial cells. Vacuolated macrophages never form papillary collections which could be confused with adenocarcinoma.

Adenocarcinoma: free cell form

About one-quarter of all effusions due to metastatic adenocarcinoma contain few aggregates of tumour cells and are composed almost entirely of discrete cells (*Plate 18*). Again, no two are exactly the same and some contain pleomorphic vacuolated cells similar to those seen in papillary populations. More commonly the discrete tumour cells are remarkably uniform, sometimes almost indistinguishable from mesothelial cells. It is this type which may be given a false negative diagnosis by all but the most experienced observer. Fortunately this type is rare. The most striking feature in these cases is the immediate appearance of cell homogeneity with an apparent pure culture of 'mesothelial cells'. This picture alone is suspect as mesothelial cells are always accompanied by leucocytes. Usually, however, the careful observer will recognize that the cells are larger than mesothelial cells and that the nuclear size is greater. The nucleocytoplasmic ratio may also be very high even when the cells are no larger than normal mesothelial cells.

Adenocarcinomas are frequently mucus-secreting and a small discrete central vacuole is present in many cells. The presence of mucin can be demonstrated by using the Periodic Acid Schiff (PAS) stain after diastase digestion of the glycogen, thereby confirming the malignant nature of the cell. Direct PAS staining will also reveal varying amounts of glycogen in tumour cells. These appear as variably sized granular masses at the periphery of the cell and never aggregate as spherical masses like mucin. Glycogen is of no value as a marker of malignancy as it can be found in both mesothelial cells and tumour cells.

Often adenocarcinoma cells display a fringe of microvilli which stain pink in Giemsa films but are not visible in Papanicolaou-stained preparations. Since mesothelial cells can also be seen to have microvilli displayed as a regular fine fringe around the cytoplasmic membrane, this is not a diagnostic finding. However, larger tufts of microvilli in localized collections are more characteristic of adenocarcinoma (*Plate 19*). These often appear as ciliated tufts projecting from the tumour cell. Extreme forms of this phenomenon resemble the guardsman's bearskin, exceeding the size of the parent cell. These cells have been seen in females with papillary cystadenocarcinoma of the ovary (*Plate 19*).

One of the limitations of cytological investigation of serous effusions is that it is not possible to distinguish between primary and metastatic adenocarcinoma in Papanicolaou or Giemsa-stained smears, nor can the primary site of the tumour be identified from the appearance of the cells in the smears. There are a few exceptions to this rule. Small uniform discrete malignant cells containing mucous vacuoles might suggest metastatic carcinoma of stomach or colon, although the malignant signet-ring cells typical of these tumours in histologic section are rarely seen. Effusions due to metastatic lobular carcinoma of breast often contain 'target' cells which have a dense globule of mucin in the cytoplasm which stains pale orange with the Papanicolaou stain and pink with Giemsa. The pseudociliated tufts often seen in cystadenocarcinomas may also be found in metastatic breast cancer. Occasionally psammoma bodies have been found in pleural and peritoneal fluids from patients with adenocarcinoma of the ovary or carcinoma of the thyroid.

Small cell undifferentiated (oat-cell) carcinoma

Next to adenocarcinoma this is the most frequent type of malignant cell seen in pleural fluids. Oat cells are among the smallest of tumour cells, and are not much larger than lymphocytes in Papanicolaou-stained smears. Characteristically oat cells have only a remnant of cytoplasm at one pole and tend to cluster tightly in a mosaic pattern. In Giemsa-stained preparations, the tumour cells appear much larger due to flattening, possibly because of the small amount of cytoplasm (*Plate 20*). The nuclei stain pale pink and the cytoplasm is often peppered with small regular lipid-containing vacuoles. The tumour cells often occur in pairs or in small clusters, therefore it is relatively easy to miss them, especially in wet-fixed Papanicolaou-stained smears. Probably well over half of the effusions associated with oat-cell carcinoma in the lungs contain these typical tumour cells.

The identification of oat cells in a smear may be difficult if there are many benign cells present. Care must be taken not to confuse them with lymphocytes which occasionally assume a mosaic appearance in air-dried smears. Very few other types of malignant cell can be confused with oat cells and this carcinoma remains the one type where the primary site can be specified. The very rare Wilms' tumour of children may produce tumour cells which resemble oat cells. Confusion should not occur if the age of the patient is taken into consideration.

Squamous cell carcinoma

Although a common cause of bronchial carcinoma, cells from squamous carcinoma are less frequently found in effusions than adenocarcinoma or oat cells. When malignant cells are present it is extremely rare to see the classical highly keratinized tumour cells with bizarre shapes found in sputum.

The cells which can be cytologically diagnosed as of squamous origin are usually single, quite uniform in appearance, and of parabasal type (*Plate 21*). They can vary in size but have abundant large homogeneous cytoplasm, usually deeply staining and without vacuolation. In Papanicolaou-stained smears, the cytoplasm stains a deep green and in Giemsa smears, royal blue. The cornified forms which are occasionally present appear orange in Papanicolaou and a pale sky-blue in Romanowsky smears. The nuclei are centrally placed and binucleate forms are frequently present. In Giemsa-stained smears, the nucleus may appear broken and have an ill-defined nuclear membrane.

Poorly differentiated squamous carcinoma cells cannot be distinguished from poorly differentiated adenocarcinoma.

Lymphoma

The highly malignant group of non-Hodgkins' lymphomas very commonly metastasize to the serous cavities. The cells are always discrete and vary in size. The nuclei show a fine chromatin pattern and often contain multiple irregular nucleoli. The nucleus almost fills the cell which may be the size of a mesothelial cell. The cytoplasm is deeply staining and in Romanowsky-stained smears is characteristically a slate blue or a deep purple-blue colour. It is extremely common for lymphoma cells to be peppered with small regular lipid-containing vacuoles. Nuclear fragments of all sizes are also commonly present.

As with other types of malignancy, the cell pattern is very variable from case to case. It is not easy to classify lymphomas on the basis of their appearance in wet-fixed or air-dried smears, as classification of these tumours depends as much on lymph node architecture as on cell morphology. However, well differentiated forms can be recognized in which the tumour cells closely resemble mature lymphocytes (*Plate 22*), and poorly differentiated forms can also be recognized by the lymphoblastoid appearance of the tumour cells. In these cases, the nuclear irregularity and enlargement is very striking (*Plate 23*). Immunocytochemical staining of air-dried smears may be of value in distinguishing between T and B cell lymphomas.

Hodgkin's lymphomas when accompanied by a malignant effusion cannot be distinguished from other high grade lymphomas and only rarely are Sternberg–Reed cells found.

Leukaemias and plasmacytoma rarely cause effusions. The malignant cells in effusions associated with plasmacytoma or multiple myelomatosis have the characteristic deep blue cytoplasm of benign plasma cells with a clear zone adjacent to the nucleus. The pattern differs from that seen in benign effusions in that the malignant plasma cells show a marked anisonucleosis and the tumour cells are the dominant cell type in the smear.

Other tumour cells

Rarer types of tumour including sarcomas can produce malignant effusions in the terminal stage. The cells from these are characteristic of the parent tumour and cannot usefully be described individually. Cells from malignant melanoma may be recognized by the presence of melanocytes (*Plate 24*) and macrophages containing phagocytosed melanin granules. For positive identification of melanomas special staining techniques are required.

Mesothelioma

This primary tumour of mesothelium may be difficult to distinguish cytologically from adenocarcinoma, sarcoma or anaplastic carcinomas.

The most distinctive cytological picture is that of many large, rounded, papillary clusters of mesothelial cells (*Plate 25*). These appear at first sight to be remarkably similar to the papillary form of adenocarcinoma, except that the balls usually appear solid. The nuclei are centrally and evenly placed within each cell. The balls are remarkably uniform and have a distinctive crenated surface outline resembling blackberries.

These large papillary clusters are always accompanied by many discrete mesothelial cells showing a marked pleomorphism with binucleate and multinucleate forms and a greatly increased cannibalistic population. All are features seen in benign mesothelial proliferation but never to such a degree as in mesothelioma. Considerable variation in cell picture may occur, and in many instances only a differential diagnosis of mesothelioma can be made.

Mesothelioma cells have numerous microvilli and are rich in glycogen; however, both features are found in benign mesothelial cells and do not contribute to the diagnosis. A rare type of mesothelioma is the keratinizing form. One case has been described by Koss (1968), and in this laboratory two cases with the typical papillary features of mesothelioma, confirmed by histology, were seen to develop into a keratinizing form before death.

Characteristic cytological patterns

Benign effusions

Purulent exudate (empyema)

Acute bacterial infection may give rise to a purulent exudate composed of degenerating neutrophils, macrophages and cell debris. A Gram stain may reveal the causative organism. Squamous carcinoma may give rise to malignant empyema; in such cases bizarre keratinized squames may be seen in the exudate. Empyema may be due to ruptured oesophagus or bronchopleural fistula. Superficial squamous cells can be seen amongst the polymorphs in smears from such cases. Rupture of the oesophagus can occur 'spontaneously' and be completely unsuspected clinically, and a cytological diagnosis can be life-saving. Mesothelial cells are rarely seen in empyema as the pleural surfaces become coated by a fibrinous exudate which prevents exfoliation of these cells.

Tuberculous effusions

Infection caused by the tubercle bacillus produces a lymphocytic response. If the pleural surfaces become coated with exudate, mesothelial cells may be absent from the fluid. A lymphocytic effusion in the complete absence of mesothelial cells is suggestive of (but not diagnostic of) tuberculosis. Small macrophages may be common but giant cells of Langhans type are almost never seen.

Lymphocytic effusions

A lymphocytic picture is not specific to tuberculosis and is often seen in sterile post-pneumonic effusions, in lymphoma and in carcinoma. In these conditions, mesothelial cells or tumour cells are also present and plasma cells are common. A lymphocytic pattern is often found in chylous effusions which can be recognized by the milky appearance of the fluid due to leakage of lymph into the pleural cavity usually because of trauma, or to obstruction of the thoracic duct either by carcinoma or lymphoma. Malignant cells are rarely if ever seen in chylous fluids except in cases associated with lymphoma.

Acute non-bacterial inflammation

Many exudates due to inflammation are sterile and give a mixed cytological picture with varying proportions of inflammatory cells. Typical effusions of this type are seen following pneumonia, influenza, pulmonary collapse and pulmonary infarction and many cases of cardiac failure. Neutrophils and lymphocytes are always present accompanied by macrophages and mesothelial cells in varying number, and occasional eosinophils, basophils and sometimes plasma cells. In pulmonary infarct, the mesothelium undergoes extreme proliferation and a very high proportion of the cells may be of active mesothelial type. Clusters of these very active mesothelial cells are not infrequently mistaken for malignant cells.

Eosinophilic effusions

Eosinophils can be present in a great many exudates (*Plate 26*). When they exceed 10% of the cell population, the findings may be significant. In those cases with an exceedingly high number of eosinophils, there is often a history of trauma (particularly pneumothorax). Eosinophils are regularly seen in effusions from patients whose effusions have been repeatedly aspirated. They are also found after thoracotomy and in patients with fractured ribs; they can be present in effusions due to Hodgkin's disease and carcinoma of the lung.

Charcot–Leyden crystals which are associated with breakdown of eosinophils in lung disease and seen regularly in sputum are rarely seen in effusions. It is probable that the effusion resolves quickly before breakdown can occur. Eosinophils seem to be very resistant to damage or degenerative changes as such fluids left for months on the bench do not contain crystals. Charcot–Leyden crystals can, however, be produced artificially by simply crushing eosinophils under a cover slip. Within 10 s of the rupture of the eosinophils, minute crystals form among the debris and rapidly grow to a large size in several minutes. Basophil polymorphonuclear cells always accompany eosinophils in small numbers but the significance of this finding is not known.

Rheumatoid effusions

Another distinctive exudate for which a cytological diagnosis can readily be made is the exudate formed in rheumatoid arthritis. The fluid is often grey and turbid in appearance with a deposit which resembles that found in empyema. Smears contain polymorphs and abundant nuclear debris. At first glance, the unwary observer may make the diagnosis of empyema but this exudate also contains large amounts of solid material in the form of small regular globules (*Plate 27*). These are apparent in Papanicolaou-stained smears as masses of regular green staining material with orange staining areas. In Romanowsky-stained smears the globules stain a very pale delicate sky blue. This material is precipitated immune complex which has been identified by immunofluorescence as containing rheumatoid factor. The combination of irregular degenerative nuclear debris and the masses of regular round amorphous globules is indicative of rheumatoid disease.

A further diagnostic feature which, when present, is absolute proof of a rheumatoid aetiology, is the presence of large multinucleate epithelioid cells. Unfortunately, these very distinctive cells which are shed from the rheumatoid nodules developing in the pleural wall are not always present. They are elongated spindle-shaped cells which develop into multinucleate forms and always retain a pronounced tail to give a 'tadpole' or 'comet' shape. Their epithelioid nature, as opposed to histiocytic giant cells, is seen in the staining reaction of the cytoplasm which is green in Papanicolaou-stained smears and blue in Romanowsky films. These extreme forms cannot easily be confused with malignant cells because of their small, uniform nuclei but occasionally the mononuclear form is the only type present and these resemble fibre cells and might be mistaken for malignant squamous cells or even sarcoma, unless the clinical history is borne in mind. Very often a rheumatoid effusion is unsuspected clinically and in some cases can precede any other clinical manifestations of the disease.

A few cases of rheumatoid arthritis have been associated with a different picture with little degeneration but many lymphoid cells, transformed cells and plasma cells. The characteristic proteinaceous globules, however, are still present and should enable a correct diagnosis to be made.

As effusions due to rheumatoid arthritis are resorbed slowly, cholesterol crystals may also be present.

Cholesterol effusions

Any longstanding 'encysted' fluid will eventually produce cholesterol crystals. Tuberculous fluids in the past were common examples of this process but at the present time have been replaced by rheumatoid disease as the commonest single cause. Occasionally, ovarian cysts are aspirated in error and mistakenly labelled ascitic fluid. The cyst fluid can be recognized as such by the presence of amorphous cellular debris and cell ghosts and phagocytic macrophages. Cholesterol crystals may also be present.

Systemic lupus erythematosus (LE)

In effusions which accompany this disease, the cytological picture may show no specific features. When there are many neutrophils present, however, LE cells may be seen. These are neutrophils which contain a large mass of nuclear material which compresses the lobed nucleus of the host cell against the cell wall. They are believed to be artefacts. Similar masses of nuclear material have been described in a few cases in small mononuclear cells resembling monocytes or plasma cells. LE cells are usually not numerous, and may be a chance finding if the clinical details are not known. A few cases of LE have been diagnosed as a result of finding this cytological change in the fluid.

Viral changes

Viral infection is usually associated with a lymphocytic effusion. The multinucleated cells associated with herpes virus have never been seen by the author. In one case of infectious mononucleosis seen at Oxford, the atypical mononuclear cells were present in large numbers in the ascitic fluid.

Malignant effusions

The presence of malignant cells in smears prepared from an effusion indicates that the tumour has involved the serosal surfaces and the tumour cells are shed into the fluid where they may undergo proliferation in the medium. However, not all effusions associated with malignant disease can be expected to contain tumour cells as the effusion may be due to obstruction of the lymphatics by the tumour or by a local inflammatory response to the tumour. Even when the tumour is growing freely in the fluid there are often numerous benign cells present in the smears. The proportion of benign mesothelial cells, leucocytes, erythrocytes and malignant cells in any smear will vary according to the method of sampling and smear preparation.

The value of cytology as a method of detecting malignant cells in effusions has been studied by Dr A. Spriggs and the author over a period of 30 years. Ten thousand effusions from 4463 patients were examined between 1952 and 1976; 3337 patients had pleural effusions and 1126 had ascites. Malignant cells were present in 59% of cytologial smears prepared from pleural effusions and 67% of ascitic fluids from patients with proven malignant disease. The percentage of effusions containing malignant cells was increased by combined cytological and histological investigation of the specimen. In our experience, a combination of biopsy and cytology increased the proportion of effusions found to contain malignant cells by about 10% compared with cytology alone. Although the accuracy of reporting is often a reflection of experience with a particular technique, we found that biopsy is of greater value than cytology for the diagnosis of tuberculosis and squamous carcinoma metastatic to the pleura.

A summary of the cytological findings made on 1746 patients with pleural effusions and 452 patients with ascites investigated between 1952 and 1964 is shown in *Table 14.1*. By far the commonest pleural effusions encountered during this time were associated with lung carcinoma and carcinoma of the breast; the commonest cause of ascites was carcinoma of ovary. Whereas about half the effusions due to carcinoma of the lung contained malignant cells, the ascites accompanying ovarian tumours were almost always malignant. Malignant cells can be found in about two-thirds of effusions due to

Table 14.1 Summary of cytological performance in serous fluids, 1952–1964

Final diagnosis	Pleural fluids				Ascitic fluids			
	Total	Cases with malignant cells	Cases with suspicious cells	% Positive or suspicious	Total	Cases with malignant cells	Cases with suspicious cells	% Positive or suspicious
Ca of lung								
Proven oat cell	49	27	1	57.1				
Proven squamous Ca	54	18		33.3				
Proven other type	57	32	4	63.1	4	3		(75.0)
Not known	115	65	5	60.8				
Ca of breast	157	93	14	68.2	30	18	3	70.0
Ca of ovary	27	17	1	66.7	110	85	3	85.4
Ca of stomach	18	7	1	44.4	45	27	1	62.2
Ca of colon	9	6		66.6	21	12	1	61.9
Ca of pancreas	7	6		85.7	14	6		42.8
Ca of corpus uteri	2	1		(50.0)	8	7		87.5
Ca of bladder	7	4	2	85.7	3	2		66.6
Ca of cervix	4	3		75.0	4	3		75.0
Other Ca	23	5		21.7	20	5	2	35.0
Lymphoma	52	28	1	55.8	15	6		40.0
Mesothelioma	6	4	1	83.3	1			
Sarcoma	13	4	1	38.5	4	2	1	75.0
Primary unknown	40	29		72.5	42	25	2	64.2
Total with malignant disease	640	349	31	59.3	321	207	13	68.5
No malignant disease	1106	3	0	0.3[a]	131	0	0	0

[a] 0.8% false positive reports

adenocarcinoma. Lymphoma is associated with pleural effusion more frequently than with ascites, but malignant cells are detectable in the fluid in about half the cases.

Reporting difficulties

The false positive rate in the series shown in *Table 14.1* was very low as a consequence of a conservative reporting policy. Reports of 'suspicious' cells were kept to a minimum. The false positive rate did not exceed 1%. In our experience, errors were most commonly due to bizarre mesothelial cells suggesting the possibility of mesothelioma. Some errors could have been avoided if relevant clinical data had been given. Examples of some of the problem cases are given below:

(1) An unusual cytological pattern of proliferating mesothelial cells was noted in an effusion on the seventh day after a lobectomy for squamous carcinoma. A diagnosis of adenocarcinoma was made without laboratory knowledge of the operation. The patient made a satisfactory recovery.

(2) A patient in intensive care with a myocardial infarct had large numbers of myeloblasts in a pleural effusion which were mistaken for oat-cell carcinoma cells. This mistake was rectified once the clinical details were revealed. although the cause of the unusual cytological pattern in the fluid was never established.

(3) An 80-year-old woman developed a transient pleural effusion during a flu-like illness. The fluid contained numerous immature 'blast' cells which constituted the vast majority of cells in the fluid. They were thought to reflect the presence of lymphoma, although karyotypes of the cells were normal. The effusion did not recur and the patient died five years later from bronchopneumonia. Autopsy revealed no tumour. On review, these cells still suggest a high grade lymphoma.

(4) A pleural fluid from a 52-year-old woman

who had had a bronchoscopy for a suspicious pulmonary shadow contained many mesothelial cells together with a separate population of discrete cells showing marked degenerative changes which were considered to be poorly differentiated adenocarcinoma cells. No carcinoma was found and the patient was alive and well several years later.

(5) A male aged 54 with chronic pancreatitis developed a pleural effusion prior to resection of a parathyroid adenoma. Many of the cells were discrete large mesothelial-like cells showing degenerative changes and considered to be a free cell form of adenocarcinoma. Repeated aspiration revealed a gradual disappearance of this cell type and their replacement by normal mesothelial cells. The serum amylase level of the fluid was known to be 15 000 units/l and the patient was found to have a pleuropancreatic fistula. After pancreatectomy, the patient remains alive and well with no evidence of malignancy. Review of the literature has revealed that others have been misled by the cytological appearance of the pleural effusions found in this rare syndrome.

(6) Finally, mention must be made of pericardial effusions. Because the mesothelium in pericarditis proliferates markedly, care must be taken in the diagnosis of these specimens. One specimen from an 86-year-old woman with massive pericarditis contained large bizarre masses of epithelioid cells, some of which showed cornification. A diagnosis of squamous carcinoma was made. Post-mortem examination revealed a typical uraemic pericarditis with fibrino-purulent exudate of the pericardium, but no evidence of malignancy.

Note on the illustrations

The illustrations in this chapter can serve to demonstrate only the salient cell types described in the text and the reader must refer to the various Atlases on the subject for a wider appreciation of cell morphology seen in serous fluids.

The photomicrographs were taken on Ektachrome 50 Professional of May-Grünwald–Giemsa-stained smears at the microscope magnification of × 540.

References and further reading

ASKIN, F. B., BRENDAN, G., MCCANN, G. and KUHN, C. (1977). Reactive eosinophilic pleuritis. A lesion to be distinguished from pulmonary eosinophilic granuloma. *Arch. Pathol. Lab. Med.*, **101**, 187–191

BODDINGTON, M. M. (1980). Mast cells in serous effusions. *Acta Cytol.*, **24**, 470

BODDINGTON, M. M., SPRIGGS, A. I., NORTON, J. A. and MOWAT, A. G. (1971). Cytodiagnosis of rheumatoid pleural effusions. *J. Clin. Pathol.*, **24**, 95–106

BOON, M. E. and DRIJVER, J. S. (1986). *Routine Cytological Staining Techniques. Theoretical Background and Practice*. London: Macmillan

BUTLER, E. B. and BERRY, A. V. (1973). Diffuse mesotheliomas: diagnostic criteria using exfoliative cytology. In *Biological Effects of Asbestos*, (Bogovski, P., ed.), pp. 68–73. Lyons International Agency for Research on Cancer

BUTLER, E. B. and STANBRIDGE, C. M. (1986). *Cytology of Body Cavity Fluids. A Colour Atlas*. London: Chapman and Hall

CAMERON, J. L. (1978). Chronic pancreatic ascites and pancreatic pleural effusions. *Gastroenterology*, **74**, 134–140

DAS, D. K., GUPTA, S. K., AYYAGARI, S., BAMBERY, P. K., DATTA, D. N. and DATTA, U. (1987). Pleural effusions in non-Hodgkin's lymphoma. A cytomorphologic, cytochemical and immunologic study. *Acta Cytol.*, **31**, 119–124

DHILLON, D. P. and SPIRO, S. G. (1983). Malignant pleural effusions. *Br. J. Hosp. Med.*, **26**, 506–510

GRUNZE, H. and SPRIGGS, A. I. (1983). *History of Clinical Cytology*, 2nd Edition. Darmstadt: Ernst Giebeler

KOSS, L. G. (1968). *Diagnostic Cytology and its Histopathologic Bases*, 2nd Edition. Philadelphia: J. B. Lippincott

MELSOM, R. D. (1979). Diagnostic reliability of pleural fluid protein estimation. *J. R. Soc. Med.*, **72**, 823–825

MURPHY, W. M. and NG, A. B. P. (1972). Determination of primary site by examination of cancer cells in body fluids. *Am. J. Clin. Pathol.* **58**, 479–488

NOSANCHUK, J. S. and NAYLOR, B. (1968). A unique cytologic picture in pleural fluid from patients with rheumatoid arthritis. *Am. J. Clin. Pathol.*, **50**, 330–335

SEARS, D. and HAJDU, S. I. (1987). The cytological diagnosis of malignant neoplasms in pleural and peritoneal effusions. *Acta Cytol.*, **31**, 85–97

SPRIGGS, A. I. (1980). Cytology of pleural fluid. In *Pulmonary Diseases and Disorders*, (Fishmann, A. P., ed.), Chapter 34. New York: McGraw-Hill

SPRIGGS, A. I. (1984). The architecture of tumour cell clusters in serous effusions. In *Advances in Clinical Cytology*, Volume 2, (Koss, L. G. and Coleman, D. V., eds), Chapter 11. New York: Masson

SPRIGGS, A. I. and BODDINGTON, M. M. (1976). Oat-cell bronchial carcinoma. Identification of cells in pleural fluid. *Acta Cytol.*, **20**, 525–529

SPRIGGS, A. I. and BODDINGTON, M. M. (1989). *Atlas of Serous Fluid Cytopathology*. Lancaster: Kluwer Academic.

SPRIGGS, A. I. and JEROME, D. W. (1975). Intracellular mucous inclusions. A feature of malignant cells in effusions in the serous cavities, particularly due to carcinoma of the breast. *J. Clin. Pathol.*, **28**, 929–936

SPRIGGS, A. I. and VANHEGAN, R. I. (1981). Cytological diagnosis of lymphoma in serous effusions. *J. Clin. Pathol.*, **34**, 1311–1325

SPRIGGS, A. I., BODDINGTON, M. M. and HALLEY, W. (1967). Uniqueness of malignant tumours. *Lancet*, **1**, 211

15

The place of special techniques in the investigation of serous effusions

Janice Guerra

Pleural and peritoneal fluids form a large part of the workload of many cytology laboratories and provide a challenging and interesting area for investigation. Many of the smears prepared from these specimens present problems of diagnosis and additional tests are needed before a confident diagnosis can be made. In this chapter we recommend a battery of tests that can be used to supplement the morphological studies of effusions described in the previous chapter, so that a high degree of diagnostic accuracy can be attained. The special tests described include histochemical, immunocytochemical, cytogenetic and ultrastructural studies.

Histochemical staining

These tests are relatively simple and inexpensive and can be carried out in most cytology laboratories. They are used most frequently to identify metastatic adenocarcinoma cells in serous effusions when there is a problem of distinguishing these cells from benign reactive mesothelial cells or mesothelioma. The single test that is of most value in this respect is the Periodic Acid Schiff (PAS)/diastase reaction although the Alcian Blue/hyaluronidase reaction is a useful adjunct. Regretfully neither test will distinguish between benign reactive mesothelial cells and mesothelioma.

Table 15.1 shows how the PAS reaction and the Alcian Blue reaction can be used to distinguish between mucin secreting malignant glandular epithelial cells and benign reactive mesothelial cells in the fluid. The same reactions will also assist the discrimination between adenocarcinoma and mesothelioma. The distinction depends on three different histochemical reactions.

(1) Both benign and malignant mesothelial cells and adenocarcinoma cells may contain a diffuse granular deposit of glycogen in their cytoplasm which stains a dark pink or

Table 15.1 Distinction between adenocarcinoma and mesothelioma in effusions using PAS and Alcian Blue stains.

	Expected results	
Histochemical technique	Benign mesothelial cells and mesothelioma	Adenocarcinoma
Periodic Acid Schiff (PAS)	+	+
Diastase/PAS	−	+
Alcian Blue	+	+
Hyaluronidase/Alcian Blue	−	+

magenta with the PAS signalling a positive reaction. Prior treatment of the cells with diastase before staining with PAS will result in a negative reaction.

(2) Mesothelial cells may also contain mucosubstances incorporating hyaluronic acid which gives a positive reaction when stained with Alcian Blue at pH 2.5. Prior exposure of the cells to hyaluronidase before staining with Alcian Blue will result in a negative reaction.

(3) Malignant glandular cells from adenocarcinoma may contain either neutral or acid muco-substances which stain positively with PAS and Alcian Blue and are resistant to both diastase and hyaluronidase digestion. Thus treatment of the tumour cells with diastase or hyaluronidase before staining with PAS or Alcian Blue will not affect the reaction. From our experience it is easier to detect mucin or glycogen deposit in air-dried smears than in wet-fixed smears.

It is worth noting that although mesothelial cells contain hyaluronic acid, it may be present in very small amounts which are difficult to demonstrate by histochemical means. This must always be taken into account when evaluating the results. Similarly, glycogen granules in mesothelial cells (particularly in malignant mesothelial cells) sometimes appear crescent shaped and may be mistaken for mucin deposits prior to diastase digestion.

Other histochemical techniques that may be useful in the cytologic investigation of serous effusions include the Masson-Fontana or Schmorl's reaction for confirmation of melanin deposition. The possibility of haemosiderin must be eliminated by using the Perl's Prussian Blue technique. When amelanotic melanoma is suspected, the dopa oxidase enzyme technique may be appropriate. In addition, perinuclear fat deposits may be present in malignant mesothelial cells and the fat stain oil red O is an ideal method of demonstrating it.

Immunocytochemical stains

The application of immunocytochemical techniques to cytological specimens is a relatively new development in laboratory practice. Early experience with this approach indicates that it has a particularly important role to play in the diagnosis of effusions (Nadji, 1980; To et al., 1981; Mason and Gatter, 1987; Ovell and Dowling, 1983). Careful selection of the appropriate antibody can assist with the discrimination between benign and malignant cells in these samples and in tumour typing (Sloane, Hughes and Ormerod, 1983; Sehested, Ralfkjaer and Rasmussen, 1983; Walts, Said and Banks-Schlegel, 1983). General aspects of the application of immunocytochemical techniques in cytology are discussed in Chapter 6. In this section we discuss special problems associated with the immunocytochemical staining of serous effusions. This includes advice on specimen collection, preparation and fixation, staining methods, selection of controls, interpretation of results and choice of antibodies.

Specimen collection and processing

Clinicians should be asked to send 100 ml of fluid whenever possible so that there is adequate material for immunocytochemical studies after smears have been prepared for Papanicolaou and May-Grünwald–Giemsa staining. Having said this, it is often possible to get adequate material for special stains with smaller volumes of fluid. A minimum of six additional smears are needed for this purpose so that control slides as well as test slides can be set up for each sample. It has become routine practice in this laboratory to make several extra smears from each sample in anticipation of immunocytochemical staining. These are readily available should further studies be indicated after the Papanicolaou and Giemsa-stained smears have been examined. Clots present in serous effusions should always be sent to the histology laboratory for processing and sectioning and staining with the same antibodies as the smears. Histological and cytological correlation of immunocytochemical staining patterns is a very useful form of quality control.

Fixation

Two factors are critical for the success of the immunocytochemical staining of smears prepared from serous fluid, namely good preservation of the antigen under investigation leaving it accessible to the primary antibody, and good presentation of the morphology of the cells in the smear. Fortunately, most antigens appear to be resistant to the cytological fixatives in routine use, i.e. 95% alcohol for wet-fixed smears and methanol or acetone for air-dried smears. However, this should not be taken for granted and a positive control should always be used.

Processing

Air-dried or wet-fixed smears can be prepared from the deposit after centrifugation although the latter is preferred in this laboratory. We have found it advisable to wash the cells harvested from the fluid in 5 ml normal saline at least once before preparing the smears as this reduces the risk of proteins in the fluid causing background staining. Cytocentrifugation may be used if the deposit is small. The latter approach is often the most economical as only a small amount of antibody is needed to cover the cells. If the fluid is bloodstained, it is essential to remove the red blood cells before preparing the smears using the density gradient method or the capillary method described in Chapter 4. By eliminating the red blood cells from the preparations, endogenous peroxidase activity is minimized. It also results in an improved tumour cell harvest with better cell morphology.

It is possible to prepare cell blocks from the cell deposit of serous fluids. However, this procedure is not recommended, since the processing is more time-consuming and the results are inferior to cytological smears as the cell morphology is less distinct. Both primary and secondary antibodies should be used at optimal dilutions and a battery of slides prepared from a large fluid sample of known specificity should be used to determine these. A store of slides should be kept at 4 °C for this purpose. Generally, higher dilutions can be used for cytological specimens than for histological sections.

Controls

A panel of slides of known reactivity should be retained as controls. These can be stored in alcohol at 4 °C. A positive and negative control should be included in every batch of immunostaining. If there is difficulty in finding positive cytological controls for the less common antibodies, it may be necessary to use histological material. Omission of the primary antibody and substitution by inappropriate antibody serves as a negative control.

Interpretation of results

Great care must be taken in the interpretation of immunocytochemical staining patterns especially when there are only a few discrete positively stained cells in the smear and the morphology of the cell must always be taken into account. Generally, two patterns of staining will be observed. Either a diffuse cytoplasmic stain which may vary in intensity from cell to cell, or a ring-like staining which is most intense at the cell periphery.

A pale or weak diffuse staining must be treated with caution and special reference should be made to the control slides. Histiocytes which have ingested proteins and antigens may show a degree of staining reaction which can cause a problem if numbers are sufficient, but generally histiocytes can be eliminated on morphological grounds.

Since staining patterns may vary from tumour to tumour, or indeed from one area to another of the same tumour, it is important to standardize the technique so that results can be considered reliable.

Choice of staining method

Any of the immunostaining methods described in Chapter 6 may be considered for use in cytology including peroxidase–antiperoxidase (PAP) and immunogold techniques. In this laboratory we use the indirect immunoperoxidase technique (Nakane and Pierce, 1966). This is currently the most widely used method, although other conjugates such as alkaline phosphatase and glucose oxidase can be used. Alkaline phosphatase is preferred for bloodstained smears such as those from bone marrow aspirate because there is less risk of non-specific staining due to endogenous enzyme activity. This is probably the only advantage of alkaline phosphatase methods. Disadvantages include a longer staining process and the need to use aqueous mountants. However, alkaline phosphatase and glucose oxidase labels are useful for double immunostaining techniques. Immunocytochemical staining of cytological specimens using the avidin–biotin peroxidase complex may increase the sensitivity of antigen detection but carries the risk of non-specific staining unless endogenous biotin is fully blocked.

Table 15.2 Antibody panel used to discriminate between benign mesothelial cells and malignant cells, and between epithelial and non-epithelial tumour cells in serous fluids, showing the staining reactions of different cell types

Cell type	Positive staining	Negative staining
Mesothelial (benign and malignant)	CAM 5.2 VIM	CEA LCA
Carcinoma	CAM 5.2 CEA	VIM LCA
Lymphoid	LCA	CAM 5.2 VIM CEA
Mesenchymal (other than mesothelial) Sarcoma	VIM	CAM 5.2 CEA LCA

Choice of antibodies

By the judicious use of a small panel of antibodies it is now possible to solve some of the longstanding problems associated with the cytodiagnosis of serous effusions (Gatter *et al.*, 1982; Ghosh *et al.*, 1983). The panel of antibodies used routinely in this laboratory is shown in *Table 15.2*. It comprises antibodies against cytokeratin (CAM 5.2), carcinoembryonic antigen (CEA), leucocyte common antigen (LCA) and vimentin (VIM).

Cytokeratin belongs to the family of the intermediate filaments which comprises also vimentin, desmin, neurofilament protein and glial fibrillary acidic protein (Gabbiani *et al.*, 1981). They are part of a series of related polypeptides with molecular weights varying between 40 and 200 kilodaltons. The cytokeratins, of which there are 19, are between 40 and 68 kilodaltons (low to high molecular weight respectively) and are present in both non-keratinizing and keratinizing epithelia.

CAM 5.2 is a murine monoclonal antibody raised against the colon carcinoma cell line HT29 and recognizes lower molecular weight intracellular cytokeratin proteins within secretory epithelia (50 000, 43 000 and 39 000 daltons) (Makin, Bobrow and Bodmer, 1984). This anticytokeratin stains normal adult epithelial cells, apart from stratified squamous epithelium. It stains all tumours arising from secretory epithelia (*Figure 15.1*) and all squamous carcinomas (Makin, Bobrow and Bodmer, 1984). CAM 5.2 is also reported to react positively with tumours of neuroendocrine origin (Lehto *et al.*, 1983).

Vimentin has a molecular weight of 57 kilodaltons and is present in cells of mesenchymal origin. It is also known as fibroblast intermediate filament (FIF) and may be prepared from transformed human fibroblasts (Gabbiani *et al.*,

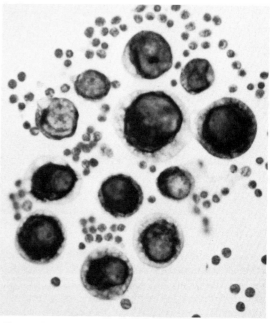

Figure 15.1 Adenocarcinoma cells from a pleural fluid showing a strong positive cytoplasmic staining with CAM 5.2. Note outer rim of unstained cytoplasm an occasional finding in adenocarcinoma cells. Indirect immunoperoxidase technique, magnification × 400, reduced to 90% in reproduction

1981). The antibody stains a wide range of tissues of mesenchymal origin, such as connective tissue. Staining of metastatic carcinoma is variable (*Figure 15.2*).

CEA is a glycoprotein with a molecular weight of approximately 200 kilodaltons which was first described by Gold and Freedman (1965a,b) as an antigen present in extract of fetal gut and carcinomas of the gastrointestinal tract (Whiteside and Dekker, 1979). Consequently the antibody stains a wide range of adenocarcinomas (*Figure 15.3*), particularly of gastrointestinal tract, lung and pancreas. It may also stain some squamous carcinomas and small cell carcinoma of lung (Said *et al.*, 1983).

Polyclonal anti-CEA has been found to react with a non-specific cross-reacting antigen (NCA) present in some epithelial cells (Von Kleist, Chavanal and Burtin, 1972). Where there is reactivity to NCA, this can be overcome by adsorbing the CEA antisera against a phenyl extract of human spleen as described by Goldenberg, Pegram and Vazquez (1975).

Leucocyte common antigen is confined to cells of haemopoietic and lymphoid origin and will therefore detect lymphocytes, granulocytes and macrophages, but will not react with cells of epithelial or mesenchymal origin.

The selective staining patterns of the cells in effusions make it possible to use this panel to discriminate between benign and malignant cells and to determine whether poorly differentiated tumour cells are of epithelial or nonepithelial origin as shown in *Table 15.2*. This table shows that benign and malignant mesothelial cells stain positively with anticytokeratin (CAM 5.2) and antivimentin but are negative for CEA. Therefore cells in an effusion which stain positively for CEA must be regarded as neoplastic, most likely from adenocarcinoma (Wang *et al.*, 1979). CEA positivity excludes tumours of mesenchymal origin, and conversely keratin negative and CEA negative tumours are most likely to be of connective tissue origin. Cytokeratin and vimentin may be co-expressed by metastatic adenocarcinoma cells in effusions (Ramaekers *et al.*, 1983). However, the same workers have shown that all non-epithelial tumour cells present in effusions were negative for keratin but positive for vimentin. Anti-leucocyte common antigen is particularly useful since its pattern of reactivity excludes all tumours other than those of leucocyte origin, e.g. lymphomas.

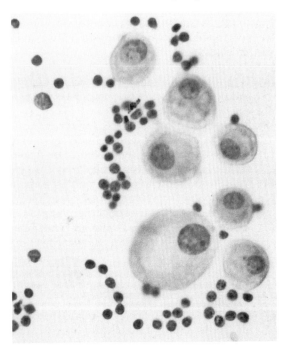

Figure 15.2 Cells from the same fluid as in *Figure 15.1*, showing a negative reaction with vimentin. Indirect immunoperoxidase technique, magnification ×400, reduced to 90% in reproduction

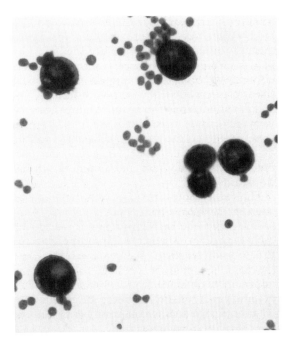

Figure 15.3 Adenocarcinoma cells from the same fluid as in *Figure 15.1*, showing strong homogeneous cytoplasmic staining. Indirect immunoperoxidase technique, magnification ×400, reduced to 90% in reproduction

Supplementary to the basic antibody panel, we have found the following antibodies useful in certain cases:

(1) T and B cell markers UCHLI (DAKO) and MBT$_2$ (Biotest), for distinction between T and B cell lymphomas.
(2) Prostatic acid phosphatase, for the identification of metastatic carcinoma of prostate.
(3) Kappa and lambda light chains to detect monoclonality of plasma cells in myeloma.
(4) Neuron-specific enolase for the identification of oat cell carcinoma, metastatic melanoma and other neuroendocrine tumours.
(5) Alpha-fetoprotein and human chorionic gonadotrophin (HCG) for the identification of germ cell tumours.
(6) Factor VIII can be used for detecting tumours of endothelial origin e.g. Kaposi's sarcoma.
(7) ERD 5, oestrogen receptor-related protein for the detection of metastatic breast carcinoma.

Cytogenetic studies

The fluids that collect in the body cavity as a result of benign and malignant disease serve as an excellent tissue culture medium for mesothelial and tumour cells which are shed into the fluid from the serosal surface. Consequently it is not unusual to see cells in mitosis in cytological smears prepared from serous effusions. In malignant effusions the mitotic figures may be very numerous indeed and the chromosome complement of the tumour cells may appear quite abnormal. In contrast, in benign effusions, mitoses are usually rare and assume a normal configuration.

The spontaneously dividing cells in malignant effusions have been the subject of intense study to determine whether karyotyping is unique for each tumour or whether a karyotope emerges that is characteristic of malignancy. If this were the case, it would be possible to utilize the information obtained from cytogenetic studies of effusions to assist with the discrimination between benign and malignant effusions and the identification of tumour type.

Several different cytogenetic patterns have been recognized in effusions. These include:

(1) Aneuploid karyotypes (*Figure 15.4*).
(2) Structural abnormality or translocation.
(3) Marker chromosomes (*Figure 15.5*).
(4) Minutes.
(5) Apparently normal diploid karyotypes.

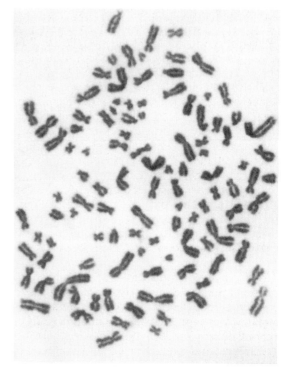

Figure 15.4 An adenocarcinoma cell in metaphase showing the very large chromosomes complement found in this case. Stain: Giemsa, magnification × 1250, reduced to 50% in reproduction

Watts *et al.* (1983) carried out cytogenetic studies on 132 specimens of serous effusions to determine whether karyotyping could be of diagnostic value. The specimens were collected fresh where possible and processed immediately or cultured for 24 h. Colchicine was added to arrest mitoses in metaphase. The cells were then exposed to hypotonic saline to spread the chromosomes. The number of chromosomes in each metaphase and the individual chromosomes were examined for structural abnormalities. Metaphases suitable for cytogenetic studies were obtained from cultured cells in 56% of the specimens and from uncultured cells in only 12% of the specimens. This relatively poor harvest from uncultured cells may have reflected the fact that the specimens were not always fresh. The cytogenetic results were compared with the cytological findings. Abnormal meta-

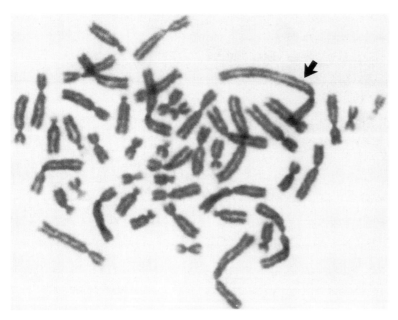

Figure 15.5 Chromosomes in metaphase from an adenocarcinoma cell in a pleural fluid. A marker chromosome is present (arrowed). Stain: Giemsa, magnification × 1250, reduced to 65% in reproduction

phases were found in 22 of the 34 specimens containing malignant cells. An unexpected finding, however, was the presence of chromosome abnormalities in five cytologically negative specimens from patients with benign effusion, including two from patients who were found to have pneumonia and three from patients who had hepatic cirrhosis. Thus cytogenetic studies are not a reliable method of discriminating between reactive mesothelial cells and carcinoma. However, it was noted that monoclonality of the tumour cell line was often a feature of non-epithelial tumours and was helpful in identifying tumour type.

Electron microscopy

Electron microscopy already has a well recognized application in special fields of diagnosis, such as in renal disease, muscle pathology, storage diseases and more significantly in tumour pathology. It can also contribute to the cytological investigation of serous effusions in selected cases.

Processing of serous fluids after electron microscopy

There are problems to be encountered when considering the use of electron microscopy in cytology. Most special investigations are carried out retrospectively and this means that there may be little, if any, material available for electron microscopy. In some cases, it may be possible to obtain more fluid from which a cell pellet can be processed after centrifugation. From this a cell block can be made which, after fixation in 4% glutaraldehyde, can be processed as described in Chapter 4. In cases where additional material is unavailable, reasonable results can be obtained by using a technique for reprocessing Papanicolaou-stained smears for electron microscopy (Coleman *et al.*, 1977); This is not an ideal solution since after reprocessing, the cell organelles are not well preserved (*see also* Chapter 4).

There are certain areas where electron microscopy may contribute to a definitive diagnosis in serous fluid cytology, and these are itemized below.

Mesothelioma

The presence of a dense covering of elongated, slender, sometimes branching surface microvilli is considered to be a useful ultrastructural marker in differentiating mesothelioma from benign mesothelial cells (*Figure 15.6*) which only have scanty microvilli. Mesotheliomas also

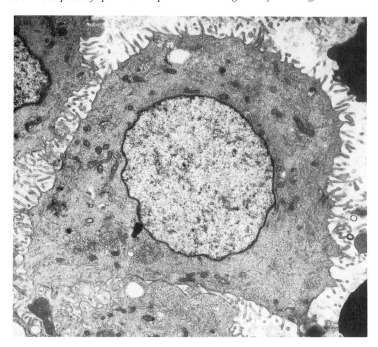

Figure 15.6 Electron micrograph (× 12 000, reduced to 50% in reproduction) of a mesothelial cell with scanty microvilli and diffuse glycogen deposits in the cytoplasm

have large aggregates of glycogen particles in the cytoplasm as well as small bundles of intermediate filaments (10 mm diameter) (Wang, 1973).

Adenocarcinoma

There may be evidence of glandular differentiation at the ultrastructural level such as microvilli, micro-lumen formation and mucin vacuoles, none of which occur in squamous cell carcinoma (Henderson and Papadimitriou, 1982).

Small cell anaplastic carcinoma

These exhibit small dense membrane-bound neurosecretory granules averaging 155 mm diameter which may be sparse or present in moderate numbers.

Bronchial carcinoid tumours

These also exhibit the characteristic neurosecretory granules, usually in greater numbers and with greater variation in size (70–500 mm) though generally in the range of 100–300 mm (Henderson and Papadimitriou, 1982).

Squamous cell carcinoma

At ultrastructural level there is an abundance of desmosomes with attached tonofilaments together with free bundles of tonofilaments in the cytoplasm (Inoue and Dionne, 1977).

Melanomas

These may be recognized by the presence of premelanosomes or melanosomes – membrane-bound vesicles with a distinctive internal structure. These may be either parallel lamellae, helical or zigzag structures with a periodicity of 8–10 nm. Electron microscopy is particularly helpful in the diagnosis of amelanotic melanoma (Henderson and Papadimitriou, 1982).

These are some examples of the role of electron microscopy in the cytological investigation of serous fluids. It is important to note that immunocytochemistry can also be applied successfully at the ultrastructural level, particularly with the immunogold–silver method (Danscher, 1981; Holgate et al., 1983) and this adds yet another dimension to this specialized technique.

In conclusion, it can be said that histochemistry, electron microscopy, cytogenetic studies and immunocytochemistry have a definite place in the routine cytology laboratory. By applying such techniques we can expand our knowledge, and improve the accuracy of cytological diagnosis.

References

COLEMAN, D. V., RUSSELL, W. J. I., HODGSON, J., TUN, P. E. and MOWBRAY, J. F. (1977). Human papovirus in Papanicolaou smears of urinary sediment detected by transmission electron microscopy. *J. Clin. Pathol.*, **30**, 1015–1020

DANSCHER, G. (1981). Localisation of gold in biological tissue. A photochemical method for light and electron microscopy. *Histochemistry*, **71**, 81–88

GABBIANI, G., KAPANCI, Y., BARAZZONE, P. and FRANKE, W. W. (1981). Immunochemical identification of intermediate sized filaments in human neoplastic cells – a diagnostic aid for the surgical pathologist. *Am. J. Pathol.*, **104**, 206–216

GATTER, K. C., ABDULAZIZ, Z., BEVERLEY, P., CORVALAN, J. R. F., FORD, C., LANE, E. B., MOTA, M., NASH, J. R. G., PULFORD, K., STEIN, H., TAYLOR-PAPADIMITRIOU, J., WOODHOUSE, C. and MASON, D. Y. (1982). Use of monoclonal antibodies for the histopathological diagnosis of human malignancy. *J. Clin. Pathol.*, **35**, 1253–1267

GHOSH, A. K., MASON, D. Y. and SPRIGGS, A. I. (1983). Immunocytochemical staining with monoclonal antibodies in cytologically 'negative' serous effusions from patients with malignant disease. *J. Clin. Pathol.*, **36**, 1150–1153

GHOSH, A. K., SPRIGGS, A. L., TAYLOR-PAPADIMITRIOU, J. and MASON, D. Y. (1983). Immunocytochemical staining of cells in pleural and peritoneal effusions with a panel of monoclonal antibodies. *J. Clin. Pathol.*, **36**, 1154–1164

GOLD, P. and FREEDMAN, S. O. (1965a). Demonstration of tumour-specific antigens in human colonic carcinoma by immunological tolerance and absorption techniques. *J. Exp. Med.*, **121**, 439–462

GOLD, P. and FREEDMAN, S. O. (1965b). Specific carcinoembryonic antigens in the human digestive system. *J. Exp. Med.*, **122**, 467–481

GOLDENBERG, D., PEGRAM, C. and VAZQUEZ, J. (1975). Identification of colon specific antigen (CSA) in normal and neoplastic tissues. *J. Immunol.*, **114**, 1008–1013

HENDERSON, D. W. and PAPADIMITRIOU, J. M. (1982). *Ultrastructural Appearances of Tumours. A Diagnostic Atlas.* Edinburgh: Churchill Livingstone

HOLGATE, C. S., JACKSON, P., COWEN, P. N. and BIRD, C. C. (1983). Immunogold–silver staining: a new method of immunostaining with enhanced sensitivity. *J. Histochem. Cytochem.*, **31**, 938–944

INOUE, S. and DIONNE, G. P. (1977). Tonofilaments in normal human bronchial epithelium and in squamous cell carcinoma. *Am. J. Clin. Pathol.*, **88**, 345–354

LEHTO, V. P., STENMAN, S., MIETTINEN, M., DAHL, D. and VIRTANEN, I. (1983). Expression of a neural type of intermediate filament as a distinguishing feature between oat cell carcinoma and other lung cancers. *Am. J. Pathol.*, **110**, 113–118

MAKIN, C. A., BOBROW, L. G. and BODMER, W. F. (1984). Monoclonal antibody to cytokeratin for use in routine histopathology. *J. Clin. Pathol.*, **37**, 975–983

MASON, D. Y. and GATTER, K. C. (1987). The role of immunocytochemistry in diagnostic pathology. *J. Clin. Pathol.*, **40**, 1042–1054

NADJI, M. (1980). The potential value of immunoperoxidase techniques in diagnostic cytology. *Acta Cytol.*, **24**, 442–447

NAKANE, P. K. and PIERCE, G. B. (1966). Enzyme labelled antibodies: Preparation and application for the localisation of antigens. *J. Histochem. Cytochem.*, **14**, 929–931

OVELL, S. R. and DOWLING, K. D. (1983). Oncofoetal antigens as tumour markers in the cytological diagnosis of effusions. *Acta Cytol.*, **27**, 625–629

RAMAEKERS, F. C. S., HAAG, D., KANT, A., MOLESKER, O., JAP, P. H. K. and VOOIJS, G. P. (1983). Coexpression of keratin- and vimentin-type intermediate filaments in human metastatic carcinoma cells. *Proc. Natl. Acad. Sci. USA*, **80**, 2618–2622

SAID, J. W., NASH, G., TEPPER, G. and BANKS-SCHLEGEL, S. (1983). Keratin proteins and carcinoembryonic antigen in lung carcinoma. *Human Pathol.*, **14**, 70–76

SEHESTED, M., RALFKJAER, E. and RASMUSSEN, J. (1983). Immunoperoxidase demonstration of carcinoembryonic antigen in pleural and peritoneal effusions. *Acta Cytol.*, **27**, 124–127

SLOANE, J. P. and ORMEROD, M. G. (1981). Distribution of epithelial membrane antigen in normal and neoplastic tissues and its value in diagnostic tumour pathology. *Cancer*, **47**, 1786–1795

SLOANE, J. P., HUGHES, E. and ORMEROD, M. G. (1983). An assessment of the value of epithelial membrane antigen and other epithelial markers in solving diagnostic problems in tumour histopathology. *Histochem. J.*, **15**, 645–654

SMITH, J., COLEMAN, D. V. (1983). Electron microscopy of cells showing viral cytopathic effects in Papanicolaou smears. *Acta Cytologica*, **27**, 605–613.

TO, A., COLEMAN, D. V., DEARNLEY, D., ORMEROD, M. G., STEELE, K. and NEVILLE, A. M. (1981). Use of antisera to epithelial membrane antigen for the cytodiagnosis of a malignancy in serous effusions. *J. Clin. Pathol.*, **34**, 1326–1332

VON KLEIST, S., CHAVANEL, G. and BURTIN, P. (1972). Identification of a normal antigen that cross reacts with the carcinoembryonic antigen. *Proc. Natl. Acad. Sci. USA*, **69**, 2492–2494

WALTS, A. E., SAID, J. W. and BANKS-SCHLEGEL, S. (1983). Keratin and carcinoembryonic antigen in exfoliated mesothelial and malignant cells: an immunoperoxidase study. *Am. J. Clin. Pathol.*, **80**, 671–676

WANG, N. S. (1973). Electron microscopy in the diagnosis of pleural mesotheliomas. *Cancer*, **31**, 1046–1054

WANG, N. S., HUANG, S. N. and GOLD, P. (1979). Absence of carcinoembryonic antigen-like material in mesothelioma – an immunohistochemical differentiation from other lung cancers. *Cancer*, **44**, 937–943

WATTS, K. C., BOYO-EKWUEME, H., TO, A., POSNANSKY, M. and COLEMAN, D. V. (1983). Chromosome studies on cells cultured from serous effusions – use in routine cytological practice. *Acta Cytol.*, **27**, 38–44

WHITESIDE, T. L. and DEKKER, A. (1979). Diagnostic significance of carcinoembryonic antigen levels in serous effusion. *Acta Cytol.*, **23**, 443–448

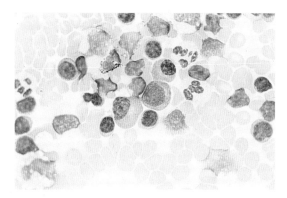

Plate 12 Plasma cells. Smear from an effusion which occured after an episode of pneumonia. The leucocytes in the smear were mostly lymphocytes. In this field are three mature plasma cells each with a clear Golgi zone adjacent to the nucleus. Giemsa stain. Magnification × 1080

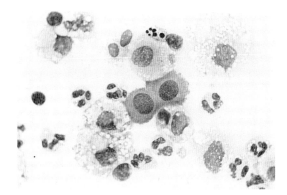

Plate 13 Macrophages and mesothelial cells. Smear prepared from a pleural effusion in a young man aged 24. This field contains two active mesothelial cells and a degenerative form above them. The other large cells with marked vacuolation of the cytoplasm are macrophages. The two macrophages on the left contain phagocytosed debris. Giemsa stain. Magnification × 1080

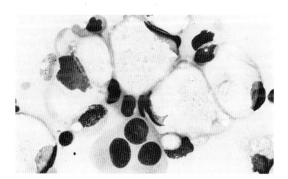

Plate 14 Signet-ring macrophages in a smear from a pleural effusion. Three degenerate mesothelial cells at 6 o'clock surrounded by macrophages. The cytoplasm contains large vacuoles giving a signet-ring appearance to the cells. Giemsa stain. Magnification × 1080

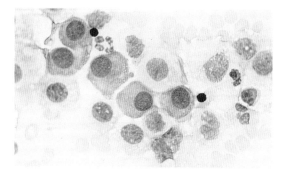

Plate 15 Mesothelial cells in serous effusion. Most of the cells in this field are mesothelial cells; the active forms have dark staining cyanophilic cytoplasm whereas the degenerative forms have pale-staining cytoplasm. Giemsa stain. Magnification × 1080

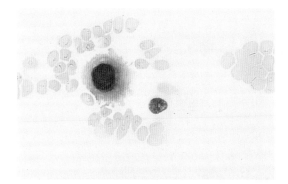

Plate 16 Microvilli on mesothelial cells in pleural effusion. A mesothelial cell showing cytoplasmic blebs and fringes of 'hairs' which stain pink in Giemsa stained films. A discrete lymphocyte can also be seen in this field. Giemsa stain. Magnification × 1080

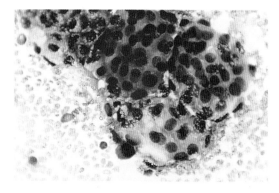

Plate 17 Papillary adenocarcinoma. Smear from a pleural effusion in a 70-year-old woman who had carcinoma of the breast resected eight years previously. The smear contained numerous typical spherical collections of adenocarcinoma cells. Note the peripheral location of the nuclei of the tumour cells in the clusters shown here. Giemsa stain. Magnification × 570

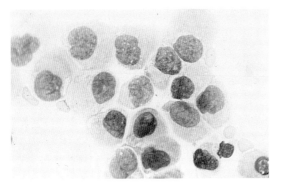

Plate 18 Adenocarcinoma, free cell form. Smear prepared from an ascitic fluid from an elderly male. The discrete tumour cells typical of adenocarcinoma differ from mesothelial cells by having pale staining opaque cytoplasm, a high nucleo-cytoplasmic ratio and an eccentric irregular nucleus. Giemsa stain. Magnification × 1080

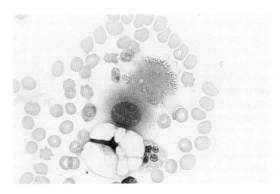

Plate 19 Microvilli on tumour cells. Malignant cells resembling mesothelial cells are present in this field. One has a large cytoplasmic spur covered in microvilli. Smear prepared from ascitic fluid from a patient with bilateral papillary cystadenocarcinoma of the ovaries. Giemsa stain. Magnification × 1080

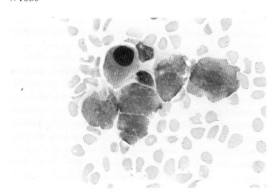

Plate 20 Oat-cell carcinoma. Smear prepared from pleural fluid from an 88-year-old woman with clinical features suggestive of bronchial carcinoma. Many cells typical of oat-cell carcinoma were present. This group of flattened pink staining nuclei form a mosaic pattern. The individual tumour cells are larger than the mesothelial cell at 11 o'clock. Giemsa stain. Magnification × 1080

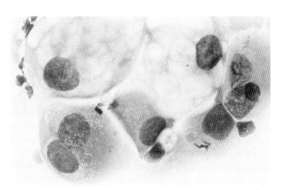

Plate 21 Squamous carcinoma. Smear from pleural effusion in a 45-year-old male with squamous carcinoma of the lung. Several large malignant squamous cells are illustrated here. These rather pleomorphic discrete cells are larger than mesothelial cells. One is binucleate – a common finding in these cases. Giemsa stain. Magnification × 1080

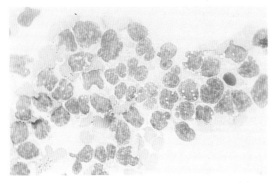

Plate 22 Lymphoblastic lymphoma. Smear prepared from pleural fluid from a male with lymphoma. The lymphoblastic cells show little cytoplasm, nuclear fragmentation and clefts. Giemsa stain. Magnification × 1080

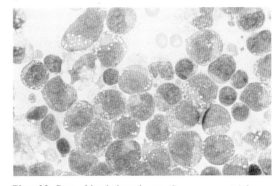

Plate 23 Centroblastic lymphoma. Smear prepared from a pleural fluid from a 47-year-old man with mediastinal lymphoma. The numerous centroblasts in this field have typical blue cytoplasm, high nuclear-cytoplasmic ratio, several nucleoli and vacuolated cytoplasm. Giemsa stain. Magnification × 1080

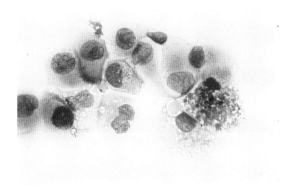

Plate 24 Melanoma. Smear of a pleural fluid from a 28-year-old male who had a history of melanoma. The effusion developed four months later and contained many macrophages packed with melanin and a monotonous population of tumour cells. Many tumour cells contained very fine melanosome granulation as illustrated. Giemsa stain. Magnification × 1080

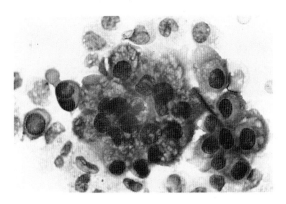

Plate 25 Mesothelioma. Smear of a pleural fluid from a male aged 72. Many mesothelial cells of benign appearance with one cell engulfing another are present. The crenated papillary formation typical of mesothelioma can be seen in this field. Giemsa stain. Magnification × 1080

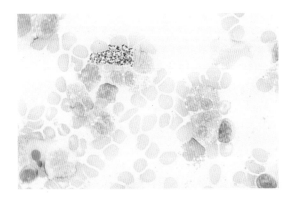

Plate 26 Eosinophils in pleural effusion. The eosinophils in this field show the bilobed nuclei and orange-staining granules in the cytoplasm which is characteristic of these cells. A basophil polymorph can be seen at 11 o'clock. Giemsa stain. Magnification × 1080

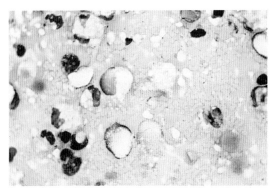

Plate 27 Rheumatoid arthritis. Smear from pleural effusion in a patient with rheumatoid arthritis. The slide is covered with an exudate composed of discrete regular globules of RA protein. Many necrotic cells are present. Giemsa stain. Magnification × 1080

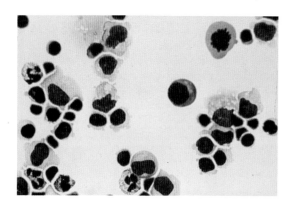

Plate 28 Reactive lymphocytes in CSF. Note variety and cell in mitosis. Stain: May-Grünwald–Giemsa. Magnification ×800

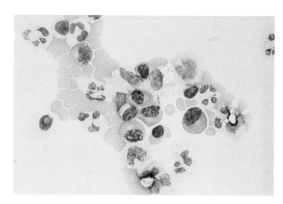

Plate 29 Reactive lymphocytes in CSF—a postoperative sample. Note plasma cell. Stain: May-Grünwald–Giemsa. Magnification ×800

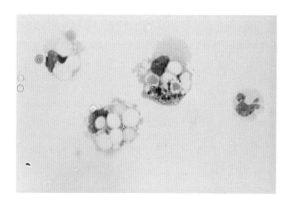

Plate 30 Macrophages in CSF containing vacuoles and ingested red cells (erythrophages). Stain: May-Grünwald–Giemsa. Magnification ×800

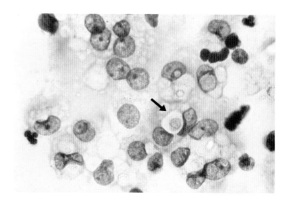

Plate 31 Macrophages in CSF containing cryptococci—round pink-staining bodies (arrowed). Note the active macrophages with nucleoli. Stain: May-Grünwald–Giemsa. Magnification ×800

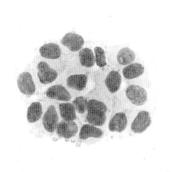

Plate 32 Clump of choroidal or ependymal cells in CSF. Stain: May-Grünwald–Giemsa. Magnification ×800

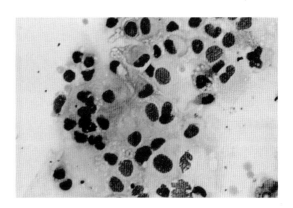

Plate 33 Brain or glial cells with some histiocytes in CSF. Stain: May-Grünwald–Giemsa. Magnification ×800

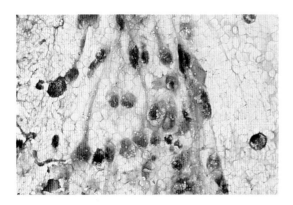

Plate 34 Astrocytoma—a smear from a cyst fluid. Well-differentiated cells showing fibrils. Stain: haematoxylin and eosin. Magnification ×800

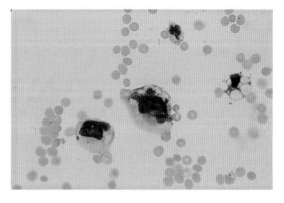

Plate 35 Astrocytoma—very undifferentiated malignant cells from a recurrent Grade IV tumour. CSF. Stain: May-Grünwald–Giemsa. Magnification ×800

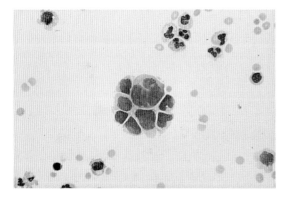

Plate 36 Medulloblastoma—characteristic clump in the form of pseudo-rosette. Note variation in size of cells. CSF. Stain: May-Grünwald–Giemsa. Magnification ×800

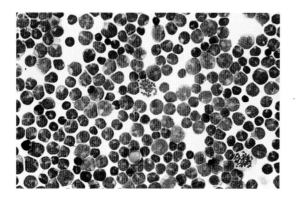 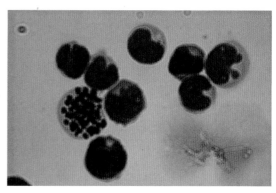

Plate 37 Lymphoblastic leukaemia. (*a*) A very dense field including mitotic figures.

Plate 37 (*b*) High power view, again with mitotic figure. CSF. Stain: May-Grünwald–Giemsa. Magnification ×800

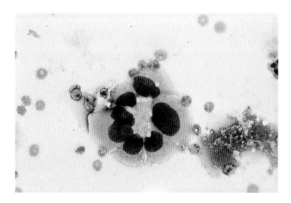 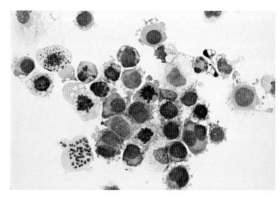

Plate 38 Breast carcinoma cells in CSF. Well-differentiated cells showing signs of glandular formation; patient had undergone mastectomy nine years previously. Stain: May-Grünwald–Giemsa. Magnification ×800

Plate 39 Breast carcinoma cells in CSF. Compare these poorly differentiated carcinoma cells with *Plate 22*. This patient had a mastectomy for an anaplastic carcinoma the previous year. The cells have extremely bizarre nuclei and a high proportion of mitotic figures. Stain: May-Grünwald–Giemsa. Magnification ×800

Plate 40 Cluster of cells from gastrointestinal tumour: PAS stain showing strongly positive staining of cells. CSF. Magnification ×800

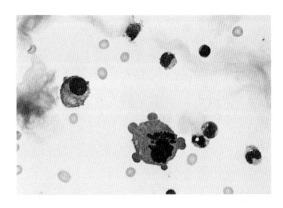

Plate 41 (*a*) Adenocarcinoma cells in CSF showing pseudopodia-like projections.

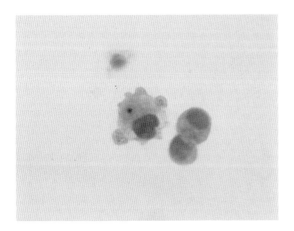

Plate 41 (*b*) Same patient—Alcian blue stain showing positive staining of projections.

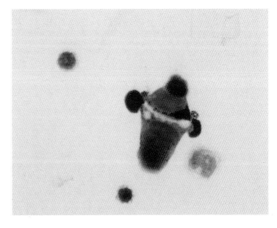

Plate 41 (*c*) PAS stain—positive staining in the pseudopodia. Magnification ×1152

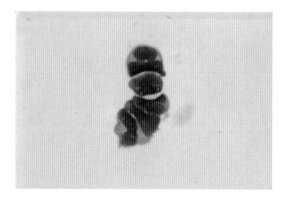

Plate 42 Lung carcinoma. Oat cell carcinoma in CSF showing characteristic 'moulding'. Stain: May-Grünwald–Giemsa. Magnification ×800

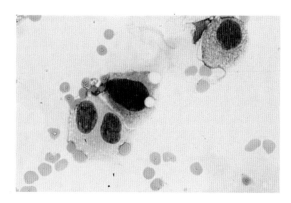

Plate 43 Melanoma cells in CSF (*a*) Stained MGG. Note one cell showing granular staining and one binucleate cell. Magnification ×800.

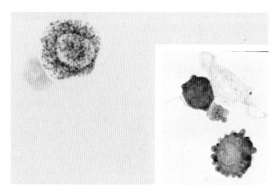

Plate 43 (*b*) Stained Masson-Fontana (specific pigment stain) strongly positive. Magnification ×1440.
Insert: S100 protein stain strongly positive. Magnification ×700

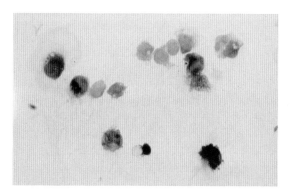

Plate 44 Macrophages in CSF staining strongly positive with non-specific esterase stain. Magnification ×800

16

Cerebrospinal fluids

Patricia M. Norman

Formation and circulation of cerebrospinal fluid

The cerebrospinal fluid (CSF) is produced by the choroid plexuses which consist of a papillary mass of cubical epithelium mounted on a simple vascular framework. The choroid plexuses are situated in all four ventricles and secrete about 0.35 ml of CSF every minute (about 500 ml/day). As some CSF is also derived from the extracellular brain fluid flowing towards the ventricular system, the total amount of CSF in the cerebral circulation is approximately 130 ml (Fishman, 1980). The CSF circulates from the lateral ventricles into the third ventricle and onwards to the fourth and then passes into the subarachnoid space via the foramina of Luschka and Majendie. Some flows down into the spinal subarachnoid space but most passes upwards over the surface of the cerebral hemispheres and cerebellum. The CSF is then mainly absorbed by the arachnoid villi which project from the superior sagittal sinus into the subarachnoid space. *Figure 16.1* shows the areas of secretion of CSF and its circulation. The main regions of the brain and spinal cord are shown and the positions of the main primary tumours are indicated. Secondary tumours are commonly found in many of the same situations.

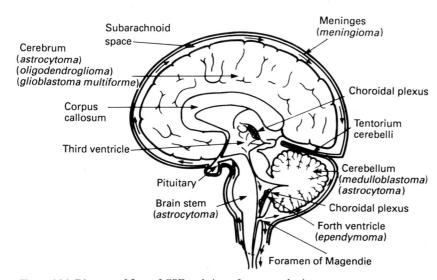

Figure 16.1 Diagram of flow of CSF and sites of common brain tumours

Aspiration of CSF

CSF for examination is obtained by aspiration using a sterile needle either by lumbar puncture at the level of the fourth/fifth lumbar vertebrae which is below the level of the spinal cord, or after a burr hole has been made in the skull which allows a needle to be passed into the ventricles. Cyst fluid may also be obtained during craniotomy. CSF can also be obtained from the cisterna magna at the upper end of the spine, the needle being inserted laterally at the junction of the head and neck. This is usually performed as part of a cervical myelogram by a radiologist. In babies CSF can be aspirated without difficulty directly from the ventricles through the fontanelle. Aspiration of CSF is indicated as a diagnostic measure in patients thought to be suffering from bacterial, viral or fungal infection. It is also of value as a diagnostic test in cases of demyelinating disease such as multiple sclerosis, sarcoidosis and in subarachnoid haemorrhage or suspected carcinomatous meningitis. The aspiration of CSF is often contraindicated in patients with large intracranial space-occupying lesions because of the risk of a major shift of vital centres in the brain stem due to alteration in pressure following the procedure (coning). Therefore, in many cases of suspected tumour CSF is not obtained.

Whenever investigation of CSF is indicated it is most important that there is the fullest possible consultation between the laboratory and the clinician as this will ensure that essential diagnostic tests on the fluid are not omitted. It is extremely important that the specimen is transferred to the laboratory as speedily as possible and that the specimen is never put in the refrigerator overnight; some delicate bacteria may not survive this treatment and cells in the CSF will begin to degenerate. The practice of adding fixative is, in general, unsatisfactory as the CSF which may not contain many cells, is further diluted.

CSF cytology

The study of cells of the spinal fluid has, with the introduction of new techniques, become an essential part of investigation of the patient with neurological disease. It should be possible for laboratories which process these fluids infrequently to produce preparations which can be forwarded for identification or special stains to specialist centres. It is important that the fluid should be correctly handled from the start and by following some simple rules good preparations can easily be made.

Cytology is particularly valuable in the detection of metastatic and recurrent tumours. Early cytological detection of involvement of the CNS may prevent surgical intervention and is essential for the correct management of patients with leukaemia. Optimal techniques and results are most likely to be developed in cytology laboratories serving neurosurgical units of large general hospitals who have a wide clinical and histological experience of the pathology of the CNS. However, it is within the competence of every cytology laboratory to analyse spinal fluid successfully should the occasion arise. In addition, most haematology departments and, in some cases, the microbiology departments can also deal with CSF specimens if necessary, if there is no cytology department within the hospital.

Methods of preparation

Collection

The CSF should be collected into three plastic bottles; sterile universal containers (e.g. Sterilin) are suitable for this purpose. It is essential to use plastic containers as cells adhere to glass and are lost to preparation. If possible, approximately 2–3 ml of CSF is allowed to drip under natural pressure into each container which must be numbered sequentially (1, 2 and 3) in order of withdrawal. A separate 0.5 ml fluoride container should also be provided for glucose and lactate analysis.

The specimen should then be taken to the laboratory as quickly as possible, preferably within one, or at most, two hours. Speed of processing is extremely important at this stage. Most laboratories are equipped with a cytocentrifuge, or have access to one in another department of their own hospital, and can process the specimen immediately. Failing this, they must send the specimen elsewhere as a matter of extreme urgency.

A cell count is undertaken using bottle No. 3. A modified Fuchs–Rosenthal counting chamber, depth 0.2 mm, should be used and red and white cells must be counted separately. If possible slides for cytology are also made from this specimen. Bottle No. 2 is used for culture, especially if there is blood contamination, and bottle No. 1 is left as a reserve for further tests, e.g. virology.

Normal CSF contains no red cells. The mean lymphocyte count in normal individuals (Tourtellotte et al., 1964), is $1.1-3.5 \times 10^9/l$. A count of $4 \times 10^9/l$ is regarded as borderline and $5 \times 10^9/l$ lymphocytes as indicative of underlying pathology. The presence of polymorphs in a CSF which is not bloodstained is also an abnormal finding.

After an initial cell count and appraisal of the clinical details, it should be possible to divide the fluid to the best advantage if the clinical condition of the patient is clearly indicated on the request forms. In some cases it may be necessary to contact the clinician as there may be several possible diagnoses, e.g. tuberculous meningitis, viral encephalitis (Dayan and Stokes, 1973) or malignant meningitis, and inadequate CSF to set up tests for them all.

Specimens with very high cell counts, up to $200 \times 10^9/l$, may need to be diluted, preferably with the supernatant from the same specimen. (Alternatively any iso-osmotic fluid can be used. Although air encephalography is very rarely performed now, in cases where specimens of CSF are obtained during this procedure the post air specimen will contain a higher number of cells. More frequently, the volume of CSF is small and the cells scanty so it must be used as efficiently as possible, occasionally lightly centrifuging the CSF before using the cytocentrifuge.)

Two slides for cytological examination should be made in all cases and stained by the May-Grünwald-Giemsa technique.

Factors to be taken into consideration in deciding whether to make further slides or to divert some of the specimens for other investigations, e.g. viral and bacterial culture, include:

(1) A high lymphocyte or polymorphonuclear count.
(2) Atypical or large cells seen in the counting chamber.
(3) Where positive cytological findings in conjunction with the clinical details indicate that further specialized stains are necessary, e.g. possible adenocarcinoma necessitates a Periodic Acid Schiff (PAS) or Alcian blue stain. Slides should also be set aside for immunocytochemistry, e.g. T and B cell markers, glial markers or non-specific esterase.

If the lymphocyte count of the CSF is raised, some of the specimen must be sent for viral studies and culture for mycobacteria. If the polymorph count is raised a pyogenic infection should be suspected and these specimens investigated accordingly.

Preparation of slides

Glass slides well cleaned in alcohol should be used. The slides may be coated with poly-L-lysine or egg albumin which is considered by some workers to give greater cell adherence, especially when immunocytochemical staining is anticipated. Bearing in mind the affinity of cells for glass this is probably, in most instances, unnecessary. In our experience using poly-L-lysine and egg albumin affords only marginal benefit.

Preparation of specimens

Several methods of preparing CSF for cytological investigations are available; these include sedimentation techniques (Sayk and Sörnäs), millipore and cytocentrifugation. It must be remembered that centrifugation of the specimen followed by smears made from the deposit, even when it is resuspended in 30% bovine albumin, is unsatisfactory and should no longer be used.

Sedimentation

This method is valued for its simplicity, and is widely used in Europe. It was pioneered by Sayk who devised the sedimentation chamber shown in *Figure 16.2* although this has never been produced commercially. Sayk's chamber and variations of it have been described by Den Hartog Jager (1969) and Kolmel (1976).

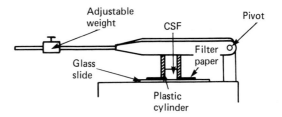

Figure 16.2 Sayk apparatus for sedimentation of CSF

The principle involved is that under the influence of gravity cells will sediment onto a glass slide while the fluid in which they are suspended will gradually be absorbed into a filter paper. The rate of sedimentation and diffusion of the liquid phase are controlled by applying pressure on the filter paper. The technique offers a very satisfactory method of preparing fluids, although there is some loss of cells into the filter paper; moreover a separate piece of apparatus is required for each slide and it is, therefore, less suitable for laboratories examining large numbers of specimens, or where several slides have to be prepared from each specimen.

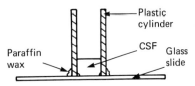

Figure 16.3 Sörnäs apparatus for sedimentation of CSF

A modified sedimentation method has been described by Sörnäs (1967) which does not require special apparatus (*Figure 16.3*) and can be constructed in any laboratory. One end of a glass or plastic cylinder 4 cm high with an inner diameter of 1 cm is immersed in liquid vaseline and applied immediately to a clean slide and kept in place until the vaseline hardens. Fresh CSF (0.7 ml) is pipetted into the chamber and the cells allowed to sediment for 40 min to 1 h. During this time most of the cells should be deposited on the slide. Sedimentation for longer periods of time increases cell degeneration. After the specified time (40 min to 1 h) the supernatant is removed with a Pasteur pipette. The last traces of fluid can be absorbed by holding the tip of a piece of blotting paper over the centre of the preparation, allowing it to dry from the periphery inwards; the vaseline is scraped off and the film air-dried as quickly as possible in an air stream or jet. The more quickly the deposit is dried the less cell shrinkage there will be.

Both the Sayk and Sörnäs procedures must be carried out in an exhaust protective cabinet.

Millipore filters

This method is rarely used now except in conjunction with another method. It provides a three-dimensional view of cells. The earlier filters could not be stained with Romanowsky stains, Papanicolaou staining being used. Nucleopore filters (Sterilin, pore size 0.45 μm) are not affected by staining techniques, allowing the cells to show up more clearly. The main advantage is that almost all the cells are collected on the matrix but may be difficult to identify. Although filter preparations are unsuitable for immunocytochemical staining, they can be used for electron microscopical studies.

Cytocentrifugation

This method is based on the principles of sedimentation and absorption but is designed to accelerate the sedimentation process by centrifugation (*see* Chapter 4).

One of the most widely used cytocentrifuges suitable for CSF cytology is the Shandon Cytospin. The head is autoclavable and can be loaded and unloaded in an exhaust protective cabinet. It can be autoclaved in cases of spillage of infected material.

In our laboratory, using the Mark II Cytospin, 0.3 ml of fluid is placed in each of two chambers and the machine is run at 1150 rev/min (G force 129) using the high acceleration mode for 5 min.

Evaluation of preparatory techniques

The two methods producing the best cytological preparations are the modified sedimentation chamber of Sayk and the Shandon Cytospin Mark II. In both methods the cells lie flat on the slide allowing the morphological details to be more fully appreciated under the microscope. Both suffer to a greater or lesser extent from the disadvantage of cell loss into the filter papers, usually depending on the rate of diffusion. In both methods some spinal fluid is available for biochemical investigations as neither method requires large volumes. Some workers (Chu, Freiling and Wassilak, 1977) found too much cellular distortion and a very poor yield with the cytocentrifuge, but in our laboratory we have not found this to be a problem. Where the cells are scanty the modified sedimentation method of Sörnäs may be useful and rapid drying of the film by air jet will minimize the otherwise inevitable shrinkage and distortion of cells (Herndon, 1971).

Two films should always be stained with

May-Grünwald–Giemsa. The staining can be done on the staining machine or in Coplin jars. In each case the stains must be made up freshly each day. The slides are fixed in methanol for 5 min, stained with May-Grünwald and diluted 50/50 with buffered water (Sorensens) for 5 min. They are then stained in a 1/10 dilution of Giemsa and buffered water pH 6.8 for 10 min and finally washed in buffered water pH 6.8. The slides are drained, allowed to dry and then mounted.

Methanol fixation is also suitable for Periodic Acid Schiff (PAS) staining. This can be carried out on slides previously stained by the May-Grünwald–Giemsa method which have been decolourized with acid alcohol (70% alcohol, 1% HCl). If the slides are to be stained by the haematoxylin and eosin method, either ethanol or methanol fixation is suitable. On the other hand, if a Papanicolaou stain is preferred, the slides must be wet-fixed in methanol.

Fixation in cold acetone is suitable if special staining techniques using monoclonal antibodies are anticipated. Air-dried films stored at $-20\,°C$ in a sealed container can also be used for immunological staining but the morphology of the cells is very poorly preserved.

Classification of cells in CSF

There is a wide variety of opinions as to functions and nomenclature of cells found in CSF. Many different names are applied to the same cell type. The classification of cells proposed by Wieczorek, Schmidt and Olischer (1974) and Oehmichen (1976) which attempts to standardize CSF cytology, appears logical and should be universally adopted. Cell types which may appear in CSF under both normal and pathological conditions can be classified as follows:

(1) Immunocompetent cells—lymphocytes.
(2) Mononuclear phagocytes—macrophages/histiocytes.
(3) Polymorphonuclear granulocytes—neutrophils, eosinophils, basophils.
(4) Erythrocytes.
(5) Cells lining the CSF space—choroid plexus cells, ependymal cells, arachnoid cells.
(6) Other cellular elements—cartilage cells, cells from bone marrow, glial cells and capillary fragments.
(7) Tumour cells.

Benign cells

Lymphocytes

Lymphocytes are the cells most frequently encountered in the CSF. A few small lymphocytes will be found in most specimens. However, numerous reactive forms (*Plate 28*) may be seen in response to antigenic stimulation. These include large 'blast' forms and plasma cells and also include a range of lymphoid cells with dark basophilic cytoplasm and an enlarged nucleus with a fine chromatin structure. Unlike the immunoblast, these intermediate forms rarely have nucleoli. The plasma cells can be recognized by their oval shape, eccentric nuclei, clear basophilic cytoplasm and distinctive clockface nuclear pattern. Reactive lymphocytes may be found in viral, bacterial or fungal meningitis, in multiple sclerosis, after radiotherapy, intrathecal administration of drugs or even post-operatively (*Plate 29*).

In tuberculous meningitis the lymphocyte count will be raised and many reactive forms including cells in mitosis may be present. The differential diagnosis must be made between this and leukaemic infiltration. Careful appraisal of the range of cell type and full clinical details should allow differentiation to be made. In a reactive state all stages of lymphocyte transformation will be present, usually including plasma cells. In lymphoma, one distinct cell line will predominate in the smears.

In multiple sclerosis the lymphocyte count is commonly elevated, rarely more than 25 cells/mm^3 and often under 10 cells/mm^3, but may be normal even during an acute exacerbation. However, even if the total cell count is normal, reactive lymphocytes or plasma cells may be present. Reactive cells are also seen in *Listeria* meningitis and in neurosyphilis.

Mononuclear phagocytes (macrophages)

Normal CSF may contain macrophages which constitute a small percentage of the mononuclear cell population. About 60% are derived from peripheral blood and resemble blood monocytes; the remainder are derived from microglial cells of mesodermal origin. The number of macrophages in CSF is invariably increased in subarachnoid haemorrhage, resolving meningitis and after intracranial surgery. When small numbers of red cells are present in the CSF the presence of erythrophages (macrophages with ingested red cells) indicate a subarachnoid haemorrhage rather than a traumatic tap (*Plate 30*). The presence of macrophages containing haemosiderin pigment together with macrophages containing ingested red cells indicates a previous as well as a recent cerebral haemorrhage. Macrophages may show considerable activity in certain infections e.g. *Cryptococcus neoformans* (*Plate 31*). The cells may be present in small or large groups and frequently have a vacuolated cytoplasm. This, together with prominent nucleoli and numerous mitoses, may mislead inexperienced observers to an erroneous diagnosis of malignancy. Only experience, caution and a full assessment of the clinical details will help to avoid this pitfall.

The distinction between macrophages and reactive lymphocytes can be very difficult. Macrophages usually have more abundant pale cytoplasm which is often vacuolated. Nuclear shape is very variable and they frequently show phagocytosis. If doubt persists, staining for non-specific esterase is helpful.

Polymorphonuclear granulocytes

Neutrophils are not present in a normal specimen unless it is contaminated with blood, but may be found after surgical intervention or infection. A differential count on sequential cytological preparations can be very helpful. In resolving meningitis and in postoperative cases, the initial high polymorphonuclear cell count will be gradually replaced by an increasing number of macrophages. The differential count can be especially useful where infection is suspected but, in fact, is not present. In these cases, the total count remains steady although the percentages of polymorphs and macrophages may be reversed. An increase in the macrophage count indicates a resolving process. An upward swing of polymorphs in a postoperative case may be the first indication of postoperative infection.

Eosinophils may be present where allergic conditions exist and in parasitic diseases. They have also been described in a variety of other conditions.

Ependymal cells and choroid plexus cells

It is difficult to distinguish between these two types of cells as they can be remarkably similar (*Plate 32*); both types can be seen in ventricular CSF but only ependymal cells will be seen in spinal CSF. Degeneration is rapid following exfoliation.

Other cellular elements (bone marrow cells)

Fragments of bone marrow may rarely be seen in CSF specimens following a traumatic puncture and can be distinguished from leukaemic infiltration by the variety of cell types present (e.g. both erythroid and myeloid types of cell).

Glial tissue (*Plate 33*) and capillary fragments may be present in ventricular CSF; the latter may also be found in lumbar puncture specimens.

Tumour cells

Most cells in spinal fluid tend to round up irrespective of their original shape and tumour cells in unfixed CSF are usually swollen. Large clumps of tumour cells are often very dense and difficult to identify. They may be few in number and numerous lymphocytes and macrophages may be present to add to the difficulty.

The criteria used in the diagnosis of malignant cells in cerebrospinal fluid are the same as for tumour cells elsewhere. The following criteria may be useful for diagnosis.

(1) The cells are foreign to the environment. They differ from any type of cell normally or accidentally found in CSF.
(2) The cells show characteristics of malignancy e.g. large size, high nucleocytoplasmic ratio, aberrant nuclear material (possibly indicating polyploidy), enlargement of nucleoli, cellular and nuclear pleomorphism, variation in the nuclear size and abnormal mitotic figures.
(3) The cells show features suggestive of a particular tumour type, e.g. cytoplasmic vacuolation and signet-ring cells suggestive of adenocarcinoma. Positive staining for mucopolysaccharide with PAS or Alcian blue or other appropriate stains will confirm this.

Primary tumours of CNS

The cytological appearance of primary tumours

in CSF preparations cannot always be related to the histological appearances in sections. Medulloblastoma cells, for example, may resemble the more primitive cells of the lymphoid series rather than the small, dark, intensely staining cells seen in histological section.

Astrocytoma

This tumour is found within a wide age group. The benign form often presents in childhood in the first and second decades. It is usually found in the cerebellum and brain stem and less commonly in the hemispheres.

Cells from such tumours are rarely seen in CSF on first presentation unless the tumour has infiltrated the wall of the ventricle and is disseminating cells into the CSF. These tumours are frequently cystic and cyst fluid may be aspirated at the time of biopsy and sent to the cytology laboratory. It may be treated in a similar fashion to CSF but if high in protein may need to be diluted in saline.

The cell type may range from well differentiated astrocytes (*Plate 34*) showing fibril formation (Grade I) to very undifferentiated cells (Grades III and IV). The fibrils are better shown using haematoxylin and eosin. Occasionally malignant astrocytoma cells may be found in the CSF if the tumour recurs (*Plate 35*).

Oligodendroglioma

These cerebral hemisphere tumours of adults can, like any others, impinge on the ventricles and cells from them can be found in the CSF but this, in practice, is rare.

Medulloblastoma

This tumour is commonly found in the midline where it may grow into the fourth ventricle; positive cytology is in practice frequently found. It is chiefly a tumour of the first two decades with a peak at 6–8 years of age. A second less sharp peak occurs in the 20–30 age group. These adult tumours are more likely to be situated laterally than in the midline and cells, therefore, are less likely to be found in the CSF. Medulloblastoma cells in the CSF are not often looked for at the time of the initial biopsy. Requests to confirm or refute their presence are most frequently made when a recurrence or postoperative spread is suspected.

Differentiation from morphologically similar cells, such as those derived from retinoblastoma or pineoblastoma, has to be made by taking the site of the tumour into account. Differential diagnosis between medulloblastoma and germinoma, which is a midline form of teratoma, and lymphoma may also be difficult. Leukaemic cells usually have clear margins and distinct outlines, and some medulloblastoma cells resemble them closely. Often, however, the cells are starting to disintegrate which gives an indistinct cell outline. They may also appear in small clumps; pseudo-rosettes, if present, are a characteristic feature and a help in diagnosis (*Plate 36*).

Lymphoma/lymphoblastic leukaemia

Primary cerebral lymphoma (microglioma) is much less common than other cerebral tumours including pineal tumours, whereas leukaemia or lymphomatous infiltration of the meninges is not uncommon (*Plate 37*).

One distinct cell line will predominate in the smear and obvious blast cells will be seen. The pattern can be distinguished from that found in tuberculous or viral meningitis by the range of reactive lymphocytes seen in the latter conditions, including atypical lymphocytes similar to those seen in infectious mononucleosis. Moreover, in lymphoma or leukaemic infiltration, plasma cells are rarely found.

The analysis of CSF from patients with leukaemia with CNS involvement who have received intrathecal cytotoxic therapy may be difficult. The cytologist may be asked to distinguish between reactive lymphocytic response to therapy and leukaemic relapse. This is really a test for the haematologist as it may be necessary to refer to bone marrow and peripheral blood smears to ascertain the nature of the cells in the CSF. Immunocytochemical staining may prove to be of value in resolving some of these cases.

Ependymoma

This is a tumour found in the fourth ventricle and spinal cord. In rare cases it can arise from other ependymal-lined areas. Unlike any other glioma these are closer in type to epithelial cells.

Choroid plexus papilloma

This tumour is usually benign and most often

arises in the fourth and left lateral ventricles; very occasionally it may be malignant. These cells also resemble epithelial cells and show characteristics difficult to distinguish from adenocarcinoma when malignant.

Meningioma and neurofibroma

Cells from these tumours are almost never found in CSF, but a choroid plexus meningioma could shed cells into the CSF.

Secondary carcinomas

Secondary carcinoma deposits may be found anywhere in the brain and spinal cord and may also involve the meninges or ependymal surfaces, resulting in a 'carcinomatous meningitis'. This condition may occur in the absence of a space-occupying lesion. Metastatic adenocarcinoma of breast (*Plates 38* and *39*) is the most common lesion although metastatic carcinomas from many other sites may also occur (*Plate 40*). Adenocarcinoma cells have pseudopodia-like projections or 'frills' (*Plate 41(a)*) which stain positively with Alcian blue and PAS (*Plate 41(b)* and *(c)*). Metastatic oat cell carcinomas (*Plate 42*) can be recognized by their characteristic moulding but are rare, adenocarcinomas being more common. Melanoma may also metastasize to the CNS. The cells may, even when stained by May-Grünwald–Giemsa, show granulation (*Plate 43(a)*), but in amelanotic forms the cytoplasm may be clear. Specific stains for melanin pigment such as Masson-Fontana (*Plate 43(b)*) can be used to confirm the diagnosis. The stain for protein S100 will also be positive (*Plate 43(c)*) in both melanotic and amelanotic forms, and in practice this may be of great value despite the fact that S100 may be found in various other conditions, e.g. histiocytosis X and certain glial tumours.

Special staining techniques as an aid to diagnosis

In some cases where a small amount of fluid has been provided or the cell count is low a positive diagnosis of malignancy may be difficult because the number of cells in the fluid is very small. A repeat specimen should be requested whenever possible to ascertain the diagnosis. In other cases, particularly where active macrophages are numerous (*Plate 44*) special stains may be of value in discriminating between tumour cells and macrophages.

Staining for non-specific esterase, which picks out macrophages and T cells, is especially useful for cyst fluids and CSF samples containing numbers of macrophages. A panleucocyte marker such as HLe 1 or DAKO LC (CD45) can also be used. These react only with leucocytes and are particularly strong markers for cells of the lymphoid series. T cell and B cell monoclonal antibodies may be helpful in differentiating reactive lymphocytic conditions from lymphoma and leukaemia (*Table 16.1*). Where a single (monotypic) light chain is demonstrated the diagnosis of a monoclonal, and thus malignant, proliferation of B cells can be made. Biclonal tumours which synthesize both kappa and lambda chains are extremely rare so that for practical purposes the finding of both chains indicates a non-malignant process.

Table 16.1 T cell and B cell monoclonal antibodies for differentiating reactive lymphocytic conditions from lymphoma and leukaemia

Antigen	Cellular distribution
T cell markers	
CD3[a]	T cells
CD4	Helper/inducer T cells
CD5	T cells, B-CLL
CD8	Cytotoxic/suppressor T cells
B cell markers	
CD19	B cells
CD10	Common ALL Antigen bearing cells

[a] CD = Cluster of differentiation

Other monoclonal antibodies, e.g. GFAP (glial fibrillary acidic protein) can help to differentiate between astrocytomas, ependymal well differentiated tumours and glioblastomas and oligodendrogliomas. The first three tumours stain positively with GFAP whereas oligodendrogliomas are negative.

However, each monoclonal antibody test requires two slides and 8–10 slides may be needed. It is not always possible to provide these with the small amounts of CSF available. The interpretation of results from monoclonal antibody tests is frequently difficult. In addition, there may be very few atypical cells on the slide and this will add to the difficulty of interpretation. Large numbers of other monoclonal antibodies are commercially available but it is neither practical nor economic to try to provide a fully comprehensive screen.

When immunocytochemical or histochemical staining is anticipated the laboratory must be forewarned and the specimen delivered there immediately after collection. Smears must be prepared at once by cytocentrifugation. The slides can be quickly air-dried and put in a sealed container (a plastic slide carrier is adequate if closed with paraffin film) and stored at −20 °C (this can be delayed for a few days). It is then possible to decide later which stains are the most appropriate. The slides can then either be fixed and stained or forwarded, still frozen, to special centres for monoclonal antibody testing.

References

CHU, J. Y., FREILING, P. and WASSILAK S. (1977). Simple method for the cytological examination of cerebrospinal fluid. *J. Clin. Pathol.*, **30**, 486–487

DAYAN, A. D. and STOKES, M. I. (1973). Rapid diagnosis of encephalitis by immunofluorescent examination of cerebrospinal fluid cells. Lancet, **1**, 177–179

DEN HARTOG JAGER, W. A. (1969). Cytopathology of the cerebrospinal fluid examined with the sedimentation technique after Sayk. *J. Neurol. Sci.*, **9**, 155–177

FISHMAN, R. A. (1980). Formation of the cerebrospinal fluid. In *Cerebrospinal Fluid in Diseases of the Nervous System*, p. 20. Philadelphia: W. B. Saunders

HERNDON, R. H. (1971). Electron microscopic slides of cerebrospinal fluid sediment. 68th Annual Meeting of American Association of Pathologists and Bacteriologists, jointly with American Association of Neuropathologists. Abstract No. 19.

KOLMEL, H. W. (1976). *Atlas of Cerebrospinal Fluid Cells*. Berlin: Springer-Verlag

OEHMICHEN, M. (1976). *Cerebrospinal Fluid Cytology*. Stuttgart: Georg Thieme and Philadelphia: W.B. Saunders

SÖRNÄS, R. (1967). A new method for the cytological examination of the cerebrospinal fluid. *J. Neurol. Neurosurg. Psychiatry*, **30**, 568–577

TOURTELLOTTE, W. E., HAERER, A. F., HELLER, G. L. and SOMERS, J. E. (1964). *Post Lumbar Puncture Headaches*. Springfield, Illinois: Charles C. Thomas

WIECZOREK, V., SCHMIDT, R. M. and OLISCHER, R. M. (1974). Zur-Standardisierung der Liquorzelldiagnostik. *Deutsche Gesundheitswesen*, **29**, 423–426

17

The urinary tract

Elizabeth Wilson

The development of urinary tract cytology

Cytological examination of urinary sediment has been used as an aid to clinical diagnosis for over 100 years. Professor Lionel Beale from King's College Hospital, London, was one of the first to recognize the value of this approach for the diagnosis of malignant lesions in the urinary tract. Beautiful hand-drawn illustrations of the tumour cells, crystals and casts which he noted in urinary sediment can be seen in his handbook *The Microscope and its Application to Clinical Medicine* which was published in 1854 (Beale, 1854). More recent reviews of urine cytology by Papanicolaou and Marshall (1945), Deden (1954), Esposti, Moberger and Zajicek (1970) and Lewis *et al.* (1976) have done much to promote the use of this method for diagnostic purposes and in this chapter the numerous applications of the procedure in modern medical practice are examined.

Anatomy and histology of the urinary tract

Anatomy

The urinary tract includes the kidneys, ureters, urinary bladder and the urethra (*Figure 17.1*). The *kidneys* are situated in the abdominal cavity, one on each side of the vertebral column, and the right kidney is slightly lower than the left. The upper pole is level with the twelfth vertebra, the lower pole with the third lumbar vertebra. Their main function is the concentration of body fluids and electrolytes as well as the excretion of excess water and the end products of metabolic activity. Each kidney measures approximately $11 \times 6 \times 3$ cm, and weighs 150 g in the male and 135 g in the female.

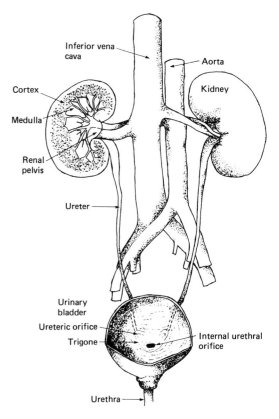

Figure 17.1 Diagram of the urinary tract

The human kidney is covered by a thin capsule, and comprises an outer cortical and an inner medullary zone, the substance of which is made up of 8–18 conical lobes or pyramids. The unit of excretion is the nephron which consists of the glomerulus, the proximal convoluted tubule, the descending and ascending loop of Henle and the distal convoluted tubule. Urine collects in the collecting tubules situated in the medulla. These form a branched system, the largest of which is the ducts of Bellini, which open through the apex of the pyramidal medulla into the renal pelvis. The medullary zone at the tip of the pyramid fits into one calyx, each calyx being an expanded portion of the proximal ureter and urine is conveyed in the ureter to the urinary bladder.

The *ureters* extend from the renal pelvis to the bladder and are two muscular tubular structures measuring 25–30 cm in length, with a diameter of 3 mm. They run in front of the psoas major muscle, into the pelvic cavity and enter the base of the bladder. The outermost fibrous coat of the ureteric wall is continuous with the fibrous capsule of the kidney and fuses with the bladder wall at its distal end. The walls of the ureter are made up of two layers of smooth muscle and peristaltic movements propel urine down the ureters to the urinary bladder. The lining of the ureter is continuous with the epithelium lining the renal pelvis and the bladder. Its surface is thrown into folds so that the ureteric lumen can expand considerably without rupturing; hence renal calculi may be passed with relatively little damage. The ureters enter the bladder at an oblique angle and have a fold of mucosa covering the orifice. The longitudinal muscles of the ureter contract during peristalsis and open the lumen allowing urine to enter the bladder. Longitudinal muscles in the bladder wall contract and close the ureteric lumen to prevent the reflux of urine into the ureters from the bladder.

The *bladder* is a reservoir situated at the base of the pelvis, and varies in size, shape and position according to the amount of urine it contains, expanding upwards and forwards into the abdominal cavity as it fills with urine. The thick muscular wall of the bladder, which is composed of three layers of innervated smooth muscle fibres, enables it to accommodate changes in the size of the cavity. The epithelium lining the bladder is continuous with that of the ureters and the urethra. The base of the bladder is a triangular area known as the trigone, at the apex of which lies the urethral opening, and at the base, the right and left ureteric orifices.

The *urethra* is a tubular structure extending from the bladder to the external orifice. In women it varies between 2 and 6 cm in length, and is 6 mm in diameter. Its inner and outer layers are continuous with that of the bladder. In the male, the urethra measures between 18 and 20 cm extending from the bladder to the external meatus at the end of the penis. It is made up of three sections: the proximal section which passes through the prostate, the membraneous region which extends to the bulb of the penis, and the distal section which is spongiosa. The lamina propria contains smooth muscle fibres, fibroelastic tissues and glands.

Histology

Kidney

The proximal and distal convoluted tubules are lined by a single layer of cuboidal cells as is the thick ascending loop of Henle. The descending loop of Henle is thinner and is lined by a single layer of squamous epithelium. The smaller collecting tubules are lined by cuboidal cells, whilst columnar epithelium lines the ducts which increase in size towards the papillae. Under certain pathological conditions, these cells may be exfoliated into the urine. Casts may also be formed in the distal parts of the nephron and the collecting tubules.

The renal pelvis, ureters, bladder and proximal part of the urethra are lined by transitional epithelium which is a specialized epithelium unique to the urinary tract (*Figure 17.2a*). It is a multilayered epithelium supported on a basal lamina which separates it from the lamina propria which is made up of connective tissue with blood vessels and nerve fibres. The white blood cells often found in the urine migrate from the blood vessels in the lamina propria through the epithelial layers to the bladder lumen.

The transitional epithelium varies in thickness. It is thickest in the bladder where up to seven layers of cells may be found; in the renal pelves, the urothelium is only 2–3 layers thick, whereas in the ureters the average number of cell layers is five. The deeper layers consist of cuboidal cells, generally with a single round or oval nucleus, whilst the superficial cells are larger and often have multiple nuclei. These

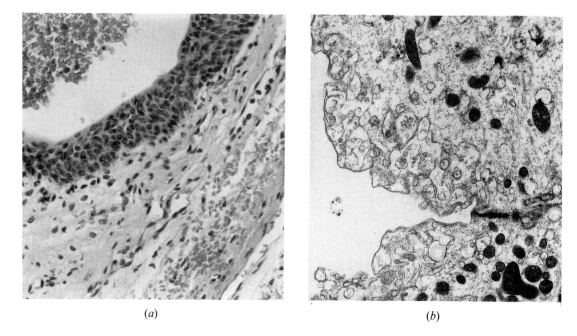

Figure 17.2 (*a*) Histological section through bladder wall showing transitional epithelium. Note the polygonal shape of the transitional cells in the contracted bladder. Haematoxylin and eosin stain. Magnification ×200. (*b*) EM appearance of section from bladder biopsy showing the asymmetric unit membrane on the luminal surface of the urothelial cells. Stain: lead acetate. Magnification ×6000

superficial cells are referred to as umbrella cells as they cover several deeper cells, and vary in shape depending on the degree of dilatation of the bladder. In a contracted bladder they are polygonal; when the bladder is fully dilated they are flat. Their luminal surface is lined by an asymmetrical unit membrane (AUM) (Koss, 1979). Transmission electron microscopy (*Figure 17.2b*) has shown that this membrane consists of two electron opaque layers, uneven in thickness, separated by an electron lucent area. It occupies three-quarters of the cell surface and is attached to filaments which are pulled taut on stretching, but prevent excess stretch during bladder dilatation. The AUM is synthesized by the Golgi apparatus, packaged into vesicles and transported to the cell surface, so it can be replaced if damaged. The transitional cell epithelium provides an avascular barrier between the urine and the underlying tissue. It is not keratinized but the AUM and the tight junctions between adjacent cells provide an adequate system to prevent the penetration of urine.

Although the main epithelial lining of the urinary tract is transitional in type, foci of squamous or glandular epithelium may also be found. Squamous metaplasia may occur in the area of the trigone in up to 50% of adult women, and a small proportion of men (Koss, 1979), and is a normal occurrence. Glandular metaplasia, another normal finding, may occur in large areas of the renal pelvis, ureters, and the bladder.

Von Brunn's nests are found in 80% of normal bladders. These are formed when groups of transitional cells extend into the lamina propria. Cysts lined by mucus-producing columnar epithelium formed as the result of metaplastic change may be found at the centre of these nests. Enlargement of the cysts due to mucus retention gives rise to the condition of cystitis cystica.

Urethra

The proximal part of the male and female urethra has a lining of transitional epithelium similar to that found in the bladder, whereas in the distal parts pseudostratified columnar epithelium is found. An abrupt change to a keratinizing squamous epithelium occurs at the external orifice. In the male, two tubuloalveolar glands known as Cowper's glands open into the membraneous urethra.

Specimen collection

There are four types of sample that are routinely sent for cytological studies: voided urine, catheter urine, ileal loop urine and bladder washings. The majority of samples sent to the laboratory are aliquots of voided urine. If there is a delay in the processing of specimens a fixative must be added.

Voided urine

Ideally a whole voided specimen should be collected in a sterile container and sent for cytological examination. However, in a busy routine department this is often impractical, and a representative sample of between 50 and 100 ml of urine is sent to the laboratory instead. Urothelial cells are often passed at the beginning and the end of voiding, and for this reason mid-stream urine specimens are unsuitable for cytological examination. Early morning urines are also inadvisable as the exfoliated cells may have been in the bladder for several hours and have degenerated (Crabtree and Murphy, 1980) making diagnosis difficult and unreliable. De Voogt, Rathert and Beyer-Boon (1977) suggest that the first morning specimen should be discarded and samples should be collected after exercise, which they believed encouraged exfoliation of cells. If delay is anticipated, the specimen should be kept at 4 °C or fixed in an equal volume of alcohol. (For other urine fixatives see Chapter 4).

Catheter urine

Koss (1979) considered urine collected after catheterization of the bladder to be the most useful specimen for urinary cytology, especially for female patients. He has recommended that three consecutive specimens, each of between 50 and 100 ml, should be examined to provide a reliable assessment of the urinary tract.

Ureteric urine is collected using a catheter at cystoscopy. It may be obtained from one or both ureters and the two samples must be processed separately (Naib, 1961).

Bladder washings

A volume of physiological saline or Ringer's solution is injected into the bladder, withdrawn, and sent for cytological examination. The National Bladder Cancer Collaborative Group (1977) recommend that 50 ml of saline be injected and aspirated five consecutive times. Generally a good cellular sample is obtained with this method (Harris *et al.*, 1971) although as instrumentation traumatizes the urothelium, large fragments of tissue may be detached. Crabtree, Jukkola and Murphy (1981) described a study comparing the efficiency of bladder washings and voided urine specimens from the same patient for the diagnosis of urinary tract pathology. They found a higher cell yield with bladder washings, but overall there was little to choose between the two methods. Other authors (Highman, 1983) consider that bladder washings are of limited value. The difficulty lies in distinguishing well differentiated transitional cell carcinoma and carcinoma *in situ* from traumatically detached hyperplastic epithelium.

Ileal loop urine

This is examined for the follow-up of patients with urinary tract diversions following cystectomy. Care must be taken in the processing of such specimens as they often contain mucoid material which traps the cells from the remaining urothelium.

Processing of cytological samples

Most specimens are unfixed when they are received by the laboratory and should be processed immediately to avoid further degeneration of the cells. If delay is anticipated, an equal amount of fixative (alcohol) should be added to the specimen. All specimens must be examined macroscopically and a note made of the volume and colour of the sample. Clots or fragments of tissue should be sent for histological sectioning. The specimen is then tested with Albustix, and when protein is present, saline added until the reaction yields only a trace. All cloudy urines should be tested with litmus. An alkaline urine is likely to contain phosphate crystals and the cloudiness can be removed by the addition of 10% glacial acetic acid drop by drop. Cloudy urine showing an acid reaction may contain urates and should be allowed to stand in hand-hot water until the cloudiness disappears. An equal volume of saline may be added to prevent the reappearance of crystals and the urine should be processed immediately. Heavily bloodstained specimens can be processed in a similar way to serous effusions to retrieve the urothelial cells.

Cell concentration methods

Efficient harvesting of cells from the urine is critical for the successful diagnosis of tumours of the urinary tract. It is not only essential to retrieve the maximum number of exfoliated cells, but also to preserve the morphology and produce a preparation covering a reasonable slide area for screening in a busy routine laboratory. The two main methods used are membrane filtration and centrifugation, either by the conventional approach or using a Cytospin. Filters and slides are routinely stained by the Papanicolaou technique but more specialized staining methods can be used to aid diagnosis. For further details of processing, cell concentration methods and staining techniques *see* Chapter 4.

Which method?

Many authors recommend the use of membrane filters for preparing diagnostic urinary specimens (Taylor, MacFarlane and Ceelen, 1963; Trott, 1967; El-Bolkainy, 1980) and suggest that the cellulose filters are more useful than the polycarbonate type (Barrett and King, 1976; Crabtree and Stine, 1983). They provide maximum cell collection and good morphological cell detail. Barrett and King (1976) suggest that Millipore filters are the most efficient filters for cytological use but, despite the excellent cell yield, several authors have found that the method is too time consuming for routine use, and recommend that smears are prepared (Beyer-Boon and Voorn-den Hollander, 1978; Marwah, Devlin and Dekker, 1978). However, much cell loss has been described from wet-fixed smears even with albuminized slides. Although Cytospin preparations ensure a good yield, the results are unpredictable, with much variation between repeat tests (Barrett and King, 1976) and this technique is not favoured for routine use. In this laboratory, Gelman filters are used routinely and drop smear preparations made if the urine is bloodstained.

Non-permanent preparations

These may be prepared for the examination of urinary sediment by phase contrast microscopy (PCM) or by toluidine staining. Preparations must be examined immediately as not only will the unfixed cellular constituents degenerate rapidly but the wet preparation will evaporate within one hour. After examination the smears or deposit may be fixed and stained to produce a permanent preparation.

Phase contrast microscopy (PCM)

The main advantage of PCM (De Voogt, Beyer-Boon and Brussee, 1975; De Voogt, Rathert and Beyer-Boon, 1977) is that it provides a rapid screening method for the urologist. It is an office procedure that can be carried out while the patient is waiting. PCM is a better method than brightfield microscopy for the examination of wet preparations, as cellular detail is clearer.

A small sample of centrifuged urine deposit is placed on a glass slide and coverslipped. Normal and abnormal cells are identified; photomicrographs taken of the latter provide a permanent record.

Wet-fresh preparations

These are also used for the identification of abnormal cells in the urine (Holmquist, 1977). The metachromatic dyes commonly used are toluidine blue and methylene blue. Several drops of dye are mixed on a glass slide with several drops of centrifuged urine deposit. The mixture is coverslipped and examined immediately.

Specialist techniques for collection of specimens from ureter, renal pelvis and kidney

One of the disadvantages of urine cytology is that it is impossible to localize tumours by examination of urinary sediment. Consequently specialized techniques have been developed for this purpose which enable the urologist to localize the lesion. These include brush, fine needle aspiration and lavage techniques all of which are particularly useful for detecting lesions in the upper urinary tract. The specimens obtained are processed in the usual way.

Brushing

Radiolucent filling defects in the ureters and renal pelves may be caused by either tumour or calculi or inflammatory processes. As a result, they present problems of differential diagnosis. The cytology of voided urine is of limited value in such cases, since it gives rise to a high false negative rate especially in the detection of low grade tumours (Blute, Gittes and Gittes, 1981). Gill, Lu and Thomsen (1973) and Gill, Lu and Bibbo (1978) have described a method for sampling material directly from a lesion, using a brushing technique under X-ray control. A small wire or nylon brush is passed through the catheter for sampling. The brush is moved from side to side to dislodge cells for examination. After the brush is withdrawn, smears are made of the cellular material attached to the bristles. Speed is essential in the processing of these samples as the cells will very quickly dry out. Blute, Gittes and Gittes (1981) found 78% of filling defects that were due to tumour were correctly diagnosed by this method. Bibbo *et al.* (1974) have recommended the use of this method for the diagnosis of lesions of the renal pelvis.

Ureteric lavage

The use of ureteric and renal pelvic lavage techniques has also increased the diagnostic accuracy of cytology in detecting pathology of the upper urinary tract. Leistenschneider and Nagel (1980) examined ureteric lavage specimens from 101 patients and obtained an overall diagnostic accuracy of 80.5% for tumours. The main advantage of lavage over the brush technique is that it enables areas to be sampled that could not be reached with a brush. However, the use of both techniques at one cystoscopic examination provides maximum material for diagnosis.

Fine needle aspiration

Fine needle aspiration of the urinary tract may also provide diagnostic material from a variety of pathological conditions. Solid and cystic renal lesions may be aspirated under fluoroscopic guidance (Harrison, Batsford and Tucker, 1951), and it is possible to diagnose a range of pathological conditions including renal cell carcinoma, oncocytoma (Rodriguez *et al.*, 1980), lesions of the renal pelvis and retroperitoneal masses by this method.

The cytology of the urinary tract

Under normal conditions, most of the cells found in voided urine are from the urothelial lining. In normal urine these are few in number but under pathological conditions, e.g. the presence of calculi or infection, they may be very numerous indeed. Other cells found in urine include contaminants from the squamous epithelium of the vulva and vagina in the female and seminiferous fluid cells in the male. Light microscopy of the urine may also reveal tumour cells, casts and crystals and a number of specific infections can be diagnosed on the basis of the smear pattern.

Transitional epithelial cells

The appearance of transitional cells in cytological preparations varies according to the method of collection of the samples. Voided urine contains cells that are exfoliated spontaneously, whereas catheter specimens contain cells which are forcibly detached.

Voided urine

Transitional cells are rarely seen in voided urine in the absence of infection. The large superficial cells, so-called umbrella cells, are flat and polygonal in shape and measure between 30 and 50 µm in diameter. They may have one or two round or oval vesicular nuclei measuring 10 µm in diameter, and the chromatin is finely granular.

Deeper layer cells may occasionally be seen and appear as rounded cells measuring 20–25 µm in diameter, with a single round or oval vesicular nucleus measuring 10 µm in diameter. The chromatin is finely granular and a small round nucleolus—a normal finding—is often present. The cytoplasm stains blue or green with the Papanicolaou stain. In the presence of infection they may show degenerative changes including cytoplasmic loss or vacuolation, variation in size, nuclear pyknosis, hyperchromasia or karyorrhexis (*Figure 17.3 (a)*). These features will disappear after treatment, and a cytological diagnosis must not be made on heavily

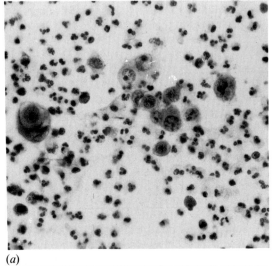

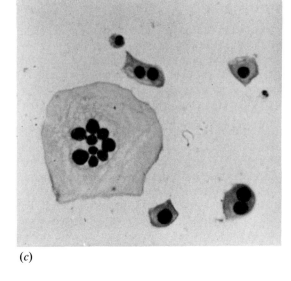

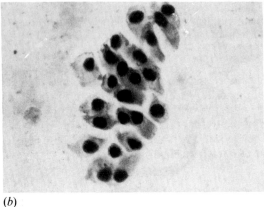

Figure 17.3 (*a*) Transitional cells in voided urine. There are numerous polymorphs in the background. The transitional cells have been exfoliated spontaneously and show degenerative changes including vacuolation and phagocytosis. They vary slightly in size and the nuclei of some cells contain prominent nucleoli. (*b*) A cluster of transitional cells in a catheter specimen of urine. Note cuboidal shape of cells which have been forcibly detached. (*c*) Large syncytial cells in urine. These are sometimes found after purgation and may be ureteric in origin. Papanicolaou stain. Magnification × 500

infected specimens where detail of transitional cell morphology is obscured or poorly preserved.

Normal exfoliated transitional cells usually occur singly in the urine and, as has been indicated, vary in size and shape. Further variation is caused by the degeneration of these cells, the extent of which will depend on the length of time they have been lying in the urine.

In an examination of DNA distribution by Levi *et al.* (1969), normal bladder epithelial cells were shown to be polyploid whereas ureteric epithelium was diploid.

Catheter specimens

These include specimens collected by catheterization of the bladder, bladder washings and ureteric specimens (Trott, 1977). Their method of collection involves forcible detachment of urothelial cells which can be distinguished from naturally exfoliated cells by the fact that they occur in sheets or clusters. They differ in shape from the spontaneously exfoliated cells seen in voided urine (*Figure 17.3(b)*). These often have irregular borders, and the individual cells in the clusters may be very large indeed. Syncytial cells 40 µm in diameter may also be seen in these samples (*Figure 17.3(c)*). Koss (1979) suggests that their appearance reflects a low turnover of the urothelium, resulting in older superficial cells lining the ureters. Holmquist (1977) considers them to be an artefact produced by the breakdown of cellular borders after a sheet of cells has been exfoliated, as these large cells have not been observed in histological preparations of the urinary tract. Another possibility is that the irritative effect of a purgation such as castor oil may cause exfoliation of these cells.

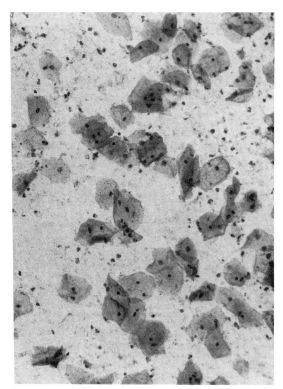

Figure 17.4 Urine sediment. Squamous cells from the vagina and vulva present in urine from a female patient. Stain: Papanicolaou. Magnification × 500

Squamous and columnar cells

Squamous cells are a normal finding in urine from female patients due to vaginal or vulval contamination (*Figure 17.4*). They may also be shed from the trigone area of the bladder if squamous metaplasia has occurred, or from the distal urethra in both the male and female.

Columnar cells derived from the glandular areas of the bladder, urethra and prostate are also a normal finding in the urine. These appear as small cells with a round vesicular nucleus eccentrically placed, and their cytoplasm stains pale blue with Papanicolaou stain. Renal tubular cells may also be exfoliated and are indistinguishable from other columnar epithelial cells.

Inflammatory cells

A few white blood cells may be present in normal voided urine and microhaematuria is commonly found (Freni and Freni-Titulaer, 1977). An infection may increase the number

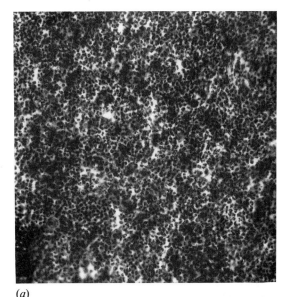

(*a*)

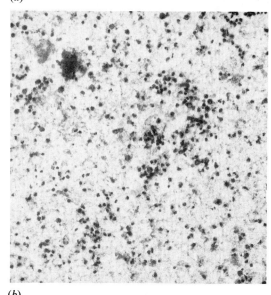

(*b*)

Figure 17.5 Urine sediment. (*a*) An inflammatory exudate obscuring urothelial cells. (*b*) A bacterial exudate obscuring urothelial cells. Stain: Papanicolaou. Magnification × 500

of exfoliated transitional cells, although their features are often obscured by the inflammatory exudate (*Figures 17.5(a)* and *(b)*).

Casts

Casts are formed in the nephrons and consist of organized elements that are shaped by the tubular lumen. They have parallel sides and may be straight or convoluted. The number, type and size of the casts found in the urine are a good indication of renal parenchymal disease (Schumann, Harris and Henry, 1978). Their basic substance is a proteinaceous material and they are generally formed in the distal portion of the nephron and collecting tubule.

Several types of cast may be seen in smears of urine sediment. Hyaline casts are made up of Periodic Acid Schiff (PAS) positive mucoprotein (*Figure 17.6*). Epithelial cell casts contain renal tubular cells, which in urine from a renal transplant patient may contain cytomegalovirus (CMV) inclusions (Schumann, Harris and Henry, 1978) (*Figure 17.7*). Granular casts are formed from the breakdown of cellular material or direct aggregation of serum proteins. Waxy casts are associated with inflammation and degeneration due to degradation of cellular material.

Crystals

Fresh urine must be examined for crystal identification. Phosphates and urates are most commonly found; these and other less common crystals are listed in *Table 17.1*. The conditions under which the crystals may be found are noted in the table.

Organisms

Bacteria are the most common organisms found in the urine and are usually identified using culture methods. Fungal elements and parasites may also be found. Virus infection may be suspected from cytopathic changes in exfoliated transitional cells.

Bacteria

The commonest cause of urinary tract infection is *E. coli*. The small rod-shaped bacilli may be seen in smears together with numerous leucocytes. However, Gram stain or culture is essential for accurate identification of the organisms prior to treatment. Although normally urine is sterile, cytological samples are frequently contaminated with commensals from the perineum. In these cases, a few leucocytes are found in the samples.

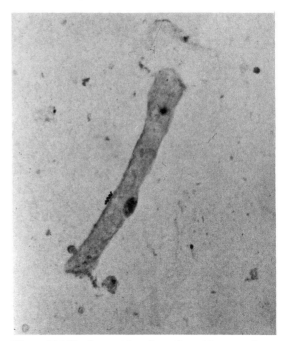

Figure 17.6 Hyaline cast in urine sediment from renal allograft recipient. Papanicolaou stain. Magnification × 1000

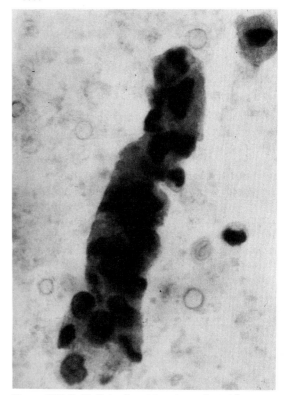

Figure 17.7 Epithelial cell cast in urine sediment from renal allograft recipient. Papanicolaou stain. Magnification × 1000

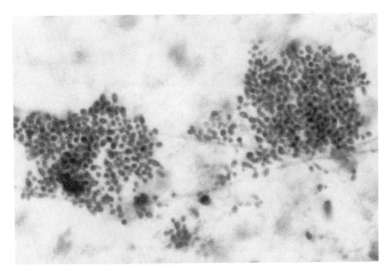

Figure 17.8 *Candida albicans.* Spores and pseudohyphae present in a urine specimen from a renal allograft recipient. Stain: Papanicolaou. Magnification ×2000

Fungi

Candida albicans is the most commonly detected fungal infection. The spores and pseudohyphae stain pink with Papanicolaou stain (*Figure 17.8*). The infection occurs in immunosuppressed patients and diabetic patients (De Voogt, Rathert and Beyer-Boon, 1977) and may form a fungal ball and block the ureter in advanced cases (Koss, 1979). *Blastosis dermatides* has also been identified in the urine (Eickenberg, Amin and Lich, 1975) although culture techniques are needed for positive identification.

Viruses

The two virus infections of the urinary tract which can be detected by cytological examination of urinary sediment are cytomegalovirus (CMV) infection and human polyoma virus (HPV) infection. These infections occur most commonly in patients whose immunity is impaired by drugs or disease e.g. renal allograft recipients (Bossen and Johnston, 1975), patients with AIDS and patients receiving chemotherapeutic agents.

Cytomegalovirus (CMV)

This is an encapsulated DNA virus measuring 120 nm in diameter which is a member of the herpes group of viruses. Infected urothelial cells are enlarged to between 25 and 35 µm in diameter, and contain an intranuclear 'bird's eye' inclusion, which is often separated from the thickened nuclear membrane by a halo. Cytoplasmic inclusions are also seen. It is important to examine multiple urine samples as only a few virus-infected cells may be present (Johnston *et al.*, 1969). Infection with CMV is common in the population, but the virus only rarely causes disease (Fenner and White 1981). Reactivation of a latent virus frequently occurs in renal allograft recipients and may cause pneumonia or pyrexia in these patients. Primary infection in pregnancy can have serious consequences for the fetus. Transplacental transmission of the infection from the mother to the fetus is often fatal, but if the baby survives, it may suffer severe mental and physical retardation.

Human polyomavirus (HPV)

This is a DNA virus with an icosahedral symmetry measuring 45 nm in diameter, which is a member of the papovavirus group. Unlike cytomegalovirus, HPV replicates exclusively in the nucleus of the cell and many infected cells are often exfoliated into the urine at any one time. Each affected cell has a cyanophilic nucleus with a basophilic inclusion. A clear halo may separate the inclusion from the thickened nuclear membrane (*Figure 17.9*) although the halo is

Table 17.1 Urinary crystals

Crystal	Conditions under which normally found	Appearance of crystals
(1) Ammonium urate (thorn apples)	Usually found in alkaline urine. Associated with ammoniacal fermentation.	
(2) Bilirubin	May be found in urine containing bile. Sometimes found in carcinoma of the bladder, pyonephrosis and haemorrhage of the urinary tract.	
(3) Calcium carbonate	Rare in man but common in herbivorous animals.	
(4) Calcium oxalate	Often found in acid urine of healthy individuals or in association with calculi. Strawberries, spinach and rhubarb are all high in oxalates.	
(5) Hippuric acid	Associated with the consumption of large amounts of fruit and vegetables.	
(6) Triple phosphate	Usually associated with ammoniacal fermentation. Common in *Proteus* infection of the bladder.	
(7) Stellar phosphate	Occurs in urines containing large amounts of calcium phosphate.	
(8) Calcium phosphate (flakes)	Occurs as a surface pellicle in some alkaline urines.	
(9) Cholesterol	Found in cases of fatty degeneration of the kidney, hydatid disease, renal calculi and severe cystitis.	

(10) Ammonium magnesium phosphate	Found in fresh wine to which ammonia has been added.
(11) Magnesium phosphate	Rarely found.
(12) Uric acid	Often found in acid urines of healthy individuals after profuse sweating. Common in leukaemia.
(13) Uric acid (Barrel and Hourglass)	As above.
(14) Cystine	Very rare. Can be a result of inherited metabolic derangement. Frequently confused with uric acid crystals.
(15) Tyrosine (A) and leucine (B)	Very rare. Associated with yellow atrophy of the liver or phosphorus poisoning.
(16) Calcium sulphate	Rarely found.
(17) Sulphonamides	Found in acid urine from patients treated with sulphonamides or sulphonamide-containing combinations.
(18) Indigo (blue)	Rarely found.

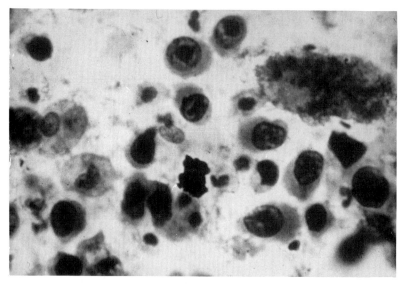

Figure 17.9 Urine sediment from renal allograft recipient showing enlarged transitional cells containing prominent basophilic intranuclear inclusions. Human polyomavirus infection was confirmed by virus isolation and electron microscopy. Papanicolaou stain. Magnification × 500

not always seen. Initial infection with these viruses occurs in childhood, the virus then remains latent in the host cell and is reactivated when the immune system is impaired (Coleman, 1975, 1979; Gardner *et al.*, 1971). Urinary tract infection may result in ureteric stenosis and obstruction in renal allograft patients, which clinically may resemble a rejection episode. There are reports of the excretion of this virus in pregnancy, although there is no evidence that the fetus is at risk (Fenner and White, 1981).

Parasites

Schistosoma haematobium is a trematode worm found commonly along the Nile delta (El-Bolkainy, 1980). In man, the infection results in the deposition of ova in the submucosa of the bladder and distal ureters. The ova are oval structures with a terminal spine, and cause inflammation and fibrosis of the bladder wall. Patients infected with this organism have a high risk of developing carcinoma of the bladder, which is often squamous in type. *Toxoplasma gondii* infection results in cysts filled with polygonal structures (De Voogt, Rathert and Beyer-Boon, 1977). The cysts are shed into the urine. *Trichomonas vaginalis* may also be detected cytologically. If a prostatic infection is suspected, prostatic massage is advised before urine collection to obtain an adequate sample (*see also* Chapter 23).

Tumour cells

The majority of tumours arising in the bladder, ureter and renal pelvis are transitional cell carcinomas, although primary and metastatic squamous carcinoma and adenocarcinoma also occur. Rarely, tumours such as lymphoma and melanoma have been reported. The tumours usually cause haematuria and cytology should be available for all patients presenting with this symptom. The tumours which may be found are listed in *Table 17.2*.

Table 17.2 Types of tumour in the urinary tract

(a) *Bladder, ureter, renal pelvis*
  (1) Transitional cell carcinoma
     i papillary non-invasive (papilloma)
     ii papillary invasive
     iii non-papillary invasive
  (2) Squamous carcinoma
  (3) Adenocarcinoma
  (4) Carcinoma *in situ*
  (5) Metastatic carcinoma
  (6) Lymphoma, melanoma, mixed mesodermal tumours—all are rare
(b) *Renal parenchyma*
  (1) Clear cell carcinoma (hypernephroma)
  (2) Nephroblastoma (Wilm's tumour)
  (3) Metastatic carcinoma

Transitional cell carcinoma

This common tumour may affect any part of the urinary tract that is lined by urothelium. They account for 4% of all male and 2% of all female malignancies. Transitional cell carcinoma is generally considered to be a disease of elderly men as more than 60% of those who develop the disease are males in their sixth or seventh decade according to the Committee on Professional Education (1973). De Voogt, Rathert and Beyer-Boon (1977) report that 96% of them occur in the bladder, 1.4% in the renal pelves, 2.1% in the ureters and 0.5% in the posterior urethra. The most commonly affected area is the trigone whereas the anterior wall and dome of the bladder are rarely affected. The tumours are often multicentric, and recur after treatment. Development of transitional cell carcinoma of the renal pelvis is often bilateral and is associated with analgesic abuse, particularly phenacetin, an aniline derivative (Rathert, Melchior and Lutzeyer, 1975). The incidence of transitional cell carcinoma of the urinary tract, particularly of the bladder, has been increasing in recent years. This may be related to several factors including an increase in cigarette smoking and exposure to carcinogenic substances.

Transitional cell carcinomas may be papillary or sessile, well differentiated or poorly differentiated, confined to the epithelial layers or invading the bladder wall. A system of grading and staging has been proposed to classify the different types of transitional cell carcinoma in histological sections. Grading refers to the degree of differentiation of the tumour cells and staging to the depth of invasion.

Papillary tumours are graded histologically according to the thickness of the epithelial lining and the morphology of the individual tumour cells. Thus papillary carcinoma (Grade I) is lined by an epithelium which is more than seven layers thick but which is composed of near normal transitional cells whereas papillary carcinoma (Grade III) is lined by a greatly thickened epithelium showing a marked nuclear pleomorphism. As may be expected, Grade I tumours are much more difficult to recognize cytologically than Grade III tumours (*Figure 17.10(a)* and *(b)*). However, Grade I lesions may be suspected if a *voided* urine contains numerous papillary clusters or rosettes of normal looking transitional cells. Care must be taken to ensure that there has not been recent catheterization or a history of urinary calculus when making a diagnosis of transitional carcinoma in such cases. Those tumours which are papillary and growing into the lumen of the bladder and are composed of well differentiated urothelium less than six cell layers thick are often designated 'papillomas'—although there is doubt as to whether these really are benign lesions, as they almost invariably recur after removal. Papilloma cannot be distinguished cytologically from papillary carcinoma Grade I. Most sessile cancers are high-grade lesions as are the majority of recurrent tumours.

Cytology cannot be used to stage urothelial cancers, although the absence of blood and pus from the urine suggests that the lesion may be confined to the surface of the bladder. Moreover, staging and grading often go together. The greater the degree of invasion, the more pleomorphic the tumour cells. Tumour cells shed from a high-grade lesion are readily recognized by their nuclear pleomorphism and nuclear hyperchromasia. Many tumour cells are often present in the smear.

Carcinoma in situ (non-papillary transitional cell carcinoma—stage 0)

Carcinoma *in situ* of the bladder, like carcinoma *in situ* of the cervix, is characterized by replacement of normal flat transitional epithelium by carcinoma cells. The lesions are often multicentric and if untreated will progress to invasive cancer. Like the preinvasive lesions of the cervix, lesser degrees of epithelial atypia (dysplasia) have been described in the bladder.

The cells shed from this lesion are usually small, discrete, irregular in shape with dark hyperchromatic nuclei (*Figure 17.11 (a)* and *(b)*). Unlike invasive tumours there is little necrotic debris or inflammatory exudate present. Cystoscopy may reveal multiple reddened patches on the bladder mucosa. The diagnosis depends on demonstration of intraepithelial carcinoma in the biopsy.

Squamous carcinoma

This is not commonly found in the urinary tract as a primary tumour, although patients with *Schistosoma heamatobium* infection often develop squamous rather than transitional carcinoma of the bladder. It may also rarely occur in the urethra. Primary squamous carcinoma

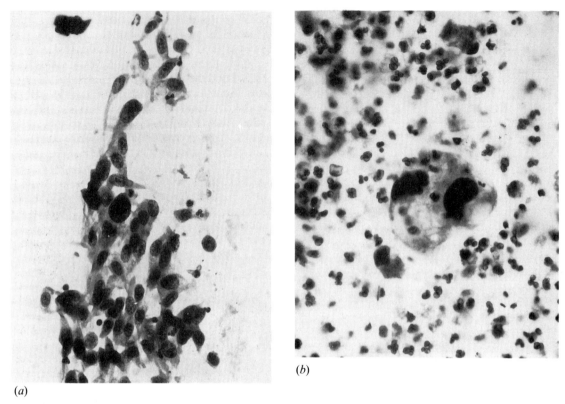

Figure 17.10 (*a*) Well differentiated transitional cell carcinoma in urine sediment. The cells show slight variation in size and shape, a coarse chromatin structure and hyperchromasia. (*b*) Poorly differentiated transitional cell carcinoma. The tumour cells have large hyperchromatic irregular nuclei. Anisonucleosis and cellular pleomorphism are prominent features of the smear. Papanicolaou stain. Magnification × 500

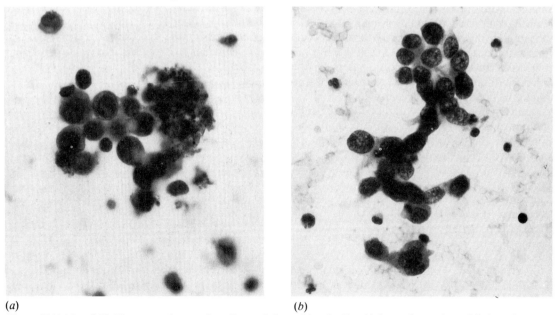

Figure 17.11 (*a*) and (*b*) Numerous clusters of small rounded transitional cells with hyperchromatic nuclei. A tendency to rosette formation was noted. Cystoscopy revealed no abnormality but random biopsy showed carcinoma *in situ*. Papanicolaou stain. Magnification × 500

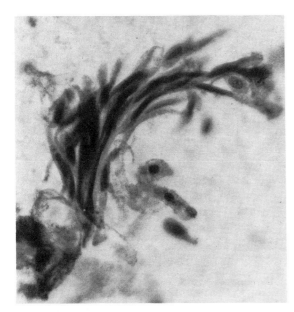

Figure 17.12 A group of elongated spindle-shaped malignant cells. Their morphology is suggestive of squamous cell carcinoma although biopsy of the tumour revealed a transitional cell carcinoma. Membrane filter preparation. Stain: Papanicolaou. Magnification × 2000, reduced to 65% in reproduction

is thought to arise in areas of squamous metaplasia, and is identified cytologically by the presence of exfoliated abnormal spindle-shaped keratinized squamous cells (*Figure 17.12*) similar to those seen in squamous carcinomas elsewhere.

Adenocarcinoma of bladder

These tumours are believed to be derived from areas of intestinal metaplasia or glanduleris cystica in the bladder wall. The tumour cells resemble adenocarcinoma cells at other sites, and are often columnar shaped, with clear vacuolated cytoplasm and large nucleoli. A distinction between primary and metastatic adenocarcinoma from colon or endometrium or prostate can only be made on the clinical history.

Renal carcinoma

Adenocarcinoma of the kidney is commonly known as clear cell carcinoma of the kidney. It is of tubular origin and arises in the renal parenchyma. It accounts for 85% of all renal neoplasms (Milsten *et al.*, 1973). It is a slow growing tumour and malignant cells are only found in the urine if it has invaded the renal pelvis. Thus, urine cytology is only likely to detect advanced lesions at a late stage of their development (Hajdu *et al.*, 1971). The exfoliated glandular cells are generally in a poor state of preservation, with finely vacuolated cytoplasm. The nuclei are large and hyperchromatic and multiple nucleoli may be present.

Cytoplasmic lipid demonstration in these cells, using an Oil Red O technique, has improved the accuracy of the cytological diagnosis in cases of renal cell carcinoma, although a positive smear is only meaningful if the patient has a known or a suspected kidney mass (Milsten *et al.*, 1973).

Metastatic carcinoma

Primary carcinoma cells from sites other than the urinary tract may be detected in urine sediment when there is invasion through the bladder wall from adjacent organs. The primary carcinomas most likely to invade are the colon, prostate (*Figure 17.13*), endometrium and the cervix by virtue of their proximity to the bladder. Metastatic squamous carcinoma from the cervix can be recognized cytologically by the presence of bizarre keratinized squamous cells in the urinary tract.

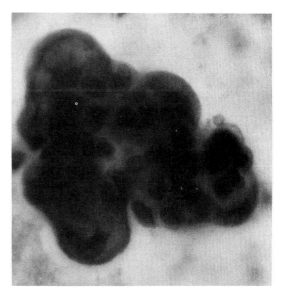

Figure 17.13 Metastatic prostatic carcinoma in a urine specimen. Membrane filter preparation. Stain: Papanicolaou. Magnification × 2000

318 The urinary tract

Other cell types

Intestinal epithelial cells in ileal conduit urine

It is advisable to examine ileal conduit urine from patients who have had urinary diversion for carcinoma of the lower urinary tract to detect recurrence of the tumour at an early stage (Malmgren *et al.*, 1971; Wolinska and Melamed, 1973). As the turnover rate of intestinal cells is much higher than that of urothelial cells, numerous single columnar cells from the intestinal epithelium are exfoliated into the urine (*Figure 17.14 (a) and (b)*). These are round and degenerate with darkly stained irregular nuclei which show pyknosis and karyorrhexis, and often cytoplasmic vacuolation. These degenerate columnar cells from the intestinal epithelium are often associated with white blood cells and mucus strands, both of which obscure cell detail. However, abnormal or neoplastic urothelial cells can be distinguished from the ileal elements in the majority of cases.

Seminal vesicle cells

The columnar cells from the seminal vesicles may be identified in the urine by the presence of yellow cytoplasmic granules of lipochrome pigment. Spermatozoa may be present in male urine and spermatocytes may be mistaken for malignant cells unless the possibility of contamination is borne in mind (*Figure 17.15*), as may corpora amylacea from the prostate.

Malakoplakia

This is a chronic inflammatory condition affecting the bladder wall resulting in the formation of yellow plaques. In this condition mononuclear cells or macrophages containing cytoplasmic inclusions termed Michaelis-Gutmann (M-G) bodies are found. The inclusions are calcified laminated lysosomes measuring 4–10 µm in diameter which are positive using the von Kossa method. They are thought to be an abnormal inflammatory response by the macrophages to bacteria, probably *Escherichia coli*. In an immunohistochemical study, Gupta, Schuster and Christian (1972) suggested that the inclusions are made up of glycolipid. However, it is unusual to find cells bearing M-G bodies in

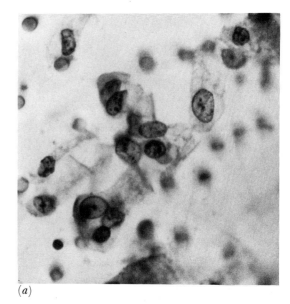
(a)

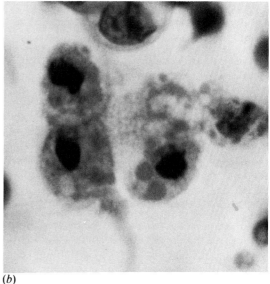
(b)

Figure 17.14 (*a*) Urinary sediment from ileal conduit urine. Several well preserved columnar cells originating from small intestine are seen. (*b*) Ileal conduit urine. Note degenerating columnar cells with nuclear pyknosis and eosinophilic cytoplasmic inclusions. Papanicolaou stain. Magnification ×400. (Photographs provided by Dr N. Stormby and Miss B. Nilsson, Cytodiagnostic Unit, Malmo General Hospital, Sweden.)

the urine as they are usually embedded in the wall of the bladder and covered by a layer of normal urothelium. Only if the urothelium is ulcerated exposing the deeper layers may they then be exfoliated.

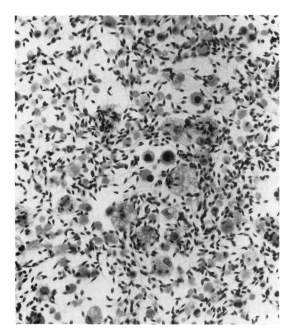

Figure 17.15 Urinary sediment containing numerous spermatozoa and spermatocytes. The large basophilic nuclei of these cells could lead to a misdiagnosis of neoplasia. Papanicolaou stain. Magnification × 500

Dyskaryotic cells

Besides containing numerous squamous cells from the vulva and vagina, urine from the female may also contain dyskaryotic cells shed from an area of CIN or squamous carcinoma of the lower female genital tract.

Use of urine cytology in the examination of the urinary tract

There are a number of important uses of urinary cytology in clinical practice. These include the diagnosis of primary and metastatic cancer, screening of high risk patients, evaluation of treatment methods, detection of recurrence of cancer, detection of infection and monitoring of renal allograft recipients.

Diagnosis of primary and metastatic bladder cancer

The major purpose of urine cytology is to detect urothelial carcinoma, particularly lesions arising in the bladder (Foot *et al.*, 1958; Umiker, 1964; Wiggishoff and McDonald, 1969; Kalnins *et al.*, 1970). Lesions may be detected at a preclinical stage, before they are evident cystoscopically (Sarnaki *et al.*, 1971), or radiologically. Cytology is specially effective for the diagnosis of high-grade tumours (De Voogt and Wielenga, 1972), with a detection rate of between 93% and 94.7%. It plays a useful role in distinguishing between transitional cell tumours and other tumour types (Brannan, Lucas and Mitchell, 1973).

Evaluation of treatment

Recurrent urothelial carcinoma may also be detected by urinary cytology in patients whose tumours have previously been treated by diathermy or radiotherapy or chemotherapy (Reichborn-Kjennerud and Hoeg, 1972; Lewis *et al.*, 1976; De Voogt, Rathert and Beyer-Boon, 1977). In fact, the detection rate is higher for recurrent than primary lesions, as recurrent tumours usually show a greater degree of pleomorphism than primary lesions. It is usual for such patients to be cystoscoped regularly (every 3–6 months), and Koss (1979) suggests that three urines examined at three-monthly intervals provide a reliable cytological follow-up sample. In these cases malignant cells are often exfoliated into the urine before the lesion is visible at cystoscopy (El-Bolkainy, 1980). Thus cytological follow-up may complement cystoscopic examination but cannot replace it as a diagnostic tool (De Voogt and Wielenga, 1972). It should be remembered that chemotherapeutic agents used to treat the neoplastic lesions may cause cellular atypia which can be misdiagnosed as malignant change.

Screening high risk populations

Urine cytology is used extensively for screening those populations at special risk of developing carcinoma of the bladder, as it is possible to detect a tumour at an early stage of development (Crabbe *et al.*, 1956; Forni, Ghetti and Armeli, 1972). It is a cheap, easily repeatable screening method, which is also highly acceptable to the patient as regular examination of voided urine is much more agreeable than repeated cystoscopy. Workers in the dyestuffs and rubber industries, who are exposed to a number of carcinogenic substances including alpha- and beta-naphthylamine and benzidine, are particularly at risk.

The examination of urine for red blood cells in order to detect occupational bladder pathology was first suggested in 1920. However, it was not until 1951 that urine cytology screening was introduced in British factories. According to statistics published by the Association of Scientific, Technical and Managerial Staffs (ASTMS), 7500 cases of bladder cancer are diagnosed annually in Britain of which the Department of Health recognize between 5 and 10% as being occupationally caused and therefore eligible for compensation (Glashan, 1982). It is generally accepted that screening checks of high risk groups should be carried out at six-monthly intervals (El-Bolkainy, 1980).

Another high risk group are those people exposed to *Schistosoma haematobium* in whom the developing carcinoma is often squamous in type.

Diagnosis of infection

Infection of the urinary tract is often caused by an obstruction to the flow of urine by calculi, a urethral stricture, prostatic enlargement, or a neoplastic growth. The cytological manifestation of infection of the urinary tract is characterized by the appearance of numerous white blood cells, red blood cells, casts and cellular debris in the urine. Bacteria may be numerous in the smear but Gram stain and culture is essential for their correct identification. There are a few specific agents that produce recognizable cytological changes, e.g. cytomegalovirus, *Schistosoma haematobium*. The M-G bodies of malakoplakia can also be detected by cytology.

Diagnosis of rejection episodes in renal allograft recipients

Renal allograft recipients are susceptible to fungal and viral infections and cytological changes characteristic of these infections may be seen in smears of urine sediment (Traystman *et al.*, 1980). In the absence of rejection, few urothelial cells are exfoliated (Bossen *et al.*, 1970; O'Morchoe *et al.*, 1976). During an episode of acute rejection the cytological pattern changes, becoming more cellular. The characteristic cytological profile demonstrated by Bossen *et al.* (1970) as indicative of acute rejection episode, is the presence of five of the seven following features: red blood cells, lymphocytes, renal tubular cells, nuclear changes in renal tubular cells, casts, 'dirty' background and mixed cell clusters.

O'Morchoe *et al.* (1976) demonstrated in a series of 57 transplant patients that the presence of an increased number of renal tubular cells, particularly small degenerate cells with an associated background of cellular debris and casts, was indicative of a rejection episode. Detection of a rejection episode by the examination of urine specimens at frequent intervals may be made before clinical signs appear. Schumann, Johnston and Weiss (1981) suggest that the shedding of renal epithelial fragments into the urine during a rejection episode is indicative of a poor prognosis for the transplanted kidney.

It is possible, using semiquantitative differential cell counting methods, to distinguish cells from the proximal and distal convoluted tubules and the small and large collecting ducts, and thus localize renal parenchymal disease (Johnston and Schumann, 1981).

An alternative approach to the problem of detecting rejection in renal allograft recipients has been adopted by Pasternack (1968), von Willebrand (1980) and Häyrey and von Willebrand (1981). These authors used fine needle aspiration of kidney substance to identify rejection with considerable success.

Problems of cytological diagnosis

There are a number of benign conditions of the urinary tract that may affect both the number and the morphology of the cells exfoliated. Beginners may encounter problems in the cytological interpretation of these cells, but these will become less frequent as he or she becomes more familiar with the cytological criteria for distinguishing each condition. A diagnosis of cellular atypia may be made on cells which, whilst not being malignant, are not absolutely normal. Such a diagnosis often occurs when a patient's urological history is not known to the cytologist, making an accurate interpretation of these cells impossible. Three of the major causes of problems in cytological interpretation are the presence of calculi, changes due to therapy and the diagnosis of virus infection.

Urinary calculi

Urinary calculi cause ulceration of the urothelium, resulting in the exfoliation of abnormal

patients. Smooth bordered clusters of transitional cells were observed in many urines. The presence of these clusters were found to be helpful in diagnosing unsuspected calculi (*Figure 17.16 (a) and (b)*). There was a small percentage of urines with epithelial changes suspicious of malignancy, but when a comparative study of urines from patients with well differentiated transitional carcinoma was made, there were clearly demonstrable differences. In patients with urinary calculi, the suspicious clusters were very few in number and although some cells had abnormal features suggestive of malignancy, this was not so in all the cells in the clusters. Also there were very few single transitional cells present with abnormal morphology. In contrast, the smears from patients with transitional cell carcinoma were composed almost entirely of tumour cells recognizable as single cells and in clusters.

Changes due to therapy

Chemotherapy

The systemic administration of cytotoxic drugs may have a marked effect on the urothelium. It has been reported that cyclophosphamide and busulphan cause enlarged, degenerative and abnormal cells to be exfoliated into the urine (Koss, 1979). The degenerative features include vacuolated cytoplasm, nuclear karyorrhexis and hyperchromasia. Koss (1979) suggests that these changes may be severe enough to imitate urothelial carcinoma (*Figure 17.17*), emphasizing the importance of an accurate history.

Radiotherapy

Ionizing radiation is used in the treatment of transitional cell carcinoma of the urinary tract (Loveless, 1973; Randwin, 1980), although it is ineffective for carcinoma *in situ*. The main cytological effect of radiation on the urothelium is an increase in cell size. This is often associated with nuclear hyperchromasia, multinucleation and cytoplasmic vacuolation. Differential diagnosis between malignant and radiation changes is often possible as the nucleocytoplasmic ratio does not change with treatment and a regular chromatin pattern remains.

Virus infection

CMV and human polyomavirus (HPV) infected

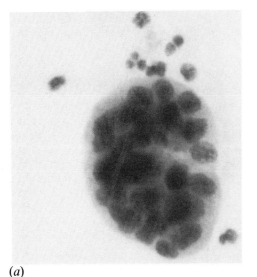

(a)

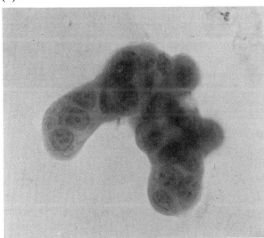

(b)

Figure 17.16 (*a*) and (*b*) Urine sediment. Smooth bordered clusters of benign transitional cells from a patient with renal calculi. Membrane filter preparation. Stain: Papanicolaou. Magnification × 2000

cells, and consequently are one of the main contributors to false positive diagnoses of the urinary tract. In a study of 15 882 urine samples, 57 false positive diagnoses were attributed to urinary calculi (Highman and Wilson, 1982) and in two other studies, the abnormal cells disappeared from the urine after removal of the calculi (Beyer-Boon *et al.*, 1978; De Ruiter, Beyer-Boon and De Voogt, 1975). A study of 154 patients with calculi by Highman and Wilson (1982) demonstrated that there were clearly recognizable benign clusters of cells in the urine from approximately half of these

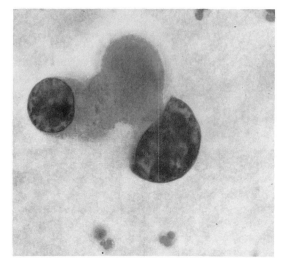

Figure 17.17 Urine sediment. Degenerate malignant transitional cell from a patient treated with methotrexate. Stain: Papanicolaou. Magnification × 2000

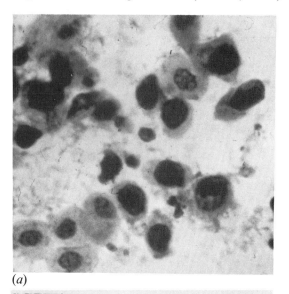

(a)

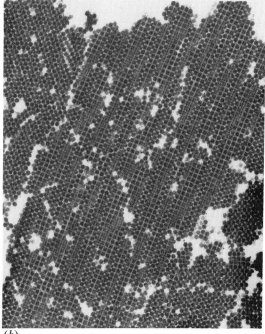

(b)

Figure 17.18 (a) Urine sediment from **immunosuppressed** patient containing numerous transitional cells with large basophilic nuclei. The nuclei are structureless and have a glassy appearance. The nuclear membranes are thickened. Papanicolaou stain. Magnification × 500. (b) Electron microscopy of the cells shown in (a) revealed the presence of virus particles in crystalline array consistent with polyomavirus infection. Uranyl acetate. Magnification × 33 900

buted to variation in the methods of analysis, and the subjectivity of interpretation of cytological and histological data (Umiker, 1964; cells have similar cytological features and need to be distinguished by virological methods. As HPV is difficult to culture, Coleman *et al.* (1977) have devised a method whereby an infected cell can be picked off a fixed smear of urine sediment and processed for electron microscopy (see Chapter 4). The results provide an excellent method for distinguishing between the two viral types.

It is important to identify the virally infected cells in the urine, since they may be mistaken for malignant urothelial cells as both may have enlarged darkly staining nuclei (*Figure 17.18(a)*). However, careful examination of the virus infected cells shows amorphous nuclear material, whilst malignant cells have coarse granular chromatin. Electron microscopy will confirm the presence of viral infection (*Figure 17.18(b)*).

Evaluation of urinary tract cytology as a diagnostic procedure

Cytological accuracy is assessed by comparing cytological and histological diagnoses (Brannan, Lucas and Mitchell, 1973). In a review of 14 studies, El-Bolkainy (1980) found the sensitivity of the method ranged from 44.7% to 97.3% with a mean of 73.8%, demonstrating the variability of the results. This can be attri-

Esposti and Zajicek, 1972). Some groups included bladder cancer only in their results, others included cancer of other urogenital organs. The population examined varied—several were reports on the screening of healthy industrial workers for bladder cancer, others based their analysis on hospitalized patients with a range of urogenital diseases. Some included papilloma in their analysis, others did not (Pearson, Kromhout and King, 1981).

Despite this variation, urinary cytology was found to be particularly effective for the diagnosis of poorly differentiated transitional cell carcinoma (Grade III) of the bladder, although it was less effective for the detection of well differentiated transitional cell carcinoma (Johnson, 1964). Accuracy has been estimated to be approximately 90% for high-grade lesions but only between 40 and 60% for low-grade lesions. This is inevitable as the cell morphology of well differentiated transitional cell carcinoma so closely resembles that of normal urothelium that a distinction is difficult to make. Urinary cytology is not effective for the diagnosis of papilloma of the urinary bladder (Umiker, 1964; Esposti and Zajicek, 1972), although it has been suggested that the presence of epithelial fronds in the urine is indicative of this condition.

Poor or inadequate sampling and processing of material may impair the accuracy of the method (Harris et al., 1971). The accuracy can be improved by examining several specimens from each patient although results are conflicting. Esposti, Moberger and Zajicek (1970) advocate the examination of more than one specimen as variable shedding of urothelial cells does occur. Foot et al. (1958) demonstrated that there were malignant cells in the first urine examined from 60% of patients, but on the examination of subsequent specimens the number of positive urines decreased. In contrast De Voogt, Rathert and Beyer-Boon (1977) state that the diagnosis of low grade tumours from urinary sediment increases from 34% if one specimen is examined to 79% if three specimens are examined.

It is important that the correct sample is investigated, as was discussed earlier, since midstream and early morning urine specimens will not provide sufficient material for a reliable diagnosis. Esposti and Zajicek (1972) and Harris et al. (1971) suggest that bladder washings obtained at cystoscopy are the most useful samples for detecting low-grade carcinoma. This must be open to question since cystoscopy will provide the opportunity to take biopsy material for a tissue diagnosis, thereby removing the need for further cytological examination.

A poorly fixed specimen, or the failure of the prepared cellular content to adhere to the glass slides, may cause diagnostic problems and inaccurate reporting. Large numbers of inflammatory cells, red blood cells or degenerate cellular material will obscure cellular detail and a repeat specimen must be examined before a diagnosis can be made.

False positive diagnoses occur in a variety of conditions where benign cells are misinterpreted as malignant. The number of false positive diagnoses of urinary tract specimens reported in the literature varies from author to author. Umiker (1964) reviewed the results of several groups and found that the false positive rate ranged from 1.3% to 11.9%, with a mean of 2.3% amongst 3609 patients. He and others (Foot et al., 1958) found that these erroneous diagnoses were made on cytological specimens from patients suffering from a variety of benign urological conditions, including chronic cystitis, prostatic hypertrophy and urinary calculi. It has been suggested that exfoliated regenerative or hyperplastic cells show features that are likely to be confused with malignancy. The cytologist must become aware of the range and variation of atypias seen in urinary sediment in order to reduce such diagnostic errors.

It is essential to follow up those patients with malignant cells in their urine but with no histological confirmation of a tumour (Heney et al., 1977). Malignant cells from an in situ carcinoma may appear in the urine before a tumour has become clinically visible. There are reports in the literature of cases where months or years have elapsed between the cytological diagnosis and the histological confirmation of a tumour (Koss, 1979). Thus Crabbe et al. (1956) noted the presence of malignant cells in the urine 46 months before a tumour was detected clinically; Highman (1983) found malignant cells in urine 33 months before tumour detection at cystoscopy and Reichborn-Kjennerud and Hoeg (1972) noted that a period of up to 23 months elapsed in some cases before the positive cytology was histologically confirmed. It should also be remembered that transitional cell carcinoma arising in a diverticulum of the bladder may be an unsuspected source of tumour cells.

Urinary cytology is less effective for diagnosing lesions of the upper urinary tract than those of the bladder (Bibbo et al., 1974). False negative rates of between 6% and 22% have been reported for tumours of the upper urinary tract when voided urine samples have been examined (Gill, Lu and Bibbo, 1978). However, the use of visualization techniques for specimen collection has notably improved the detection of tumours arising in the ureters and renal pelves. It is advisable to examine a voided urine specimen prior to instrumentation, so that a cellular baseline can be established to aid the interpretation of the collected cell sample (Highman, 1983).

Clear cell carcinoma of the kidney is rarely diagnosed by urinary cytology (O'Morchoe et al., 1976), as tumours of the renal parenchyma do not exfoliate into the urine unless the mass has grown into the renal pelvis and has breached the urothelium (Soloman et al., 1958). Results are improved if fat stains are carried out on urine sediments to identify the malignant cells (Hajdu et al., 1971). Similarly urine cytology is not a good method for the diagnosis of prostatic carcinoma where a detection rate of only 15% was reported by Foot et al. (1958). This percentage rises to between 88% and 94.7% when prostatic massage specimens are examined (Holmquist, 1977). Koss (1979) reports that the accuracy of a clinical diagnosis of prostatic cancer is 50%, whilst in a comparative study of fine needle aspiration and biopsy from 162 cases of prostatic carcinoma, the cytology agreed with the histology in every case.

The future

The search for an automated system for preparation and screening of urinary sediment to both speed up the process and eliminate human diagnostic error has occupied researchers over the last 15 years (Bales, 1981). The number of transitional cells within each specimen varies greatly and each sample should contain between 1000 and 10 000 cells to provide sufficient material for a reliable diagnosis. The results of computerized image analysis of urinary sediment (Koss, Sherman and Adams, 1983) are encouraging since it is now possible to differentiate between normal and malignant cells, and between two grades of atypia. If a classifier is used with this method, degenerated cells and multinucleated cells and cell clusters may be eliminated from the assessment thus improving accuracy.

Urine specimens are, by their nature, ideally suited for examination with a flow cytometer where cells are examined in suspension. A recent study (Melamed and Klein, 1984) using the metachromatic fluorescent dye acridine orange to stain DNA and RNA, has shown that flow cytometry has the potential to identify malignant cells. Although automated systems are still in an early stage of development, they may become an important technique in the future.

References

BALES, C. E. (1981). A semi-automated method for preparation of urine sediment for cytologic evaluation. *Acta Cytol.*, **25**, 323–326

BARRETT, D. L. and KING, E. B. (1976). Comparison of cellular recovery rates and morphologic detail obtained using membrane filter and cytocentrifuge techniques. *Acta Cytol.*, **20**, 174–180

BEALE, L. (1854). *The Microscope and Its Application to Clinical Medicine.* London: Samuel Highley

BEYER-BOON, M. E. and VOORN-DEN HOLLANDER, M. J. A. (1978). Cell yield obtained with various cytopreparatory techniques for urinary cytology. *Acta Cytol.*, **22**, 589–594

BEYER-BOON, M. E., DE VOOGT, H. J., VAN DER VELDE, E. A., BRUSSEE, J. A. M. and SCHABERG, A. (1978). The efficacy of urinary cytology in the detection of urothelial tumours. *Urol. Res.*, **6**, 3–12

BIBBO, M., GILL, W. B., HARRIS, M. J., LU, C., THOMSEN, S. and WIED, G. L. (1974). Retrograde brushing as a diagnostic procedure of ureteral, renal pelvic and renal calyceal lesions. A preliminary report. *Acta Cytol.*, **18**, 137–141

BLUTE, R. D., GITTES, R. R. and GITTES, R. F. (1981). Renal brush biopsy: survey of indications, techniques and results. *J. Urol.*, **126**, 146–149

BOSSEN, E. H. and JOHNSTON, W. W. (1975). Exfoliative cytopathologic studies in organ transplantation, IV. *Acta Cytol.*, **19**, 415–419

BOSSEN, E. H., JOHNSTON, W. W., AMATULLI, J. and ROWLANDS, D. T. (1970). Exfoliative cytopathologic studies in organ transplantation. *Acta Cytol.*, **14**, 176–181

BRANNAN, W., LUCAS, T. A. and MITCHELL, W. T. (1973). Accuracy of cytologic examination of urinary sediment in the detection of urothelial tumours. *J. Urol.*, **109**, 483–485

COLEMAN, D. V. (1975). The cytodiagnosis of human polyomavirus infection. *Acta Cytol.*, **19**, 93–96

COLEMAN, D. V. (1979). Cytological diagnosis of virus-infected cells in Papanicolaou smears and its application in clinical practice. *J. Clin. Pathol.*, **32**, 1075–1089

COLEMAN, D. V., RUSSELL, W. J. I., HODGSON, J., TUN PE and MOWBRAY, J. F. (1977). Human papovavirus in Papanicolaou smears of urinary sediment detected by transmission electron microscopy. *J. Clin. Pathol.*, **30**, 1015–1020

COMMITTEE ON PROFESSIONAL EDUCATION. (1973). *Clinical Oncology*. Berlin: Springer-Verlag.

CRABBE, J. G. S., CRESDEE, W. C., SCOTT, T. S. and WILLIAMS, M. H. C. (1956). The cytological diagnosis of bladder tumours amongst dyestuff workers. *Br. J. Ind. Med.*, **13**, 270–276

CRABTREE, W. N. and MURPHY, W. M. (1980). The value of ethanol as a fixative in urinary cytology. *Acta Cytol.*, **24**, 452–455

CRABTREE, W. N. and STINE, C. R. (1983). The polycarbonate disc versus the cellulose membrane filter in urinary cytology. *Acta Cytol.*, **27**, 577

CRABTREE, W. N., JUKKOLA, A. F. and MURPHY, W. M. (1981). Urine versus bladder washing in urinary cytology. *Acta Cytol.*, **25**, 441–442

DEDEN, C. (1954). Cancer cells in urinary sediment. *Acta Radiol.*, Supplementum, 115, Esselte Aktiebolag, Stockholm

DE RUITER, F., BEYER-BOON, M. E. and DE VOOGT, H. J. (1975). The effect of calculi on transitional epithelium. *Urol. Res.*, **3**, 67–72

DE VOOGT, H. J. and WIELENGA, G. (1972). Clinical aspects of urinary cytology. *Acta Cytol.*, **16**, 349–351

DE VOOGT, H. J., BEYER-BOON, M. E. and BRUSSEE, J. M. (1975). The value of phase contrast microscopy for urinary cytology; reliability and pitfalls. *Acta Cytol.*, **19**, 542–546

DE VOOGT, H. J., RATHERT, P. and BEYER-BOON, M. E. (1977). *Urinary Cytology*. Berlin: Springer-Verlag

EICKENBERG, H., AMIN, M. and LICH, R. (1975). Blastomycosis of the genitourinary tract. *J. Urol.*, **113**, 650–652

EL-BOLKAINY, M. N. (1980). Cytology of bladder carcinoma. *J. Urol.*, **124**, 20–22

ESPOSTI, P. L. and ZAJICEK, J. (1972). Grading of transitional cell neoplasms of the urinary bladder from smears of bladder washings. *Acta Cytol.*, **16**, 529–537

ESPOSTI, P. L., MOBERGER, G. and ZAJICEK, J. (1970). The cytologic diagnosis of transitional cell tumours of the urinary bladder and its histological basis. *Acta Cytol.*, **14**, 145–155

FENNER, J. J. and WHITE, D. O. (1981). *Medical Virology*, 2nd Edition. London: Academic Press

FOOT, N. C., PAPANICOLAOU, G. N., HOLMQUIST, N. D. and SEYBOLT, J. F. (1958). Exfoliative cytology of urinary sediments. *Cancer*, **11**, 127–137

FORNI, A., GHETTI, G. and ARMELI, G. (1972). Urinary cytology in workers exposed to carcinogenic aromatic amines: a six year study. *Acta Cytol.*, **16**, 142–145

FRENI, S. C. and FRENI-TITULAER, L. W. J. (1977). Microhematuria found by mass screening of apparently healthy males. *Acta Cytol.*, **21**, 421–423

GARDNER, S. D., FIELD, A. M., COLEMAN, D. V. and HULME, B. (1971). New human papovavirus (BK) isolated from urine after renal transplantation. *Lancet*, 1253–1257

GILL, W. B., LU, C. T. and THOMSEN, S. (1973). Retrograde brushing: a new technique for obtaining histologic and cytologic material from ureteral renal pelvic and renal caliceal lesions. *J. Urol.*, **109**, 573–578

GILL, W. B., LU, C. T. and BIBBO, M. (1978). Retrograde ureteral brushing. *Urology*, **XII**, 279–283

GLASHAN, R. W. (1982). Industrial bladder cancer. *Br. Med. J.*, **284**, 614

GUPTA, R. K., SCHUSTER, R. A. and CHRISTIAN, W. D. (1972). Autopsy findings in a unique case of malacoplakia. *Arch. Pathol.*, **93**, 42–48

HAJDU, S. I., SAVINO, A., HAJDU, E. O. and KOSS, L. G. (1971). Cytologic diagnosis of renal cell carcinoma with the aid of fat stain. *Acta Cytol.*, **15**, 31–33

HARRIS, M. J., SCHWINN, C. P., MORROW, J. W., GRAY, R. L. and BROWELL, B. M. (1971). Exfoliative cytology of the urinary bladder irrigation specimen. *Acta Cytol.*, **15**, 385–399

HARRISON, J. H., BOTSFORD, T. W. and TUCKER, M. R. (1951). The use of the smear of the urinary sediment in the diagnosis and management of neoplasm of the kidney and bladder. *Surg. Gynecol. Obstet.*, **92**, 129–139

HÄYREY, P. and VON WILLEBRAND, E. (1981). Monitoring of human renal allograft rejection with fine needle aspiration cytology. *Scand. J. Immunol.*, **19**, 87–97

HENEY, N. M., SZYFELBEIN, W. M., DALY, J. J., PROUT, G. R. and BREDIN, H. C. (1977). Positive urinary cytology in patients without evident tumour. *J. Urol.*, **117**, 223–224

HIGHMAN, W. J. (1983). The role of urine cytology in urological practice. *Acta Urol. Belg.*, **51**, 7–11

HIGHMAN, W. J. and WILSON, E. (1982). Urine cytology in patients with calculi. *J. Clin. Pathol.*, **35**, 350–356

HOLMQUIST, N. D. (1977). *Diagnostic Cytology of the Urinary Tract*. Monographs in Clinical Cytology. Basle: Karger

JOHNSON, W. D. (1964). Cytopathological correlations in tumours of the urinary bladder. *Cancer*, **17**, 867–880

JOHNSTON, J. L. and SCHUMANN, G. B. (1981). Differential renal epithelial cell counts in urine sediment. *Acta Cytol.*, **25**, 709

JOHNSTON, W. W., BOSSEN, E. H., AMATULLI, J. and ROWLANDS, D. T. (1969). Exfoliative cytopathologic studies in organ transplantation. *Acta Cytol.*, **13**, 605–610

KALNINS, Z. A., RHYNE, A. L., MOREHEAD, R. P. and CARTER, B. J. (1970). Comparison of cytologic findings in patients with transitional cell carcinoma and benign urologic diseases. *Acta Cytol.*, **14**, 243–248

KOSS, L. G. (1979). *Diagnostic Cytology and its Histopathologic Bases*. Philadelphia: J. B. Lippincott

KOSS, L. G., SHERMAN, A. B. and ADAMS, S. E. (1983). The use of hierarchic classification in the image analysis of a complex cell population. Experience with the sediment of voided urine. *Anal. Quant. Cytol.*, **5**, 159–166

LEISTENSCHNEIDER, W. and NAGEL, R. (1980). Lavage cytology of the renal pelvis and ureter with special reference to tumours. *J. Urol.*, **124**, 597–600

LEVI, P. E., COOPER, E. H., ANDERSON, C. K., PATH, M. C. and WILLIAMS, R. E. (1969). Analysis of DNA content, nuclear size and cell proliferation of transitional cell carcinoma in man. *Cancer*, **23**, 1074–1085

LEWIS, R. W., JACKSON, A. C., MURPHY, W. M., LEBLANC, G. A. and MEEHAM, W. L. (1976). Cytology in the diagnosis and follow-up of transitional cell carcinoma of the urothelium: a review with a case series. *J. Urol.*, **116**, 43–46

LOVELESS, K. J. (1973). The effects of radiation upon the cytology of benign and malignant bladder epithelia. *Acta Cytol.*, **17**, 335–360

MALMGREN, R. A., SOLOWAY, M. S., CHU, E. W., DEL VECCHIO, P. R. and KETCHAM, A. S. (1971). Cytology of ileal conduit urine. *Acta Cytol.*, **15**, 506–509

MARWAH, S., DEVLIN, D. and DEKKER, A. (1978). A comparative cytologic study of 100 urine specimens processed by the slide centrifuge and membrane filter techniques. *Acta Cytol.*, **22**, 431–434

MELAMED, M. R. and KLEIN, F. A. (1984). Flow cytometry of urinary bladder irrigation specimens. *Hum. Pathol.*, **15**, 302–305

MILSTEN, R., FRABLE, W. J., TEXTER, J. H. and PAXSON, L. (1973). Evaluation of lipid stain in renal neoplasms as adjunct to routine exfoliative cytology. *J. Urol.*, **110**, 169–171

NAIB, Z. M. (1961). Exfoliative cytology of renal pelvic lesions. *Cancer*, **14**, 1085–1087

NATIONAL BLADDER CANCER COLLABORATIVE GROUP (1977). Cytology and histopathology of bladder cancer cases in prospective longitudinal study. *Cancer Res.*, **37**, 2911–2915

O'MORCHOE, P. J., EROZAN, Y. S., COOKE, C. R., WALKER, W. G., O'MORCHOE, C. C. C., TRAYSTMAN, M. D., COWLES, L . T., DORSCH, R. F. and FROST, J. K. (1976). Exfoliative cytology in the diagnosis of immunologic rejection in the transplanted kidney. *Acta Cytol.*, **20**, 454–461

PAPANICOLAOU, G. N. and MARSHALL, V. F. (1945). Urine sediment smears as a diagnostic procedure in cancers of the urinary tract. *Science*, **101**, 519–520

PASTERNACK, A. (1968). Fine needle aspiration biopsy of human renal allografts. *Lancet*, **ii**, 82–85

PEARSON, J. C., KROMHOUT, L. and KING, E. B. (1981). Evaluation of collection and preservation techniques for urinary cytology. *Acta Cytol.*, **25**, 327–333

RANDWIN, H. M. (1980). Radiotherapy and bladder cancer: a critical review. *J. Urol.*, **124**, 43–46

RATHERT, P., MELCHIOR, H. and LUTZEYER, W. (1975). Phenacetin: a carcinogen for the urinary tract? *J. Urol.*, **113**, 653–657

REICHBORN-KJENNERUD, S. and HOEG, K. (1972). The value of urine cytology in the diagnosis of recurrent bladder tumours. A preliminary report. *Acta Cytol.*, **16**, 269–272

RODRIGUEZ, C. A., BUSKOP, A., JOHNSON, J., FROMOWITZ, F. and KOSS, L. G. (1980). Renal oncocytoma. Preoperative diagnosis by aspiration biopsy. *Acta Cytol.*, **24**, 355–359

SARNAKI, C. T., MCCORMACK, L. J., KISER, W. S., HAZARD, J. B., MCLAUGHLIN, T. C and BELOVICH, D. M. (1971). Urinary cytology and the clinical diagnosis of urinary tract malignancy: a clinicopathologic study of 1400 patients. *J. Urol.*, **106**, 761–764

SCHUMANN, G. B., HARRIS, S. and HENRY, J. B. (1978). An improved technic for examining urinary casts and a review of their significance. *Am. J. Clin. Pathol.*, **69**, 18–23

SCHUMANN, G. B., JOHNSTON, J. L. and WEISS, M. A. (1981). Renal epithelial fragments in urine sediment. *Acta Cytol.*, **25**, 147–152

SOLOMAN, C., AMELAR, R. D., HYMAN, R. M., CHAIBAN, R. and EUROPA, D. L. (1958). Exfoliated cytology of the urinary tract: a new approach with reference to the isolation of cancer cells and the preparation of slides for study. *J. Urol.*, **80**, 374–382

TAYLOR, J. N., MACFARLANE, E. W. E. and CEELEN, G. H. (1963). Cytological studies of urine by millipore filtration technique: second annual report. *J. Urol.*, **90**, 113–115

TRAYSTMAN, M. D., GUPTA, P. K., SHAH, K. V., REISSLING, M., COWLES, L. T., HILLIS, W. D. and FROST, J. K. (1980). Identification of viruses in the urine of renal transplant recipients by cytomorphology. *Acta Cytol.*, **24**, 501–510

TROTT, P. A. (1967). Cytological examination of urine using a membrane filter. *Br. J. Urol.*, **39**, 610–614

TROTT, P. A. (1977). Ureteric urine examination. *Br. J. Hosp. Med.*, May, 493–497

UMIKER, W. (1964). Accuracy of cytological diagnosis of cancer of the urinary tract. *Acta Cytol.*, **8**, 186–193

VON WILLEBRAND, E. (1980). Fine needle aspiration cytology of human renal transplants. *Clin. Immunol. Immunopathol.*, **17**, 309–322

WIGGISHOFF, C. C. and MCDONALD, J. H. (1969). Urinary exfoliative cytology in tumours of the kidney and ureter. *J. Urol.*, **102**, 170–171

WOLINSKA, W. H. and MELAMED, M. R. (1973). Urinary conduit cytology. *Cancer*, **32**, 1000-1006

18

Semen analysis, and the cytology of hydrocoeles, spermatocoeles, and the prostatic gland

Margaret Haddon

The male reproductive system includes the testes, the epididymis, seminal vesicles, prostate and external genitalia (*Figure 18.1*). In the adult male, the testis has two main functions—spermatogenesis (the production of spermatozoa) and steroidogenesis (the synthesis and subsequent secretion of androgens). The adult testes are ovoid organs suspended by the spermatic cords within the scrotum. Each testis is approximately 4–5 cm long, 2.5 cm in breadth and weighs 12.5 g. Except on its posterior surface where the epididymis is located, the testis is covered by a fold of mesothelium—the tunica vaginalis. The mass of the testis comprises highly coiled seminiferous tubules which are lined predominantly by germ cells (spermatocytes) whose proliferation and maturation result in the production of spermatozoa (*Figure 18.2*). Interspersed between the germ cells are Sertoli cells which have a supportive function and may control spermatogenesis. In the intertubular spaces, closely related to the capillaries, are Leydig cells which are the primary source of testosterone. The epididymis contributes to the

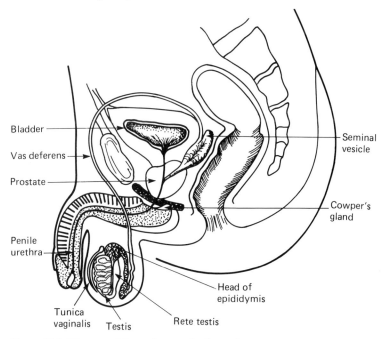

Figure 18.1 Diagram of the male reproductive system

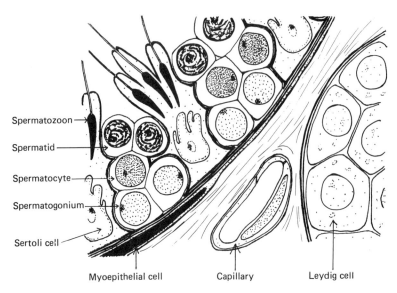

Figure 18.2 Diagram of normal testis to show relationship of Sertoli cells, germ cells and Leydig cells

maturation and reabsorption of degenerating spermatozoa, and the accessory organs provide most of the volume of seminal plasma which makes up the ejaculate; 65% of seminal fluid is derived from the seminal vesicles and 30% from the prostate.

In 1677 van Leeuwenhoek wrote to the Royal Society of London reporting the demonstration of motile spermatozoa in semen. He followed this with a further report in 1685 associating the existence of spermatozoa with fertility in the male (MacLeod, 1951). This stimulated great interest in the significance of abnormalities of spermatozoa and male reproductive potential and semen analysis soon became an integral part of the investigation of the infertile couple.

It has been shown that normal semen analysis does not necessarily mean that the man is fertile, nor does the presence of abnormal forms inevitably equate with infertility, and in this chapter we discuss the significance of this long established investigative procedure.

Sample collection and delivery

The patient is instructed both verbally and in writing regarding specimen collection and delivery to the laboratory. The following information should be given:

(1) The semen should be collected after a three-day period of abstinence from sexual intercourse.

(2) Masturbation is the preferable method of specimen production. Contraceptive sheaths (either rubber or plastic) are not acceptable as methods of collection as they have spermicidal properties which will invalidate results. The first portion of the ejaculate contains the highest concentration of spermatozoa and for this reason coitus interruptus is also unacceptable as a means of collection.

(3) The whole of the ejaculate must be collected directly into a sterile non-toxic (plastic) container. Any loss of a portion of the specimen must be recorded.

(4) The time of the specimen collection must be recorded.

(5) The specimen must be delivered to the laboratory within two hours of production. It is preferable for the specimen to be produced at the clinic/laboratory as this avoids any delay of analysis, allowing far more consistent results. Hafez (1977) noted that changes in pH, increased sperm death and alterations of biochemical composition occur very quickly and affect the results.

(6) During transportation the specimen must be protected from extremes of temperature (not higher than 40 °C or lower than 20 °C) as these will cause damage to the spermatozoa and enhance lytic processes.

The need to repeat the analysis on at least two or three occasions is well recognized.

Semen analysis

Freshly produced semen plasma made up of various secretions from the male accessory glands has a pH normally in the region of 7.2–7.6. The ejaculate coagulates within one minute, trapping most spermatozoa. Liquefaction occurs some 10–20 minutes later as a result of enzymes in the seminal plasma and must be complete before analysis can commence. Gentle agitation will often speed up the process of liquefaction and ensure the specimen is well mixed. The time should be noted when analysis begins as any undue delay in processing may affect the result. The following tests should be carried out on the specimen in the following order: inspection, viscosity, volume, motility, viability, spermatozoa density (concentration) and spermatozoa morphology.

Inspection

The specimen should be held against a white background. A normal fresh semen is milky grey and slightly opalescent. Seminal vesicle secretions may give it a yellow tinge. The presence of white blood cells results in a turbid, white appearance; red blood cells will give a reddish tinge. Certain drugs may also colour the semen.

Viscosity

Following complete liquefaction the viscosity may be assessed. The specimen is drawn up into a Pasteur pipette. Normally, semen is relatively fluid and easily drawn into the pipette with a few thread formations being present. The more viscous the semen, the harder it is (sometimes impossible) to pipette the specimen. The viscosity of semen may be described as:

(a) Normal
(b) Hyperviscid—the semen is thick, mucoid and stringy. It may be impossible to pipette. Hyperviscosity is associated with poor sperm mobility.
(c) Hypoviscid—the semen is watery and very pale in colour. This may be due to drug reactions (Toovey et al., 1981). Post-vasectomy semen usually falls into this category.

Abnormalities of semen liquefaction and viscosity are difficult to quantitate and are often the result of enzymatic defects which may be due to inflammatory changes in the seminal vesicles or prostate.

Volume

The volume of the semen is important for it neutralizes the high acidity of the vagina and determines the total number of spermatozoa in the ejaculate. Sample volume may be easily measured by pouring the sample into a 5 ml or 10 ml measuring cylinder, taking care that no loss of specimen occurs. Mean volumes of 2.8 ml (Kinloch, Nelson and Bunge, 1974) and 3.5 ml (Hafez, 1977) have been reported. Low volumes may be due to spillage, pathological conditions such as obstruction of the ejaculatory duct, or incomplete specimen collection. If the first ejaculatory portion is missed, the semen will have a severely decreased spermatozoa concentration and a low acid phosphatase content due to a low prostatic fluid content. If the last portion of the ejaculate is lost, the spermatozoa concentration may be normal but the semen will be lacking seminal vesicle secretions and thus be low in fructose and liquefy poorly.

Volumes greater than 6 ml may be due to the patient pooling several samples or to overproduction of the prostatic secretions resulting in a decreased spermatozoa concentration which may cause decreased fertility due to a 'washing out' effect in the vagina. A long period of sexual abstinence may also result in a high semen volume.

Motility

As motility is temperature dependent the semen sample must be brought to its optimal temperature (37 °C). One drop of semen should be placed on a warm glass slide, and a coverslip applied. This should be examined immediately as air drying may occur. A ×40 lens should be used and at least four microscopic fields examined ideally on a temperature-controlled microscope plate. Lowering the substage condenser will improve clarity.

Motility can be expressed in two ways, quantitatively or qualitatively. Quantitative motility is the percentage of motile spermatozoa present. This is determined by counting both non-motile and motile forms in randomly selected fields and expressing the total as a percentage of motile spermatozoa counted. The percentage of motile spermatozoa observed will decrease with advancing age of the specimen. Hafez quotes figures of 70% motility for specimens less than one hour old, and 60% motility for specimens up to three hours old.

Qualitative motility is a sub-division whereby the percentage of motile spermatozoa showing forward progression is assessed. Forward progression of the spermatozoa can be graded as absent, poor, moderate or excellent. MacLeod's system of qualitative assessment uses a grading of 0–4 (total immobility to good forward progression). A normal sample of seminal fluid half an hour to three hours after ejaculation will show 60% or more motile spermatozoa with most showing good forward progression. Those showing 40% or less motile spermatozoa and less than good forward progression must be reanalysed. Circular movements of spermatozoa should also be reported. The evaluation of motility is subjective but when assessed by a skilled observer gives valuable information on spermatozoa quality. Methods using photomicrography and image analysis are currently being investigated (Katz and Dott, 1975; Makler, 1980; Schmassman et al., 1982) as a means of providing a more accurate assessment of spermatozoa motility.

Viability

Supravital staining techniques enable live motile and non-motile spermatozoa to be distinguished from the dead spermatozoa. Many techniques have been tried including trypan blue, neutral red, eosin–opal blue, and eosin–nigrosin. Studies by Dougherty et al. (1975) show eosin-nigrosin
to be superior and this is the technique recommended by Belsey et al. (1980) in the World Health Organization manual.

Eosin–nigrosin technique
Method

(1) Mix one drop of semen with two drops of a 5% aqueous eosin Y in a small tube; leave for 15 seconds.
(2) Add three drops of 10% aqueous nigrosin. Mix thoroughly.
(3) Prepare a thin smear in the manner of a blood film, and air dry—a warming plate may be used.

Results

The eosin will penetrate the membranes of dead spermatozoa staining them red, whilst the live spermatozoa will remain unstained against the blue/black nigrosin background. At least 200 spermatozoa should be counted and the percentage of live and dead spermatozoa calculated. Using this supravital staining technique it is possible to differentiate between immotile live spermatozoa and those which are dead. It can also be used as a check on the motility assessment.

Spermatozoa density

This is one of the most frequently used tests in the investigation of the infertile male, although variability in consecutive spermatozoa counts and poor correlation with fertility has led to dissatisfaction with the results. Spermatozoa density (concentration) is usually determined by diluting the seminal fluid and counting the number of spermatozoa present using a haemocytometer, the improved Neubauer being the most popular counting chamber (*Figure 18.3*).

Method

(1) Using a micropipette dilute the well mixed semen specimen 1 : 10, 1 : 20, 1 : 50 and 1 : 100 with diluent in a small tube. The diluent consists of 5 g sodium bicarbonate ($NaHCO_3$), 1 ml 35% formalin, 99 ml distilled water and 0.5 ml of saturated gentian violet (this can be excluded if phase contrast microscopy is available).
(2) Thoroughly mix the diluted sample and transfer one drop to a coverslipped haemocytometer (*Figure 18.3*).
(3) Place in a moist chamber for 15 min to allow the cells to sediment.
(4) Using a light microscope with the condenser lowered (or phase contrast) and a magnification of ×400, count the numbers of spermatozoa (morphologically mature germinal cells with tails) present. When spermatozoa cross the

lines of the grid, count only those that cross the lines at the top and right hand sides.

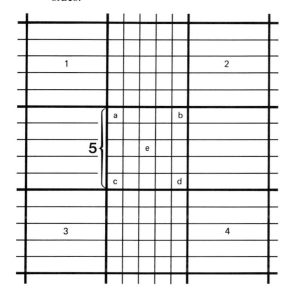

Figure 18.3 Improved Neubauer counting chamber. Three methods of counting have been described using this chamber: (1) Count all cells in the four corner squares 1, 2, 3 and 4. The multiplication factor is 2000. (2) Count all cells in square 5. The multiplication factor is 10 000. (3) Count all cells in small squares 5a, 5b, 5c, 5d and 5e. The multiplication factor is 50 000. Both sides of the haemocytometer should be counted. N.B. These multiplication factors apply only to the improved Neubauer haemocytometer.

Result

The spermatozoa concentration is normally expressed as the number of spermatozoa/ml seminal fluid. This is calculated by multiplying the number of spermatozoa counted by the multiplication factor for the haemocytometer and the dilution of the semen sample. This can be expressed simply in the following equation:

spermatozoa concentration/ml =
  number of spermatozoa counted ×
  multiplication factor ×
  dilution factor.

Occasionally a total spermatozoa count may be needed and this is obtained by multiplying spermatozoa concentration by semen volume.

The definition of a 'normal' spermatozoa count varies greatly ranging from 48 million/ml (Kinloch, Nelson and Bunge, 1974) to 107 million/ml (MacLeod and Gold, 1951). MacLeod and Gold suggested that men with spermatozoa counts above 20 million/ml or total spermatozoa counts above 100 million/ml should be considered fertile. The spermatozoa density from a normal healthy male may vary widely from week to week, month to month, being influenced by stress, disease or drugs (Driff, 1982). In a study of a fit healthy male, Paulsen (Belsey et al., 1980) demonstrated that a count may vary from below 20 million/ml to 170 million/ml over a period of 120 weeks. It is impossible to say with certainty that a male is infertile whilst he produces a few spermatozoa in his ejaculate, as this oligospermia may be a temporary event which reverses at a later date. Infertility is not absolute until the patient is azoospermic.

Various counting chambers have now been designed for semen analysis, for example the Horwell 'Fertility' Semen Counting Chamber* and the Makler Sperm Counting Chamber (Makler, 1978). Their advantage is that they have been developed specifically for semenology and dilution errors have been minimized. A monolayer of spermatozoa is established for simpler, more accurate counting and because spermatozoa are free to move in a horizontal plane only, the motility may be simultaneously assessed. The Makler Counting Chamber has also been used in conjunction with supravital staining and multiple exposure photography for evaluation of viability and motility (Makler, 1979).

Various electronic counters have been employed, e.g. the Coulter counter suitably calibrated (Iverson, 1963), but they are more expensive than a haemocytometer and also possess the disadvantage of counting other cellular elements that are present, such as white blood cells, epithelial cells and immature spermatocytes, in addition to the spermatozoa.

Morphology

Spermatozoa morphology is an essential part of a full semen analysis, as a spermatozoa count may fall within normal limits and yet the individual may have a high percentage of abnormal spermatozoa forms. Human semen usually contains between 21% (MacLeod and Gold, 1951) and 30–40% (Hafez, 1977) abnormal forms. Independent studies carried out by Kinloch,

---

* Arnold R Horwell Ltd, 73 Maygrove Rd, West Hampstead, London NW6 2BP

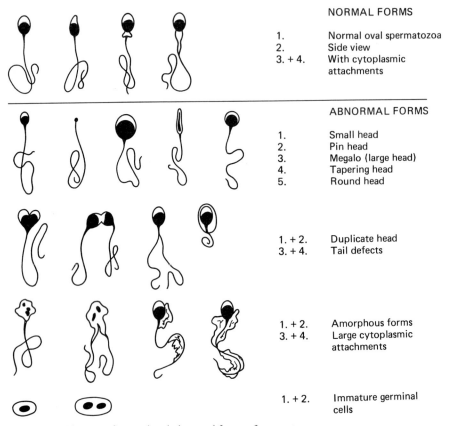

Figure 18.4 Diagram of normal and abnormal forms of spermatozoa

Nelson and Bunge (1974) and Rehan, Sobrero and Fertig (1975) report abnormal forms at 26% and 27% respectively.

Semen morphology is generally assessed microscopically using various techniques to demonstrate various spermatozoa constituents. Results are subjective, therefore methods using photomicrography and image analysis are currently being investigated (Toovey et al., 1981; Schmassman et al., 1982).

Smear preparation for morphology studies

Clean grease-free slides are essential. If the spermatozoa count is greater than 20×10^6/ml the smear may be directly prepared in the manner of a blood film. If the density is less than 20×10^6/ml the semen must be concentrated to obtain sufficient spermatozoa for complete assessment. This is achieved by centrifuging the sample at 2000 rev/min for 10 min, removing the supernatant and smearing one drop of the deposit as above.

Fixation will depend on the staining procedures to be employed, but for routine purposes alcohol is the fixative of choice. A major problem in assessing morphology is that background staining often masks the spermatozoa. This can be reduced by partial air-drying of the smears before immersion in alcohol. Spermatozoa do not seem prone to air-drying artefact when prepared in this manner. Background staining will also be reduced if the smear is kept relatively thin and even.

Staining methods

Several staining methods are available, the method of choice depending on the specific morphological requirements to be investigated. Papanicolaou or haematoxylin/eosin will give good nuclear detail with adequate cytoplasmic staining for evaluation of gross abnormalities. For more detail of cytoplasmic attachments, haematoxylin/rose bengal staining is recommended. Rose bengal is a homologue of eosin and is used as a 1% aqueous solution. All smears for the above staining methods may be fixed in alcohol. For a full assessment of immature spermatocytes and mature spermatozoa Bryan's stain (Bryan, 1970) will provide greater detail. Inflammation of the genital tract may cause difficulties in distinguishing white blood cells from immature germ cells. Couture et al. (1976) overcame this problem by combining Bryan's stain with Leishman's to produce a superior staining technique.

Microscopy

The morphology of normal spermatozoa, their variants and abnormal forms are shown in *Figure 18.4*. Normal spermatozoa exhibit a regular oval-shaped head measuring 3–5 µm in length, 2–3 µm in width, together with a mid portion and tail of at least 45 µm in length (*Figure 18.5*) (Belsey et al., 1980). Small cytoplasmic attachments or collars may be present. Abnormal forms of spermatozoa may be grouped into head defects (small, pin, megalo, tapering and round head forms), duplicate head/tail defects, tail defects, amorphous forms and those with large cytoplasmic attachments. Immature germ cells do not have tails, appearing as single or multinucleated cells, ranging from 4–9 µm in diameter (Belsey et al., 1980). At least 200 spermatozoa should be assessed, using different microscopic fields. Only the mature spermatozoa should be counted, with a separate count of mature/immature forms performed. The abnormal forms may be reported either as a percentage of the total number of spermatozoa seen, e.g. 50% abnormal forms present, or subdivided into their respective abnormalities, e.g. 20% pin head, 10% tapering head, 20% amorphous forms.

Cervical mucus penetration and postcoital tests

Whilst semen analysis provides evidence of potential fertility, the interaction of the spermatozoa with cervical mucus is also relevant to the outcome. Ovarian hormones regulate cervical mucus secretion, with cyclic alterations influencing receptivity, migration, penetration and survival. Evaluation of cervical mucus properties include assessment of ferning (crystallization) patterns, viscosity and pH (the optimal being between 7 and 8.5).

The cervical mucus penetration test is performed *in vitro* and is a measurement of spermatozoa penetration and migration into cervical mucus. Two different techniques have been described, one a capillary tube method (Kremer, 1965), the other a slide method (Moghissi, 1966).

Postcoital tests are *in vitro* investigations, providing information of spermatozoa survival and behaviour on the endocervix. The test should be performed 6-10 h after intercourse, and involves the sampling of mucus from the ectocervix and endocervical canal (Moghissi, 1976).

Semen immunology

Autoimmune reactions against spermatozoa are now recognized as a primary cause of infertility. Antibody formation appears to be related to a permanent (vasectomy) or temporary obstruction of the ejaculatory duct such as inflammation or testicular trauma, when the blood/testis barrier of the reproductive system is breached, bringing sperm and the immune system into contact.

Anti-sperm antibodies can be detected in the seminal plasma as locally produced IgA which coats the spermatozoa and impairs cervical mucus penetration, and as circulating serum IgG. If an immunological disorder is suspected, a mixed antiglobulin reaction test (MAR) should be performed. The principle of the MAR reaction is the use of sensitized red cells to detect immunoglobulins attached to the spermatozoa. It is a primary screening test and does not give an indication of the antibody titre. A positive MAR reaction should be followed by tray agglutination test (TAT) and gelatin agglutination test (GAT) of both serum and seminal plasma so that antibody titres may be obtained. For details of these techniques see Hafez (1977). A serum antibody titre of 32 or over usually indicates a significant impairment of fertility especially when seminal plasma antibodies are present as well.

Other methods of seminal fluid analysis include spermatozoa cytochemistry, electron microscopy, cytogenetics, microbiology and biochemistry. In all these cases, great care must be exercised to ensure that the whole semen sample is well mixed and the volume measured before the specimen is split and despatched to the various laboratories. In the majority of

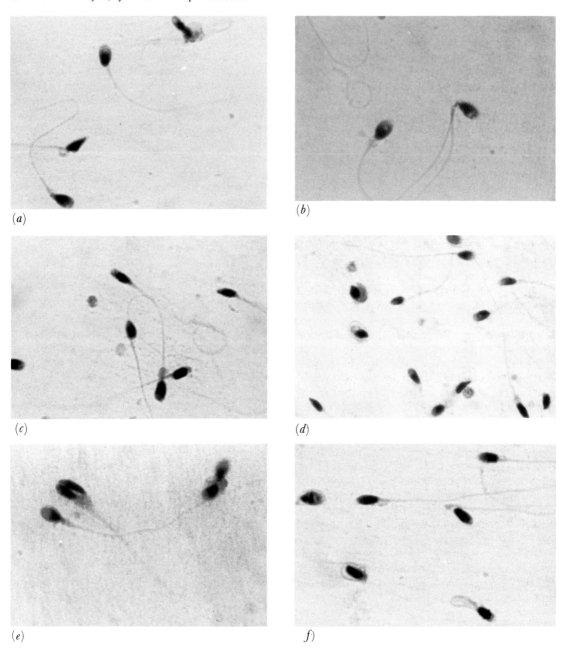

Figure 18.5 Normal (*a*) and abnormal (*b–f*) spermatozoa in seminal fluid. (*a*) Normal forms; (*b*) tail defect; (*c*) tapering head defect; (*d*) small head defect; (*e*) duplicate head defect; (*f*) cytoplasmic attachment. Stain: Papanicolaou. Magnification ×980

cases, however, the tests described in the previous sections will provide sufficient information for the clinician to counsel his patients who present with problems of infertility.

Place of semen analysis in clinical practice

There are many causes of infertility in the male, and the investigation of this condition can be very complex indeed. It will include extensive history taking and meticulous physical examination, hormone estimation and microbiological studies, as well as semen analysis. The semen examination will focus not only on spermatozoa cells but also allow biochemical evaluation of seminal fluid for carnitine GPC, fructose and prostatic acid phosphatase for assessment of function of the epididymis, seminal vesicle and prostate respectively.

The causes of azoospermia can be categorized by (a) spermatogenic failure, (b) obstruction at any level of the genital tract, and (c) ejaculative failure leading to retrograde ejaculation. Ectopic testes, Klinefelter's syndrome and irreversible tubular damage following mumps orchitis are common causes of azoospermia. The causes of oligospermia and the presence of abnormal or poorly motile forms are more difficult to identify and toxic drug therapy or genetic factors must be considered as possible aetiological factors. In the long run, the cause of the defect may never be identified and a diagnosis of idiopathic infertility has to be made.

Hydrocoele and spermatocoele

Hydrocoeles and spermatocoeles are commonly occurring lesions of the male reproductive system. Fluid is withdrawn in a similar way to aspiration of any palpable lesion but great care must be taken to avoid damage to the underlying testis.

The specimen is transferred to a tube containing anticoagulant and for preparatory purposes in the laboratory is treated as any other serous fluid (*see* Chapter 14). Alcohol-fixed smears should be stained by the Papanicolaou and/or haematoxylin and eosin method and air-dried smears by the May-Grünwald–Giemsa technique.

Hydrocoele

Hydrocoeles are caused by an accumulation of serous fluid between the mesothelial layers of the tunica vaginalis and it has been suggested that delayed absorption rather than excess production is the main factor causing this accumulation. The incidence of hydrocoeles is uncertain, but they occur mainly in males over 20 years of age and are commoner in tropical regions than temperate zones. Hydrocoeles usually pursue a chronic clinical course and may be associated with inflammatory conditions including tuberculosis, gonorrhoea and filariasis as well as neoplastic conditions.

Hydrocoeles, when uninfected, yield a translucent amber-coloured fluid on aspiration. Occasionally, the aspirate is haemorrhagic (haematocoele) if bleeding due to trauma, torsion or tumour has occurred in the sac; or turbid if there is an inflammatory reaction.

Cytology

The fluid from hydrocoeles is often acellular, or shows only a few mesothelial cells and lymphocytes. In inflammatory conditions, numerous leucocytes may be found along with red cells. Parasites have also been identified in hydrocoele fluids (Vassilakos and Cox, 1974). Markedly atypical mesothelial cells have also been observed (Koss, 1979). These constitute a diagnostic pitfall cytologically as the cells show many of the features associated with malignancy. Large metaplastic cells have also been described. Malignant testicular tumours such as a seminoma, may exfoliate cells into a hydrocoele fluid (Koss, 1979). The reason for a hydrocoele fluid having a high mesothelial cell content is usually due to repeated aspirations. High cell counts are also found if the testis or the epididymis (spermatozoa, spermatocytes or columnar cells) are mistakenly aspirated.

Spermatocoele

Spermatocoeles occur after puberty and are most commonly found in patients aged between 40 and 60 years. The aetiology is unknown.

A spermatocoele is a cyst (either uni- or multilocular) occurring in the rete testis or the head of the epididymis. The cyst walls are normally lined by cuboidal or flattened epithelium on a supporting fibrous layer. On aspiration the fluid appears milky or turbid. Spermatocoeles often contain large numbers of spermatozoa though these may be absent in elderly patients. The fluid may also contain occasional lymphocytes, epithelial cells and immature spermatocytes. These germ cells can look quite active and arouse concern in the inexperienced cytologist (*Figure 18.6*). Another possible pitfall described by Wentworth, Wager and Unitt (1971) is the presence of exfoliated ductal epithelial cells into the spermatocoele fluid. These appear as bizarre eosinophilic-orangeophilic cells with angular hyperchromatic nuclei. Similar cells appear in smears obtained by prostatic massage. Various inflammatory processes may also result in the appearance of these bizarre cells. In cases of tuberculosis, epithelioid cells, granulomatous and caseous necrotic material (possibly containing tubercle bacilli) may be found (Cardozo, 1976). Sometimes a spermatocoele ruptures, producing a spermatic hydrocoele. In all cases of fluid aspiration from cystic lesions careful follow-up examination of the affected testis must be made, with re-aspiration of any palpable residual mass to exclude possible underlying tumour.

 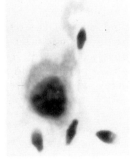

Figure 18.6 Spermatocytes (composite) in a spermatocoele. Stain: Papanicolaou. Magnification × 400

The prostate and seminal vesicles

Anatomy and histology

The prostate is a compound tubulo-alveolar gland, the size of a chestnut measuring 4 × 3 × 3 cm, weighing approximately 8 g in the average adult male. It surrounds the urethra and is covered by a capsule of connective tissue. There are three types of glands in the prostate: the smallest mucosal glands which are situated in the peri-urethral tissue and may form adenomatous nodules in older men; the submucosal glands which surround these, and the main prostatic glands which are in the largest outer part of the prostate. The mucosal glands open into the lumen of the urethra, the submucosal and prostatic glands open into the urethral sinuses. The two seminal vesicles, situated between the posterior surface of the bladder and the rectum, are elongated coiled structures 5 cm long, 3–4 mm in diameter. They are continuous with the ejaculatory ducts which pass through the prostate, dividing it into three uneven lobes.

The prostatic ducts and the seminal vesicles are lined by tall columnar epithelium. The prostatic secretion contains acid phosphatase, the level of which is measured in the bloodstream to detect prostatic tumours. The lumina of the prostatic glands often contain corpora amylacea, calcified prostatic concretions which may also be exfoliated.

Specimen collection

Fine needle aspiration

This is often used as it is an easily repeatable test which provides a speedy diagnosis of prostatic pathology. The Franzen needle is used to aspirate the prostate. This is a 22 gauge needle, approximately 20 cm in length with a specialized handle and needle guide. The needle is guided into the rectum and prostatic material aspirated under negative pressure. Smears are prepared and fixed immediately in alcohol or air-dried for Romanowsky staining.

In a series of approximately 17 000 transrectal aspirates of the prostate reported by Zajicek (1979), complications were encountered in only 0.4% cases following the procedure. A second series demonstrated that it is inadvisable to carry out fine needle aspiration on patients with prostatitis, as widespread infection may subsequently develop. There is a risk of local tumour cell dissemination if needles with a gauge larger than 18 or 22 are used.

Prostatic massage

Prostatic massage may be carried out to obtain diagnostic material from the prostate. Massage is performed manually and transrectally and the secretions produced are smeared directly onto glass slides. Alternatively a post-massage urine is collected, which will contain the prostatic secretions from the urethra, as well as exfoliated cells from the bladder and ureters.

Cytology of benign conditions

Prostatic glandular cells may appear singly or in small clusters. The nuclei are round and centrally placed, and the cytoplasm is pale and granular. In prostatic massage specimens corpora amylacea may also be present, as well as cells from the seminal vesicles which are identified by the presence of yellow cytoplasmic granules. Spermatozoa and their precursors are also found in these specimens.

Prostatic hyperplasia is a glandular proliferation, commonly occurring in elderly men and results in an enlarged prostate. Material collected using a fine needle is usually made up of sheets of regular, round or polygonal cells, with centrally placed nuclei. These sheets display a honeycomb pattern, with cell borders readily visible. There is often a thin film of blood present with this condition.

Prostatitis is a condition in which numerous inflammatory cells are found and the glandular cells show degenerative features, including cytolysis and changes in nuclear size.

Cytology of prostatic neoplasia

The cytological features present in prostatic carcinoma reflect the degree of differentiation of the tumour. Cells from a poorly differentiated adenocarcinoma vary in size, with malignant nuclei and large nucleoli. Well differentiated adenocarcinoma of the prostate sheds clusters of cells with well preserved features, and is often found in the posterior lobe. There may be coexisting carcinoma of the bladder. Prostatic duct carcinoma is cytologically similar to transitional cell carcinoma of the bladder causing diagnostic difficulties.

Prostatic neoplasia causes raised blood acid phosphatase levels, although these levels are only raised in 20% of well differentiated adenocarcinoma. The peak incidence of this disease is in the seventh and eighth decade, and it rarely occurs in men aged less than 40 years. The presence of benign hyperplasia is associated with the disease, but not related to it.

References

ALEXANDER, N. J., FREE, M. J., ALUIN-PAULSEN, C., BUSCHBOM, R. and FULGHAM, D. L. (1980). A comparison of blood chemistry, reproductive hormones and the development of anti-sperm antibodies after vasectomy in man *J. Androl.*, **1**, 40–50

BELSEY, M., ELIASSON, G., GALLEYUS, A., MOGHISSI, K., PAULSEN, C. and PRASAD, M. (1980). Laboratory manual for the examination of human semen and semen–cervical mucus interaction. Press Concern, Singapore. WHO Publication

BRYAN, J. (1970). An eosin-fast, green naphthol yellow mixture for differential staining of cytoplasmic components in mammalian spermatozoa. *Stain Tech.*, **45**, 231–236

CARDOZO, P. L. (1976). *Atlas of Clinical Cytology*. Targ b.v.'s Hertogenbusch

COUTURE, M., ULSTEIN, M., LEONARD, J. and PAULSEN, C. (1976). Improved staining method for differentiating immature germ cells from white blood cells in human seminal fluid. *Andrologica*, **8**, 61–66

DOUGHERTY, K., EMILSON, L., COCKETT, A. and URRY, R. (1975). A comparison of subjective measurements of human sperm motility and viability with two live-dead staining techniques. *Fertil. Steril.*, **26**, 700–703

DRIFF, J. (1982). Drugs and sperm. *Br. Med. J.*, **284**, 844–845

HAFEZ, E. S. E. (1977). *Techniques in Human Andrology. Human Reproductive Medicine I.* Amsterdam: North Holland Publishing Company

IVERSON, S. (1963). Coulter counting sperm. *J. Agric. Sci.*, **62**, 219–223.

JONES, R. C. (1973). Preparation of spermatozoa for electron and light microscopy. *J. Reprod. Fertil.*, **33**, 145–149

KATZ, D. F. and DOTT, H. M. (1975). Methods of measuring swimming speed of spermatozoa. *J. Reprod. Med.*, **45**, 263–272

KINLOCH NELSON, C. and BUNGE, R. (1974). Semen analysis—evidence of changing parameters of male fertility potential. *Fertil. Steril.*, **25**, 503–507

KOSS, L. (1979). *Diagnostic Cytology and Its Histopathologic Bases*, 3rd Edition. Philadelphia: J. B. Lippincott

KREMER, J. (1965). A simple sperm penetration test. *Int. J. Fertil.*, **10**, 209–215

MACLEOD, J. (1951). Sperm quality in one thousand men of known fertility and eight hundred cases of infertile marriages. *J. Fertil. Steril.*, **2**, 115

MACLEOD, J. and GOLD, R. Z. (1951). The male factor in fertility and infertility. II. Spermatozoon counts in 1000 men of known fertility and 1000 cases of infertile marriages. *J. Urol.*, **66**, 436

MAKLER, A. (1978). A new chamber for rapid sperm count and motility estimation. *Fertil. Steril.*, **30**, 313–318

MAKLER, A. (1979). Simultaneous differentiation between motile, non-motile, live and dead human spermatozoa by combining supravital staining and MEP procedures. *Int. J. Androl.*, **2**, 32–42

MAKLER, A. (1980). Distribution of normal and abnormal forms among motile, non-motile, live and dead human spermatozoa. *Int. J. Androl.*, **3**, 620–628

MOGHISSI, K. S. (1966). Cyclic changes of cervical mucus in normal and progesterone treated women. *Fertil. Steril.*, **17**, 663–675

MOGHISSI, K. S. (1976). Postcoital test: physiologic basis, technique and interpretation. *Fertil. Steril.*, **27**, 117–129

REHAN, N. E., SOBRERO, A. J. and FERTIG, J. (1975). The semen of fertile men. Statistical analysis of 1300 men. *Fertil. Steril*, **26**, 492–502

ROSE, N. R., HARPER, M., HJORT, R., VYAZOV, O. and RUMKE, P. (1976). Techniques for detection of iso and auto antibodies to human spermatozoa. *Clin. Exp. Immunol.*, **23**, 175–199

RUMKE, P. (1974). The origins of immunoglobulins in semen. *Clin. Exp. Immunol.*, **17**, 287–297

SCHMASSMAN, A., MIKUZ, G., BARTSCH, G. and ROHR, H. (1982). Spermiometrics: objective and reproducible methods for evaluating sperm morphology. *Eur. Urol.*, **8**, 274–279

TOOVEY, S., HUDSON, E., HENDRY, W. and LEVI, A. J. (1981). Sulphasalazine and male infertility. *Gut*, **22**, 445–451

TUNGE, K. S. K. (1975). Human sperm antigens and antisperm antibodies. *Clin. Exp. Immunol.*, **20**, 93–104

VAN LEEUWENHOEK (1685). Letter to the Royal Society of London, dated March 30th 1685. Reproduced in the Collected Letters of Antoni Van Leevwenhoek, edited by a Committee of Dutch Scientists, Swets and Zeitlinger, Amsterdam 1941

VASSILAKOS, P. and COX, J. N. (1974). Filariasis diagnosed by cytological examination of hydrocele fluid. *Acta Cytol.*, **18**, 62–64

WENTWORTH, P., WAGER, S. and UNITT, M. (1971). Atypical cells in spermatocele fluid. *Acta Cytol.*, **15**, 210–211

ZAJICEK, J. (1979). In *Diagnostic Cytology and its Histo pathologic Bases*, (Koss, L. G., ed.), pp. 1001–1008; 1093–1097. Philadelphia: J. B. Lippincott

19

Fine needle aspiration cytology

David Melcher, M. Reeves and Russell Smith

Disabuse your minds of much of that which you have read or been told. In books tumours are 'innocent' or 'malignant' but in Nature such distinctions do not exist. In a little while I shall tell you of tumours which seemed to be innocent and yet were malignant, and of tumours which seemed malignant and yet were innocent. Do not be misled by the positive air wherewith some assert the nature of tumours which they cannot see and have only felt with difficulty. The errors they fall into would be ludicrous were they not fraught with tragedy.

C. B. Lockwood
Lancet, 29 June 1904

These eloquent phrases are taken from a lecture by Mr C. Lockwood, a prominent surgeon at St Bartholomew's Hospital, London at the turn of the century. Mr Lockwood was addressing a group of medical students on the subject of early microscopical diagnosis of tumours and he was advising them of the need for accurate pathological diagnosis before the commencement of treatment. What is so fascinating about his advice is that his attitude to the diagnosis and management of tumours is as valid today as it was 80 years ago.

At the turn of the century when Lockwood was lecturing, the favoured method for rapid tissue diagnosis was frozen section. However, within a few years alternative methods of obtaining tumour tissue for rapid diagnosis by fine needle aspiration cytology were reported (Dudgeon and Patrick, 1927; Martin and Ellis, 1930) and surgeons were soon presented with a choice. Should they select open biopsy which was traumatic for the patient or should they employ fine needle aspiration cytology which was better tolerated by the patient but less reliable than surgical biopsy? The dilemma persists to this day, although there is good evidence that in the last decade many more surgeons are turning to fine needle aspiration cytology for tumour diagnosis than ever before.

It is probable that fine needle aspiration (FNA) was first used by pathologists for the investigation of enlarged lymph nodes. One of the earliest descriptions of the technique is provided by Grieg and Gray (1904) who identified trypanosomes in material aspirated from enlarged nodes in patients with sleeping sickness using a hypodermic needle and syringe. According to Koss, Woyke and Olszewski (1984) FNA was introduced into the Memorial Hospital, New York in the 1930s to allay the fears of many of the clinicians and pathologists employed there who believed that surgical biopsy and manipulation of cancerous tissue *in vivo* increased the risk of spread of tumours. Once fears of tumour spread were allayed, the technique fell into disuse and its potential for rapid diagnosis remained unappreciated for many years.

The recent emergence of FNA as one of the key diagnostic procedures in clinical practice today reflects the work and enthusiasm of Professor Josef Zajicek and his associates at the Karolinska Hospital in Sweden who succeeded in convincing their medical colleagues throughout the world and the thousands of students who passed through their hands, that the technique had much to offer. In their hands it was proven simple, safe, quick and accurate. In addition it was shown to be acceptable to the patient and free from haemorrhage and sepsis. Unlike biopsy, general anaesthesia is rarely required for FNA, so that the technique is particularly appropriate for use in an outpatient setting. Moreover, aspiration could be repeated without distressing the patient if a second specimen was needed to further the diagnosis. Clinical experience has shown that FNA is particularly useful for the diagnosis of metastatic tumour or recurrence of tumour, thereby avoiding the need for further investigation of the patient, particularly the need for exploratory laparotomy or thoracotomy. It should not be forgotten that microbiological studies can also be carried out on the samples.

The limitations of FNA are few. One of the main drawbacks of the technique is the limited amount of information that can be gleaned from a cytological specimen compared with tissue biopsy where, in histological section, the architecture of the tissue is preserved intact. Another drawback is that lesions that contain abundant fibrous tissue or are very vascular do not lend themselves to aspiration. Risk of tumour spread along the needle track is very low. It has been recorded but only in cases where a wide bore needle has been used.

FNA was first used in clinical practice for the investigation of palpable lumps such as enlarged cervical or inguinal glands, breast lumps and thyroid or parotid tumours. It was also used for the rapid diagnosis of thoracic or abdominal tumours exposed at operation. However, with the development of computer assisted tomography and ultrasonography for tumour imaging, it is now possible to sample deep seated lesions at almost any body site by percutaneous FNA. In this chapter we describe the principles of FNA and discuss the advantages and limitations of the technique as it is applied to each site of the body. We have outlined the use of FNA for the investigation of breast lumps, lymph nodes and some intrathoracic and intra-abdominal lesions, bearing in mind the fact that this book is intended for laboratory scientists and cytotechnicians rather than pathologists. In consequence the role of the technologist in the collection and preparation of the samples is emphasized at the expense of details of the cytomorphology of the aspirated cells.

Fine needle aspiration technique

The basic technique of specimen collection through a fine needle is the same regardless of the site of the lesion being sampled. This is shown diagrammatically in *Figure 19.1(a)–(j)* and a general description of the technique for aspiration of palpable lumps is given in this section. Modifications are necessary for deep seated lesions and these are also discussed.

(1) Palpate the lump and define its boundaries. Assemble needle and syringe. The preference in this laboratory is for a 20 ml syringe with a Luer lock although some centres prefer a smaller syringe. A 21 gauge needle of the type used for venepuncture is appropriate for the aspiration of palpable lumps. Clean the overlying skin with appropriate antiseptic.

(2) Immobilize lump between thumb and forefinger (*Figure 19.1a*). Insert the needle into the tumour and induce negative pressure in the needle by withdrawing the plunger (*Figure 19.1b*). The operator can often detect a change in consistency in the tissue as the needle enters the lesion.

(3) While maintaining the negative pressure, gently move the needle tip from side to side so that the cells are dislodged and sucked into the lumen of the needle (*Figure 19.1c*). Try to ensure that the whole aspirate is contained within the lumen of the needle and the barrel of the syringe contains only air. Fluid may appear in the syringe if the lesion is cystic or contains pus. Blood will be aspirated if the lesion is very vascular. Gently release the plunger of the syringe to equalize the pressure before withdrawing from the lesion.

(4) Immediately detach needle from syringe and withdraw plunger to fill syringe with air (*Figure 19.1d*).

(5) Replace needle and direct it toward the upper end of a clean glass microscope slide prelabelled with the patient's name. Gently depress plunger expelling contents of the needle onto the slide (*Figure 19.1e*).

(6) Spread material evenly over slide with the

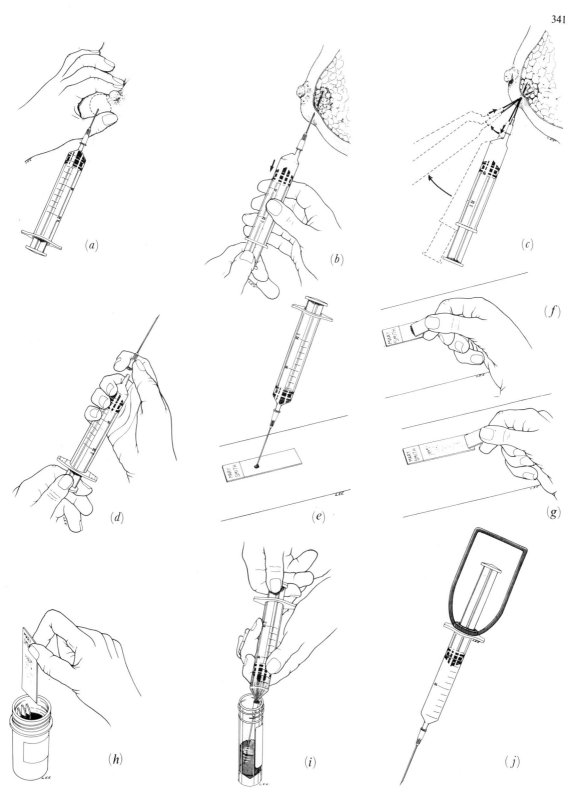

Figure 19.1 (a)–(j) Diagrammatic description of technique of specimen collection through a fine needle. From *Operative Surgery* 4th edition 1982, General Principles, Breast and Extracranial Endocrines (Eds Hugh Dudley and Walter Pories) London: Butterworths

edge of a second slide as for a blood film (*Figure 19.1f* and *g*). Allow to air dry, and stain by May-Grünwald–Giemsa (MGG) method. In this laboratory, MGG staining is the preferred method of staining as the amount of material available for staining is usually limited and air drying of the smears before fixation is a common occurrence.

(7) If adequate material is available prepare a second smear in the same way. Fix immediately in alcohol before air drying occurs and stain by the Papanicolaou method (*Figure 19.1h*). Prepare additional alcohol fixed slides if sufficient material is available.

(8) Rinse out needle and syringe with sterile saline or Hank's medium in order to salvage any material remaining in the needle or syringe for further studies of immunocytochemistry (*Figure 19.1i*). Collect the washings in an Universal container and cytocentrifuge.

(9) Some operators prefer to use the Cameco syringe holder (Cameco A6, Tabyvagen 71-180, Enebybang, Sweden) which permits the aspiration to be performed with one hand, thereby freeing the other hand for fixing the tumour (*Figure 19.1j*).

(10) The preparation of satisfactory smears from fine needle aspirates takes considerable skill and experience. The pathologist should make every attempt to train the clinicians in these skills if he or she is not taking the aspirate personally. In the event of failure the clinician should be advised to collect the aspirate in a suspension of Hank's fluid although it should be emphasized that this is 'second best'.

(11) Arrangements should be made in advance if microbiological study of the aspirated material is required. If the needle and syringe are still sterile the samples described in section 5 can also be sent for microbiological studies. If the needle has been contaminated by touching the slide, aspiration may have to be repeated to obtain an appropriate sample.

(12) Dispose of needle and syringe in accordance with safety guidelines.

(13) Ensure specimens are cleary labelled and that you are in possession of all the clinical information necessary for accurate interpretation of the smears.

Special aspects of aspirating deep seated lesions

The percutaneous fine needle aspiration of intra-abdominal or intrathoracic lesions requires several modifications of the general aspiration technique described in *Figure 19.1*. The operation is usually carried out by the radiologist who has the special skill needed to localize the lesion on ultrasound or tomographic scan and to guide the needle into the tumour without damaging adjacent organs or tissue. Aspiration is usually performed under local anaesthesia as larger needles are required and more than one attempt at aspiration may be needed to get an optimal sample. Moreover, it is essential to have the full cooperation of the patient as the positioning of the needle may be quite time consuming and technically difficult (Husband and Golding, 1983; Livraghi *et al.*, 1983).

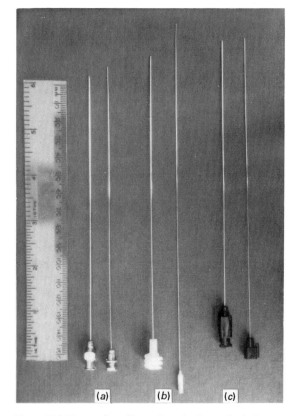

Figure 19.2 Types of needle used to obtain cytological material. (*a*) 22 gauge spinal needle; (*b*) 20 gauge Rotex screw tip needle; (*c*) 22 gauge Cook's Franzen needle

The choice of needle is in the hands of the operator and many different needles are now available. The needle most frequently used at our hospital is the spinal needle which was originally designed for lumbar puncture (*Figure 19.2a*). A 22 g spinal needle, 10–15 cm long, is appropriate for most situations (Pagani, 1983). If the lesion is particularly difficult to access, a 22 g or 23 g Chiba or Okuda needle (a 'skinny' needle) can be used. These skinny needles were designed for transhepatic cholangiography and have the advantage of being thinner and more flexible than spinal needles. Consequently they can be passed through bowel or other viscera as well as arteries and veins without causing damage. As the needle is light it does not transfix the tissue but moves with it, thereby minimizing local trauma to the tissue at the needle tip. The main disadvantage of this needle is that it is more difficult than the rigid spinal needles to position correctly into the lesion.

Some needles intended for aspiration have been modified in such a way that a small core of tissue can be obtained to provide a mini-biopsy. Of these modified aspiration needles the authors have experience of the Rotex screw tip needle (*Figure 19.2b*) which is principally used for lung biopsies (Nahman *et al.*, 1985), and the Franzen needle (*Figure 19.2c*) for prostatic aspiration biopsy. In order to facilitate location of the needle in the lesion, some needles are coated with Teflon. Others have undulated shafts which can be readily visualized if the plane of the needle is in the same plane as the tomographic slice (Porter, Karp and Forsberg, 1981; Linsk and Franzen, 1986).

Rapid cytological diagnosis

An important aspect of cytological investigation in terms of its value as a diagnostic tool is the speed and accuracy of reporting. As a routine, FNA specimens should be processed and reported within 24 h of receipt in the laboratory, the limiting factors being availability of staff to prepare the smears, stain them and examine them under the microscope. A great advantage of FNA is that it is possible to make a cytological diagnosis in a much shorter time than this, often within minutes of taking the sample should the need arise. This facility for rapid diagnosis is of particular value for the investigation of patients attending an outpatient clinic or for the investigation of tumours exposed at operation. It is also useful for the diagnosis of intrathoracic or intra-abdominal organs using ultrasound or tomographic imaging techniques. In these situations a diagnosis of malignancy has an immediate effect on patient management, e.g. early admission to hospital in the case of the patient attending the outpatient clinic or intra-peritoneal chemotherapy for the patient in the operating theatre. Moreover, an on-the-spot diagnosis in an out-patient clinic or in the X-ray department eliminates the need to recall the patient for repeat tests should the specimens prove to be inadequate for diagnosis, e.g. heavily bloodstained or very scanty.

Unfortunately some of the advantages of rapid diagnosis are offset by the demands it makes on technician and consultant time, but it is possible to organize FNA sessions to minimize the time spent by the laboratory staff in attendance at the clinic. A portable trolley and microscope should be available in the cytology laboratory for this purpose. The optimal arrangement, as evidenced by the Scandinavian system, is for the pathologist to take the specimen whenever feasible.

There are several staining procedures suitable for rapid cytological diagnosis. Most centres favour the Dif Quick stain, a modified Giemsa method which is used on air-dried smears. Methylene blue which is a vital stain is also appropriate for rapid diagnosis. The smears can be decolourized by immersing them in alcohol which serves as a fixative so that they can be restained by the Papanicolaou method. Methylene blue is particularly useful as only nucleated cells take the stain up so that they can be quickly identified in a heavily bloodstained smear.

If rapid diagnosis is requested by the surgeon it is important for the laboratory to have all the necessary equipment to hand. It is convenient to have a trolley or tray prepared for the occasion in advance. The following items should be included on the trolley:

(a) Glass slides with frosted ends.
(b) Glass coverslips (22 mm × 40 mm).
(c) Pencil.
(d) Alcohol fixative (Spray fix).
(e) 5 ml ampoules of Hank's fluid.
(f) Sterile universal container for samples requiring microbiological study.
(g) Hank's fluid for needle washing.
(h) Dif Quick stain.
(i) Blotting paper for removing excess stain.

(j) Protective gloves and sterile pack.
(k) Portable microscope.

Cytological features of benign and malignant lesions

The cytological patterns seen in smears prepared from material obtained by FNA are very variable. They may show characteristics which correspond very clearly with the histological features of a particular tumour and in such cases a very precise diagnosis is possible. In many specimens, however, there is only sufficient information for the pathologist to give a diagnosis of benign or malignant. The features which assist the cytopathologist with his diagnosis are shown in *Table 19.1* They relate mainly to the diagnosis of epithelial tumours.

Table 19.1 Cytological features of benign and malignant lesions

Benign features	Malignant features
Scanty cells in the smear	Very cellular smears
Epithelial cells in tight clusters	Loss of cohesiveness of epithelial cells
Uniformity of cell type and shape	Pleomorphism of cells and cell nuclei
Regular chromatin and small nucleoli (if any)	Irregular chromatin and multiple nucleoli

Reporting fine needle aspirates

It is the author's practice to use several categories for the reporting of FNA cytology. These are defined below.

(1) *Unsatisfactory*. A report of an unsatisfactory sample should indicate that essential epithelial or other tissue elements are absent from the smear and that the specimen is therefore unsuitable for analysis.
(2) *Benign*. A benign report indicates that adequate tissue elements are present for study and that no malignant cells are seen. This report does not exclude malignancy completely as the sample may not be representative of the lesion. However, in most cases a report indicating that there is no evidence of malignancy is acceptable when allied to other investigations, e.g. clinical examination and mammography.
(3) *Malignant*. A malignant report indicates that there is no doubt in the cytologist's mind that malignant cells are present and that the clinician can proceed accordingly with definitive treatment.
(4) *Suspicious*. A suspicious report should be issued when, for example, the specimen is very scanty but contains a few abnormal cells or when the specimen contains benign material in combination with abnormal cells. Approximately 4% of breast aspirates in the author's laboratory fall into this category. Of these 80% prove to be malignant.

FNA of specific target organs

Breast

Fine needle aspiration of the breast has an important part to play in the investigation of the women presenting at the outpatient clinic with a palpable lump in the breast (Gardecki *et al.*, 1981). It also has a useful role in the investigation of non-palpable lesions detected by mammography in women who are participating in breast cancer screening programmes. In these cases, the lesion is located by stereotactic imaging and aspiration performed under radiological control (Kopans *et al.*, 1985).

Fine needle aspiration of the breast enables the clinician to ascertain whether the lump is cystic or solid, benign or malignant so that the management of each case can be carefully planned on an individual basis. For example, if the lesion is cystic, aspiration of the fluid should result in disappearance of the lump. Blood-stained cyst fluid or a residual lump after aspiration should raise a suspicion of malignancy and hospital admission for biopsy planned. On the other hand frank malignant cells in an aspirate should be an indication for prompt admission for biopsy. Thus there should never be any need for the situation to occur where a woman submits to surgery without being informed of the nature of the lesion in her breast and the probable outcome of the operation.

Several studies have shown that the accuracy of a cytodiagnosis of breast cancer is very high and in skilled hands the false positive rate is less than 1% (Zajicek, 1974). In centres which can claim a high degree of diagnostic accuracy, it is not uncommon for the surgeons to start treatment without resorting to surgical biopsy for confirmation of the diagnosis, especially if the clinical examination is consistent with the cytological findings. It must be remembered, however, that the false negative rate of FNA of the breast is high (usually in the region of 10%) mainly because of inadequate sampling, and a clinical suspicion of malignancy should always override a negative cytology report. In the author's experience the percentage of unsatisfactory samples is in the region of 25% and is a direct reflection of the ability of the clinician who is taking the sample. Experience has shown that cytology reporting is at its most accurate when the pathologist takes the sample and examines it immediately so that a repeat specimen can be taken as necessary. Moreover, the pathologist can obtain additional clues to the diagnosis by palpation of the lump (cancerous lesions are often stony hard and feel gritty when penetrated by the needle).

Several methods of preparing the aspirates have been described. These include direct smears, cytocentrifuged preparations from a suspension of the aspirated material, and millipore preparations. The direct smear probably provides the most useful information as the relationship of the tumour cells to each other is not distorted during processing. Air drying often occurs before the smear can be placed in fixative as the number of cells and amount of mucus in the sample may be scanty. Such samples should always be stained by the May-Grünwald–Giemsa method.

Structure of the breast

The breast is a modified sweat gland whose secretions drain into the lactiferous ducts which open into the lactiferous sinus beneath the nipple. In the adolescent the lactiferous ducts are short and are embedded in scanty connective tissue. At puberty, in the non-pregnant female, the breast enlarges and the lactiferous ducts subdivide into smalller and smaller branches which terminate in secretory acini. In the adult breast the lactiferous ducts and their complement of secretory acini are separated by fibroblastic connective tissue and abundant adipose tissue, thereby dividing the breast into 15–20 lobes. Within each lobe the gland is further divided into lobules by a fine network of connecting tissue surrounding the terminal ducts and the secretory acini. In pregnancy the structure of the breast changes again as the secretory acini proliferate to form a greatly expanded secretory surface.

The larger ducts are lined by a pseudostratified columnar epithelium which in histological section appears as a double layer of cells. The smaller ducts and alveoli are lined by a single layer of cuboidal epithelium beneath which a low flattened layer of myoepithelial cells may be seen. These cells contain cytoplasmic filaments similar to those found in smooth muscle and have a role in the contraction of the gland and the secretion of milk from the alveoli into the lactiferous ducts. The ductal epithelium of the breast retains the potential for metaplastic change and it is not unusual to see evidence of apocrine metaplasia in histological sections reflecting the embryonic origin of this exocrine gland.

Cytology of benign lesions

The benign lesions which are found most commonly in younger women are single or multiple breast cysts, fibroadenosis or fibrocystic hyperplasia and fibroadenomas. They result from hormonal imbalance and hormone fluctuation during the menstrual cycle and pregnancy. Fat necrosis is another common lesion but occurs mainly in older women.

Cystic disease

Breast cysts vary in size and are often multiple. Needle aspiration is particularly useful in this situation as the technique is often curative and the patient's obvious worry disappears on the spot. Any residual lump left behind when the cyst contents are aspirated should be re-aspirated or biopsied. The cytology of cyst contents is variable, the common cellular constituents being:

(a) Leucocytes: usually neutrophils and often abundant.
(b) Red blood cells: these may reflect the trauma of sampling but altered blood indicates old haemorrhage.
(c) Foam cells: histiocytes (*Figure 19.3*).
(d) Duct epithelial cells: these epithelial cells from the cyst lining may show the changes of apocrine metaplasia which reflect the fact that the breast is a modified sweat gland. The ductal cells may also form papillary clusters but rarely cause a problem of cytodiagnosis.

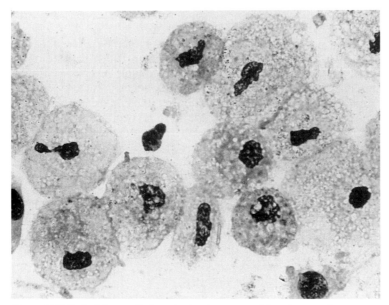

Figure 19.3 Typical foam cells from benign breast cyst. Note degenerative changes. Stain: May-Grünwald–Giemsa. High power view

The fluid aspirated from a benign cyst has a typical grey/green opalescent appearance and only merits cytological examination if time permits. In contrast bloodstained cyst aspirates should always be sent for cytological study.

The authors have not seen malignant cells in breast cyst fluid unless heavily bloodstained, and in these cases, which are rare, it is likely that the bloody fluid is originating from a necrotic or haemorrhagic area within a carcinoma rather than in a cyst. Any residual lump left behind when the cyst contents are aspirated should be re-aspirated or biopsied.

Fibroadenosis

This lesion is characterized by intralobular fibrosis and proliferation of ductules and acini. On aspiration this lesion typically yields a dual cell population, i.e. groups of ductal epithelial cells surrounded by frequent stromal cells (*Figures 19.4* and *19.5*).

Occasionally myoepithelial cells can also be identified within the groups of ductal cells. The ductal cells are regular, oval, tightly adherent with a rather coarse chromatin pattern. The stromal cells are scattered singly or in pairs around the groups of ductal cells. These stromal cells are smaller than the ductal cells and are often devoid of cytoplasm and appear as stripped or naked nuclei which have a homogeneous chromatin pattern.

Fibroadenoma

Fibroadenoma is a benign tumour consisting of both glandular and fibrous elements; it tends to occur in younger women during the reproductive years. On aspiration numerous groups of hyperplastic duct cells are seen together with many scattered stromal cells (*Figures 19.6* and *19.7*).

It is not always possible to distinguish fibroadenoma from fibroadenosis in cytological preparations but aspirates from a fibroadenoma tend to be very much more cellular. However, the mixed cell population, cohesiveness of the cellular clusters and regular population of duct cells indicate that the lesion is benign.

Fat necrosis

This lesion is believed to be caused by trauma to the breast. As its name indicates, this lesion manifests as an area of necrotic adipose tissue usually with accompanying haemorrhage. The lesion is self-limiting but can simulate a tumour. The smear may contain numerous neutrophils, lipid-filled histiocytes, adipose cells, fat and a few giant cells. Duct epithelial cells and stromal

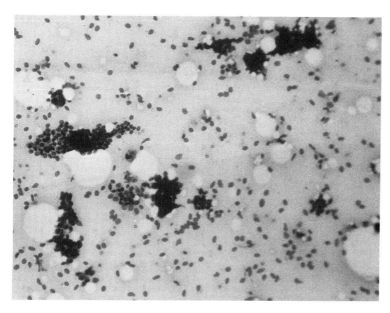

Figure 19.4 A low power view of fibroadenosis of the breast. Tightly adherent groups of duct epithelial cells with single stromal cells scattered throughout. Stain: May-Grünwald–Giemsa

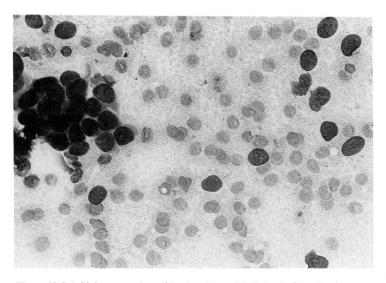

Figure 19.5 A high power view of benign duct epithelial cells from the breast together with stromal cells. Stain: May-Grünwald–Giemsa

cells are not usually present and the chronic inflammatory nature of the lesion confirms its benign nature. It may be difficult to distinguish fat necrosis from other inflammatory conditions such as plasma cell mastitis and duct ectasia on cytologic grounds. It is unlikely that fat necrosis will be confused with an acute breast abscess.

The lactating breast

Breast tissue aspirated during lactation can present a worrying picture to the inexperienced cytologist as the epithelial cells tend to increase in size and lose their cohesiveness. The presence of frequent small lipid droplets in the smear is

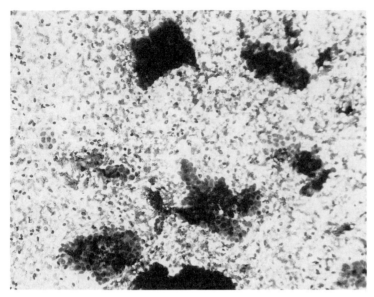

Figure 19.6 A low power view of fibroadenoma. Frequent large groups of hyperplastic duct cells with many scattered stromal cells. Stain: May-Grünwald-Giemsa

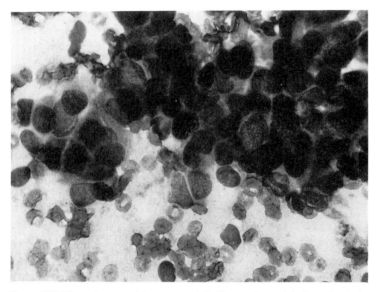

Figure 19.7 Fibroadenoma. A large group of adherent duct epithelial cells. Stain: May-Grünwald-Giemsa. High power view

a useful marker of the nature of the aspirate (*Figure 19.8*) and hopefully the cytologist will be forewarned by his clinical colleague of the patient's condition.

Cytology of malignant lesions

By far the most frequently encountered malignant lesions of the breast are carcinomas arising from the duct epithelium. Connective tissue tumours or lymphomas are rarely encountered. Attempts to identify the different histological types of carcinoma found on the breast, e.g. medullary, scirrhous or ductal carcinoma, in cytological specimens are unrewarding and classification and staging of breast tumours is

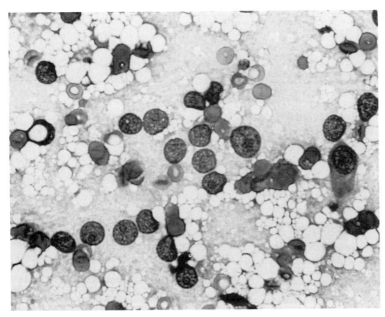

Figure 19.8 Lactation. Frequent scattered active nuclei and numerous lipid droplets. Stain: May-Grünwald–Giemsa. High power view

far better left to the histologist. Breast carcinomas exhibit several characteristic features in smears prepared from FNA, regardless of tumour type. These descriptions are based on May-Grünwald staining morphology:

(1) The specimen usually contains numerous epithelial cells. This in itself is a suspicious finding in the older woman.
(2) The tumour cells show loss of cohesiveness which is often marked. In those cases where cell aggregates are apparent, the clusters are very loose compared with benign duct epithelium (*Figure 19.9*).
(3) The cells are usually of a single type, i.e. tumour cells only. If, for example, stromal cells or foam cells are also present, then a malignant report should only be issued when the malignant cells show obvious pleomorphism and nuclear enlargement (*Figure 19.10*). This malignant mixed picture presumably occurs because the needle tip encounters both benign and malignant tissue during aspiration.
(4) Large tumour cells are obviously a useful marker of malignancy but size alone is not a reliable criterion as many malignant breast carcinomas, particularly in old patients, are composed of cells which differ very little in size from normal ductal cells.
(5) The chromatin of the normal ductal cells is rather coarse but regular, whereas many of the tumour cells show a fine chromatin pattern. Chromatin patterns are not reliable as a single marker for malignancy.
(6) When large and multiple, nucleoli are a useful feature of malignancy.

Paget's disease of the nipple

Paget's disease of the nipple is usually suspected on clinical grounds as the nipple appears eczematous and crusted. The disease can be recognized cytologically by the presence of the so-called 'Paget cells' in scrapings from the areola of the nipple or in smears prepared from the nipple discharge (*Figure 19.11*). These are large cells with abundant clear cytoplasm. The prominent nucleus often contains a large nucleolus. Paget's disease of the nipple occurs in association with intraduct carcinoma.

Cytology of the male breast

Male patients may present with diffuse enlargement of the breast (gynaecomastia) or with a discrete breast lump. Smears from patients with gynaecomastia show the cytological features of fibroadenosis with a typical dual cell popu-

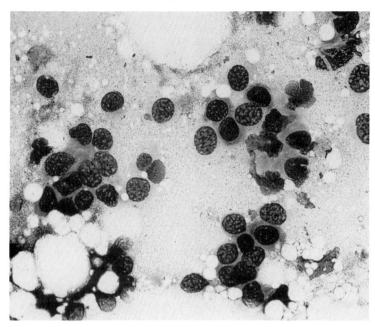

Figure 19.9 Cytologically differentiated carcinoma of the breast. Although the tumour cells are small and regular, they show obvious loss of adhesion and stromal cells are absent. Stain: May-Grünwald–Giemsa. High power view

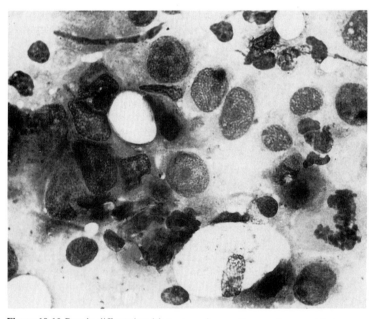

Figure 19.10 Poorly differentiated breast carcinoma. Large, pleomorphic tumour cells with prominent nucleoli. Stain: May-Grünwald–Giemsa. High power view

lation, i.e. duct epithelial cells and stromal cells. Carcinoma of the male breast shows similar characteristics to those seen in the female. Gynaecomastia is a frequent finding in males with Klinefelter syndrome. A single Barr body (sex chromatin) may be seen in the cells in the aspirate reflecting the patient's abnormal karyotype (47 XXY) in PAP stained preparations.

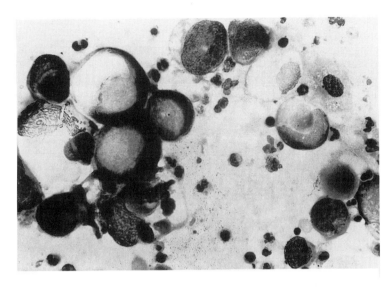

Figure 19.11 Paget's disease of the nipple. Large tumour cells, some showing obvious cytoplasmic clearing. Stain: May-Grünwald–Giemsa. High power view

Lymph nodes

Structure and function

Lymph nodes are found at many sites in the body—in superficial locations in the fascia of the neck and groin and in deep locations along the great vessels and in the layers of the omentum. They vary in size but in the absence of disease are rarely more than a few millimetres in diameter. Superficial lymph nodes cannot be palpated unless they are enlarged. Their function is to serve as a filtration unit for lymphatic fluid but they also have a role in phagocytosis and antibody production.

The structure of a lymph node is shown diagrammatically in *Figure 19.12*. They are classically described as bean-shaped. Lymphatic fluid is delivered to the node via afferent lymphatics which drain into the subcapsular sinus and passes through a meshwork of channels to exit at the hilum. Three regions of the node can be distinguished: the cortex which is situated beneath the capsule, the medulla adjacent to the hilum, and the paracortical region.

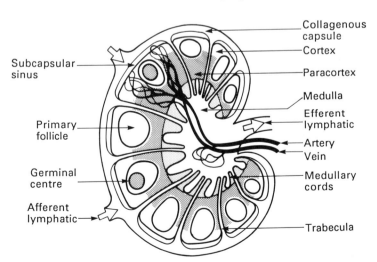

Figure 19.12 Diagram of structure of lymph node

The nodes contain many different cell types. The sinuses are lined by a discontinuous layer of endothelial cells and phagocytic cells, while the cortex and medulla are populated with many different classes of lymphocytes. In the absence of disease the distribution of T and B lymphocytes follows a regular pattern; B lymphocytes predominate in the cortex and medulla whereas T cells are found mainly in the paracortical zone. The germinal centres in the cortex are the sites of B cell activity and a range of cell types may be found there. Thus cytological interpretation of FNA of lymph nodes requires a thorough knowledge of the immunological responses of the various component cells present in the node.

Indications for FNA of lymph nodes

Fine needle aspiration of lymph nodes is particularly valuable for the investigation of persistent lymphadenopathy where nodes have remained enlarged for a period of several weeks. It is not advised for the investigation of the lymphadenopathy associated with acute infections.

It has a special place in the differential diagnosis of chronic lymphadenopathy due to granulomatous diseases such as tuberculosis and sarcoidosis and malignant disease, particularly metastatic carcinoma. In these latter cases the carcinoma cells can be readily distinguished from the cells of the lymphoid series one may expect to find in the smear. The metastatic cancer cells in the FNA are remarkably similar to those seen in the primary tumour and metastatic squamous carcinoma (*Figure 19.13*), oat cell carcinoma, adenocarcinoma (*Figure 19.14*), and melanoma can be readily recognized. A cytodiagnosis of metastatic cancer in patients with a history of carcinoma of breast, colon, pharynx, etc. is particularly rewarding as it obviates the need for further investigation of the patient and permits treatment to be started without delay.

The cytological diagnosis of primary lymphoma from FNA biopsy specimens can be rewarding in skilled hands. It is important for the cytologist interpreting the smear to be familiar with the cellular patterns associated with the different types of lymphoma as many of the morphological criteria applied to histological specimens apply equally well to cytological material. Thus the monomorphic patterns that characterize the non-Hodgkin's lymphoma can be recognized in smears (*Figures 19.15, 19.16 and 19.17*) and morphology can be supplemented by marker studies. Experts in this field claim that much information can be gained from imprint cytology which preserves the relationship of the cells in the cortex and the medulla. Readers are advised to consult Koss *et al.* (1984) for detailed descriptions of the cytology of Hodgkin's and non-Hodgkin's lymphomas (*Figures 19.15–19.19*).

Problems of differential diagnosis abound.

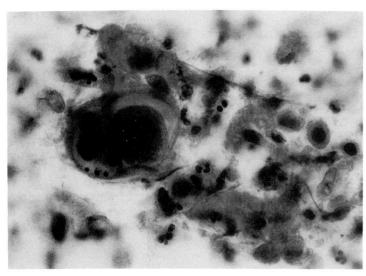

Figure 19.13 Lymph node aspirate. Metastatic squamous cell carcinoma. Stain: May-Grünwald–Giemsa. High power view

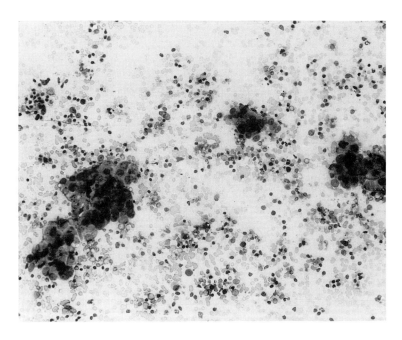
(a)

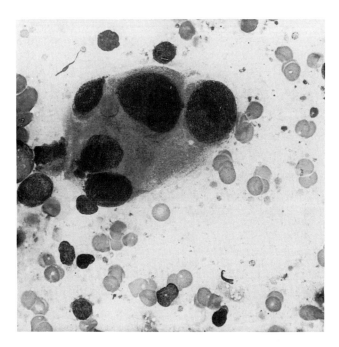
(b)

Figure 19.14 (a) Metastatic adenocarcinoma in a lymph node. This low power view shows clusters of tumour cells together with scattered lymphoid cells.
(b) Metastatic adenocarcinoma in a lymph node. A high power view of the same tumour as shown in (a). Stain: May-Grünwald–Giemsa. Reduced to 80% in reproduction

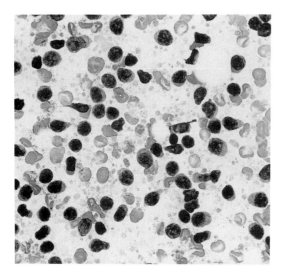

Figure 19.15 Low grade malignant lymphoma (lymphocytic). A population of small lymphoid cells, many showing nuclear irregularity. Stain: May-Grünwald–Giemsa

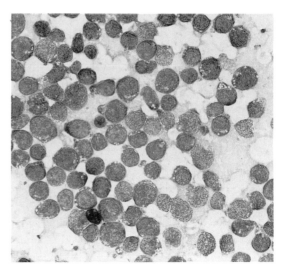

Figure 19.17 Malignant lymphoma (Burkitt's type). Note monomorphic picture. Many of the cells exhibit small vacuoles which are a feature of the tumour. Stain: May-Grünwald–Giemsa

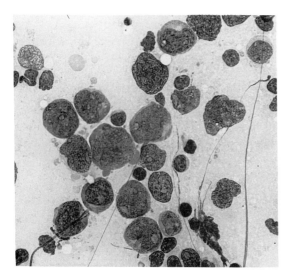

Figure 19.16 High grade malignant lymphoma (centroblastic). Large irregular malignant lymphoid cells show multiple nucleoli. Stain: May-Grünwald–Giemsa

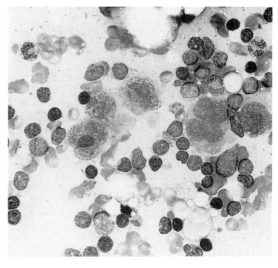

Figure 19.18 Hodgkin's disease. Mixed lymphoid cells together with Sternberg-Reed cells. Note the coarse nuclear chromatin pattern. Stain: May-Grünwald–Giemsa

Reactive hyperplasia of the nodes may occur in response to bacterial or viral infection and biopsy may be the only way of distinguishing chronic lymphadenitis from malignant lymphoma. In general, however, the aspirates from a node draining an area of inflammation contain lymphocytes in various stages of maturation and there is a diversity of cell types in the reactive node which is not usually seen in lymphoma (*Figures 19.20* and *19.21*). Another pitfall stems from the assumption that all swellings in the cervical region are enlarged lymph nodes. An aspirate from a branchial cyst may be mistaken for metastatic carcinoma in a node unless the possibility of this lesion is borne in mind.

Recently FNA has been used to investigate

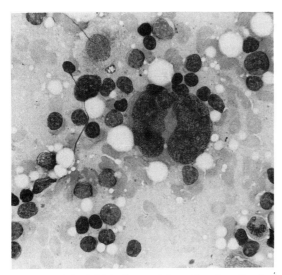

Figure 19.19 Hodgkin's disease. A large Sternberg-Reed cell together with mixed lymphoid cells. Stain: May-Grünwald–Giemsa

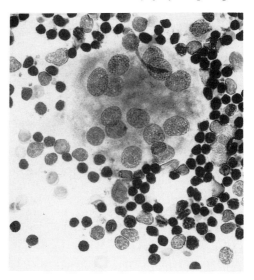

Figure 19.21 Lymph node aspirate. Large giant cells with scattered lymphoid cells from a tuberculous node. Stain: May-Grünwald–Giemsa

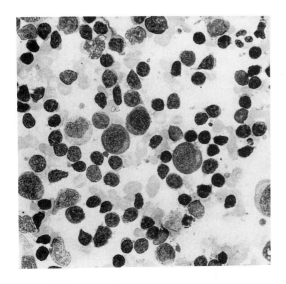

Figure 19.20 Lymph node aspirate showing reactive hyperplasia. Note mixed picture of lymphoid cells. Stain: May-Grünwald–Giemsa

the lymphadenopathy associated with AIDS. These patients are susceptible to viral and fungal infection and tuberculosis as well as malignant lymphoma. Aspirates from these patients should be submitted to microbiological investigation as well as to cytological studies.

Lung

Transthoracic percutaneous FNA can be used to investigate the benign or malignant nature of any lesion of the lung parenchyma which is visible on a standard chest radiograph and which is suspected of being a tumour. In practice it is usually reserved for the investigation of peripheral lung lesions which are clinically suspicious and for which pathological confirmation of malignancy is being sought. In such cases, FNA is a logical 'next step' when routine investigation by fibreoptic bronchoscopy, endobronchial biopsy and sputum cytology is negative.

In 1978 Wang, Terry and Marsh described an alternative to percutaneous FNA of lung lesions, namely aspiration of lung tissue via a transbronchial or transtracheal route. The needle was attached to a hollow shaft 55 cm in length and manipulated into position through a bronchoscope. Suction was provided by a 20 ml syringe attached to the shaft. Aspiration of carinal lymph nodes or lung parenchyma by this method has proved to be very successful especially as complications of the procedure are rare.

Complications

(1) *Pneumothorax*: this is the commonest complication and is more likely in patients with

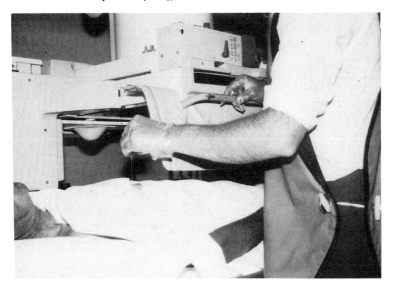

Figure 19.22 Performing a percutaneous lung biopsy under X-ray image intensification

obstructive airways disease or diffuse interstitial lung disease. Lesions which are close to the surface or at the apex are less likely to result in pneumothorax than deeper lesions. If an X-ray is taken after the FNA, the incidence of pneumothorax may be found to be as high as 40% but very few patients require treatment for this condition. Pneumothorax is rare in transbronchial FNA.

(2) *Haemoptysis*: this occurs in approximately 8% of patients. It is nearly always minor and self-limiting but the patient must be warned in advance as it can be frightening. It is mandatory to have full resuscitation facilities within the department when performing FNA of the lung.

Contraindications

FNA of pulmonary lesions should not be carried out under the following circumstances:

(1) When there is a risk of chronic obstructive airway disease, pneumothorax is a common complication and the ensuing reduction in respiratory reserve could prove fatal.
(2) When there is a bleeding diathesis due to haemophilia, anticoagulant therapy, etc. FNA should not be attempted in those cases where a vascular lesion such as haemangioma or aneurysm is suspected.
(3) If the patient has pulmonary hypertension as the risk of fatal haemorrhage is too great.
(4) When an echinococcal cyst is suspected as there is a risk of anaphylactic shock in these cases if the cyst ruptures.
(5) Aspiration of mesothelioma has been associated with spread of tumour along the needle track, but this is even more likely with open biopsy.

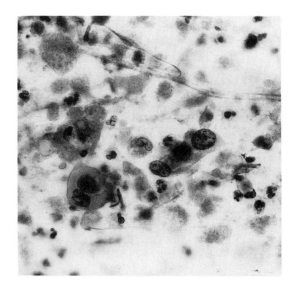

Figure 19.23 Keratinizing squamous cell carcinoma of the lung. Note inflammation and necrosis are also present. Stain: May-Grünwald–Giemsa

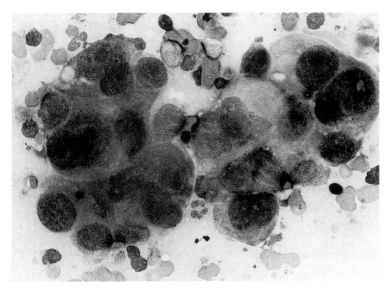

Figure 19.24 Large cell groups aspirated from an adenocarcinoma of the lung. Stain: May-Grünwald–Giemsa. High power view

Special technical notes

It is preferable for the patient to be fasting for approximately 4 h in case general anaesthesia is needed for the emergency treatment of complications such as haemothorax or pneumothorax. It is advisable to perform lateral tomography before proceeding to aspiration to establish the distance of the lesion from the anterior or posterior chest wall. The shortest route in an antero-posterior plane is chosen and the patient lies on the X-ray screening table either prone or supine, depending on whether the lesion lies closer to the anterior or posterior chest wall (*Figure 19.22*). The patient must be prepared to cooperate throughout the procedure for the correct positioning of the needle (Melvin Stevens and Jackman, 1984).

Cytology

Aspirates from normal lung tissue contain numerous macrophages and normal bronchial epithelial cells which can readily be recognized by cytologists familiar with the exfoliative cytology of the respiratory tract. Aspirates from tuberculous lesions, bacterial or viral infection contain a large complement of inflammatory cells (Michel, Lushipan and Ahmed, 1983).

The primary malignant tumours of the lung which are encountered most frequently include squamous carcinoma (*Figure 19.23*) and adenocarcinoma (*Figure 19.24*). The latter are usually peripherally situated and therefore suitable targets for FNA. The tumour cells in smears prepared from the aspirates differ slightly in appearance from those seen in smears prepared from sputum samples. Oat cells for example do not appear in 'streaks' and adenocarcinoma cells are rarely vacuolated.

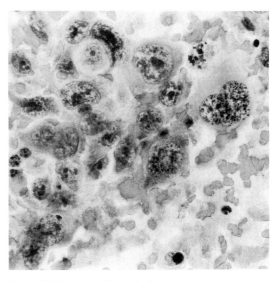

Figure 19.25 Large cell anaplastic carcinoma of the lung. Stain: Papanicolaou

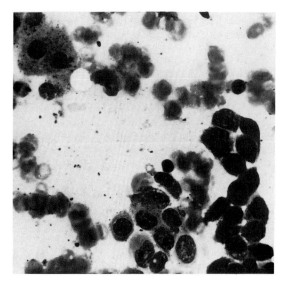

Figure 19.26 Small cell carcinoma of the lung (intermediate cell type). Note pigmented macrophages are also present. Stain: May-Grünwald–Giemsa

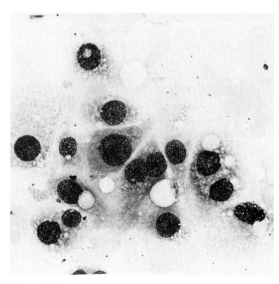

Figure 19.27 Cells from a metastatic clear cell carcinoma of the kidney aspirated from the lung. Stain: May-Grünwald–Giemsa

Some examples of FNA cytology of primary and metastatic tumours in the lung are shown in *Figures 19.25–19.29*. Immunocytochemical techniques may be used to identify the source of a metastatic carcinoma. Transbronchial biopsy has a special place in the diagnosis of *Pneumocystis carinii* in patients whose respiratory reserve is slightly limited as the risk of pneumothorax is small.

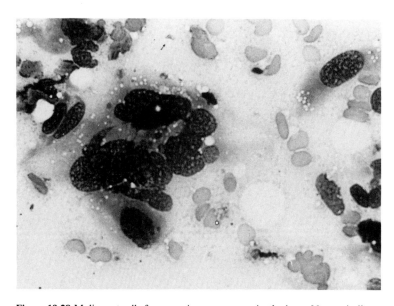

Figure 19.28 Malignant cells from a primary sarcoma in the lung. Note spindle shaped nuclei and coarse nuclear chromatin. Stain: May-Grünwald–Giemsa. High power view

FNA of specific target organs 359

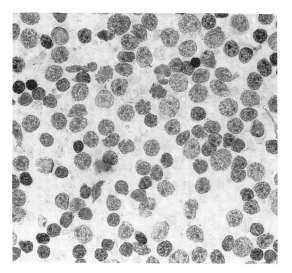

Figure 19.29 A low grade centrocytic malignant lymphoma metastatic to the lung. The sputum sample from this case also showed numerous tumour cells. Stain: May-Grünwald–Giemsa.

Liver

Fine needle aspiration is indicated for the diagnosis of focal lesions of the liver which are suspected of being due to primary or metastatic carcinoma or liver abscess. It should not be used for the evaluation of diffuse disorders of the liver parenchyma (Bell *et al.*, 1986). In these cases a core of liver tissue is needed for accurate diagnosis. The lesion can be localized for FNA by computed tomography or ultrasound (*Figures 19.30–19.32*). The latter is quicker and cheaper. If an abscess is suspected, antibiotic cover is advised.

Complications

Liver biopsy with cutting needles has been in use for over 50 years and complications are rare. Nevertheless they do occur, and haemorrhage is a real risk of the procedure. A platelet count and a prothrombin should be obtained before FNA is performed. Peritonitis either due to leakage of bile or spread of infection is also a rare complication.

Aspiration of cysts should be undertaken with special care as a rupture of a hydatid cyst caused by *Echinococcus* may cause anaphylactic shock.

Cytology

The predominant cell types in a FNA which pass through normal liver tissue are hepatocyte and bile duct cells. The hepatocytes have abundant cytoplasm which may contain blue-black lipofuschin granules in MGG-stained smears (*Figure 19.33*). The cells appear discrete or in loose clusters and the nuclei vary in size. Tetraploid or octoploid nuclei may be seen in elderly patients. However, the nuclear outline is

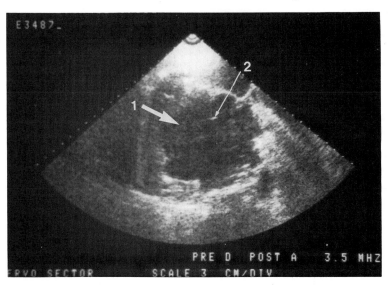

Figure 19.30 An ultrasound scan, transverse section through the liver of a 70-year-old female. This shows a large echo poor mass (1). The biopsy needle tip echo (2) is shown in the mass

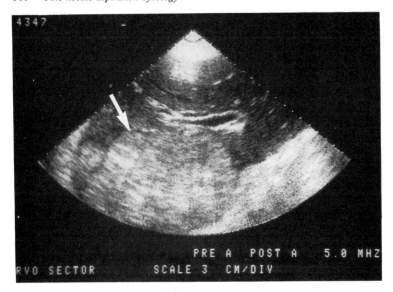

Figure 19.31 A longitudinal section of an ultrasound scan through the liver in a 60-year-old man. The echogenic mass arrowed in the right lobe of the liver proved to be a hepatoma

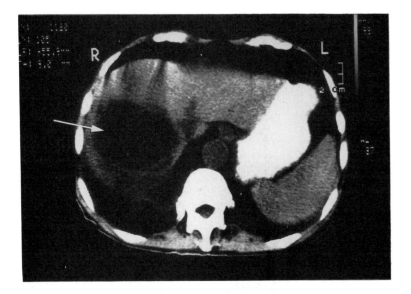

Figure 19.32 A CT scan of a 72-year-old male showing a fluid collection in the right lobe of the liver. This proved to be a liver abscess on aspiration. It was drained percutaneously

smooth and rounded and the chromatin content bland.

The epithelial cells of the bile duct are smaller than the hepatocytes and have a cubical shape with small round nuclei. They often appear in sheets. Kupfer cells, which are the macrophages of the liver, are occasionally seen in smears.

The commonest lesion to be aspirated is metastatic carcinoma of the liver (*Figure 19.34*). The aspirate often contains a pure population of cancer cells. Identification of malignant cells follows the guideline set out earlier in this chapter. Identification of tumour type is often possible, e.g. squamous, adenocarcinoma even if the site of the primary tumour is not known. The primary site of a tumour can rarely be

FNA of specific target organs   361

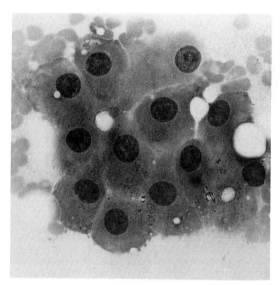

Figure 19.33 A group of normal liver cells (hepatocytes). Note low nuclear/cytoplasmic ratio. Stain: May-Grünwald–Giemsa.

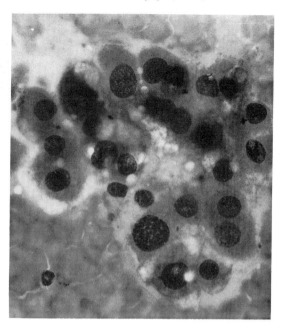

Figure 19.35 Primary liver carcinoma (hepatoma). Note pleomorphism and coarse nuclear chromatin. Stain: May-Grünwald–Giemsa

malignant lesion of the liver. Hepatocellular carcinoma (*Figure 19.35*) can be particularly difficult to distinguish from normal or regenerating liver in those cases where the tumour is well differentiated. In less well differentiated cases the tumour cells exhibit an obvious pleomorphism. The distinction between this type of

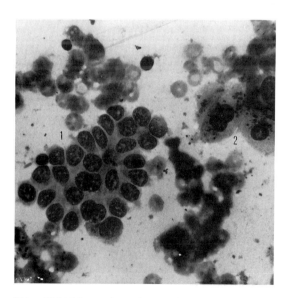

Figure 19.34 Metastatic carcinoma of the breast (1) together with hepatocytes containing lipofuscin pigment (2). Stain: May-Grünwald–Giemsa

determined from the appearance of the metastatic tumour cells. One of the most common tumours to metastasize to the liver is gastric carcinoma and this possibility should always be borne in mind.

FNA is frequently used to ascertain the diagnosis in patients suspected of having a primary

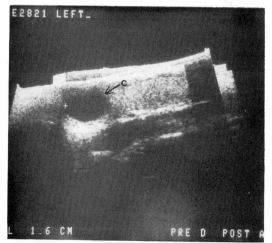

Figure 19.36 Thyroid ultrasound scan of a 25-year-old female showing a cyst (C) in the left lobe of the thyroid. This is shown as a transonic region with increased throughput of sound. Aspiration produced chocolate coloured fluid

tumour and poorly differentiated adenocarcinoma can be difficult. The presence of lipofuschin in the cytoplasm of the tumour cells may help clarify the diagnosis.

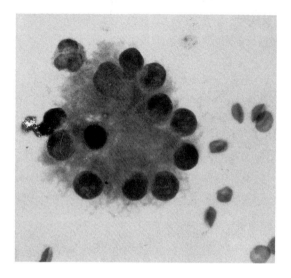

Figure 19.37 Thyroid. Cell group from a benign follicular adenoma. Stain: May-Grünwald–Giemsa. High power view

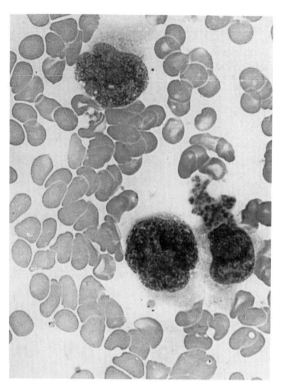

Figure 19.38 Large, irregular tumour cells from an anaplastic carcinoma of the thyroid. Stain: May-Grünwald–Giemsa. High power view

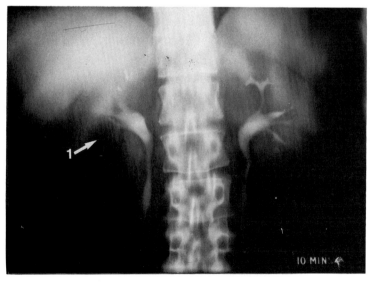

Figure 19.39 An IVU series tomograph of a 68-year-old female with haematuria. There is a space-occupying lesion in the right lower pole (1)

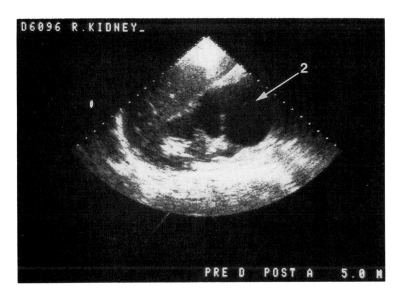

Figure 19.40 A longitudinal ultrasound scan of the patient discussed in *Figure 19.39* showing a bilocular cyst (2). This was aspirated in view of the haematuria and proved to be a simple cyst

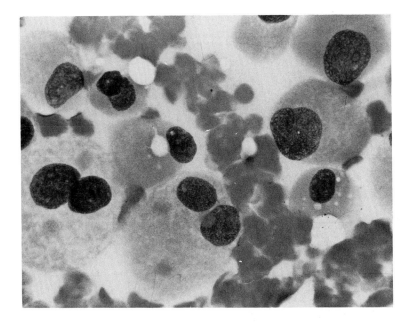

Figure 19.41 Renal aspirate. Pleomorphic malignant cells from a renal clear cell carcinoma. Note abundant cytoplasm. Stain: May-Grünwald–Giemsa

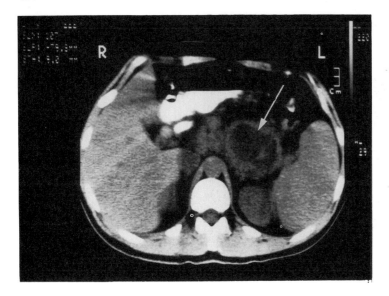

Figure 19.42 A CT scan through the upper abdomen of a 52-year-old man. This shows a pseudocyst (arrowed) in the tail of the pancreas

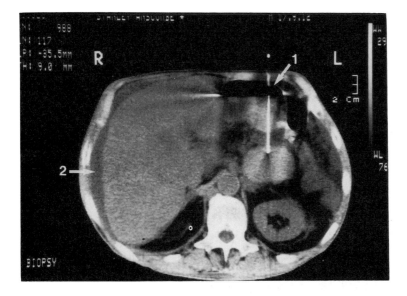

Figure 19.43 A CT scan of the pancreas of a 72-year-old male with a mass in the tail of the pancreas. The needle has been placed with its tip in the mass. The needle has traversed the stomach en route (1). Ascitic fluid (2) is noted adjacent to the liver

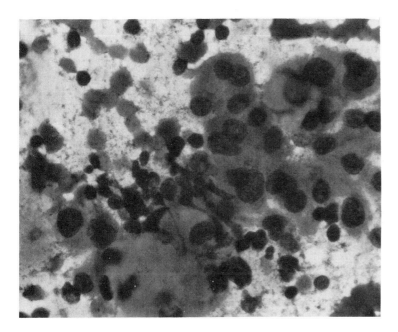

Figure 19.44 Aspirated carcinoma of the pancreas. Pleomorphic tumour cells seen here with abundant cytoplasm. Stain: May-Grünwald–Giemsa

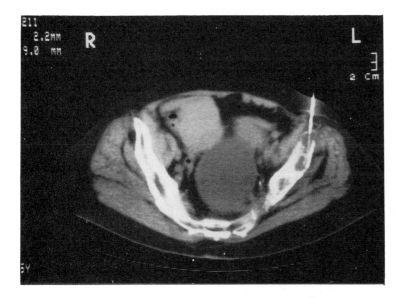

Figure 19.45 A CT scan of a destructive lesion in the left iliac bone of a 65-year-old female. This proved to be a metastatic adenocarcinoma. The needle is seen in the lesion. An incidental ovarian cyst is noted

Conclusion

Fine needle aspiration of tumours at almost any body site is now possible using imaging techniques. In most centres aspiration of palpable lumps in the breast far exceeds aspiration of deep seated tumours. Nevertheless we have been called on to interpret aspirates from many sites other than those discussed in this chapter and examples of smears prepared from thyroid (*Figures 19.36–19.38*), kidney (*Figures 19.39–19.41*), pancreas (*Figures 19.42–19.44*) and bone (*Figures 19.45–19.47*) are shown as a point of interest. For excellent accounts of the cytology of these organs readers are advised to consult the following references: An-Foraker and Fong-Mui (1982); Franzen and Brehmer-Andersson (1982); Hall-Craggs and Lees (1986); Lowhagen and Sprenger (1974); Nosker *et al.* (1982); Schwartz *et al.* (1982); Solbiat *et al.* (1985); Stormby and Akerman (1973). Chapters on FNA of abdominal and thoracic organs are also to be found in the following books: Melcher, Linehan and Smith (1984) *Practical Aspiration Cytology*; Orell *et al.* (1986) *Manual and Atlas of Fine Needle Aspiration Cytology*; Zajicek (1974, 1979) *Aspiration Biopsy Cytology*; and Koss, Woyke and Olszewski (1984) *Aspiration Biopsy—Cytologic Interpretation and Histologic Bases*.

Success with this technique depends on the combined skills of the surgeon, radiologist, cytologist and cytotechnologist. The surgeon and radiologist must develop good sample col-

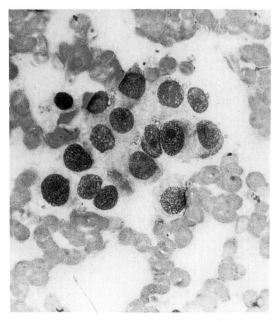

Figure 19.46 Aspirate from lytic lesion of bone showing cells from metastatic carcinoma of the breast. Stain: May-Grünwald–Giemsa

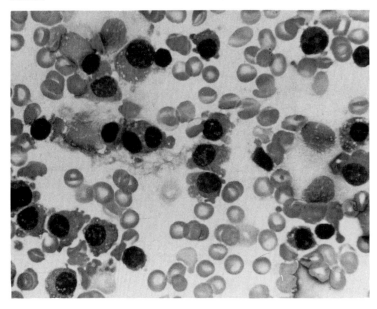

Figure 19.47 Cells aspirated from a lytic lesion in the pubic ramus. Atypical plasma cells demonstrated from a plasmacytoma. Stain: May-Grünwald–Giemsa

lection technique and, with the help of the cytotechnologist, learn to make satisfactory smears. The cytotechnologist has a critical role to play in specimen preparation and choice of stain. Provided the cytologist has a sound knowledge of the histopathology of the organ and a good understanding of the differential diagnosis, a reliable report can be given in most cases.

References

AN-FORAKER, S. H. and FONG-MUI, K. K. (1982). Cytodiagnosis of lesions of the pancreas and related areas. *Acta Cytol.*, **26**, 814–818

BELL, D. A., CARR, C. P. and SZYFELBEIN, W. M. (1986). Fine needle aspiration cytology of focal liver lesions. *Acta Cytol.*, **30**, 397–402

DUDGEON, L. S. and PATRICK, C. V. (1927). A new method for the rapid microscopical diagnosis of tumours, with an account of 200 cases so examined. *Br. J. Surg.*, **15**, 250–261

FRANZEN, S. and BREHMER-ANDERSSON, E. (1982). Cytologic diagnosis of renal cell carcinoma. *Prog. Clin. Biol. Res.*, **100**, 425–432

GARDECKI, T. I., HOGBIN, B. M., MELCHER, D. H. and SMITH, R. S. (1981). Aspiration cytology in the pre-operative management of breast cancer. *Lancet*, **ii**, 790–792

GRIEG, E. D. W. and GRAY, A. C. H. (1904). Note on the lymphatic glands in sleeping sickness. *Br. Med. J.*, **i**, 1252

HALL-CRAGGS, M. A. and LEES, W. R. (1986). Fine-needle aspiration biopsy: pancreatic and biliary tumours. *AJR*, **147**, 399–403

HUSBAND, J. E. and GOLDING, S. J. (1983). The role of computed tomography-guided needle biopsy in an oncology service. *Clin. Radiol.*, **34**, 255–260

KOPANS, D. B., LINDFORS, K., MCCARTHY, K. A. and MEYER, J. E. (1985). Spring hookwire breast lesion localiser: use with rigid-compression mammographic systems. *Radiology*, **157**, 537–538

KOSS, L. G., WOYKE, S. and OLSZEWSKI, W. (1984). *Aspiration Biopsy—Cytologic Interpretation and Histologic Bases*. New York and Tokyo: Igaku Shoin

LINSK, J. N. and FRANZEN, S. (1986). *Fine Needle Aspiration for the Clinician*. New York: J. B. Lippincott

LIVRAGHI, T., DAMASCELLI, B., LOMBARDI, C. and SPAGNOLI, I. (1983). Risk in fine-needle abdominal biopsy. *J. Clin. Ultrasound*, **11**, 77–81

LOWHAGEN, T. and SPRENGER, E. (1974). Cytologic presentation of thyroid tumours in aspiration biopsy smears. *Acta Cytol.*, **18**, 192–197

MARTIN, H. E. and ELLIS, E. B. (1930). Biopsy by needle puncture and aspiration. *Ann. Surg.*, **92**, 169–181

MELCHER, D., LINEHAN, J. and SMITH, R. (1984). *Practical Aspiration Cytology*. Edinburgh: Churchill Livingstone

MELVIN STEVENS, G. and JACKMAN, R. J. (1984). Outpatient needle biopsy of the lung: its safety and utility. *Radiology*, **151**, 301–304

MICHEL, R. P., LUSHIPAN, A. and AHMED, M. N. (1983). Pathologic findings of transthoracic needle aspiration in the diagnosis of localized pulmonary lesions. *Cancer*, **51**, 1563–1672

NAHMAN, B. J., VAN AMAN, M. E., MCLEMORE, W. E. and O'TOOLE, R. V. (1985). Use of the Rotex needle in percutaneous biopsy of pulmonary malignancy. *AJR*, **145**, 97–99

NOSKER, J. L. AMOROSA, J. K., LEIMAN, S. and PLAFKER, J. (1982). Fine needle aspiration of the kidney and adrenal gland. *J. Orol.* **128**, 895–899

ORELL, S. R., STERRETT, G. F., WALTERS, M. N. I. and WHITAKER, D. (1986). *Manual and Atlas of Fine Needle Aspiration Cytology*. Edinburgh: Churchill Livingstone

PAGANI, J. (1983). Biopsy of focal hepatic lesions: comparison of 18 and 22 gauge needles. *Radiology*, **147**, 673–675

PORTER, B., KARP, W. and FORSBERG, L. (1981). Percutaneous cytodiagnosis of abdominal masses: ultrasound guided fine-needle aspiration biopsy. *Acta Radiol. Diagn*, **22**, 663–668

SCHWARTZ, A. E., NIEBURGS, H. E., DAVIES, T. F., GILBERT, P. L. and FRIEDMAN, E. W. (1982). The place of fine needle biopsy in the diagnosis of nodules of the thyroid. *Surg. Gynecol. Obstet.*, **155**, 54–58

SOLBIAT, L., BOSSI, M. C., BELLOTTI, E., RAVETTO, C. and MONTALI, G. (1985). Focal lesions in the spleen; sonographic patterns and guided biopsy. *AJR*, **144**, 471–474

STORMBY, N. and AKERMAN, M. (1973). Cytodiagnosis of bone lesions by means of fine needle aspiration biopsy. *Acta Cytol.*, **17**, 166–172

WANG, K. P., TERRY, P. B. and MARSH, B. (1978) Bronchoscopic needle aspiration biopsy of paratracheal tumours. *Am. Rev. Respir. Dis.*, **118**, 17–21

ZAJICEK, J. (1974). *Aspiration Biopsy Cytology Part I: Cytology of Supradiaphragmatic Organs*. Basel: S. Karger

ZAJICEK, J. (1979). *Aspiration Biopsy Cytology Part II: Cytology of Infradiaphragmatic Organs*. Basel: S. Karger

20

Gastrointestinal tract

Michael Drake

There are four areas of the gastrointestinal tract that are accessible to the techniques of diagnostic cytology, namely the oesophagus, stomach, duodenum and large bowel. Although cytological investigation of the liver is considered frequently in association with gastrointestinal cytology, this investigation involves needle aspiration techniques, usually by the percutaneous route, and hence will not be dealt with in this chapter. Conversely, the biliary tract and pancreas, although not strictly anatomical components of the gastrointestinal system, drain directly into that system and hence will be mentioned briefly in relation to duodenal cytology. At the outset it is emphasized that diagnostic cytology is of greatest value in the investigation of the oesophagus and stomach, particularly of the latter organ.

General principles of specimen collection, preparation and evaluation

Specimen collection

In all areas of the alimentary tract two methods of specimen collection may be employed. The method initially used was that of blind lavage whereby the oesophagus or stomach was intubated, the mucosal surfaces vigorously washed using a suitable physiological fluid, and the washings were then retrieved and examined for malignant cells. With the development of flexible fibreoptic endoscopes, however, it has become possible to visualize most areas of the gastrointestinal tract and to brush, wash, or biopsy the mucosa under direct vision. These latter methods have increased greatly the accuracy and sensitivity of the diagnostic procedures and are less tedious, although not necessarily more comfortable, for the patient. However, it is important not to discard totally the blind lavage techniques as they may be extremely useful in the investigation of specific patients, particularly those with a suspected lesion not visible endoscopically. They also provide a technique for cytological screening of carefully selected high risk population groups.

Specimen preparation

In some laboratories the cytological material, whether derived from the endoscopic brush or by the centrifugaton of the lavage specimen, is allowed to air-dry and is subsequently stained by the May-Grünwald–Giemsa technique. However, it is strongly recommended that all specimens be fixed while still wet and stained by the Papanicolaou method. Excellent and reproducible results are obtainable by this method and a more direct correlation with subsequent histological material is possible, the latter being stained usually by the haematoxylin and eosin method. If sufficient material is available there is some merit in leaving one or two slides unstained to use if special staining techniques are indicated by the initial evaluation. Stains for mucin may be of particular value but other special stains, such as those for melanin, may be indicated occasionally.

Specimen evaluation

Detection

Although the initial screening procedures involve similar skills to those employed in the cytological examination of other body systems, gastrointestinal cytology does make special demands upon the cytotechnologist. Obviously there is a need for a thorough knowledge of the cells and other structures that may be encountered in specimens from the various organs, both in the normal state and in association with the many disease processes that may occur. In addition, the cytotechnologist must appreciate the elusiveness of the cells shed from some malignant processes. For example, as will be pointed out subsequently, anaplastic carcinoma of the stomach usually sheds relatively few cells often in association with a considerable amount of necrotic debris and inflammatory exudate. The carcinoma cells are small and relatively inconspicuous and may easily be overlooked by the technologist inexperienced in the evaluation of gastric material.

Before systematically screening each slide, a rapid and random examination is desirable. This enables the technologist to obtain an overall impression of both the epithelial cells present and also the background of inflammatory exudate, blood, contaminants such as meat fibres and vegetable cells, and any other non-cellular material which may be present such as barium. In addition specimens from patients with a malignant lymphoma often have a curious pale blue background. Although the nature of this is uncertain, its presence should alert the technologist to search most diligently for small single malignant cells which are often missed by the inexperienced observer or mistaken for lymphocytes and ignored. Indeed an examination of lymphocytes under high power may be necessary to ensure that they are normal lymphocytes, such as may frequently accompany gastritis or peptic ulceration, and do not have criteria indicative of a malignant lymphoma.

After making this preliminary assessment the specimens must be screened systematically to ensure that all cells on every slide are examined. It is preferable to screen in a vertical manner and all abnormal cells should be marked for further evaluation. The method of marking varies considerably amongst different laboratories. The use of rings to mark the slide is undesirable as they tend to obscure too much cellular detail. Small dots placed to one side of the abnormal cells are preferred and it is strongly recommended that white marking ink is used. It is surprising how many laboratories continue to use a black or dark coloured marking ink which is quite invisible against the black background of the microscope stage. Where a group of cells is considered to be of particular diagnostic significance the relevant dot may be highlighted by a small line or arrow.

Interpretation

The final evaluation depends on a detailed knowledge of the disease processes that may be encountered and their cytological manifestations. However, in making this evaluation it is essential for the cytotechnologist and the cytopathologist to be aware of the way in which the specimen was collected and to be familiar with the very significant differences between cells obtained by direct brushing and those by washing—particularly by the blind lavage techniques.

Usually direct brushings yield large sheets of cells or tissue fragments in contrast to the single cells or small groups of cells that are commonly seen in lavage specimens. In addition, in the latter specimens the cells tend to round up in the fluid medium—a phenomenon of surface tension. They are thus seen as three-dimensional structures or 'balls' of cells rather than the flatter sheets of cells seen in brushings. These features are particularly relevant to the problem of differentiating between the regenerative changes that may accompany chronic gastritis or a chronic peptic ulcer, and those changes indicating malignancy. In lavage specimens the tendency for tissue showing regenerative activity only to appear as large sheets contrasts with the small groups of cells so typical of a carcinoma. This diagnostic aid may be lost in brushing specimens although the poor cohesion of malignant cells is usually manifest even in material collected by the brushing techniques.

There are also differences in the appearances of the two methods. The cytoplasm of cells in a gastric brushing is usually more abundant and has an uneven outline, whereas in washings the cells tend to round up, the cytoplasm appears less, and the cell outline is smoother. There is some disagreement as to variations in nuclear detail. Thus Shida and Ishioka (1988) state that the cells collected by the lavage technique have an extremely abnormal chromatin pattern with uneven condensation and irregularity of chromatin arrangement, thickening of nuclear borders and prominent nucleoli. They claim that this contrasts with the features seen in direct smears where, as a rule, the cell pattern appears less malignant. In our experience the comparative features of specimens collected by the two techniques are very variable. Whilst it is true that in some cases the nuclear malignant criteria are accentuated in lavage specimens, in other cases the reverse is true.

Finally, drying artefact is more likely to be seen in the specimen collected by the brushing technique than by lavage methods. Thus artefact is particularly noticeable at the periphery of the smear where drying commences. To avoid this problem it may be necessary to have a cytotechnologist attend the endoscopy to personally supervise the immediate fixation of the smear.

It is important to note that, regardless of the method of collection and preparation employed, the basic principles of diagnosis of malignancy are always applicable. Indeed it is stressed that identification of malignant criteria is essential if false diagnoses of malignancy are to be avoided.

With these general principles in mind each organ will now be considered individually.

Oesophagus

Anatomy and histology

The oesophagus is a hollow muscular tube which measures approximately 25 cm or 10 inches in length. It extends from the pharynx, beginning at the level of the sixth cervical vertebra, to the cardia of the stomach joining that structure 2.5 cm to the left of the midline opposite the eleventh thoracic vertebra. The oesophagus pierces the diaphragm at the level of the tenth thoracic vertebra and has an intra-abdominal portion which measures 1.25 cm in length and is conical in shape, the base of the cone being continuous with the cardiac orifice of the stomach. It is constricted at four sites, namely, at its commencement, where it is crossed by the aortic arch, where it is crossed by the left main bronchus, and where it pierces the diaphragm. These constrictions are of considerable importance when instruments are being passed along the oesophagus and they also appear to be of importance in the localization of various disease processes.

The histological structure of the oesophagus follows the same general pattern as the rest of the alimentary tract having four layers or coats. Of greatest importance to the diagnostic cytologist is the mucosa which is composed of stratified squamous non-keratinizing, or non-hornifying, epithelium. This squamous epithelium changes abruptly to glandular epithelium of gastric type usually, although not always, at the gastric orifice. The stratified squamous epithelium of the lower oesophagus may be replaced by metaplastic glandular epithelium of gastric type. If this process is extensive it is referred to as Barrett's syndrome, a condition of considerable importance in the genesis of some oesophageal malignancies.

Cytology

Oesophageal cytology is, for the most part, relatively straightforward. The normal cytology is extremely simple whilst abnormal cytology is concerned mostly with the distinction between inflammatory conditions and carcinoma, the latter being almost exclusively squamous cell carcinoma and adenocarcinoma.

Normal cytology

The epithelial cell component of both washings and brushings of the normal oesophagus is composed of squamous cells of superficial and intermediate cell type. Cells from the deeper layers, the parabasal cells, may occasionally be seen in a normal brushing, presumably reflecting the vigour of that procedure. More often, however, their presence indicates an undue friability of the mucosa with exposure of the deeper layers. In addition to these normal cellular components, other structures such as swallowed respiratory elements and food particles may be found in oesophageal specimens, particularly if there is any delay at the cardio–oesophageal junction.

Inflammatory conditions

Oesophagitis may be due to an identifiable microorganism or, alternatively, may be non-specific. Perhaps the commonest causes now of primary infection of the oesophagus are the herpes simplex virus and *Candida* or *Monilia albicans*. Both are most commonly seen in immunosuppressed patients but both may also infect otherwise healthy people.

The changes in herpetic oesophagitis are both distinctive and most striking. Histologically the earliest lesion is seen in the squamous epithelium, the surface epithelial cells showing abnormalities characteristic of the viral infection. These epithelial cell abnormalities are clearly manifest in the cytological specimen collected from the infected oesophagus (*Figure 20.1*). The cells may be mononuclear but more commonly are multinucleate. The most characteristic feature is the margination of the nuclear chromatin material, the bulk of the nucleus having a striking refractile or 'ground-glass' appearance. Within the multinucleate cells the enlarged nuclei show conspicuous moulding. Intranuclear inclusions are characteristic. These may be extremely small but much more commonly are large, eosinophilic and surrounded by a clear halo.

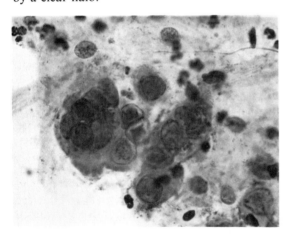

Figure 20.1 Cells from herpetic oesophagitis. Oesophageal brushings. Stain: Papanicolaou. Magnification × 790, reduced to 80% in reproduction

The cytological features of monilial oesophagitis are less specific being due to coexistent degenerative and regenerative activity within the infected epithelium. Both processes are evident within the desquamated cells whilst inflammatory exudate and fungal elements are prominent.

Non-specific oesophagitis is a relatively common condition and is most commonly due to a reflux of gastric content into the lower oesophagus. Again the inflammatory reaction may be intense whilst changes due to both degeneration and regeneration are evident in the cytological specimen. Parabasal cells are shed frequently and there is often nuclear enlargement, clumping of the nuclear chromatin material and prominent nucleoli. In addition, degenerative changes are manifest by cytoplasmic vacuolation and fragmentation of both cytoplasm and nucleus of the more affected cells.

Neoplasms

Benign neoplasms of the oesophagus are rare and of no cytological importance. Conversely malignant neoplasms are an important form of gastrointestinal malignancy and are particularly amenable to cytological diagnosis. Although a relatively large number of malignant oesophageal neoplasms may occur it is important to realize that only two occur with sufficient frequency to be of major concern to the diagnostic cytologist. These are the squamous cell carcinoma and adenocarcinoma, the former predominating.

The squamous cell carcinomas of the oesophagus, as in other parts of the body, may be graded as well, moderately well, and poorly differentiated. In addition, undifferentiated forms may occur. It is important to distinguish this very poorly differentiated squamous cell, or 'large cell undifferentiated', carcinoma from the much less common 'small cell undifferentiated' or 'oat cell' carcinoma more characteristic of the bronchus.

Examination of the brushings or washings from a well differentiated squamous cell carcinoma reveals cells which show considerable variation in size and shape and which are frequently quite bizarre in appearance (*Figure 20.2*). Nuclear malignant criteria are usually clearly recognizable although increased hyperchromasia of the nucleus may tend to obscure the finer details of nuclear structure. Evidence of squamous differentiation may be prominent with obvious cytoplasmic keratinization, malignant epithelial 'pearls' and spindle or fibre cells (*Figure 20.3*). Less well differentiated forms may require careful examination to determine their squamous origin. Finally, undifferentiated or anaplastic carcinomas yield cells which vary in size and shape, being round, oval or polygonal, usually have good nuclear malignant criteria, and rather scanty non-descript cytoplasm.

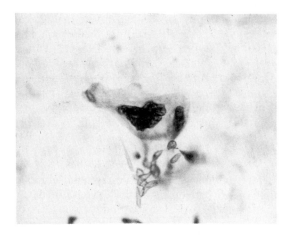

Figure 20.2 Bizarre malignant cell from squamous cell carcinoma of oesophagus. Oesophageal brushings. Stain: Papanicolaou. Magnification × 790, reduced to 80% in reproduction

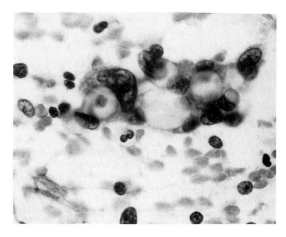

Figure 20.4 Vacuolated cells from oesophageal adenocarcinoma. Oesophageal brushings. Stain: Papanicolaou. Magnification × 790, reduced to 80% in reproduction

Stomach

Anatomy and histology

The stomach is a J-shaped organ which extends from the lower end of the oesophagus at the level of the eleventh thoracic vertebra, about 2.5 cm to the left of the midline, to end in the duodenum just to the right of the lower border of the first lumbar vertebra. It is divided rather arbitrarily into four parts. The cardia is the region about 2–3 cm in width immediately distal to the oesophagus. The fundus is that part of the stomach which lies above a line drawn horizontally through the gastro–oesophageal junction. The body comprises roughly the proximal two-thirds of the remainder, and the antrum the distal one-third which leads, by way of the pyloric sphincter, into the duodenum. A more rational division may be made on the basis of the mucosal glands, three regions being distinguished in this way. Thus the narrow ring-shaped area around the cardia is called the cardiac area and contains the cardiac glands. The fundus and proximal two-thirds of the stomach, which contain the gastric glands proper, form the second zone. The third zone is called the pyloric region being characterized by the presence of the pyloric glands. It occupies the most distal portion of the stomach and extends further along the lesser curvature than along the greater. It should be noted that the zones are not sharply demarcated, the gland

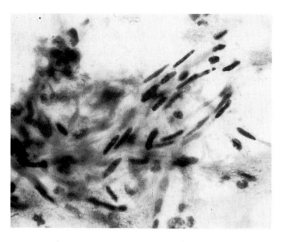

Figure 20.3 Spindle cells from squamous cell carcinoma of oesophagus. Oesophageal brushings. Stain: Papanicolaou. Magnification × 790, reduced to 80% in reproduction

Adenocarcinomas involving the oesophagus are usually located near the cardio–oesophageal junction and again all degrees of differentiation may occur. The cytological appearances reflect these variations in histological pattern. Papillary fragments may be seen as may acinar structures, these indicating a well differentiated lesion. More commonly, however, an oesophageal adenocarcinoma is manifested by groups or sheets of cells which show good malignant criteria and varying degrees of cytoplasmic vacuolation (*Figure 20.4*).

types intermingling to some extent at the sites of transition.

Again the wall consists of four layers, the most important to the cytologist being the innermost layer or mucosa. This is a complex tissue the surface of which exhibits numerous small depressions—the gastric pits. The gastric glands open into the bottom of these pits, three types of glands being recognized. The main gastric glands, which occupy the mucosa of the body and fundus of the stomach, are of most importance and contain two specialized cell types, the parietal or oxyntic cells which secrete hydrochloric acid and the chief or zymogenic cells responsible for the digestive enzyme pepsinogen.

The epithelium that lines the gastric pits or foveolae and covers the free surface between the pits is uniformly of the same structure. The cells are tall columnar, mucus-secreting cells with their nuclei towards the base of the cell. The cytoplasm may appear eosinophilic and finely granular but, more frequently, is vacuolated. The vacuoles are small. Cells with large secretory vacuoles—the so-called goblet cells—do not occur within the normal gastric mucosa.

Specimen collection

Although gastric cytological specimens are usually collected by endoscopically directed brushing or washing techniques, this organ does lend itself to blind lavage procedures as indicated in the introductory comments. These procedures are described in considerable detail in a number of texts (Drake, 1980). In summary the patient is examined in the fasting state, a tube being passed into the stomach. The fasting gastric content is aspirated and then successive aliquots of isotonic saline are forcefully injected, aspirated, and re-injected. During the procedure the patient is positioned in different planes and the abdomen firmly palpated to maximize contact of the injected fluid with the gastric mucosa. The washings are finally aspirated and centrifuged and slides are prepared from the centrifuged deposit.

Cytology

Normal cytology

Within the normal gastric specimen the only type of mucosal cell seen commonly is the surface epithelial cell. These are rarely seen as single cells but almost invariably appear in small sheets. When viewed from above a sheet of cells has a characteristic honeycomb appearance. In profile the cells are columnar and sometimes have a tapering tail. Not uncommonly small clusters of cells are seen arranged in a circular fashion to produce a 'cartwheel' or rosette appearance. These cells may be joined by their tails or by their lumenal borders. The nuclei of the normal gastric mucosal cells are rounded or oval with uniform granular chromatin, a distinct nuclear membrane, and often a small nucleolus. The cytoplasm is relatively abundant, rather delicate in appearance, and granular or finely vacuolated.

Gastric washings are contaminated frequently by material from the oral cavity and oesophagus and also by swallowed sputum. The oral cavity and oesophagus yield squamous cells whilst swallowed sputum may contain ciliated columnar cells from the respiratory epithelium, Curschmann's spirals and histiocytes or macrophages, the latter often containing ingested dust particles. Food particles, such as vegetable cells and meat fibres, may be seen as may red blood cells, inflammatory cells and a variety of microorganisms.

Non-neoplastic abnormalities

Gastritis and peptic ulceration

A variety of non-neoplastic abnormalities may be seen but of greatest importance are gastritis and peptic ulceration. The term gastritis is a general one used to denote any inflammatory disorder of the stomach but should be restricted to those cases where pathological changes are demonstrable. Two broad categories of gastritis are recognized—acute and chronic—the former, by definition, being a relatively transient event whilst the latter is a condition of long standing with a rather complex pattern of histological changes.

Since acute gastritis is of short duration, histological and cytological studies are few in number. Perhaps the most cytologically significant form of this condition is that which follows ingestion of various therapeutic substances, notably analgesic/anti-inflammatory drugs such as aspirin (acetylsalicylic acid), Naprosyn (naproxen) and Indocid (indomethacin). As indicated by Koss (1979), the cell and nuclear abnormalities consequent upon aspirin ingestion may be considerable and indistinguishable from gastric carcinoma. That this diagnostic problem is not limited to aspirin is shown in *Figure 20.5* which depicts cells from a patient receiving high doses of indomethacin as an anti-arthritic measure.

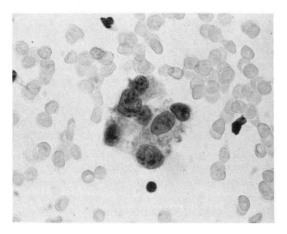

Figure 20.5 Abnormal cells from patient receiving indomethacin. Gastric brushings. Stain: Papanicolaou. Magnification × 790, reduced to 80% in reproduction

The entity chronic gastritis is a complex one but essentially three main categories are recognized. These are chronic superficial gastritis, chronic atrophic gastritis and gastric atrophy. The cytological abnormalities are most prominent in chronic superficial gastritis, an entity in which the lamina propria of the gastric mucosa becomes congested and oedematous and densely infiltrated by inflammatory cells. In association with this inflammatory reaction there is a progressive destruction of the surface and glandular epithelial cells, the degeneration and subsequent necrosis of these cells being accompanied by intense regenerative activity.

The cytological consequences of this combination of degeneration and regeneration are considerable. There is usually nuclear enlargement, the enlarged nuclei varying in staining capacity although most often being hyperchromatic. The nuclear chromatin material is often granular or clumped and nucleoli are prominent and often irregular in shape. Whilst the lesser degrees of abnormality are readily recognizable, the more marked changes may closely mimic malignancy (*Figure 20.6*). However, despite the wide variation in cytological presentation there is usually a preservation of tissue fragments and a general uniformity of cell size and shape within these fragments. This contrasts with the tendency of carcinoma cells to separate from each other due to poor cell adhesion. In addition, where groups of carcinoma cells are seen there is pleomorphism and loss of cell polarity with these groups.

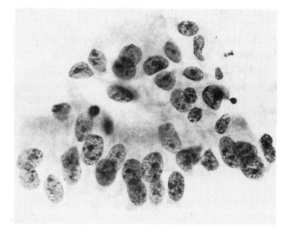

Figure 20.6 Abnormal cells from case of chronic superficial gastritis. Gastric brushings. Stain: Papanicolaou. Magnification × 790, reduced to 80% in reproduction

Peptic ulcers may also be acute or chronic but it is the chronic variant that is of greatest interest to the diagnostic cytologist. Thus at the periphery of the chronic ulcer there is an ongoing process of epithelial cell degeneration and necrosis accompanied by intense regenerative activity. Regenerating cells may also extend as a single layer across the floor of the ulcer. As chronic peptic ulcers are subject to exacerbations and remissions of activity so too do the processes of necrosis and regeneration wax and wane. However, as the patient harbouring the ulcer is usually investigated during a period of activity the cells derived from the edges and floor of the ulcer, which are usually shed in strips and fragments, can cause considerable diagnostic difficulty. The cytological presentation is virtually identical to that of chronic superficial gastritis and hence the problems of diagnosis and differential diagnosis are as already described.

Intestinal metaplasia

Although chronic atrophic gastritis and gastric

atrophy are said to have characteristic cytological manifestations, such manifestations are non-specific and variable. Of greater importance, however, is that these conditions are associated frequently with intestinal metaplasia. In this condition the normal gastric mucosa is replaced by intestinal-type mucosa. Goblet cells and non-secretory columnar cells with microvilli, producing the so-called brush border, so characteristic of small bowel, are prominent. The cytological recognition of this condition is of considerable importance as it is commonly seen in association with gastric cancer and is regarded by many as a 'precancerous' condition (Morson, 1955; Kawachi, 1977). Its cytological manifestations are somewhat variable. Cells with a characteristic brush border may be seen in association with intestinal metaplasia, particularly if the specimen is collected by the brushing technique. However, there is no doubt that the mucus-secreting or goblet cells are the most common manifestation of intestinal metaplasia in both washings and brushings. The cells may be seen singly, in small groups or, occasionally, in small strips, the latter presumably torn intact from the mucosal surface (*Figure 20.7*). Occasionally cells with a brush border and goblet cells may be seen in the same microscopic field (*Figure 20.8*). A very characteristic feature is the presence of polymorphonuclear leucocytes within the cytoplasmic vacuole of the goblet cell.

Figure 20.8 Prominent 'brush border' and 'goblet' cell from patient with intestinal metaplasia. Gastric washings. Stain: Papanicolaou. Magnification × 790, reduced to 80% in reproduction

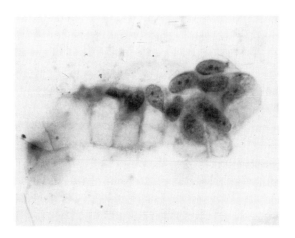

Figure 20.7 Strip of 'goblet cells' from patient with intestinal metaplasia. Gastric washings. Stain: Papanicolaou. Magnification × 790, reduced to 80% in reproduction

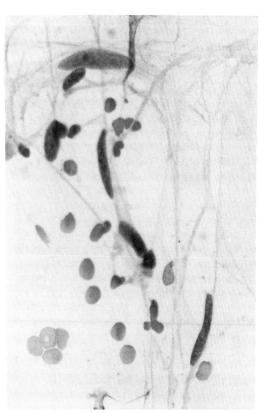

Figure 20.9 Smooth muscle cells from surface of ulcerated leiomyoma of stomach. Gastric brushings. Stain: Papanicolaou. Magnification × 630.

In evaluating the goblet cells shed from an area of intestinal metaplasia, particularly when they are seen as single cells, it is extremely important to distinguish them from the so-called 'signet-ring' cells characteristic of one form of adenocarcinoma. This distinction relies upon an accurate assessment of nuclear morphology and of the relationship of the nucleus to the cytoplasmic vacuole.

Benign neoplasms

Benign neoplasms of the stomach are relatively common. Although mucosal polyps represent the most common benign tumours they are of little cytological significance. More important, perhaps, is the leiomyoma which arises from the smooth muscle component of the gastric wall. If large these neoplasms may project from the gastric mucosa in a pedunculated manner and the overlying mucosa may ulcerate. Should this occur abnormal epithelial cells, as already described in association with a chronic peptic ulcer, often appear in the gastric washings or brushings. In addition, brushing of the exposed tumour tissue may yield characteristic elongated smooth muscle cells (*Figure 20.9*).

Malignant neoplasms

Carcinoma

Carcinoma is the most common malignant tumour occurring in the stomach. A bewildering variety of histological patterns may be seen and indeed the histological classification of gastric carcinoma has been described as 'hopeless and unrewarding' (Stout, 1953). Nevertheless many classifications have been proposed, most conforming to the traditional descriptive approach utilizing the presence or absence of recognizable tissue patterns. Thus the World Health Organization Classification of Malignant Epithelial Gastric Tumours (Oota and Sobin, 1977) includes as its major group, adenocarcinoma, with a subdivision into papillary, tubular, mucinous and signet-ring cell forms. By contrast with these 'differentiated' forms of carcinoma, an entity 'undifferentiated' carcinoma is also included in the WHO classification. In the cytological sense, the term 'undifferentiated' or 'anaplastic' should only be used to designate those cells that have relatively scanty cytoplasm and in which there is no evidence of specialized cytoplasmic function.

In 1965 Lauren proposed a classification that departed from the classical descriptions and allocated gastric carcinoma into two main groups—intestinal type and diffuse type. In general terms intestinal-type carcinomas have a glandular pattern, and are usually well demarcated. The diffuse type of carcinoma is mainly characterized by scattered single cells or small clusters of cells, the latter showing poor cell cohesion. This type of neoplasm is poorly demarcated with a tendency to spread widely within the gastric wall. This classification would appear to be of some importance since the two major types of carcinoma of the stomach, the intestinal and the diffuse, and the less common mixed type, show differences in epidemiological factors, clinical behaviour and natural history. However, for cytological purposes the more traditional system of nomenclature, as exemplified by the WHO classification, is preferred.

As already emphasized, it is important to note that malignant criteria are usually present and clearly recognizable. Indeed gastric cancers are usually characterized by a number of malignant criteria present in numerous cells. Evidence of cytoplasmic differentiation is also usually prominent, the various cell patterns being manifest in the cytological preparations (*Figure 20.10*). The majority of carcinomas, however, are characterized by small groups or sheets of cells with abnormal nuclei and variable degrees of cytoplasmic differentiation (*Figure 20.11 (a) and (b)*).

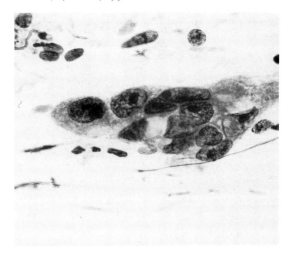

Figure 20.10 Papillary fragment from well differentiated adenocarcinoma of stomach. Gastric brushings. Stain: Papanicolaou. Magnification ×790, reduced to 80% in reproduction

The completely undifferentiated or anaplastic carcinoma may present diagnostic difficulties because of the frequent paucity of cells in the preparation and the relatively inconspicuous nature of these cells. However, when detected their evaluation is straightforward, the cells being characterized by the relatively large nucleus with good malignant criteria and a narrow rim of cytoplasm (*Figure 20.12*). When seen singly they may be difficult to distinguish from the cells of malignant lymphoma but the almost invariable presence of small clusters of cells displaying a so-called epithelial grouping is diagnostic (*Figure 20.13*).

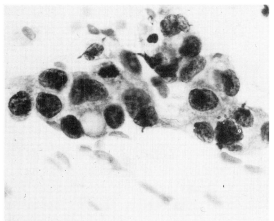

(a)

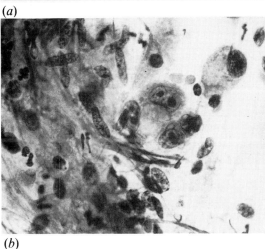

(b)

Figure 20.11 (*a*) and (*b*) Vacuolated cells from moderately well differentiated adenocarcinoma of stomach. Gastric brushings. Stain: Papanicolaou. Magnification × 790, reduced to 80% in reproduction

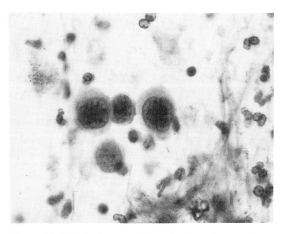

Figure 20.13 Cells from undifferentiated carcinoma of stomach showing so-called epithelial grouping. Gastric washings. Stain: Papanicolaou. Magnification × 790, reduced to 80% in reproduction

Malignant lymphoma

Although uncommon, primary gastric lymphoma occurs sufficiently frequently to be an important diagnostic problem to those who practise gastric cytology.

The nomenclature applied to malignant lymphomas is immensely complex and subject constantly to modification. For the gastric cytologist, however, the following simplified classification is adequate depending, as it does, on cell identification and descriptive morphology.

(1) Hodgkin's lymphoma—no subdivision is desirable or necessary.
(2) Non-Hodgkin's lymphoma:
    (a) Small cell (lymphocytic lymphoma or lymphosarcoma)

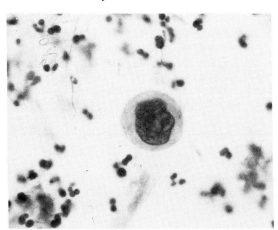

Figure 20.12 Single cell from undifferentiated carcinoma of stomach. Gastric washings. Stain: Papanicolaou. Magnification × 790, reduced to 80% in reproduction

(b) Large cell (histiocytic lymphoma or reticulum cell sarcoma).

On the basis of this simplified classification there are three main cell types that may be recognized in a cytological preparation:

(1) The small cell, or 'lymphocytic' cell, is a round cell that is usually larger than the normal lymphocyte. It has a round nucleus, a prominent nucleolus, and a narrow rim of cytoplasm (*Figure 20.14*).

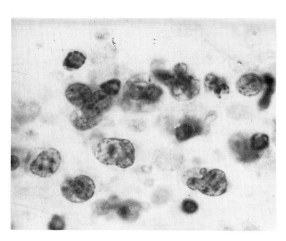

Figure 20.15 Cells from large cell lymphoma of stomach. Gastric brushings. Stain: Papanicolaou. Magnification × 1250, reduced to 80% in reproduction

The cytological diagnosis of malignant lymphoma of the stomach depends largely on the recognition and interpretation of these various cell types. However, in addition, the following morphological features are emphasized:

(1) The nuclei of malignant lymphoma cells quite commonly show small or large, often multiple, nipple-like protrusions. This feature is apparent in *Figure 20.15*.
(2) Cells with cleaved nuclei, whilst not necessarily indicative of malignancy, may assist in the recognition that a malignant cell is from a lymphoma.
(3) Nuclear fragmentation is quite commonly seen in both gastric washings and brushings from both treated and untreated patients, and indeed is seen quite frequently also in histological sections.

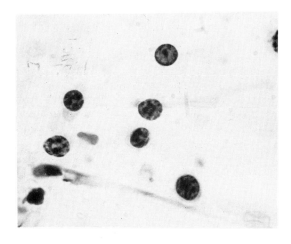

Figure 20.14 Cells from small cell lymphoma of stomach. Gastric brushings. Stain: Papanicolaou. Magnification × 1250, reduced to 80% in reproduction

(2) The large cell ('histiocyte or reticulum cell') is, as the name implies, a larger cell although considerable size variation is common. The cell is usually round but may be irregular in shape. The nucleus is often irregular in outline, there are usually one or more prominent nucleoli, and the cytoplasm, although variable in amount, is more abundant than in the small cell (*Figure 20.15*). Often the cytoplasm is fragmented.
(3) The Reed-Sternberg cell is a large cell with a nucleus that is large in proportion to the size of the cell. Usually the cell contains two or more nuclei and in the classical 'mirror-image' cell two similar nuclei are opposed to one another in the same cell. The nuclear membrane is thick and the nucleus tends to be rather vesicular in appearance. A large nucleolus is a prominent feature of each nucleus.

A number of problems of differential diagnosis arise in relation to malignant lymphoma but undoubtedly the most important is that already referred to, namely, its distinction from the single cells of undifferentiated or anaplastic carcinoma.

The problem can be resolved by immunocytochemical staining using a panel of the appropriate monoclonal antibodies, (see Chapter 6).

Early or superficial gastro-oesophageal carcinoma

Before leaving this consideration of gastro-oesophageal cytology, brief mention will be made of the cytological diagnosis of early or superficial carcinoma of these two organs.

Although preinvasive lesions of the oesophagus have been recognized for some years reports of their diagnosis have been infrequent. Recently, however, limited screening programmes have been initiated in high risk populations notably in Iran (Crespi et al., 1979), China (Co-ordinating Group for the Research of Esophageal Carcinoma, 1973) and South Africa (Berry, Baskind and Hamilton, 1981). Using an abrasive balloon technique cells have been obtained from large numbers of asymptomatic people and examination of these cells has revealed a spectrum of abnormalities ranging from mild dysplasia to severe dysplasia/carcinoma in situ. In addition, cases of microinvasive and frankly invasive squamous cell carcinoma have been detected. In these various studies it has become apparent that the genesis of oesophageal squamous cell carcinoma is identical cytologically to that which occurs in the uterine cervix.

It is generally stated that early cancer of the stomach and advanced cancer are not distinguishable cytologically or that the differences, when present, are relatively insignificant. Whilst it is true that many cases of superficial gastric cancer have cytological features identical to those of the more advanced lesions, cases of the former condition are frequently distinguished by the clean background with a conspicuous absence of tumour debris, and small compact cell clusters. In addition, the cells from these superficial cancers quite frequently display abnormalities that appear to lie between those of extreme regenerative activity, such as may be seen in a florid chronic superficial gastritis in the active phase and unequivocal malignancy. Thus there is a tendency to increased pleomorphism, some irregularity of nuclear membranes and an increasing disorder in the relationship of one cell to another.

Duodenum, biliary tract and pancreas

Anatomy and histology

The duodenum is a C-shaped structure which forms the first or most proximal portion of the small intestine. It is almost entirely retroperitoneal and has a fairly constant length of 20 cm. It is in direct continuity with the stomach, commencing at the pylorus, and it terminates in the jejunum with no recognizable line of demarcation. The duodenum embraces the head of the pancreas and the pancreatic and common bile ducts open into its second part. This opening is usually by way of a common orifice at the ampulla of Vater but sometimes the ducts open separately.

The duodenal mucosa is characterized by the presence of numerous fingerlike processes or villi. These are outgrowths of the mucous membrane which impart a velvety appearance to the mucosa. Between the bases of the villi are the openings of many intestinal glands—the crypts of Lieberkuhn. The epithelium covering the free surface of the mucous membrane is simple columnar, three cell types being distinguished. These are the columnar absorptive cells, the goblet cells and the argentaffin cells. The former are of greatest importance to the diagnostic cytologist being tall columnar in shape and having a distinctive brush border. The common bile duct and the pancreatic duct are both lined by a single layer of columnar cells.

Specimen collection

As in other parts of the gastrointestinal tract material may be collected from the duodenum and the pancreatico–biliary system either by lavage techniques or by endoscopic means. The lavage method, originally evolved by Raskin and his colleagues (1958), has been used with considerable success by a number of workers. A double lumen tube is used so that the gastric secretions may be drained separately from those within the duodenum. The initial aspiration is usually followed by the intravenous injection of secretin which results in a brisk flow of pancreatic secretions. The successful intubation of the duodenum and the subsequent procedure is tedious and time consuming for both patient and operator alike and hence has never been used widely. That it can be an effective method, however, is witnessed by the excellent results reported by several units, and by the case depicted in *Figure 20.16*.

The more recent availability of the duodenal fibreoptic endoscope has facilitated a direct inspection and sampling of the duodenal mucosa and ampulla of Vater and, by retrograde intubation, the collection of pancreatic secretions and bile and brushing of the pancreatic duct.

Finally, material may be collected from the pancreas by needle aspiration, either by percutaneous puncture or by direct puncture intraoperatively.

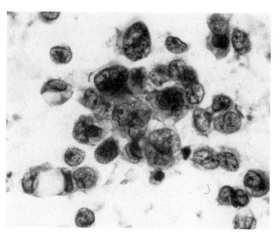

Figure 20.17 Adenocarcinoma of pancreas. Percutaneous fine needle aspiration. Stain: Papanicolaou. Magnification ×790, reduced to 80% in reproduction

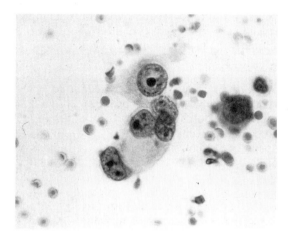

Figure 20.16 Cells from adenocarcinoma of duodenal mucosa. Duodenal washings. Stain: Papanicolaou. Magnification ×790, reduced to 80% in reproduction

Cytology

Normal cytology

In normal duodenal and pancreatic washings relatively few cells are seen. These are columnar in type and it is difficult to determine their precise site of origin. Like columnar cells elsewhere they have a honeycomb appearance when viewed 'from above' whilst in profile they have a columnar configuration, usually with a visible brush border. The nuclei are usually basally placed within the cell, are round and vesicular, and frequently have a small nucleolus.

Malignant neoplasms

Although inflammatory processes do occur the only conditions of cytological significance in this site are malignant neoplasms, notably carcinoma. Carcinomas may arise within the duodenal mucosa, the biliary tracts, and the pancreas, the latter being by far the commonest site. The neoplasms are usually carcinomas of varying degrees of differentiation as seen in *Figure 20.17*.

Large intestine

Anatomy and histology

The large intestine extends from the termination of the ileum to the anus and measures approximately 150 cm in length. It comprises the caecum, the ascending, transverse, and descending colon, the sigmoid colon, rectum and anal canal. With the exception of the anus, the whole of the large bowel is lined by a simple columnar epithelium, the transition to stratified squamous epithelium occurring along an irregular line about 2 cm above the anal opening. The epithelial component of the colonic and rectal mucosa consists of straight perpendicular tubules or crypts. The surface epithelium around the openings of the crypts comprises simple columnar cells with occasional goblet cells only, whereas the tubules are lined predominantly by goblet cells.

Specimen collection

Again material may be collected by a blind lavage technique or under direct vision using the fibreoptic colonoscope. The latter allows inspection of the colonic mucosa and also selective washing, brushing, or biopsy under direct vision.

The washing technique is again time consuming and requires a considerable degree of cooperation on the part of the patient. Adequate preparation of the patient is of paramount

Cytology

Normal cytology

The normal cytology is relatively uncomplicated. The colonic epithelial cells are usually seen in small sheets which, when viewed from above, have the characteristic honeycomb appearance of columnar epithelium. Both cell types are seen, columnar absorptive cells and goblet cells being present often within the same sheet. Goblet cells, with their cytoplasmic vacuoles and basally placed nuclei, predominate in most preparations.

Inflammatory conditions

Although a variety of inflammatory disorders may affect the large bowel, there is no doubt that the one of greatest importance to the diagnostic cytologist is chronic ulcerative colitis. This disease process, the cause of which remains obscure, is characterized by episodes of severe colitis with intervening periods of remission. During the acute or active phase there is an inflammatory reaction, which is confined largely to the mucosa and submucosa, and is associated with extensive destruction of the mucosal epithelium. The mucosal ulceration is followed by intense regenerative activity.

Patients with this disorder have a significantly increased risk of developing carcinoma. This increased risk may be correlated with the frequent development of foci of dysplasia and carcinoma *in situ* in patients with long standing ulcerative colitis. This is characterized by enlargement and stratification of nuclei, nuclear hyperchromasia and increased mitotic activity. Goblet cells are decreased in number and the tubules become irregular in outline and exhibit a process of budding.

The cytological manifestations of chronic ulcerative colitis may pose considerable diagnostic problems. Frequently cells are seen which appear to be larger than normal and have large rather pale nuclei. These cells, referred to as 'bland' cells, do not suggest malignancy. However, cells may also be seen which have more hyperchromatic nuclei and prominent irregular nucleoli—the so-called 'active' cells. Atypia may be considerable, presumably reflecting the presence of mucosal dysplasia or carcinoma *in situ*. There is no doubt that it is extremely difficult to predict cytologically the onset of malignant transformation although cytological surveillance is recommended frequently for patients with long standing ulcerative colitis.

Neoplastic conditions

Mucosal polyps occur frequently within the large bowel. They may be pedunculated or sessile and are divided histologically in a variety of ways, perhaps the most useful classification being into tubular, villous and tubulovillous forms. Cytology has little place in the assessment of polyps although occasionally a brushing done at the time of colonoscopy may indicate a focus of dysplasia or malignancy.

Carcinoma of the colon and rectum is one of the commonest malignancies and hence its diagnosis, preferably at an early stage, is of considerable importance. Approximately 65% of large bowel cancers arise in the sigmoid colon and rectum, the majority occurring within the rectum or at the rectosigmoid junction. They are thus in an area that can be readily visualized endoscopically but also in the area most effectively examined by lavage techniques.

Most carcinomas of the large intestine are adenocarcinomas. Varying degrees of differentiation may be seen but well differentiated forms predominate. The cytological features are similar to adenocarcinoma elsewhere. Malignant criteria are usually clearly evident with, in particular, nuclear enlargement and prominent, often irregular and frequently multiple, nucleoli. Tissue structures such as acini and papillary fragments are sometimes seen, but, more frequently the carcinoma is manifest by sheets or clumps of malignant cells with a varying degree of cytoplasmic vacuolation (*Figures 20.18* and *20.19*)

Results of gastrointestinal cytology

Analyses of performance in gastro-oesophageal cytology have now been published by many laboratories throughout the world. It is difficult to compile such analyses and even more difficult to assess their validity. To a large extent they suggest a black and white situation and do not

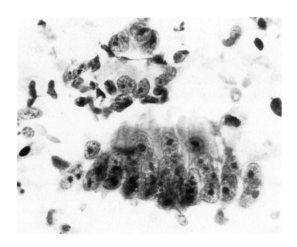

Figure 20.18 Cells from adenocarcinoma of colon. Colonic brushings. Stain: Papanicolaou. Magnification × 790, reduced to 80% in reproduction

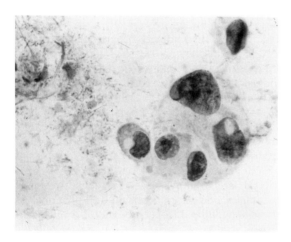

Figure 20.19 Cells from adenocarcinoma of rectum. Colorectal washings. Stain: Papanicolaou. Magnification × 790, reduced to 80% in reproduction

take into account the subtle nuances of communication between the pathologist and clinician. It is important to have a clear understanding of the method of reporting used within a particular laboratory and also the diagnostic philosophy of the laboratory. In the practice of diagnostic cytology there is a constant balancing of sensitivity and accuracy. Thus the diagnostic criteria can be set at such a level that 'false positive' diagnoses will virtually never occur. Conversely, however, in such circumstances the incidence of 'false negative' reports may rise to an unacceptable level. In particular there will be a tendency to miss some cases of surface or intramucosal carcinoma where, as already indicated, the cell yield may be small and the abnormalities difficult to distinguish from those of dysplasia or even florid gastritis.

The basis for the 'final' or definitive diagnosis is also important in the evaluation of results. Traditionally histopathology has been regarded as the final arbiter, the accuracy or otherwise of the cytological diagnosis being determined by the histological investigation. Whilst this may be a reasonable approach when a gastrectomy or autopsy specimen is available for examination, it is not necessarily so when the only tissue available is a minute endoscopically directed biopsy—particularly if taken, for example, from the inflamed or necrotic tissue comprising the floor of an ulcer. Even when an operative specimen is available the way in which this specimen is examined may be of crucial importance. Whilst most carcinomas are readily observed and sampled, a minute intramucosal carcinoma may be missed if the specimen is not subjected to meticulous examination and extensive sampling. A parallel situation exists in gynaecological cytology where an inadequate examination of a cone biopsy specimen from the uterine cervix may deny the cytological diagnosis. Similarly a minute colposcopically directed cervical biopsy may not necessarily sample accurately the abnormality present.

It would be tedious and unnecessary to reproduce the many reports of individual laboratory results that have been published. Indeed it could well be counter-productive since, as already indicated, a knowledge of the method and philosophy of reporting for each laboratory is desirable if an analysis of their results is to be meaningful. Hence reference to the original publications is desirable. The following reviews are recommended to those seeking further information on this subject: Kasugai and Kobayashi (1974); Pilotti et al. (1977); Qizilbach et al. (1980) and Young, Hughes and Hole (1982).

What is evident from a study of these many publications is that excellent results are obtainable from all areas of the gastrointestinal tract. If a particular laboratory is not achieving good results then it is not the techniques that are at fault but the way in which they are being applied. Further it would appear that the methods used in obtaining and evaluating the cytological material are of secondary importance. What is of paramount importance is the skill and dedication of the person or persons carrying out the investigation, meticulous attention to detail being essential at every stage of the procedure if optimal results are to be achieved.

References

BERRY, A. V., BASKIND, A. F. and HAMILTON, D. G. (1981). Cytologic screening for esophageal cancer. *Acta Cytol.*, **2**, 135–141

CO-ORDINATING GROUP FOR THE RESEARCH OF ESOPHAGEAL CARCINOMA, CHINESE ACADEMY OF MEDICAL SCIENCES AND HONAN PROVINCE (1973). The early detection of carcinoma of the esophagus. *Sci. Sin.*, **16**, 457–463

CRESPI, M., MUNOZ, N. and GRASSI, A. et al. (1979). Oesophageal lesions in Northern Iran: a premalignant condition? *Lancet*, **2** 217–221

DRAKE, M. (1980). Gastric cytology. In *Compendium on Diagnostic Cytology*, (Wied, G. L., Koss, L. G. and Reagan, J. W., eds), p. 468. Chicago: Tutorials of Cytology.

KASUGAI, T. and KOBAYASHI, S. (1974). Evaluation of biopsy and cytology in the diagnosis of gastric cancer. *Am. J. Gastroenterol.*, **62**, 199–203

KAWACHI, T. (1977). Intestinal metaplasia and its relation to gastric cancer. *Taisha*, **14**, 919–926

KOSS, L. G. (1979). *Diagnostic Cytology and its Histopathologic Bases*, 3rd Edition. Philadelphia: J. B. Lippincott

LAUREN, P. (1965). The two histological main types of gastric carcinoma: diffuse and so-called intestinal type carcinoma. *Acta Pathol. Microbiol. Scand.*, **64**, 31–49

MORSON, B. C. (1955). Carcinoma arising from areas of intestinal metaplasia in the gastric mucosa. *Br. J. Cancer*, **9**, 377–385

OOTA, K. and SOBIN, L. H. (1977). *Histological Typing of Gastric and Oesophageal Tumours*. International Histological Classification of Tumours, No. 18. Geneva: World Health Organization

PILOTTI, S., RILKE, F., CLEMENTE, C., ALASIO, H. and GRIGIONI, M. (1977). The cytologic diagnosis of gastric carcinoma related to histologic type. *Acta Cytol.*, **21**, 48–59

QIZILBACH, A., CASTELLI, M., KUWALSKI, M. and CHURLY, A. (1980). Endoscopic brush cytology and biopsy in the diagnosis of cancer of the upper gastrointestinal tract. *Acta Cytol.*, **24**, 313–318

RASKIN, H. F., KIRSNER, J. B. and PALMER, W. L. et al. (1958). *Arch. Int. Med.*, **101**, 731–740

SHIDA, S. and ISHIOKA, K. (1988). Gastric cytology: its evaluation for the diagnosis of early gastric cancer. In *Compendium on Diagnostic Cytology*, (Wied, G. L., Koss, L. G. and Reagan, J. W., eds); pp. 382–388. Chicago: Tutorials of Cytology

STOUT, A.P. (1953). Tumours of the stomach. In *Atlas of Tumour Pathology*, Section 6, Fasc. 21. Washington DC: Armed Forces Institute of Pathology

YOUNG J., HUGHES, H. and HOLE, D. (1982). Morphological characteristics and distribution patterns of epithelial cells in the cytological diagnosis of gastric cancer. *J. Clin. Path.*, **35**, 585–590

21

Synovial fluids

Margaret Haddon

The structure of synovial joints, or diarthroses, permits a wide range of movement between bones. Where each bone moves over the other a special type of hyaline cartilage, the articular cartilage, is present. Elsewhere the joint space is lined by a folded membrane, the synovium. The articular surfaces are held together and the joint given stability by the joint capsule and surrounding ligaments (*Figure 21.1*). The small amount of synovial fluid within the joint space acts as a lubricant, decreasing friction and increasing joint efficiency. The synovial membrane consists of a special fibrovascular connective tissue with an incomplete layer of synovial lining cells. For the most part these resemble mesothelial cells, but some behave more like fibroblasts in that they secrete a mucopolysaccharide, hyaluronic acid.

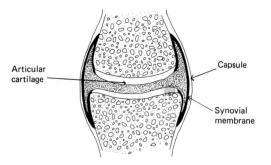

Figure 21.1 Diagram of a synovial joint

Synovial fluid is essentially an ultrafiltrate of plasma with added hyaluronic acid. Although its method of production is not clearly understood, it is generally assumed to be synthesized by the synovium. Analysis of synovial fluid can provide a rapid and accurate method for identifying a variety of joint conditions including those caused by trauma, infections, idiopathic disorders and tumours. Perhaps the most useful test performed in the cytology laboratory is the identification of crystals by compensated polarized light microscopy. All results, however, must be considered in the light of clinical findings and other laboratory tests because considerable overlapping occurs in a number of conditions.

Collection of synovial fluid

Arthrocentesis must be performed using an aseptic procedure. The need for local anaesthetic depends on the size and tenderness of the joint to be aspirated but it must be remembered that urate crystals are water-soluble and a large quantity of anaesthetic may cause dissolution of crystal in small samples of fluid. Once collected the fluid is then divided according to the laboratory tests required:

(1) A plain tube for gross appearance, viscosity and, if preferred, crystal studies. Opinion differs as to whether a sample for crystal examination should be treated with anticoagulant or not. Powdered anticoagulants such as oxalates are themselves crystalline and may mask the presence of uric acid or pyrophosphate crystals.

(2) A tube with suitable anticoagulant e.g. EDTA or 5% tri-sodium citrate for total

Table 21.1 Summary of findings on synovial fluid

	Normal	Degenerative; non-inflammatory	Inflammatory; aseptic	Inflammatory; septic	Crystal synovitis
Colour	Clear	Clear	Cloudy	Cloudy	Cloudy
Viscosity	High	High	Low	Low	Low
WBC ($\times 10^9$/l)	<0.2	0.2–2.0	2.0–50.0	20.0–100.0	20.0–100.0
Polymorphs (%)	<25	<25	<50	>75	>50
Organisms	None	None	None	May detect organisms using special stains and culture	None
Crystals	None	None	None	None	Urate, pyrophosphate and others

cell count, differential white cell count, cytological studies and crystal studies. The author recommends the use of an anticoagulant because small crystal deposits and cells may be bound into clots and give false negative results.

(3) A sterile plain tube for microbiological investigations which are essential to exclude infection.
(4) A potassium or fluoride oxalate tube for glucose determination.
(5) Appropriate additional tubes may be required for serological, biochemical and electron microscopical analysis.

Routine examination of synovial fluid

Tests on the extracted fluid should be performed within a few hours of collection; however, samples can be left overnight in the refrigerator at 4 °C. Cytologically, the presence or absence of crystals and the cellular content are the most important observations. Other useful information includes the gross appearance and the viscosity. The findings in a variety of diseases are summarized in *Table 21.1*.

Gross appearance

Initially the specimen should be observed for colour and clarity. A non-inflammatory fluid will be clear; the turbidity increases with the cellular content. A traumatic tap may have blood unevenly distributed throughout the specimen in the form of small clots. In contrast, a true bloody synovial effusion will have a uniform colour with no clots, and upon centrifugation leaves a xanthochromic supernatant. Bloody synovial effusions may be found in cases of trauma, tumour or haemophilia. Fluids from patients with rheumatoid arthritis may contain small fibrin clots or appear milky due to the presence of cholesterol crystals.

Viscosity

The viscosity may be simply assessed either while decanting the fluid into a disposable centrifuge tube or by drawing an orange stick through the specimen. Normal synovial fluid is viscous and will form tenacious strings several inches in length before breaking. If the fluid pours easily or the strings break short, the viscosity is considered to be abnormally low. Low viscosity is found in both septic and aseptic inflammatory joint diseases due to the enzymatic breakdown of hyaluronic acid. Hyaluronic acid may be quantitatively assayed biochemically or qualitatively assessed with the mucin clot test. This is a simple qualitative viscosity test based on the polymerization of hyaluronic acid. Various methods have been published (Cohen, 1967; Naib, 1973) but they are all based on the addition of acetic acid to the synovial fluid resulting in the coagulation of hyaluronate and formation of a homogeneous clot.

Method

(1) To 1 ml of synovial fluid add 1 ml 5% acetic acid.
(2) Mix well and allow to stand for 1 min.

Clots are graded as good, fair or poor (*Figure 21.2*). Fluids from patients with osteoarthritis clot readily, whereas those from patients with acute infections or crystal synovitis clot poorly.

Figure 21.2 Mucin clot test. (*a*) 'Good' mucin clot–a tightrope-like clot that does not disperse on agitation in a clear fluid; (*b*) 'fair' mucin clot–a softer clot with some threads in solution giving a turbid appearance; (*c*) 'poor' mucin clot–a few small clots which easily disperse on agitation

Total cell count

It is not possible to specifically categorize joint disease from the total count, but it provides useful diagnostic information. Synovial fluid that has been collected into anticoagulant must be used.

Method

(1) Mix the specimen thoroughly and using a white blood cell diluting pipette make a 1:20 dilution of the sample with normal saline (0.85%) containing a few drops of methylene blue. The addition of methylene blue aids cellular differentiation.
(2) Load a standard haemocytometer e.g. a Neubauer, and leave for 5 min to allow the cells to settle.
(3) Make a count of the red and white cells in the usual manner using a ×40 objective. Express the answer as number of cells $\times 10^9/l$. Alternatively the cells may be counted electronically using a Coulter counter.

Preparation techniques for the examination of crystals and cells

Method for synovial fluid that has been collected into anticoagulant:

(1) Centrifuge the sample at 3000 rev/min for 10 min.
(2) Decant the supernatant and resuspend the deposit.
(3) Make the following preparations:

   (a) Two wet preparations for crystal studies. (A small amount of the specimen is placed onto a clean glass slide and a coverslip gently dropped over it. The expanding edge of the compressed fluid should just reach the edges of the coverslip. The edges of the coverslip are then sealed with nail polish or wax to prevent rapid dehydration during examination.)
   (b) Two alcohol-fixed smears for staining by the Papanicolaou method.
   (c) One air-dried smear for staining by the May-Grünwald–Giemsa technique.

Alcohol-fixed and air-dried smears should be examined under the light microscope for cell content and differential count if required.

Method for synovial fluid specimens that have clotted:

(1) Make one wet preparation of squashed clot for crystal studies.
(2) Centrifuge the sample at 3000 rev/min for 10 min.
(3) Decant the supernatant and resuspend the deposit.

Table 21.2 Table summarizing the optical properties of various crystals

Crystals	Morphology	Size (µm)	Birefringence	Intracellular or extracellular
Monosodium urate	Needles	5–10	Strongly negatively birefringent	Intracellular and extracellular
Calcium pyrophosphate	Brickets	2–6	Weakly positively birefringent	Intracellular and extracellular
Cholesterol	Notched pane	Varied	Varied	Extracellular
Hydroxyapatite	Round or star	0.02–0.08	Visualized using electron microscopy	Intracellular and extracellular
Starch	Maltese cross	8–10	Varied	Extracellular
'Plastic'	Ellipsoid	10–30	Varied	Extracellular
Other miscellaneous particles	Varied	Varied	Varied	Intracellular and extracellular

(4) Make the following preparations:

(a) A wet preparation of the deposit (making two wet preparations in total).
(b) Two alcohol-fixed smears for staining by the Papanicolaou method.
(c) One air-dried smear for staining by the May-Grünwald–Giemsa technique.

As an alternative method the clot can be extracted, fixed in alcohol and treated as a histological specimen. Tissue sections can then be examined for crystals and cells. N.B. Aqueous fixatives such as formal saline cannot be used because the crystals of interest are water-soluble and will dissolve.

Polarizing microscopy

The polarizing microscope has made it possible to distinguish different types of crystals found in synovial fluids. In practice the only two of significance are crystals of monosodium urate monohydrate and calcium pyrophosphate dihydrate. Other crystals which may be identified include cholesterol and a variety of contaminants.

A wet preparation of the synovial fluid is placed on a light microscope stage. Polarized light is induced by placing a polarizing filter between the light source and the condenser, and a further polarizing filter (analyser) between the objective and the eyepiece. Each filter allows through only that light which is polarized in one particular plane. When the planes of these two filters are at right angles to each other (i.e. crossed), no light will reach the observer; only when the specimen between the filters rotates the plane of light (i.e. is birefringent) will an image be seen. The relationship of the planes to each other can be varied by rotating one of the polarizers, thereby illuminating or darkening the field of view. These filters may be added to any standard microscope so that birefringent material can be seen.

Crystals in synovial fluid are birefringent and they may be detected using crossed polarized filters appearing light on a dark background. The size and shape of the crystals are characteristic and a preliminary identification may be made (*Table 21.2*).

For further crystal identification a first order red compensating filter is used. It is placed between the two polarizing filters with its optical axis at 45° to the axis of transmission between the filters. This filter absorbs light of wavelength 565 nm (green/yellow band) giving red light. The absorption of various light wavelengths differs according to the birefringent material introduced, resulting in colour changes which aid crystal identification (*Figure 21.3*). The specimen is rotated until the long axis of the crystal is parallel to the axis of the filter (which will be marked on the filter holder).

A positive birefringent crystal causes part of the red end of the spectrum to be absorbed and the crystal appears blue on a red background. A negatively birefringent crystal has the reverse effect and makes the blue light in phase, therefore the crystal appears yellow on a red background.

The slide should be carefully rotated through 90° and the colour changes of any crystals observed during rotation. The long axis of the crystal will now be at 90° to the axis of the filter. Negative birefringent material will now be blue and positive birefringent material yellow (*Table 21.3*).

As a reference and to make crystal identification easier for the beginner it is recommended that a control slide of urate (e.g. a tissue section from a gouty tophus fixed in alcohol) and calcium pyrophosphate (obtainable from the chemist) are kept in the laboratory.

Zaharopoulas and Wong (1980) described how a standard microscope may be converted into a polarizing microscope complete with red compensating filter by the addition of two polarizing filters and substitute red filter made from the application of Cellophane tape on a glass slide.

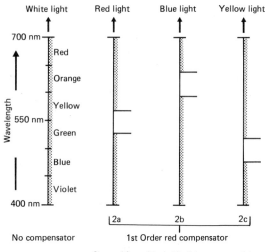

Figure 21.3 The effect produced on transmitted light with the introduction of a first order red compensation filter and birefringent material placed between two polarizing filters

Crystal identification

Monosodium urate crystals

Monosodium urate crystals tend to be thin, needle-shaped structures, although occasionally rod-shaped crystals are observed (*Figure 21.4*). The size ranges between 5 and 10 µm and the crystals may be both intracellular or extracellular (*Figure 21.5*). Urate crystals exhibit strong negative birefringence, and for this reason they are easier to identify than calcium pyrophosphate crystals.

Sir Alfred Garrod first identified uric acid crystals in association with joint inflammation in the mid-nineteenth century, although it was not until the early 1960s that the diagnostic importance of crystal identification in synovial material was fully appreciated. Uric acid is the end product of purine catabolism in man and

Table 21.3 Determination of birefringence using a first order red compensating filter

Axial position of crystal	Angle of extinction	Crystal type	Colour	Birefringence
Long axis	Parallel //	Monosodium urate	Yellow	– – –
		Calcium pyrophosphate	Blue	+
Long axis	At 90° ⊥	Monosodium urate	Blue	– – –
		Calcium pyrophosphate	Yellow	+

Crystal identification 389

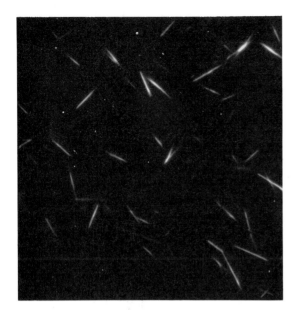

Figure 21.4 Urate crystals. Polarizing microscope. Magnification ×160

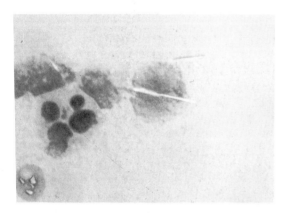

Figure 21.5 Intracellular urate crystals. May-Grünwald–Giemsa smear. Magnification ×650

elbows, hands, feet and heels may be affected. Deposition of urates in the kidneys can cause renal impairment (urate nephropathy) and stone formation. Fortunately, the medical treatment of the condition is usually successful.

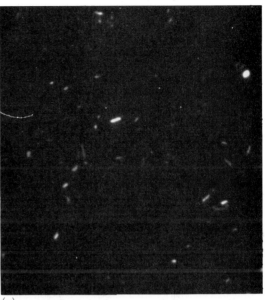

(a)

(b)

Figure 21.6 Pyrophosphate crystals. Polarizing microscope. (a) Magnification ×160; (b) Magnification ×400

there are many biochemical or physiological derangements that may cause hyperuricaemia in a patient, though the exact mechanism of crystal deposition in joints is not fully understood (Dieppe, 1981). The deposition of uric acid crystals leads to the condition of gout, classically described as affecting the first metatarsophalangeal joint in middle-aged males. However, the disease is not confined to the male population and almost any joint may be affected. The onset of symptoms may be sudden and severe, or the disease may follow a more chronic course. Urates may be deposited in and around the joints producing gouty tophi. The ears,

Calcium pyrophosphate crystals

Calcium pyrophosphate crystals are in general smaller than urate crystals, ranging between 2 and 6 µm in length. They usually present as rhomboids or parallelepipeds and sometimes as rods with blunt ends. They are weakly birefringent (*Figure 21.6(a)* and *(b)*) and like urates they may be both intracellular and extracellular.

Calcium pyrophosphate crystal deposition in the synovium results in pyrophosphate arthropathy ('pseudogout'). The radiological appearance of flecks of calcification within articular cartilage gives the disease its other name—chondrocalcinosis. A clinical diagnosis of pyrophosphate arthropathy is not always supported by the identification of crystals using polarizing microscopy, as the crystals may be too small for detection by this method. Their identification by scanning electron microscopy may be helpful in these circumstances (Dieppe *et al.*, 1979). A large number of conditions have been associated with pyrophosphate deposition including other forms of acute arthritis, diabetes mellitus, hypertension and uraemia, but sporadic crystal deposits frequently occur without obvious metabolic disturbances. Unlike gout, there is no satisfactory medical treatment for this condition.

Cholesterol crystals

Cholesterol crystals are easily identified with their flat plate or window pane appearance that has a notched corner (*Figure 21.7*). They vary from quite small discrete single crystals, to irregularly stacked multiple crystalline structures. Cholesterol crystals are found more frequently in effusions of long standing and their presence in joint fluids was first described by Ropes and Bauer (1953).

A study of synovial effusions containing cholesterol crystals was later carried out by Ettlinger and Hunder in 1979. Their work suggests local factors are most important in the development of cholesterol crystals but the exact mechanisms are unknown. As cholesterol is dissolved in alcohol, the crystals themselves cannot be demonstrated in alcohol-processed material using polarizing microscopy, but the spaces the crystals occupied (their ghosts) remain (*Figure 21.8*).

Contaminants

Starch granules with their striking 'Maltese cross' appearance (*Figure 21.9*) are commonly

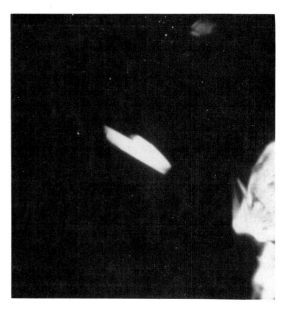

Figure 21.7 Cholesterol crystals. Polarizing microscope. Magnification × 160

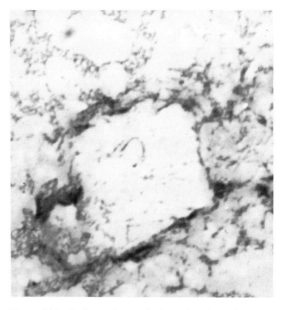

Figure 21.8 Cholesterol crystal 'ghost' in a May-Grünwald–Giemsa smear. Magnification × 650

seen. These contaminants are introduced at arthrocentesis from surgical gloves. Talcum was used in surgical gloves for many years but its use discontinued after talcum crystals were found to elicit a foreign body inflammatory reaction (Kirshen, Naftolin and Benirschke, 1974).

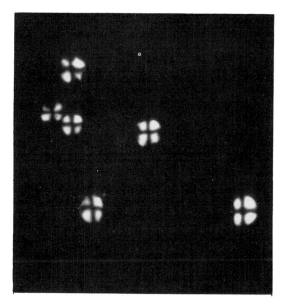

Figure 21.9 Starch granules. Polarizing microscope. Magnification ×650

Various unusual particulate and crystalline substances have been described in synovial fluid. Fragments of cartilage and collagen fibrils are frequent observations. Metal, plastic, plant thorns, amyloid fibrils, crystals of lithium heparin, calcium oxalate and granules of corticosteroids (following intraarticular injection of these drugs) may all cause confusion; but their size, shape and variable patterns of birefringence should allow their distinction from urates or pyrophosphates.

Diagnostic pitfalls

Possible diagnostic pitfalls include mixed crystal deposits. All particulate matter must be carefully analysed, and the possibility of crystal formation after aspiration minimized by prompt examination. Synovial fluid that has been left standing overnight in polythene containers may reveal elongated crystalline structures upon polarizing microscopy. Although this material exhibits birefringence it can be morphologically distinguished from urates or pyrophosphate by size (10–30 μm), generally ellipsoid shape and tendency to aggregate (*Figure 21.10(a) and (b)*).

Other crystals
Hydroxyapatite crystals

Hydroxyapatite crystals are too small for identi-

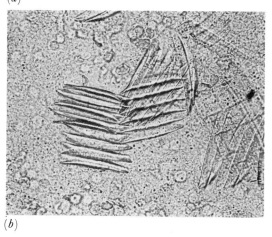

Figure 21.10 'Polythene crystalline structures' in a synovial fluid. (*a*) Polarizing microscope, magnification ×160, reduced to 50% in reproduction; (*b*) unstained smear, magnification ×650, reduced to 50% in reproduction

fication by polarizing microscopy and it was with the said of electron microscopy that their existence was recognized (Dieppe *et al.*, 1976; Schumacher *et al.*, 1977). Preparatory techniques have been described by Crocker *et al.* (1976). Hydroxyapatite crystals appear round or star-shaped, varying between 0.02 μm and 0.08 μm in size (*Figures 21.11 (a) and (b)*). The relevance of hydroxyapatite crystals remains unknown, but their presence in occasional cases of acute synovitis raises the possibility that they may be another cause of crystal-induced synovitis (Dieppe *et al.*, 1979).

Hydroxyapatite is a natural mineral of teeth and bone. It may be found in numerous pathological, extra-articular calcifications. Deposits in lung, lymph nodes or vessel walls are often an unexpected radiological finding.

(a)

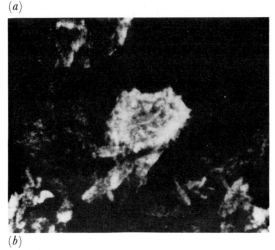

(b)

Figure 21.11 Hydroxyapatite crystals. (a) Scanning transmission electron micrograph, total magnification ×150 000; (b) scanning electron micrograph, total magnification ×75 000, reduced to 50% in reproduction. Courtesy of Dr Dieppe, Consultant Rheumatologist, University of Bristol

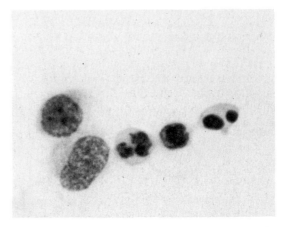

Figure 21.12 Two synovial cells and three leucocytes. Papanicolaou stain. Magnification ×650, reduced to 80% in reproduction

Synovial fluid cytology

Normal synovial fluid contains remarkably few cells (between 10 and 200 cells/ml), comprising histiocytes (65%), lymphocytes (25%) and polymorphs (10%). The histiocytes have foamy phagocytic cytoplasm which occasionally contains azurophilic granules and a central nucleus which may be round, oval or crescent in shape. Prominent nucleoli are common. A few synovial cells may be present. These resemble reactive mesothelial cells with abundant cytoplasm and round or oval eccentric nuclei which have a fine regular chromatin pattern (*Figure 21.12*). To obtain a differential count, count 200 cells and express the total of each cell type as a percentage.

During acute stages of inflammatory disorders, white blood cell counts frequently exceed $50 \times 10^9/l$ with polymorphs constituting over 50% of the white cell population. In chronic inflammatory disorders the synovial fluid shows a mixed population of polymorphs, histiocytes and lymphocytes together with synovial cells either singly or in sheets. In traumatic fluids histiocytes containing haemosiderin or forming foreign body giant cells are a common finding and occasionally in these fluids cartilage cells may also be present. These appear as large, oval mononuclear or binuclear cells with a thick nuclear membrane and an encapsulated appearance of the matrix. Chondrocytes may also be found in degenerative joint diseases.

Reiter's syndrome

In 1967 Pekin, Malinin and Zvaifler described large macrophages which had phagocytosed one or more polymorphonuclear leucocytes in the synovial fluid of patients suffering from Reiter's disease, but Spriggs, Boddington and Mowat (1978) could not confirm that their presence was a specific feature of this disease because these cells were seen in other inflammatory conditions.

Naib (1973) described single, swollen, hyperchromatic synovial cells of which 20% appeared to contain chlamydial inclusions, but cytological features alone were unreliable for diagnosis. Monoclonal antibody techniques may be a valuable diagnostic asset. Schachter *et al.* (1966) isolated chlamydial agents from infected synovial membranes using yolk sac isolation techniques. The electron microscope has also proved useful for the identification of chlamydial-infected cells. *Figure 21.13* demonstrates a large subnuclear, intracytoplasmic inclusion in a McCoy cell in tissue culture containing many dense secondary or elementary bodies.

Reiter's syndrome produces symptoms of polyarthritis, conjunctivitis and urethritis.

that they were complexes of the abnormal gamma globulins collectively known as 'rheumatoid factor'.

The precise role of chronic inflammatory cells in the pathogenesis of rheumatoid arthritis is not fully understood (Meijer *et al.*, 1976). DeVere Tyndall *et al.* (1983) have demonstrated a relationship between dendritic 'veiled' cells (*Figure 21.14*) and a subgroup of lymphocytes. The possibility that these veiled cells are actively presenting antigen to lymphoid cells is being investigated.

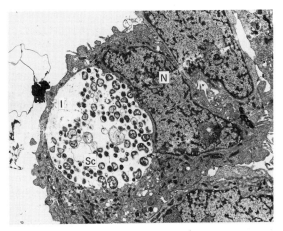

Figure 21.13 Chlamydial inclusion body in McCoy cell. N, nucleus; I, inclusion body; Sc, secondary bodies. Electron micrograph, total magnification × 6384, reduced to 50% in reproduction

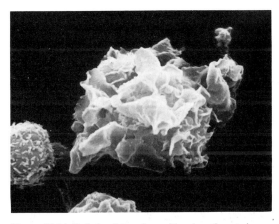

Figure 21.14 Veiled cell together with a lymphocyte in a synovial fluid. Electron micrograph, total magnification × 6250, reduced to 75% in reproduction

Rheumatoid disease

In 1966 Hollander, Reginato and Torralba described the existence of polymorphs containing between 12 and 15 round, basophilic, cytoplasmic inclusions in synovial fluid. These cells were termed 'ragocytes' and their identification led to much research into their diagnostic value in rheumatoid disease. Naib (1973) reported that the percentage of ragocytes varies between 5% and 80% and that their presence is significant only when exceeding 10%.

Coimbra and Lopez-Vaz (1967) with the use of the electron microscope and histochemistry demonstrated that some of these dense bodies were phosphatase active whilst others contained lipid and glycogen. In addition, they found that one group of granules resembled antibody complexes. Fallet, Boussina and Fellman (1968) using immunofluorescent techniques, showed

Rheumatoid disease may affect patients of all ages. It has an unknown aetiology and the diagnosis is usually made clinically or immunologically and not by cytological criteria.

Lupus erythematosus (LE)

In 1953 Ropes and Bauer described the appearance of LE cells in joint fluid and emphasized the importance of diagnosis of lupus erythematosus involving the joints. LE cells are easily recognized due to their large size and distinctive appearance (*Figure 21.15*). They are mature polymorphs with large, hyaline, pink, homogeneous inclusion bodies which distend the cytoplasm. The nuclei are crescent-shaped and appear compressed against the periphery of the inclusion. These LE cells are not always present, and immunological tests are a far more reliable method of diagnosing this disease.

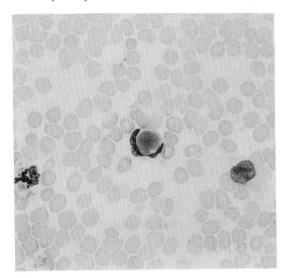

Figure 21.15 LE cell with its characteristic crescent-shaped nucleus and homogeneous cytoplasmic inclusion. May-Grünwald–Giemsa smear. Magnification × 650, reduced to 90% in reproduction

Malignant tumours

Cytological diagnosis of primary or metastatic disease involving the joints is rare. Osteogenic sarcoma has been described in synovial fluid (Meisels and Berebichez, 1961; Naib, 1973) as have cases of metastatic tumour to the joints (Naib, 1973).

Acknowledgements

Many thanks to the Medical Illustration Department, Clinical Research Centre for photographic assistance and illustrations; to Dr P. Dieppe, Consultant Rheumatologist, University of Bristol and Mr M. Broderick, Olympus Microscopes; and to Mrs C. L. Robinson for secretarial help.

References

COHEN, A. S. (1967). *Laboratory Diagnostic Procedures in the Rheumatic Diseases*, 1st Edition. London: J. A. Churchill

COIMBRA, A. and LOPEZ-VAZ, A. (1967). Acid phosphatase-positive cytoplasmic bodies in leucocytes of rheumatic synovial fluid. *Arthritis Rheum.*, **10**, 337–342

CROCKER, P. R., DIEPPE, P. A., TAYLOR, T., CHAPMAN, S. K. and WILLOUGHBY, D. A. (1976). The identification of particulate matter in biological tissues and fluids. *J. Pathol.*, **121**, 37–40

DE-VERE TYNDALL, A., KNIGHT, S. C., EDWARDS, A. J. and CLARKE, J. (1983). Veiled (dendritic) cells in synovial fluid. *Lancet*, **i**, 472–473

DIEPPE, P. A. (1981). Crystal-induced arthropathies and osteoarthritis. In *Recent Advances in Rheumatology*, No. 2, (Watson-Buchanan, W. and Carson-Dick, W., eds), pp. 1–18. Edinburgh: Churchill Livingstone

DIEPPE, P. A., CROCKER, P., HUSKINSON, E. C. and WILLOUGHBY, D. A. (1976). Apatite deposition disease. *Lancet*, **i**, 266–269

DIEPPE, P. A., CROCKER, P. R., CORKE, C. F., DOYLE, D. V., HUSKINSON, E. C. and WILLOUGHBY, D. A. (1979). Synovial fluid crystals. *Quart. J. Med.*, New Series **XLVIII** (192), 533–553

EDWARDS, J. C. W. (1982). The origin of type A synovial lining cells. *Immunobiology*, **161**, 227–231

ETTLINGER, R. E. and HUNDER, G. G. (1979). Synovial effusions containing cholesterol crystals. *Mayo Clin. Proc.*, **54**, 366–374

FALLET, G. H., BOUSSINA, I. and FELLMAN, N. (1968). Diagnostic value and possible role of synovial ragocytes in rheumatology. *Rev. Rhum. Mal. Osteoartic.*, **35**, 590–600

HOLLANDER, J. L., REGINATO, A. and TORRALBA, T. P. (1966). Examination of synovial fluid as a diagnostic aid in arthritis. *Med. Clin. North Am.*, **50**, 1281–1293

KIRSHEN, E. J., NAFTOLIN, F. and BENIRSCHKE, K. (1974). Starch glove powders and granulomatous peritonitis. *Am. J. Obstet. Gynecol.*, **118**, 799–804

LINNE, J. and RINGSRUD, K. (1979). *Basic Techniques for the Medical Laboratory*, 2nd Edition. New York: McGraw-Hill

MALININ, T. I., PEKIN, T. J. and ZVAIFLER, N. J. (1967). Cytology of synovial fluid in rheumatic arthritis. *Am. J. Clin. Pathol.*, **47**, 203–208

MEIJER, C. J., VAN DER PUTTE, L. B., EULDERINK, R., KLEINJAN, R., LAFEBER, G. and BOTS, G. (1976). Characteristics of mononuclear cell populations in chronically inflamed synovial membranes. *J. Pathol.*, **121**, 1–8

MEISELS, A. and BEREBICHEZ, M. (1961). Exfoliative cytology in orthopedics. *Can. Med. Assoc. J.*, **84**, 957–959

NAIB, Z. M. (1973). Cytology of synovial fluid. *Acta Cytol.*, **17**, 299–309

PEKIN, T. J., MALININ, T. I. and ZVAIFLER, N. J. (1967). Unusual synovial fluid findings in Reiter's syndrome. *Ann. Intern. Med.*, **66**, 677–684

ROPES, M. W. and BAUER, W. (1953). *Synovial Fluid Changes in Joint Disease*. Cambridge, Massachusetts: Harvard University Press

SCHACHTER, J., BARNES, M. G., JONES, P. J., ENGLEMANN, E. P. and MEYER, K. F. (1966). Isolation of bedsoniae from the joints of patients with Reiter's syndrome. *Proc. Soc. Exp. Biol. Med.*, **122**, 283–285

SCHUMACHER, H. R., SMOLYO, A. P., TSE, R. and MAURER, K. (1977). Arthritis associated with apatite crystals. *Ann. Intern. Med.*, **87**, 411–416

SPRIGGS, A. I., BODDINGTON, M. M. and MOWAT, A. G. (1978). Joint fluid in Reiter's disease. *Ann. Rheum. Dis.*, **37**, 557–560

ZAHAROPOULOS, P. and WONG, J. Y. (1980). Identification of crystals in joint fluids. *Acta Cytol.*, **24**, 197–202

22

Tumour cells in blood and bone marrow

A. J. Salsbury

Tumour cell metastasis

Cancer is largely feared because of its ability to metastasize or spread, often to sites far removed from its primary site of origin. When they are localized, most malignant tumours can be surgically removed. Even when disseminated, cancer may be contained and its spread delayed by surgery, radiotherapy and chemotherapy.

What is metastasis? Briefly, it consists of a malignant cell or cells capable of leaving the primary tumour and proliferating locally or more distantly along the lymphatic channels and lymph nodes, or by the bloodstream, and of implanting, developing and establishing a further growth. Such behaviour characterizes the majority of carcinomas, sarcomas and lymphomas.

No one really knows why some malignant tumours frequently metastasize to bone marrow and others do not. There is considerable evidence that venous invasion, and involvement of lymph nodes with subsequent probable passage to the bloodstream, is more likely to give rise to distant metastases. But metastasis is not predictable. Some tumours are known to have a predilection for the bone marrow, such as carcinomas of the breast, kidney, thyroid, prostate and lung; malignant melanoma; neuroblastoma; many sarcomas and the majority of the lymphomas. In contrast, many carcinomas of the head and neck, and of the lower abdomen produce few bone marrow metastases.

It seems likely that the primary site of origin, its blood supply and the ultimate destination of any metastasizing tumour cells are largely a matter of pure chance, although there is some agreement that certain organs, such as the spleen, seem unfavourable fields for tumour cell metastasis. Most recent reviews suggest that the ability to metastasize is not a major factor in the biology of cancer (Alexander, 1984) and that spontaneous metastasis is determined more by random survival of tumour cells than by the selection of pre-existing metastatic cell variants (Milas, Peters and Ito, 1983). This view is borne out by a recent paper (Tarin et al., 1984) on patients with malignant ascites treated with peritoneovenous shunts. They found that metastases did not necessarily develop even when large numbers of viable tumour cells regularly entered the blood.

Bone marrow examination in malignant disease

In all suspected cases of malignant disease, bone marrow should be aspirated and a trephine biopsy taken. With modern needles both samples can be taken under a local anaesthetic and the need for a general anaesthetic, often needed in the past for trephine samples, is obviated. Bone marrow aspiration and trephine biopsy tend to complement one another. Aspiration is useful for detecting any malignant cells and for differentiating them from non-malignant cells, whereas trephine biopsy helps one to gain an idea of the nature and structure of any malignant cells found.

A suitable site should be selected, usually the anterior superior iliac spine, the posterior iliac spine or the iliac crest, according to the build of the patient. The sternum should not be chosen as, although it is suitable for aspiration biopsies, it is not usually thick enough for trephine biopsies.

After skin preparation, approximately 1–2 ml of local anaesthetic should be injected with a fine needle, making sure that an intradermal bleb is raised and that anaesthetic is also injected under the periosteum. Sufficient time must be left for the anaesthetic to have effect before anything further is done.

Bone marrow aspiration

This is best done with a Klima or similar screw-guard needle, both because one can use the rotating guard as a pivot with one hand whilst pressing and rotating the needle with the other, and because this type of needle will not yield if suddenly penetrating to softer bone marrow. Bone marrow aspirates should be used to make smears of any marrow particles as soon as possible. If a 'dry trap' is experienced—not uncommon in cases of malignant disease—the marrow needle should be fitted with a 10–20 ml syringe and the marrow withdrawn under suction. In this way valuable material may often be found in the lumen of the syringe. The bone marrow aspirate films are fixed in methanol and stained by a Romanowsky method—this is very useful for identifying the different types of non-malignant bone marrow cells. After keeping further films for spares and for special stains, the remainder of the marrow should be centrifuged and sections cut as for the trephine biopsy specimen.

Bone marrow trephine biopsy

Samples can be taken from the same skin site, slightly varying the area where the bone is penetrated. The Jamshidi and Sultan needles both give excellent results with little distortion of marrow morphology. Both are inserted through the outer table of dense bone and then the inner stylet removed so that the outer trephine can take a biopsy of bone marrow. A further penetration of 1–2 cm is desirable, although much depends on whether the patient experiences discomfort. The Jamshidi needle is slightly withdrawn and advanced slightly in another direction to free the biopsy sample; the Sultan needle is simply rotated a few times to free the sample. The needle is then withdrawn and the specimen removed. Some authorities recommend aspirating a sample through the trephine needle; in my experience it is better to aspirate first and then take a trephine biopsy in a slightly different spot. The specimen should be fixed in Bouin's fluid rather than 10% formaldehyde as this preserves better nuclear architecture (a factor important in the identification of malignant cells). The specimen is then well decalcified, paraffin sections cut and the latter stained with haematoxylin and eosin, van Gieson's stain and a silver stain for reticulin fibrils. Other stains can also be performed where appropriate. Specimens can be embedded and cut in methacrylate resin if desired. The van Gieson and reticulin stains are important: firstly, they may be very helpful in marrow infiltration by some of the lymphomas; secondly, they may direct one's attention to areas with a high connective tissue content (as happens sometimes in carcinomatous involvement of bone marrow) which are less prominent on the haematoxylin and eosin sections.

Tumour cells are likely to be found in marrow aspirates in approximately 10% of cases of advanced malignant disease. The incidence tends to be higher in certain carcinomas such as oat cell carcinoma of the bronchus where a detection rate of around 40% is common. The detection rate is increased if trephine biopsies are also examined and also if multiple samples are taken. As an example, in cases of neuroblastoma Franklin and Pritchard (1983) had an overall detection rate of about 50% in children whose marrow they studied. They found that trephine biopsies were more effective than aspirates for detection of tumour cells in 20% of combined aspirate/trephine examinations, while the converse applied in 7%. A further improvement in detection rate of 10% was achieved by taking samples from two sites (in this case, both posterior iliac crests). Similar results have been reported in other types of malignant disease such as non-Hodgkin's lymphomas (Brunning et al., 1975).

Identification of tumour cells in bone marrow

Tumour cells can vary very widely in appearance. It is often easier to describe what characteristics they are not likely to possess, rather than those that they do. The appearance of malignant cells also depends, to some extent, on their type and site of origin. The appearance of carcinoma cells will first be described, and points of difference in other malignant tumours subsequently.

Carcinoma

Generally carcinoma cells in bone marrow possess the following characteristics:

(1) They are often found in clusters (*Figures 22.1* and *22.2*).
(2) They usually vary very markedly in size and shape (*Figure 22.3*).

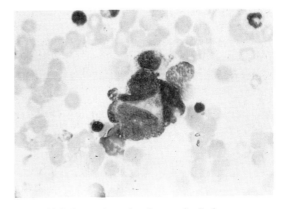

Figure 22.3 A very irregular cluster of cells from a carcinoma of the prostate. Bone marrow aspirate. Stain: May-Grünwald–Giemsa. Magnification × 500, reduced to 65% in reproduction

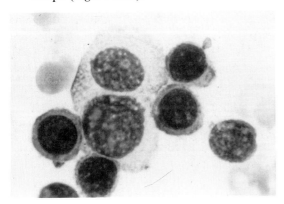

Figure 22.1 Carcinoma of the breast cells in bone marrow. The cells vary in size and in the amount of cytoplasm that they possess. Bone marrow aspirate. Stain: May-Grünwald–Giemsa. Magnification × 1200, reduced to 65% in reproduction

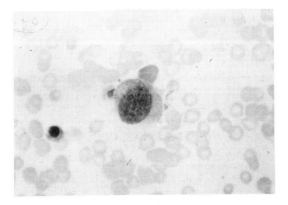

Figure 22.4 A single large malignant cell from an undifferentiated carcinoma of the bronchus. The nucleus is primitive and the cytoplasm has some pseudopodial projections. Bone marrow aspirate. Stain: May-Grünwald–Giemsa. Magnification × 500, reduced to 65% in reproduction

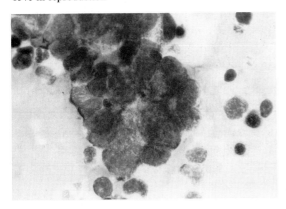

Figure 22.2 Cluster of malignant cells from a renal carcinoma. The cells are very densely packed. Bone marrow aspirate. Stain: May-Grünwald–Giemsa. Magnification × 500, reduced to 65% in reproduction

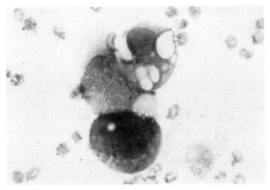

Figure 22.5 A cluster of cells from a carcinoma of the ovary. The upper cell is vacuolated and shows evidence of secretory activity. Bone marrow aspirate. Stain: May-Grünwald–Giemsa. Magnification × 1200, reduced to 65% in reproduction

(3) In spite of the variation in size, carcinoma cells tend to be large cells (*Figure 22.4*).
(4) The cytoplasm is usually scanty, sometimes vacuolated (*Figures 22.5* and *22.6*) and usually deeply basophilic, with a clearer area around the nucleus. Sometimes the cytoplasm contains azurophil granules.
(5) The nucleus is large and usually possesses an abnormal chromatin network.
(6) The nucleoli tend to be large, prominent and often multiple.
(7) Mitotic figures are usually prominent, but this is not a constant feature. In general, the more poorly differentiated the carcinoma the more numerous the mitoses. If mitoses are present in any number, atypical or bizarre mitotic figures are likely to be seen.
(8) Necrotic and degenerative changes are often found.
(9) Occasionally polyploid or multinucleate cells are seen.

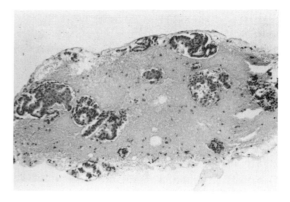

Figure 22.7 Section of clotted bone marrow aspirate to show adenocarcinoma of the stomach. Stain: haematoxylin and eosin. Magnification ×60, reduced to 65% in reproduction

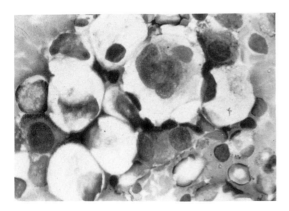

Figure 22.6 Carcinoma of the stomach. The cytoplasm of most of the malignant cells is distended by mucin production. One multinucleate cell is present. Bone marrow aspirate. Stain: May-Grünwald–Giemsa. Magnification ×1200, reduced to 65% in reproduction

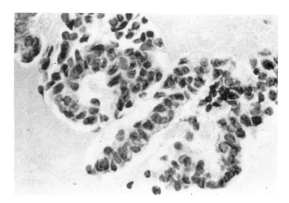

Figure 22.8 Higher power view of *Figure 22.7*. The cells have a tendency towards a tubular arrangement. Stain: haematoxylin and eosin. Magnification ×250, reduced to 65% in reproduction

These features can usually lead to a diagnosis of carcinoma in bone marrow aspirations. In sections of clots (*Figures 22.7* and *22.8*) or in bone marrow trephine biopsy sections, further information may be obtained which can lead to speculation as to the primary site of origin. Firstly, information can be gained on the arrangement of carcinoma cells: whether they tend to occur individually, in sheets, or in secretory tubules (*Figure 22.9*). Secondly, the extent of the fibrous tissue stroma can be assessed (the van Gieson and reticulin stains are of help). Thirdly, special staining techniques

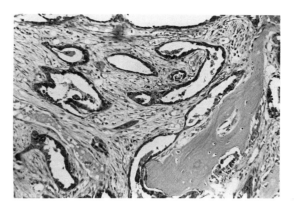

Figure 22.9 Trephine biopsy section of a carcinoma of the prostate. Tubule formation is marked with some secretion in the lumina, and there is a good deal of fibrous tissue stroma. Stain: haematoxylin and eosin. Magnification ×100, reduced to 65% in reproduction

may well be of value, particularly in adenocarcinomas where a stain for mucin or for secretory products may prove to be essential in giving a likely source of origin of the carcinoma.

In embryonal carcinomas such as teratoma, neuroblastoma, nephroblastoma and retinoblastoma, spread to the bone marrow is highly likely, particularly in advanced cases (*Figure 22.10*). The tumour cells are usually arranged in clumps. A common finding in neuroblastomas is the presence of 'rosettes' i.e. a group of malignant cells oriented towards the centre, which may be open and consist of a network of fibrillary material.

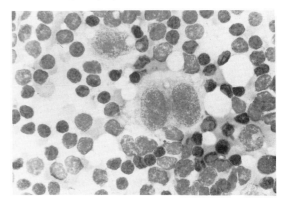

Figure 22.11 Hodgkin's lymphoma of mixed cell type. Sternberg-Reed cells are prominent, with very large nuclei and prominent nucleoli. Trephine biopsy. Stain: haematoxylin and eosin. Magnification × 250, reduced to 65% in reproduction

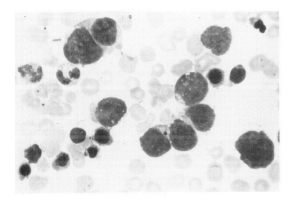

Figure 22.10 Neuroblastoma cells in bone marrow. The cells are arranged in small clusters and show some vacuolation, but there is no 'rosette' formation in this case. Bone marrow aspiration. Stain: May-Grünwald–Giemsa. Magnification × 500, reduced to 65% in reproduction

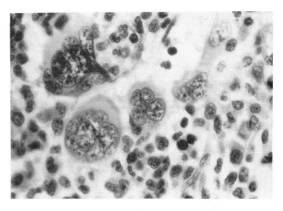

Figure 22.12 Multinucleate Sternberg-Reed cells in Hodgkin's lymphoma. Trephine biopsy. Stain: haematoxylin and eosin. Magnification × 250, reduced to 65% in reproduction

Sarcoma

The above remarks apply to sarcoma. Mitotic figures tend to be numerous. Trephine biopsy sections may give a clue to the type of sarcoma, e.g. in osteogenic sarcoma osteoid elements may be present. The sections may also show the close relationship of tumour cells to vascular channels, with no evidence of a connective tissue stroma.

Melanoma

Malignant melanoma frequently metastasizes to bone marrow. The cells, in general, are heavily pigmented but, for some unknown reason, deposits in bone marrow are often amelanotic and are difficult to differentiate from any other anaplastic tumour.

Hodgkin's lymphoma

In Hodgkin's lymphoma the characteristic cell is the Sternberg-Reed cell (*Figure 22.11*). This is found in almost all the bone marrows infiltrated by the disease. It is a large cell, sometimes multinucleated (*Figure 22.12*) with a large nucleus possessing a fine, but occasionally rather coarse, chromatin pattern. The nucleolus is typically large and prominent and in Romanowsky-stained preparations is a clear bright blue. The amount of cytoplasm is variable, but usually clear and basophilic. In nodular sclerosing Hodgkin's disease there is much connective tissue and any aspiration or trephine biopsy material containing fibrous tissue or clusters of fibroblasts should be carefully examined for the presence of Sternberg-Reed cells; the presence of young fibrous tissue is strong presumptive evidence for Hodgkin's disease (*Figure 22.13*). In contrast, in mixed cell Hodgkin's disease the infiltrate consists of histiocytes, plasma cells, lymphocytes, eosinophils and fibroblasts, in addition to a variable number of Sternberg-Reed cells.

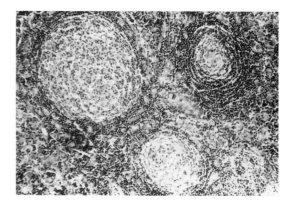

Figure 22.14 Lymph follicles in the bone marrow from a patient with a well differentiated nodular lymphoma. Trephine biopsy. Stain: haematoxylin and eosin. Magnification × 100, reduced to 65% in reproduction

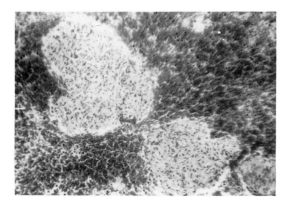

Figure 22.13 Young granulation tissue in a section of clotted bone marrow from a patient who subsequently was shown to have Hodgkin's lymphoma. Bone marrow aspirate. Stain: haematoxylin and eosin. Magnification × 100, reduced to 65% in reproduction

Non-Hodgkin's lymphoma

The classification of the non-Hodgkin's lymphomas is at present in a somewhat confused state, as we pass from a purely histological to a functional point of view. For the purposes of this section the histological classification of Rappaport (1966) will be used.

Lymphocytic lymphoma

Both well and poorly differentiated and both nodular or diffuse forms can be easily recognized in bone marrow aspirates (*Figure 22.14*). Poorly differentiated lymphoma cells (or lymphosarcoma cells) resemble lymphoblasts, but often possess vacuolated cytoplasm or cytoplasmic fragmentation. Cleaved nuclei and mitotic figures are common (*Figure 22.15*).

Histiocytic lymphomas

These usually show abundant and slightly basophilic cytoplasm. Nuclei are large and often in an eccentric position. Nuclear irregularity is often pronounced and nuclear indentation com-

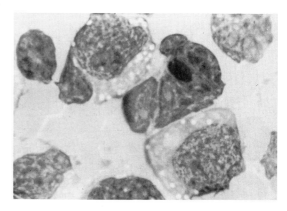

Figure 22.15 Malignant cells in a poorly differentiated diffuse lymphoma. The middle cell shows nuclear cleaving. The other cells show cytoplasmic vacuolation. Bone marrow aspiration. Stain: May-Grünwald–Giemsa. Magnification × 1200, reduced to 65% in reproduction

mon (*Figures 22.16* and *22.17*). There are one or two nucleoli, often sited towards the margin of the nucleus. Mitotic figures are usually prominent. There is usually a diffuse increase in reticulin fibrils (*Figure 22.18*) and multinucleate cells are frequently present, so leading to a possible confusion with Hodgkin's disease.

Mixed cell lymphomas

As their name suggests, these are a mixture of lymphoid cells and histiocytes. Undifferentiated lymphomas can be very hard to distinguish from any other undifferentiated tumours, but may contain a vague and disrupted network of reticulin fibrils.

Other lymphomas

Burkitt's lymphoma cells are often found in the bone marrow and in general resemble lymphoblasts. Some rare lymphomas, such as mycosis fungoides and Sezary's syndrome rarely spread to the bone marrow. In cases of 'hairy' cell leukaemia (leukaemic reticuloendotheliosis) the cells are almost always present in the bone marrow. The 'hairy' cells are similar to prosmonocytes or prolymphocytes, but may have cytoplasmic projections. They are frequently positive for tartrate-resistant acid phosphatase. Electron microscopy may be of considerable help. In histiocytic medullary reticulosis (or malignant histiocytosis), a very rare condition, the bone marrow is usually involved. The histiocytes in the infiltration are capable of phagocytosing large quantities of red blood cells and nuclear material.

Do not forget that the presence of more than one type of malignant tumour in the same patient occasionally occurs. Probably the most common is the presence of chronic lymphatic leukaemia or lymphoma in a bone marrow which is also infiltrated by oat cell carcinoma of the bronchus (*Figure 22.19*). One must be very wary of the protean manifestations of malignant disease.

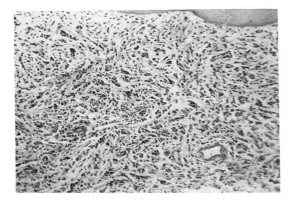

Figure 22.16 Bone marrow infiltration by a histiocytic lymphoma. The appearances are very pleomorphic. Trephine biopsy. Stain: haematoxylin and eosin. Magnification ×100, reduced to 65% in reproduction

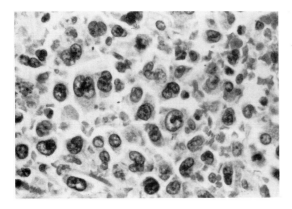

Figure 22.17 A higher power view of *Figure 22.16*. There is a considerable degree of nuclear irregularity and indentation. Trephine biopsy. Stain: haematoxylin and eosin. Magnification ×250, reduced to 65% in reproduction

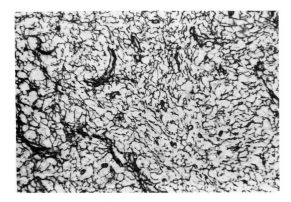

Figure 22.18 Dense and disordered reticulin pattern in a histiocytic lymphoma. Trephine biopsy. Silver stain for reticulin fibrils. Magnification ×100, reduced to 65% in reproduction

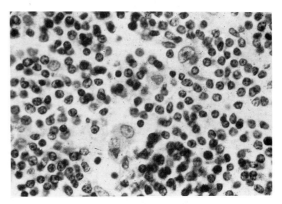

Figure 22.19 Scanty oat cell carcinoma of the bronchus cells and abundant small lymphocytes in a patient with chronic lymphocytic leukaemia and carcinoma of the bronchus. Trephine biopsy. Stain: haematoxylin and eosin. Magnification ×250, reduced to 65% in reproduction

Non-malignant cells in bone marrow

Cells in bone marrow films or sections which may be confused with tumour cells are:

(1) Blood cell precursors.
(2) Atypical white blood cells.
(3) Non-haemopoietic marrow elements.
(4) Cells foreign to the bone marrow.

Blood cell precursors

Myeloblasts, lymphoblasts and early normoblasts, which are present in normal bone marrow, can usually be identified as non-malignant with relative ease. Reticulum cells may possess some of the attributes of malignant cells but are almost always discrete, with little variation in size and appearance (*Figure 22.20*). Megakaryocytes and their precursors present much more of a problem. The precursors are large cells with nuclear appearances sometimes closely similar to malignant cells (*Figure 22.21*). Their most pronounced points of difference from malignant cells are:

(1) Discrete cells, with little tendency to cluster.
(2) A tendency to nuclear cleaving. In any multinucleate cell the nuclei are generally uniform in size and appearance.
(3) Nucleoli are generally less prominent than in tumour cells.
(4) The cytoplasm stains an even basophilic blue.
(5) If there is any azurophilic cytoplasmic granulation, this is usually separated by areas of oxyphilic staining (at times, this is difficult to distinguish from the azurophil cytoplasmic granulation sometimes seen in carcinomas).
(6) Evidence of platelets or platelet budding.

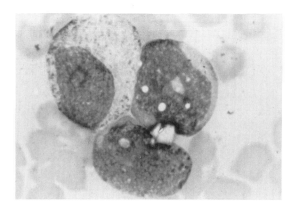

Figure 22.21 Two megakaryocyte precursors (at right) showing a tendency to nuclear cleaving and some azurophil cytoplasmic granulation. The cell at upper left is a promyelocyte. Bone marrow aspiration. Stain: May-Grünwald–Giemsa. Magnification × 1200, reduced to 65% in reproduction

Atypical white blood cells

Atypical mononuclear cells, Türk cells or immunoblasts tend to be increased in patients with malignant disease. Although they can be large cells, their nuclear structure is unlike that of a tumour cell. The nuclear chromatin pattern is very dense and the nucleoli are usually not prominent. The cytoplasm tends to be abundant and very deeply basophilic, often staining more deeply towards the periphery of the cell.

Non-haemopoietic marrow elements

Capillaries are frequently seen in trephine biopsies and endothelial cells can often be present in marrow aspirates. The cells can be recognized by their regular appearance in size and shape, by their elongated appearance and by the lack of most of the characteristics of tumour cells (*Figure 22.22*).

Osteoblasts may resemble malignant cells. They are large cells and tend to be present in small clusters, but usually have abundant clearly-defined basophilic cytoplasm and their nuclei are eccentrically placed like those of plasma cells (*Figure 22.23*). The nucleoli tend to be small, with two to four to a nucleus.

Osteoclasts are multinucleate 'giant' cells. The nuclei are fairly uniform in size with, usually, one rather indistinct nucleolus (*Figure 22.24*). The cytoplasm is blue, abundant and may contain azurophilic granules of varying size.

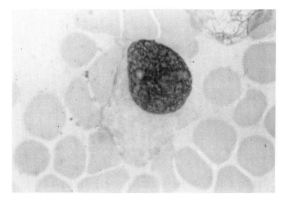

Figure 22.20 Reticulum cell. The nucleus is large but the cell is individual and the cytoplasm is abundant. Bone marrow aspiration. Stain: May-Grünwald–Giemsa. Magnification × 1200, reduced to 65% in reproduction

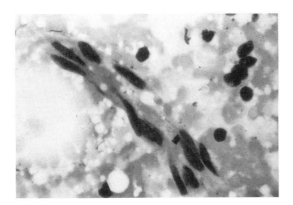

Figure 22.22 Endothelial cells. Their elongated appearance is characteristic. Bone marrow aspiration. Stain: May-Grünwald–Giemsa. Magnification × 500, reduced to 65% in reproduction

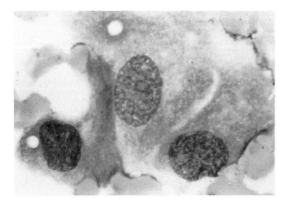

Figure 22.23 Osteoblasts. The cells have abundant cytoplasm and their nuclei tend to be eccentrically placed. Bone marrow aspiration. Stain: May-Grünwald–Giemsa. Magnification × 1200, reduced to 65% in reproduction

Cells foreign to the bone marrow

Epithelial cells can easily be introduced to the marrow specimen by skin puncture, as can fat, muscle and cartilage cells. These are all easily recognizable. Other cellular elements may be introduced as artefacts, e.g. *Paramecium* contaminating distilled water used for staining. These have only a superficial resemblance to tumour cells.

Plasma cells in malignant disease

Plasma cells tend to be increased in the bone marrow in cases of malignant disease. For example, in oat cell carcinoma of the bronchus their incidence is usually 5–10% of the total nucleated count. Moreover, an increase in plasma cells should encourage one to search the bone marrow films very carefully for evidence of malignant cells.

Plasma cells can normally be recognized quite readily. They are smaller than malignant cells and have abundant deep blue cytoplasm with eccentric nuclei and often a 'clock face' nuclear chromatin pattern. An Unna–Pappenheim stain may be of help in detecting them and determining their numbers. However, in cases of myelomatosis the plasma and myeloma cells may be so atypical that there is a risk of confusing them with carcinoma cells (*Figure 22.25*); all information must be taken into account including, for example, the presence or absence of paraproteins and a monoclonal immunoglobulin band. Correct identification may well depend on the trephine biopsy.

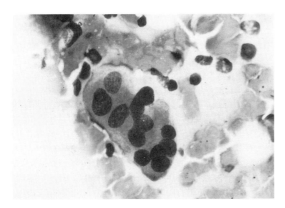

Figure 22.24 Osteoclast. A multinucleate 'giant' cell with fairly uniform nuclei. Bone marrow aspiration. Stain: May-Grünwald–Giemsa. Magnification × 500, reduced to 65% in reproduction

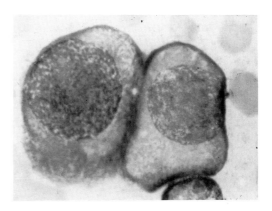

Figure 22.25 Anaplastic myeloma cells. At times these can be difficult to differentiate from carcinoma cells. Bone marrow aspirate. Stain: May-Grünwald–Giemsa. Magnification × 1200, reduced to 65% in reproduction

Significance of tumour cells in bone marrow

If tumour cells are found in bone marrow it is a sign of generalized disease and carries a worse prognosis than if tumour cells are not found. Moreover, tumour cells in the bone marrow are more likely to be found with a poorly differentiated tumour which carries a worse prognosis.

More important from a practical point of view is that if bone marrow metastases are found, very careful consideration should be given to the possibility of cytotoxic chemotherapy, since there is little hope of local measures such as surgery and radiotherapy controlling the disease. Occasionally, a bone marrow biopsy may provide the first evidence of malignancy; usually in these cases the bone marrow biopsy cannot supply a full histological diagnosis and every effort must be made to ascertain and biopsy the primary site of origin. It is possible that, in the future, antimetastatic drugs may be developed which will prevent blood-borne metastasis from a malignant tumour. Encouraging results have already been obtained in animals (for a review, see Hellmann, 1984).

The role of electron microscopy

On the whole, electron microscopy has only a limited place in the routine examination of marrow for tumour cells. Occasionally, electron microscopy may yield useful information on doubtful cells, both malignant and non-malignant, and special stains may be of help.

Carcinoma cells may be identified by evidence of secretory activity. Cytoplasmic protrusions in 'hairy' cell leukaemia can often be visualized better by electron microscopy than by light microscopy. Malignant histiocytic cells may be very difficult to identify and here the electron microscope may be of assistance. Such cells tend to contain ingested material which may actually deform the nucleus. Nuclear chromatin may show some peripheral condensation along the nuclear membrane. The Golgi apparatus is well developed and the cytoplasm contains numerous strands of endoplasmic reticulin. The number and size of ribosomes and mitochondria are small.

As regards non-malignant cells, megakaryocyte precursors are large cells with indented or lobed nuclei and an even distribution of chromatin. The cytoplasm adjacent to the nucleus is rich in free ribosomes, short channels of endoplasmic reticulin, mitochondria, vacuoles, vesicles and specific granules. The peripheral cytoplasm has a few ribosomes but no cellular organelles. As the megakaryocyte matures, the vesicles and tubules merge to provide large communicating systems of cisternae in the perinuclear zone and eventually form cell membranes for the developing platelets.

Plasma cells have a characteristic appearance. The cytoplasm is largely occupied by sacs of endoplasmic reticulin arranged in concentric and parallel patterns and lined by ribosomes. The interiors of the sacs are frequently distended by immunoglobulins. The Golgi apparatus is well developed and mitochondria are large. The nucleus may contain, apart from nuclei, 'nuclear bodies' i.e small round bodies of dense granules surrounded by a clear zone.

New developments

In recent years the emergence and considerable proliferation of monoclonal antibodies as a means to diagnose, classify and indeed, at times, to treat malignant disease has progressed rapidly. Their principal use so far has been in the classification of the lymphomas, where several functional classifications have been proposed (Holgate et al., 1983). Monoclonal antibodies have been prepared that are specific for differentiation antigens on normal and malignant myeloid cells and have now been used to treat myeloblastic leukaemia (Ball et al., 1983).

The Ca1 antibody has also been used in the identification of malignant cells. The Ca antigen was originally extracted from a segregant malignant hybridoma produced by fusion of diploid fibroblasts and malignant cells from a human cervical carcinoma (Ashall, Bramwell and Harris, 1982). The monoclonal antibody to this antigen was reputed to identify malignant cells, but it has now become clear that the Ca antigen is not exclusive to malignancy (Paradinas, Boxer and Bagshawe, 1984) and is probably a secretory product. There seems little doubt that the use of monoclonal antibodies to detect tumour cells will be a fruitful field in the future.

A somewhat different, and very promising, approach to the problem of tumour cells in bone marrow has recently been described (Dearnaley et al., 1983; Redding et al., 1983). Marrow was taken from multiple sites at primary surgery for carcinoma of the breast. After concentration, marrow cells were immunocytochemically stained for epithelial membrane antigen (EMA). This technique uses rabbit anti-EMA raised against human milk fat globule membranes and subsequently a sheep anti-rabbit alkaline phosphatase conjugate. The development of a colour reaction with a Brentamine fast-red substrate demonstrated tumour cells. Their numbers ranged from over 500 to *only one* per specimen, none being detected in conventionally stained smears. Tumour cells were found in 28% of marrows at primary surgery and in 57% of marrows at first relapse. The significance of so sensitive a technique is that it not only gives some idea of prognosis and of the patients who might benefit from high dose chemotherapy and marrow rescue, but that it offers the hope that it might lead to a method of removing tumour cells from the bone marrow. If sufficient bone marrow can be taken and any tumour cells removed, the patient could receive potentially lethal radiotherapy and cytotoxic chemotherapy and subsequently receive a graft of his own bone marrow free from tumour cells. A recent paper (Treleaven *et al.*, 1984) describes the use of monoclonal antibodies bound to polystyrene microspheres containing magnetite to achieve such a purpose.

Tumour cells in the blood

It should be emphasized that tumour cells are most unlikely to be found in an ordinary blood film. On the whole, certain special circumstances such as blood taken from a local vein draining a carcinoma during nailing of a pathological fracture are necessary before there is any hope of finding circulating tumour cells.

Detection and identification of circulating tumour cells

To increase the chances of finding circulating tumour cells the nucleated cells in the blood must be concentrated. This can be achieved in several ways:

(1) Centrifugation.
(2) Accelerated sedimentation using agents like fibrinogen or dextran to promote rouleaux formation.
(3) Lysis of red blood cells using a substance such as streptolysin O and subsequent centrifugation.

All these methods result in a suspension containing many white blood cells, in addition to any malignant cells. To remove the majority of white blood cells several methods have been used, including centrifugation over albumin, filtration through a fine filter, incubation with a fine suspension of iron and subsequent removal of phagocytic cells with a strong magnet (for a review of methods and results *see* Salsbury, 1975). More recent methods have usually involved density separation employing such agents as 'Ficoll' and sodium metrizoate.

Whatever the method of concentration, the preparation is probably best stained with a Romanowsky stain as this is more likely to result in correct identification of any non-malignant cells. Alternatively, a Papanicolaou stain may be used by those experienced with the method.

The problems of detection and identification of tumour cells in the blood are the same as in the bone marrow. Similar criteria apply. Similar nonmalignant cells may be encountered (*Figures 22.26* and *22.27*), with the addition of mesothelial cells (*Figure 22.28*) if the sampling needle traverses the pleural, peritoneal or pericardial cavities, trophoblast cells in the blood of pregnant women and tissue cells from various organs, e.g. liver cells in blood from an inferior mesenteric cannula which has been misdirected (*Figure 22.29*). The fact that the specimen has been concentrated means that the chances of

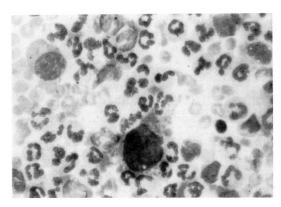

Figure 22.26 Megakaryocyte precursor (towards bottom) and Türk cell (upper left) in a blood concentrate. Stain: May-Grünwald–Giemsa. Magnification × 500, reduced to 65% in reproduction

finding atypical non-malignant cells are far higher than in normal blood film.

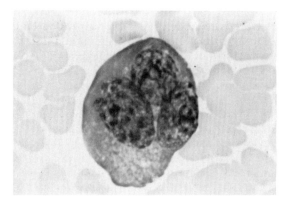

Figure 22.27 Unidentified cell (although probably an osteoclast) in a blood concentrate. Stain: May-Grünwald–Giemsa. Magnification × 1200, reduced to 65% in reproduction

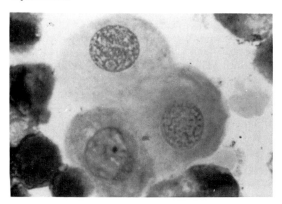

Figure 22.28 Mesothelial cells in a blood concentrate. Stain: May-Grünwald–Giemsa. Magnification × 1200, reduced to 65% in reproduction

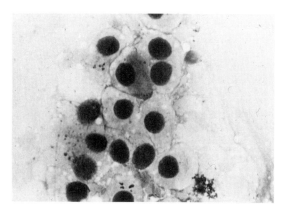

Figure 22.29 Hepatic cells in a blood concentrate. Stain: May-Grünwald–Giemsa. Magnification × 500, reduced to 65% in reproduction

Incidence of tumour cells in the blood

The results of examination of blood for malignant cells have been conflicting. Between 1955 and 1962 various reports on the incidence of tumour cells in peripheral blood ranged from 1% to 57%. In subsequent years, the incidence reported rose to 80% in serial samples and 90% in single blood samples of 100 ml. This wide variation is probably due to a number of factors:

(1) The method of concentration. All methods have their drawbacks but methods involving lysis are particulary prone to partial disintegration of cells and their subsequent misidentification.
(2) The correct identification of cells as malignant.
(3) The correct identification of non-malignant cells. As mentioned above, a Romanowsky stain is probably the best single stain; multiple staining methods may produce even better results.
(4) The number and size of blood samples. As might be expected the incidence of circulating tumour cells rises with the number of blood samples and with their volume.
(5) The site from which the blood was taken. The incidence and numbers of tumour cells increase when blood is taken from a vein draining the tumour, rather than a peripheral vein.
(6) The type of malignant tumour. Reports of the incidence of circulating tumour cells tend to be higher in sarcoma where malignant cells are generally closely approximated to vascular channels, than in carcinoma where there is often a fibrous tissue stroma separating malignant cells and blood vessels. On the whole, the less differentiated a carcinoma the more likely it is that circulating tumour cells will be found. The incidence of circulating tumour cells and the occurrence of blood-borne metastases is higher in patients with demonstrable tumour invasion of veins than in those without venous invasion.
(7) External factors may also affect the incidence and numbers of circulating tumour cells. The likelihood of finding cancer cells in the blood during nailing of a pathological fracture has already been noted. There is good evidence that any interference with a malignant tumour, including manipulation,

biopsy and operative removal, is likely to increase the incidence of circulating tumour cells.

Significance of tumour cells in the blood

If tumour cells are found in the blood, what is their prognostic significance? In the state of present knowledge, we simply do not know. Some well-documented reports suggest that the presence of circulating malignant cells carries a somewhat worse prognosis than their absence. Moreover, numerous animal experiments and a few experiments in humans suggest that some, at least, of circulating malignant cells are viable.

In contrast, many animal experiments have shown that metastases are reduced if anticoagulants or fibrinolytic agents are given, and this is borne out in humans to some extent. It may be that any tumour cells detected in the blood are linked to some form of fibrinolytic process and that they cannot implant and form metastases. Investigations into this question are currently in progress.

Rather paradoxically, it may be that the tumour cells which one does *not* find in the blood are the most dangerous, i.e. those which have already implanted and have the capacity to form a metastasis. One hopes that future research may provide the answer.

References

ALEXANDER, P. (1984). The biology of metastases. *Cancer Topics*, **4**, 116–117

ASHALL, F., BRAMWELL, M. E. and HARRIS, H. (1982). A new marker for human cancer cells. 1. The Ca antigen and the Ca1 antibody. *Lancet*, **2**, 1–6

BALL, E. D., BERNIER, G. M., CORNWELL, G. G., MCINTYRE, O. R., O'DONNELL, J. F. and FANGER, M. W. (1983). Monoclonal antibodies to myeloid differentiation antigens: *in vivo* studies of three patients with acute myelogenous leukaemia. *Blood*, **62**, 1203–1210

BRUNNING, R. D., BLOOMFIELD, C. D., MCKENNA, R. W. and PETERSON, L. (1975). Bilateral trephine bone marrow biopsies in lymphoma and other neoplastic diseases. *Ann. Intern. Med.*, **82**, 365–366

DEARNALEY, D. P., SLOANE, J. P., IMRIE, S., COOMBES, R. C., ORMEROD, M. G., LUMLEY, H., JONES, M. and NEVILLE, A. M. (1983). Detection of isolated mammary carcinoma cells in marrow of patients with primary breast cancer. *J. Roy. Soc. Med.*, **76**, 359–364

FRANKLIN, I. M. and PRITCHARD, J. (1983). Detection of bone marrow invasion by neuroblastoma is improved by sampling at two sites with both aspirates and trephine biopsies. *J. Clin. Pathol.*, **36**, 1215–1218

HELLMANN, K. (1984). Antimetastatic drugs: laboratory to clinic. *Clin. Exp. Metastasis*, **2**, 1–4

HOLGATE, C. S., JACKSON, P., LAUDER, I., COWEN, P. N. and BIRD, C. C. (1983). Surface membrane staining of immunoglobulins in paraffin sections of non-Hodgkin's lymphomas using immunogold–silver staining technique. *J. Clin. Pathol.*, **36**, 742–746

MILAS, L., PETERS, L. J. and ITO, H. (1983). Spontaneous metastasis: random or selective? *Clin. Exp. Metastasis*, **1**, 309–315

PARADINAS, F. J., BOXER, G. and BAGSHAWE, K. D. (1984) Distribution of the Ca (Oxford) antigen in lung neoplasms and non-neoplastic lung tissues. *J. Clin. Pathol.*, **37**, 1–5

RAPPAPORT H. (1966). *Tumours of the Hematopoietic System*, Section III, Fascicle 8. Washington DC: Armed Forces Institute of Pathology

REDDING, W. H., COOMBES, R. C., MONAGHAN, P., CLINK, H. MCD., IMRIE, S., DEARNALEY, D. P., ORMEROD, M. G., SLOANE, J. P., GAZET, J.-C., POWLES, T. J. and NEVILLE, A. M. (1983). Detection of micrometastases in patients with primary breast cancer. *Lancet*, **2**, 1271–1274

SALSBURY, A. J. (1975). The significance of the circulating cancer cell. *Cancer Treatment Rev.*, **2**, 77–96

TARIN, D., PRICE, J. E., KETTLEWELL, M. G. W., SOUTER, R. G., VASS, A. C. R. and CROSSLEY, B. (1984). Clinicopathological observations on metastasis in man studied in patients treated with peritoneovenous shunts. *Br. Med. J.*, **288**, 749–751

TRELEAVEN, J. G., UGELSTAD, J., PHILIP, T., GIBSON, F. M., REMBAUM, A., CAINE, G. D. and KEMSHEAD, J. T. (1984). Removal of neuroblastoma cells from bone marrow with monoclonal antibodies conjugated to magnetic microspheres. *Lancet*, **1**, 70–73

23

Parasitic infections

George W. Wikeley

In western communities enjoying a high standard of living endemic parasitic infections are relatively rare, but with the increasing use of air travel former residents of areas with endemic parasitic diseases are found in almost all communities while tourists may be exposed to disease with which they have had no previous contact. Today, when almost 90% of travellers take to the air, people may be infected by a parasite and return to base within a few hours where they may remain undiagnosed and untreated for many years. The aeroplane must be added to the list of mechanical vectors, i.e. a vector which is not essential to the life cycle of a parasite, involved in the transmission of parasitic infection.

It is also evident that the widespread use of cytotoxic therapy which impairs the host immune response results in the reactivation of latent parasitic infections, nor should it be overlooked that the impaired immunity associated with AIDS will have the same effect.

Often the first intimation of parasitic disease comes as a result of cytological investigation and cytologists should be aware that such disease may be asymptomatic or that the symptoms may frequently be misinterpreted. Cytological evidence is mostly found in Papanicolaou-stained smears and cell samples from the genital, urinary and gastrointestinal tracts as well as from serous fluids and fine needle aspiration biopsy.

Classification of human parasites of cytological interest

As a general rule the parasites with which the cytologist can be concerned can be divided into two groups: Protozoa and Helminths (*Tables 23.1* and *23.2*).

Protozoa are unicellular (acellular) organisms in which all the functions of metabolism and reproduction are performed within the individual cell. They are classified according to their motile system and include Mastigophora which have flagella, Rhizopoda which have pseudopodia, Ciliophora which are ciliated and Sporozoa which form spores.

Helminths (worms) are wormlike animals parasitic to man and the group is divided into Platyhelminths or flatworms and Nematodes which are roundworms. The Platyhelminths are divided into two classes: Trematoda (flukes) and Cestoda (tapeworms). Nematodes, the roundworms, include parasites of the Filaria group.

Many of these are cosmopolitan whilst others have, in the past, been confined to distinct geographical areas.

Specimen collection and preparation

Parasites more often than not present as unexpected findings in cytological material.

Table 23.1 Classification of human protozoan parasites of cytological interest

Mastigophora	Rhizopoda	Ciliophora	Sporozoa
Giardia lamblia	Entamoeba coli	Balantidium coli	Toxoplasma gondii
Trichomonas vaginalis	Entamoeba histolytica		Plasmodium falciparum
Trichomonas hominis	Entamoeba gingivalis		Plasmodium vivax
Leishmania sp.			Plasmodium ovale
Trypanosoma sp.			Plasmodium malariae

Table 23.2 Classification of human helminth parasites of cytological interest

Platyhelminths (flatworms)		Nematodes (roundworms)
Trematoda (flukes)	Cestoda (tapeworms)	
Schistosoma haematobium	Taenia saginata	Enterobius (Oxyuris) vermicularis (pinworm)
Schistosoma mansoni	Taenia solium	Trichuris trichiura (whipworm)
Schistosoma mattheei	Echinococcus granulosus (T. echinococcus)	Ascaris lumbricoides
Schistosoma intercalatum	Hymenolepis nana	Strongyloides stercoralis
Schistosoma japonicum	Dipilidium caninum	Ancylostoma duodenale (hookworm)
Paragonimus westermani (lung fluke)	Diphyllobothrium latum	Nector americanus
Paragonimus kellikotti		Ancylostoma braziliensis (sandworm)
Clonorcis sinensis (liver fluke)		
Opisthorcis felineus		
Fasciola hepatica (liver fluke)		

Standard Papanicolaou and May-Grünwald–Giemsa techniques show parasites adequately. Specialized techniques to confirm the identity of the parasite may occasionally be required. These may be found in any standard work, or by consultation with a specialist parasitology laboratory.

Protozoa

Mastigophora (flagellates)

Flagellates may be found in cell samples from the urogenital and gastrointestinal tracts, in peripheral blood, tissues and body fluids. Two specimens are often incriminated in inflammatory lesions of the genitourinary and gastrointestinal tracts, *Trichomonas* sp. and *Giardia lamblia* (syn. *Giardia intestinalis*), both of which are found worldwide. *Leishmania* sp. and *Trypanosoma* sp. are found in blood and tissues.

Genitourinary and gastrointestinal tract flagellates (T. vaginalis and G. lamblia)

Trichomonas will already be familiar to the cytologist as a pyriform (pear-shaped) organism, rounded anteriorly and pointed at the posterior, bearing a number of anterior flagella, usually three to five. A further flagellum may be attached to an undulating membrane spiralling down the side of the body. Two main species may be recognized: *Trichomonas vaginalis*, the largest, inhabits the female genital tract whilst *Trichomonas hominis* normally inhabits the caecum and is usually regarded as non-pathogenic. A further species *Trichomonas tenax* may be found in the mouth where oral hygiene is poor. It is also non-pathogenic. In vaginal smears *T. vaginalis* may vary in size and shape, ranging from 10–30 μm. The identifying features are the nucleus, often referred to as the 'mongol eye', and the axostyle which is a dark rod running down the centre of the body and protruding slightly through the posterior tip. *T. vaginalis*

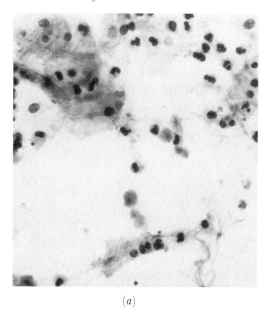 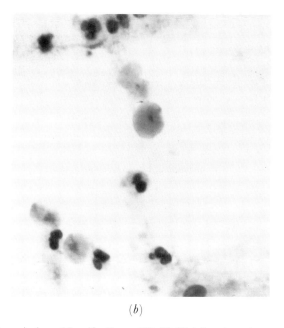

Figure 23.1 (*a*) *Trichomonas vaginalis*, cervical smear. Stain: Papanicolaou. Magnification × 500, (*b*) Slightly enlarged area of (*a*) to show pleomorphism

may also inhabit the male urogenital tract and in refractory cases both the patient and her partner should be investigated and, if necessary, treated (*Figures 23.1(a)* and *(b)*).

Giardia lamblia inhabits the duodenum, jejunum, bile duct and gallbladder and may be encountered in aspirates where neoplasia is suspected. The trophozoite is flat and pear-shaped with two nuclei, each containing a karyosome, in the widest portion (*Figure 23.2*). A darkly staining bar is seen across the midline and four pairs of flagella can be demonstrated. It ranges from 9–16 µm long and 9–12 µm wide.

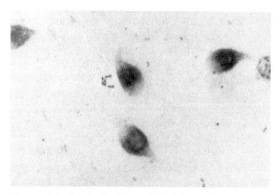

Figure 23.2 *Giardia lamblia*, gall bladder aspirate. Stain: Papanicolaou. Magnification × 500

Blood and tissue flagellates (Leishmania *and* Trypanosoma)

These flagellate protozoa depend upon an intermediate host in their life cycle, usually a blood sucking insect. The parasites develop in the blood sucking invertebrate from which the infection is transferred to the vertebrate host. *Leishmania tropica*, *L. braziliensis* and *L. donovani* parasitize man as do *Trypanosoma rhodesiense*, *T. gambiense*, *T. cruzi* and *T. rangeli*.

Leishmaniasis (*Figure 23.3*)

Leishmaniasis has two clinical manifestations. Some varieties may be confined to lesions of the skin (*L. tropica* and *L. braziliensis*) whilst others, including *L. donovani*, infect the reticuloendothelial system. There is some geographical separation of the two diseases although the countries around the Mediterranean support both varieties. Typically the organism appears as an ovoid body, Leishman Donovan (L.D.) bodies measuring 2–5 µm × 1–3 µm. External flagella can only be demonstrated in the insect vector or in tissue culture. L.D. bodies are found most frequently in the cells of the reticuloendothelial system.

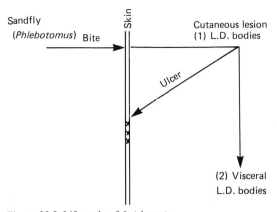

Figure 23.3 Life cycle of *Leishmania*

Trypanosomiasis (*Figure 23.4*)

The intermediate host in this case is a biting fly, the tsetse fly (*Glossina*). The organisms are elongated (12–30 μm long) and slender (1.5–3 μm wide). There is a single nucleus which contains a karyosome. A single flagellum arises at the posterior end and runs along the periphery attached to an 'undulating membrane'. Movement is forward with the flagellum leading. Trypanosomes may be found in the blood and tissues of the majority of vertebrates across a wide belt of tropical Africa and America. The two main diseases of man are sleeping sickness (African) and Chagas' disease (South American). *T. gambiense* and *T. rhodesiense* are associated with African sleeping sickness and *T. cruzi* with South American Chagas' disease. *T. gambiense* and *T. rhodesiense* live in the blood and lymph glands where, in the early stages of the infection, extensive multiplication occurs. This can result in lymph gland enlargement. Fine needle aspiration of the gland may reveal the organism unexpectedly if a full history of the patient is not available.

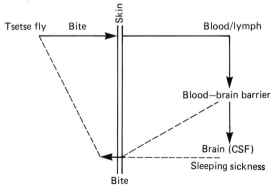

Figure 23.4 Life cycle of *Trypanosoma*

Three forms of the organism may be recognized. A slender form at the initial infection progressing through an intermediate stage to a shorter stumpy form. This pleomorphism is a distinctive feature and should not be confused with infection by more than one type of organism.

Rhizopoda (amoebae)

Amoebae may be parasitic or commensal in man. Infection occurs worldwide and in endemic areas 30–80% of the population may harbour the parasites. Many people who harbour these parasites remain asymptomatic, in a state of equilibrium, until some other disease intervenes.

Rhizopoda move by means of pseudopodia and reproduce by mitotic division. The mature trophozoite averages about 25 μm in size and is about the size of a parabasal cell, although smaller and larger forms are not uncommon. The cytoplasm is clear and finely granular. The pathogenic forms of the organism ingest red blood cells which are an aid to correct identification. The nucleus is spherical with a central karyosome. A cystic form may be encountered in intestinal amoebiasis. Three members of the genus *Entamoeba* are the concern of the cytologist: *E. histolytica*, *E. coli* and *E. gingivalis*.

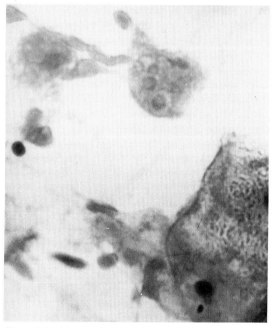

Figure 23.5 *Entamoeba histolytica*, vaginal vault. Stain: Papanicolaou. Magnification × 500

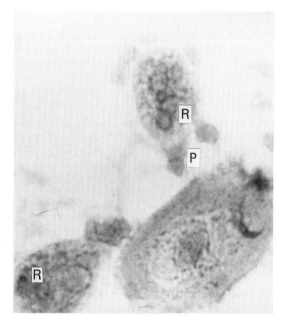

Figure 23.6 *Entamoeba histolytica*, cervical smear. Note projecting pseudopodia (P) and ingested red cell (R). Stain: Papanicolaou. Magnification × 500

E. histolytica (*Figures 23.5* and *23.6*) has been found in liver aspirates, sputum, pleural fluid, genital tract and in cell samples from many other sites. The primary site of infection is the gastrointestinal tract but extra-intestinal amoebiasis has been demonstrated in practically all organs of the body, and it is these manifestations of the disease which may be encountered in cytological practice. Infection is acquired by ingestion of faecally contaminated fluid or water which contains the cystic form of the parasite. All patients with extra-intestinal amoebiasis must, at some time or other, have suffered from a bowel infection, subsequent spread being blood-borne to the liver and from there the organisms may be disseminated throughout the body. Liver aspiration in cases of suspected neoplasia may reveal an amoebic abscess. Pinkish 'anchovy paste' pus is withdrawn on aspiration. An immediate wet preparation should reveal motile trophozoites but frequently smears are made and sent to cytology for investigation. The finding of amoebae in sputum is not uncommon, although great care must be taken in distinguishing it from *E. gingivalis* (*Figure 23.7*) which is normally a mouth commensal associated with poor oral hygiene. It can be distinguished from it by its smaller size and the presence of several small pseudopodia

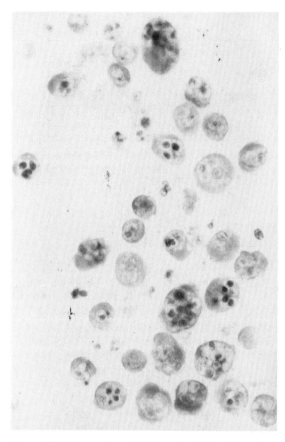

Figure 23.7 *Entamoeba gingivalis*, saliva. Stain: Papanicolaou. Magnification × 500

Table 23.3 Characteristics of amoebae in stained preparations

	E. histolytica	E. coli	E. gingivalis
Size (μm)	10–40 (average 25)	15–30 (sometimes elongated to 40–50)	10–25
Pseudopodia	Rounded tongue-like	Short and blunt	Up to 5 finger-like projections
Inclusions	Red blood cells, bacteria and cell debris	Bacteria and debris	Green debris
Staining (Papanicolaou)	Amphophilic	Amphophilic	Amphophilic or basophilic
Nuclei	Small central karyosome	Coarse chromatin, large karyosome	Central karyosome

Shape and size are variables dependent upon the motility of the trophozoite at time of fixation.

(*Figure 23.8*). Being a commensal rather than a pathogen it never contains ingested red blood cells (*Table 23.3*).

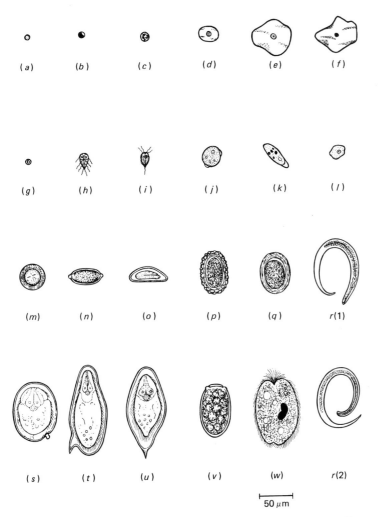

Figure 23.8 Comparative sizes of some common parasites. *a*, erythrocyte; *b*, lymphocyte; *c*, polymorph; *d*, parabasal cell; *e*, intermediate squamous cell; *f*, superficial squamous cell; *g*, *Toxoplasma*; *h*, *Giardia lamblia*; *i*, *Trichomonas vaginalis*; *j*, *Entamoeba coli*; *k*, *Entamoeba histolytica*; *l*, *Entamoeba gingivalis*; *m*, *Taenia* species; *n*, *Trichuris trichiuris*; *o*, *Enterobius vermicularis*; *p*, *Ascaris lumbricoides*; *q*, *Ascaris lumbricoides* decorticated ovum; *r* (1), *Strongyloides stercoralis* rhabditiform larva; *r* (2), *Strongyloides stercoralis* filariform larva; *s*, *Schistosoma japonicum*; *t*, *Schistosoma mansoni*; *u*, *Schistosoma haematobium*; *v*, *Paragonimus*; *w*, *Balantidium coli*

Amoebic infection may also co-exist with carcinoma. An elderly lady who had been treated for bronchogenic carcinoma with pleural instillations of a cytostatic drug subsequently developed a pleural effusion. Thoracentesis yielded purulent fluid. Papanicolaou-stained smears showed both trophozoites of *E. histolytica* and malignant cells. A middle-aged patient with biliary peritonitis yielded an aspirate containing vegetative amoebae as did cell samples from an infected hydrocoele in an elderly man.

The finding of amoebae in genital tract smears is not uncommon arising from poor hygiene or from a rectovaginal fistula following surgery and irradiation for genital carcinoma. Amoebae resembling *E. gingivalis* have been described in a cervical smear from a patient fitted with an intrauterine contraceptive device. 'Free living' amoebae, contaminants from water and soil, may also be found.

Ciliophora (ciliates)

Balantidium coli (*Figure 23.9*) is the largest protozoan found in man. It measures 50–200 μm in length and 40–70 μm in breadth. It was thought

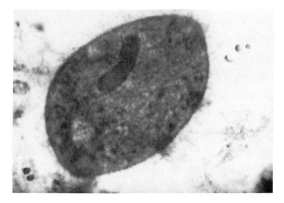

Figure 23.9 *Balantidium coli* trophozoite form, vaginal smear. Note kidney-shaped macronucleus and small micronucleus within the curvature. Stain: Papanicolaou. Magnification × 500, reduced to 65% in reproduction

to be exclusively parasitic to the digestive tract but it has been incriminated in fatal peritonitis and in chronic vaginitis. It has also been found as a contaminant of sputum.

In appearance *Balantidium coli* is a large, round to ovoid parasite. The cytoplasm contains food vacuoles and two prominent nuclei, a bean-shaped macronucleus and a small round micronucleus which can be seen in the inner curvature of the macronucleus. The organism is ciliated and the cilia beat with a ripple effect along the surface and the beat may be reversed. At the posterior end is a small opening through which the cytoplasmic vacuoles are emptied. Towards the more pointed anterior end is a curved depression, the gullet. As encystment takes place the organism begins to round off forming, without retraction of the cilia, a tough cyst wall (*Figure 23.10*).

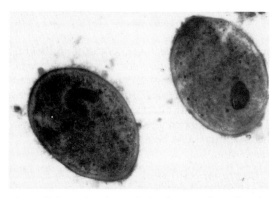

Figure 23.10 *Balantidium coli* showing rounding off to the cystic form. Stain: Papanicolaou. Magnification × 500, reduced to 65% in reproduction

A 45-year-old woman living in a remote rural community presented at the nearest hospital complaining of bloodstained discharge and postcoital bleeding. A cervical scrape smear, stained by the standard Papanicolaou technique, showed numerous trophozoites and a number of smaller precystic forms of *Balantidium coli*. The background picture was inflammatory and showed polymorphonuclear leucocytes, eosinophils and histiocytes. The epithelial cells showed inflammatory changes.

A male patient was being investigated for pulmonary tuberculosis and *Balantidium coli* was detected in smears from his sputum. Infection was accidentally acquired by the ingestion of contaminated vegetables.

Sporozoa

The members of this genus are obligate intracellular parasites of body fluid and tissues and include toxoplasmas and malarial parasites found in vertebrates and invertebrates.

Toxoplasma gondii

Toxoplasmosis presents as a chronic infection in many organs of the body such as liver, spleen, uterus and brain. The identification of both trophozoite and cystic forms of *Toxoplasma gondii* in cytological material requires great care. Reservoir hosts are domestic animals and the cat is implicated in the sexual cycle. Trophozoites are ovoid structures 5 μm × 2.5 μm with a polar nucleus. They are often found as intracellular organisms in monocytes and histiocytes presenting as thin walled vacuoles called pseudocysts in the cytoplasm of the cell. Workers in Spain (Dominguez and Girou, 1976; San Cristobal, Roset and Bloy, 1976) reported the presence of such cysts in vaginal smears. Subsequent correspondence (Naylor, 1977; Frenkel, 1977) has indicated the need for care and confirmatory investigations by fluorescent antibody or complement fixation tests although these are outside the scope of the cytology laboratory.

Malarial parasites (Plasmodium)

Malarial parasites have, in the past, been of little significance in cytological investigations. However, relaxation of malarial control in parts of Africa have resulted in an awareness that cytology may have a part to play in diagnosis.

In routine practice the parasites may be seen in both Papanicolaou and May-Grunwald–Giemsa stained preparations. Malaria is a tropical disease and can only thrive in areas where mosquitoes are found. *Plasmodium falciparum*, *P. vivax*, *P. malariae* and *P. ovale* are parasites of man requiring the *Anopheles* mosquito for the sexual phase of development. In the intermediate host, man, sporozoites injected by the female *Anopheles* mosquito invade the liver parenchyma where they become active undergoing binary fission forming schizonts each of which contain merozoites. This is called the pre-erythrocytic phase. Merozoites burst from the liver cells and invade the bloodstream to participate in the erythocytic phase. This is the characteristic ring form. Further asexual reproduction within the red cell sees the formation of more merozoites. As the red cell disintegrates it releases merozoites into the bloodstream where they invade more red cells. This cycle is repetitive and eventually some of the merozoites undergo change to produce gametes which mature only in the stomach of a mosquito. Details of the life cycle of the parasite can be found in any of the definitive works listed in the References. The incubation period following a mosquito 'bite' may vary from one week to as long as four, depending upon type and host response. In non-endemic countries malaria is easily overlooked or misdiagnosed. It is important to know where the patient has been. The cytologist should be aware that both the gamete and ring forms may be present in a cell sample.

Helminths

Platyhelminths (flatworms)

Ova of helminths may be found in a variety of sites. In smears from the genital tract they are often incidental contaminants, the result of poor personal hygiene or evidence of rectovaginal fistulae following surgical and/or radiation treatment. In patients on immunosuppresive therapy lack of host response often enables the worm to increase rapidly. Two subgroups are recognized: the trematodes and cestodes.

Trematodes (flukes)
Schistosoma

Amongst the flatworms possibly the most important species is *Schistosoma*. The life cycle of *Schistosoma* (*Figure 23.11*) involves man and a freshwater snail. Man disposes of excreta containing fertile ova into rivers which almost immediately hatch to produce miracidia. If suitable snails are present they may be infected with subsequent development of sporocysts. Each sporocyst will produce numerous cercariae. The snail cycle is roughly one month long. Cercariae, discharged into the water, are capable of penetrating the unbroken skin of persons bathing or washing in the water. They invade blood vessels which carry them to the lungs from where they migrate to the liver and into the excretory system.

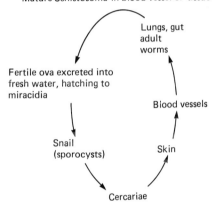

Figure 23.11 Simplified life cycle of *Schistosoma* sp.

Schistosomiasis (bilharzia) was probably endemic in the Sinai peninsula and Egypt at the time of the Exodus. It is a widespread, debilitating disease of man, insidious and damaging. Distribution of the three main invaders, *Schistosoma haematobium*, *S. japonicum* and *S. mansoni* has tended to be geographical with *S. japonicum* confined to the Far East. *S. mansoni* and *S. haematobium* are endemic through most of Africa and there are pockets in South America (*S. mansoni*) and in India and Asia (*S. haematobium*). Berry (1971) has reported the finding of *S. mansoni*, *S. mattheei* and *S. intercalatum* which is endemic in the Congo basin. Identification of ova may be difficult as shape and size variation within the species is not uncommon.

Freshly passed ova are oval with a chitinous shell. *S. haematobium* (*Figures 23.12* and *23.13*) has a terminal spine whilst *S. mansoni* has a lateral one (*Figure 23.14*). The length and apparent sharpness of the spine is variable

416 *Parasitic infections*

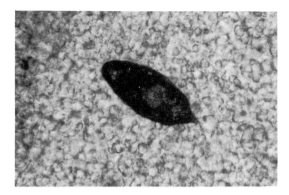

Figure 23.12 Ovum of *S. haematobium* in wet preparation of sputum. The terminal spine is clearly seen. Magnification × 120

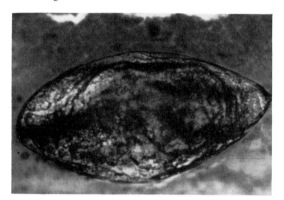

Figure 23.13 Fertile ovum of *S. haematobium* in cervical smear. Stain: Papanicolaou. Magnification × 500, reduced to 65% in reproduction

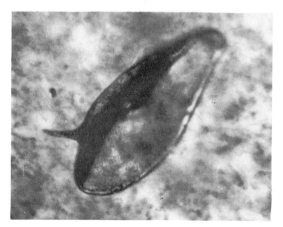

Figure 23.14 *S. mansoni* lateral spine is clearly seen. Stain: Papanicolaou. Magnification × 500

amongst ova in the same specimen. Ova of *S. haematobium* range from 80–170 µm long and from 30–70 µm broad whilst those of *S. mansoni* are marginally larger and *S. mattheei* longer and narrower (*Figure 23.15*). Inside the ovum an embryo miracidium may be seen in an undeveloped or well developed stage and emergent miracidia are not uncommon.

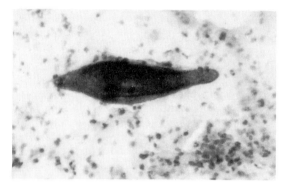

Figure 23.15 Long, narrow ovum of *S. mattheei*, cervical smear. Stain: Papanicolaou. Magnification × 500, reduced to 85% in reproduction

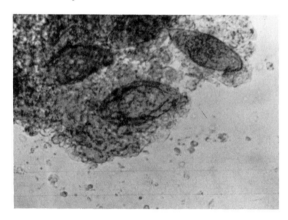

Figure 23.16 Ova of *S. haematobium* in semen during infertility investigation. Wet preparation. Magnification × 500, reduced to 70% in reproduction

In cell samples from the genital tract some or all of the following may be encountered:

(1) Empty shells (*Figure 23.16*): refractile, folded and crumpled with no visible internal structure.
(2) Degenerative ova: variable appearance dependent upon the length of time they have been dead. Partially blackened and completely black and opaque forms are frequently seen, particularly in post-irradiation smears.
(3) Immature ova: they have a granular appearance inside the refractile, chitinous shell. The staining reaction with Papanicolaou

technique ranges from frank basophilia to pale mauve. Immature ova with scanty granular contents may be derived from unfertilized female worms (Berry, 1971).

(4) Mature ova with visible miracidium (*Figure 23.17*). The cytoplasm of the miracidium tends to be eosinophilic whilst the nuclear material takes up haematoxylin. Ova that are about to hatch are larger than the immature ones and the miracidium is fixed in a contorted position.

(5) Emergent and free lying miracidia (*Figures 23.17, 23.18* and *23.19*): oval, round and elongated shapes are encountered and as they are often buried in inflammatory detritus it is this form which is most often missed by inexperienced screeners. However, the presence of empty shells and giant histiocytes should be a pointer to the presence of miracidia.

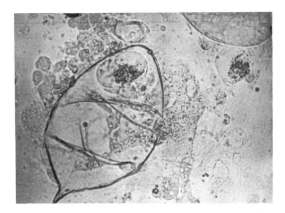

Figure 23.19 Miracidium of *S. haematobium* breaking free of grossly swollen ovum in semen during infertility investigation. Stain: Papanicolaou. Magnification × 500, reduced to 65% in reproduction

Finally it should be emphasized that ova may be present in any tissue as may adult worms (*Figure 23.20*).

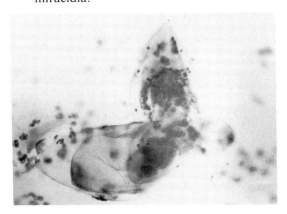

Figure 23.17 Miracidium of *S. haematobium* increasing in size as it emerges from the ovum. Stain: Papanicolaou. Magnification × 500, reduced to 70% in reproduction

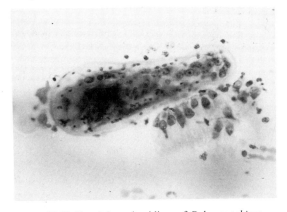

Figure 23.18 Free lying miracidium of *S. haematobium*, cervical smear. Stain: Papanicolaou. Magnification × 500, reduced to 65% in reproduction

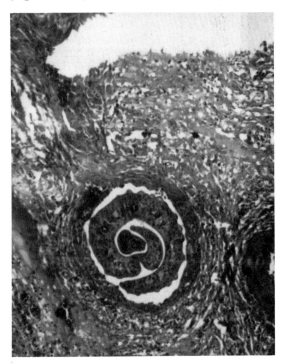

Figure 23.20 Section of Fallopian tube showing adult worm of *S. haematobium in situ*. Numerous ova were also present. Stain: haematoxylin and eosin. Magnification × 500

Paragonimus

During the past ten years only two reports of the finding of *Paragonimus* species have been published in *Acta Cytologica* (McCallum, 1975; Willie and Snyder, 1977). This lung fluke is widely spread and may be found in West, Central and Southern Africa where the reservoir hosts include domestic dogs and cats. The eggs are broad, oval and operculate, i.e. they possess a lid-like structure at one extremity. They measure 100 μm × 50 μm.

Cestodes (tapeworms)

Ova of *Taenia* species are not infrequently found in genital tract cell samples and are commonly the result of poor personal hygiene. Human infection by larvae of *Taenia solium* is an ever-present threat and the examination of cerebrospinal fluid (CSF) for evidence of cerebral cysticercosis is a frequent request in some areas. The significance of an eosinophil reaction in CSF has been discussed in a number of publications (Spriggs, 1968; Wilber, King and Howes, 1980).

Taenia sp. ova may be found in genital smears. The ovum is round, about 40 μm in diameter, with a thick wall carrying radial striations which contain the embryo; hooklets may be visible in well preserved specimens. The life cycle of *Taenia* is shown in *Figure 23.21*.

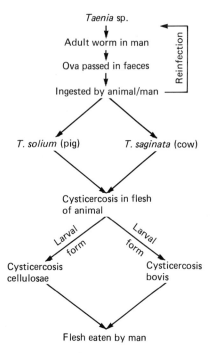

Figure 23.21 Life cycle of *Taenia* sp.

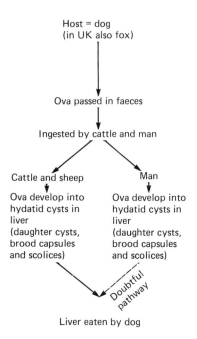

Figure 23.22 Life cycle of *Echinococcus* sp.

Hydatid cysts

Echinococcus granulosus (*Taenia echinococcus*) hosts are carnivorous animals including domestic dogs. The adult worm is small, less than 1 cm long, and is found attached to the intestinal mucosa. Man is infected by ingestion of ova which hatch in the intestines and the embryos migrate through the lymphatics and blood to the liver where cysts develop (*Figure 23.22*). These are slow growing and it may be some years before they become clinically evident. Rupture of the cysts may lead to pulmonary and cerebral hydatid disease. Cysts, which may reach a considerable size, contain scolices, free-lying hooklets and 'brood capsules' giving the fluid a characteristic granular or sandy appearance. The condition is often difficult to diagnose clinically and may be encountered unexpectedly as the result of aspiration of cyst fluid. Where cysts have ruptured into the respiratory, urinary or intestinal tract, evidence of the infection may be seen in sputum, stool, urine or body cavity fluids in the form of free-lying hooklets and/or colices (*Figures 23.23* and *23.24*).

Nematodes (roundworms)

Roundworms are elongated, unsegmented cylindrical parasites; all the reproductive organs

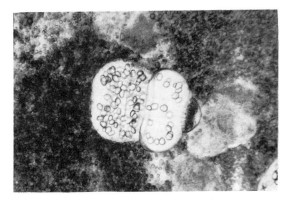

Figure 23.23 *Echinococcus granulosus*, scolex in cyst fluid (wet preparation). Magnification × 125, reduced to 65% in reproduction

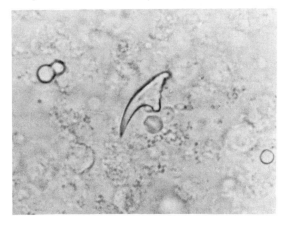

Figure 23.24 *Echinococcus granulosus*, free lying hooklet (wet preparation). Magnification × 400

lie within the body cavity and the sexes are separate. Whilst the human intestinal tract is the common habitat visceral migration does occur within the human host (*Figure 23.25*). *Ascaris lumbricoides* may be found in the lungs whilst *Ancylostoma* and *Strongyloides*, which enter the body through the skin, are bloodborne to the lungs and from there migrate to the intestines. Ova of the roundworms may also be found in genital tract smears as the result of poor standards of cleanliness.

Strongyloides stercoralis

S. stercoralis is a parasite of tropical and subtropical regions where a large percentage of the lower socioeconomic strata may be infected. Carriers may be symptomless until, as a result of repeated autoinfection, the onset of other debilitating diseases or immunosuppressive therapy, symptoms occur.

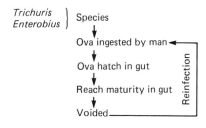

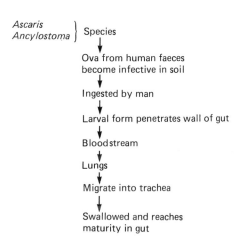

Figure 23.25 Life cycle of Nematoda. (*a*) *Trichuris, Enterobius* and (*b*) *Ascaris, Ancylostoma*.

Infection occurs at the skin where the filariform larvae enter leaving papular eruptions and an oedematous rash. They enter the cutaneous blood vessels and travel via the heart to the lungs where considerable development takes place. They then spread through the respiratory tree and thence into the gastrointestinal tract. Poor personal hygiene results in autoinfection when rhabditiform larvae emerge from the anus and, changing to the filariform state, the cycle starts again. During pulmonary migration a productive cough often develops with consequent requests for sputum cytology. Larval forms of the parasite may be present in considerable numbers often as a mixture of rhabditiform and filariform larvae. Rhabditiform larvae are about 250 μm long and are frequently coiled, with a short buccal cavity. Filariform larvae are shorter with longer oesophagus and a small notch in the tail (*Figures 23.26, 23.27* and *23.28*).

A 50-year-old man was referred to one of our skin clinics with a history of urticaria, generalized itching and chronic cough over a period of

420 Parasitic infections

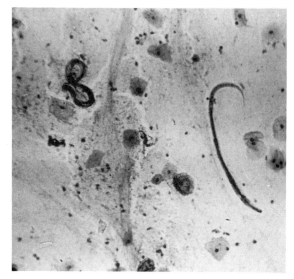

Figure 23.26 Rhabditiform and filariform larvae of *Strongyloides stercoralis* in sputum. Stain: Papanicolaou. Magnification × 250

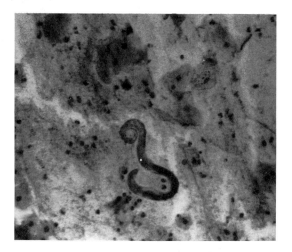

Figure 23.27 Rhabditiform larva of *S. stercoralis* in vaginal smear. Stain: Papanicolaou. Magnification × 250, reduced to 80% in reproduction

about four years. As he had been employed by a carpet manufacturing concern, with consequent exposure to an atmosphere of floating fibres, a tentative diagnosis of an allergic reaction was made. Within a few weeks he developed respiratory distress and was admitted to hospital where, a few days later, diarrhoea of 'alarmingly explosive proportions with blood and mucus' occurred together with a productive cough. X-rays showed an opacity at the left apex and at this stage a diagnosis of bronchial neoplasm and fulminating amoebiasis was entertained. Sputum specimens sent for the investigation of the suspected neoplasm showed rhabditoid larvae of *Strongyloides stercoralis* averaging about 35 parasites in each smear. Malignant tumour cells were not demonstrated in the sputum. The patient died 15 days after admission and at autopsy the nematode infection was noted with malignant disease and healed amoebiasis.

Figure 23.28 Rhabditiform larva of *S. stercoralis* in sputum. Stain: Papanicolaou. Magnification × 500, reduced to 65% in reproduction

Ascaris lumbricoides

This parasite is frequently encountered in cytological material in the form of ova. As it infects the lungs and gastrointestinal tract the ova may be found in sputum, faeces or occasionally as an incidental contaminant in genital smears or urine. *Ascaris lumbricoides* is a tapering worm; the female is about 35 cm long and the male about 20 cm. In the male the posterior end is curved. The fertilized egg is surrounded by a thick, transparent layer and enclosed by a rough mamillated shell. The whole measures $70 \times 45\,\mu m$. Unfertilized eggs are slightly longer and narrower containing a mass of granules.

Trichuris trichiura

The whipworm is a tropical parasite of man and infection is direct without an intermediate host. The distinctive ovum may be encountered in cytological material from the genital tract. It measures about $50\,\mu m \times 20\,\mu m$ and a mucoid plug at each end gives it an unforgettable 'tea-tray' appearance.

Enterobius (Oxyuris) vermicularis

The pinworm is widely distributed with a greater incidence in temperate zones than in the tropics and subtropics. The male worm which is about 2–5 mm long is tightly curled at the posterior whilst the female has a long tapering tail and may reach a total length of 10–12 mm. The eggs are flattened on one side and are usually embryonated (*Figure 23.29*). Although the worms usually inhabit the terminal ileum they may migrate through the rectum to the genital tract. There have been numerous reports of the finding of *Enterobius* in the vagina (San Cristobal and de Mundi, 1976).

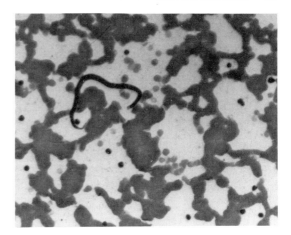

Figure 23.30 *Loa loa* in peripheral blood smear. Stain: May-Grünwald–Giemsa. Magnification × 125

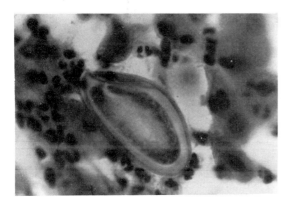

Figure 23.29 Ovum of *Enterobius vermicularis* in vaginal smear. Stain: Papanicolaou. Magnification × 500, reduced to 65% in reproduction

Other common parasites

Microfilaria

Microfilaria have been reported in a number of cell samples including cells from the urinary tract and the endometrium. Filariasis is endemic in many tropical areas and may be found in the subtropics. It is probably better known as 'elephantiasis'. Diagnosis is usually made in peripheral blood samples. *Wuchereria bancrofti* is noctural and samples should be collected in the late evening. *Loa loa* appear in the blood during the day (*Figure 23.30*). Both microfilaria are approximately the same length, 250–300 µm, tapering towards the tail. They are 8–10 µm broad at the widest part and *Loa loa* tends to be marginally more slender. Definitive descriptions can be found in the texts referred to in the References.

Pneumocystis

Pneumocystis carinii was considered for many years to be a yeast; however, morphological studies showed both trophozoite and cystic forms of the organism which is responsible for pneumocystic pneumonia, particularly in undernourished children (Kim, Hughes and Feldman, 1972; Kim and Hughes, 1973). It is a ubiquitous organism which appears to be present in the environment. However, in healthy adults it remains quiescent until lowered resistance, malnutrition, cystostatics and/or immunosuppressive therapy or disease such as AIDS ensues. Then opportunistic infection occurs with rapid proliferation of the organism.

There are a number of morphological variables; a trophozoite stage, a thick walled stage, early cyst form and cysts containing sporozoites. The various forms range in size from 4–10 µm. They are best demonstrated by the Gomori methenamine–silver technique although Giemsa preparations are usually adequate for identification of the organism (*see also* p. 258).

Contaminants and look-alikes

A variety of water-borne contaminants may confuse the inexperienced screener. *Vorticella* (protozoa) were reported in genital tract smears (San Cristobal, Roset and Bloy, 1976), unidentified ciliates in the rubber tubes of a staining machine transferred to genital tract smears (Bourne, 1979) and *Acanthamoeba* causing a similar problem (Rubel, 1979).

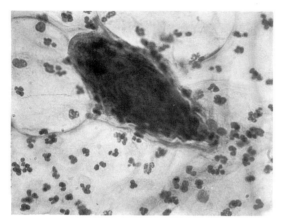

Figure 23.31 Tight cluster of squamous cells which may be mistaken for an ovum of *S. mansoni*. Stain: Papanicolaou. Magnification × 400, reduced to 75% in reproduction

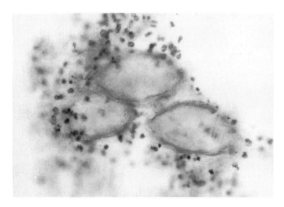

Figure 23.32 Food debris in sputum wrongly diagnosed as ova. Stain: Papanicolaou. Magnification × 500, reduced to 65% in reproduction

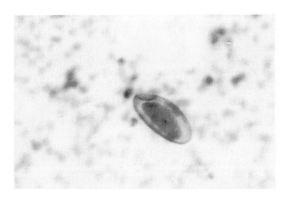

Figure 23.33 Vegetable seeds found in a cervical scrape smear. Seeds of this type may be confused with ova of *Enterobius vermicularis*. The dense internal mass bears no similarity to the coiled embryo which can be seen in *Figure 23.29*. Stain: Papanicolaou. Magnification × 100

Food particles, vegetable cells and pollen are frequently encountered whilst parasites from nonhuman sources may be occasionally encountered (*Figures 23.31, 23.32* and *23.33*). The *Atlas of Medical Helminthology and Protozoology* (Jeffrey and Leach, 1975) is of considerable help with the identification of the latter.

The value of cytological techniques in the diagnosis of non-neoplastic disease is often overlooked. The identification of parasitic infestations is particularly rewarding.

Acknowledgements

The author is indebted to Dr A. V. Berry for permission to use *Figure 23.13* and to Mrs S. M. McCallum for *Figures 23.23* and *23.24*.

References

BERRY, A. (1971). Evidence of gynecologic bilharziasis in cytologic material. *Acta Cytol.*, **15**, 482–498

BOURNE, R. (1979). A cervical smear contaminant. *Acta Cytol.*, **23**, 174

KIM, H. K. and HUGHES, W. T. (1973). Comparison of methods of identification of *Pneumocystis carinii* in pulmonary aspirates. *Am. J. Clin. Pathol.*, **60**, 462–466

KIM, H. K., HUGHES, W. T. and FELDMAN, S. (1972). Studies of morphology and immunofluorescence of *Pneumocystis carinii*. *Proc. Soc. Exp. Biol. Med.*, **141**, 304–309

MCCALLUM, S. M. (1975). Ova of the lung fluke *Paragonimus kellicotti* in a fluid from a cyst. *Acta Cytol.*, **19**, 279–280

RUBEL, L. R. (1979). Smear contaminants. *Acta Cytol.*, **23**, 260

SAN CRISTOBAL, A., ROSET, M. and BLOY, C. C. (1976). Finding of ciliated protozoa genus *Vorticella* on cervical and endocervical smears. *Acta Cytol.*, **20**, 387–389

SPRIGGS, A. (1968). *The Cytology of Effusions*, 2nd Edition, pp. 17–20. London: Heinemann

WILBER, R. D., KING, E. B. and HOWES, E. L. (1980). Cerebrospinal fluid cytology in five patients with cerebral cysticercosis. *Acta Cytol.*, **24**, 421–426

WILLIE, S. M. and SNYDER, R. N. (1977). The identification of *Paragonimus westermanii* in bronchial washings. *Acta Cytol.*, **21**, 101

Selected reading from *acta cytologica*
Amoebae

HOFFLER, A. S. and RUBEL, S. R. (1974). Free-living amoebae identified by cytologic examination of gastrointestinal washings. *Acta Cytol.*, **18**, 59–61

KAPILA, K. and VERMA, K. (1982). Cytologic detection of parasitic disorders. *Acta Cytol.*, **26**, 359–362

MCNEILL, R. E. and MORAES-RUEKSEN, M. DE (1978). Amoebae trophozoites in cervico-vaginal smear of a patient using an intrauterine device. *Acta Cytol.*, **22**, 91–92; 413–420

SAYGI, G. (1982). Protozoal organisms in cervico-vaginal smears. *Acta Cytol.*, **26**, 750

WALSH, T. J. et al. (1983). Cytologic diagnosis of extra-colonic amoebiasis. *Acta Cytol.*, **27**, 671–675

WALTER, A. (1982). Diagnosis of amoebic cervicitis from cervico-vaginal smears. *Acta Cytol.*, **26**, 378–379

Ascaris

GARUD, M. A. et al. (1980). Vaginal parasites. *Acta Cytol.*, **24**, 34–35

Balantidium

LAHIRI, V. L., ELHENCE, B. R. and AGARWAL, B. M. (1977). *Balantidinum* peritonitis diagnosed in cytological material. *Acta Cytol.*, **21**, 123

Echinococcus

GARRETT, M., HERBSMAN, H. and FIERST, S. (1977). Cytologic disease of *Echinococcus*. *Acta Cytol.*, **21**, 553–554

KAPILAN, K. and VERMA, K. (1982). Cytologic detection of parasitic disorders. *Acta Cytol.*, **26**, 359–362

WESLEY I. et al. (1979). Cytologic diagnosis of *Echinococcus*. *Acta Cytol.*, **23**, 134–136

Enterobius

BAK, M., BAK, M. JR. and BODE, M. (1982). Vaginal enterobiasis. *Acta Cytol.*, **26**, 264–265

GARUD, M. A. et al. (1980). Vaginal parasites. *Acta Cytol.*, **24**, 34–35

MASCOLO, G., PIZZANATO, V. and NOVELLI, G. G. (1979). Colpocytologic observation of eggs of *Enterobius vermicularis*. *Acta Cytol.*, **23**, 425–426

SAN CRISTOBAL, A. and DE MUNDI, A. (1976). *Enterobius vermicularis* larvae in vaginal smears. *Acta Cytol.*, **20**, 190–192

SARASWATHI, T. R., VENI, C. and KHAN, J. A. (1982). *Enterobius vermicularis* in a vaginal wet smear. *Acta Cytol.*, **26**, 97

WONG, J. V. and BECKER, S. N. (1982). *Enterobius vermicularis* in routine cervico-vaginal smears. *Acta Cytol.*, **26**, 484–487

Microfilariae

AFFANDI, M. Z. (1980). Microfilariae in the endometrial smear. *Acta Cytol.*, **24**, 173–174

AGARWAL, P. K., SRIVASTAVA, A. N. and AGARWAL, N. (1982). Microfilariae in association with neoplasms. *Acta Cytol.*, **26**, 488–490

DE BRUX, J. A., BAUP, H. F. C. and KAEDING, H. (1983). *Loa loa* microfilariae in an endometrial smear. *Acta Cytol.*, **27**, 547–548

KAPILAN, K. and VERMA, K. (1982). Cytologic detection of parasitic disorders. *Acta Cytol.*, **26**, 359–362

RANI, S. and BESHAR, P. C. (1981). Microfilariae in bone marrow aspirates. *Acta Cytol.*, **25**, 425–426

WALTER, A., KRISHMASWAMI, H. and CARIAPPA, A. (1983). Microfilariae of *Wuchereria bancrofti* in cytologic smears. *Acta Cytol.*, **27**, 432–436

WEBBER, C. A. and EVELAND, L. K. (1982). Cytologic detection of *Wuchereria bancrofti* microfilariae in urine collected during a routine workup for haematuria. *Acta Cytol.*, **26**, 837–840

Paragonimus

MCCALLUM, S. M. (1975). Ova of the lung fluke *Paragonimus kellicotti* in fluid from a cyst. *Acta Cytol.*, **19**, 279–280

WILLIE, S. M. and SNYDER, R. N. (1977). The identification of *Paragonimus westermanii* in bronchial washings. *Acta Cytol.*, **21**, 101–102

Schistosoma

BERRY, A. (1976). Multispecies schistosomal infections of the female genital tract detected in cytology smears. *Acta Cytol.*, **20**, 361–365

EL-BOKAINY, M. N. et al. (1982). Cytologic detection of bladder cancer in a rural Egyptian population infested with schistosomiasis. *Acta Cytol.*, **26**, 303–310

HUMBERTO, L., CARVAHLO, G. and CARVAHLO, J. M. (1979). Carcinoma *in situ* and invasive squamous cell carcinoma associated with schistosomiasis of the uterine cervix. *Acta Cytol.*, **23**, 45–48

Strongyloides

AVAGNINA, M. A. et al. (1980). *Strongyloides stercoralis* in Papanicolaou-stained smears of ascitic fluid. *Acta Cytol.*, **24**, 36–39

HUMPHREYS, K. and HIEGER, L. R. (1979). *Strongyloides stercoralis* in routine Papanicolaou-stained sputum smears. *Acta Cytol.*, **23**, 471–476

KENNEY, M. and WEBBER, C. A. (1974). Diagnosis of *Strongyliosis* in Papanicolaou-stained smears. *Acta Cytol.*, **18**, 270–273

PERAETA, N. R. and RODRIGUES, A. (1978). *Strongyloides stercoralis* in gastric and duodenal aspirates. *Acta Cytol.*, **22**, 61–63

WONG, T. et al. (1980). Diagnosis of *Strongyloides stercoralis* in sputum cytology. *Acta Cytol.*, **24**, 40–43

YASSIM, S. M. A. and GARRET, M. (1980). Parasites in cytodiagnosis. *Acta Cytol.*, **24**, 539–544

Toxoplasma

CHRIST, M. L. and FELTES-KENNEDY, M. (1982). Fine needle aspiration cytology of toxoplasmic lymphadenitis. *Acta Cytol.*, **26**, 425–428

DOMINGUEZ, A. and GIROU, J. J. (1976). Toxoplasma cysts in vaginal smears. *Acta Cytol.*, **20**, 269–271

FRENKEL, J. K. (1977). *Toxoplasma* cysts in vaginal smears. *Acta Cytol.*, **21**, 492 (letter)

NAYLOR, B. (1977). Regarding cyanophilic bodies, *Toxoplasma* cysts and ferruginous bodies. *Acta Cytol.*, **21**, 490 (letter)

Further reading

FAUST, E. C., BEAVER, P. C. and JUNG, R. C. (1975). *Animal Agents and Vectors of Human Disease*, 4th Edition. Philadelphia: Lea and Febiger

FRIPP, P. J. (1983). *An Introduction to Human Parasitology with Reference to Southern Africa*, 2nd Edition. Johannesburg: Macmillan South Africa

JEFFREY, H. C. and LEACH, R. M. (1975). *Atlas of Medical Helminthology and Protozoology*, 2nd Edition. Edinburgh: Churchill Livingstone

SPENCER, F. M. and MONROE, L. S. (1961). *The Colour Atlas of Intestinal Parasites*. Springfield: Charles C. Thomas

24

Iatrogenic changes

Tadao K. Kobayashi

Introduction

One of the most perplexing problems in modern medical practice is the amount of damage and disease which is caused as a result of treatment. This 'physician-induced' or iatrogenic disease is often seen in cytological specimens and may take the form of cell damage and destruction, dysplasia, hyperplasia or other disorders of growth. Alternatively, treatment may result in the patient having an increased risk of infection so that the cytological specimens exhibit evidence of viral or fungal infection. The aim of this chapter is to describe the iatrogenic changes that are most likely to be encountered in cervical smears and other samples, and discuss them in relation to the treatment modalities that cause them. The cytological patterns are grouped as follows:

(1) Changes in cervical smears associated with cryotherapy, electrocautery and laser ablation of cervical intraepithelial neoplasia (CIN).
(2) Radiation change.
(3) The effects of cytotoxic drug therapy.
(4) The intrauterine contraceptive device.
(5) The effect of exogenous hormones on cervical smear pattern.
(6) The effect of instrumentation, e.g. catheterization and endoscopic examination.

Changes associated with cryotherapy, electrodiathermy, 'cold' coagulation diathermy

The techniques of cryotherapy, high temperature electrodiathermy, 'cold' coagulation diathermy and carbon dioxide (gas molecular) laser therapy have largely replaced traditional cone biopsy for the ablation of small areas of CIN (Singer and Walker, 1985). All three methods of treatment are valuable since they are safe, effective and acceptable to the patient. They can be performed in an outpatient setting without anaesthesia, and complications are unusual. Destruction of tissue to a depth of 5 mm is achieved without distress to the patient and risk of recurrence of the lesion after therapy is low.

Laser therapy

The laser principle was first proposed by Schawlow and Townes (1958); the term is an abbreviation for 'light amplification by stimulated emissions of radiation'. The development of laser surgery has been a major advance in clinical practice and the use of the laser for treatment of CIN is just one of many therapeutic uses of this instrument. The use of the laser for the treatment of cervical and vaginal lesions was pioneered by Kaplan, Goldman and Ger (1973),

Townsend (1976), Stafl, Wilkinson and Mattingly (1977) and Masterson et al. (1981). Tissue destruction is achieved by vaporization of the cells exposed to the beam. The light energy is instantly converted to heat as the laser beam is absorbed by the cells, leaving a crater where the cells were previously located.

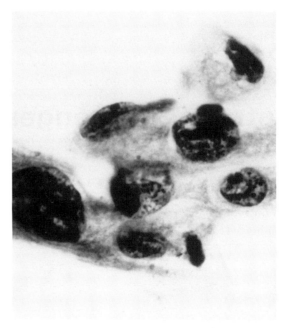

Figure 24.2 Atypical repair cells. Smear taken 2–4 weeks after laser treatment. Tissue repair-like cells show hyperchromasia, anisokaryosis and pronounced nucleoli with cyanophilic cytoplasm. Stain: Papanicolaou. Magnification × 400. Photograph provided by Drs S. Ono and S. Sekimoto, Department of Obstetrics and Gynecology, Fukushima Medical College, Fukushima, Japan

Figure 24.1 Extremely elongated cells in the cervical smear immediately after laser irradiation from patient with cervical intraepithelial neoplasia. Stain: Papanicolaou. Magnification × 200

Smears taken within one week of carbon dioxide laser ablation show very characteristic changes. They contain abundant necrotic debris together with extremely elongated cells of epithelial and connective tissue origin (*Figure 24.1*) due to dehydration and shearing of the cells adjacent to the area of destruction. Within two weeks of treatment, cellular changes consistent with repair can be seen. Smears taken at this time contain regenerating parabasal cells with enlarged hyperchromatic nuclei and large nucleoli. There may be some anisonucleosis (*Figure 24.2*) which may make it difficult to distinguish the repair process from residual CIN (Ono et al., 1982). Epithelial regeneration is usually complete within a month, but in view of the difficulty in interpreting the smear it is advisable to wait at least two months before taking a follow-up smear to assess the effectiveness of therapy.

Electrodiathermy, 'cold' coagulation diathermy and cryotherapy

Many of the changes seen in cervical smears taken after electrodiathermy, 'cold' coagulation diathermy or cryotherapy are similar to those seen in smears taken after laser therapy (Holmquist, Bellina and Danos, 1976). Within the first few days, necrosis and degenerative changes are very much in evidence. Elongated cells may be seen in the smears and a frequent finding is cytoplasmic vacuolation and nuclear pyknosis and karyorrhexis (Gondos, Smith and Townsend, 1970). Similar elongated cells can be seen in bladder biopsies taken immediately after electrocautery (*Figure 24.3*).

One week after therapy many parabasal cells may be found in the smear which have enlarged hyperchromatic nuclei and prominent nucleoli. As with laser therapy they reflect epithelial regeneration. Some anisonucleosis may be present which may be misinterpreted as dyskaryotic change (Hasegawa, Tsutui and Kurihara, 1975) and it is suggested that, as with laser therapy, at least two months should elapse before a follow-up smear is taken (Townsend, 1976).

The regenerating epithelium of the cervix in women who have been treated with electrodiathermy has been studied histologically (González-Merlo et al., 1973) and by electron microscopy (Norum et al., 1969). These studies reveal a two stage process of regeneration. Initially, a single layer of cells grows from the margin of the damaged epithelium over the injured area. The epithelium becomes multilayered and within four weeks the epithelium stratifies and is restored to normality. Similar findings have been reported after cryotherapy (Hasegawa, Tsutui and Kurihara, 1975) and laser therapy.

epithelial cells in cervical smears in the first few days after exposure to ionizing radiation differ from those seen at a later stage. The early changes include swelling of the cytoplasm and nucleus so that the cells appear balloon-shaped and there is a loss of morphological detail. Other early changes include vacuolation of the cytoplasm and nucleus (*Figure 24.4*), wrinkling of the nuclear membrane, multinucleation and condensation of the nuclear chromatin. These changes, which reflect cellular degeneration and death, affect benign and malignant epithelial cells alike.

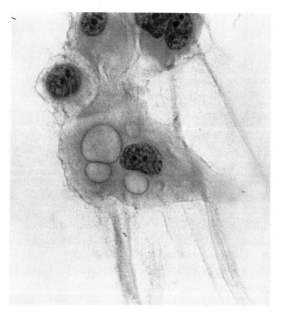

Figure 24.4 Acute radiation changes on squamous cells of the uterine cervix. Note the cellular enlargement with cytoplasmic vacuolation. Stain: Papanicolaou. Magnification × 400

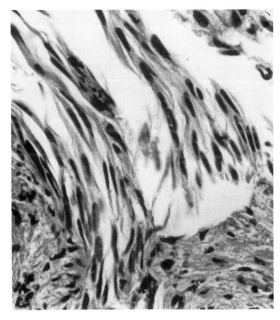

Figure 24.3 Biopsy from a patient with bladder tumour removed by electrocautery. Many elongated cells can be seen. Stain: haematoxylin and eosin. Magnification × 200

Changes due to ionizing radiation

The effect of ionizing radiation on the epithelium of the body varies according to the dose, period of exposure and interval between treatments. Unlike the epithelial changes after laser therapy, cryotherapy or electrodiathermy, the changes following radiotherapy are long-lived and may be seen in cytological and histological samples for many months or even years after the initial exposure. The changes seen in

Changes seen in cervical epithelial cells six months after radiotherapy differ from those described above. Nuclei may be enlarged, hyperchromatic and irregular, reflecting a post-radiation dysplasia which may be difficult to distinguish from neoplastic change. Multinucleate cells and bizarre cell shapes may also be seen adding to the problem of differential diagnosis. However, the abundant cytoplasm and 'smudging' of the nuclear chromatin pattern are distinguishing features of radiation change not found in neoplasia.

Changes similar to those described above can be seen in other epithelial cells. They have been observed in urothelial cells in urine from patients who have received pelvic irradiation (Forni, Koss and Geller, 1964; Cowen, 1975) and in the epithelium of the respiratory tract after radiotherapy. Koss (1979) reports that sputum samples from patients who have had radiotherapy for a malignant tumour of the lung may contain greatly swollen columnar cells with multiple nuclei, cilia and elongated cytoplasm. Should acute radiation pneumonitis supervene, the sputum may contain necrotic cellular debris, numerous degenerating cells, strands of smeared nuclear material and many polymorphs. The long-term result of irradiation of pulmonary tissue is squamous metaplasia or the dysplastic change described above.

The most detailed studies of radiation changes have been made on cervical smears (Graham and Graham, 1953; Zimmer, 1959). The degenerative and reactive changes which can be seen in cervical smears after radiotherapy of a cervical neoplasm were believed by Graham to be a reliable predictor of therapeutic outcome. However, these observations have not been confirmed and the extent of radiation change seen in a smear is no indication of the risk of recurrence of malignant disease.

Patten *et al.* (1963), Wentz and Reagan (1970), von Hamm (1979) and Koss (1979) all comment on the difficulty in distinguishing between post-radiation dysplasia in cervical smears and recurrence of carcinoma. The presence of hyperchromatic nuclei showing marked anisonucleosis and nuclear crowding suggests neoplastic change. The presence of mitotic figures in these abnormal cells can be taken as firm evidence of malignancy.

Changes due to chemotherapy

One of the most important new developments in therapeutic medicine in the last decade has been the introduction of a range of cytotoxic drugs for the control of malignant disease and the prevention of rejection in patients who are recipients of an organ transplant. The efficacy of cytotoxic drugs depends on their ability to inhibit the growth of dividing cells. Those drugs administered for the control of malignant disease act mainly by inhibiting the growth of poorly differentiated cancer cells although other dividing cells in the body, e.g. those involved in gametogenesis or haemopoiesis are also susceptible to their action. Those drugs administered to organ transplant recipients act mainly by inhibiting the proliferation of T lymphocytes and therefore have an immunosuppressive action.

The iatrogenic changes induced by cytotoxic drugs are of two types: they induce dysplastic change in the epithelial cells of the host and they increase the patient's risk of infection with bacteria, viruses and fungal agents. Some of the immunosuppressive drugs used in organ transplantation also increase the host's susceptibility to malignant disease (Penn, 1978).

The dysplastic epithelial changes induced by cytotoxic drugs have been described in histological section of respiratory tract, urinary tract, breast, skin, pancreas, thyroid, alimentary tract and liver and in a variety of cytological specimens including urine, sputum, cervical smears, cerebrospinal fluid, gastric and oesophageal brushings (*Figure 24.5*). The changes mimic those found in cancerous epithelium and great care must be taken in the interpretation of cytological and histological findings in patients who are on a treatment regime which involves the use of these drugs. A complete clinical history must be provided with each sample if misdiagnosis is to be avoided. A list of cytotoxic agents in common use is shown in *Table 24.1*. The iatrogenic changes associated with their use is discussed below.

Cyclophosphamide

Cyclophosphamide is a synthetic alkylating agent widely used for the treatment of malignant disease, especially lymphoproliferative and myeloproliferative disorders. The drug is related to nitrogen mustard but is inactive until metabolized in the liver. The metabolites are excreted by the kidney and are probably responsible for the haemorrhagic cystitis associated with the administration of this drug. Cytological alterations associated with therapy have been demonstrated in experimental animals (Forni, Koss and Geller, 1964; Koss, 1967; Bellin, Cherry and Koss, 1974) and in clinical practice. They are to be found in urogenital epithelium and are very similar to those seen in radiation change. They are characterized by nuclear and cytoplasmic enlargement of exfoliated urothelial cells which may have marked cytoplasmic vacuolation (*Figure 24.6*). No relation between the degree of cellular atypia and dosage has

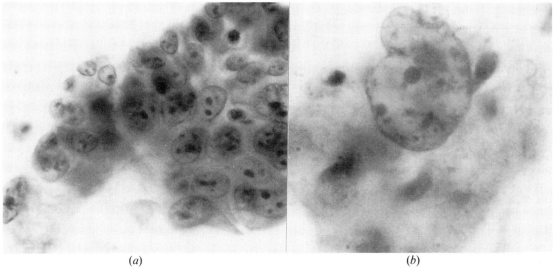

Figure 24.5 Epithelial atypia in an oesophageal brush specimen from a patient receiving cytotoxic drug therapy. (*a*) Note the crowded and overlapped cells, the variation in nuclear size, increased nucleocytoplasmic ratio, cleared, clumped and irregular chromatin and prominent multiple nucleoli. Stain: Papanicolaou. Magnification ×950. (*b*) Note high nucleocytoplasmic ratio, irregular nuclear margin and cleared and clumped chromatin. Stain: Papanicolaou. Magnification ×2400. Reproduced from O'Morchoe, Lee and Kozak (1983) with permission

been observed but the changes are more likely to be seen in patients who have had long-term therapy (Forni, Koss and Geller, 1964).

Table 24.1 Partial list of chemotherapeutic agents and related drugs

Alkylating agents:
Cyclophosphamide
Busulphan
Cisplatin
Thiotepa
Melphalan
Carmustine
Semustine

Antibiotics:
Bleomycin
Doxorubicin
Mitomycin
Toyomycin
Mithramycin

Antimetabolites:
Cytarabine
Methotrexate
Mercaptopurine
Fluorouracil

Immunosuppressive agents:
Azathioprine
Cyclosporin

Other biological response modifiers
Interferon
OK-432
Tumour necrotizing factor

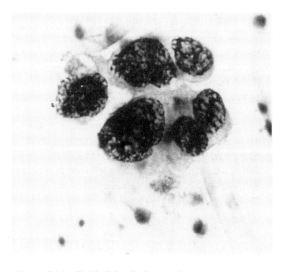

Figure 24.6 Epithelial cell changes due to cyclophosphamide therapy in urinary sediments. The nuclear chromatin shows the so-called 'salt and pepper' pattern. Stain: Papanicolaou. Magnification ×400

Busulphan

Busulphan was introduced for the treatment of chronic myeloid leukaemia in 1953 (Haddow and Timmis, 1953). The complications associated with its use include bone marrow aplasia, pulmonary fibrosis, hyperpigmentation and amenorrhoea (Oliner *et al.*, 1961). The drug also has a marked effect on the epithelial cells of the lung, uterine cervix and urinary tract, inducing cellular and nuclear enlargement and nuclear hyperchromasia which, in cervical smears, is indistinguishable from dyskaryosis and in sputum samples closely mimics bronchial carcinoma (Nelson and Andrews, 1964; Koss, Melamed and Mayer, 1965; Heard and Cooke, 1968; Feingold and Koss, 1969). A full case history is essential with each specimen to avoid the pitfalls of diagnosis associated with this drug.

Thiotepa, adriamycin, bleomycin, melphalan and methotrexate

The efficacy of thiotepa, adriamycin and mitomycin for the treatment of superficial bladder cancer is well documented (Drew and Marshal, 1968; Mitchell, 1971; Banks *et al.*, 1977; Yagoda *et al.*, 1977; Sloway *et al.*, 1981). Clinical and experimental studies (Rasmussen *et al.*, 1980) have shown that treatment is associated with swelling and vacuolation of the urothelial cells (*Figure 24.7*). Drug-induced atypias have also been described in sputum due to bleomycin (Bedrossian and Corey, 1978); in breast due to thiotepa (Moore, 1958); and in pulmonary tissue due to melphalan (Taetle, Dickman and Feldman, 1978). Sostman *et al.* (1976) reported methotrexate-induced pneumonitis. Personal experience with methotrexate (Kobayashi *et al.*, 1984b) has shown that when this drug is introduced intrathecally for carcinomatous meningitis, nuclear enlargement and cytoplasmic vacuolation can occur in the cells in the cerebrospinal fluid (*Figure 24.8*).

Azathioprine and cyclosporin

These are potent immunosuppressive agents used mainly to prevent transplant rejection but also used for the treatment of certain autoimmune diseases. Azathioprine blocks lymphocyte proliferation by inhibiting DNA synthesis but also acts against other haemopoietic cells and its administration may result in depression of bone marrow function. Cyclosporin is more selective and acts by inhibiting the production of interleukin-2, a T cell growth factor. It does not cause generalized bone marrow depression. Both these drugs reduce the ability of the body to fight infection and opportunistic infection

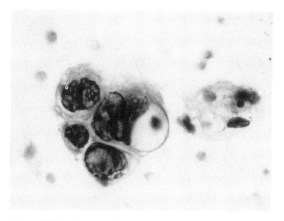

Figure 24.7 Changes in transitional cells due to adriamycin therapy. Urinary sediment. Note the marked hyperchromasia, irregular nuclear margin and vacuolated cytoplasm. Stain: Papanicolaou. Magnification × 400

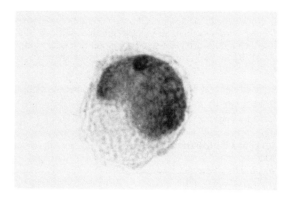

Figure 24.8 Cellular changes due to methotrexate therapy in a cerebrospinal fluid specimen. Cellular enlargement with homogeneous nuclear chromatin distribution and tiny cytoplasmic vacuoles. Stain: Papanicolaou. Magnification × 1000

with viruses, fungi and parasites is common. Cytopathic changes due to infection with herpes simplex and herpes zoster, cytomegalovirus and human polyomavirus can be seen in histological and cytological samples from these patients (Coleman, 1981). Schramm (1970) and Kay, Frable and Hume (1970) reported a susceptibility to cervical dysplasia in renal allograft recipients receiving azathioprine. Gupta, Pinn and Taft (1969) confirmed this finding.

Interferons, OK-432 and tumour necrotizing factor

These agents all have an antitumour effect and are known to possess biological response modifiers (Carter, 1980). Hsu *et al.* (1984) injected interferon into the cervices of patients with cervical intraepithelial neoplasia and claimed that the drug had a cytocidal effect on the dyskaryotic cells (*Figure 24.9*). A report by Katano and Torisu (1982) indicated that intraperitoneal injection with OK-432 (a substance derived from group A *Streptococcus pyogenes*) in a case of malignant ascites resulted in the disappearance of the adenocarcinoma cells in the fluid and an increase in the number of neutrophils. The author's experience with OK-432 for the treatment of a malignant pleural effusion due to metastatic squamous carcinoma has been similar (*Figure 24.10*).

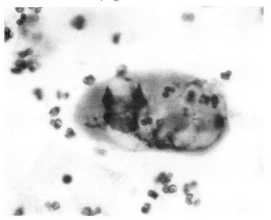

Figure 24.9 Selective cytocidal effect on a large dyskaryotic cell with bizarre nucleus following fifth local injection of interferon. Karyolysis and condensation of nuclear membrane are obvious. A distinct cell border and intracytoplasmic vacuoles infiltrated by neutrophils are present. Stain: Papanicolaou. Magnification × 575. Reproduced from Hsu *et al.* (1984) with permission

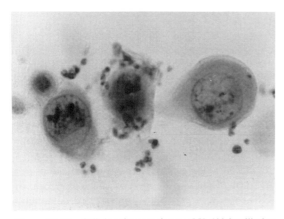

Figure 24.10 Cellular changes due to OK-432 instillation in pleural cavity. Note the cellular enlargement and vacuolated cytoplasm infiltrated by polymorphonuclear leucocytes. Stain: Papanicolaou. Magnification × 400, reduced to 80% in reproduction

Cytopathological changes associated with the intrauterine contraceptive device

Numerous light microscope and electron microscope studies of the effect of the intrauterine device (IUD) on the endometrium (Israel and Davis, 1966; Parr, Schaedler and Hirsch, 1967; Potts and Pearson, 1967; Wynn and Sawaragi, 1969; Ober, 1977; Gawad *et al.*, 1977) have shown that introduction of an IUD into the uterine cavity results in a low grade inflammatory response which is believed by some authors to account for the anti-fertility effect of the device (Parr, Schaedler and Hirsch, 1967). The cytological changes associated with the presence of an IUD have been studied in cervical smears and in uterine fluid.

Cytological changes in cervical smears

Studies of the cytological pattern in cervical smears from IUD users have been carried out by many groups including Ayre (1965), Piver, Whiteley and Bolognese (1966), Sağiroğlu and Sağiroğlu (1970), Ishihama *et al.* (1970), Gupta *et al.* (1976, 1978), Gupta (1982) and Kobayashi *et al.* (1982). Several different cell types have been noted in the smear:

(1) Macrophages and other inflammatory cells reflecting the foreign body reaction in the uterus.
(2) Normal and abnormal endometrial cells shed from the uterus probably reflecting a chronic endometritis.
(3) Metaplastic squamous cells from the cervical epithelium due to local reaction of the cervix due to the IUD.
(4) *Actinomyces* and *Entamoeba gingivalis*-like organisms.

Inflammatory response

The histological changes in the endometrium which result from IUD insertion are characterized by local infiltration of the stroma with polymorphs, lymphocytes and macrophages. The stroma itself may show decidual change. These changes are reflected in the cervical smear by the presence of polymorphs, macrophages and foreign body giant cells of uterine origin.

Endometrial cells

Prolonged use of an IUD is often associated with the exfoliation of endometrial cells. These cells have been found in 58% of smears from IUD users (Ashton and Johnston, 1975) and can be found throughout the menstrual cycle. The cells may show a marked nuclear enlargement (*Figures 24.11* and *24.12*) and unless care is taken, may be mistaken for cells shed from an adenocarcinoma. The abundant cytoplasm frequently found in some of these atypical endometrial cells has occasionally resulted in their being mistaken for dyskaryotic squamous cells shed from an area of carcinoma *in situ*.

Atypical multinucleate cells have been described in cervical smears (*Figure 24.13*) which have been interpreted as histiocytes (Bibbo, 1979) or endocervical cells (Koss, 1979). However, by consensus, they are now regarded as endometrial in origin (Herting and Tauber, 1978; Gupta *et al.*, 1978; Ng and Reagan, 1979; Kobayashi, Casslén and Stormby, 1983).

It has been pointed out that there is some similarity between the atypical endometrial cells seen in smears from IUD users and the endometrial cells seen in smears from patients who have had a miscarriage (Kobayashi *et al.*, 1980), or ectopic pregnancy (Albukerk and Gnecco, 1977; Kobayashi *et al.*, 1983). Clearly the atypical cells associated with IUD usage may result in an erroneous diagnosis of cancer and the correct interpretation of these cells in cervicovaginal smears is of great importance.

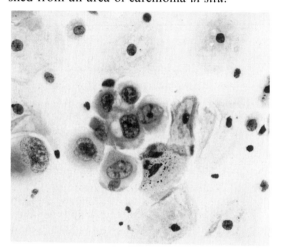

Figure 24.11 Atypical endometrial cells in cervicovaginal smear. The nuclei are large showing irregularly distributed chromatin and recognizable nucleoli. These cells can mimic adenocarcinoma in routine smear. Stain: Papanicolaou. Magnification × 400. Reproduced from Kobayashi, Casslén and Stormby (1983) with permission

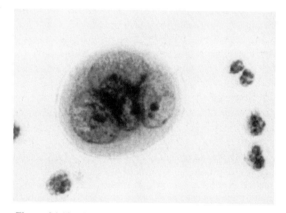

Figure 24.13 Atypical multinucleated giant cell with enlarged nuclei and nucleoli in cervicovaginal smear. Stain: Papanicolaou. Magnification × 1000

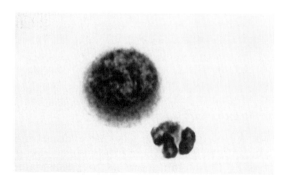

Figure 24.12 Atypical endometrial cell in cervicovaginal smear. This cell has a high nucleocytoplasmic ratio and simulates a dyskaryotic squamous cell. Note hyperchromasia and pronounced nucleoli. Stain: Papanicolaou. Magnification × 1000

Changes in cervical epithelial cells

Ayre (1965) suggested that IUD usage causes transformation of the cervical epithelium from normal to dysplastic. Piver, Whiteley and Bolognese (1966) and Ishihama *et al.* (1970) were unable to confirm this. It is now agreed that although the presence of an IUD may result in metaplastic change, patients using these devices are not at increased risk of cervical neoplasia. However, it is worth noting that the incidence of pelvic inflammatory disease is reported to be higher in IUD wearers than in women using oral contraceptives (Tatum *et al.*, 1975; Faulkner and Ory, 1976; Eschenbach, Harnisch and Holmes, 1977).

Table 24.2 Prevalence of *Actinomyces*-like organisms in cervicovaginal smears with and without IUD

	Positive for Actinomyces-*like* organisms (%)	Average duration (years)	Average age (years)
Tailed IUD ($n = 350$)	47 (13.4)	5.7	39.1
Tail-less IUD ($n = 170$)	6 (3.5)	3.2	34.2
Non-IUD users ($n = 140$)	–	–	40.4

Actinomyces *associated with the use of an IUD*

It is well known that *Actinomyces* is associated with IUD usage. A more detailed description of this bacterium is given in Chapter 11.

We have recently had the opportunity to assess the prevalence of *Actinomyces*-like organisms in routine cervical smears from IUD users (Kobayashi and Tara, 1983). Of 520 patients using the Ohta ring, 53 (10.2%) were positive for *Actinomyces*-like organisms (*Table 24.2*). Of the 53 positive cases, 47 were fitted with a tailed IUD and 6 were fitted with a tail-less IUD. In no instance were *Actinomyces*-like organisms observed in women without an IUD. The duration of IUD insertion averaged 5.7 years in the tailed and 3.2 years in the tail-less groups respectively. Our results and those of others (Valicenti *et al.*, 1982) indicate that infection is related to duration of use of IUD.

Cellular and cytochemical studies of uterine fluid

Uterine fluid is the term used to describe the fluid found in the uterine cavity and its cellular composition has been extensively studied by Casslén, Kobayashi and colleagues (Casslén, 1981; Casslén, Kobayashi and Stormby, 1981, 1982; Kobayashi, Casslén and Stormby, 1982). The cells in the fluid were found to be similar to those found in cervical smears. Macrophages and polymorphs constitute the dominant cell type in the fluid, although mast cells may also be seen. Cytochemical studies of the macrophages show a varying degree of activity of both non-specific esterase and acid phosphatase which suggests that they may be derived from blood monocytes. Casslén, Kobayashi and Stormby (1981) found that macrophages constitute less than one-third of the cells in the uterine fluid surrounding the devices, and polymorphonuc-

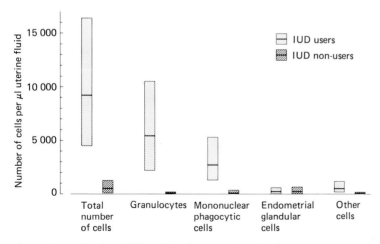

Figure 24.14 Total and differential cell counts/µl uterine fluid in 75 IUD users and 48 non-users. Samples from all menstrual phases are included. Results are given as medians with upper and lower quartiles. Reproduced from Casslén, Kobayashi and Stormby (1981) with permission

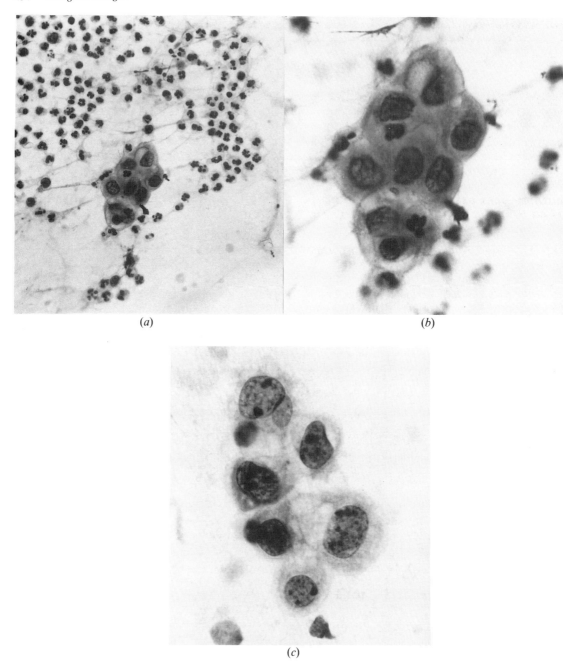

Figure 24.15 Atypical cell cluster in a uterine fluid smear obtained from an IUD user. (a) Atypical glandular cells with granulocytes and macrophages. Stain: Papanicolaou. Magnification ×400. (b) High magnification (×1000) of the same cluster. (c) Atypical glandular cell cluster with enlarged nuclei and pleomorphism. The chromatin is clumped and nucleoli pronounced. The cytoplasm shows a vacuolated, foamy appearance. Stain: Papanicolaou. Magnification ×1000. Reproduced from Kobayashi, Casslén and Stormby (1983) with permission

lear leucocytes constitute the remainder (*Figure 24.14*).

Kobayashi, Casslén and Stormby (1983) showed that atypical endometrial cell clusters, similar to those found in cervical smears, could be seen in uterine fluid in clusters of 3–40 cells (*Figure 24.15*). A study of the DNA content of these atypical cells (Kobayashi *et al.*, 1984a) indicated that these cells had a diploid to octoploid DNA content with a major peak in the tetraploid region.

Effect of hormone administration

The epithelial cells lining the urogenital tract are sensitive to oestrogen, progesterone and other steroid hormones. The administration of these hormones for the purpose of contraception, replacement therapy or tumour control, results in cytologcial changes in cervical and vaginal smears and in smears of urinary sediment.

Cellular changes due to contraceptive drugs

Clinical trials of contraceptive steroid therapy began in the late 1950s. The cervicovaginal epithelium is affected by the action of the steroid contained in oral contraceptive drugs and often reflects a progesterone pattern (Ober, 1977). The presence of giant endocervical cells showing multinucleation and prominent nucleoli has been reported as an occasional finding in cervical smears from patients receiving oral contraceptive therapy (Kline, Holland and Wemple, 1969; Koss and Melamed, 1970). These changes may reflect the presence of microglandular hyperplasia commonly found in cervical biopsies from women taking these drugs. Occasionally, an Arias-Stella-like phenomenon comprises part of these lesions and this may be reflected in the smear as papillary fragments. Despite the presence of occasional atypical cells in the smears of oral contraceptive users, there is no evidence to suggest that these women have an increased risk of cervical neoplasia (Jordan and Singer, 1980).

Hormone replacement therapy (HRT)

Exogenous hormones are frequently administered to women at the menopause to counter the effects of hot flushes, dyspareunia or dysarthria associated with reducing ovarian function after the fourth decade of life. They are usually administered in tablet form although oestrogen cream is sometimes used. Whatever the form of therapy, cervical smears may appear well oestrogenized in women who have not menstruated for many years. Although a proliferative phase smear pattern in a post-menopausal woman is most commonly due to HRT, it is important to bear in mind the differential diagnosis in such cases. The rare oestrogen-producing tumours such as granulosa cell tumours must be considered as a possible cause of the abnormal smear pattern. Occasionally, HRT may result in endometrial proliferation and hyperplasia so that atypical endometrial cells are found in the smears. Curettage is advised to exclude the possibility of coincidental endometrial carcinoma in these cases.

Smear patterns in diethylstilboestrol (DES) exposed women

An association between prenatal exposure to DES and the development of clear cell carcinoma of the cervix was first proposed by Herbst, Ulfelder and Poskaner (1971). Subsequent studies have shown that the risk of this tumour is not high; nevertheless it should be borne in mind whenever smears from daughters of women who had taken DES in pregnancy are examined.

More recent studies of females who have been exposed to DES *in utero* show that they have an increased risk of several non-neoplastic conditions. These include the development of cervical hoods and cervicovaginal ridges and various types of structural abnormalities of the cervix. In addition, they may have extensive ectropion extending well beyond the external os onto the vaginal wall. The presence of ectopic foci of glandular epithelium in the vagina—a condition which has been designated 'vaginal adenosis'—has also been noted.

The large expanse of glandular epithelium in these patients undergoes metaplastic change in the normal way. Stafl and Mattingly (1974) suggested that in view of the greatly extended transformation zone in these women, they may be at increased risk of CIN. However, this has not been borne out by experience. Nevertheless, vaginal adenosis may be suspected cytologically whenever glandular cells or immature metaplastic cells are seen in a vaginal smear.

Cellular changes due to oestrogen therapy for prostatic cancer

Oestrogen therapy is frequently given to patients with well differentiated carcinoma of the prostate. Cytological changes reported in these patients include the presence of glycogen-

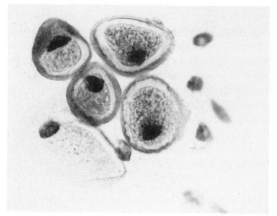

Figure 24.16 Cellular changes due to oestrogen therapy for prostatic cancer. Glycogen-bearing squamous-like cells in urinary smear. Note numerous glycogen deposits in the cytoplasm. Stain: Papanicolaou. Magnification × 400

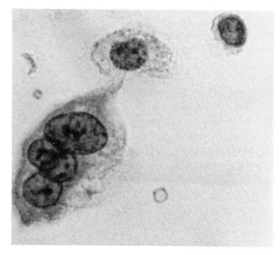

Figure 24.17 Catheterized uretheral urine. Multinucleated cell. Note the large nuclei and pronounced chromocentres. The cytoplasm is somewhat vacuolated. Stain: Papanicolaou. Magnification × 400

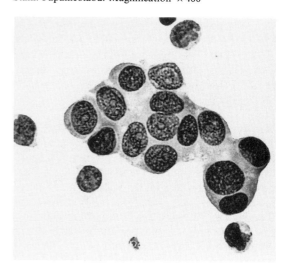

Figure 24.18 Peritoneal dialysis fluid preparation. Cluster of activated mesothelial cells. Note hyperchromatic nuclei with pronounced nucleoli. Stain: May-Grünwald–Giemsa. Magnification × 400, reduced to 80% in reproduction

bearing squamous cells in the urinary sediment (*Figure 24.16*).

Effect of instrumentation on the cells

Many of the instruments used to evaluate, diagnose or treat disease coincidentally cause damage to the epithelial cells. Cytologists must familiarize themselves with the cellular changes which result from catheterization, aspiration or brushings of mucosal surfaces, as well as the changes associated with the peritoneal dialysis and ventriculoperitoneal or ventriculopleural shunts. These changes are discussed in the paragraphs below.

Changes following catheterization

Immediately after catheterization of the bladder, large sheets of urothelial cells appear in the urine. Single cells with columnar configuration and hyperchromatic nuclei may occasionally be noted and may be mistaken for adenocarcinoma. The most striking cell change is multinucleation (*Figure 24.17*); the urothelial cells contain 2–50 or more nuclei of variable size with prominent chromocentres, and probably originate from the umbrella cell layer of the urothelium. In patients with prolonged catheter placement, the urine often contains crystalline debris which probably represents various calcium salts (Tweedale, 1977).

Changes due to peritoneal dialysis

Changes in the cells in peritoneal dialysis fluid specimens have been described by Koss (1979) and Carlon and Giustina (1983). Atypical mesothelial cells may be present in cohesive sheets or as isolated cells (*Figure 24.18*). The nuclei are slightly enlarged and show a coarse, granular chromatin structure, and sometimes have a high nucleocytoplasmic ratio. Biopsies of

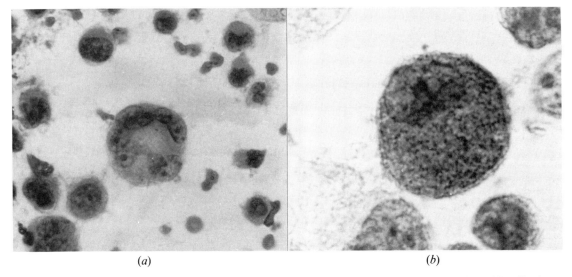

Figure 24.19 Smear from ascitic fluid. (a) Markedly atypical tumour cells are seen. Stain: Papanicolaou. Magnification ×400. (b) Immunoperoxidase staining. Note GFAP-positive granules in the cytoplasm of tumour cells. Stain: anti-GFAP-peroxidase antiperoxidase with haematoxylin counterstain. Magnification ×1000

the peritoneum of these patients show a fibrinous peritonitis and proliferation of mesothelial cells (Koss, 1979). The presence of erythrophagocytosis has also been observed in cells recovered from dialysis fluid specimens.

Extracranial spread of brain tumour via ventriculoperitoneal or ventriculopleural shunt

The value of a shunt in the management of a patient with a posterior fossa tumour prior to craniotomy has been well established (Abraham and Chandy, 1963). Extracranial metastasis of central nervous system neoplasms via the shunt is a rare complication. However, a few cases of the passage of brain tumour cells through a ventriculoperitoneal or ventriculopleural shunt to the pleural or peritoneal cavity have been reported (Brust, Moiel and Rosenberg, 1968; Kessler, Dugan and Concannon, 1975; Hoffman, Hendrick and Humphreys, 1976). The author has personal experience of extracranial spread of a glioblastoma via a ventriculoperitoneal shunt. The metastatic glioblastoma cells were detected in ascitic fluid (*Figure 24.19a*). Immunoperoxidase staining of these malignant cells showed that they were strongly positive for glial fibrillary acidic protein (GFAP) (*Figure 24.19b*) which indicated their astrocytic lineage. A similar observation has been made by Kimura *et al.*, (1984) of a pineal endodermal sinus tumour which metastasized to the peritoneal cavity through a ventriculoperitoneal shunt and caused ascites. Subsequently, α-fetoprotein-containing tumour cells were demonstrated by immunocytochemistry.

Acknowledgements

I am deeply indebted to Michael Dictor, MD for reviewing the manuscript and for his helpful comments. I also thank Toshiyuki Tsujioka, CT (IAC), Akira Inagaki, CT (IAC), Takashi Okumoto, CMIAC, Iwao Matsushita, CMIAC, Tarumi Yamaki, MD, Hyoe Yamada, MD, Isamu Sawaragi, MD, MIAC and Nils Stormby, MD, FIAC for their invaluable contributions to this chapter.

References

ABRAHAM, J. and CHANDY, J. (1963). Ventriculoatrial shunt in the management of posterior fossa tumours. *J. Neurosurg.*, **20**, 252–253

ALBUKERK, J. and GNECCO, C. A. (1977). Atypical cytology in tubal pregnancy. *J. Reprod. Med.*, **19**, 273–276

ASHTON, P.R. and JOHNSTON, W. W. (1975). Cytopathologic alterations associated with intrauterine contraceptive devices. *Acta Cytol.*, **19**, 583.

AYRE, J. (1965), Human precarcinogenic cell manifestations associated with polyethylene contraceptive device. *Industrial Med. Surg.*, **34**, 393–403

BANKS, M. D., PONTES, J. E., IZBICKI, R. M. and PIERCE, J. M. (1977). Topical instillation of doxorubicin hydrochloride in the treatment of recurring superficial transitional cell carcinoma of the bladder. *J. Urol.*, **118**, 757–760

BEDROSSIAN, C. W. M. and COREY, B. J. (1978). Abnormal sputum cytopathology during chemotherapy with bleomycin. *Acta Cytol.*, **22**, 202–207

BELLIN, H. J., CHERRY, J. M. and KOSS, L. G. (1974). Effects of a single dose of cyclophosphamide: V. Protective effect of diversion of the urinary stream on dog bladder. *Lab. Invest.*, **30**, 43–47

BIBBO, M. (1979). Look-alikes in cytology of the female genital tract. In *Compendium on Diagnostic Cytology*, 4th Edition, (Wield, G. L., Koss, L. G. and Reagan, J. W., eds), pp. 194–197. Chicago: Tutorial of Cytology

BRUST, J. C. M., MOIEL, R. H. and ROSENBERG, R. N. (1968). Glial tumour metastases through a ventriculo-pleural shunt: resultant massive pleural effusion. *Arch. Neurol.*, **18**, 649–653

CARLON, G. and GIUSTINA, D. D. (1983). Atypical mesothelial cells in peritoneal dialysis fluid. *Acta Cytol.*, **27**, 706–708

CARTER, S. K. (1980). Biologic response-modifying agents: what is an appropriate phase I–II strategy?. *Cancer Immunol. Immunotherapy*, **8**, 207–210

CASSLEN, B. (1981) Proteinases and proteinase inhibitors in uterine fluid, with special reference to IUD-users. *Acta Obstet. Gynecol. Scand.*, Supplement 98

CASSLEN, B., KOBAYASHI, T. K. and STORMBY, N. (1981). The cellular composition of uterine fluid in IUD users: a quantitative study. *Contraception*, **24**, 685–694

CASSLEN, B., KOBAYASHI, T. K. and STORMBY, N. (1982). Cyclic variation of the cellular components in human uterine fluid. *J. Reprod. Fert.*, **66**, 213–218

COLEMAN, D. V. (1981). The cytologic diagnosis of human polyomavirus infection and its value in clinical practice. In *Advances in Clinical Cytology*, (Koss, L. G and Coleman, D. V., eds), pp. 136–159. London: Butterworths

COWEN, P. N. (1975). False cytodiagnosis of bladder malignancy due to previous radiotherapy. *Br. J. Urol.*, **47**, 405–412

DREW, J. E. and MARSHAL, V. F. (1968). The effects of topical thiotepa on the recurrence rate of superficial bladder cancers. *J. Urol.*, **99**, 740–743

ESCHENBACH, D. A., HARNISCH, J. P. and HOLMES, K. K. (1977). Pathogenesis of acute pelvic inflammatory disease: role of contraception and other risk factors. *Am. J. Obstet. Gynecol.*, **128**, 838–850

FAULKNER, W. L. and ORY, H. W. (1976). Intrauterine devices and acute pelvic inflammatory disease. *J. Am. Med. Ass.*, **235**, 1851–1853

FEINGOLD, M. L. and KOSS, L. G. (1969). Effects of long term administration of busulfan. *Arch. Intern. Med.*, **124**, 66–71

FORNI, A. M., KOSS, L. G. and GELLER, W. (1964). Cytologic study of the effect of cyclophosphamide on the epithelium of the urinary bladder in man. *Cancer*, **17**, 1348–1355

GAWAD, A. H. A., TOPPOZADA, H. K., SAWI, M. E., SALEH, F. and SAHWI, S. E. (1977). Study of the uterine environment in association with intrauterine contraceptive devices. *Contraception*, **16**, 469–485

GONDOS, B., SMITH, L. R. and TOWNSEND, D. E. (1970). Cytologic changes in cervical epithelium following cryosurgery. *Acta Cytol.*, **14**, 386–389

GONZALEZ-MERLO, J., AUSIN, J., LEJARCEQUI, A. and MARQUEZ, M. (1973). Regeneration of the ectocervical epithelium after its destruction by electrocauterization. *Acta Cytol.*, **17**, 366–371

GRAHAM, R. M. and GRAHAM, J. B. (1953). A cellular index of sensitivity to ionizing radiation: the sensitization response. *Cancer*, **6**, 215–223

GUPTA, P. K. (1982). Intrauterine contraceptive devices: vaginal cytology, pathologic changes and clinical implications. *Acta Cytol.*, **26**, 571–613

GUPTA, P. K., PINN, V. M. and TAFT, P. D. (1969). Cervical dysplasia associated with azathioprine (Imuran) therapy. *Acta Cytol.*, **13**, 373–376

GUPTA, P. K., HOLLANDER, D. H. and FROST, J. K. (1976). Actinomycetes in cervico-vaginal smears: an association with IUD usage. *Acta Cytol.*, **20**, 295–297

GUPTA, P. K., BURROUGHS, F., LUFF, R. D., FROST, J. K. and EROZAN, Y. S. (1978). Epithelial atypias associated with intrauterine contraceptive device (IUD). *Acta Cytol.*, **22**, 286–291

HADDOW, A. and TIMMIS, G. M. (1953). Myleran in chronic myeloid leukemia: chemical constitution and biological action. *Lancet*, **i**, 207–208

HASEGAWA, T., TSUTUI, F. and KURIHARA, S. (1975). Cytomorphologic study on the atypical cells following cryosurgery for the treatment of chronic cerivicitis. *Acta Cytol.*, **19**, 533–537

HEARD, B. E. and COOKE, R. A. (1968). Busulfan lung. *Thorax*, **23**, 187–193

HERBST, A. L., ULFELDER, H. and POSKANER, D. C. (1971). Adenocarcinoma of the vagina; association of maternal stilboestrol with tumour appearance in young women. *N. Engl. J. Med.*, **284**, 878–881

HERTING, V. M. and TAUBER, P. F. (1978). Endometrium-Zytologie bei Kupferhaltigen Intrauterinpessaren. *Fortschr. Med.*, **96**, 311–314

HOFFMAN, H. J., HENDRICK, E. B. and HUMPHREYS, R. B. (1976). Metastasis via ventriculoperitoneal shunt in patients with medulloblastoma. *J. Neurosurg.*, **44**, 562–566

HOLMQUIST, N. D., BELLINA, J. H. and DANOS, M. L. O. (1976). Vaginal and cervical cytologic changes following laser treatment. *Acta Cytol.*, **20**, 290–294

HSU, C., CHOO, Y-C., SETO, W-H. *et al.* (1984). Exfoliative cytology in the evaluation of interferon treatment of cervical intraepithelial neoplasia. *Acta Cytol.*, **28**, 111–117

ISHIHAMA, A., KAGABU, T., IMAI, T. and SHIMA, M. (1970). Cytologic studies after insertion of intrauterine contraceptive devices. *Acta Cytol.*, **14**, 35–41

ISRAEL, R. and DAVIS, H. J. (1966). Effect of intrauterine contraceptive devices on the endometrium. *J. Am. Med. Ass.*, **195**, 144–148

JORDAN, J. A. and SINGER, A. (1980). Effect of oral contraceptive steroids upon epithelium and mucus. In *The Cervix*, (Jordan, J. A. and Singer, A., eds), pp. 192–209. Philadelphia: W. B. Saunders

KAPLAN, I., GOLDMAN, J. and GER, R. (1973). The treatment of erosion of the uterine cervix by use of the carbon dioxide laser. *Am. J. Obstet. Gynecol.*, **41**, 795–800

KATANO, M. and TORISU, M. (1982). Neutrophil-mediated tumour cell destruction in cancer ascites. *Cancer*, **50**, 62–68

KAY, S., FRABLE, W. J. and HUME, D. M. (1970). Cervical dysplasia and cancer developing in women on immunosuppression therapy for renal homotransplantation. *Cancer*, **26**, 1048–1052

KESSLER, L. A., DUGAN, P. and CONCANNON, J. P. (1975). Systemic metastases of medulloblastoma promoted by shunting. *Surg. Neurol.*, **3**, 147–152

KIMURA, N., NAMIKI, T., WADA, T. and SASANO, N. (1984). Peritoneal implantation of endodermal sinus tumour of the pineal region via a ventriculoperitoneal shunt: cytodiagnosis with immunocytochemical demonstration of alpha-fetoprotein. *Acta Cytol.*, **28**, 143–147

KLINE, T. S., HOLLAND, M. and WEMPLE, D. (1969). Atypical cytology with contraceptive hormone medication. *Am. J. Clin. Pathol.*, **53**, 215–222

KOBAYASHI, T. K. and TARA, K. (1983). *Actinomyces*-like organisms in cervicovaginal cytology from women wearing an IUD. *Jpn. J. Med. Technol.*, **32**, 622

KOBAYASHI, T. K., CASSLEN, B. and STORMBY, N. (1982). Enzyme cytochemical studies of the mononuclear phagocytic cells in human uterine fluid. *Anal. Quant. Cytol.*, **4**, 221–224

KOBAYASHI, T. K., CASSLEN, B. and STORMBY, N. (1983). Cytologic atypias in the uterine fluid of intrauterine contraceptive device users. *Acta Cytol.*, **27**, 138–141

KOBAYASHI, T. K., YUASA, M., FUJIMOTO, T., FUKUDA, M., HARAMI, K. and HAYASHI, K. (1980). Cytologic findings in post-partum and postabortal smears. *Acta Cytol.*, **24**, 328–334

KOBAYASHI, T. K., FUJIMOTO, T., OKAMOTO, H., HARAMI, K. and YUASA, M. (1982). The cells from intrauterine contraceptive devices, with special reference to the occurrence of mast cells. *Acta Cytol.*, **26**, 614–617

KOBAYASHI, T. K., FUJIMOTO, T., OKAMOTO, H., HARAMI, K. and YUASA, M. (1983). Cytologic evaluation of atypical cells in cervicovaginal smears from women with tubal pregnancies. *Acta Cytol.*, **27**, 28–32

KOBAYASHI, T. K., UENO, T., TANAKA, N., CASSLEN, B. and STORMBY, N. (1984a). Nuclear DNA content of atypical glandular cells in the uterine fluid of IUD users. *Acta Cytol.*, **28**, 192–194

KOBAYASHI, T. K., YAMAKI, T., YOSHINO, E., HIGUCHI, T. and KAMACHI, M. (1984b). Immunocytochemical demonstration of carcinoembryonic antigen in cerebrospinal fluid with carcinomatous meningitis from rectal cancer. *Acta Cytol.* **28**, 430–434.

KOSS, L. G. (1967). A light and electron microscopic study of the effects of a single dose of cyclophosphamide on various organs in the rat: I. The urinary bladder. *Lab. Invest.*, **16**, 44–65

KOSS, L. G. (1979). *Diagnostic Cytology and its Histopathologic Bases*, 3rd Edition, pp. 510–530, 590–601, 718–745, 902, 962–964. Philadelphia and Toronto: J. B. Lippincott

KOSS, L. G. and MELAMED, M. R. (1970). Epithelial abnormalities on the ectocervix during pregnancy. *J. Reprod. Med.*, **4**, 13–14

KOSS, L. G., MELAMED, M. R. and MAYER, K. (1965). The effect of busulfan on human epithelia. *Am. J. Clin. Pathol.*, **44**, 385–397

MASTERSON, B. J., KRANTZ, K. E., CALKINS, J. W., MAGRINA, J. F. and CARTER, R. P. (1981). The carbon dioxide laser in cervical intraepithelial neoplasia: a five-year experience in treating 230 patients. *Am. J. Obstet. Gynecol.*, **139**, 565–567

MITCHELL, R. J. (1971). Intravesical thio-tepa in treatment of transitional cell bladder carcinoma. *Br. J. Urol.*, **43**, 185–188

MOORE, G. E. (1958). Clinical experience with triethylene thiophosphoramide, with special reference to carcinoma of the breast. *Ann. N. Y. Acad. Sci.*, **68**, 1074–1080

NELSON, B. M. and ANDREWS, G. A. (1964). Breast cancer and cytologic dysplasia in many organs after busulfan (Myleran). *Am. J. Clin. Pathol.*, **42**, 37–44

NG, A. B. P. and REAGAN, J. W. (1979). Changes simulating adenocarcinoma. In *Compendium on Diagnostic Cytology*, 4th Edition, (Wied, G. L., Koss, L. G. and Reagan, J. W., eds), pp. 173–175. Chicago: Tutorial of Cytology

NORUM, M. L., MOYER, D. L., TOWNSEND, D. E. and HIROSE, F. M. (1969). Ultrastructural changes in normal human ectocervical epithelium immediately following cryosurgery. *Lab. Invest.*, **21**, 11–18

OBER, W. B. (1977). Effects of oral and intrauterine administration of contraceptives on the uterus. *Human Pathol.*, **8**, 513–527

OLINER, H., SCHWARTZ, R., RUBIO, F. and DAMESHEK, W. (1961). Interstitial pulmonary fibrosis following busulfan therapy. *Am. J. Med.*, **31**, 134–139

O'MORCHOE, P. J., LEE, D. C. and KOZAK, C. A. (1983) Esophageal cytology in patients receiving cytotoxic drug therapy. *Acta. Cytol.*, **27**, 630–634

ONO, S., SAKAMOTO, K., NIITSUMA, K., SEKIMOTO, S., KONNO, K. and KIMURA, K. (1982). The carbon dioxide laser therapy for cervical intraepithelial neoplasia: cytological findings according to time following irradiation. *J. Jpn. Ass. Clin. Cytol.*, **21**, 559–568

PARR, E. L., SCHAEDLER, R. W. and HIRSCH, J. G. (1967). The relationship of polymorphonuclear leukocytes to infertility in uteri containing foreign bodies. *J. Exp. Med.*, **126**, 523–537

PATTEN, S. F., REAGAN, J. W., OBENAUT, M. and BALLARD, L. A. (1963). Post-irradiation dysplasia of uterine cervix and vagina: an analytical study of the cells. *Cancer*, **16**, 173–182

PENN, I. (1978). Malignancies associated with immunosuppressive or cytotoxic therapy. *Surgery*, **83**, 492–502

PIVER, M. S., WHITELEY, J. P. and BOLOGNESE, R. L. (1966). Effect of an IUD upon cervical and endometrial exfoliative cytology. *Obstet. Gynecol.*, **28**, 528–531

POTTS, M. and PEARSON, R. M. (1967). A light and electron microscope study of cells in contact with intrauterine contraceptive devices. *J. Obstet. Gynaecol. Br. Commonw.*, **74**, 129–136

RASMUSSEN, K., PETERSON, B. L., JACOBO, E., PENICK, G. D. and SALL, J. (1980). Cytologic effects of thiotepa and adriamycin on normal canine urothelium. *Acta Cytol.*, **24**, 237–243

SAGIROGLU, N. AND SAGIROGLU, E. (1970). The cytology of intrauterine contraceptive devices. *Acta Cytol.*, **14**, 58–64

SCHAWLOW, A. L. and TOWNES, C. H. (1958). Infrared and optical lasers. *Phys. Rev.*, **112**, 1940–1946

SCHRAMM, G. (1970). Development of severe cervical dysplasia under treatment with azathioprine (Imuran). *Acta Cytol.*, **14**, 507–509

SINGER A. and WALKER, P. (1985). The treatment of CIN: conservative methods. *Clin. Obstet. Gynaecol.*, **12**, 121–132

SLOWAY, M. S., MURPHY, W. M., DEFURIA, M. D., CROOKE, S. and FINEBAUM, P. (1981). The effect of mitomycin C on superficial bladder cancer. *J. Urol.*, **125**, 646–648

SOSTMAN, M. S., MATTHAY, R. A., PUTMAN, C. E. and SMITH, G. J. (1976). Methotrexate-induced pneumonitis. *Medicine*, **55**, 371–388

STAFL, A. and MATTINGLY, R. F. (1974). Vaginal adenosis: a pre-cancerous lesion? *Am. J. Obstet. Gynecol.*, **120**, 666–673

STAFL, A., WILKINSON, E. J. and MATTINGLY, R. F. (1977). Laser treatment of cervical and vaginal neoplasia. *Am. J. Obstet. Gynecol.*, **128**, 128–136

TAETLE, R., DICKMAN, P. S. and FELDMAN, P. S. (1978). Pulmonary histopathologic changes associated with melphalan therapy. *Cancer*, **42**, 1239–1245

TATUM, H. J., SCHMIDT, F. H., PHILLIPS, D., MCCARTY, M. and O'LEARY, W. M. (1975). Microbial migration in the thread attached to an IUD as a possible factor in infectious complications. *J. Am. Med. Ass.*, **231**, 711–717

TOWNSEND, D. E. (1976). The management of cervical lesions by cryosurgery. In *The Cervix*, (Jordan, J. A. and Singer, A., eds), pp. 305–313. London: W. B. Saunders

TWEEDALE, D. N. (1977). *Urinary Cytology*, 1st Edition, pp. 33–63. Boston: Little Brown

VALICENTI, J. F., PAPPAS, A. A., GRABER, C. D., WILLIAMSON, H. O. and WILLIS, N. F. (1982). Detection and prevalence of IUD-associated *Actinomyces* colonization and related morbidity: a prospective study of 69 925 cervical smears. *J. Am. Med. Ass.*, **247**, 1149–1152

VON HAMM, E. (1979). Radiation cell changes. In *Compendium on Diagnostic Cytology*, 4th Edition, (Wied, G. L., Koss, L. G. and Reagan, J. W., eds), pp. 242–267. Chicago: Tutorial of Cytology

WENTZ, W. B. and REAGAN, J. W. (1970). Clinical significance of post-irradiation dysplasia of the uterine cervix. *Am. J. Obstet. Gynecol.*, **106**, 812–817

WYNN, R. M. and SAWARAGI, I. (1969). Effects of intrauterine and oral contraceptive on the ultrastructure of the endometrium. *J. Reprod. Fert. (Supplement)*, **8**, 45–57

YAGODA, A., WATSON, R. C., WHITMORE, W. F., GRABSTALD, H., MIDDLEMAN, M. P. and KRAKOFF, I. H. (1977). Adriamycin in advanced urinary tract cancer: experience in 42 patients and a review of the literature. *Cancer*, **39**, 279–285

ZIMMER, T. S. (1959). Late irradiation changes: cytological study of cervical and vaginal smears. *Cancer*, **12**, 193–196

25

Automation

Keith C. Watts and O. A. N. Husain

In recent years cellular pathology has benefited from the advances made in biophysical and computer technology. By computer analysis, the morphological and functional properties of cells can be quantified enabling objective reproducible descriptions of specimens to be recorded. Analytical studies for leucocyte differentiation, cell counting, cell cycle kinetics, automated karyotyping and morphometry are well established in routine laboratory practice.

Other applications of this approach have been in the development of automated systems for cervical cancer screening. In the UK screening programmes for cervical cancer have generated an ever increasing workload (currently estimated at 3 million smears per annum) for routine cytology laboratories. The manual examination of these smears has proved to be a time-consuming, cost-ineffective task. Quality control studies have revealed a false negative rate as high as 14% (Husain et al., 1974a); cytotechnologists miss an average of 7–10% of all cancer cases (Wied et al., 1981). For these reasons the concept of automation has received widespread attention since the early 1950s.

Automated analytical systems have developed via two different technical principles:

(1) Static image cytometry (where cells are prepared on glass microscope slides), which is based upon microspectrophotometry.
(2) Flow cytometry where cells in suspension are forced through a focused beam of monochromatic light (usually a laser).

Microspectrophotometry

Microspectrophotometry or microdensitometry is the measurement of the concentration or mass of a chromophore in microscopically defined regions of cells or tissues. It was first used in the measurement of DNA using the Feulgen reaction but has since been applied to the measurement of cellular biochemical activity by using stoichiometric chromogenic reactions (Bitensky, 1980).

Static cytometry instrumentation

High resolution image analysis is achieved using a combination of a microscope with a stepping motor driven stage, an optical sensor, e.g. television camera or linear array diode, and a computer. The stage moves the slide bearing 'stained cells' through the scan line of the optical sensor generating light absorbance or transmission values which are processed in the computer thus creating a digitized impression of the cells. The instrumentation of high resolution systems including the development of laser scanning microscopes is described by Bartels, Koss and Weid (1980).

Early developments in systems designed for cervical screening

One of the first instruments for cytology automation was the Cytoanalyser (Bostrom, Sawyer and Tolles, 1959). This used a 'Nipkow disc' image plane scanner consisting of a rotating disc with a series of small holes arranged in spiral formation round the periphery. The light passing through each hole in turn as it traversed the image was converted into a corresponding electrical signal by a photomultiplier tube. This signal was then thresholded to find dark image areas corresponding to cell nuclei. The area and density of each nucleus was measured and the results passed to classification circuits. Cells of various types were then grouped and counted, and false measurements from objects such as cell clusters and debris eliminated or cancelled. However, clinical trials with the Cytoanalyser gave disappointing results (Spencer and Bostrom, 1962).

A new interactive approach was developed in the UK by the Vickers Company which featured a scanner which would prescreen slides and recognize suspicious objects (susps); the final classification of the specimen was completed by human visual inspection of the susp (Spriggs, Diamond and Meyer, 1968). The most serious problem with this system was the large number of false positive alarms caused by artefacts such as clumps of polymorphonuclear leucocytes, cell overlaps and debris. Clearly a means of reducing these false positive signals was required. Subsequent projects dedicated to the development of automated cervical prescreening retained the principle of interactivity but also included computer algorithms for pattern recognition and artefact rejection which could distinguish between positive signals due to abnormal cells and those caused by artefacts (Husain et al., 1974b; Zahniser et al., 1980; Erhardt et al., 1980; Tucker and Husain, 1981; Ploem et al., 1986; Husain and Watts, 1987). Other projects such as the CYBEST Model 3 have been more adventurous and have attempted direct machine specimen classification without human interaction (Tanaka et al., 1982).

Cytological recognition of abnormal cells

The detection of abnormal cells is based upon the measurement of morphometric and densitometric parameters which will identify the features of neoplasia. These parameters are derived from the fundamental cytological criteria for malignancy. In many instances the nuclei of neoplastic cells which have been stained with a basic dye will appear relatively enlarged and hyperchromatic; compared to the benign cell the neoplastic nucleus shows an increased chromatin condensation in the form of coarse and irregular granulation, especially beneath the nuclear membrane. The cytoplasm of the neoplastic cells may also be reduced in overall size so that malignant cells exhibit an abnormally high nucleocytoplasmic (N/C) ratio.

Morphometric parameters

Planimetry was used to measure the nuclear and cytoplasmic areas and calculate the N/C ratio of 1000 normal and malignant epithelial cells (Johnston, 1952). From these measurements it was found that this parameter gave good but imperfect separation between normal and malignant cell populations.

Reagan and his colleagues carried out a more extensive study of 20 000 malignant cells and 11 000 normal cells, in which the nuclear and cytoplasmic areas were computed. From this study a number of significant differences were noted between populations of normal cells and those from cases of dysplasia, carcinoma *in situ* and invasive carcinoma, particularly in the nuclear shape and N/C ratio of cells (Reagan, Hamonic and Budd Wentz, 1957).

In 1961 Tolles, Horvath and Bostrom measured the nuclear area and N/C ratio from 140 000 randomly selected cells from normal and neoplastic cervical scrape specimens. The correlation of these measurements produced differences between normal and neoplastic cell populations.

DNA content

The value of quantitative DNA analysis in the diagnosis and prognosis of neoplasia has been studied extensively since the 1950s (Friedlander, Hedley and Taylor, 1984). Cytogenetic studies have demonstrated that many malignant human tumours and many benign and premalignant neoplastic conditions feature structural and/or numerical chromosome anomalies. As early as 1955 Manna described his findings on the number of chromosomes in the cells from 29 cervical carcinomas. In a later study of six carcinomas of the cervix, karyotyping demonstrated that each tumour had its individual structural as well as numerical chromosomal characteristic (Tonomura, 1959). However, at that time techniques for karyotyping were not well developed and another method of assessing the karyology of neoplastic cells was pursued, namely Feulgen cytophotometry. Atkin subsequently established the fact that the quantitative DNA analysis of cells can be used to reflect the total chromosome content (Atkin and Richards, 1956; Atkin and Kay, 1979). Studies of normal, malignant, and premalignant cells from the uterine cervix showed that DNA microspectrophotometry can be used to distinguish between neoplastic and normal cells (Atkin, 1964; Caspersson, 1964). Many programmes dedicated to the development of automated cervical screening systems have adopted the nuclear integrated optical density (IOD) parameter (Ploem *et al.*, 1986; Husain and Watts, 1987).

Quantitative cytochemistry

Cellular biochemical activity can be measured using a suitable chromogenic reaction and the microspectrophotometer. The following cytochemical studies have shown a quantitative difference in biochemical activity between normal and neoplastic tissues which could be used in automated screening (Bitensky, Chayen and Husain, 1984).

Changes in chromatin: DNA hydrolysis profiles

The Feulgen reaction can be used to demonstrate nuclear DNA by acidic hydrolysis which liberates the purines and exposes the aldehyde groups of the deoxyribose moieties which can be coloured by reaction with Schiff reagent. The rate of hydrolysis in this reaction is determined by the molarity of the acid and the temperature at which the reaction is done. By using a relatively slow rate of hydrolysis, i.e. 5 mol/l HCl at a temperature of 20 °C, different species of DNA can be identified. These species may be visualized by plotting hydrolysis profiles where multiple slides from the same sample are prepared and reacted with Schiff reagent after varying times of acid hydrolysis. The relative absorption is plotted against hydrolysis time to achieve the profile.

Millet *et al.* (1982) have applied this technique to cervical scrape specimens from patients with invasive cancer and intraepithelial neoplasia and measured both structurally normal and abnormal cells present in each specimen. Some of the invasive cancer specimens could be distinguished from the normal controls by the presence of acid-labile DNA after a relatively short hydrolysis time (5–10 min). However, the results on cases of intraepithelial neoplasia were far more variable and the sensitivity is too low to be used as a routine test.

Proteolytic enzymes

In 1980 Millet and co-workers compared the activity of lysosomal naphthylamidase in cervical scrape material from patients with intraepithelial neoplasia and invasive cancer with normal and inflammatory controls. Many of the neoplastic cases showed a high activity as did specimens collected after cryosurgery or those known to be infected with herpes simplex virus. The test had the advantage of being sufficiently robust to be done on air-dried smears several days after collection of the sample; the specificity was poor, precluding its use in routine cervical screening.

Glutathione

Increased levels of glutathione (GSH) have been found in malignancy. Millet (1979) found that the concentration of GSH in cytologically abnormal cervical epithelial cells was approximately 2.5 times greater than in normal cells from both normal and neoplastic cases. However, GSH is a small very diffusible molecule and some slides also appeared to show extracellular staining which can lead to errors in measurement by microspectrophotometry.

Pentose-shunt dehydrogenase activity

An increase in the activity of two dehydrogenases, namely glucose-6-phosphate dehydrogenase (G6PD) and 6-phosphogluconate dehydrogenase (6PGD), has been demonstrated in malignancy (Bitensky, Chayen and Husain, 1984).

Cytochemical studies demonstrated an increase in the activity of 6PGD in cells which had undergone proliferative or neoplastic change (Chayen *et al.* 1962). Bonham and Gibbs (1962) subsequently presented 'a new biochemical test for gynaecological cancer' which involved measuring the 6PGD activity in vaginal fluid. The authors claimed that all cases of invasive cancer and carcinoma *in situ* were detected and that the false positive rate was only 3%. Cameron and

Husain (1965) showed that although this test detected all of the 36 invasive cervical cancers, only 6 of 11 cases of carcinoma *in situ* gave a positive result. The false positive rate was also disappointing: 30% for premenopausal cases, rising to 70% in women over 45.

Gibbs subsequently modified the method for determining the cell content of the sample. He discovered a linear relationship between the number of cells and the potassium content of the vaginal fluid. This proved to be a more consistent and accurate method of estimating the cell content than measuring the freeze-dried mass of the sample (Gibbs, Labrum and Stagg, 1968).

Rees *et al.* (1970) made further improvements to the method and used saponin to lyse the cells rather than freeze drying. The 6-phosphogluconate dehydrogenase activity was expressed as international units per milliequivalent of potassium. Vaginal fluids from 24 women with cervical neoplasia were studied; all 12 cases of invasive squamous carcinoma were positive but 3 of the 12 cases of carcinoma *in situ* assayed were not detected. The specificity had certainly improved with a false negative rate of only 7.3% in samples collected from pregnant women.

Other workers (Cohen and Way, 1966) demonstrated increased activity of G6PD in malignant cells by cytochemical methods. Altman *et al.* (1970) suggested that the increased G6PD activity found in neoplastic cells as compared to normal controls was enhanced if the reaction was done in an atmosphere of oxygen.

The cytochemical demonstration of dehydrogenase activity involves the reduction of $NADP^+$ to $NADPH_2$ and the transfer of the 'hydrogen' or reducing equivalents via phenazine methosulphate (PMS), an intermediate electron acceptor, to a soluble tetrazolium salt which is reduced to an insoluble formazan (Chayen, Bitensky and Butcher, 1973). In all these studies neotetrazolium chloride (NT) was used as the final hydrogen acceptor. If the reaction is performed on normal cells in an atmosphere of oxygen, competition for the reducing equivalents takes place between the oxygen and the NT and very little formazan is produced. However, when the reaction in oxygen is done on malignant cells, there appears to be some mechanism which prevents oxygen from competing with NT for reducing equivalents derived from the NADPH. This test was used on 15 benign and 18 malignant cytological smears prepared from gastric brushings; in an atmosphere of oxygen the residual activity of G6PD in the malignant cases was 31% as opposed to only 0.02% in the smears from benign cases (Ibrahim *et al.*, 1983).

Immunocytochemistry

Advances made in the development of immunocytochemical and hybridoma techniques have led to a growing number of highly specific monoclonal antibodies and an expanding spectrum of detectable antigens in human tissues. Attempts to raise antibodies are in progress to produce an antibody which would identify and distinguish between neoplastic and normal cells. Immunocytochemical techniques have been developed which can be adapted to both flow and image cytometry (Husain *et al.*, 1984; Moncrieff, Omerod and Coleman, 1984).

Texture features and malignancy-associated changes

With the introduction of high resolution instrumentation diagnostic clues offered by cytologically normal intermediate squamous cells have been demonstrated. In 1981 Burger *et al.* identified a pattern using monochromatic light (Burger, Jutting and Rodenacker, 1981). Other workers (Wied *et al.*, 1980; Bibbo *et al.*, 1981) demonstrated colour based features that distinguish between cytologically normal intermediate cells in patients with neoplastic lesions, from similar cells in women without disease. This observation is probably a reflection of the so-called malignancy-associated changes first described by Nieburgs in 1959. Malignancy associated changes were observed using an oil immersion lens in the buccal squamous cells of patients with malignant tumours (Nieburgs, Herman and Reisman, 1962). The obvious advantage of this technique is that searching the sample for neoplastic cells would not be necessary; a statistical test could be applied to determine whether the sample should be referred for human review or classified as normal. A distinction between lesions which are likely to regress and those which are at risk of progression to invasive cancer could serve to prevent unnecessary referrals for colposcopy. Measurements of the texture features in Papanicolaou-stained smears from women with CIN I–II have been

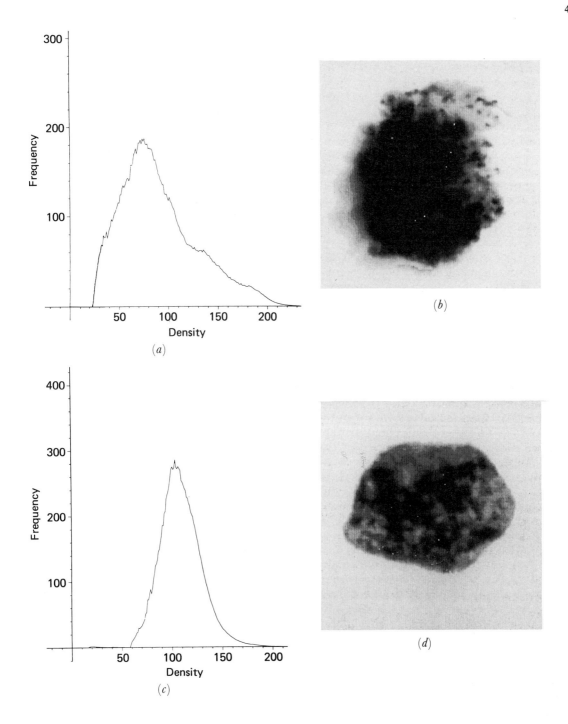

Figure 25.1 (a) Pixel density histogram (texture) of artefact, shown in (b); (c) pixel density histogram (texture) of a cell nucleus, shown in (d). Reproduced by permission of Dr J. Tucker, Pattern Recognition Section, MRC Clinical and Population Cytogenetics Unit, Edinburgh

made but it has not been possible to demonstrate any correlation with subsequent progression or regression of the lesion (Timmers, 1987).

Pattern recognition and artefact rejection

Most of the artefact rejection techniques incorporated in automated screening instruments have been designed to test morphological features of detected objects. These differentiate between the relatively smooth, round/elliptical outline of the cell nucleus and the more irregular shape of the artefact. The features used have generally been those which can be easily extracted from digitized images of microscope fields. Although the details and implementations of artefact rejection techniques vary considerably from system to system, many of the concepts are basically similar. For example the height/width ratio of an object has been used to eliminate long thin objects such as dust fibres, mucus and fungal hyphae. Another algorithm of this type is the box test in which the measured area of the object is compared with the area of the enclosing rectangle calculated from vertical and horizontal projections and chords. Other techniques make use of local features of the object outline. Indentation tests have been developed to detect large irregularities in object outline. The corner test removes artefacts exhibiting outlines with sharp angles or corners. These tests can provide a substantial reduction in the number of false positive signals due to artefacts. However, they may also cause misclassification of abnormal, highly irregular, elongated nuclei as artefactual (Husain, Tucker and Page Roberts, 1976). Other shape and contour measurements of the nucleus and cytoplasm have been employed to reduce false positive alarms (Bowie and Young, 1977a,b; Meyer, 1979). Texture features have also shown some ability to separate high IOD objects into cells and artefacts (*Figure 25.1*) (Tucker et al., 1989).

Flow cytometry

In 1956 Coulter introduced the concept of volume measurement of cells in suspension. This analysis was based upon the passage of non-conducting cells suspended in an electrolyte solution through a narrow aperture. By applying a potential difference to the aperture a resistance increment could be measured as an electrical pulse. Theoretical equations relating the pulse amplitude and cell volume were also developed but inaccurate results were often recorded. These artefacts were largely due to differences within the electrical field of the aperture entrance and exit. Subsequent modifications of this principle were introduced which overcame the problems and improved the resolution of the system. One of the most useful advances was the introduction of a coaxial flow of cells suspended within a sheath of cell-free solution. Later developments included the introduction of a system which measured more than one parameter, namely light scatter, and absorption of unstained cells illuminated with ultraviolet light (Kamentsky, Derman and Melamed, 1963).

Light scatter

When a cell crosses a light beam through a flow chamber some of the original light is scattered and falls outside the original beam. Modern flow systems include apparatus for the detection or analysis of this scattered light using one or more detectors at selected angles to the original beam. The signals received by the light scatter are used to assess the granularity and/or size of the cells under analysis (Brunsting, 1974; Loken, Sweet and Herzenberg, 1976). These parameters are of value in the interpretation of DNA histograms since an indication of size can distinguish between different cell populations with equal DNA content. They can also identify some false positive signals due to cell aggregates.

Fluorescence

Many of the more recently developed flow systems use fluorescence to measure cellular parameters. Fluorochromes can be cytochemically or immunocytochemically bound to cells and used to measure an ever increasing number of parameters and cellular features (*Table 25.1*). A beam of exciting monochromatic light is generated usually by a laser and focused upon a stream of cells flowing through an analysis chamber. The emission quanta are detected and recorded. Barrier and longpass filters are utilized to exclude the exciting light from the sensor. As with total absorption measurements, fluorescence emission quanta at several wave-

Table 25.1 Cellular parameters studied by cytometry

Feature	Technique	Reference
DNA	Feulgen: Acriflavine	Tanke et al. (1979)
	Thionine	Oud et al. (1981)
	DNA in situ hybridization	Baumann et al. (1984)
DNA/RNA[a]	Ethidium bromide	Le Pecq and Paoletti (1967)
	Propidium iodide	Crissman and Steinkamp (1973)
	Mithramycin	Crissman et al. (1978)
	Hoechst benzimide: 33258	Hillwig and Gropp (1972)
	Indole compounds: DAPI/DIPI	Stohr and Goerttler (1979)
	Acridine Orange	Adams (1974)
RNA	Pyronine Y	Tanke et al. (1980)
Cell cycle S-phase	Anti-bromodeoxyuridine	Dean et al. (1984)
Protein	Sulphorhodamine 101	Stohr et al. (1980)
	Fluorescamine	Stohr et al. (1977)
Antibody labelling	Fluorescein isothiocyanate	Crissman and Steinkamp (1973)
	Rhodamine isothiocyanate	Crissman and Steinkamp (1982)
	Phycoerythrin	Vernon et al. (1982)
Enzymes	Fluorogenic substrates	Watson (1980)
		Dolbeare and Smith (1979)
Volume	Electrical conductivity	Kachel (1976)
Size	Small angle light scatter	Brunsting (1974)
Granularity	Large angle light scatter	Loken et al. (1976)
Oestrogen receptors	Fluorescent oestrogen probe	Kute et al. (1983)
Oncogenes	c-myc-oncoprotein	Watson et al. (1985)

[a] Ribonuclease should be used for specific DNA measurements

lengths can be measured simultaneously using sensors with dichroic filters.

Cell sorting

Another facility offered by the flow cytometer is the electronic cell sorting technique (Melamed, Mullaney and Mendelsohn, 1979) which enables the cells to be separated into discrete populations. After passing through the laser beam the stream is converted into tiny droplets by a piezoelectric crystal oscillator attached to the orifice outlet assembly. The droplets are generated at a rate of approximately 30 000–40 000/s. The average rate of particle analysis is about 1000/s as only 3–4% of the droplets contain a cell. Droplets that contain cells within certain defined limits according to electronic thresholds can be electrostatically charged. This analysis of selected parameter values is known as 'windowing'. Traversing downstream through a strong electric field, these droplets are deflected to the left or the right, depending on the polarity of the charge applied to them. The deflected droplets can be collected in tubes or deposited directly into 'cytobuckets' for monolayer preparation and subsequent visual assessment or image analysis (*Figure 25.2*) (Tanke et al., 1983).

Cervical screening using flow analysis

The speed, accuracy and statistical reliability offered by flow systems has led to the idea of applying it to the automation of cervical cancer screening. Initial trials using cell volume as a discriminatory parameter were not successful (Ladinsky, Sarto and Peckham, 1964) but the measurement of DNA content has shown more promise for identifying neoplastic specimens. In

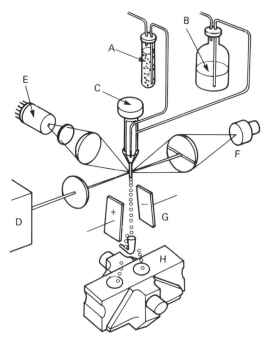

Figure 25.2 Schematic representation of a flow sorter used for sorting cells for subsequent image analysis. A, Cell suspension; B, sheath fluid; C, nozzle with piezoelectric crystal; D, laser; E, fluorescence detector; F, light scatter detector; G, deflection plates; H, centrifugation bucket. Reproduced with permission from Tanke et al. (1983)

a study of 264 cervical smears stained with acridine orange, using hypertetraploidy as the criterion of neoplasia, two-thirds of the false positive results were due to cell and bacterial aggregates (Taft and Adams, 1979). In 1979 Sprenger and Witte reported 15% false negative and 39% false positive rates using this approach. Other workers have evaluated a multiparametric approach which includes the use of DAPI and sulphorhodamine (SR) 101 stain for DNA and protein respectively (Stohr and Goerttler, 1979; Stohr et al., 1980), but the number of false alarm signals was still unacceptably high. Other studies have shown that although flow analysis can be used to detect abnormal cervical scrape specimens, a significant reduction in the number of false positive signals is essential before flow analysis can be considered suitable for automated cervical cancer screening (Fowlkes, Herman and Cassidy, 1976; Jensen, Mayall and King, 1979; Gill et al., 1979; Barrett et al., 1979; Habberset et al., 1979; Linden et al., 1979).

Specimen preparation for static and flow cytometry

Special preparatory methods are necessary before analysis of cells and tissues can be achieved. Some of the preparation techniques are applicable to both flow and static image analysis systems, and many have been adopted for the preparation of cervical scrape material (Wheeless and Onderdonk, 1974). Obviously, whole tissue samples cannot be analysed on an individual cell basis without separating the cells; but even specimens collected by aspiration or exfoliation may contain sheets of cells, syncytia or three-dimensional primary clusters. Primary clusters and sheets are those cells grown together possessing desmosome links; the secondary clusters are due to reaggregation in solution. Mechanical and biochemical techniques have been used to disaggregate cell clusters and produce single cell suspensions. However, this can only be achieved to a limited extent since excess mechanical and/or biochemical treatment will cause cell damage and debris (Husain, Page Roberts and Millet, 1978; Wooley et al., 1979). Mechanical methods include the use of ultrasonification probes and baths (Garcia and Tolles, 1977) or pumps. The latter involves the sample being pumped through a narrow gauge hypodermic needle (19 g) or vortex whistle connected to a syringe or a peristaltic pump. Chemical techniques range from the reduction of disulphide bonds using dithiothreitol to enzyme digestion using pepsin, trypsin, collagenase and other proteases; it is even possible to examine paraffin-processed material (Watts et al., 1987).

To some extent the nature of the tissue or cells can influence the likelihood of achieving good homogeneous suspensions with a high proportion of single cells. For example, good suspensions can be obtained from cell cultures and haemopoietic tumours which tend to exist as single cells when in suspension (Thornthwaite, Sugarbaker and Temple, 1980). It has also been shown that the number of single abnormal cells in smears from patients with CIN increases with the severity of the lesion (i.e. highest number for patients with CIN III) (Bibbo et al., 1975, 1976a,b). However, it is generally agreed that a certain fraction of all samples cannot be dispersed but sufficient numbers of single cells may be present to satisfy the diagnostic requirement (Bahr et al., 1979).

The conventional smear prepared by the clinician is not suitable for automated analysis due to the variation in cell thickness and cell overlap. The sample collected by scraping the uterine cervix may contain in addition to squamous and glandular epithelial cells, mucus, polymorphonuclear leucocytes, erythrocytes, various bacteria, fungi and protozoa. Most preparation procedures for automated screening are based upon a cell suspension. This is accomplished by transferring material from the cervical spatula collection instrument into a specimen bottle containing a small volume of preservative fluid. The fluid must provide a stabilizing environment for the cells and must remain effective for at least seven days. This enables samples to be collected at clinics which are not within the vicinity of the preparation laboratory. Collection fluids in common use include buffered salt solution, with or without the addition of antibiotics, low concentrations (10–25%) of ethanol, glutaraldehyde, paraformaldehyde and formaldehyde. Other additives include mucolytic agents such as dithiothreitol or 1% glacial acetic acid. Cells exfoliated from the cervix can be disaggregated to a limited extent using the syringing method described previously.

One major consideration in the preparation of cervical scrape for automated analysis is that the cells must be presented as far as possible as a monolayer so that individual cell boundaries remain distinct and do not overlap. This has been effected by the use of slides coated with various adhesives such as poly-L-lysine hydrobromide (Husain, Page Roberts and Millet, 1978). The rationale of the method is based on the electrostatic charge interaction between anionic sites present on the cell membrane and protonated groups present along the length of the polymer. Cells are suspended in buffered salt solution and are settled onto the slide surface under the force of gravity (Watts et al., 1984) or in combination with a cytocentrifuge (Van Driel-Kulker et al., 1980). For overview see Rosenthal and Manjikian (1987).

Staining protocols

Automated screening of monolayer smears stained by the Papanicolaou method may be possible, but only if the scanning system is capable of multispectral analysis. There is also the problem of standardization due to wide variation between different dye solutions and the lack of stoichometry of Harris's haematoxylin. Other methods have been sought which would not only overcome the problems of standardization (Wittekind, 1985) and stoichiometry, but also provide a stain which visually resembles the Papanicolaou stain. All of these methods are based upon stoichiometric cytochemical methods for nuclear DNA and/or cytoplasmic proteins. Double staining methods are suitable for absorption providing the absorption spectra of the dyes are sufficiently distinct. Alternatively filters which transmit light in specific regions of the visible spectrum can be used; cytoplasmic staining with eosin can be made 'invisible' by illuminating the slide with 'red' (600–620 nm) light. The traditional Feulgen pararosaniline method has been modified by many workers and used in conjunction with protein stains such as Acriflavin SITS (Tanke, van Ingen and Ploem, 1979) and Congo Red (Oud et al., 1981). Other methods for nuclear staining include the rapid gallocyanin chrome alum technique (Husain and Watts, 1984) and thionine (Wittekind, 1985).

Urinary tract

Cytometric studies on cells exfoliated from the urinary tract have been performed for two reasons: to supplement cytological and cystoscopical assessment of patients with suspected urothelial neoplasia (Freni et al., 1975) and to try to establish a method for screening high risk individuals working in the aniline dye industry. The cells used for such studies have been collected from voided urine or from bladder washing (Collste et al., 1979, 1980; Koss, Sherman and Adams, 1983). High-grade malignant tumours from the bladder often have a high DNA content which is significantly different from normal urothelial cells. In contrast cells from many low-grade neoplasms are near diploid, thus detection by DNA measurement may not be possible. Quantitative analysis is further limited by the fact that bladder neoplasms may be multicentric and cells collected and measured may not be representative of the true diagnosis. Furthermore the presence of binucleate normal transitional cells can cause an excessive number of false positive signals (Cambier et al., 1976).

Gastrointestinal tract

Sprenger and Witte (1980) used flow cytometry

to supplement the cytological study of gastroscopic samples from 161 patients. The false negative and false positive rates were 18% and 39% respectively. Weiss et al. (1980) analysed gastric specimens from patients with chronic atrophic gastritis and concluded that DNA quantitation and cell cycle studies could be used as a prognostic indicator to identify patients most likely to develop carcinoma.

Respiratory tract

In many cases of carcinoma of the lung and bronchus, malignant cells exfoliated from the neoplastic epithelium can be detected by examination of sputum. Routine cytology has been correlated with flow cytometric analyses using acridine orange staining. Frost et al. (1979a,b) were able to identify small numbers of abnormal aneuploid cells in the sample by the simultaneous measurement of green and red fluorescence and light scatter.

Fine needle aspiration specimens and serous effusions

Static and flow cytometry have both been used to supplement the cytodiagnosis of other cytological specimens including fine needle aspiration of the prostate (Tribukait, Ronstrom and Esposti, 1980), thyroid (Boon, Lowhagen and Willems, 1980) and breast (Cornelisse et al., 1983), and in the evaluation of serous effusions (Kwee et al., 1982; Evans et al., 1983).

Chromosome analysis and karyotyping

Cytogenetic studies have benefited from the introduction of automated techniques namely metaphase finding and quantitative karyotyping. High resolution pattern recognition has been used to locate metaphases with minimal chromosome overlaps and produce a karyotype using the conventional Denver classification (Piper, 1982). Flow analysis has also been used for quantitative karyotyping which can detect minute chromosome aberrations (Young et al., 1981). This form of analysis is of value in antenatal diagnosis, karyological studies of tumours and in detecting cells which have been subjected to mutagenic damage (Otto and Oldiges, 1980). Other clinical applications have been reviewed by Laerum and Farsund (1981).

Prospects for the future

In recent years significant advances in the development of instrumentation for cytometric analysis have been made; it is well recognized that machines can detect cellular features which cannot be perceived by visual assessment. These features have proved to be capable of the detection, diagnosis and classification of neoplasia. In some cases prognostic information can also be obtained. Static scanners are also capable of identifying neoplastic cells within a large population of normal cells. This automated detection of the 'rare event' has been a particularly valuable development in the search for a technique for the automated screening for cervical cancer and has shown some promise in the detection of micrometastases in bone marrow smears from patients with breast or prostate cancer.

Improved specimen preparation techniques have been developed which provide far more suitable specimens for automated screening than previous methods. However, problems requiring further work are apparent; the techniques for preparation are often laborious and time-consuming and considerable engineering ingenuity is required to integrate these into a cost-effective, automated cytology system. There is also an urgent need for a definition of realistic requirements for an automated screening programme which will be an improvement on the present manual system. Criteria such as acceptable false negative and positive rates, the ability of the machine to detect borderline cases and the number of cells to be analysed before specimen adequacy can be assumed are yet to be established and universally agreed.

Active research is in progress into the development of higher resolution instrumentation for rapid scanning and for new parameters which will increase the specificity and sensitivity of cytometric analysis even further.

References

ADAMS, L. R. (1974). Acridine orange staining of epithelial cells in strong salt solution. *J. Histochem. Cytochem.*, **22**, 492–494

ALTMAN, F. F., BITENSKY, L., BUTCHER, R. G. and CHAYEN, J. (1970). In *Cytology Automation*, (Evans, D. M. D., ed.), pp. 82–99. Edinburgh: Livingstone

ATKIN, N. B. (1964). The DNA content of malignant cells in cervical smears. *Acta Cytol.*, **8**, 68–72

ATKIN, N. B. and KAY, R. (1979). Prognostic significance of modal DNA value and other factors in malignant tumours, based on 1465 cases. *Br. J. Cancer*, **40**, 210

ATKIN, N. B. and RICHARDS, B. M. (1956). Deoxyribonucleic acid in human tumours as measured by microspectrophotometry of Feulgen stain. A comparison of tumours arising at different sites. *Cancer*, **10**, 769–789

BAHR, E. F., BARTELS, F. H., WIED, G. L. and KOSS, L. G. (1979). Automated cytology. In *Diagnostic Cytology and its Histopathologic Bases*, (Koss, L. G., ed.), p. 1152. Philadelphia: J. B. Lippincott

BARRETT, D. L., JENSEN, F. H., KING, E. B., DEAN, F. N. and MAYALL, E. H. (1979). Flow cytometry of human gynecologic specimens using log chromomycin A3 fluorescence and log 90 light scatter signals. *J. Histochem. Cytochem.*, **27**, 573–578

BARTELS, F. H., KOSS, L. G. and WIED, G. L. (1980). Automation in cytology: Computerised high resolution scanning of cervical smears. In *Advances in Clinical Cytology*, (Koss, L. G. and Coleman, D. V., eds), pp. 314–342. London: Butterworths

BAUMANN, J. G. J., VAN DER PLOEG, M. and VAN DUIJN, P. (1984). Fluorescent hybridocytochemical procedures DNA/RNA hybridisation *in situ*. In *Investigative Microtechniques in Medicine and Biology*, Volume 1, (Chayen, J. and Bitensky, L. eds), pp. 41–47. New York: Marcel Decker

BIBBO, M., BARTELS, P. H., CHEN, M., HARRIS, M. J., TRUTTMANN, B. and WIED, G. L. (1975). The numerical composition of cellular samples from the female reproductive tract. I. Carcinoma *in situ*. *Acta Cytol.*, **19**, 438–447

BIBBO, M., BARTELS, P. H., CHEN, M., HARRIS, M. J., TRUTTMANN, B. and WIED, B. L. (1976a). The numerical composition of cellular samples from the female reproductive tract. II. Cases with invasive squamous carcinoma of uterine cervix. *Acta Cytol.* **20**, 249–254

BIBBO, M., BARTELS, P. H., CHEN, M., HARRIS, M. J., TRUTTMANN, B. and WIED, G. L. (1976b). The numerical composition of cellular samples from the female reproductive tract. III. Cases with mild and moderate dysplasia of uterine cervix. *Acta Cytol.*, **20**, 565–572

BIBBO, M., BARTELS, P. H., SYNCHRA, J. J. and WIED, G. L. (1981). Chromatin appearance in intermediate cells from patients with uterine cancer. *Acta Cytol.*, **25**, 23–28

BITENSKY, L. (1980). Microdensitometry. *Trends Enzyme Histochem. Cytochem.*, **May**, 181–202

BITENSKY, L., CHAYEN, J. and HUSAIN, O. A. N. (1984). Cytochemical detection of cancer: a review. *J. Roy. Soc. Med.*, **77**, 677–681

BONHAM, D. G. and GIBBS, D. F. (1962). A new enzyme test for gynaecological cancer: 6-phosphogluconate dehydrogenase activity in vaginal fluid. *Br. Med. J.*, **2**, 823–824

BOON, M. E., LOWHAGEN, T. and WILLEMS, J. S. (1980). Planimetric studies on fine needle aspirates from follicular adenoma and follicular carcinoma of the thyroid. *Acta Cytol.*, **24**, 145–148

BOSTROM, F. C., SAWYER, H. S. and TOLLES, W. E. (1959). The cytoanalyser—an automatic prescreening instrument for cancer detection. In *Automation in Medicine: Medical Electronics*, Proceedings 2nd International Conference, Paris, p. 479.

BOWIE, J. E. and YOUNG, I. T. (1977a). An analysis technique for biological shape. II. *Acta Cytol.*, **21**, 455–464

BOWIE, J. E. and YOUNG, I. T. (1977b). An analysis technique for biological shape. III. *Acta Cytol.*, **21**, 739–746

BRUNSTING, A. (1974). Can light scattering techniques be applied to flow through cell analysis? *J. Histochem. Cytochem.*, **22**, 607–615

BURGER, G., JUTTING, U. and RODENACKER, K. (1981). Changes in benign cell populations in cases of cervical cancer and its precursors. *Anal. Quant. Cytol.*, **3**, 261–271

CAMBIER, M. A., CHRISTY, W. J., WHEELESS, JR. L. L. and FRANK, I. N. (1976). Slit scan cytofluorometry. Basis for automated prescreening of urinary tract cytology. *J. Histochem. Cytochem.*, **24**, 305

CAMERON, C. B. and HUSAIN, O. A. N. (1965). 6-phosphogluconate dehydrogenase activity in vaginal fluid: limitations as a screening test for genital cancer. *Br. Med. J.*, **1**, 1529–1530

CASPERSSON, O. (1964). Quantitative cytochemical studies of normal, malignant, premalignant and atypical cell populations from the human uterine cervix. *Acta Cytol.*, **8**, 45–60

CHAYEN, J., BITENSKY, L., AVES, E. K., JONES, G. R. N., SILCOX, A. A. and CUNNINGHAM, G. J. (1962). Histochemical demonstration of 6-phosphogluconate dehydrogenase in proliferating and malignant cells. *Nature*, **195**, 714–715

CHAYEN, J., BITENSKY, L. and BUTCHER, R. (eds) (1973). *Practical Histochemistry*. New York: John Wiley

COHEN, S. and WAY, S. (1966). Histochemical demonstration of pentose-shunt activity in smears from the uterine cervix. *Br. Med. J.*, **1**, 88–89

COLLSTE L. G., DARZYNKIEWICZ, Z., TRAGANOS, F., SHARPLESS, T. K., DEVONEC, M., CLAPS, M. L. K., WHITMORE, JR. W. F. and MELAMED, M. R. (1979). Cell-cycle distribution of urothelial tumour cells as measured by flow cytometry. *Br. J. Cancer*, **40**, 872

COLLSTE, L. G., DEVONEC, M., DARZYNKIEWICZ, Z., TRAGANOS, F., SHARPLESS, T. K., WHITMORE, JR. W. F. and MELAMED, M. R.(1980). Bladder cancer diagnosis by flow cytometry. Correlation between cell samples from biopsy and bladder irrigation fluid. *Cancer*, **45**, 2389

CORNELISSE, C. J., TANKE, H. J., DE KONING, H., BRUTEL, G. and RIVIERE DE LA, G. B. (1983). DNA ploidy analysis and cytologic examination of sorted cell populations from human breast tumours. *Anal. Quant. Cytol.*, **5**, 173–183

COULTER, W. A. (1956). High speed automatic blood cells counter and cell size analyser. *Proc. Nat. Electron Conf.*, **12**, 1034

CRISSMAN, H. A. and STEINKAMP, J. A. (1973). Rapid, simultaneous measurement of DNA, protein and cell volume in single cells from large populations. *J. Cell Biol.*, **59**, 766–771

CRISSMAN, H. A. and STEINKAMP, J. A. (1982). Rapid, one step staining procedures for analysis of cellular DNA and protein by single and dual laser flow cytometry. *Cytometry*, **3**, 84–90

CRISSMAN, H. A., STEVENSON, A. P. and KISSANE, R. J. (1978). Flow microfluorometric and spectrophotometric analysis of DNA staining in mammalian cells. In *Pulse-Cytophotometry*, (Lutz, D., ed.), pp. 251–265. Ghent: European Press Medikon

DEAN, F. N., DOLBEARE, F., GRATZNER, H., RICE, G. C. and GRAY, J. W. (1984). Cell cycle analysis using a monoclonal antibody to Brd Urd. *Cell Tissue Kinet.*, **17**, 427–436

DOLBEARE, F. A. and SMITH, R. E. (1979). Flow cytoenzymology: rapid enzyme analysis of single cells. In *Flow Cytometry and Sorting*, (Melamed M. R., Mullaney, F. F. and Mendelsohn, M. L., eds), pp. 317–333. New York: John Wiley

ERHARDT, R., REINHARDT, E. R., SCHLIPF, W. and BLOSS, W. H. (1980). FAZYTAN A system for fast automated cell segmentation, cell image analysis and feature extraction based on TV-image pickup and paralleled processing. *Anal. Quant. Cytol.*, **2**, 25

EVANS, D. A., THORNTHWAITE, J. T., NG, A. E. P. and SUGARBAKER, E. V. (1983). DNA flow cytometry of pleural effusions: comparison with pathology for the diagnosis of malignancy. *Anal. Quant. Cytol.*, **5**, 19–27

FOWLKES, E. J., HERMAN, C. J. and CASSIDY, M. (1976). Flow microfluorometric system for screening gynaecologic specimens using propidium iodide-fluorescein isothiocyanate. *J. Histochem. Cytochem.*, **24**, 322

FRENI, S. C., REIJNDERS WARNER, O., DE VOOGT, H. J., BEYER BOON, M. E. and BRUSSE, J. A. M. (1975). Flow fluorophotometry on urinary cells compared with conventional cytology. In *Pulse-Cytophotometry*, I. (Haanen, O. A. M., Hillen, H. F. P. and Wessels, J. M. C., eds), Ghent: European Press Medikon

FRIEDLANDER, M. L., HEDLEY, D. W. and TAYLOR, I. W. (1984). Clinical and biological significance of aneuploidy in human tumours. *J. Clin. Pathol.*, **37**, 961–974

FROST, J. K., TYRER, H. W., PRESSMAN, N. J., ALBRIGHT, C. D., VANSICKEL, M. H. and GILL, G. W. (1979a). Automatic cell identification and enrichment in lung cancer. I. Light scatter and fluorescence parameters. *J. Histochem. Cytochem.*, **27**, 545

FROST, J. K., TYRER, H. W., PRESSMAN, N. J., ADAMS, L. A., VANSICKEL, M. H., ALBRIGHT, C. D., GILL, G. W. and TIFFANY, S. M. (1979b). Automatic cells identification and enrichment in lung cancer. III. Light scatter and two fluorescence parameters. *J. Histochem. Cytochem.*, **27**, 557

GARCIA, G. L. and TOLLES, W. E. (1977). Ultrasonic disaggregation of cell clusters. *J. Histochem. Cytochem.*, **25**, 508–512

GIBBS, D. F., LABRUM, A. H. and STAGG, B. H. (1968). Vaginal fluid enzymology. A new method with enzyme/potassium ratios. *Am. J. Obstet. Gynec.*, **102**, 982–988

GILL, J. E., WHEELESS, JR. L. L., HANNA-MADDEN, C. and MARISA, R. J. (1979). Cytofluorometric and cytochemical comparisons of normal and abnormal human cells from the female genital tract. *J. Histochem. Cytochem.*, **27**, 591

HABBERSET, M. C., SHAPIRO, M., BUNNAG, B., NISHIYA, I. and HERMAN, C. (1979). Quantitative analysis of flow fluorometric data for screening gynaecologic cytology specimens. *J. Histochem. Cytochem.*, **27**, 536

HILLWIG, I. and GROPP, A. (1972). Staining of constitutive heterochromatin in mammalian chromosomes with a new fluorochrome. *Exp. Cell Res.*, **75**, 122–126

HUSAIN, O. A. N. and WATTS, K. C. (1984). The rapid demonstration of nucleic acids using 'oxidised' gallocyanin and chromic potassium sulphate: methods and applications. *J. Clin. Pathol.*, **37**, 99–101

HUSAIN, O. A. N. and WATTS, K. C. (1987). The development of computerised cell scanners for cancer screening. In *The Computer in Obstetrics and Gynaecology*, (Dalton, K. J. and Fawdry, R. D. S., eds), Oxford: IRL Press

HUSAIN, O. A. N., BLANCHE BUTLER, E., EVANS, D. M. D., MACGREGOR, J. E. and YULE, R. (1974a). Quality control in cervical cytology. *J. Clin. Pathol.*, **27**, 935–944

HUSAIN, O. A. N., ALLEN, R. W. B., HAWKINS, E. J. K. and TAYLOR, J. E. (1974b). The quantimet cytoscan and interactive approach to cancer screening. *J. Histochem. Cytochem.*, **22**, 678–684

HUSAIN, O. A. N., TUCKER, J. H. and PAGE ROBERTS, B. A. (1976). Automation in cervical cancer screening. *Biomedical Engineering*, May and June

HUSAIN, O. A. N., PAGE ROBERTS, B. A. and MILLET, J. A. (1978). A sample preparation for automated cervical cancer screening. *Acta Cytol.*, **22**, 15–21

HUSAIN, O. A. N., WATTS, K. C., FRAY, R. E. and TO, A. C. W. (1984). Immunocytochemical markers in cervical screening. *Lancet*, **1**, pp. 338–339

IBRAHIM, K. S., HUSAIN, O. A. N., BITENSKY, L. and CHAYEN, J. (1983). A modifed tetrazolium reaction for identifying malignant cells from gastric and colonic cancer. *J. Clin. Pathol.*, **36**, 133–136

JENSEN, F. H., MAYALL, E. H. and KING, E. B. (1979). Multiparameter flow cytometry applied toward diagnosis of cervical carcinoma. In *The Automation of Cancer Cytology and Cell Image Analysis*, (Pressmann, N. J. and Wied, G. L. eds), pp. 95–102. Chicago: Tutorials of Cytology

JOHNSTON, D. G. (1952). Cytoplasmic: nuclear ratios in the cytological diagnosis of cancer. *Cancer*, **5**, 945

KACHEL, C. V. (1976). Basic principles of electrical sizing of cells and particles and their realisation in the new instrument 'Metricell'. *J. Histochem. Cytochem.*, **24**, 211–230

KAMENTSKY, L. A., DERMAN, H. and MELAMED, M. R. (1963). Ultraviolet absorption in epidermoid cancer cells. *Science*, **142**, 1580–1583

KOSS, L. G., SHERMAN, A. B. and ADAMS, S. E. (1983). The use of hierarchic classification in the image analysis of a complex cell population. Experience with sediment of voided urine. *Anal. Quant. Cytol.*, **5**, 159–166

KUTE, T. E., LINVILLE, C. and BARROWS, G. (1983). Cytofluorometric analysis for estrogen receptors using fluorescent estrogen probes. *Cytometry*, **4**, 132–140

KWEE, W. S., VELDHUIZEN, R. W., ALONS, C. A., MORAWETZ, F. and BOON, M. E. (1982). Quantitative and qualitative differences between benign and malignant mesothelial cells in pleural fluid. *Acta Cytol.* **26**, 401–406

LADINSKY, J. L., SARTO, E. G. and PECKHAM, B. M. (1964). Cell size distribution patterns as a means of uterine cancer detection. *J. Lab. Clin. Med.*, **64**, 970.

LAERUM, O. D. and FARSUND, T. (1981). Clinical application of flow cytometry: a review. *Cytometry*, **2**, 1–13

LE PECQ, J. B. and PAOLETTI, C. (1967). A fluorescent complex between ethidium bromide and nucleic acids. *J. Mol. Biol.*, **27**, 87–106

LINDEN, W. A., OCHLICH, K., BAISCH, H., SCHOLZ, K. L., MAUSS, H. J., STEGNER, H. E., JOSHI, D. S., WU, C. T., KOPROWSKA, I. and NICOLINI, C. (1979). Flow cytometric prescreening of cervical smears. *J. Histochem. Cytochem.* **27**, 529

LOKEN, M. R., SWEET, R. G. and HERZENBERG, L. A. (1976). Cell discrimination by multiple light scattering. *J. Histochem. Cytochem.*, **24**, 284–291

MANNA, G. K. (1955). Chromosome number of human cancerous cervix uteri. *Naturwissenschaften*, **9**, 253

MELAMED, M. R., MULLANEY, F. F. and MENDELSOHN, M. L. (eds) (1979). *Flow Cytometry and Sorting.* New York: John Wiley

MEYER, F. (1979). Iterative image transformations for an automated screening of cervical smears. *J. Histochem. Cytochem.*, **27**, 512–519

MILLET, J. A. (1979). PhD Thesis. University of London

MILLET, J. A., CHIN, Y., BITENSKY, L., CHAYEN, J. and HUSAIN, O. A. N. (1980). Lysosomal naphthylamidase activity as a possible aid to cytological screening. *J. Clin. Pathol.*, **33**, 684–687

MILLET, J. A., HUSAIN, O. A. N., BITENSKY, L. and CHAYEN, J. (1982). Feulgen hydrolysis profiles in cells exfoliated from the cervix uteri: a potential aid in the diagnosis of malignancy. *J. Clin. Pathol.*, **35**, 345–349

MONCRIEFF, D., OMEROD, M. G. and COLEMAN, D. V. (1984). Tumour marker studies of cervical smears. Potential for Automation. *Acta Cytol.*, **28**, 407–410

NIEBURGS, H. E., HERMAN, B. E. and REISMAN, H. (1962). Buccal changes in patients with malignant tumours. *Lab. Invest.*, **2**, 80–88

OTTO, F. J. and OLDIGES, H. (1980). Flow cytogenetic studies in chromosomes and whole cells for the detection of clastogenic effects. *Cytometry*, **1**, 99

OUD, P. S., ZAHNISER, D. J., RAAIJMAKERS, M. C. T., VOOIJS, P. G. and VAN DE WALLE, R. T. (1981). Thionine–Feulgen–Congo Red staining of cervical smears for the BIOPEPR image analysis system. *Anal. Quant. Cytol.*, **3**, 289–294

PIPER, J. A. (1982). Interactive image enhancement and analysis of prometaphase chromosomes and their band patterns. *Anal. Quant. Cytol.*, **4**, 233–240

PLOEM, J. S., VAN DRIEL-KULKER, A. M. J., GOYARTS-VELDSTRA, L., PLOEM-ZAAIJER, J. J., VERWOERD, N. P. and VAN DER ZWAN, M. (1986). Image analysis combined with quantitative cytometry: results and instrumental developments for cancer diagnosis. *Histochemistry*, **84**, 549–555

REAGAN, J. W., HAMONIC, M. J. and BUDD WENTZ, W. (1957). Analytical study of the cells in cervical squamous cell cancer. *Lab. Invest.*, **6**, 241

REES, K. R., SLATER, T. F., GIBBS, D. F. and STAGG, B. (1970). A modified method for 6-phosphogluconate dehydrogenase in samples of vaginal fluid from women with and without gynecologic cancer. *Am. J. Obstet. Gynecol.*, **107**, 857–864

ROSENTHAL, D. L. and MANJIKIAN, V. (1987). Techniques in the preparation of a monolayer of gynecologic cells for automated cytology. *Anal. Quant. Cytol.*, **9**, 55–59

SPENCER, C. C. and BOSTROM, R. C. (1962). Performance of the cytoanalyser in recent clinical trials. *J. Nat. Cancer Inst.*, **23**, 267

SPRENGER, E. and WITTE, S. (1979). The diagnostic significance of deoxyribonucleic acid measurement in automated cytology. *J. Histochem. Cytochem.*, **27**, 520

SPRENGER, E, and WITTE, S. (1980). The diagnostic significance of flow cytometric nuclear DNA measurements in gastroscopic diagnosis of the stomach. *Path. Res. Pract.*, **169**, 269

SPRIGGS, A. I., DIAMOND, R. A. and MEYER, F. W. (1968). Automated screening for cervical smears. *Lancet*, **1**, 359

STOHR, M. and GOERTTLER, K. (1979). The Heidelberg flow analyser and sorter (HEIFAS) approach on the prescreening of uterine cancer. *J. Histochem. Cytochem.*, **27**, 564

STOHR, M., EIPEL, H. and GOERTTLER, K. (1977). Extended application of flow microfluorometry by means of dual laser excitation. *Histochemistry*, **51**, 305–313

STOHR, M., FROHBERG, S., SANDER, H., HEBERLIN, D. and GOERTTLER, K. (1980). Mulitparameter flow cytometry of past irradiation cervical smears. In *Flow Cytometry*, IV, (Laerum, O., Lindmo, T. and Thoruds, E., eds), pp. 424–426. Bergen-Oslo-Trondheim: Universitetsforlaget

TAFT, P. D. and ADAMS, L. R. (1979). Flow cytofluorometry in routine diagnostic cytology. A pilot study. *Am. J. Clin. Path.*, **72**, 533

TANAKA, N., UENO, T., IKEDA, H., ISHIKAWA, A., KONOIKE, K., SHIMAOKA, Y., YAMAUCHI, K., HOSOI, S., OKAMOTO, Y. and TSUNEKAWA, S. (1982). CYBEST Model 3. Automated cytologic screening system for uterine cancer utilising image analysis processing. *Anal. Quant. Cytol.*, **4**, 279–285

TANKE, H. J., VAN INGEN, E. M. and PLOEM, J. S. (1979). Acriflavine Feulgen Stilbene (AFS) staining: a procedure for automated cervical cytology with a television based system (LEYTAS). *J. Histochem. Cytochem.*, **27**, 84–86

TANKE, H. J., NIEUWENHUIS, L. A. B., KOPER, G. J. M., SLATS, J. C. M. and PLOEM, J. S. (1980). Flow cytometry of human reticulocytes based on RNA fluorescence. *Cytometry*, **1**, 313–320

TANKE, H. J., VAN DRIEL-KULKER, A. M. J., CORNELISSE, C. T. and PLOEM, J. S. (1983). Combined flow cytometry and image cytometry of the same sample *J. Microsc.*, **130**, 11–22

THORNTHWAITE, J. T., SUGARBAKER, E. V. and TEMPLE, W. J. (1980). Preparation of tissue for DNA flow cytometric analysis. *Cytometry*, **1**, 229–237

TIMMERS, T. (1987). PhD Thesis. Free University of Amsterdam

TOLLES, W. E., HORVATH, W. J. and BOSTROM, R. C. (1961). A study of the quantitative characteristics of exfoliated cells from the female genital tract. I. Measurement methods and results. II. Suitability of quantitative cytological measurements for automatic prescreening. *Cancer*, **14**, 437–468

TONOMURA, A. (1959). A chromosome survey in six cases of human uterine cervix carcinomas. *Jap. J. Genetics*, **34**, 401–406

TRIBUKAIT, B., ESPOSTI, P. L. and RONSTROM, L. (1980). Tumour ploidy for characteristic of prostatic carcinoma: flow cytofluorometric DNA studies using aspiration biopsy material. *Scand. J. Urol. Nephrol. (Suppl.)*, **55**, 59

TUCKER, J. H. and HUSAIN, O. A. N. (1981). Trials with the Cerviscan experimental prescreening device on poly-lysine-prepared slides. *Anal. Quant. Cytol.*, **3**, 117–120

TUCKER, J. H., RODENACKER, K., JUETTING, U., NICKOLLS, P., WATTS, K. AND BURGER (1989). Interval coded texture features for artifact rejection in automated cervical cytology. *Cytometry*, **9**, 418–425

VAN DRIEL-KULKER, A. M. J., PLOEM-ZAAIJER, J. J., VAN DER ZWAN, M. and TANKE, H. J. (1980). A preparation technique for exfoliated and aspirated cells allowing staining procedures. *Anal. Quant. Cytol.*, **2**, 243–247

VERNON, T. O., GLAZER, A. N. and STRYER, L. (1982). Fluorescent phycobiliprotein conjugates for analysis of cells and molecules. *J. Cell Biol.*, **93**, 981–986

WATSON, J. V. (1980). Enzyme kinetics studies in cell populations using fluorogenic substrates and flow cytometric techniques. *Cytometry*, **1**, 143

WATSON, J. V., SIKORA, K. and EVAN, G. I. (1985). A simultaneous flow cytometric assay for c-myc oncoprotein and DNA in nuclei from paraffin embedded material. *J. Immunol. Methods*, **83**, 179–192

WATTS, K. C., HUSAIN, O. A. N., TUCKER, J. H., STARK, M., EASON, P., SHIPPEY, G., RUTOVITZ, D. and FROST, G. T. B. (1984). The use of cationic polyelectrolytes in the preparation of cell monolayers for automated cell scanning and diagnostic cytopathology. *Anal. Quant. Cytol.*, **6**, 272–278

WATTS, K. C., CAMPION, M. J., BUTLER, E. BLANCHE, JENKINS, D., SINGER, A. and HUSAIN, O. A. N. (1987). Quantitative DNA analysis of patients with mild cervical atypia; a potentially malignant lesion? *Obstet. Gynecol.* **70**, 205–207

WEISS, H., GUTZ, H. J., WILDNER, G. P., EBELING, K. and SCHIMDT, W. (1980). Characterisation of chronic atrophic gastritis by means of pulse cytophotometry. In *Flow Cytometry*, IV, (Laerum, O. D., Lindmo, T. and Thorud, E., eds), pp. 473–477. Bergen-Oslo-Trondheim: Universitetsforlaget

WHEELESS, JR. L. L. and ONDERDONK, M. A. (1974). Preparation of clinical gynaecologic specimens for automated analysis. An overview. *J. Histochem. Cytochem.*, **22**, 522

WIED, G. L., BARTELS, P. H., BIBBO, M. and SYNCHRA, J. J. (1980). Cytomorphic markers for uterine cancer in intermediate cells. *Anal. Quant. Cytol.*, **2**, 257–263

WIED, G. L., BARTELS, P. H., BIBBO, M. and KEEBLER, C. M. (1981). Frequency and reliability of diagnostic cytology of the female genital tract. *Acta Cytol.*, **25**, 543–550

WITTEKIND, D. (1985). Standardization of dyes and stains for automated cell pattern recognition. *Anal. Quant. Cytol. Histol.*, **7**, 6–31

WOOLEY, R. C., HERZ, F., DEMBITZER, H. M., SCHREIBER, K. and KOSS, L. G. (1979). The monodisperse cervical smear. Quantitative analysis of cell dispersion and loss with enzymatic and chemical agents. *Anal. Quant. Cytol.*, **1**, 43–49

YOUNG, E. D., FERGUSON SMITH, N. A., SILLAR, R. and BOYD, E. (1981). High resolution analysis of human peripheral lymphocyte chromosomes by flow cytometry. *Proc. Nat. Acad. Sci., USA*, **78**, 7727–7731

ZAHNISER, D. J., OUD, F. S., RAAIJMAKERS, M. C. T., VOOYS, G. P., WALLE, R. T. and VAN DE WALLE (1980). Field test results using the BioPEPR cervical smear prescreening system. *Cytometry* **1**, 2000–2003

26

Laboratory organization and management

John Sims and Phillipa C. Carroll

The aim of this chapter is to advise the manager of a cytology laboratory how to develop an efficient, friendly, safe and reliable laboratory service. This calls for a thorough knowledge of the hospital and community needs, good communication with other health care workers, a well designed and conveniently located laboratory, careful planning and sound decision-making skills, sufficient resources for teaching and research, and a dedicated and highly motivated staff. Unfortunately, such ideal conditions rarely exist and the laboratory manager must do the best he or she can with the resources available. The laboratory is not always in the best geographical position for good communication and the internal structure is often inherited rather than designed. Trained staff are hard to find and resources are scarce. These are all hurdles which must be overcome if the laboratory is to be well managed.

Despite the frustrations, the rewards of good management skills are high. They include job satisfaction coupled with good patient care. In addition there is the comforting knowledge that one is making a major contribution to the development of a new and useful branch of medical laboratory science which has an increasingly important role to play in clinical diagnosis and primary health care.

Workload

The size and type of the workload of a cytopathology department depends upon several factors: the population served, the location of the laboratory, the range of services provided and the interests of the pathologist and clinical staff.

Most cytology laboratories undertake both cervical screening and clinical diagnostic work. These represent very different types of work which require different approaches. The cervical screening service undertakes the investigation of healthy women of whom only a small proportion are found to have abnormal smears. The diagnostic service, on the other hand, involves the examination of non-gynaecological and gynaecological specimens from patients with clinical symptoms suggestive of malignancy. In these cases the proportion of abnormal specimens is high. This distinction is not absolute and a screening service may be extended to workers exposed to aniline dyes or asbestos for whom the risk of cancer is higher than in the general population. The ratio of diagnostic work to screening depends upon whether the laboratory is situated within a histopathology department serving a hospital population, or whether the cytology laboratory is independent of histology and has a large cervical screening commitment.

The manager's responsibility starts with the arrival of specimens at the laboratory. Cervical smears arrive already fixed whereas most non-gynaecological specimens are unfixed. For the best possible result, non-gynaecological specimens should be received in the laboratory as fresh as possible, and certainly within 24 h. Cerebrospinal fluid should be received within

2 h of being taken. Advice should be given at regular intervals to the wards, out-patient departments and clinics on specimen collection and transport. Close cooperation and liaison with the consultants, junior medical staff, nurses, porters and other hospital personnel is recommended to promote good working relationships between departments

Specimens should be accompanied by a distinctive cytology request form. The specimen should be labelled with enough details for patient identification and an unlabelled slide or container must not be accepted. The form should include all the relevant information required for processing the specimen: surname, forename, identification number, sex, date of birth, sender's name and address, date of request, type of specimen, investigation required and a brief synopsis of the clinical history. Additional information is required when gynaecological cytology is requested; this includes menstrual status, obstetric history and hormone and contraceptive details. The name and address of the general practitioner is also necessary for follow-up and recall of patients for cervical screening. It is a good policy to take particular notice of the details in the request form to ascertain whether the specimen is a biohazard, and that the correct type of specimen has been sent for the investigation requested. Similarly, if more than one specimen has been sent on a patient, the specimens must be labelled clearly and correlated with the information on the request form.

The organization and flow of work through the department should follow a logical progression and the working day should be structured in such a way that work finished by one member of staff can be continued by another. The workload should be organized so that specimens can be easily located and their position reflects the stage of processing. The layout of the laboratory will partially dictate how the work runs through the department, but by the division of work and coordination of working groups, wasted effort can be eliminated and result in the smooth running of the department.

The non-gynaecological specimens should be reported within 24 h of being received. However, urgent samples such as fine needle aspirates should be processed much more quickly and the system should allow this fast throughput. Cervical smears generally represent the bulk of the workload and turnover of these samples need not be so rapid unless urgently requested. However, a backlog of more than two weeks is unsatisfactory, and a backlog of more than a month contravenes DHSS instructions (DHSS, 1988). Smears from gynaecological departments, from patients with a history of abnormal smears and from patients with symptoms such as postmenopausal or intermenstrual bleeding, should be given high priority and screened before routine work.

The mode by which the screening work flows through the department must be clearly defined. The consultant cytopathologist is ultimately responsible for all reports that leave the laboratory regardless of whether the report is negative or abnormal. Abnormal cervical smears, problem cases and smears from symptomatic patients attending a gynaecological clinic, should be funnelled to senior or supervisory personnel. All abnormal cervical smears must then be directed to the consultant cytopathologist for signing out. It is advisable for all diagnostic samples (both gynaecological and non-gynaecological) to be seen by a pathologist, especially fine needle aspiration biopsy specimens. In many laboratories, negative cervical cytology reports are signed out by non-medical scientists. This is quite acceptable as long as the competence of the screener has been tested and standards of quality control are high.

It is important to maintain a consistent reporting policy. A uniformity in reporting style should exist in any one laboratory. This can exist without loss of flexibility and personal expression. The employment of a computerized system often enforces a level of conformity due to the coding system it utilizes. Terminology for the diagnosis of cervical smears can be found in the document prepared by the British Society for Clinical Cytology (Evans, 1986) and is strongly recommended.

The size of the workload should equate with staff levels (Penner, 1982). Independent departments will benefit from their own hierarchical staff structure with their own pathologist and laboratory manager whereas cytology laboratories sited within histopathology departments will be more dependent upon shared staff. In order to equate staff levels to workload, a time and motion study can be employed to provide figures for both inter- and intra-laboratory comparisons. This method has proved to be a useful management tool in the face of rapid change in laboratory technology. It can identify changing patterns and give rational projections of staff needs and improve efficiency.

In estimating workload, separate consideration must be given to the different types of work undertaken, namely clinical diagnostic samples and cervical screening. The cervical screening commitment can be calculated from the female population served by the laboratory and controlled by preventing opportunistic screening. The number of diagnostic samples, both gynaecological and non-gynaecological, depends to some extent upon the population served, but is also influenced by the working practices of the clinicians and specialized departments within the hospital.

As a result of time-engineered studies by the Canadian Association of Pathologists, with a development grant from the Canadian National Health and Welfare funds, the 'Canadian Units' system of workload measurement in pathology was introduced. This was based on the principle that one unit was to represent one minute's work, the concept of 'productive' and 'non-productive' time was introduced and the system was 'dynamic', that is, subject to amendment as new or modified techniques became recognized. The first manual of 'unit values' allocated to lists of specific jobs calculated from actual timed measurements was published in 1969 and although there have been several up-dated editions, notably 1983 (Statistics, Canada Health Division, 1982–83) and 1987 (Statistics, Canada Health Division, 1986–87), the basic principles remain unaltered.

The Canadian Association is responsible for development and maintenance of the laboratory system and most recently the emphasis on data collection has been to establish a proper relationship between unit-producing activities and other activities necessary to maintain an acceptable level of laboratory service; in other words, to calculate the correct ratio of unit-producing to non unit-producing activities and thus to establish realistic targets for performance indicator units per paid hour. Activities such as education, administration, research and method development have not yet been taken into account as measurable parameters.

The values have been arrived at by averaging the results from numerous laboratories and comprise a range of laboratory size, complexity and geographical distribution (throughout Canada).

Canadian unit values may be permanent (P) or temporary (T), depending on whether a sufficient number of time-studies at various locations have been carried out on a particular procedure. The unit values are shown in *Table 26.1*. (*Note* this table is prepared using data from Canada Health Division 1986–87 and these time allocations are being revised in the light of UK experience.)

In summary, a 'Canadian Unit' is defined as one minute of productive time of technical, clerical and laboratory aide staff. Note that medical staff involvement is not included, and

Table 26.1 Units accorded to cytology specimens in the Canadian system

Laboratory procedure	Item for count	Unit value
Initial identification, gynaecological*	specimen	10
Initial identification, non-gynaecological*	specimen	10
Cytohormonal evaluation	specimen	10
Preparation of fluids by centrifugation†	specimen	T7
Preparation of fluids by membrane filter technique	membrane filter	T8
Preparation of smears from fine-needle aspiration	specimen	T10
Preparation of sputa by pick and smear technique	specimen	T6
Screening, technical gynaecological	slide	5
Screening, technical non-gynaecological	slide	5
Despatch of biological materials to outside laboratories	specimen	6
Handling and reporting of slides received from referring lab	specimen	5
Travel time-trips outside the department for the procurement of specimens	round trip	8
Procurement of smears for cytopathology‡	patient	T6

Specific units are allocated for specimen handling; these are additional to the unit value for the procedure
* comprises clerical functions, staining (including daily preparation), reporting, slide-filing and follow-up
† also includes Cytospin technique. Although this procedure is more time consuming, this is offset by shorter screening time per slide
‡ only members of laboratory and clerical staff to be counted

productive staff are referred to as 'Canadian unit producers' (CUPS). It should also be noted that time that is 'not productive' is not wasted.

British variants of this scheme include: Welsh Workload Measurement System for Pathology (WWMS) (Welsh Office, 1987–89), Southwest Regional Health Authority Canadian Unit Working Party and Histopathology Performance Indicators Group (Personal Communications).

Potential data could be applied for staffing requirements, cost-effective equipment purchasing, space utilization, monitoring laboratory functions, projecting staff and space requirements, identifying areas of increased demand and budget preparation and monitoring. WWMS has the additional advantage of assigning units to individual 'request' categories, for example, in-patients, out-patients, general practitioners, other hospitals, private patients.

The British Society for Clinical Cytology recommends that no more than 7000 slides should be screened per annum by a productive staff member. When these workload figures per staff member were calculated, training and supervisory grades were not included (BSCC, 1986). The figures were based upon the time involved in screening one slide (5–10 min or 8 slides/h), the time spent in one working day looking down the microscope and the time taken to prepare and process the specimens. It was calculated that a screener should read up to 32 smears per day and suggested that one-fifth of the day should be free for other working activities such as training, quality control and assessment, meetings and seminars. It also recommends that no more than 4 h/day is spent at the microscope and that a break is taken after 2 h; beyond this it is feared that the screener's attention is severely diminished.

Sustaining a screener's attention so that an occasional abnormality is detected in such repetitive tasks as cervical screening is termed 'vigilance' (Fowkes, 1986). Deterioration in performance is known as 'vigilance decrement'. Several factors have been shown to affect vigilance. These include background noise, screening experience, smear presentation, the observer's perception of the benefits to be derived from correct identification and their expectation of the frequency of the abnormality (Fowkes, 1986). The diagnostic sensitivity is worse with lowered prevalence of abnormal smears. Since the prevalence of abnormal smears in cervical screening is quite low, it is important to maintain optimal conditions; thus the screening area must be kept as quiet as possible, away from noisy telephones, and it should be carpeted for sound proofing. Ergonomically designed chairs should be used, so the seating position does not promote screener fatigue. Above all, it is essential to have optimally prepared samples, well fixed and stained for maximum clarity. More recently these conditions have also been applied to visual display unit users (BSCC, 1986).

In small laboratories, a minimum number of specimens per annum has been suggested to ensure vigilance is maintained (BSCC, 1986). It has been suggested that a cytology laboratory should process a minimum of 15 000 specimens annually so that the screeners encounter enough abnormal smears to maintain a high level of vigilance.

One aspect of the workload which is rarely considered is coverslip size. In the UK there are three different sizes in regular use in cytology laboratories (32×22 mm, 40×22 mm and 50×22 mm). A survey carried out in the department of cytology at St Mary's Hospital, Manchester (Proctor, 1988) showed that abnormal cells were more likely to be missed when the small or medium sized cover slips are used. Thus only the large size coverslips should be used for coverslipping cervical smears. However, increasing the size of the coverslip results in a significant increase in the workload and a requirement for more staff.

Workforce

A cytology laboratory is made up of a complex group of medical, scientific and clerical staff who must cooperate and harmonize with a joint sense of purpose for optimal performance (Fowkes, 1986). This can only be achieved if the laboratory is well organized and under good management. Managerial skills must be learned. This is not always appreciated by senior laboratory workers who arrive in a managerial position through academic achievement and at-the-bench experience (Singer, 1987a). Few are promoted due to their expertise in organization, supervision and communication. Most will be more efficient at analytical thinking by logical deduction than creative, lateral or divergent thinking (Singer, 1987b). Thus many senior laboratory personnel who find themselves in managerial positions may benefit from a training course in laboratory management.

Staff time should be used efficiently and productively (Singer, 1987c). The manager should identify quiet and busy periods in the laboratory and redistribute staff accordingly. Managers must have a good imagination and a flexible way of handling individuals. They must recognize individual ambitions, capabilities and give new challenges to avoid repetitiveness and boredom. Each member of staff is unique with his or her own personality, perception and potential. Managers must be aware of this and allow individuals to develop their skills to the best interest of the department. It is important to motivate and stimulate staff, to maintain morale by making work more interesting and rewarding, so staff gain a sense of achievement and job satisfaction. Job rotation, delegation and promotion are ways in which this can be done. People are the most valuable resources available to managers and their welfare must be considered at all times (Singer, 1987d).

Senior laboratory staff should be familiar with current legislation which affects their workforce (Fowkes, 1986), in particular that which affects the recording and investigation of all accidents in the laboratory, procedures for handling high-risk specimens, inoculations, grievance and disciplinary procedures.

A high calibre of personnel is required to work in a busy cytology laboratory and generally the tone is set by the calibre of the manager. When a vacancy does arise and a job specification is drawn up, selection of a candidate follows. Selection depends upon the job specification and obviously the candidate must be suitably qualified. However, the candidate with the highest qualifications is not always the best person for the job. Relevant experience and general qualities are also important; dedication, motivation, communication skills and ability to work with the existing staff should be considered. Placing the advertisement in the correct medium can help find the right person for the right job. Similarly interview structure and technique can affect the way the candidates perform (Singer, 1987e).

The staffing structure of the department depends upon the type of laboratory and its workload. As a general rule, the structure of the department should remain balanced with a well defined hierarchy. The head of the department is the consultant pathologist (BSCC, 1986) who is ultimately responsible for the quality of the work and all reporting. There must be one member of the senior scientific staff who assumes a supervisory role and is responsible for the day-to-day manning of the department and reports directly to the consultant pathologist. Other supervisory roles include training and monitoring junior staff, checking and quality control, statistical analysis and computerization, which can be delegated to other senior scientific staff members. There must be a core of productive staff who complete the primary screening. They may be engaged in advanced courses leading to a more senior position. Junior members of staff in training grades are not regarded as productive members despite their contribution to the practical work. Their numbers should be carefully calculated to equate with staff turnover. This is obviously a dynamic process and the training grades must act as a source of manpower which can be fed into the department, but at the same time not be a drain on it. A cytology laboratory requires efficient clerical back-up and a ratio of one clerical officer to 2–3 scientific staff has been recommended (BSCC, 1986). Experience with computerization and data storage systems is an advantage.

Untrained and unqualified staff have a limited place in the cytology department. Unskilled tasks do exist, such as washing up and filing, but generally do not amount to a job specification in themselves and can be carried out by junior members through job rotation schemes.

Planning

The effective management of resources depends on careful planning. It involves the determination of objectives and the ability to pursue a course of action which will lead to the attainment of a prespecified goal. It therefore relies upon adequate statistical information, continual reassessment of resources and close communication with colleagues, so realistic objectives can be set. There is little point in setting targets well above the capability of the department; this will result in disappointment and dissatisfaction.

A workload measuring system can be a useful tool when making projections about future requirements, such as numbers of staff (Penner,

1982), space allocation and level of automation. It can reveal unseen trends and gives a quantifiable figure to what might otherwise be interpreted as mere opinion or impression. Each department should have a short, medium and long-term plan covering improvement in the quality of the service, training programmes, succession of staff and replacement of equipment (Singer, 1987b).

Improvements in the quality of the service may require changes and reorganization (Singer, 1987b). Developments in techniques and cytological diagnosis are taking place at a rapid rate and it is essential that laboratories ensure patients benefit from these advances. Similarly, through research and development, improvements can be made to well established techniques to increase cell harvest or afford better preservation. Changes may also be necessary to the organization of the laboratory to accommodate clinics or other hospital departments. These changes need strategic and tactical planning for implementation. It is necessary to evaluate and review at each stage of the plan and to modify it accordingly.

Each laboratory should have a formal training scheme for its members, not only for those in junior grades but one which is suited to each position and personality. The schemes should be planned well ahead with early consultation. They should progress along with the employee giving new activities and responsibilities at a time when they are ready for them. To plan and formalize training, a training manual or personal record may be helpful. Alternatively, staff appraisal can be conducted at specified intervals (as in industry) giving an opportunity to review performance, monitor progress, counsel and to listen.

As staff members progress and become qualified, they will succeed through the department to higher grades. It is important to have a long-term plan to account for this and to maintain a balance of junior and senior grades. There will inevitably be staff losses which cannot always be foreseen. However, when a vacancy does arise it should be regarded as an opportunity for change and reappraisal. It allows the possibility of internal promotion with the re-appointment of a junior grade.

There are many sound reasons why a laboratory may wish to replace equipment; for example, the present equipment may be unreliable with increasing operating costs and deteriorating performance. However, before planning the purchase of new equipment both existing and predicted requirements must be considered, and if major expenditure is to be incurred, each step should be planned with consultation of medical and non-medical staff. Purchasing and operating costs must be acceptable, it must harmonize with existing laboratory apparatus, meet safety requirements and be flexible and reliable. Reliability should be considered not only of the instrument but of its marketing company and the standard of their after sales service.

Budgeting

Budgeting is the efficient management of resources and in a well managed department, planning and budgeting are interdependent. Each laboratory manager should be cost conscious, continually examining cheaper alternative consumables such as alternative suppliers, bulk purchases and energy saving measures.

Although the head of the department is the budget holder, it is very difficult to control or limit expenditure in a demand-led service such as cytopathology. Clinicians and general practitioners are free to investigate and test their patients in any way they feel is in the best interest of the patient and obviously by providing a screening service there is an open invitation to well women to have a smear taken at any time. Thus, opportunistic screening (defined as taking smears as and when the patient presents at a clinic or surgery) and increasing patient demand all contribute to the uncontrollable stretching of resources. Where resources are limited it has been proposed to allocate cost to individual units of medical health care (Korner et al., 1981), a unit being one request. This gives the budget to the clinician who can then be 'charged' for each individual request made. Indeed, one practical way of cutting costs is by restricting the screening of unnecessarily repeated smears within the recall period unless clinical details suggest otherwise.

The cytopathology department is a labour intensive one where the largest percentage of the budget is spent on staff salaries. It follows therefore that the measure of efficiency and performance of the workforce is an essential part of laboratory management. Performance indicators have been produced as indicators of efficiency and are measured as a ratio of inputs to outputs such as requests per staff member.

However, the assessment of performance indicators is not without difficulty and their use has been criticized (Stenton, Hyde and Hoctor, 1987). They do not take into account capital expenditure, the quality of the service provided, the speed at which the report is generated or clinical specialities that might be served. Nevertheless, costing data can help the laboratory manager in showing how money is spent and pinpointing where savings might be made. One system which has been designed called CATER (Stenton, Hyde and Hoctor, 1987) uses a measure of workload for individual tests based on the units derived from the Canadian System of Workload Measurement and relates staff time to staff costs. It enables the laboratory to compare its activities from one period to another and enables direct comparisons between hospitals.

During times of tight financial constraints, there is a move towards making the service more cost effective and there is no doubt that major cost savings could be made by integration of small departments and centralization. However, in the fight for the most efficient form of laboratory service, we must not lose sight of the function of the laboratory, namely that of providing a high quality service and maintaining a high standard of patient care.

Data storage

A comprehensive data storage system, whether manual or computerized, should be maintained within the cytology department. Records should be challenged every time a request is received and the results of previous cytological investigations noted for each patient. While it is advisable to keep all records and slides it is not always feasible in departments where storage space is minimal, and negative reports and smears over 10 years old may be discarded. Space can be a problem even when records are on computer and culling of negative reports may have to be done. Alternatively microfilm or microfiche can be used for archiving.

A laboratory specimen number is generated for each request; this is then recorded in a daybook in chronological order. This provides proof of receipt of the specimen. Most cytological request forms are also the report form and, once reported, a copy is sent to the sender and one is kept for filing. In gynaecological cases a further copy is sent to the general practitioner if he is not the original sender. The copy kept in the cytology department is generally filed in alphabetical order and subsequent tests are attached together, thereby building a patient's report profile.

Patient index cards are an alternative means of storing data, especially for positive cases. The master patient index is challenged for each new request and the index card attached to the request form. The report is recorded on the index card which is refiled in the master index file (Proctor, 1988). In a busy cytology laboratory the manual filing, refiling and follow-up of smear reports can be a complex clerical problem. Furthermore, statistical analysis, reminder letters and cross referencing with histological reports can be awkward, cumbersome and extremely time consuming, in a manual system (Walsh, 1981). As demands on cytology departments continue to grow, the wastage of scientific staff on this necessary clerical work should be reduced to a minimum. This problem has eased considerably with the introduction of sophisticated computer hardware and software into the cytology laboratory. This is further discussed in Chapter 27.

The computer system employed must be flexible and adaptable to the consumers' needs. It must be able to handle large volumes of information and provide quick and easy access to the data stored for statistical analysis, workload measurement and costing purposes. The search routines should be comprehensive to identify patients' previous specimens including histological reports for cross referencing. There should be provision to accommodate updates and new developments and it should be 'user friendly' with the option of producing ad hoc reports through free-text statements. Daily printouts can be requested to equate with a daybook, although with adequate peripherals, visual display units and printers this may not be necessary. There should be limitation of access to confidential data stored for security reasons. It is also recommended that back-up systems are used to avoid data loss (Walsh, 1981).

The data are usually entered in a coded form to save space. There are the international systems SNOMED (Systematized Nomenciature in Medicine; SNOMED, 1984) and SNOP (Systemized Nomenclature in Pathology; SNOP, 1965) which use topographical and morphological codes; or a unique user code system can be generated for in-house use that can be tailor made. Entries should be available for special requests and special investigations since they may represent considerable staff time and can affect workload figures. Call and recall commitment lies beyond the sphere of the laboratory and at best the laboratory can only recall. It is taken that the responsibility for informing the patient of the results and initiating action on an abnormal smear report lies with the smear taker. However, the laboratory should, if appropriate action has not been taken, remind the sender that a repeat smear is overdue. A complete computerized system can generate the reminder letter and record whether a further smear has been received and action taken. If no action has been taken, the computer will generate a reminder.

Since all the information is stored on computer it makes it unnecessary to keep a manual filing system and, indeed, with a comprehensive computerized system the request form and report form can be quite separate. The computer can generate the whole report, printed and sealed with requests not only going to the sender and general practitioner but also, if directed, to the patient in a letter form.

Perhaps in the not too distant future all specimens will arrive in the laboratory with magnetic bar code labels eliminating the need for any paper work and providing complete patient identification.

Teaching and training

Junior members of staff require a basic grounding in science as an entrance requirement to a career in cytology. Their training can then follow a prespecified course through day release or block release courses. Complete training, however, is a blend of formal educational courses and in-service training, which should correlate and be monitored at regular intervals, to assess the progress of the trainee. In-service training not only shows how to do the work but also lays the foundations of the trainee's attitude towards the job. It gives an opportunity for the trainee to appreciate how the work relates to patient care, leading to a greater understanding of the whole patient care process.

The exact practical procedures of preparing specimens should be demonstrated giving clear instructions at each stage of the procedure with reasons why it is performed that way. It may take some time before the trainee is proficient, and duplicate numbers of smears from non-gynaecological specimens may be made during this time. The normal day to day duties should be outlined with special reference to how they affect the running of the laboratory and why they are performed in that manner. Fixing and staining criteria should be explained and a laboratory manual can help, though should not replace a personal and verbal introduction. Close supervision is required throughout this induction period.

The training programme should also include safety procedures and ideally laboratories should have a safety handbook that can be issued to new members to be read and signed (Singer, 1987f). This should then be reinforced by a tour to show the location of fire points, fire extinguishers, fire blankets, eye wash bottles, first aid kits and collecting points after evacuation.

For screening, individual coaching is invaluable and time spent on a multiheaded microscope is time well spent. A selection of teaching slides should be examined at specified stages in the training programme, interspersed with routine screening sessions where the trainee records all of their findings. These slides can then be reviewed on the multiheaded microscope for discussion especially when there are discrepancies. During this time the trainee develops a tolerance to the microscope and builds a library of morphological features. This training period usually takes a minimum of 12 months to build up a library sufficient enough for competent cervical screening. The analysis of non-gynaecological specimens takes longer to learn. Nothing can replace time at the microscope to create a competent screener.

A variety of atlases of cytopathology should be available together with text books on pathology, histology, histopathology and medical sciences. The trainee should also be encouraged to keep up to date by reading scientific journals. Several teaching aids are recommended, such as closed circuit colour television with video recording facilities, adapted to a high quality microscope and an audio visual teaching machine with synchronized recorded commentary (Singer, 1987f).

Training courses in cytopathology at both

introductory and advanced levels exist and provide a high standard of tuition in theoretical and practical aspects of cytology. It is often recommended for trainees to attend one of these courses during their training programme, especially if the amount and type of material at their place of work is limited (Singer, 1987f). Seminars, lectures, update courses and workshops can be equally beneficial to staff in non-training grades. They often stimulate staff, who, given the opportunity, respond positively to new ideas. Exchange schemes and group sessions can also stimulate interest such as discussion groups, journal clubs, case studies and user group meetings. Attendance should be encouraged. Our science is continually developing and it can be difficult to keep pace with new advances.

Training is therefore not just a mere list of instructions but the development of an individual's behaviour, knowledge, attitude and skills in order to achieve a desired standard of performance. Training should commence but never cease. Indeed once a high level of competence to screen has been achieved, there is the avenue of management training to pursue!

Safety

In the organization and management of a laboratory, safety must be regarded as high priority. The responsibility for safety within the laboratory rests with the head of the department (Howie *et al.*, 1978), who may delegate particular functions and the authority to carry them out. It is recommended that a Safety Officer is nominated (Howie *et al.*, 1978) who has sufficient training and experience not only to understand the control of laboratory infection but also to cope with fire prevention, the safe maintenance of electrical equipment, the safe storage of flammable and corrosive substances, monitoring of toxic fume levels and accident prevention.

All new members of staff must be made aware of the hazards of dealing with infectious material, and in their training should be instructed in the correct behaviour, attitude and procedures necessary for working in a hazardous environment.

New members should be effectively immunized before working with contaminated material. This is a condition of employment unless there are contraindications and in such cases work with infected material should be prohibited (Howie *et al.*, 1978).

Protective clothing should be worn at all times when working in the laboratory. It should be removed when leaving and placed on the pegs provided. Disposable gloves should be worn when working with specimens and these should also be removed when leaving the laboratory. Hands should be washed and hand basins with elbow taps should be available together with disposable hand towels. Obviously food, drink and cosmetics are banned from the laboratory (Howie *et al.*, 1978).

Special accommodation should be made for processing specimens and the processing room must be kept separate from the screening room. A warning note 'Danger of infection' within the international biohazard symbol must be displayed prominently on the outside of the door of the processing room and restricted access enforced. A separate area must be provided for the receipt of specimens, which has an impervious surface resistant to disinfectants. Specimens labelled 'danger of infection' should not be unpacked by reception staff but by laboratory staff only and specimens which have leaked should be dealt with by the Safety Officer (Howie *et al.*, 1978). Each cytology laboratory should have or have access to a Class I open fronted safety cabinet, for the processing of specimens potentially infected with tubercle bacilli or hepatitis. The air flow of the cabinet should register 150 ft/min and regular anemometer readings should be made. It should be fitted with a high efficiency particulate air filter to British Standard 3928; it should not recirculate air and must maintain negative pressure when in use (Howie *et al.*, 1978). After use it should be wiped with 2% glutaraldehyde, a non-corrosive disinfectant. The apparatus and equipment kept within the cabinet should be fumigated with formaldehyde vapour and this must also be done before servicing and changing the filter. Servicing should be done at intervals not longer than six months.

When dealing with potentially infected material, good microbiological technique must be observed. Specimens should be centrifuged in sealed centrifuge buckets to avoid aerosols and transported in these to and from the cabinet. A refrigerator should be available in the processing room for the safe storage of specimens.

Specimens from patients infected with the

human immunodeficiency virus (HIV) should be dealt with under Category 3 containment level. It is regarded by the Advisory Committee on Dangerous Pathogens (ACDP) as a Group 3 pathogen and it is advised that such specimens are handled in a room set aside for this purpose (ACDP, 1984). It is not sufficient to set apart a section of the ordinary laboratory. This room must have restricted access with only authorized personnel allowed entry. Gowns and gloves should be provided in an ante room. Category I safety cabinet and autoclaving facilities must be available. Disinfectants at their working concentrations must be placed at each working station and changed regularly. Articles should be totally immersed in disinfectant which should be in contact with all the inner surfaces. Phenolic disinfectants are used for organic material and tubercle bacilli; hypochlorites should be used for blood and viruses. Aldehyde-based disinfectants, such as 2% glutaraldehyde, are used for metal surfaces as they are non-corrosive.

Infected or potentially infected waste material must be treated by autoclaving or incineration and laboratory staff should render their own refuse safe for other people to handle. It must not leave the laboratory unless it has been effectively autoclaved or placed in a safe and secure container and transported to the incinerator in a safe manner. Infected sharps should be placed in a 'CINBIN', then autoclaved and/or incinerated. The sterilization and disposal of all laboratory waste must be supervised by the Safety Officer and infected waste should not be placed where the general public have access. Strong plastic bags should be provided and a colour coded system can help identify infected and non-infected waste for disposal purposes.

Other risks within the laboratory which should be minimized include fire, electric shock, risks of physical injuries from centrifuges and other equipment and exposure to poisonous chemicals, carcinogens and vapours.

Fire precautions taken should be in accordance with those recommended by the Fire Officer. Indeed the Safety Officer must liaise with the Fire Officer to make the correct arrangements and assess standards. Fire extinguishers of the correct type must be installed; they should be tested and kept up to date. Regular fire drills should be conducted and fire doors kept shut and not propped open. Special provision must be made for the storage of flammable substances; ideally a lockable metal cupboard or trunk should be used. It is advisable to store only small quantities of flammable substances in the laboratory with a main store elsewhere. Flammable material should not be stored in the refrigerator. Escape routes should be clearly marked and kept clear of furniture or equipment.

Similarly, corrosive substances should be stored in a separate cupboard which is labelled 'danger of corrosion' with the international symbol displayed on the containers. Plastic containers are available for bottles and are ideal for such storage. The solvents xylene, toluene, etc. should also be kept in minimum quantities within the laboratory and the levels of their fumes monitored. It may be necessary to ventilate the working area or install a filtration fume cabinet if safety levels are exceeded. This is especially necessary if carcinogens are going to be handled, such as diaminobenzidine used in immunocytochemistry.

All the mechanical equipment in the cytology laboratory should be serviced regularly. Similarly, electrical equipment must be installed and serviced by qualified electricians. While working in the laboratory, maintenance personnel should be supervised at all times.

In conclusion, it is advised that all clinical laboratories have a safety policy. Each policy should contain a general statement of intent and specify the objectives, thus giving a basic plan of the aims of the laboratory in health and safety, describing who is responsible for various measures and how they should be carried out. It is important that the Safety Officer monitors the implementation of the policy by a strict system of documentation, recording all accidents and carrying out regular safety audits. The Safety Officer should be kept up to date with new developments in safety and help to develop new measures to reduce or eliminate risks. Effective hazard control is of prime consideration to the Safety Officer and areas such as occupational hygiene, fire prevention and safety training should be given special attention. However, it is impossible to promote safety without the full cooperation and commitment of all employees and they should be encouraged to participate in preventive measures and given a say in developing an approach to accident prevention.

Quality control

Quality control schemes provide objective measures of performance and offer a good opportunity for the profession to explore the difficult territory of professional self scrutiny. Unfortunately because of its subjective nature, cytopathology has not lent itself well to quality control. Cytological diagnoses are greatly dependent upon personal experience, judgment and skill and there are no rigid measurable criteria as in other medical laboratory disciplines, such as clinical chemistry (Carney, 1973). The wide variety of systems that have been proposed only emphasize the difficulty in finding anything satisfactory and the rescreening of 10% of negative cervical smears, often recommended as a quality control exercise, has proved most unsatisfactory and statistically non-productive (Melamed et al., 1973). However, this does not eliminate the need for quality control but rather intensifies the search for an acceptable system.

The purpose of quality control is to ensure consistent high quality performance. In the cytology laboratory, a high quality of performance requires more than periodic testing. It requires good basic training, continuing education, good working conditions, proper collection of specimens, technically good preparations and good record keeping. It requires consistent attention to follow-up and a continuing programme of review of negative specimens whenever there is evidence of neoplasia on later surgical or cytological material. Comparison of the performance of individual screeners in terms of abnormal or unsatisfactory smears reported may be of value, assuming screeners are examining smears from similar sources.

Correlation of cytological and surgical reports is mandatory and must be given special attention. The cytological specimen should be reviewed with the surgical specimen, either with a multihead microscope or in a seminar with video microscopy facilities. The conference may be brief or detailed, depending on the complexity of the case or how unusual or interesting it is. This can frequently stimulate constructive critical comment and interprofessional discussion. The information gained from such discussion is particularly useful; many will involve diagnoses that are rare or difficult and the surgical diagnosis may equally be difficult or equivocal. Errors are regretted and learned from and those present find that it exposes them to educational, cytological and histological material. Usually the result is that quality is maintained, correlation is good and the value of such reinforcement is obvious.

External quality assessment

This refers to a system of checking laboratory results by means of an external agency. It includes comparisons of a laboratory's results at intervals with those of other laboratories and the main object is to establish comparability between laboratories (Whitehead and Woodford, 1981).

The reference point to which each laboratory's result is compared must be acceptable and consensus is virtually the only available means of arriving at a generally acceptable reference point in cytopathology. Alternatively, an expert in a particular field can be consulted or peer review of a selection of test slides (proficiency testing) can be carried out (Collins and Patacsil, 1986). Generally, consensus techniques are satisfactory if a large enough number of laboratories participate. However, in cytology a large number of identical test specimens are almost impossible to obtain and sending the same stained smear through a succession of laboratories for comparison of results is time consuming and cumbersome. Several small clusters of laboratories presently take part in slide exchange schemes and five or six laboratories in one scheme seems optimal (Melamed et al., 1973; Whitehead and Woodford, 1981). Afterwards the results are circulated and the participants meet to discuss the slides, especially where there were discrepancies. Such an exercise can be criticized on the grounds that it is largely educational and not a true reflection of laboratory standards.

A system of laboratory accreditation can be a strong force in establishing and maintaining high standards of laboratory performance (Melamed et al., 1973). It reflects peer judgment and in that respect will be a measure of the influence of their professional and scientific evaluation. The criteria for laboratory accreditation should be flexible and the accrediting board should be free to revise regulations as experience dictates. One system that has been developed involves a site visit during which the laboratory staff are presented with a tray of smears which must be screened and reported within a specified time. Although accreditation systems are not compulsory and carry no legal weight, sanctions can be introduced into quality assurance systems. This has happened in New York (Carney, 1973; Collins and Patacsil, 1986) where, if laboratories fail the proficiency testing programme repeatedly, their licence is removed.

Ideally, assessment of a laboratory's performance would be based upon its total output of results and would take into account the correctness of the result and its interpretation for every investigation, the speed of response and the absence of administrative and clerical errors. Clearly only a small fraction of these features can be assessed externally and external quality assessment is not an alternative method to internal quality control. However, coupled with a further series of education and training schemes concerned not only with specimen examination but also with specimen collection, transport and handling, a reasonable system of assessing standards can be developed. The sum of all these procedures can be called a quality assurance programme. This does not imply that perfectly reliable results can be assured at all times, but that a controlled systematic effort is being made towards achieving the highest possible quality of performance in patient care.

Conclusion

In conclusion, there is no universal system for the effective management and organization of a cytology laboratory. However, with careful planning, sensible budgeting, good communications, friendly team work, training and education and a programme of quality assurance, a smooth functioning laboratory should develop. Participation in research and attendance at meetings should be encouraged and senior staff should be free to explore new techniques and introduce new ideas of scientific value. Emphasis must be placed upon the safety and welfare of all the laboratory staff who should, in turn, never be allowed to forget their important role in patient care.

Enthusiasm and interest for the subject should be transmitted to junior members who will respond in a positive way. Continuing education will improve the quality of the work, and lead to greater insight and knowledge of this rapidly expanding field of clinical cytology.

References

ACDP (1984). Categorisation of pathogens according to hazards and categories of containment. Health and Safety Executive, Bootle, Merseyside

BSCC (1986). Recommended code of practice for laboratories providing a cytopathology service. British Society for Clinical Cytology, London

CARNEY, C. (1973). Quality control in cytopathology. *Acta Cytol.*, **28**, 535–540

COLLINS, D. and PATACSIL, D. (1986). Proficiency testing in cytology in New York. *Acta Cytol.*, **30**, 634–644

EVANS, D. M. D., HUDSON, E. A., BROWN, C. L., BODDINGTON, M. M., HUGHES, H. L., MACKENZIE, E. F. D. AND MARSHALL, T. (1986). Terminology in gynaecological cytopathology: report of the working party of the British Society for Clinical Cytology. *J. Clin. Path.*, **39**, 933–944

DHSS (1988). DHSS Health Circular HC(88)1 HC(FP)(88)2

FOWKES, F. G. R. (1986). Diagnostic vigilance. *Lancet*, **i**, 493–494

HOWIE, J. et al. (1978). Code of practice for the prevention of infection in clinical laboratories and post mortem rooms. London: HMSO

KORNER, E. et al. (1981). Report of the steering group on health service information. London: DHSS

MELAMED, M. et al. (1973). Presidential address: 20th annual scientific meeting. *Acta Cytol.*, **17**, 285–288

PENNER, D. (1982). The workload recording method: a management tool for the clinical laboratory. *Hum. Path.*, **13** 393–398

PROCTOR, D. T. (1988). Letters to the Editor: Replies to questions on quality assurance measures in cytopathology. *Acta Cytol.*, **32** (6), 926–927

SINGER, R. (1987a). Management in the NHS. *Med. Lab. World*, March, 37–40

SINGER, R. (1987b). Managing change. *Med. Lab. World*, June, 37–40

SINGER, R. (1987c). Managing resources. *Med. Lab. World*, April, 39–43

SINGER, R. (1987d). Managing people. *Med. Lab. World*, April, 49–53

SINGER, R. (1987e). Personnel selection. *Med. Lab. World*, August, 43–47

SINGER, R. (1987f). Training appraisal and counselling. *Med. Lab. World*, July, 53–57

SNOMED (1984). Systematized nomenclature of medicine. College of American Pathologists

SNOP (1965). Systemized nomenclature of pathology. College of American Pathologists

STATISTICS, CANADA HEALTH DIVISION (1982–83). Canadian schedule of unit values for clinical laboratory procedures. Institutional Statistics Section, Ministry of Supply and Service, Ottawa, Canada

STATISTICS, CANADA HEALTH DIVISION (1986–87). Canadian schedule of unit values for clinical laboratory procedure

STENTON, P., HYDE, K. and HOCTOR, D. (1987). Pathology costings (CATER). *IMLS Gaz.*, **31**, 281–284

WALSH, S. (1981). Cytology tracking system. *Med. Technol.*, October, 55–58

WELSH OFFICE (1987–88). Welsh workload measurement system for pathology; manual with schedule of unit values

WHITEHEAD, T. P. and WOODFORD, F. P. (1981). External quality assessment of clinical laboratories in the UK. *J. Clin Pathol.*, **34**, 947–957

27

Computerization

Noel R. Padley

Computers are used in cytology in two main ways. They are used to analyse microscopic images in attempts to automate diagnosis and they are used to provide record systems. The former remains a research application; the latter is now commonplace in routine laboratories and it is this which is the subject of this chapter.

Advantages of computerized records

Computerized records have many advantages over paper files. The more important ones are considered below.

Economies of space

Paper files are very bulky. To keep alphabetical and numerical files of results requires a large amount of space and numerous filing cabinets. To save space many laboratories discard normal results after seven years. This is undesirable when women are being screened throughout their lifetime and an accurate record of their negative tests is important both for quality control and for regulating the interval for screening. Modern digital storage systems using hard disks, laser written disks and magnetic tape allow for extended storage of records in a small space with duplication of records and multiple indexing. They do not suffer from degradation by misfiling, loss of records or fading. Copies of records can be kept in more than one place, which makes them more secure than paper.

Legibility of records

In the United Kingdom most cytology units have used the nationally agreed form which combines request and report. This is multi-part non-carbon copy and is handwritten. This has proved satisfactory for providing a result in an individual case but has caused problems where reports are to be used on a cumulative basis for reference at a later date or for recalling patients for further tests or for use as statistics. Handwritten patient details which are easily read by the doctor or nurse who wrote them are often illegible or ambiguous to others. The copies are also often faint and they fade with age. To make this data reliable and usable it needs to be typewritten, which requires a considerable increase in secretarial resources. This is beyond the scope of all but a few laboratories. The use of computers to transcribe, store, sort and print this data requires only a modest increase in staff and makes typewritten records available to all laboratories.

Referring to previous results

The patient file on a computer system allows each request to be checked quickly as it is received to see if the patient has been tested before. This has proved impractical in the past with paper files which are often grouped in years, and usually only those with abnormal results or those of special interest could be checked. With a computer system, a cumulative report of the results of each patient's previous biopsies and cytology is available when the slides are read. This historical information puts the current test in its proper context, which improves the standard of reporting. When deciding to advise a repeat smear, further investigations or suitable treatment, this cumulative summary presents the necessary data quickly and conveniently.

Laboratory records extend over many years and large numbers of people are tested. To identify individuals it is necessary to make use of numerous personal details. In practice the use of some of these can be difficult. For example, in the UK the National Health Service (NHS) number is a very accurate discriminant. Unfortunately few people know what their number is; on the other hand, although everyone can easily provide their date of birth, it is often only their age which is entered on the record. The NHS number and the date of birth are the only identifiers in common use which remain constant; all others, such as names, addresses, etc. are subject to change. Computer systems cope with this by allocating a unique number to each patient and by providing for the other parameters to be updated. Acquiring details of such changes relies on the subject notifying the laboratory either directly or indirectly through a new test, or via an outside agency such as the Family Practitioner Committee (FPC).

At present, notification of such changes is haphazard and incomplete; because of this it is important to devise comprehensive patient recognition procedures. When designing or choosing a system it is necessary to have multiple key fields as well as combinations of these key fields for use as an index to the patient file. This index should be available in all the areas of the programs for quick reference. Only in this way will individuals be identified and errors of double entry and confusion between patients be eliminated. These problems are not unique to mechanical systems, but although errors are frequent on manual systems using paper files this is not apparent because they remain largely undetected. It is only when such files are transferred to a computer that these errors come to light.

Cross-reference with histology

The use of a single computer for cytology and histology makes efficient use of the patient file and of staff and resources, but has the major advantage of providing automatic cross-reference between cytology and histology, with advantages to both. Not only are past results available for comparison, but the existence of current samples undergoing examination is automatically notified. This can be provided as a patient history. Statistical analyses of correlations can also be performed.

Checks on abnormal results

If a cervical smear is inadequate, it is usual to ask for it to be repeated. By recording a recall period of, for example, three months, it is possible to use the computer to generate a reminder if a further smear is not received within the allotted time. If a smear is received, that recall is automatically cancelled. The names of those patients who have not been followed up can be printed as a monthly list, together with their details and the reasons for recall.

Abnormal results can be flagged in a similar manner to ensure appropriate action. Each smear can be allotted a suitable default period and if no further smear or biopsy is received within that period, attention is drawn to this by its appearance on the monthly abnormal recall list. A variety of automated letters may be used as reminders to general practitioners, clinics or patients to make sure that abnormal results receive appropriate action. These arrangements can be handled manually, but they can be extended and made more comprehensive on a computer, while at the same time simplifying the operation and saving considerable clerical time. It is comparatively easy to build routines into a mechanical system which perform several different checks to ensure that no patient has been overlooked. These additional safeguards are too cumbersome to use on manual systems.

Recall of normal results

The laboratory file of adequate normal smears is used as the basis for recalling well women at regular intervals for screening. Whatever the interval and age range which is agreed upon, the laboratory computer can identify these women automatically. It can provide lists to general practitioners and clinics and FPCs and can automatically generate letters to patients inviting them to have a smear. The smear may be taken by the general practitioner or, if preferred, at a local clinic.

Recruitment

In the past many ways have been used to recruit

women into a cervical screening programme. These have included a general mixture of education and advertisements, together with more specific measures through family planning clinics, well-women clinics and general practitioners. Campaigns using mobile clinics have also been successful. However, call systems in which all women at risk are personally invited to have a smear are thought to offer a more systematic approach. Lists of names and addresses could be fed into the laboratory computer, those who are already on it as a recall or who have been retired from the system could be identified, and the remainder written to automatically by the system. Ideally, the list would contain names, addresses and dates of birth of all women and would be stored on magnetic media. However, no such list exists at present in the UK. The electoral roll is stored on a magnetic medium and is up-to-date, but does not distinguish age or sex. Nevertheless, pilot studies have shown that it can be used successfully (Cook and Wald, 1985). Lists from FPCs are incomplete and not automated. There are many other lists which could be used, such as consumer mailing lists, payroll lists from large employers, union membership lists and lists from other screening projects. Provided the lists can be obtained in a suitable form they can be read into the system in turn and combined automatically to build up a comprehensive list of women who are at risk in the area. Retirement from a screening system may be by choice, by old age or death, or by hysterectomy or (occasionally) some other medical condition. It is necessary to record the reason for retirement on the system so that deliberate retirements are not later confused with errors and omissions.

Statistics for research

The amount of detail recorded about the result will vary from system to system, but with modern storage devices most systems can store the entire report indefinitely. Statistical analysis of these records is a useful epidemiological research tool.

An extension of the conventional report is to use a systematic nomenclature to code the type of test and the result. This forces categorization, improves precision of diagnosis and translates the data into a form in which it can be easily sorted and analysed. Several systems have been devised for this purpose; the most universal of these is SNOMED (Systematized Nomenclature of Medicine) which has been developed by the American Institute of Pathology in cooperation with many international bodies. It is an expansion of their earlier (1965) SNOP (Systematized Nomenclature of Pathology) and now supersedes it. The tumour part of SNOMED is compatible with ICD-O (International Classification of Diseases for Oncology) which was devised in 1976. Many systems contain a SNOMED dictionary and have automatic encoding. Coded diagnoses can be used for research in the laboratory and also form a useful input to cancer registries and other information systems.

Statistics for laboratory management

More detailed statistics can easily be produced for quality control and management. Comparing rates of diagnosis in specific age groups between laboratories, and for different time periods within the same laboratory, provides data on the quality of diagnosis. Automatic costings can be provided for clinical budgeting and management of the laboratory. Statutory statistical returns can also be generated.

Practical considerations

It is important that the computer system should be large enough to be comprehensive. Small systems are better avoided because although their initial low cost is attractive, they are less efficient in their use of staff, are unable to support all the functions required and may be limited in size of storage and speed of response. Most of these small systems are based on desktop microcomputers. In a department which has a catchment area with a population of 250 000, a suitable machine would be a minicomputer with 10–15 terminals, a central processor of one or two megabytes, four or five printers of various types and hard disk or similar storage of about 400 megabytes. A system of this size can run special software which integrates a relational database with a comprehensive word processor to generate records and produce reports. It also has enough storage to be able to give immediate access to records which span ten or more years and rapid access to older records. Production of daybooks, worksheets, lists of patients for individual clinicians or clinics, recall letters and reminders to doctors and patients, statutory statistical returns and statistical analyses is automatic. Cost analysis of laboratory activities and clinical budgeting is possible. Although many small systems are advertised as offering similar features, in practice these promises are usually incompletely fulfilled.

Commercial systems are more effective, reliable and economic than home built systems. Only a well established software house is able to provide the variety of professional, trained experts which are needed to maintain a system. It is better to buy this expertise on a contract basis than to employ full-time computer staff in the laboratory. A maintenance contract is supported by a full range of engineers, systems analysts and programmers. Companies design their systems to be easy to operate and they will teach all members of the laboratory staff to be competent computer users with only a few days training. An alternative approach has been to retrain laboratory staff to design and run a system. However, this wastes the expertise that they have already acquired and is often unsatisfactory in the long run as the system which they produce is idiosyncratic and is dependent for its continued operation on the individuals who created it. If one or more of them is absent for any length of time the system may fail.

Modern minicomputers will operate in normal office conditions. A clean power supply is usually required. To protect the system from losing data if the electrical supply is interrupted there are devices which monitor the supply and shut down the system automatically. Most systems will not work on hospital emergency generators, but this is not usually a problem as prolonged power cuts are infrequent. Sudden large fluctuations in temperature are best avoided and, as the equipment may give off several kilowatts of heat, a separate room with modest air conditioning is ideal but not strictly necessary. The introduction of a computer with its associated changes in working practice is stressful to staff and, if not arranged carefully, may be very disruptive. Adequate time should be given for staff training with additional temporary staff to cover the laboratory routine. Attention should be given to lighting in the office and to the siting of visual display units. Desks and chairs which have been designed for computer operation usually prevent the aches and pains which some people experience when they first start operating computer terminals for long periods. Impact printers produce considerable noise and should be housed in a separate room or, if this is not practicable, in sound deadening cabinets. When the layout of the system is arranged care should be taken to ensure the security of the system, its equipment and its contents. Care is also needed to arrange that the system and its operation comply with local statutes for the protection of data.

Conclusions

Computerizing cytology records makes major improvements in the quality of service provided by a laboratory. It provides the basis for a screening programme and enables data previously buried in paper files to be used for research and management. Its use of staff and resources is highly efficient and well worth the initial cost and training. These advantages all confer significant benefits to patients and make an effective cervical screening programme a practical proposition.

Reference

COOK, G. A. and WALD, N. J. (1985). Can the coverage of screening for cancer of the cervix be improved using the electoral register? A pilot study. *Cancer Letters*, **28**, 307–310.

Index

Abscess, lung, 249
Accurette endometrial sampler, 221
Acetic-orcein technique, 93
Acid and alcohol fast bacillus (AAFB), see *Mycobacterium tuberculosis*
Acid haemolysis, 68
Acquired immune deficiency syndrome (AIDS), 31, 41
  lymphadenopathy associated with, FNA, 355
  opportunistic infections, 250
  Pneumocystis carinii pneumonia, 238, 239, 250
  sputum collection, 238
Actinomyces israelii, IUCDs and, 178–179, 226–227, 433
Acute bronchitis, 249
Acyclovir, 183
Adenocarcinoma, see *individual organs*; Carcinoma
Adenosquamous carcinoma of endometrium, 229
Adenoviruses, 188
Adrenal cortical stimulating hormone (ACTH), 43
Adriamycin, iatrogenic changes, 430
Agar, 69
Air-dried smears, 55, 74
Albuminized slide method, 70
Alcian blue/PAS staining technique, 90
Alcohol-fixed smears, 55, 73
Alcoholic fixatives, 52–53, 54
Alkaline phosphatase, enzyme marker, 111–112
Alkylating agent, 428
Allergic reaction, 39, 298
Allergic rhinitis (hay fever), 248, 255
  nasopharyngeal smears, 237, 255
Alveolar cell (bronchoalveolar) carcinoma, 254, 264

Alveolitis
  diffuse fibrosing (Hamman-Rich syndrome, idiopathic diffuse pulmonary fibrosis, interstitial pneumonia), 253, 261
  extrinsic allergic, 253
Alveolus, 237
3-Amino,9-ethylcarbazole, 111
Amoebic infection, 408, 409, 411–413
Anaphase, 22, 23
Ancylostoma sp., 419
Anisonucleosis, 168
Ankylosing spondylitis, HLA and, 30, 31
Anti-carcinoembryonic antigen, 286, 287
Anticoagulant, in serous effusion, 272
Anti-cytokeratin (CAM 5.2), 286
Anti-leucocyte common antigen, 287
Anti-sperm antibodies, 333
Anti-vimentin, 286–287
Antibody, 35
  markers, 108–113
    enzyme, 109–112
    fluorescent, 109
    other, 112–113
  monoclonal
    in CSF studies, 300
    fixative, 115
    production, 107–108
  polyclonal
    fixative, 115
    production, 107
  purification
    affinity, 108
    tissue powder absorption, 108
Antigens, 35
Antisera, 107–108
Apocrine cells, in breast aspirates, 345
Argentaffin, staining, 95
Arias Stella reaction, 146

Artefacts, 31, 52, 75–76
  automated recognition, 446
  'carbowax', 81, 103
Arthritis
  HLA and, 30, 31
  rheumatoid, see Rheumatoid arthritis
Arthrocentesis, 384
Asbestos
  bodies (ferruginous), 57, 245, 261
  staining, 95
Asbestosis, 252–253
Ascaris lumbricoides, 190, 419, 420
Ascitic effusions, see Serous effusions
Aspergillus fumigatus
  opportunistic lung infection, 250
  vaginal infection, 180
Aspiration pneumonia, 249
Asthma, 251
  Creola bodies, 256
  histology, 255
  sputum, 245–246, 256
Astrocytoma, 299
Asymmetrical unit membrane (AUM), 304
Atheroma, 42
Atherosclerosis, 42
Atrophy, 43
Autoimmune disease, 31
Automation
  abnormal cell recognition, 442–444
  artefact recognition, 446
  cervical screening, early developments, 441–442
  coverslipping, 101
  cytometry
    flow, 446–448
    specimen preparation, 448–449
    staining protocols, 449–450
  immunocytochemistry, 444
  malignancy-associated changes, 444–446

Automation (cont.)
  pattern recognition, 446
  staining, 79
  technical principles, 441
  texture features, 446
Ayre spatula, 4, 149
Azathioprine, iatrogenic changes, 430
Azoospermia, aetiology, 335

B cells (B lymphocytes), 35
  monoclonal antibodies, 300
Bacilli, 174
Bacteria
  cellular damage, 31–32, 33
  normal (commensals), 31–32, 157
  staining reactions, 88
  urinary, 310
  vaginal, 174–179 passim
    coitus effect, 177–178
  see also specific organisms
Bacteroides sp., 175, 177
Balantidium coli, 189, 413–414
Barr bodies, see Chromatin bodies
Barrett's syndrome, 370
Bartholin's gland, 137
Basement membrane (basal lamina), 8
Beale, Professor L. S., 2, 302
Best's carmine technique, 86
Bile, 379
Bilharzia, see Schistosomiasis
Biliary tract, 379
Biotin–avidin, antibody marker, 112–113
Bird fancier's lung, 253
Bladder
  anatomy, 303
  catheterization, urothelial changes, 435
  histology, 304–305
  malakoplakia, 318
  washings, 305
  see also Urinary tract; Urine
Bladder carcinoma
  adeno-, 317
  cytology, evaluation, 322–324
  diagnosis, 319
  in situ (non-papillary transitional cell), 315
  metastatic, 317
  occupational-related, 319–320
  screening, 54
  squamous, 315
  transitional cell, 315
Blast cells in cerebrospinal fluid, 299
Blastomyces sp., 180
Blastosis dermatides, 313
Bleaching techniques, 95–96
Bleomycin, iatrogenic changes, 430
Blood
  cells, 27
  immunocytochemistry, 120
  tumour cells in, 405–407
    detection, 405–406
    identification, 405–406
    incidence, 406–407
    significance, 407

Bloodstained specimens
  fixative, 53
  preparation, 55, 56, 66–67, 67–68
Bone
  description, 27
  fine needle aspiration, 365–366
Bone marrow
  in CSF, 298
  electron microscopy, role of, 404
  fixative, 54
  immunocytochemical methods, 120, 404–405
  non-malignant cells in, 401–403
    blood cell precursors, 402
    endothelial cells, 402, 403 (fig.)
    foreign, 403
    osteoblasts, 402, 403 (fig.)
    osteoclasts, 402, 403 (fig.)
    plasma cells, 403
    white blood cells, atypical, 402
  tumour cells in
    aspiration, 395–396
    Burkitt's lymphoma, 401
    carcinoma, 397–399
    detection rate, 396
    Hodgkin's lymphoma, 399–400
    melanoma, 399
    non-Hodgkin's lymphoma, 400–401
    sarcoma, 399
    significance, 404
    trephine biopsy, 395–396
Bouin's fluid, 54
Bradykinin, 34
Brain
  tissue, preparation, 70
  tumours
    extracranial spread via shunt, 437
    main sites, 293, see also Cerebrospinal fluid
Breast
  cancer
    fine needle aspiration, 344, 348–349
    genetic factors, 49
    metastases, CNS, 300
    unopposed oestrogen and, 50
  fine needle aspiration, 344–345
    benign lesions, cytological features, 344
    cystic disease, 345–346
    fat necrosis, 346–347
    fibroadenoma, 346
    fibroadenosis, 346
    lactating, 347–348
    male, 349–350
    malignant lesions, 344, 348–349
    Paget's disease of the nipple, 349
  structure, 345
Bronchial biopsy, cytology compared to, 49
Bronchial brushings, 239
  cytology
    evaluation, 268
    normal, 244–245

Bronchial brushings (cont.)
  non-cellular elements, 245
  specimen preparation, 56, 242
Bronchial carcinoma, 253–254
Bronchial epithelium
  busulphan-induced changes, 260
  metaplasia, 21, 22, 254, 260
  radiation-induced changes, 260
Bronchial secretions
  collection of, 238
  non-cellular elements, 245
  normal cytology, 244–245
  specimen preparation, 56, 57, 241–242
  see also Sputum
Bronchial washings, 238
  cytology, evaluation, 268
  non-cellular elements, 245
  normal cytology, 244–245
  specimen preparation, 56, 241–242
Bronchiectasis, 39
Bronchioles, 236–237
Bronchitis
  acute, 249
  chronic, 260, 260–261
Bronchoalveolar (alveolar cell) carcinoma, 254, 264
Bronchoalveolar lavage, 238–237
  cell count, 242
  non-cellular elements, 56
  normal cytology, 245
  smear preparation, 56, 242
Bronchopneumonia, 249
Bronchopulmonary dysplasia of the newborn, 261
Bronchoscope
  fibreoptic, 238
  rigid, 238
  sampling, 238–239
Bronchus
  anatomy, 235
  cytology, 244
  histology, 236
Brush biopsy, bronchial, see Bronchial brushings
Brush border, gastric, 375
Bryan's sperm stain, modified, 99
Buccal smear, preparation, 92–93
Buffy layer technique, 67
Burkitt's lymphoma, 49, 50, 401
Busulphan
  iatrogenic changes, 429–430
  urothelium affected by, 321

Ca antigen, 404
Calcium pyrophosphate deposition (chondrocalcinosis, 'pseudogout'), 389–390
Calcospherites (Psammoma bodies), 173
Canadian System of Workload Measurement, 457–458
Cancer, see Carcinoma
Candida albicans
  oesophagitis, 371

Candida albicans (*cont.*)
  opportunistic lung infection, 250
  in urine, 313
  vaginal infection, 179–180
Carazzi's haematoxylin stain, 80
Carbol chromotrope 2R technique, Lendrum's 85
Carbon, staining, 95
Carcinoembryonic antigen (CEA) 83, 119
Carcinogens, 29
Carcinoid, 266
Carcinoma
  adenocarcinoma
    free cell form, 276–277
    papillary and acinar, 275–276
  adenosquamous, of epithelium, 299
  alveolar cell, 254, 264
  of bladder, see Bladder carcinoma
  blood, see, Blood, tumour cells
  bone marrow, see, Bone marrow, tumour cells
  brain, see, Brain tumours; Meningioma
  breast, see Breast, cancer
  bronchial, 253–254
  Burkitt's lymphoma, 49, 50, 401
    see also Non-Hodgkin's lymphoma
  cell neoplasia, 44
    see also Neoplasia; Neoplastic cells; Tumours
  cervix, see Cervical carcinoma; Cervical smears
  in CSF, see Cerebrospinal fluid
  CIN, see Cervical intraepithelial neoplasia
  choroid plexus papilloma, 298–299
  of colon, see Colonic carcinoma
  duodenal, 380
  of endometrium, see Endometrial carcinoma
  enzyme activity in malignant cells, 443
  gastric, see Stomach; Gastrointestinal
  growth, see Neoplasia; Neoplastic cells; Tumours
  Hodgkin's lymphoma, 399–400
  in immunocompromised patients, 50
  intraepithelial neoplasia, 45–46
  kidney, see Kidney; Renal
  large intestine, 381, 382
  of liver, see Liver
  of lung, see Lung; Pulmonary
  lymphoma, see Lymphomas
  melanoma, in serous effusion, 278
  meningioma, 300
  mesothelioma, 278
  nasal, 248
  nasopharyngeal, 248
  oat cell, see under Lung
  oesophageal, see Oesophagus
  oligodendroglioma, 299

Carcinoma (*cont.*)
  ovary, 215–216
  pancreatic, see Pancreas
  prostatic, see Prostate
  rectal carcinoma, 381, 382
  in serous effusion, see under Serous effusions
  skin, see Skin cancer
  squamous, 277, 290
    of stomach, 376–377
  in synovial fluid, 394
  transitional cell, 315, 316 (figs), 319
  of upper respiratory tract, 248
  urinary tract, 314–319
    see also Bladder; Kidney
  urothelial cells, 315, 316 (figs), 319
  vulval, 216
  see also under individual organs; Tumours
Carnoy's fluid, 53
Carnoy's haemolysis, 67
Cartilage, description, 27
Catheterization, see Iatrogenic changes, instrument-induced
Cautery, see Iatrogenic changes, cryotherapy
Cell blocks
  fixative, 53, 54
  frozen, 69
  preparation, 57–58, 69
  staining, 83
  see also specific tissues
Cell-mediated immunity, 35, 36
Celloidin procedure for minimizing cell loss, 77
Cells
  actin filaments, 12
  atrophy, 43
  biochemical activity, see Cytochemistry
  centrioles, 20
  cycle, 22–23
  cytoplasm, 12
  desmosomes, 11
  differentiation and control mechanisms, 24–25, 42
  endoplasmic reticulum, 14–16
  environmental damage, 32–33
  ferritin granules, 21
  filaments, 19–20
    actin (fine), 19
    intermediate, 19–20
    myosin (thick), 20
  glycogen granules, 20
  Golgi apparatus (body), 17–18
  growth disorders, 42–46
  hyperplasia, 42, 43
  hypertrophy, 42
  junctions, 11
  labile, 37
  lipid droplets, 20
  lysosomes, 18–19
  membrane, environmental damage, 32
  metaplasia, 43–44
  microtubules, 19, 20

Cells (*cont.*)
  mitochondria, 13
  necrosis, 32, 33
  neoplasia, 44, see also Neoplastic cells; Tumours
  nucleus, 21–22
  organelles, 13–20
  permanent, 37
  plasma membrane, 7–12
  ribosomes, 14
  secretory vesicles, vacuoles, 17–18
  specimen preparation, see Specimen preparation
  stable, 37
  universal features, 7
Cellularity, assessment, 63
Cellularity Comparison Card (CCC), 62–63
Centrifugation, 58–59
Centrioles, 20
Centromere, 23
Cerebrospinal fluid (CSF)
  aspiration of, 294
  benign cells
    bone marrow cells, 298
    choroid plexus cells, 298
    classification, 297
    ependymal cells, 298
    lymphocytes, 297
    macrophages, 297–298
    neutrophils, 298
  cell count, 294
  cytology, 294–297
  formation and circulation, 293
  monoclonal antibody tests, 300
  non-specific esterase stain, 298, 300
  slide preparation, 295
  special staining techniques, 300–301
  specimen collection, 294–295
  specimen preparation, 56, 295–296
    cytocentrifugation, 59, 296
    evaluation of techniques, 296–297
    fixation, 297
    Millipore filters, 296
    sedimentation, 296–297
  tumour cells
    astrocytoma, 299
    choroid plexus papilloma, 299–300
    criteria of malignancy, 298
    ependymoma, 299
    lymphoma/lymphoblastic leukaemia, 299
    medulloblastoma, 299
    meningioma, neurofibroma, 300
    oligodendroglioma, 299
    secondary, 300
Cervical canal, see Endocervical canal
Cervical carcinoma
  adenocarcinoma, 211, 212
    clear cell, 212, 215
    cytological diagnosis, 214–215
    endocervical type, 212, 213 (fig.), 215
    endometrial type, 212, 215

Cervical carcinoma (*cont.*)
  in situ, 212
  cytology, 213–215
    criteria of malignancy, 213
    herpes simplex virus and, 183
    HPV infection and, 36, 50, 184, 188, 209–210
    immunosuppression and, 50
    mixed, 212–213
    rare, 213
    secondary (metastatic), 213, 215, 216 (fig.)
    squamous cell, 211–212
      cytological diagnosis, 213, 214
      keratinizing, 212
      large cell non-keratinizing, 212, 214
      small cell non-keratinizing, 212, 214
    in urine, 319
Cervical cone biopsy, interpretation, 209
Cervical epithelium
  fibrosis, 191
  hormonal effects on, 140–141
  iatrogenic changes, 425–428
    cryotherapy, diathermy, 426–427
    IUCD, 431–432
    laser therapy, 427, 428
  keratinization (leukoplakia), 141
  metaplasia, 43, 141–144
    immature squamous, parabasal cells, 155
    mature squamous, 142, 144
  regenerating, 190–191
  see also Cervix, squamous epithelium
Cervical intraepithelial neoplasia (CIN), 45–46, 201–211
  aetiology, 202
  cytology
    diagnosis, 90, 206, 209–211
    errors, 181–183
    interpretation, 179–180
    terminology, 206–207
    see also Cervical smears
  dyskaryotic cells, 206–207
    differential diagnosis, 201, 209–211
    in urine, 309
  epidemiology, 201–202
  histology
    features, 203–206
    terminology, 202
  HPV infection and, 46, 181, 184
  'male factor', 202
  malignant cells, 206
  pathogenesis, 202
CIN I, 203, 204 (fig.), 207
CIN II, 203, 204 (fig.), 207
CIN III, 203–206, 207, 208
Cervical mucus penetration test, 333
Cervical scrape, 4, 137, 148, 149
Cervical screening, 46, 164, 216
  automation, 441–450 *passim*
  data storage, 461–462

Cervical screening (*cont.*)
  computerized, 461–462, 467–468
  flow cytometry in, 447–448
  recall, computerized, 468
  recruitment, computerized, 469
  reporting policy, 456
  training, 462–463
  workload, 455–456
    Canadian measurement system, 457–458
    recommended, 458
    vigilance, 458
Cervical smears, 137
  abnormal
    interpretation, 207–208
    management, 161–162, 217
  acute inflammatory changes, 167–169
  adenovirus infection, 188
  amoebae and IUCD, 413
  anisonucleosis, 169
  apparatus, 150
  at menstruation, 158–159
  atrophic, 155, 156 (fig.) 172–173, 175, 211
  bacterial flora, 175–179 *passim*, 210
    coitus effect, 177–178
  'blue blobs', 172, 173
  chlamydial infection, 181
  chronic inflammatory changes, 169–173
  ciliocytophthoria, 169
  cytolytic, 211
  cytoplasmic staining, 168–169
  differential diagnosis, 209–211
  dyskaryotic cells, 201, 206–207, 209–211
  dyskeratocytes, 185
  endocervical cells, 169
  endometrial cells, 220, 222–223
  'epithelioid' cells, 171, 177 (fig.)
  fungal infections, 179–180 *passim*
  granulomatous changes, 171
  histiocytes ('exodus'), 158–159, 160 (fig.)
  HPV infection, 185–188 *passim*
  hyperkeratosis, 170, 211
  iatrogenic changes, 210–211
  immature metaplasia, 201, 210
  immunological staining, 114, 120–121
  infection, 210
  interpretation, 152
  intraepithelial neoplasia, *see under* Cervical intraepithelial neoplasia
  invasive carcinoma, *see under* Cervical carcinoma
  IUCD-induced changes, 431
  karyopyknosis, 168
  karyorrhexis, 168
  koilocytes, 185
  Langhan's giant cells, 171, 177, 178 (fig.)
  management, 216–217
  metaplastic cells, 169, 200–201

Cervical smears (*cont.*)
  compared to dyskaryotic cells, 201
  negative, management, 217
  normal, 152–159
    atrophic/immature metaplasia, 155–156
    commensal organisms, 157
    endocervical, 155
    endometrial, 156–157
    hormonal status, effect of, 158–159
    other components, 157
    squamocolumnar junction, 159
    squamous epithelial, 152–155
  parabasal cells, 200
  parakeratosis, 170, 171 (fig.)
  parasitic infections, 188–189 *passim*
  plasma cell cervicitis, 172
  postcoital, 157
  in pregnancy, 159
  Psammoma bodies (calcospherites), 173
  report, 159–164
    principles, 160–161
    validity of, 162–163
  reserve cells, 210
  spider cells, 200–201
  transformation zone, 159
  treatment, monitoring after, 217
  unsatisfactory, 163, 217
  uses of, 163–164
  vacuolation, 168, 169 (fig.)
  viral changes, 210
    152–157 *passim, see also individual viruses*
Cervical spatulas, 4, 149, 150
Cervicitis
  acute, 167–169
  atrophic, 172–173
  chronic, 169–174
    granulomatous, 171
    hyperkeratosis, 170
    leukoplakia, 170
    lymphocytic (follicular), 171–172
    parakeratosis, 198
  plasma cell, 172
Cervix
  acute inflammation, 167–169
  anatomy, 139–144
  bacterial flora, 174–179
  chronic inflammation, 169–170
  CMV infection, 184
  diethylstilboesterol (DES) and, 195
  ectopy, 141, 195–197
  HPV infection
    and cervical carcinoma, 184
    endophytic (inverted) warts, 187
    exophytic warts (condyloma acuminata), 185, 186
    laboratory diagnosis, 187–188
    subclinical (SPI), 185
  inflammation, *see* Cervicitis
  neonatal, 195, 196 (figs)
  postmenopausal, 196, 197 (fig.)

Cervix (cont.)
  reserve cell hyperplasia, 195
  cytology, 199
  specimen collection, 148–151
  squamocolumnar junction, 141–142, 195, 197 (fig.)
    cervical smears, 159
  squamous epithelium
    atrophic, 140 (figs), 142
      smear pattern, 155, 156 (fig.)
    intermediate cells, 153
    navicular cells, 153, 154 (fig.)
    parabasal cells, 153–155
    superficial cells, 152–153
  squamous metaplasia, 195, 197–201
  transformation zone, 142, 144, 198
    smear pattern, 155–156
Cestodes (tapeworms), 418
Chagas disease, 411
Charcot-Leyden crystals
  in eosinophilic effusions, 279
  in sputum, 245–246
Chemical agents
  carcinogenesis, 50
  cellular damage, 32
Chemotherapy, iatrogenic changes, 428–431
Chief cells in gastric mucosa, 373
Chlamydia psittaci, 180
Chlamydia trachomatis
  follicular cervicitis and, 172
  genital infection, 208–209
  lymphogranuloma venereum, 209
Chlamydial infections
  electron microscopy and, 71
  immunofluorescent staining, 120
Chlorate bleaching method, 96
4-Chloro-1-naphthol, 111
Cholesterol crystals, 390
Cholesterol effusions, 280
Chondrocalcinosis (calcium pyrophosphate), 390
Chondrocytes, 27
Choroid plexus cells, 298
Choroid plexus papilloma, 298–299
Chromaffin, staining, 80, 95
Chromatids, 23
Chromatin, 21
Chromatin bodies (Barr bodies)
  in sex-linked disorders, 93
  staining, 31, 82, 92–93
Chromosomes, 21
  abnormalities, 29–30
  duplication, division, 23–24
  Philadelphia, 49
  sex, 22, 23, 30
  see also Deoxyribonucleic acid
Chronic bronchitis, 250–251, 260
Chylous effusions, 278–279
Cilia, 10–11
Ciliocytophthoria (CCP), 169, 258
Ciliphora (ciliates), 408, 409, 413–414
Clara cells in respiratory tract, 236
Clinical cytology
  future of, 5
  historical review, 1–4

Clitoris, 138
Clue cells, 176
  see also Gardnerella
Coal worker's pneumoconiosis, 252
Coating fixatives (polyethylene glycol, PEG), 53
Cocci, 174
Colonic carcinoma, 381, 382 (fig.)
Colposcopy, 217
Commensals, 31–32, 157
Complement, 34–35 *passim*
Computerization, 467–470
  advantages, 467–469
  data storage, 461–462
  practical points, 469–470
Coning, of medulla, 294
Connective tissues, description, 27
Contaminants, 102, 103–104, 175, 421–422
Contraceptive device, *see* IUCD
Corpora amylacea of prostate, 337
Corynebacterium vaginalis, *see* *Gardnerella vaginalis*
Coverslipping, 101–102
  automated, 101
Creola bodies, 256
Cresyl fast violet technique, 93
Cristae, *see* Mitochondria
Crowded cell index, 164
Cryotherapy, iatrogenic changes, 425, 426–427
Cryptococcus sp., vaginal, 180
Cryptococcus neoformans, CSF, 298
Crystals
  refractile, 103
  synovial, *see under* Synovial fluids
  urinary, 310
Curschmann's spirals, 245
Cushing's syndrome, 15
Cyclophosphamide
  iatrogenic changes, 428–429
  urothelium affected by, 186
Cyclosporin, iatrogenic changes, 430
Cystic fibrosis, 30
Cystoscopy, 319
Cysts
  bladder, 304
  breast, 345
  liver, 359
Cytoanalyser, 441–442
Cytocentrifugation, 59–62
  CSF preparation, 60, 296
  haemolysis of smears prepared by, 67
  quantitative cell suspension, 62–63
Cytochemistry, 443–444
  summary, 447
Cytoclair, 58
Cytogenetic studies, 442–443
  automation, 450
  serous effusions, 288–289
  see also Chromosomes; Deoxyribonucleic acid
Cytomegalovirus (CMV)
  cervical infection, 184
  opportunistic lung infection, 250

Cytomegalovirus (cont.)
  'owl's eye', 257
  respiratory, 257–258
  urinary infection, 311, 320, 321–322
Cytometry
  automation, 450
  chromosome analysis, karyotyping, 450
  fine needle aspiration, 450
  flow, 446, 467
    cell sorting, 447
    cervical screening using, 447–448
    fluorescence, 446–447
    light scatter, 446
    urine, 324
  gastrointestinal tract, 449–450
  respiratory tract, 450
  serous effusions, 450
  specimen preparation, 448–449
  staining protocols, 449
  static, 441
  urinary tract, 449
Cytopathic effect, 33
Cytoplasm, 7
Cytosieve filtration capsule, 65
Cytospin, 59–62
Cytotoxic therapy, DNA damage, 32, 33

Davson and Danielli, cell membrane model, 8
Decidual reaction in pregnancy, 147 (fig.)
Deoxyribonucleic acid (DNA), 21–22
  cytometric studies, 449
  environmental damage, 32–33
  hydrolysis profiles, 443
  and neoplasia, 442–443
  staining, 21, 93–94
Desmosome, 11
Destaining, 99–100
3,3-Diaminobenzidine (DAB), 110–111
Diarthroses, *see* Synovial joints
Diastase digest technique, 86
Diathermy, iatrogenic changes, 425, 426–427
Diazo reaction, 96
Dieterle spirochaete stain, modified, 88–89
Diethylstilboestrol (DES), 195, 435
Diffuse interstitial lung diseases, 252–253, 260–261
Direct smears
  blood, 57
  centrifugation, 58–59
  criss-cross, 57
  fish-tail, 57
  large lake, 57
  preparation, 57–58
  squash-spread, 57
DNA viruses, oncogenic, 50
Doderlein bacillus (*Lactobacillus* sp.), 157, 159, 175
Donovania granulomatis, 176–177

DOPA oxidase technique, 96–97
Down's syndrome, 30, 49
Dry slide method, 70
Dubosq Brazil sputum fixative, 54
Ductal cells in breast, 345
Duodenum
　anatomy, 379
　cytology
　　adenocarcinoma, 380
　　malignant neoplasms, 380
　　normal, 380
　histology, 379
　specimen collection, 379–380
Dyskaryosis, see Cervical intra-epithelial neoplasia
Dysplasia, 202

Echinococcus granulosa (Taenia echinococcus), 418
Ectocervix, 141
Ectoplasm, 12
Ectopy, 141, 195–197
Electrodiathermy, iatrogenic changes, 425, 426–427
Electron microscopy
　preparation
　　fresh cells, 71
　　Papanicolaou-stained smears, reprocessing, 71–72
　serous effusions
　　adenocarcinoma, 290
　　bronchial carcinoid tumour, 290
　　melanomas, 290
　　mesothelioma, 289–290
　　processing, 289
　　small cell anaplastic carcinoma, 290
　　squamous cell carcinoma, 290
Embolism, disease process, 42
Emphysema, 249
Empyema (purulent exudate), 278
Endocervical canal, 139, 141
　brush specimen, 151
Endocervical cells
　in cervical smear, 155, 169
　inflammatory changes, 169
　reactive columnar, 196, 198 (fig.)
Endocervical glands, 139–141, 144, 197
　acute inflammation, 169
　hyperplasia (adenomatous hyperplasia), 196
　mature squamous metaplasia overlying, 144
Endocervix, acute inflammatory changes, 169
Endocytosis, 9, 11–12
Endometrial aspirates, 157
Endometrial carcinoma, 228–231
　cytological features, 230
　direct sampling/curettage, 232
　oestrogen and, 50, 229
　prognosis, 230
　types of, 229–230

Endometrial cells
　in cervical smears, 156–157, 220
　IUCD-induced changes, 225, 227, 431, 432
Endometrial fragments, fixative, 54
Endometrial hyperplasia, 227–228, 232
Endometrial sampling
　abnormal exfoliation, causes, 224
　direct, 220–222
　　clinical role, 232–233
　　compared to curettage, 232
　　interpretation, 223–224
　　menstrual cycle effect on, 223–224
　　postmenopausal (atrophic), 224
　indirect (exfoliated), 220
　　endometrial plaque, 222, 223 (fig.)
　　histiocytes, 'exodus', 222–223
　　menstrual cycle and, 222
　　normal appearance, 222–223
　　reporting, 231–232
　　samplers, 231–232
Endometritis, acute, 225
Endometrium, 144
　atrophic postmenopausal, 145, 224
　hormonal effects on, 224–225
　menstrual cycle
　　degenerative, 146, 147 (fig.)
　　proliferative phase, 145
　　secretory phase, 145–146
　in pregnancy, 146
Endoplasmic reticulum, 14–16
Endoscopic brushes
　acetate fibre, 155
　nylon fibre, 155
　preparation, 155
Entamoeba coli, 411–413
Entamoeba gingivalis, 179, 411–413
Entamoeba histolytica, 179, 189, 411–413
Enterobius (Oxyuris) vermicularis (pinworm), 190, 419, 421, 422
Environmental agents, cellular effects of, 31–33
Enzyme bridge method, 110
Enzymes
　activity, in malignant cells, 443
　Cytoclair digestion of mucus, 58
　digestion, in immunoenzyme staining, 113
　environmental damage, 32, 33
　lysing agents, 68
Eosin–nigrosin technique, sperm viability, 330
Eosinophilic effusions, 279
Eosinophilic index (EI), 165
Eosinophils, 34
　staining, 85
Ependymal cells, 298
Ependymoma, 299
Epithelia
　cuboidal/columnar, 26
　description, 25–26
　pseudostratified, 26

Epithelia (cont.)
　squamous, 25–26
　stratified squamous, 26
　transitional, 26
Epithelial membrane antigen (EMA), 120, 405
Epithelial regeneration, 38
Epstein–Barr virus, Burkitt's lymphoma and, 50
Erlich's haematoxylin stain, 79
Escherichia coli
　urinary tract infection, 308, 310
　vaginal, 177
Esposti's fixation fluid, 54
Esterase, non-specific
　CSF, 298, 300
　staining methods, 91–92
Ethanol (ethyl alcohol, absolute alcohol), 52–53
Euchromatin, 21
Exocytosis, 9, 11–12
Exodus, 159
Exudates, 36–37
　in cervix, 168
　in serous fluid, 274

Fallopian tube (oviduct), 147
Farmer's lung, 253
Fat stain, 86
Ferritin granules, 20
Feulgen reaction, 93–94, 441, 443
Feyter cells, 236
Fibrinous clots
　fixative, 53, 54
　preparation, 56, 69
　staining, 83
Fibroadenoma of breast, 346
Fibroblast intermediate filament (FIF), see Vimentin
Fibrocyte, 26–27
Fibrosis, 37–38
Filaments, 19–20
Filariasis, 421
Filters, 63–64
Filtration
　capsules, 65–66
　membrane, 63–66
　positive pressure, 65–66
　vacuum, 63–65
Fine needle aspiration (FNA), 55, 56, 59, 380
　bone, 365–366
　breast, 344–345
　　benign lesions, cytological features, 344
　　cystic disease, 345–346
　　fat necrosis, 346–347
　　fibroadenoma, 346
　　fibroadenosis, 346
　　lactating, 347–348
　　male, 349–350
　　malignant lesions, 344, 348–349
　　Paget's disease of the nipple, 349
　clinical role, 340
　cytometric studies, 449

Fine needle aspiration (cont.)
  historical review, 4–5, 339–340
  kidney, 362–365 (figs), 366
  limitations, 340
  liver, 359–362
  lung
    complications, 355–356
    contraindications, 356
    cytology, 357–358
    special notes, 357
    see also Transbronchial;
      Transthoracic percutaneous
  pancreas, 364–365 (figs), 366
  prostate, 336
  reporting, 344
  specimen
    immunocytochemical methods, 115, 118–119
    preparation, 55, 56, 59
    staining, 82
  technique
    deep-seating lesions, 342–343
    general procedure, 340–342
    needle choice, 342–343
    rapid diagnosis, 343–344
  thyroid, 361–363 (figs)
  transbronchial, 239, 355
    complications, 355–356
    contraindications, 356
    cytology, 357–358
    diagnostic accuracy, 268
    non-cellular elements, 245
    normal cytology, 245
    special notes, 357
    specimen preparation, 243
    touch preparations, 239
  transthoracic percutaneous, 239–240, 355
    complications, 355–356
    contraindications, 356
    cytology, 357–358
    diagnostic accuracy, 268
    non-cellular elements, 245
    normal cytology, 245
    special notes, 357
    specimen preparation, 243
  urinary tract, 307
Fistula, cytology of tumour cells in ventricular peritoneal shunt, 437
Fixation artefact, 52
Fixatives, 52–54
  alcoholic, 52–53, 54
  coating (polyethylene glycol), 53
  immunocytochemical, 88
  special purpose, 53–54
Flagella, 11
Flagellates (Mastigophora), 409–411
Flatworms (Platyhelminths), 415–418
Flotation technique, 67–68
Flow cytometry, see Cytometry, flow
Fluid mosaic membrane model, 8
Flukes (Trematodes), 415–418
Foam cells in nipple secretions, 345
Folded cell index, 164
Follicle, 148

Follicular cervicitis, see Cervicitis, lymphocytic
Food particles, in sputum, 246
Formal calcium fixation, 54
Formal saline fixation, 53–54
Formal vapour fixation, 54
Formaldehyde, alcoholic, 54
Fornices, 139
Frozen slide method, 70
Fungal infections, 82
  female genital tract, 179–180
  respiratory, 257–258
  staining reactions, 88
  urinary infection, 313

Gallocyanin–chrome alum technique, 94
Gap junction, 11
Gardnerella vaginalis, 157, 175–176
  'clue cells', 176
Gastric atrophy, 374
Gastric carcinoma
  adenocarcinoma, 376, 377
  classifications, 376
  diffuse type, 376
  early (superficial), 378–379
  intestinal type, 376
  undifferentiated (anaplastic), 376, 377
  see also Stomach
Gastric metaplasia, 44, 374–375
Gastritis
  acute, 373–374
  chronic
    atrophic, 374
    gastric atrophy, 374
    superficial, 374
Gastrointestinal tract
  cytology, clinical role, 381–383
  cytometric studies, 449
  specimen
    collection, 368
    detection, 369
    interpretation, 369–370
    preparation, 56, 57, 368
  see also Duodenum; Oesophagus; Stomach
Gelatin agglutination tests (GAT), 333
Gelman filters, 63
  mounting and coverslipping, 102
  urinary specimen, 306
Gene mutations, and disease, 30
Genetic diseases, 29–31
Genetic factors, tumour development, 49
Genital herpes, 183
Genital tract
  female
    anatomy, 137–138
    fungal infections, 179–180
    parasitic infections, 188–190
    tuberculosis, 177
    viral infections, 93, 182–188

Genital tract (cont.)
  see also specific organs; Cervical; Endometrial
  male
    description, 327–328
    see also Prostate; Spermatozoa
Geotrichum candidum, 180
GERL membrane, 16, 18
Ghost cells, sputum, 262
Giardia lamblia (intestinalis), 409, 410
Gill's haematoxylin stain, 80
Glacial acetic acid (2 M urea) preparation, 66
Glial fibrillary acidic protein (GFAP), 300
Glucose oxidase, as enzyme marker, 112
Glucose-6-phosphate dehydrogenase (G6PD), in malignancy, 443–444
Glutaraldehyde-fixed smears, electron microscopy, 71–72
Glutathione (GSH) levels, in malignancy, 443
Glycogen
  cell granules, 20
  fixative, 53
  removal, diastase digest technique, 86
  staining, 85–86, 95
Goblet cells, 375
Gold, antibody labelling, 113
Golgi apparatus (body), 17–18
Gonorrhoea, 176
Gordon and Sweet's stain, 98
Gout (uric acid), 389
Gram–Weigert technique, 88
Gram's stain, 87–88
Granulation tissue, formation of, 38
Granulomatous disease, chronic, 40–41
Gravlee jet wash, of endometrium, 221

H & E (haematoxylin and eosin) stain, 83–84
Haematoxylin stains, 79–80
Haemolysis
  acid, 68
  Carnoy's method of, 67
  enzyme lysing agents, 68
  glacial acetic acid (2 M urea) method, 66
Haemophilia, 30
Haemophilus vaginalis, see *Gardnerella vaginalis*
Haemosiderin, staining, 95
Hageman factor (clotting factor), 34
Hamman-Rich syndrome, 253, 261
Harris's haematoxylin stain, 79–80
Hay fever, see Allergic rhinitis
Heart failure cells in pulmonary oedema, 259
Helminthic infections
  classification, 408, 409

478 Index

Helminthic infections (cont.)
  specimen collection, preparation,
      408–409
  see also individual species
Hepatitis, fixatives, 53
Hepatocytes, 359
Hepatoma, 361
Herpes simplex virus (HSV)
  and cervical cancer, 183
  as genital infection, 120, 182–183
  as opportunistic lung infection, 250
  as respiratory infection, 257
Herpes zoster, genital infection, 184
Herpetic oesophagitis, cytology, 371
Heterochromatin, 21–22
Hexamine silver technique, modified,
    89–90
Hexazonium salt technique, 91–92
Histamine, 35
Histiocytes, 35
Histochemical staining
  CSF, 300, 301
  serous effusions, 283–284
Histocompatibility complex (HLA),
    diseases associated, 30–31, 40
Histopaque flotation technique, 68
HIV, see Human immune deficiency
    virus
Hodgkin's lymphoma, 399–400
  gastric, 377–378
  lymph nodes, fine needle
      aspiration, 352, 354–355
  serous effusions, 277–278
  see also Lymphomas;
      Non-Hodgkin's lymphoma
Hookworm, see Ancylostoma
Hormonal effects
  carcinogenesis, 50
  diethylstilboesterol, 435
  oral contraceptives, 435
  prostatic cancer oestrogen therapy,
      435
  replacement therapy, 435
Hormone cytology, 164–165
Human immune deficiency virus
    (HIV), 250
  safety procedures, 463–464
Human papillomavirus (HPV), 31,
    50, 184
  cervical neoplasia and, 36, 46, 50,
      184, 188, 209, 210
  genital infection, 185
    endophytic (inverted) warts, 187
    exophytic warts (condyloma acu-
        minata), 185, 186
    in female, 185–188
    in male, 185
    subclinical (SPI), 185
  laboratory diagnosis, 187–188
  staining, 91, 120
Human polyomavirus (HPV), urinary
    infection, 313–314, 321–322
Human T lymphotropic virus type III
    (HTLV III), see Human
    immune deficiency virus
Husain's preservation fluid, 54

Hyaluronic acid in mesothelioma, 283
Hydatid cyts, 418
Hydrocoele
  cytology, 335
  description, 335
  specimen collection, preparation,
      335
  spermatic, 336
Hydroxyapatite crystals, 391–392
Hyperplasia, 42, 43
Hypersensitivity reactions, 36, 39, 40
Hypertrophy, 43

Iatrogenic changes
  bronchial epithelium, 260
  cervical epithelium, 425–428
  chemotherapy-induced, 428–431
  cryotherapy, diathermy-induced,
      425, 426–427
  hormone-induced, 435–436
    contraceptives, 435
    in DES-exposed women, 435
    oestrogen therapy for prostatic
        cancer, 435–436
  instrument-induced, 437
  IUCD-associated, 431–435
  laser-induced, 425–426
  radiation-induced, 427–428
Idiopathic diffuse pulmonary fibrosis,
    253, 261
Ileal loop urine
  intestinal epithelial cells in, 318
  specimen collection, 305
Immune complex-induced hyper-
    sensitivity, 39
Immune deficiency diseases, 40–41
Immune response, 35–37
Immunity, cell-mediated, 35, 36
Immunocompromised patients
  cancer risk in, 50
  opportunistic lung infections, 250
  reactivation of infection in, 31
  viral urinary tract infections, 313
Immunocytochemistry
  antibody markers
    enzyme, 109–112
    fluorescent, 109
    other, 112–113
  antisera
    monoclonal, 107–108
    polyclonal, 107–108
  applications, 118–121
  automation, 444
  blood/bone marrow smears, 93
  cervical smears, 114, 120–121
  CSF, 300–301
  fine needle aspirates, 115
  fixation, 115
  historical development, 106
  impression smears, 114, 118
  principles, 106–107
  schedules, 117–118
  serous effusions, 114–115, 119–120,
      284–288
    antibodies, choice of, 286–288
    controls, 285

Immunocytochemistry (cont.)
  serous effusions (cont.)
    fixation, 284–285
    interpretation, 285
    processing, 285
    specimen collection, 284
    staining method, choice of, 285
  specificity, 116–117
  specimen preparation, 114–115
  sputum, 114, 119
  touch preparations, 118
Immunoenzyme staining, 109–113
Immunofluorescence, 106, 108–109,
    115
Immunoglobulin-gold probes, 113
Immunoglobulins, 35–36
Immunological surveillance, 50
Immunoperoxidase systems, 107,
    109–111
Immunosuppressive drugs, iatrogenic
    changes, 430
Impression smears, 114
Imprint preparations, 70, 83
Infectious agents, cell damage due to,
    31–32
Infertility, male, 335
Inflammatory disease, 36–39
  acute, 36–37
  chronic, 4, 38–39, 169
    'active', 169
  granulomatous, 171, 178
Inflammatory response, 33–37, 167
  cellular components, 34
  chemical mediators, 34–37
  fibrosis, 37–38, 167
  regeneration and repair, 37, 167,
      190–191
Intercellular bridges, see Desmosome
Interferons, iatrogenic changes, 430
Intermediate cells, see Cervix,
    squamous epithelium
Interphase, 22
Interstitial pneumonia, 253, 261
Intestinal metaplasia, 44, 374–375
Intestinal parasitic worms, see
    Parasitic infections
Intrauterine contraceptive device
    (IUCD)
  Actinomyces israelii and, 178–179,
      226–227, 433
  cytopathological changes
    aetiology, 431
    cervical smears, 431–432
    endometrial cells, 226–227, 431,
        432
    uterine fluid, 433
  Entamoeba gingivalis and, 179
Ionizing radiation
  as carcinogenic agent, 33, 50
  effect on DNA, 33
Irradiation fibrosis, 32
Isaacs endometrial sampler, 221
Ischaemic infarction, disease process,
    41
Isopropanol (propan-2-ol, isopropyl
    alcohol), 53

Karyopyknosis, 169 (fig.)
Karyopyknotic index, (KPI), 164–165
Karyorrhexis, 168
  see also Cervical smear
Karyotyping, 442, 450
Keratin, staining, 91
Kidney
  adenocarcinoma (clear cell), 317, 324
  anatomy, 302–303
  fine needle aspiration, 362–363 (figs), 366
    bilocular cyst, 363
    clear cell carcinoma, 363
  histology, 303–304
  see also Renal
Kinetochore, see Centromere
Kinocilia, 11
Klinefelter's syndrome, 30, 350
Kohler illumination, 135–136
Koilocytes, 185, 186–187 (figs)
Kreyberg's stain, modified, 91

Laboratory
  budgeting, 460–461
  data storage, 461–462
  planning, 459–460
  quality control, 464–465
    computerization, 469
    external assessment, 465–466
  safety, 463–464
  training, 460, 462–463
  waste disposal, 464
  workforce, 458–459
  workload, 455–458
    Canadian measurement system, 457–458
Lactation, fine needle aspiration during, 347–348, 349 (fig.)
Lactobacillus sp. (Doderlein bacillus), 157, 159, 175
Langhan's giant cell, 40, 171, 178 (fig.)
Large intestine
  anatomy, histology, 380
  carcinoma, 381, 382 (fig.)
  chronic ulcerative colitis, 381
  mucosal polyps, 381
  normal cytology, 381
  specimen collection, 380–381
Larynx
  singer's nodes, 248
  specimen collection, 237, 238
Laser therapy
  cellular damage, localization of, 32
  iatrogenic changes, 425–426
LE (lupus erythematosus) cells, 280, 393
Lebert, Henri, 1
Legionella pneumophila, 249
Legionnaires' bacilli, modified Dieterle spirochaete stain, 88–89
Legionnaire's disease, 249

Leishmania braziliensis, L. donovani, L. tropica, 410
Leishmaniasis, 410–411
Lendrum's carbol chromotrope 2R technique, 85
Leptothrix, 175
Leucocyte common antigen (LCA), 287
Leukoplakia, 141
Lillie's performic acid bleaching method, 96
Lipid
  cell droplets, 20
  fixatives, 53, 54
  pneumonia, 249
  staining, 20, 86–87
Lipofuscin, staining, 95
Listeria meningitis, CSF, 297
Listeria monocytogenes, vaginal, 177
Liver
  carcinoma
    metastatic, 360–365
    primary, 36
  fine needle aspiration, 359–362
    complications, 359
    cytology, 359–362
    ultrasound guidance, 359, 360 (figs)
  normal cells, 359–360, 361
  ultrasound scan
    hepatoma, 360 (fig.)
    liver abscess, 360 (fig.)
Loa loa, 421
Lung
  abscess, pathology, 249
  carcinoma
    adeno-, 254, 263, 264
    carcinoid tumours, 254–255, 266–267
    classification, 253
    giant cell carcinoma, 267
    in situ, 263
    large cell anaplastic, 254, 264–265
    metastatic, 267
    mucoepidermoid, 267
    others, 255
    small cell anaplastic, ('oat cell'), 254, 265–266
    squamous (epidermoid), 253–254, 261–263
  fine needle aspiration
    complications, 355–356
    contraindications, 356
    cytology, 357–358
    special notes, 357
    transbronchial, see under Fine needle aspiration
    transthoracic, see under Fine needle aspiration
Lupus erythematosus (LE) cells, 280, 393
Lymph nodes
  description, 351–352
  fine needle aspiration
    indications, 352

Lymph nodes (cont.)
    interpretation, 352–355
    reactive hyperplasia, 354–355
Lymphadenopathy, fine needle aspiration, 352
Lymphadenopathy associated virus (LAV), see Human immune deficiency virus
Lymphocytic effusions, 279–280
Lymphogranuloma venereum, 181
Lymphomas
  cerebral, 299
  Hodgkin's, see Hodkin's lymphoma
  non-Hodgkin's, see Non-Hodgkin's lymphoma
  serous effusions, 277–278
  staining, 82
Lysosomes, 18–19, 33

Macrophages, 34, 167
  'epithelioid', 40, 171, 177 (fig.)
  staining, 86, 89
Macropinocytosis (phagocytosis), 12
Malakoplakia, 318
Malarial parasites (*Plasmodium* sp.), 414–415
'Malignant diathesis', 214
Masking technique, paraffin wax, 100
Masson–Fontana silver reduction stain, 97
Mast cells, 35
Mastigophora (flagellates), 408, 409–411
Maturation index (MI), 165
Mayer's haematoxylin stain, 80
Medulloblastoma, 299
  pseudo-rosettes, 299
Meiosis, 23–24
Melanin, staining, 89, 95
Melanomas
  metastases
    bone, 399
    CNS, 300
  serous effusions, 278, 290
Melphalan, iatrogenic changes, 430
Membrane filtration, 63–66
  mounting and coverslipping, 102
  urinary specimens, 306
Meningioma, 300
Menstrual cycle, 144–145
  endometrial cells, 222–223
Menstruation, cerival smear at, 158–159
Mesodermal mixed tumour, 230
Mesothelioma, 278
  see also under Serous effusions
Metabolism, inborn error, 30
Metachromatic stain, 90
Metaphase, 22, 23
Metaplasia, 43–44, 139
  bronchial, 254, 260
  cervical, 43, 141–144, 155
  intestinal, 44, 374–375
  mucous, 44
  squamous, 43

Methanol (methyl alcohol), 53
Methotrexate
  iatrogenic changes, 430
  urothelium affected by, 320, 322 (fig.)
Methyl green–pyronin technique, 94
Methylated spirit (alcohol, meths), 53
Metronidazole, for *G. vaginalis*, 176
Mi-Mark endometrial sampler, 221
Michaelis-Gutmann (M-G) bodies, 318
Microdensitometry (microspectrophotometry), 441
Microfilaria, 421
Microglioma, 299
Microhaematocrit technique, 67
Microorganisms
  commensal, 31–32, 157, 173
  pathogenicity, 173–174
  staining, 87–90
  see also individual organisms
Micropinocytosis (pinocytosis), 12
Micropolyspora faeni, 253
Microscopy
  advanced, 132–134
  bright field illumination, 125–126
  care, cleaning, 132
  components, 125, 130–131
  condensers, 131
  dark field illumination, 126
  differential interference contrast, 126–127
  eyepieces, 130–131
  fluorescence, 127–128
  Kohler illumination, 135–136
  objectives
    achromats, 130
    fluorite, 130
    planapochromats, 130
  phase-contrast (PCM), 126
  polarized light, 126
  procedure, 128–129
  tumours, 48–49
Microspectrophotometry (microdensitometry), 441
Microtubules, 19–20
Microvilli, 9–10
Millipore filters, 63
  CSF preparation, 296
  mounting and coverslipping, 102
Mitochondria, 13–14
Mitosis, 23, 24
Miracidia, 416
Mixed antiglobulin reaction test (MAR), 333
Mixed mullvian tumours, 230
Molluscum contagiosum, 188
Monoclonal antibodies, *see under* Antibody
Monocytes, *see* Macrophages
Monosodium urate crystals, 388–389
'Mosaicism', 30
Mounting, 100–101
  media, 101
  membrane filters, 102
Mouth, benign lesions, 255

Mucin, staining, 90–91
Mucin clot test, 385–386
Mucinous samples, smear preparation, 57–58
Mucus
  in bronchoscopic specimens, 245
  liquefaction methods, 58
  in sputum, 245
Mucus-secreting cells, 244
Muller, Johannes, 1–2
Multinucleation in inflammatory smears, 171, 172
Multiple myelomatosis, serous effusions, 278
Multiple sclerosis, CSF lymphocyte count, 297
Muscle cells, 27–28
Mycobacterium tuberculosis (acid and alcohol fast bacillus, AAFB), 177, 250
Mycotic infections, *see* Fungal infections
Myelomatosis, 278
Myoepithelial cells in FNA of breast, 346

Nasal carcinomas, 248
Nasal polyps, 248
Nasal secretions, collection, 237
Nasopharynx, squamous carcinoma, 248
Navicular cells, 153, 154
Neisseria gonorrhoeae (gonorrhoea), 176
Nematodes (roundworms), 418–421
Neonate
  bronchopulmonary dysplasia, 261
  cervix, 195, 196 (figs)
Neoplasia 44, *see also* Tumours
  intraepithelial, 45–46
Neoplastic cells
  cytological recognition, 442
  DNA content, 442–443
  glutathione levels, 443
  malignancy-associated changes, 444–446
  morphometric parameters, 442
  nucleocytoplasmic (N/C) ratio, 442
  pentose-shunt dehydrogenase activity, 443–444
  proteolytic enzymes, 443
  texture features, 444–446
Neotetrazolium chloride (NT), 444
Nervous tissue, description, 28
Neurilemma, 7
Neurofibroma, 300
Neutrophils (polymorphonuclear leucocytes), 34
Nexuses, 11
Nile blue technique, 86–87
Non-Hodgkin's lymphoma, 399–400
  gastric, 377–378
  lymph nodes, fine needle aspiration, 352, 354–355
  serous effusions, 277

Non-Hodgkin's lymphoma (*cont.*)
  see also Hodgkin's lymphoma; Lymphomas
Nucleic acids, staining, 93–94, *see also* Deoxyribonucleic acid; Ribonucleic acid
Nucleocytoplasmic (N/C) ratio, 442
Nucleolus, 21
Nucleopore filters, 63, 102
Nucleus, 22–25

Oat cell carcinoma, *see under* Lung
Oesophageal rupture, empyema due to, 278
Oesophagitis
  cytology, 370–371
  herpetic, 371
  monolial, 371
  non-specific, 371
Oesophagus
  anatomy, 370
  Barrett's syndrome, 370
  cytology
    adenocarcinomas, 372
    early (superficial) carcinoma, 378–379
    normal, 370
    oesophagitis, 370–371
    squamous cell carcinomas, 371–372
  histology, 370
  see also Gastrointestinal tract
Oestrogen effect
  cervical smear, 158–159, 435
  ectocervix, upper vagina, 227
  endometrium, 224–225
  cancer, 229
  hyperplasia, 227–228
  prostatic cancer therapy, 435–436
Oil red O technique, 87
OK-432, iatrogenic changes, 430–431
Oligodendroglioma, 299
Oligospermia, aetiology, 335
Oncogenes, 49
Oral contraceptives, cervical cells and, 435
Organs, definition, 25
Ovary, 148
  carcinoma and cervical smear, 215–216
Oviduct (fallopian tube), 147

Paget, Sir James, 4
Paget cells, 349, 351 (fig.)
Paget's disease of the nipple, FNA, 349
Pancreas
  cytology
    malignant neoplasms, 380
    normal, 380
  fine needle aspiration, 364–365 (figs), 366, 380
    adenocarcinoma, 380
    carcinoma, 365 (fig.)

Pancreas (*cont.*)
    pseudocyst, 364 (fig.)
    specimen collection
        duodenal endoscope, 379
        lavage, 379–380
PAP (peroxidase/anti-peroxidase) system, 107, 110, 117–118
Papanicolaou, G. N., 2–4, 206
Papanicolaou stain, 80–82
    alcohol-fixed smears and, 55, 73
    'carbowax' artefact, 81
    electron microscopy, reprocessing, 55, 71–72
    technical aspects, 73
    urinary speciemn, 306
Papillary lesions of bladder, 315
Papovaviruses, 184–188; *see also* Human papillomavirus; Human polyomavirus
Parabasal cells, *see under* Cervix
Paraffin wax masking technique, 100
Paragonimus sp., 417–418
Parasitic infections
    classification, 408, 409
    comparative size of organisms, 413
    contaminants, 421–422
    CSF eosinophils, 298
    female genital tract, 188–190
    specimen
        collection, preparation, 408–409
        staining reactions, 88
    urinary, 314
    see also individual species
Parietal cells of gastric mucosa, 373
PAS stain, 85–86
PCP, *see Pneumocystis carinii* pneumonia
Pemphigus in Tzanck test, 184
Pentose-shunt dehydrogenase levels, in malignancy, 443–444
Peptic ulceration, cytology, 374
Pericardial effusions, *see* Serous effusions
Peritoneal dialysis, cellular changes following, 435–436
Peritoneal fluids, *see* Serous effusions
Perls' prussian blue reaction, 97
Permanganate bleaching method, 96
Peroxidase
    activity, 110–111
    endogenous, blocking techniques, 111
Peroxidase/anti-peroxidase (PAP) system, 107, 110, 117–118
Phagocytosis (macropinocytosis), 12
Pharynx
    benign lesions, 255
    carcinoma, 248
    specimen collection, 237
Phase-contrast microscopy, *see* Microscopy, phase-contrast
Philadelphia chromosome, 49
Physical agents
    carcinogenesis, 50
    cellular damage, 32
Pigments, staining, 95–98

Pinocytosis (micropinocytosis), 12
Pinworm (*Enterobius vermicularis*), 419, 421, 422
Plasma cells, 21, 35
Plasma membrane, 7–12
    actin filaments, 9
    carbohydrate, 8
    cilia, 10–11
    endocytosis, exocytosis, 11–12
    flagella, 11
    lipid/protein models, 7–8
    microvilli, 9–10
    transport
        active, 8
        facilitated diffusion, 8
        passive, 8
Plasmacytoma, serous effusions, 278
Plasmodium falciparum, *P. vivax*, *P. malariae*, *P. ovale*, 414–415
Platyhelminths (flatworms), 415–418
Pleural fluids, *see* Serous effusions
Pneumoconiosis, 252–253
Pneumocystis carinii, 421
    cytological diagnosis, 70, 89, 258–260
    pneumonia, 238, 239, 250
Pneumonia
    aspiration, 249
    broncho-, 249
    cytology, 256
    interstitial, 253, 261
    Legionnaire's disease, 249
    lipid, 249
    lobar, 249
    Pneumocystis carinii, 238, 239, 250
Pneumothorax
    as cause of eosinophilia in effusion, 279
    as complication of FNA of lung, 240
Podocytes, 10
Polyclonal antibodies, *see under* Antibody
Polyethylene glycol (coating fixative, PEG), 53
Polyomavirus infection of urinary tract, 321–322
Positive pressure filtration, 65–66
Postcoital tests, 333
Posterior fornix
    aspirate, method, 150–151
    smears, endometrial cells, 220, 222–223
Poxviruses, 188
Pregnancy changes
    in endometrium, 146
    in cervical smear, 159
Prekeratin, staining, 91
Preparatory techniques, *see* Specimen preparation
Progesterone, effect on endometrium, 224–225
Prophase, 22, 23
Prostate
    benign conditions, 337
    carcinoma, 337

Prostate (*cont.*)
    oestrogen therapy and, 435
    urinary tract metastatic, 317
    description, 327, 336
    hyperplasia, 337
    specimen collection
        fine needle aspiration, 336
        massage, 336–337
Prostatitis, 337
Protein A, as antibody marker, 112
Protein S100 stain, 300
Proteolytic enzymes, in immuno-enzyme staining, 113
Protozoan infections, 409–415
    classification, 408, 409
    specimen collection, preparation, 408–409
    see also individual species
Psammoma bodies (calcospherites), 173
'Pseudogout' (calcium pyrophosphate crystals), 389–390
Pseudomelanin, staining, 95
Pulmonary fibrosis, idiopathic diffuse, 253, 261
Pulmonary hypertension, 252, 260
Pulmonary oedema, 251–252, 260
Pulmonary thromboembolism, 251
Pus, 37
Pyrophosphate arthropathy, 390

Radiation therapy
    bronchial epithelium, 260
    effect on DNA, 33
    fibrosis induced by, 32
    iatrogenic changes, 427–428
Radiotherapy, cellular damage, 32
Ragocytes in synovial fluid of rheumatoid arthritis, 393
Rectal carcinoma, cytology, 381, 382 (fig.)
Reed-Sternberg cell, 378
Reiter's syndrome, 392–393
Renal
    allograft, rejection cytology, 320
    aspirate, 363 (fig.)
    carcinoma, 324
    see also Kidney
Reporting cervical smears, 159–163
Reserve cells, 139, 198, 199 (fig.), 210
    hyperplasia, 142, 143 (figs), 198, 199
Respiratory tract
    cytology
        bronchial biopsy compared to, 49
        normal, 243–246
        tumour typing by, 49–50
    cytometric studies, 449
    description, 235–237
    functions, 236, 237
    lower, pathology, 248–255
    physiology, 237
    specimen
        collection, 237–240

Respiratory tract (cont.)
  preparation, 240–243
  unsuitability, 246
  upper
    infections, 248
    inflammatory conditions, 248
    normal cytology, 243–244
    pathology, 248
    specimen collection, 237–238
    tumours, 248
  see also Lung; Pulmonary
Reticulin, staining, 98–99
Rheumatoid arthritis
  HLA and, 30, 31
  serous effusions, 160–161
  synovial fluid, 393
  viral agent, 31
Rhizopoda (amoebae), 408, 409, 411–413
Ribonucleic acid (RNA), 22–23
  staining, 21, 93–94
Ribosomes, 14
RNA retroviruses, oncogenic, 49, 50
Romanowsky (ROM) stain, 55, 74, 82–83
  smear preparation, 74
Roundworms (nematodes), 418–419

Saccharomyces cerevisiae, 180
Saccomanno's fixative, 54
Saccomanno's technique, 58
Saline fixation, 53–54
Saponin, 68
Sarcoidosis, 40, 253
Scar tissue, formation of, 38
Schistosoma haematobium, 314, 315, 320, 415–417
Schistosoma mansoni, 415–417
  contaminant misidentified as, 422
Schistosoma matthei, 415–417
Schistosomiasis (bilharzia), 415–417
  genital, 189–190
Schmorl's reaction, 97–98
Schwann cell, 28
Secretory vesicles and vacuoles, 17–18
Semen analysis
  clinical role, 335
  collection and delivery, 328–329
  eosin-nigrosin technique, 330
  fertility and, 328
  immunology, 333–334
  inspection, 329
  liquefaction, 329
  spermatozoa
    abnormal/normal, 332 (figs), 333, 334 (figs)
    density, 330–331
    microscopy, 333
    morphology, 331–332
    motility, 329–330
    smear preparation, 332
    staining methods, 332–333
  viability, 330
  viscosity, 329
  volume, 329
Seminal vesicles, description, 327, 336

Sensitivity of cervical smears, 162
Serous effusions
  acute non-bacterial inflammation, 279
  air-dried smears, 55, 272, 273
  benign cells
    leucocytes, 274
    macrophages, 274
    mast cells, 275
    mesothelial cells, 275
    plasma cells, 274
  bloodstained, clotted, 272–273
  cell block technique, 273
  cholesterol, 280
  collection, 271–272
  cytogenetic studies, 288–289
  cytometric studies, 449
  electron microscopy, 289–290
  eosinophilic effusions, 279
  exudates, 274
  histochemical staining, 283–284
  immunocytochemical staining, 114–115, 119–120, 284–288
    antibodies, 118–120, 286–288
    cell type/staining, 285
    controls, 285
    fixation, 115, 284–285
    interpretation, 285
    processing, 114–115, 285
    specimen collection, 114, 284
    staining method, 285
  lymphocytic, 278–279
  malignant cells, 275, 280–281
    acinar adenocarcinoma, 275–276
    free cell adenocarcinoma, 276–277
    lymphoma, 277–278
    melanomas, 278
    mesothelioma, 278
    papillary adenocarcinoma, 275–276
    sarcomas, 278
    small cell undifferentiated ('oat cell') carcinoma, 277
    squamous cell carcinoma, 277
  preparation, 56, 272
    centrifugation, 58–63
  purulent exudate (empyema), 278
  reporting difficulties, 16, 281–282
  rheumatoid, 279–280
  SLE, 280
  transudates, 274
  tuberculous, 278, 280
  viral infection, 280
  wet-fixed smears, 272, 273
Serous membranes, description, 271
Sex chromosomes, 22, 23, 30
Sex-linked disorders, 30, 93
Shunts, extracranial spread of brain tumour via, 437
Sickle cell disease, 30
Silicosis, 252
Singer and Nicholson, cell membrane model, 8
Skin cancer
  human papillomavirus and, 50
  ultraviolet light and, 50

Sleeping sickness, 411
Smears, see Specimen preparation
Soderstrom, Nils, 5
Southgate's mucicarmine technique, 90–91
Specificity of cervical smears, 162
Specimen preparation, 54–57
  air-dried smears, 55, 73
  bloodstained smears, 66–68
  cell block method, 69
  cell loss
    adhesiveness, 75
    atmospheric pressure, 75
    electrostatic charge, 77
    fixative, 76–77
    minimizing, 77
    sedimentation rate, 75
    shape, 76
    surface tension, 75–76
    viscosity, 75
  centrifugation, 58 59
  cross-contamination, 103
  cytocentrifugation, 59–63
  direct smears, 57–58
  electron microscopy, 70–72
  endoscopic brushes, 70
  enzyme lysing agents, 68
  fibrinous clots, 69
  'floaters', 75, 103
  imprint preparations, 70
  membrane filtration, 63–66
    positive pressure, 65–66
    vacuum, 63–65
  mucinous samples, 57–58
  tissue fragments, 70
  unfixed (fresh) samples, 55–57, 72–74
    technical aspects, 72–74
  wet-fixed smears, 55, 73
Spermatocoele
  cytology, 336
  description, 335–336
  specimen collection, preparation, 335
Spermatozoa
  cervical mucus penetration test, 333
  in cervical smear, 157, 159 (fig.)
  immunology, 333–334
  postcoital tests, 333
  production, 327
  semen analysis, see Semen analysis
  staining, 99
Sporozoa, 408, 409, 414–415
Sputum
  asbestos (ferruginous) bodies, 245
  bronchial epithelial cells, 244
  carbon-laden histiocytes, 243 (fig.), 244
  Charcot-Leyden crystals, 245–246
  chronic bronchitis, 260
  ciliocytophthoria, 258
  collection, 238, 240
  Curschmann's spirals, 245
  cytology
    evaluation, 267–268
    normal, 243 (fig.), 244
  fixatives, 54

Sputum (cont.)
  food particles, 246
  herpes-infected cells in, 257
  immunological staining, 114, 119
  mucus in, 244, 245
  non-cellular elements, 245
  pollen grains, 246
  preparation
    cell blocks, 57–58, 241
    direct smear, 56, 57, 240
    liquefaction, 240–241
Squamocolumnar junction, 141–142, 159, 195, 197 (fig.)
Squamous carcinoma, *see individual organs*; Carcinoma
Squamous cells, *see under* Cervix
Staff
  management, 458–459
  training, 460, 462–463
Staining, 79–105
  artefacts, 31, 102–103
  contaminants, 102, 103–104
  mounting and coverslipping, 101–102
  multiple, 100
  objectives, 79
  paraffin wax masking technique, 100
  rapid, 84
  secondary, 99–100
  special, 84–99
  see also individual substances; Stains
Stains
  acetic–orcein technique, 93
  alcian blue/PAS technique, 90
  Best's carmine technique, 86
  bleaching techniques, 95–97
  Bryan's sperm stain, modified, 99
  buccal, 93
  chlorate bleaching method, 96
  cresyl fast violet technique, 93
  diastase digest technique, 86
  diazo reaction, 96
  Dieterle spirochaete, modified, 88–89
  DOPA oxidase technique, 96–97
  Feulgen reaction, 93–94
  gallocyanin–chrome alum technique, 94
  Gordon and Sweet's technique, 98–99
  Gram–Weigert technique, 88
  Gram's 87–88
  haematoxylin, 79–80
  haematoxylin and eosin (H & E) stain, 83–84
  hexamine silver technique, modified, 89–90
  hexazonium salt technique, 91–92
  Kreyberg's, modified, 91
  Lendrum's carbol chromotrope 2R technique, 85
  Lillie's performic acid bleaching method, 96
  Masson–Fontana silver reduction, 97
  metachromatic, 90

Stains (cont.)
  methyl green–pyronin technique, 94
  Nile blue sulphate technique, 86–87
  oil red O technique, 87
  Papanicolaou, 80–82
  PAS technique, 85–86
  Perls' prussian blue reaction, 97
  permanganate bleaching method, 96
  'reference', 83
  Romanowsky, 82–83
  Schmorl's reaction, 97–98
  Southgate's mucicarmine technique, 90–91
  Sudan black B technique, 87, 98
  Ziehl–Neelsen (ZN) technique, 90
Staphylococcus sp., 177
Starch granules, synovial fluids, 390–391
Static cytometry, *see* Cytometry, static
Sternberg cells, *see* Hodgkin's lymphoma
Stomach
  anatomy, 372
  benign neoplasms, 376
  brush border, 375
  carcinoma, 376–377
  gastritis, 373–374
  goblet (mucus-secreting) cells, 375
  histology, 373
  intestinal metaplasia, 374–375
  large (histiocyte, reticulum) cell, 378
  malignant lymphoma, 377–378
  normal, 373
  peptic ulceration, 374
  Reed-Sternberg cell, 378
  signet-ring cells, 376
  small (lymphocytic) cell, 378
  specimen collection, 373
  see also Gastric; Gastrointestinal
Storage diseases, 30
Streptococcus sp., 177
Streptolysin, 68
Strongyloides stercoralis, 420–421
  in cervical smear, 190
  respiratory infection, 260
Subarachnoid haemorrhage, CSF, 294, 297–298
Sudan black B technique, 87, 98
Superficial cells, *see under* Cervix
Swin-lok filtration capsule, 65
Swinnex filtration capsule, 65
Synovial fluids
  collection, 384–385
  crystals
    calcium pyrophosphate, 389–390
    cholesterol, 390
    contaminants, 390–391
    diagnostic pitfalls, 391
    hydroxyapatite, 391–392
    monosodium urate, 388–389
    optical properties, 387 (fig.)
    polarizing microscopy, 387–388
    preparation for, 386–387

Synovial fluids (cont.)
  cytology
    malignant tumours, 394
    preparation for, 386–387
    Reiter's syndrome, 392–393
    rheumatoid disease, 393
    SLE, 393
  gross appearance, 385
  mucin clot test, 385–386
  routine examination, 385–386
  summary of findings on, 385 (fig.)
  total cell count, 386
  viscosity, 385
Synovial joints (diarthroses), 385
Synovial membrane, 384
Syphilis, 176, 297
Systemic lupus erythematosus (SLE)
  serous effusions, 280
  synovial fluid, 393–394

T cells (T lymphocytes), 35
  monoclonal antibodies, 300
Taenia echinococcus (*Echinococcus granulosa*), 418
Taenia solium, 418
Tapeworms (cestodes), 418
Target cells in serous effusions, 276
Telophase, 22, 23
Terminology
  histology, 202
  cytology, 206
Testis, 327–328
Tetramethyl benzidine, 111
Thiotepa, iatrogenic changes, 430
Thrombosis, 41–42
Thyroid
  fine needle aspiration, 361–363 (figs)
    anaplastic carcinoma, 362 (fig.)
    benign follicular adenoma, 362 (fig.), 366
  ultrasound scan, cyst, 361 (fig.)
Tissues
  connective
    blood, 27
    bone, 27
    cartilage, 27
    fibrous, 26–27
  definition, 25
  epithelia, 25–26
  fragments
    fixative, 53
    preparation, 56, 70
  muscle, 27–28
  origin, 25
  see also individual tissues
Toluidine staining, 306
Torulopsis glabrata, vaginal infection, 180
Touch preparations, 118, 239
Toxic shock syndrome, 177
Toxoplasma gondii, 314, 414
Toxoplasmosis, 414
Trachea, specimen collection, 237
Transbronchial fine needle aspiration, *see under* Fine needle aspiration

Transformation zone, 142, 144, 202
  smear pattern, 155–156, 159
Transitional cell carcinoma, 315, 316
  (figs), 319
Transitional epithelium, 303–304,
  307–308
Transthoracic percutaneous fine
  needle aspiration, see under
  Fine needle aspiration
Transudates, 274
Tray agglutination test, (TAT), 333
Trematodes (flukes), 415–418
Treponema pallidum (syphilis), 176
Trichomonas hominis, 409
Trichomonas tenax, 409
Trichomonas vaginalis, 409–410
  in cervical smear, 159
  genital infection, 175, 188–189
  and *Leptothrix*, 175
  in urine, 314
Trichomoniasis, 188–189
Trichosporon sp., 180
Trichuris trichiura (whipworm), 190,
  419, 420
*Trypanosoma cruzi, T. gambiense, T.
  rhodesiense*, 411
Trypanosomiasis, 411
Tuberculosis, 249–250
  caseation, 256, 257 (fig.)
  cytology, 256–257
  disease process, 259
  epithelioid cells, 256, 257
  fixatives, 53
  Ghon focus, 250
  granulomata, 171
  Langhan's giant cells, 256, 257
  lymph node aspirate, 352, 355
  meningitis, lymphocyte count, 297
  pulmonary infection, 250
  serous effusions, 61, 278
Tumour necrotizing factor, iatrogenic
  changes, 430–431
Tumours
  aetiology, 49–50
  connective tissue, 48
  cystic, cavitating, 47
  differentiation, 44
  encephaloid, 46
  epithelial, 48
  from fetal structures, 49
  germ cell, 48–49
  growth, 44
  immunological surveillance, 50
  morphology
    gross, 46–47
    microscopic, 48–49, 70
  neuroectodermal, 48
  neuroendocrine, 48
  polyp, papillomas, papillary carci-
    nomas, 46–47
  reticuloendothelial, 48
  scirrhous and sclerosing, 46
  spread of, 44–45
  staging, 45
Turner's syndrome, 30
Tween 80 technique, 58

Tzanck test, 184

Ulcerative colitis, chronic, cytology,
  381
Ultraviolet light
  effect on DNA, 32–33
  and skin cancer, 50
Unfixed samples, smear preparation,
  72–74
Ureter
  anatomy, 303
  catheter urine collection, 305
  lavage, 307
Urethra
  anatomy, 303
  histology, 304–305
Uric acid crystals (gout), 388–389
Urinary tract
  anatomy, 302–303
  brushing, 307
  calculi, 320–321
  casts, 303, 309–310
  columnar cells, 303, 309
  crystals, 310, 311
  cytology
    automation, 324
    development, 302
    evaluation, 322–324
    iatrogenic changes, 321
    preparatory techniques, 56, 306
  cytometric studies, 324, 449
  fine needle aspiration, 307
  histology, 303–304
  infection, 310, 313–314
  inflammatory cells, 309
  malakoplakia, 318
  organisms, 310, 313–314
  squamous cells, 304, 309
    carcinoma, 315, 317
    metaplasia, 304
  transitional cells, 303–304, 307–308
    non-papillary (*in situ*) carcinoma,
      315, 316 (fig.)
    papillary carcinoma, 315, 316
      (fig.)
  tumours, 314–319
    bladder, 317
    diagnosis, 306–307, 319
    evaluation, 319
    metastatic, 317
    renal carcinoma, 317
    screening, 319–320
    types of, 314
  viral infection, 321–322
  see also Bladder; Kidney; Urine
Urine
  catheterization, changes following,
    436
  centrifugation, 56, 306
  cloudy, 305
  collection, 305
  dyskaryotic cells in, 319
  fixation, 54, 305
  flow cytometry, 324, 449
  intestinal epithelial cells in, 318

Urine (*cont.*)
  membrane filtration, 56, 306
  non-permanent preparations, 306
  phase-contrast microscopy, 306
  processing, 305–306
  seminal vesicle cells in, 318
  spermatozoa in, 318, 319 (fig.)
  see also Urinary tract
Urothelial cells, 303–304, 307–308
  carcinomas, 315, 316 (figs), 319
  therapy-induced changes, 321, 435
Uterus, corpus, 144–146, see also
  Endometrium

Vacuum filtration, 63–65
Vagina
  acute inflammation, 167–169
  anatomy, 138–139
  bacterial flora, 174
  candidiasis, 179–181
  chronic inflammation, 169–173
  hyperkeratosis, 170
  oestrogen effect on, 139, 151
  parakeratosis, 170
  smears, 151
    endometrial cells, 220, 222–223
    historical review, 2–4
    hormone cytology, 164
    vault, 151
  wall, scrape of upper third, 151
Vascular disease processes, 41–42
Vegetable cells, see Contaminants
Vimentin, 119
Viral infections, 182
  cellular damage, 31
  ciliocytophthoria, 248
  cytological diagnosis, 88, 182
  electron microscopy and, 71
  genital, 182, 188
  lymphocytic effusion, 280
  respiratory, 257–258
  urinary, 313, 321–322
Viral replication, 33
Vogel, Sir Julius, 2
Von Brunn's nests, 304
Von Gierke's disease, 30
Vulva
  anatomy, 139
  carcinoma, 216

Wart virus infection, see Human
  papillomavirus
Warty atypia, 185
Wegener's granulomatosis, 248
Wet-fixed smears, 54–55, 73
Whipworm (*Trichuris trichiura*), 419,
  420
Wound healing, fibrosis, 37–38
Wuchereria bancrofti, 421

Zajicek, Josef, 5
Zaponin, 68
Ziehl–Neelsen (ZN) technique, 90